Theory of Recursive Functions and Effective Computability

Theory of Recursive Functions and Effective Computability

Hartley Rogers, Jr.

The MIT Press
Cambridge, Massachusetts
London, England

Second printing, 1988

First MIT Press paperback edition 1987

Original edition published by McGraw-Hill Book Company, 1967.

This book was printed and bound in the United States of America.

Library of Congress Cataloging-in-Publication Data

Rogers, H. (Hartley), 1926–
Theory of recursive functions and effective computability.

Bibliography: p.
Includes indexes.
1. Recursive functions. 2. Computable functions.
I. Title.
QA9.615.R64 1987 511.3 86-33764
ISBN 0-262-68052-1 (pbk.)

TO MY PARENTS

It may be we shall touch the Happy Isles,
And see the great Achilles whom we knew.
Tennyson: *Ulysses*

Preface to Paperback Edition (1987)

The theory of recursive functions (more commonly known as *recursion theory*) emerged, as a distinct branch of mathematics, from formal studies in the 1930's by Kleene, Church, and Turing. In the 1940's, Post showed, by example, that the intuitive simplicity and naturalness of the concept of *general recursive function* permitted discourse and proof at a level of informality comparable to that occurring in more traditional mathematics. In the 1950's, recursion theory began a remarkable period of growth and of interaction with other areas of logic and other branches of mathematics. The informal approach of Post was carried forward and elaborated by Dekker, Myhill, and others. (The present book follows and builds on this approach.) In the middle and late 1950's, Friedberg, Spector, Sacks, and others discovered new methods and combinatorial principles, such as the *priority method* and the *infinite injury method,* which provided important tools for further exploration and development of the theory. The decade of the 1950's was a time of excitement, enthusiasm, and promise for students of recursion theory. The month-long Summer Institute for Symbolic Logic at Cornell in 1957 marked an epoch for the theory (as well as for other areas of logic) at a moment when a significant number of independent and largely self-taught researchers came together for the first time to share approaches and ideas.

The present book was written during the period 1957–1967 and began as a set of mimeographed notes published by the M.I.T. Mathematics Department in 1957. The growth of recursion theory during this time was so rapid that, even at its publication in 1967, the book could not claim to be a comprehensive survey. Its most significant omission, perhaps, was the theory of recursion for finite types as initiated by Kleene.

The development of recursion theory during the two decades since 1967 has been extraordinary both in depth (where we have witnessed the solution of numerous open problems and the uncovering of new and unexpected combinatorial structure) and in breadth (where new areas of theory have emerged and interactions with other parts of logic, mathematics, science, and technology have continued to grow). It is impossible in the few paragraphs of this preface to do more than indicate some main features of the development since 1967.

The related theories of *recursively enumerable sets,* of *degrees of unsolvability,* and of *Turing degrees* in particular, are a central concern of this book. Since 1967, these topics have continued as a rich field of study—richer and more active, perhaps, than might have been predicted in 1967. Many individuals have made substantial contributions, and it is invidious to mention names and results in the limited scope of these comments. Nonetheless, by way of example, one may note the work of Lachlan on the *monstrous injury method,* the work of Soare on *maximal sets* and *automorphisms,* the work of Shore in settling a number of interesting global questions about degrees, and the comprehensive study of degrees by Lerman.

A second group of topics, in later chapters of the book, has to do with generalizations of recursion theory. Extensive development in this area has also occurred since 1967, with work by Sacks and others on *α-recursion theory* and with exploration of *higher-type recursion theory* as defined and studied by Gandy, Normann, Moschovakis, Slaman, and others. A variety of fruitful applications of generalized recursion theory have been made to set theory, to model theory, and to proof theory. Moreover, the theory of *hyperarithmetical computability* has emerged as a branch of recursion theory in its own right. One may note the separation theorem of Louveau in this area.

A third group of topics, mentioned only tangentially in the book, has to do with *subrecursive computability* and *subrecursive hierarchies*. Discovery by Cobham, in the middle 1960's, of the natural and invariant notion of *computability in polynomial time* has led to the study of computational complexity, a field of active interest in theoretical computer science with applications to number theory, cryptography, and other areas. A further field of computer science, also related to recursion theory, has arisen from the exploration by Scott and others of sets of *continuous functionals* as bases for a semantics of programming languages.

Computer scientists have told me that republication of this book will be welcomed by them and by their students. A basic reason, perhaps, for such a welcome may lie in the naturalness and appropriateness of the mathematical concept of partial function for the purpose of modelling the action of a computer. A partial function with a given argument may or may not be defined, just as a computer with a given input may or may not produce an output in a specified number of steps.

Apart from the correction of several typographical errors, the text of this 1987 republication remains unchanged from the 1967 printing. Further updating or modification (beyond this preface) would be a formidable task, and I have not ventured to attempt it. Since the original publication in 1967, researchers and students have informed me from time to time of statements or questions in the text (chiefly in the Exercise material) which are incorrect or now out of date. I am grateful for the interest and help of those who have communicated with me. Emendations are indicated following the Introduction.

I am especially grateful to The MIT Press, to its Director, Frank Urbanowski, and to its Science Editor, Laurence Cohen, for their initiative and encouragement in bringing about this republication. From 1974 to 1981, I had a close association with The MIT Press as chairman of its Editorial Board. I am honored that the Press should now wish to publish a work of my own.

Hartley Rogers, Jr.

Contents

Introduction: Prerequisites and Notation

This book is intended for use as a senior undergraduate or first-year graduate text. It assumes a knowledge of basic set-theoretical terminology and techniques such as might be obtained in an undergraduate course in modern algebra. In most parts of the book, no knowledge of logic is assumed, but the reader will find some knowledge of logic helpful. We shall use a few notations from elementary logic, as described below.

The literature of recursive function theory unfortunately lacks a single, commonly used terminology and notation. We present our choice of basic terminology and notation in this introductory section. The reader is urged to give this section a careful, preliminary reading and then to return to it from time to time as may later prove necessary. Initial effort by the reader here will, hopefully, be rewarded with clarity and facility in the main text.

For the most part, we deal with nonnegative integers, sets of nonnegative integers, and mappings from nonnegative integers to nonnegative integers. Unless specifically indicated otherwise, we use the words *number* and *integer* to mean nonnegative (rational) integer. N is the set of all integers. A, B, C, . . . (Latin capitals, early in the alphabet) denote subsets of N. $\emptyset$ is the empty set. x, y, z, . . . (Latin lowercase, late in the alphabet) denote members of N, i.e., integers.

We use the following set-theoretical notations. $A = B$ means that A and B are identical as sets, i.e., have the same members. $x \in A$ means that x is a member of A. $\{\ |\ \}$ is the notation to indicate set formation. $\{x|\ \cdots x \cdots\}$ is the set of all x such that the expression $\cdots x \cdots$ is true when "x" is interpreted as the integer x. The universe from which a set is formed is indicated by the style of symbol appearing before the vertical bar. Thus $\{x|\ \cdots\}$ must be a set of integers.

$A \cup B$ is the *union* of A and B, that is, $\{x|x \in A \textit{ or } x \in B \textit{ or both}\}$. $A \cap B$ is the *intersection* of A and B, that is, $\{x|x \in A \textit{ and } x \in B\}$. $\bar{A}$ is the complement of A, that is, $\{x|\textit{not } x \in A\}$. $A \subset B$ means that A is a *subset of B;* that is to say, for all x, if $x \in A$ then $x \in B$. $A \supset B$ means that $B \subset A$. A is a *proper* subset of B if $A \subset B$ and not $A = B$. (Thus $A \subset B$ and $B \subset A$ imply $A = B$.)

We occasionally denote a finite set by an expression in braces listing its members, in any order. For example $\{2,5,3\}$ is the set of the first three primes. We sometimes suggest certain infinite sets by a "listing" in braces. For example, $\{0,2,4, \ldots ,2n, \ldots\}$ is the set of even integers.

Given x and y, $\langle x,y\rangle$ is the *ordered pair* consisting of x and y in that order. Similarly $\langle x_1,x_2, \ldots ,x_k\rangle$ is the *ordered k-tuple* consisting of $x_1, \ldots , x_k$ in that order. $A \times B$ is the *cartesian product* of A and B, that is, $\{\langle x,y\rangle | x \in A \text{ and } y \in B\}$. Similarly

$$A_1 \times A_2 \times \cdots \times A_k = \{\langle x_1, \ldots ,x_k\rangle | x_1 \in A_1 \text{ and } \cdots x_k \in A_k\}.$$

The cartesian product of A with itself k times is denoted as A^k. $P, Q, R, \ldots$ (Latin capitals, late middle alphabet) will stand for *relations on N*, i.e., subsets of N^k for some $k > 0$. If $R \subset N^k$, R is called a *k-ary relation.*

Let R be any k-ary relation. We say that R is *single-valued* if, for every $\langle x_1, \ldots ,x_{k-1}\rangle$, there exists at most one z such that $\langle x_1, \ldots ,x_{k-1},z\rangle \in R$. If R is single-valued, we say that its *domain* is $\{\langle x_1, \ldots ,x_{k-1}\rangle | \text{there is a } z \text{ such that } \langle x_1, \ldots ,x_{k-1},z\rangle \in R\}$, and we call this collection *domain R.* Clearly a single-valued k-ary relation may be viewed as a mapping from its domain into N.

For this reason, instead of saying that R is a single-valued k-ary relation, we shall sometimes say, synonymously, that R is a *partial function of k − 1 variables.* (Here the word "partial" suggests that the domain of R may not be all of N^{k-1}.) We shall use $\varphi, \psi, \ldots$ (Greek lowercase, late alphabet) to denote partial functions, and we shall often use ordinary functional notation with these symbols; thus $\varphi(x,y) = z$ will mean that $\langle x,y,z\rangle \in \varphi$. The reader should keep in mind that, fundamentally, a partial function is to be construed as a relation. (We thus identify a partial function with its "graph.") We shall most often be concerned with the case $k = 2$, i.e., partial functions of one variable, and *partial function* will mean partial function of one variable unless otherwise indicated. With partial functions, the functional notation can result in ambiguity. For example, the assertion that $\varphi(x)$ *not* $= y$ might mean that $\langle x,y\rangle$ *not* $\in \varphi$, or it might mean that there is a z such that $\langle x,z\rangle \in \varphi$ and z *not* $= y$. Our intended meaning in such situations will always be clear. If φ is a partial function, we say that φ is *defined* (or *convergent*) at x if $x \in$ *domain* φ; otherwise φ is *undefined* (or *divergent*) at x. (Similarly for partial functions of more than one variable.)

In case a partial function of k variables has all of N^k as its domain, we call it a *function*, or, occasionally for emphasis, a *total function.* We use $f, g, h, \ldots$ (Latin lowercase, early middle alphabet) to denote functions. As before, $f(x) = y$ will mean $\langle x,y\rangle \in f$.

$\mathcal{A}, \mathcal{B}, \mathcal{C}, \ldots$ (script capitals, early alphabet) denote either sets of subsets of N or sets of relations on N.

The *range* of a partial function φ of k variables is $\{z | \text{there exist } x_1, \ldots , x_k \text{ such that } \langle x_1, \ldots ,x_k,z\rangle \in \varphi\}$, and we denote it *range* φ. Members of *range* φ are called *values* of φ. If $\varphi(x_1, \ldots ,x_k) = y$, y is called the *value* of φ corresponding to *argument* $\langle x_1, \ldots ,x_k\rangle$. A partial function is *onto* if its range $= N$. A partial function φ is *one-one* if, for every y, there is at most one k-tuple $\langle x_1, \ldots ,x_k\rangle$ such that $\varphi(x_1, \ldots ,x_k) = y$. c_A will be the *characteristic function* of the set A; hence $c_A(x) = 1$ if $x \in A$, and $c_A(x) = 0$ if x not $\in A$. Occasionally, a set A is represented by a (nonunique) function f such that $A = \{x | f(x) = 0\}$. Such an f is called a *representing function* for A.

Let [—x—] be an expression such that given any integer in place of "x," the expression defines at most one corresponding value. (For example, the expression "$x^2 + x$" defines a value for every integer; while the expression "least proper prime divisor of x" defines a value for integers which are not prime and are different from 1.) Then λx[—x—] denotes the partial function: $\{<x,y> |$[—x—] *defines the value* y *when* "x" *is interpreted as the integer* $x\}$. This is Church's *lambda notation* for defining partial functions. For example, given φ_1 and φ_2, then $\lambda x[\varphi_1(x) + \varphi_2(x)]$ is the partial function ψ such that *domain* ψ = *domain* $\varphi_1 \cap$ *domain* φ_2 and $\psi(x) = \varphi_1(x) + \varphi_2(x)$ for all x in *domain* ψ. We also use the lambda notation for partial functions of k variables, writing $\lambda x_1 x_2 \cdots x_k$ in place of λx.

If ψ and φ are partial functions, $\psi\varphi$ indicates their *composition*, i.e., the partial function $\{<x,y> |$*there is a* z *such that* $<x,z> \in \varphi$ *and* $<z,y> \in \psi\}$. (Note the reversed order of φ and ψ so that $\psi\varphi(x)$ can be expressed as $\psi(\varphi(x))$.) Other common notations include $\psi^{-1} = \{<y,x> | <x,y> \in \psi\}$; $\psi(A) = \{y |$*for some* x, $x \in A$ *and* $\psi(x) = y\}$; $\psi^{-1}(A) = \{x |$*for some* y, $y \in A$ *and* $\psi(x) = y\}$.

For binary (that is, 2-ary) relations, the terms *transitive, reflexive, equivalence, linear ordering, partial ordering* will be given their usual meanings (which we do not define here). *Partial orderings* will sometimes be strict ($<$) and sometimes nonstrict ($\leq$).

We use certain notations and conventions from elementary logic: "and" will sometimes be abbreviated as "&"; "or" will be used in the inclusive (and/or) sense and will sometimes be abbreviated as "$\vee$"; "$\Rightarrow$" will abbreviate "only if"; "$\Leftrightarrow$" will abbreviate "if and only if"; "$\neg$" will abbreviate "not" and will be placed before the statement which it negates. We sometimes combine "$\neg$" with "$\in$" or "$=$" as "$\notin$" or "$\neq$." Brackets will be used to indicate grouping of statements, except that (contrary to logical usage but following common mathematical usage) "(1) $\Rightarrow$ (2) $\Rightarrow \cdots \Rightarrow$ (n)" (where (1), (2), . . . , (n) are statements) will abbreviate "[[(1) $\Rightarrow$ (2)] & [(2) $\Rightarrow$ (3)] & $\cdots$ [(n − 1) $\Rightarrow$ (n)]]," and "(1) $\Leftrightarrow$ (2) $\Leftrightarrow \cdots \Leftrightarrow$ (n)" will abbreviate "[[(1) $\Leftrightarrow$ (2)] & [(2) $\Leftrightarrow$ (3)] & $\cdots$ [(n − 1) $\Leftrightarrow$ (n)]]." "$\forall$" and "$\exists$" are called *universal* and *existential quantifier symbols*, respectively. "$(\forall x)$" is read: "*for all* x." "$(\exists x)$" is read: "*there exists an* x *such that* . . .". Groups of symbols such as "$(\forall x)$" and "$(\exists x)$" are called, respectively, *universal* and *existential quantifiers*.

The above logical symbols serve as convenient abbreviations of ordinary mathematical language. For example, the meaning of $A \subset B$ can be expressed $(\forall x)[x \in A \Rightarrow x \in B]$, and the definition of *domain* φ can be expressed $\{x | (\exists y)[\varphi(x) = y]\}$.

$\aleph_0$ is the cardinality of N. 2^N is the set of all subsets of N; we sometimes call this collection $\mathfrak{N}$. $2^{\aleph_0}$ denotes the cardinality of 2^N, i.e., the cardinality of the continuum. $\mu x[\cdots x \cdots]$ is the least integer x such that the expression $\cdots x \cdots$ is true when "x" is interpreted as the integer x, if this least integer exists.

Various other general and special notations will be introduced as the book progresses.

Both logic and recursive function theory lack a universally accepted system of notation. Our choice of logical abbreviations is not uncommon. A choice of notation for recursive function theory presents some difficulties, especially in a treatment that covers a variety of areas. Part of the current literature uses Greek lowercase letters for sets. We do not do so because of the almost universal use of these symbols for ordinals. The reader should note that by *number-theoretic predicate*, Kleene and others mean *relation on N*. The reader should also note that some writers use $f, g, \ldots$ to denote partial functions in general, rather than only total functions, and that in a significant part of the literature, Greek lowercase letters early in the alphabet are used for total functions.

In the assertion of a theorem or lemma, unless otherwise specified, we follow the usual mathematical convention that all unquantified variable symbols (representing integers, sets, or relations) are to be taken as operated on by unexpressed universal quantifiers standing at the beginning of the entire assertion.

In the course of a proof, certain variable symbols may appear as *universal variables*, e.g., "Let x be any integer such that $\cdots$." Other variable symbols may be introduced as *temporary names*, e.g., "Choose x_0 to be some fixed integer greater than x . . . ."† In general, but not always, we shall use subscripted symbols in the latter case. (Subscripted symbols may also be used in the former case when sufficiently many distinct variable symbols are required.) Occasionally, for emphasis, we shall also use letters in the middle of the Latin lowercase alphabet, $i, j, k, m, n, \ldots$, with or without subscripts, as temporary names of integers. Occasionally, for emphasis, we shall use letters late in the Latin capital alphabet, X, Y, Z, as universal variables for sets. In this respect our usage will not always be consistent, but it will be clear from the context whether we adhere to these conventions or deviate from them.

Chapters are divided into sections. Thus §7.4 is the fourth section of Chapter 7. Theorems receive roman numerals, beginning anew in each chapter. Theorem 7-VI is the sixth theorem in Chapter 7. Exercises receive arabic numerals. Exercise 7-14 is the fourteenth exercise at the end of Chapter 7. When reference is made to a theorem or exercise occurring

† In the terminology of elementary logic, a *universal variable* is a variable symbol introduced by *universal specification*, and a *temporary name* is a variable symbol introduced by *existential specification* (see Suppes [1957]).

in the same chapter, the chapter number may be omitted. The conclusion of a proof is indicated by the symbol ☒.

Reference to entries in the bibliography is made by giving the author's name and a bracketed date. The item will be found under that name and beside that date in the bibliography.

Emendations

This section lists some substantive corrections to the text. It also mentions, for some questions posed in the text, answers which have been communicated to the author but which have not, to the author's knowledge, appeared in the literature. It does not consider questions, like those on pages 228 and 229 and in Chapter 13, which have been treated (in some cases, extensively) in the literature since 1967.

In the references that follow, "page 228, line 12," for example, refers to the twelfth line on page 228, *not counting* the line of the running head at the top of the page, while "page 301, line 4b" refers to the fourth line from the bottom on page 301 *including* the lines of the footnote on that page.

Page 41.	Exercise 2–10 (*c*) is incorrect. Condition 2 is not a sufficient condition as claimed. (R. Byerly, G. Riccardi.)
Page 43.	In exercise 2–14 (*f*), the characterization is too narrow. The class a_1 must also satisfy the closure rule: if g and h are in a_1, and M is a periodic set of natural numbers, and if $f = g$ on M, and $f = h$ elsewhere, then f is in a_1. (D. Siefkes,)
Page 55, line 12.	A partial answer to this question has been obtained by Y. Vladik.
Page 122.	Exercise 8–7 (*c*) may be incorrect.
Page 129, lines 13b–5b.	A simpler oracle-machine for this purpose can be defined: $q_0 1 B q_0$, $q_0 B q_1 q_2$, $q_1 B 1 q_2$, $q_2 1 R q_3$, $q_2 B R\, q_3$, $q_3 1 B q_2$.
Page 155.	Exercise 9–5 is incorrect. (D. Hickerson, D. MacQueen, D. Morris.) See also lines 5–7 on page 132.

Page 156. Exercise 9–38(ii) is incorrect. One can find a recursive f and a hyperimmune A, where $d(A,f)>0$, such that $(\forall x)[d(\{x\},f) = 0]$. (W. Jackson.)

Page 228, line 12. For "an element" read "a non-unit element".

Page 291, lines 12b–8b. The proof of Theorem XXV requires modification as follows. Replace step (g) of the construction with the following:
"(g) For each x which was thrown out of some z′-list during (e) and (f) but which did not receive a *plus*, add x to the z-list of *each* z such that:
(i) $\pi_1(x)>z$,
(ii) x is used negatively in $\phi_z^{A_n}(z)$,
(iii) the computation $\phi_z^{A_n}(z)$ has not become obsolete,
(iv) if $z_x = \mu z'$ [*x is in the z'-list after completion of step* (f)] and if any such z_x exists, then $z < z_x$. (Note that z_x varies with x.)"

Page 301, line 4b. For "is recursive" read "is A-recursive."

Page 325. The answer to the question in the second footnote is negative. (Y. Vladik.)

Page 362, line 20. For "where k is chosen . . . less than n" read "where k is chosen so that there is no index less than n for any function equal to g on inputs $\{0,1, \ldots ,n\}$". (H. Katseff.)

Page 394, line 17b. For "every vertex" read "every non-limit vertex".

Page 450. In Exercise 16–56, replace the hints in (a) and (b) by the single hint: "Show that if $A \in \Sigma_1^1$ then T_α is also Σ_1^1, by considering the possible existence of order-preserving maps from trees into A." (H. Enderton.)

Page 451. Exercise 16–60 is incorrect. It is easy to find a z such that $z \in T^2$ but $(\forall f)$ $[z \in T^{\tau(f)}]$.

1 *Recursive Functions*

§1.1 THE INFORMAL NOTION OF ALGORITHM

In this chapter we give a *formal* (i.e., mathematically exact) characterization of *recursive function.* The concept is basic for the remainder of the book. It is one way of making precise the *informal* mathematical notion of function computable "by algorithm" or "by effective procedure." In this section, as a preliminary to the formal characterization, we discuss certain aspects of the *informal* notions of *algorithm* and *function computable by algorithm* as they occur in mathematics.

Roughly speaking, an algorithm is a clerical (i.e., deterministic, bookkeeping) procedure which can be applied to any of a certain class of symbolic *inputs* and which will eventually yield, for each such input, a corresponding symbolic *output.* An example of an algorithm is the usual procedure given in elementary calculus for differentiating polynomials. (The name *calculus,* of course, indicates the algorithmic nature of that discipline.)

In what follows, we shall limit ourselves to algorithms which yield, as outputs, integers in some standard notation, e.g., arabic numerals, and which take, as inputs, integers, or k-tuples of integers for a fixed k, in some standard notation. Hence, for us, an algorithm is a procedure for computing a *function* (with respect to some chosen notation for integers). For our purposes, as we shall see, this limitation (to numerical functions) results in no loss of generality. It is, of course, important to distinguish between the notion of *algorithm,* i.e., procedure, and the notion of *function computable by algorithm,* i.e., mapping yielded by procedure. The same function may have several different algorithms. We shall occasionally refer to functions computable by algorithm as *algorithmic functions.*†

Here are several examples of functions for which well-known algorithms exist (with respect to the usual denary notation for integers).

a. $\lambda x[x\text{th } \textit{prime number}]$. (The method of *Eratosthenes' sieve* is an algorithm here.) (We are assuming Church's lambda notation. To say that $f = \lambda x[x\text{th } \textit{prime number}]$ is to say that for all x, $f(x) = x$th *prime number.*‡)

b. λxy[*the greatest common divisor of* x *and* y]. (The *Euclidean algorithm* serves here.)

c. λx[*the integer* ≤ 9 *whose arabic numeral occurs as the* xth *digit in the decimal expansion of* $\pi = 3.14159 \cdots$]. (Any one of a number of common approximation methods will give an algorithm, e.g., quadrature of the unit circle by Simpson's rule.)

Of course there are even simpler and commoner examples of functions computable by algorithm. One such function is

d. $\lambda xy[x + y]$. Such common algorithms are the substance of elementary school arithmetic.

Several features of the informal notion of algorithm appear to be essential. We describe them in approximate and intuitive terms.

*1. *An algorithm is given as a set of instructions of finite size.* (Any classical mathematical algorithm, for example, can be described in a finite number of English words.)

*2. *There is a computing agent, usually human, which can react to the instructions and carry out the computations.*

*3. *There are facilities for making, storing, and retrieving steps in a computation.*

*4. *Let P be a set of instructions as in* *1 *and L be a computing agent as in* *2. *Then L reacts to P in such a way that, for any given input, the computation is carried out in a discrete stepwise fashion, without use of continuous methods or analogue devices.*

*5. *L reacts to P in such a way that a computation is carried forward deterministically, without resort to random methods or devices, e.g., dice.*§

Virtually all mathematicians would agree that features *1 to *5, although inexactly stated, are inherent in the idea of algorithm. The reader will note an analogy to digital computing machines: *1 corresponds to the

† Beginning in §1.5, we shall extend our use of the word *algorithm* to include procedures for computing nontotal partial functions.

‡ As we proceed, we shall assume, without further comment, the conventions of notation and terminology set forth in the Introduction. In addition to the lambda notation, the restriction of *function* and *partial function* to mean mappings on (nonnegative) integers is important for Chapter 1.

§ In a more careful discussion, a philosopher of science might contend that *4 implies *5. Indeed, he might question whether there is any real difference between *4 and *5.

program of a computer, *2 to its *logical* elements and circuitry, *3 to its storage *memory*, *4 to its *digital* nature, and *5 to its *mechanistic* nature.

A straightforward approach to giving a *formal* counterpart to the idea of algorithm is, first, to specify the symbolic expressions that are to be accepted as sets of instructions, as inputs, and as outputs (we might call this the *P-symbolism*), and, second, to specify, in a uniform way, how any instructions and input determine the subsequent computation and how the output of that computation is to be identified (we might call this the *L-P specifications*).

Once we begin a search for a useful choice of P-symbolism and L-P specifications, *1 to *5 serve as a helpful intuitive guide. There are, however, several features of the informal idea of algorithm that are less obvious than *1 to *5 and about which we might find less general agreement. We discuss them briefly here, formulating them as questions and answers. Later, after we have settled on a particular formal characterization, we shall return and see how our answers accord with our chosen formal characterization. There are five questions. They are closely interrelated, as will be evident, and all have to do with the role of arbitrarily large sizes and arbitrarily long times.

The first three questions are:

*6. *Is there to be a fixed finite bound on the size of inputs?*

*7. *Is there to be a fixed finite bound on the size of a set of instructions?*

*8. *Is there to be a fixed finite bound on the amount of "memory" storage space available?* (For each of *6, *7, and *8, size could be measured by the number of elementary symbols (or English words) used.)

Most mathematicians would agree in answering "no" to *6. They would assert that a general theory of algorithms should concern computations which are possible *in principle*, without regard to practical limitations. For the same reason, they would agree in answering "no" to *7. However, *7 raises an issue that is already implicit in *6, namely, what sort of intellectual "capacity" do we require of L? If instructions are to be unbounded in size, will not this require unbounded "ability" of some kind on the part of L in order that L may comprehend and follow them? We consider this further under *9 below.

Question *8 is interesting in that physically existing computing machines *are* bounded in their available storage space. One might at first suppose that a negative answer to *8 is implied by our negative answers to *6 and *7, since arbitrarily large inputs and sets of instructions would, in themselves, require arbitrarily large amounts of space for storage. We can interpret *8, however, as referring to that storage space which is necessary over and above the space needed to store instructions, input, and output. Under this interpretation, *8 becomes of interest, apart from our answers to *6 and *7. We might conceive, for instance, of an ordinary computing machine of fixed finite size and fixed finite memory where the instructions P take the form of a finite printed tape fed into the machine, where the input is fed in on a second tape which (unlike the instruction tape) moves in only one direction, and where the output is printed, digit by digit, on a third tape which moves in only one direction. It is not difficult to show that a number of simple functions, including $\lambda x[2x]$, can be computed by an arrangement of this kind.† It is possible, however, to make a rather convincing and general argument that the function $\lambda x[x^2]$ cannot be computed by any such arrangement; as input x increases, larger and larger amounts of space for "scratch work" are required. On account of this narrowness, most mathematicians would answer "no" to any form of question *8. We therefore take "no" as our answer to questions *6, *7, and *8.

Our comments on *7 lead us to a fourth question about the informal notion of algorithm.

*9. *Is there to be, in any sense, a fixed finite bound on the capacity or ability of the computing agent L?* Let the reader imagine the following situation: he is given unlimited supplies of ordinary paper and pencil; he is given two tapes upon each of which is written a 1-million digit integer; and he is asked to apply the Euclidean algorithm to these integers and to write the result on a third tape. After some reflection, the reader will find it credible that he could work out a bookkeeping and cross-reference system whereby he could keep track of his progress and mark his place at various stages of the computation, and whereby he could indeed carry out the computation satisfactorily, given enough time. Indeed, the reader could doubtless find a uniform system that would work for input integers of arbitrary size. By such a system, he would, in effect, transfer excessive demands on his own mental capacities as L into additional demands on his (unlimited) paper-and-pencil memory storage. Similar "place-marking" systems can be introduced when the set of instructions P is of great length and complexity, provided that P is sufficiently well organized and detailed. Such a system would serve to "mark one's place" in P as well as in the input, output, and computation. In fact, we would expect that such a place-marking system could, in some sense, be made a part of P itself, if the P-symbolism is sufficiently flexible. We therefore answer "yes" to question *9.

When we later present and discuss our formal characterization, we shall see that these rather vague plausibility arguments can be substantiated (see §1.8). Indeed, once the P-symbolism and computation symbolism are given in sufficiently detailed form, it is possible to limit L to the following (without otherwise limiting the notion of algorithm): (*a*) a few simple clerical operations, including operations of writing down symbols, operations of moving one symbol at a time backward or forward in the

† Such functions are sometimes called *functions computable by finite-state machine*. (What functions are so computable depends, in part, on the choice of symbolism for inputs and outputs.) See Exercise 2-14.

computation to or from symbols previously written, operations of moving one symbol at a time backward or forward in P to or from symbols previously examined, and operations for writing the output; (*b*) a finite short-term memory of fixed size which at any point preserves symbols written or examined in various of the preceding steps; and (*c*) a fixed finite set of simple rules according to which the clerical operation next to be performed and the next state of the short-term memory are uniquely determined by the contents of the short-term memory together with the symbol written or examined last. (This remark will become clearer after §§1.5 and 1.8.)

We now turn to a final and somewhat deeper question about the informal notion of algorithm. It is a question upon which considerable disagreement can exist.

*10. *Is there to be, in any way, a bound on the length of a computation? More specifically, should we require that the length of a particular computation be always less than a value which is "easily calculable" from the input and from the set of instructions P? To put it more informally, should we require that, given any input and given any P, we have some idea, "ahead of time," of how long the computation will take?*

The question is vague. If one is to give an affirmative answer without begging the question, one must define "easily calculable" with care. Nevertheless, an affirmative answer to *10 is an essential feature of the notion of algorithm for many mathematicians.

We propose, however, to make no such affirmative answer to the question, arguing that it is simpler and more natural to accept such a restriction only if it proves to be a consequence of our other assumptions. We thus require only that a computation terminate after *some* finite number of steps; we do not insist on an a priori ability to estimate this number. As we shall see, this attitude toward *10 will accord with the formal characterization we select. To the extent that a reader can make *10 precise and can give an affirmative answer to *10 which is not a consequence of our formal characterization—to that extent will his informal notion be narrower than our formal characterization.

As we shall see (Theorem XI in §1.10), our position on *10 is fundamental. The absence of any such a priori requirement is a distinctive feature of the discipline developed in the remainder of this book.

§1.2 AN EXAMPLE: THE PRIMITIVE RECURSIVE FUNCTIONS

One method for characterizing a class of functions is to take, as members of the class, all functions obtainable by certain kinds of *recursive definition.* A recursive definition for a function is, roughly speaking, a definition wherein values of the function for given arguments are directly related to values of the same function for "simpler" arguments or to values of "simpler" functions. The notion "simpler" is to be specified in the chosen characterization—with the constant functions, among others, usually taken as the simplest of all. This method of formal characterization is useful for our purposes, in that recursive definitions can often be made to serve as algorithms.

Recursive definitions are familiar in mathematics. For instance, the function f defined by

$$\begin{aligned} f(0) &= 1, \\ f(1) &= 1, \\ f(x+2) &= f(x+1) + f(x), \end{aligned}$$

gives the Fibonacci sequence: 1, 1, 2, 3, 5, 8, 13, ... (The study of *difference equations* concerns the problem of going from recursive definitions to algebraic definitions. The Fibonacci sequence is given by the algebraic definition

$$f(x) = \frac{\sqrt{5}}{5}\left(\frac{1+\sqrt{5}}{2}\right)^{x+1} - \frac{\sqrt{5}}{5}\left(\frac{1-\sqrt{5}}{2}\right)^{x+1}.)$$

The *primitive recursive functions* are an example of a broad and interesting class of functions that can be obtained by such a formal characterization.

Definition The class of *primitive recursive functions* is the smallest class $\mathcal{C}$ (i.e., intersection of all classes $\mathcal{C}$) of functions such that

(i) All *constant functions*, $\lambda x_1x_2 \cdots x_k[m]$, are in $\mathcal{C}$, $1 \leq k$, $0 \leq m$;

(ii) The *successor function*, $\lambda x[x+1]$, is in $\mathcal{C}$;

(iii) All *identity functions*, $\lambda x_1 \cdots x_k[x_i]$, are in $\mathcal{C}$, $1 \leq i \leq k$;

(iv) If f is a function of k variables in $\mathcal{C}$, and $g_1, g_2, \ldots, g_k$ are (each) functions of m variables in $\mathcal{C}$, then the function $\lambda x_1 \cdots x_m[f(g_1(x_1, \ldots, x_m), \ldots, g_k(x_1, \ldots, x_m))]$ is in $\mathcal{C}$, $1 \leq k,m$;

(v) If h is a function of $k+1$ variables in $\mathcal{C}$, and g is a function of $k-1$ variables in $\mathcal{C}$, then the unique function f of k variables satisfying

$$\begin{aligned} f(0,x_2, \ldots, x_k) &= g(x_2, \ldots, x_k), \\ f(y+1, x_2, \ldots, x_k) &= h(y, f(y,x_2, \ldots, x_k), x_2, \ldots, x_k) \end{aligned}$$

is in $\mathcal{C}$, $1 \leq k$. (For (v), "function of zero variables in $\mathcal{C}$" is taken to mean a fixed integer.)

It follows directly from the definition that, for any f, f is primitive recursive if and only if there is a finite sequence of functions $f_1, f_2, \ldots, f_n$ such that $f_n = f$ and for each $j \leq n$, either f_j is in $\mathcal{C}$ by (i), (ii), or (iii), or f_j is directly obtainable from some of the f_i, $i \leq j$, by (iv) or (v). (To show this, let $\mathfrak{D}$ be the class of all functions f for which such a sequence $f_1, \ldots, f_n$ exists. $\mathfrak{D}$ is evidently contained in every class $\mathcal{C}$ that is closed under (i) to (v); furthermore $\mathfrak{D}$ is itself closed under (i) to (v). It follows that $\mathfrak{D}$

coincides with the intersection of all such $\mathcal{C}$.) If such a sequence for f is described, together with a specification of how each f_j is obtained for $j \leq n$, we say that we have a *derivation* for f as a primitive recursive function.

For an example, consider the function f given by the derivation

$f_1 = \lambda x[x]$	by (iii)	(a function of 1 variable)
$f_2 = \lambda x[x+1]$	by (ii)	(1 variable)
$f_3 = \lambda x_1x_2x_3[x_2]$	by (iii)	(3 variables)
$f_4 = f_2 f_3$	by (iv)	(3 variables)
f_5 to satisfy		
$f_5(0,x_2) = f_1(x_2)$		
$f_5(y+1, x_2) = f_4(y,f_5(y,x_2),x_2)$	by (v)	(2 variables)
$f = f_6 = f_5(f_1,f_1)$	by (iii)	(1 variable).

It is easy to verify that f_6 is the function $\lambda x[2x]$ (and, incidentally, that f_5 is $\lambda xy[x+y]$). Hence we can conclude that the function $\lambda x[2x]$ is primitive recursive.

A derivation can be written down in any one of a number of standard symbolic forms. A written derivation can serve as a set of instructions for effectively computing the function which it defines. For instance, to compute $f(2)$ in the preceding example, the derivation leads us to the computation

$$
\begin{aligned}
f(2) &= f_6(2) \\
&= f_5(f_1(2),f_1(2)) \\
&= f_5(2,f_1(2)) \\
&= f_5(2,2) \\
&= f_4(1,f_5(1,2),2) \\
&= f_4(1,f_4(0,f_5(0,2),2),2) \\
&= f_4(1,f_4(0,f_1(2),2),2) \\
&= f_4(1,f_4(0,2,2),2) \\
&= f_4(1,f_2(f_3(0,2,2)),2) \\
&= f_4(1,f_2(2),2) \\
&= f_4(1,3,2) \\
&= f_2(f_3(1,3,2)) \\
&= f_2(3) \\
&= 4.
\end{aligned}
$$

We obtain the computation uniquely by working from the inside out and from left to right.

All this suggests that we include the precise notion of primitive recursive function within our informal notion of function computable by algorithm. How does this accord with our discussion in §1.1? The computing agent is human (and *not* formally defined); nevertheless, the computation depends on the derivation in so simple and direct a way and via such obviously mechanical steps, that *1 to *5 are evidently satisfied. We can choose a standard P-symbolism for expressing derivations, and the L-P specifications are the simple substitution rules according to which a derivation and input determine a computation. Note, in passing, that questions *6 to *8 receive the same answers for primitive recursive functions as were given in §1.1. Question *9 must remain vague, since the computing agent is not formally defined. Question *10, as we shall indicate in a moment, can be given the answer "yes."

How inclusive is the class of primitive recursive functions? Perhaps it is broad enough to include all desired algorithms, and perhaps, in consequence, one can contend that it is an accurate formal counterpart to the informal notion of *function computable by algorithm.* Although the defining rules for primitive recursive functions might at first seem limited, one can supply an impressive body of evidence to support this contention. Virtually all the algorithmic functions of ordinary mathematics can be shown to be primitive recursive. (All the examples so far mentioned in this chapter are primitive recursive.) Ways to illustrate and demonstrate the breadth of primitive recursiveness are found in Péter [1951, pp. 1–67].

Unfortunately, it is possible to construct functions, with obvious algorithms, which are not primitive recursive. One such is the *Ackermann generalized exponential,* a function f of three variables such that

$$\begin{aligned}
f(0,x,y) &= y + x,\\
f(1,x,y) &= y \cdot x,\\
f(2,x,y) &= y^x,\\
&\cdots\cdots\\
f(z+1,\ x,\ y) &= \text{result of applying } y \text{ to itself } x-1 \text{ times}\\
&\quad\ \text{under the } z\text{th level operation } \lambda uv[f(z,u,v)].
\end{aligned}$$

A more formal (and "recursive") definition for this f is given by the conditions:

$$\begin{aligned}
f(0,0,y) &= y,\\
f(0,\ x+1,\ y) &= f(0,x,y) + 1,\\
f(1,0,y) &= 0,\\
f(z+2,\ 0,\ y) &= 1,\\
f(z+1,\ x+1,\ y) &= f(z, f(z+1,\ x,\ y), y).
\end{aligned}$$

There is no primitive recursive derivation for this function (see Péter [1951, p. 68]). Indeed, as Péter in effect shows, a function similar to f can be used to obtain an "easily calculable" function that gives an affirmative answer to *10 for the primitive recursive functions (if we take "easily calculable" to mean having simple defining conditions, like those for f above).

Since the generalized exponential would be almost universally accepted as a function computable by algorithm, and since it is not primitive recursive, we must reject the primitive recursive functions as an accurate formal counterpart to the informal notion of algorithmic function.†

§1.3 EXTENSIONALITY

As was remarked in §1.1, it is important to distinguish between the notion of *algorithm* and the notion of *algorithmic function.*‡ We now give several examples further to emphasize this distinction. In particular, we define a function g for which we can prove that an algorithm exists but for which we do not know how to get a specific algorithm. Consider the functions f and g defined by

$$f(x) = \begin{cases} 1, & \text{if a consecutive run of \textit{exactly} } x \text{ 5's occurs in the decimal expansion of } \pi; \\ 0, & \text{otherwise;} \end{cases}$$

and
$$g(x) = \begin{cases} 1, & \text{if a consecutive run of \textit{at least} } x \text{ 5's occurs in the decimal expansion of } \pi; \\ 0, & \text{otherwise.} \end{cases}$$

At the present time, no algorithm is known for computing f. Indeed, it may be that no algorithm exists for f. (Once our formal characterization is given, the notion of a *function having no algorithm* will become precise. We shall see that such functions exist.) In contrast to our ignorance about f, we do have the knowledge that g is primitive recursive. For either g must be the constant function $\lambda x[1]$, or else there must exist some fixed k such that

$$g(x) = 1, \qquad \text{for } x \leq k,$$

and

$$g(x) = 0, \qquad \text{for } k < x.$$

In either case, a primitive recursive derivation exists (see Exercise 2-1), but no one knows, at the present time, how to identify the correct derivation.

For an even simpler example, take an unsettled conjecture of mathematics, e.g., Goldbach's conjecture that every even number greater than 2 is the sum of two primes, and define a function h by

$$h(x) = \begin{cases} 1, & \text{if conjecture true;} \\ 0, & \text{if conjecture false.} \end{cases}$$

† The question naturally arises: does there exist a function of *one variable* which is algorithmic but not primitive recursive? It can be shown that $\lambda x[f(x,x,x)]$, where f is the Ackermann generalized exponential, is such a function. We shall see another example in §1.4.

‡ Note that one and the same primitive recursive function can have an infinite number of different derivations, i.e., algorithms. One trivial way of obtaining such derivations is to insert additional appearances of $\lambda x[x]$ in a given derivation.

h is evidently a constant function. Hence it is primitive recursive, though again we do not know how to identify its correct derivation.†

We shall be concerned both with functions and with algorithms. Our chief emphasis will be on functions. In traditional logical terminology, our emphasis will be *extensional*, in that we shall be more concerned with objects *named* (functions) than with objects *serving as names* (algorithms).

§1.4 DIAGONALIZATION

In §1.2 we gave an example of an (intuitively) algorithmic function that is not primitive recursive. We now look at a method which can be applied to a variety of formally characterized classes of algorithmic functions and which, in each case, produces an algorithmic function falling outside of the given formally characterized class. We call this method *diagonalization* and describe it through an example. In the example, we apply the method to the primitive recursive functions.

Consider all possible primitive recursive derivations. It is easy to set up a precise formal symbolism for derivations which uses only a finite number of basic symbols. These symbols would include a function symbol; several symbols for variables; digits for subscript numerals; digits for ordinary numerals; parentheses; the comma; plus and equals signs; several special symbols for indicating constant, successor, and identity functions; and a special symbol to mark the end of a line. Any derivation could then be represented as a single finite string of these basic symbols. Furthermore, an obvious effective (i.e., algorithmic) test would exist for determining, given any string of basic symbols, whether or not that string constituted a legitimate primitive recursive derivation. Hence we could list, in sequence, all possible primitive recursive derivations by first examining all strings of length 1, then examining all strings of length 2, etc. Indeed, we could give a definite, if informal, algorithmic procedure for making this list. (The list is infinite, but each derivation is reached at some finite point.) From this, we could, in turn, devise an algorithmic procedure which would list just the derivations for primitive recursive functions of one variable. Let Q_x be the $(x + 1)$st derivation in this latter list. Let g_x be the function determined by Q_x. Define h, by

$$h(x) = g_x(x) + 1.$$

† The proofs that g and h are primitive recursive use the logical principle of *the excluded middle*. Such nonconstructive methods are qualified or rejected in various "constructive" reformulations of mathematics, such as that of the *intuitionists*. Throughout this book we allow nonconstructive methods; we use the rules and conventions of classical two-valued logic (as is the common practice in other parts of mathematics), and we say that an object exists if its existence can be demonstrated within standard set theory. We include the axiom of choice as a principle of our set theory.

Evidently, we have an algorithm for computing h; namely, to get $h(x)$ for given x, generate the list of derivations out to Q_x, then employ Q_x to compute $g_x(x)$, then add 1. On the other hand, h cannot be primitive recursive. If it were, we would have $h = g_{x_0}$ for some x_0. But then we would have $g_{x_0}(x_0) = h(x_0) = g_{x_0}(x_0) + 1$, a contradiction. (The reader will note an analogy to Cantor's *diagonal proof* of the nondenumerability of the real numbers, in classical set theory.)

It is evident that the diagonalization method has wide scope, for it is applicable to any case where the sets of instructions in the P-symbolism can be effectively (i.e., algorithmically) listed. At first glance, it is difficult to see how a formal characterization can avoid such effective listing and still be useful. The diagonal method would appear to throw our whole search for a formal characterization into doubt. It suggests the possibility that no single formally characterizable class of algorithmic functions can correspond exactly to the informal notion of algorithmic function. Perhaps, no matter what P-symbolism and L-P specifications we choose, that symbolism and those specifications can be augmented by stronger symbolism and more complex "effective" operations to yield new functions. Even if we use the entire English language as P-symbolism, it may be that there are more complex clerical operations that demand new names. Perhaps, indeed, the algorithmic functions form a nondenumerable class, and perhaps there is a spectrum of algorithmic computability upon which *all* functions fall.

These are some of the considerations and difficulties, albeit vague and informal, that surround the problem of getting a satisfactory characterization of algorithm and of algorithmic function. They had to be faced by mathematicians who first addressed themselves to that problem in the 1930's, mathematicians who were stimulated in their work by recent successes of formal logic and its methods.

§1.5 FORMAL CHARACTERIZATION

We can avoid the diagonalization difficulty by allowing sets of instructions for nontotal partial functions as well as for total functions. Of course, situations may then arise where there is no evident way to determine whether a set of instructions yields a total function or not. Assume, for example, that we can have an expression of the P-symbolism which embodies the instructions: "To compute $f(x)$, carry out the decimal expansion of π until a run of at least x consecutive 5's appears; if and when this occurs, give the position of the first digit of this run as output." Or, for a simpler example, take: "To compute $g(x)$, examine successive even numbers greater than 2 until one appears which is not the sum of two primes; if and when this occurs, give the output $g(x) = 0$." In each example, unlike the illustrations in §1.3, where we had nonconstructive definitions for specific functions (but no algorithms), we have a specific computing procedure but do not know whether this procedure gives a function, i.e., whether it always terminates and yields an output. What we *can* conclude is that each procedure gives a *partial function.* If it should happen to be true that there are runs of eight 5's but none of greater length in π, then the first example would give a set of instructions for a partial function whose domain consisted of the first nine integers. If Goldbach's conjecture is true, then the second example would give the empty partial function; if the conjecture is false, then the second example would give the constant function $\lambda x[0]$. In any case, each example provides specific calculating instructions which determine a specific partial function.

With a formal characterization for a class of *partial functions* we are not immediately subject to the diagonalization difficulty. For let ψ_x be the partial function determined by the $(x + 1)$st set of instructions Q_x, and let x_0 be chosen so that ψ_{x_0} is the partial function φ defined by the following instructions: to compute $\varphi(x)$, find Q_x, compute $\psi_x(x)$, and if and when a value for $\psi_x(x)$ is obtained, give $\psi_x(x) + 1$ as the value for $\varphi(x)$. The equation $\psi_{x_0}(x_0) = \varphi(x_0) = \psi_{x_0}(x_0) + 1$ does not yield a contradiction, since $\varphi(x_0)$ does not need to have a value. We might perversely hope to reinstate diagonalization by effectively selecting just those sets of instructions which do yield total functions; however, as we have noted, there may be no evident way to do this. Indeed, if we are to avoid diagonalization, it must be the case that no algorithm for such a selection procedure can exist. (These comments are related to the basic *incompleteness theorems* of mathematical logic. We discuss this further in Chapter 2.)

The approach by way of partial functions is, in essence, the approach taken by Kleene [1936], Church [1936], Turing [1936], and others in the 1930's. Each obtained a formal characterization for a wide class of partial functions. The characterizations differed both in outline and in detail. They had the common features, however, *first,* of giving (through a P-symbolism) a formal counterpart to the notion of *algorithm* (for partial functions) and *second,* in consequence (and via L-P specifications), of giving a counterpart to the notion of *partial function computable by algorithm.*†,‡

† Virtually all the discussion and terminology of §§1.1 and 1.3 can be applied, *mutatis mutandis,* to the problem of characterizing algorithm for a *partial* function and *partial* function computable by algorithm.

‡ Historically, in a number of instances, e.g., that of Kleene, the investigator did not give an explicit first treatment of partial functions but rather gave a single, more complex characterization for total functions computable by algorithm. In retrospect, each of these more complex characterizations can be analyzed into two steps: first, characterization of the algorithmic partial functions; and second, identification of the algorithmic functions as those algorithmic partial functions which happen to be total. Our discussion of the Kleene characterization below will make this retrospective modification, although it will detract, in certain respects irrelevant to our purposes, from the simplicity of Kleene's original formulation.

The Turing Characterization

We look at Turing's characterization first (as presented in Davis [1958]). The Turing characterization will be taken as basic in this book. It is convenient and instructive to approach the Turing characterization by way of a physical picture, although the ultimate characterization is entirely mathematical.

Consider a finite mechanical device which has associated with it a paper tape of infinite length in both directions. The tape is marked off, along its length, into spaces of equal size. We refer to these spaces as *cells*. The device is arranged so that the tape runs through it and so that there is room for one tape cell to lie within it. We refer to the tape cell lying within the device as the cell being *examined* by the device. We refer to the two directions along the tape as *right* and *left*. The device is equipped to perform any of four basic operations: (1) it can write a "1" on the cell it is examining, if no "1" already appears in that cell; (2) it can erase the cell it is examining and thus make it blank, if that cell is not blank already; (3) it can shift its attention one cell to the right along the tape (by shifting the tape one cell to the left); (4) it can shift its attention one cell to the left along the tape (by shifting the tape one cell to the right). The device, when active, performs basic operations at a rate of one operation per unit time. We sometimes refer to the performance of a basic operation as a *step* in the working of the device. At the conclusion of any step, the device itself (as distinct from the tape) takes on one of a fixed finite set of possible internal (mechanical) configurations. We refer to these internal configurations as *internal states*. We use the symbols q_i, $i = 0, 1, 2, \ldots$, to denote distinct internal states. Finally, the device is so constructed that it behaves according to a certain finite list of deterministic rules. These rules determine, from the current internal state and the condition of the cell being examined, the operation next to be performed and the internal state next to be taken on (at the end of that next operation).† Let 1 and B denote the possible conditions (B for *blank*) of any cell. Let 1, B, R, and L denote the basic operations (1), (2), (3), and (4), respectively. (Operation (1) makes no change on the tape if the cell being examined already has a "1" in it; and operation B makes no change on the tape if the cell being examined already is blank.) Then the set of rules determining the behavior of the device may be expressed as a set of *quadruples*. Each quadruple consists of symbols (in order) for (i) an internal state, (ii) a possible tape-cell condition, (iii) an operation, (iv) an internal state. A quadruple (i, ii, iii, iv) expresses the rule that given (i) and (ii), the device performs (iii) and takes on (iv). We allow any set of such quadruples as a set of rules for a device, subject only to the restriction that any two distinct quadruples must differ

† We assume that the device is equipped to "read" the condition of the tape cell lying within it.

at (i) or (ii). We call this the *consistency restriction;* it prevents a set of rules from requiring two or more different courses of action at the same time. We do not, however, ask that every combination of (i) and (ii) be provided for in a set of rules; thus we permit a device to perform *no* operation under certain circumstances. In such circumstances we say that the device *stops.*

As an illustration, consider a device with states q_0 and q_2 whose behavior is determined by the two quadruples $q_0 1 B q_2$ and $q_2 B R q_0$. If such a device is given a tape with a finite run of consecutive 1's and is started in state q_0 on the leftmost cell of the run, it will erase all the 1's and then stop. If such a device is started on a tape consisting entirely of 1's, it will never stop.

A device of the general kind described above is called a *Turing machine.* For our purposes, a Turing machine can be identified with the set of quadruples defining its behavior. Any set of quadruples, using any (finite) number of internal states, constitutes a Turing machine, provided only that the consistency requirement is satisfied.

It is easy to define *Turing machine* in more orthodox mathematical language. Let $T = \{0,1\}$ and $S = \{0,1,2,3\}$. Then a Turing machine can be defined as a mapping from a finite subset of $N \times T$ into $S \times N$. Here T represents conditions of a tape cell, S represents operations to be performed, and N gives possible labels for internal states.

Given a Turing machine, a tape, a cell on that tape, and an initial internal state, the Turing machine carries out a uniquely determined succession of operations, which may or may not terminate in a finite number of steps. We can associate a partial function with each Turing machine in the following way. To represent an input integer x, take a string of $x + 1$ consecutive 1's on the tape. Start the machine in state q_0 on the leftmost cell containing a 1. As output integer, take the total number of 1's appearing anywhere on the tape when (and if) the machine stops. Under this convention for inputs, outputs, and initial conditions, each Turing machine yields a partial function.† For example, the following machine yields the function $\lambda x[2x]$, as the reader can verify.

$$
\begin{gathered}
q_0 1 B q_1 \\
q_1 B R q_2 \\
q_2 1 B q_3 \\
q_3 B R q_4 \\
q_4 1 R q_4 \\
q_4 B R q_5 \\
q_5 1 R q_5
\end{gathered}
$$

† If q_0 does not explicitly occur in the set of quadruples, then the machine takes no action, and the partial function determined is $\lambda x[x + 1]$. Although the convention for inputs, outputs, and initial conditions is rather arbitrary, we shall see that the same *class* of partial functions is determined under a wide variety of such conventions.

$$q_5B1q_6$$
$$q_61Rq_6$$
$$q_6B1q_7$$
$$q_71Lq_7$$
$$q_7BLq_8$$
$$q_81Lq_1$$
$$q_11Lq_1.$$

This machine terminates either in state q_2 (for input 0) or in state q_8 (for input other than 0). The reader should carry through by hand the computation of this machine for input 2, i.e., for input tape $\cdots B111B \cdots$. He will see that the quadruples can be grouped according to various *subroutines* for (*a*) erasing an input digit, (*b*) moving right to the output digits, (*c*) adding two new output digits, (*d*) moving left to remaining input digits, etc.

We can also associate a partial function of k variables with each Turing machine by representing an input $<x_1, \ldots, x_k>$ as a tape with a string of $x_1 + 1$ consecutive 1's followed by a B followed by $x_2 + 1$ consecutive 1's followed by a $B \cdots$ followed by $x_k + 1$ consecutive 1's; and by again starting the machine in state q_0 on the leftmost cell containing 1. The reader can verify that, for a function of k variables, the particular machine described in the preceding paragraph yields $\lambda x_1 \cdots x_k[2x_1 + x_2 + \cdots + x_k + k - 1]$.

Each stage in a Turing-machine calculation can be described by giving (i) the condition of the tape, (ii) the internal state of the Turing machine, and (iii) the location of the cell being examined. We say that such information determines an *instantaneous description*. This information can be expressed in the form

$$\cdots q_i \cdots,$$

where $\cdots\cdots$ is a string of adjacent 1's and B's from the tape which includes all that portion of the tape that is not blank, where q_i is the current state, and where "q_i" is inserted immediately to the left of the symbol for the cell currently being examined.

For example, if the above machine for $\lambda x[2x]$ is applied to the input tape for 2, then the following instantaneous descriptions give the first few steps in the computation.

$$q_0111$$
$$q_1B11$$
$$q_211$$
$$q_3B1$$
$$q_41$$
$$1q_4B$$
$$\cdots.$$

At first glance, the class of algorithmic partial functions given by Turing machines might appear rather limited. Nevertheless, the definition does provide us with P-symbolism and L-P specifications. Each set of quadruples defining a machine may be viewed as a set of instructions P. As basic symbols for our P-symbolism we need only q, 1, B, R, L, and digits for numeral subscripts. We may take the computing agent L to be human. The L-P specifications are the simple rules according to which a succession of machine-tape configurations is determined from the initial tape and from P. The relation between the Turing-machine definition and the discussion of §1.1 will be considered further in §§1.6 and 1.8.

The Kleene Characterization

We next look at Kleene's formal characterization. Consider recursive relations of the general kind employed in §1.2 to define the Ackermann exponential. A set of instructions P consists of a set of such "recursion equations." A *computation* is a finite sequence of equations, beginning with P, where each equation after P is obtained from preceding equations *either* by the substitution of a numeral expression for a variable symbol throughout an equation *or* by the use of one equation to substitute "equals for equals" at any occurrence in a second equation *or* by the evaluation of an instance of the successor function $\lambda x[x + 1]$.

In P, we allow *auxiliary function symbols*, in addition to the *main function symbol* in whose evaluation we are interested. Thus the set of equations

$$\begin{aligned} f(0) &= 0, \\ g(x) &= f(x) + 1, \\ f(x + 1) &= g(x) + 1, \end{aligned}$$

with g as auxiliary function symbol and f as main symbol, determines the function $\lambda x[2x]$, as can be verified.

Three related difficulties arise in connection with this notion of computation: (1) the course of a computation is not uniquely determined by the input and by P; (2) it is possible that two different outputs may be obtained from the same input (by different computations); (3) it is possible that *no* output may be obtainable from a given input.†,‡

We avoid difficulties (1) and (2) in the following way. We say that an equation is *deducible* from a given P if it is obtainable in the course of *some* computation from P. It is possible to describe a uniform procedure

† Kleene, in his original characterization, simply accepts difficulty (1) and goes on formally to define algorithmic functions as those functions which are obtainable from sets of instructions which happen not to be subject to difficulties (2) and (3). The fact that there is no evident way of identifying which those sets of instructions are is the price paid for avoiding diagonalization trouble.

‡ If, in the Turing characterization, the consistency restriction were dropped, then the Turing characterization (thus modified) would be subject to a difficulty similar to (2).

according to which, given any P, we can effectively generate a list of all equations deducible from P. (The procedure makes, in effect, an exhaustive list of all possible computations and is similar to the procedure discussed in §1.4 for listing all possible primitive recursive derivations.) Formal details of such a procedure would be complex, and we do not give them. Assume that we have made a fixed and permanent selection of such a procedure. Then given any P and given any input, we define the *principal output* for that input as the *first* output associated with it in the standard list of deducible equations generated from P. Each input can yield at most one principal output; thus the relation between inputs and principal outputs defines a partial function. We can hence associate a partial function with every set of equations P; and we have a formal characterization for a class of algorithmic partial functions.

The P-symbolism in this case can be described in greater detail as follows. Take as basic symbols f, g, x, $+$, $=$, parentheses, comma, digits for ordinary numerals, and digits for numeral subscripts (for use with g and x).

$x, x_0, x_1, \ldots$ are called *variables.*

$f, g, g_0, g_1, \ldots$ are called *function symbols*, and f is the *principal* function symbol.

Terms are defined inductively as follows:

A variable is a term;

An ordinary numeral is a term;

If τ is a term, τ followed by "$+ 1$" is also a term;

Let σ consist of a function symbol followed by parentheses containing a string of terms which are separated by commas; then σ is also a term.†

The expression formed by placing "$=$" between two terms is called an *equation.*

Any finite set of such equations constitutes a set of instructions P.

The *L-P* specifications describe how the uniform list of all derivable equations (from any given P) is to be generated. We omit details.

The reader will note, incidentally, that every primitive recursive derivation is expressible in the P-symbolism in an obvious way. The derivation for $\lambda x[2x]$ in §1.2 would become a set of equations beginning

$$\begin{aligned} g_1(x) &= x, \\ g_2(x) &= x + 1, \\ g_3(x_1,x_2,x_3) &= x_2, \\ &\ldots\ldots\ldots\ldots \end{aligned}$$

It is easy to show that the function determined by these equations in the new formalism is identical with the function determined by the original primitive recursive derivation according to the procedures of §1.2.

† An expression such as "$x + 2$" can, in effect, be used as a term but must be written as "$x + 1 + 1$".

§1.6 THE BASIC RESULT

In §1.1 we raised the question of whether or not a characterization can be found which supplies a satisfactory formal counterpart to the informal notions of algorithm and algorithmic function. In §1.4 we indicated difficulties that any such characterization must face; in particular, we mentioned (§1.4) the possibility that there might be no single, maximal, formally characterizable class of algorithmic functions. In §1.5 we gave the formal characterizations of Turing and of Kleene. During the 1930's and since then, separate formal characterizations have been proposed by Church [1936], Post [1936], Markov [1951], and others. These characterizations have varied widely in form; each, however, can be represented as a certain choice of P-symbolism and a certain choice of L-P specifications.

How satisfactory are these various characterizations? What is their relation to each other? How successfully do they avoid the difficulties of §1.4? Extensive work has been done on these questions. We summarize this work in the following *Basic Result*, which is fundamental for the remainder of the book.

Basic Result

Basic Result, Part I *By means of detailed combinatorial studies* (see, for example, Turing [1937] and Kleene [1936*a*],) *the proposed characterizations of Turing and of Kleene, as well as those of Church, Post, Markov, and certain others, were all shown to be equivalent; that is to say, exactly the same class of partial functions (and hence of total functions) is obtained in each case.*

Definition The functions falling within this class are called *recursive functions.* The partial functions of this class might, naturally, be termed "recursive partial functions." It has become standard usage, however, to call them *partial recursive functions.*

These equivalence demonstrations can be generalized to show that over certain very broad families of enlargements of these formal characterizations the class of partial functions obtained remains unchanged. (For example, if we allow more than one tape, or other symbols than 1 and B, in the definition of Turing machine, the partial functions obtainable are still partial recursive functions; see Turing [1936]. For a development based on programming techniques, see Shepherdson and Sturgis [1963].) *In fact, if certain general (and reasonable) formal criteria are laid down for what may constitute a P-symbolism and L-P specifications, it is possible to show that the class of partial functions obtained is always a subclass of the "maximal" class of all partial recursive functions.*

Basic Result, Part II *A wide variety of particular partial functions, each agreed to be intuitively algorithmic, have been studied. Each has been* *demonstrated to be a partial recursive function; that is to say, a set of instructions for it has been found within one of the standard formal characterizations. A variety of useful principles and techniques have been developed for making these demonstrations.* (Parts I and II thus provide strong empirical evidence that the formal characterizations are sufficiently inclusive.)

Basic Result, Part III *The proofs for the results in Part I have the following common structure. In every instance, the fact that one formally characterized class of partial functions is contained in another is demonstrated by supplying and justifying a uniform procedure according to which, given any set of instructions P from the first characterization, we can find a set of instructions P from the second characterization for the same partial function. Although it does not operate on integers, this uniform procedure itself happens, in each instance, to be* algorithmic (*in the unrestricted informal sense of that word, with no restriction to numerical inputs and outputs*).

Comment. In Parts I and II, emphasis was extensional, i.e., on the class of algorithmic partial functions defined rather than on the class of "algorithms" defined. Parts I and II show that there is a sense in which each standard formal characterization appears to include all possible algorithmic *partial functions.* Part III, taken together with Part I, now shows that there is a sense in which each standard characterization appears to include all possible *algorithms* (for partial functions). For, given a formal characterization of the kind mentioned at the end of Part I, there is a uniform effective way to "translate" any set of instructions (i.e., algorithm) of that characterization into a set of instructions of one of the standard formal characterizations. (We discuss this further in §1.7; see also Exercise 2-11.)

Some of the detailed work upon which the Basic Result rests will be found in Davis [1958] and in Kleene [1952]. (These references include several principles and techniques of the kind mentioned in Part II of the Basic Result.) In work on the Basic Result, the Turing definition has been especially useful as a standard to which other characterizations can be reduced.

From a mathematical point of view, the formal characterizations of §1.5 are *noninvariant,* i.e., they are dependent on arbitrary choices. This is true of all known characterizations. The Basic Result shows that the characterizations nevertheless provide a natural and significant class of partial functions. It has been remarked that this class is one of the few *absolute* mathematical concepts to originate in work on foundations of mathematics.†

† Where *absolute* means "existing apart from and largely independent of particular symbolic formulations." Quantificational provability (for classical two-valued logic) is another such concept.

§1.7 CHURCH'S THESIS

The claim that each of the standard formal characterizations provides satisfactory counterparts to the informal notions of *algorithm* and *algorithmic function* cannot be proved. It must be accepted or rejected on grounds that are, in large part, empirical. (That the claim for one characterization is equivalent to the claim for another follows from Parts I and III of the Basic Result.) The Basic Result provides impressive evidence that the class of partial functions defined is a natural one (Part I) and that it is sufficiently inclusive (Parts I and II). The Turing characterization provides convincing evidence that every partial function in the class is computable by a procedure that is, intuitively, "mechanical." (In §1.10 we shall discuss further the possibility that the formal class is too inclusive; see question *10 in §1.1.) On the basis of this evidence, many mathematicians have accepted the claim that the standard characterizations give a satisfactory formalization, or "rational reconstruction," of the (necessarily vague) informal notions. This claim is often referred to as *Church's Thesis*. Church's Thesis may be viewed as a *proposal* as well as a claim, a proposal that we agree henceforth to supply certain previously intuitive terms (e.g., "function computable by algorithm") with certain precise meanings.

In recent theoretical work, the phrase "Church's Thesis" has come to play a somewhat broader role than that indicated above. In Parts II and III of the Basic Result, we noted that a number of powerful techniques have been developed for showing that partial functions with informal algorithms are in fact partial recursive and for going from an informal set of instructions to a formal set of instructions. These techniques have been developed to a point where (*a*) a mathematician can recognize whether or not an alleged informal algorithm provides a partial recursive function, much as, in other parts of mathematics, he can recognize whether or not an alleged informal proof is valid, and where (*b*) a logician can go from an informal definition for an algorithm to a formal definition, much as, in other parts of mathematics, he can go from an informal to a formal proof. Recursive-function theory, of course, deals with a precise subject matter: the class of partial functions defined in §1.5. Researchers in the area, however, have been using informal methods with increasing confidence. We shall rely heavily on such methods in this book. They permit us to avoid cumbersome detail and to isolate crucial mathematical ideas from a background of routine manipulation. We shall see that much profound mathematical substance can be discussed, proved, and communicated in this way. *We continue to claim, however, that our results have concrete mathematical status as results about the class of partial functions formally characterized in* §1.5. *Of course, any investigator who uses informal methods and makes such a claim must be prepared to supply formal details if challenged.*

Proofs which rely on informal methods have, in their favor, all the evidence accumulated in favor of Church's Thesis. Such proofs will be called *proofs by Church's Thesis.*

We meet our first examples of such informal methods in the remaining sections of this chapter. Almost all the proofs in this book will use Church's Thesis to some extent. The analogy to informal methods of proof in other parts of mathematics is instructive. In both cases, the use of informal methods is a matter not of extremes but of degree. The degree of formalization of a proof usually depends upon the complexity and abstraction (what might be called the "danger") of the argument. The degree of formal detail we employ in this book will similarly vary with circumstances.

The beginning reader, who does not possess first-hand knowledge of the evidence for Church's Thesis, may be troubled by our arguments. To whatever extent he experiences doubt, we urge him to use the books of Davis and Kleene, in which he will find the tools needed to formalize our arguments fully.

§1.8 GÖDEL NUMBERS, UNIVERSALITY, *s-m-n* THEOREM

We have adopted the Turing-machine characterization as basic. We saw in §1.5 that a set of instructions is a set of quadruples satisfying the consistency restriction. It is possible to list all sets of instructions by a procedure similar to that indicated in §1.4 for listing all primitive recursive derivations. This procedure is itself algorithmic (in our first, unrestricted, informal sense of that word). It can be viewed as a procedure which associates with each integer x the set of instructions falling at the $(x + 1)$st place in the list of all sets of instructions. We assume now that we have selected one such listing procedure. We keep it fixed for the remainder of the book. We do not give formal details.

Definition P_x is the set of instructions associated with the integer x in the fixed listing of all sets of instructions. x is called the *index* or *Gödel number* of P_x.

$\varphi_x^{(k)}$ is the partial function of k variables determined by P_x. x is also called an *index* or *Gödel number* for $\varphi_x^{(k)}$. (We shall drop the superscript (k) when its value is clear from context or when $k = 1$. We shall be most often concerned with functions of one variable.†)

Clearly the listing procedure gives us *both* (*a*) an algorithm for going from any x to the corresponding P_x, and (*b*) an algorithm for going from any consistent set of quadruples P to a corresponding integer x such that P is P_x.

† The notation $\{x\}$ (for our φ_x) also appears in the literature. We use the φ_x and P_x notations to emphasize further the distinction between extension and name, i.e., between partial function and set of instructions.

Our fixed choice of Gödel numbering will be used throughout the book. It appears to be a rather noninvariant feature of our theory. We shall see, however (Chapter 4 and Exercise 2-10), that our results possess an invariant significance independent of this choice. In this respect, our use of a particular Gödel numbering is much like the use of a particular coordinate system to establish coordinate free results in geometry.

Several facts, already implicit in the Basic Result, can now be stated.

Theorem I *There are exactly* $\aleph_0$ *(a countable infinity of) partial recursive functions, and there are exactly* $\aleph_0$ *recursive functions.*

Proof. All constant functions are recursive, by Church's Thesis. Hence there are at least $\aleph_0$ recursive functions. The Gödel numbering shows that there are at most $\aleph_0$ partial recursive functions. ☒

Theorem II *There exist functions which are not recursive.*

Proof. By Cantor's theorem, there are $2^{\aleph_0}$ (continuously many and therefore nondenumerably many) functions. The theorem follows. ☒

The next fact appears to depend, for its proof, upon our particular characterization and Gödel numbering.

Theorem III *Each partial recursive function has* $\aleph_0$ *distinct indices.*

Proof. We need only show that there are at least $\aleph_0$ indices. Let a partial recursive function φ_{x_0} be given. Let m be an integer greater than the subscript integer of any internal-state symbol occurring in P_{x_0}. Then if P_{x_0} is modified by adding to it the quadruples $q_m 1 1 q_m$, $q_{m+1} 1 1 q_{m+1}$, . . . , $q_{m+k} 1 1 q_{m+k}$, the partial function determined remains unchanged, since none of the states q_m, . . . , q_{m+k} can be entered. As k varies, this gives $\aleph_0$ distinct sets of instructions for φ_{x_0}. ☒

Theorem III is not an accident of our chosen formal characterization and Gödel numbering. Exercise 2-10 will give it invariant significance.

Observe that the following is, informally, an algorithm for a partial function ψ: given any input $<x,y>$, find P_x (by generating the standard effective list of all sets of instructions until P_x is reached), then apply P_x to input y to calculate $\varphi_x(y)$; if and when $\varphi_x(y)$ gets an output, take this as output value for $\psi(x,y)$. Therefore,

$$\psi(x,y) = \begin{cases} \varphi_x(y), & \text{if } \varphi_x(y) \text{ convergent;} \\ \text{divergent}, & \text{if } \varphi_x(y) \text{ divergent.} \end{cases}$$

By appeal to Church's Thesis, we conclude that ψ is partial recursive and that $\psi = \varphi_{x_1}^{(2)}$ for some x_1. (Formal details of P_{x_1} can be found in Davis [1958].) Stating this as a theorem, we have Theorem IV.

Theorem IV *There exists a* z *such that for all* x *and* y, $\varphi_z(x,y) = \varphi_x(y)$ *if* $\varphi_x(y)$ *is defined, and* $\varphi_z(x,y)$ *is undefined if* $\varphi_x(y)$ *is undefined.*

The $\varphi_z^{(2)}$ of Theorem IV is called a *universal partial function* for the partial recursive functions of one variable. P_z is a single Turing machine which can be used to duplicate *any* partial recursive function of one variable. Theorem IV, sometimes called the *enumeration theorem*, was a main result of the early (and more formal) work on recursive function theory. Clearly, the proof of Theorem IV can be generalized to yield, for each $k \geq 1$, a function of $k + 1$ variables which can serve as a universal partial function in the obvious sense for the partial functions of k variables. Theorem IV is the case $k = 1$.

Theorem IV has a nontrivial practical significance. It shows that, for computing partial functions of one variable, there is a critical degree of "mechanical complexity" (that of P_z) beyond which all further complexity can be absorbed into increased size of program and increased use of memory storage. (In Exercise 2-5, we shall see that the restriction to partial functions of a particular fixed number of variables is not essential.) P_z is called a *universal machine*.

If we compare our formal characterization with questions *6 to *10 given in §1.1, Theorem IV shows that the computing agent L need not be human, and it substantiates our claim in §1.1 that L can be severely limited in its "abilities." Thus our formal Turing-machine concept is compatible with our answers to questions *6 to *9. It is also compatible with our response to *10 in that it makes no restriction of the kind that an affirmative answer to *10 would require. We shall look further at *10 in Theorem XI and in Exercise 2-8.

Theorem V, now to be given, also appears to depend on the particular characterization and Gödel numbering used. Like Theorem III, however, it has invariant significance. In conjunction with Church's Thesis, it will be basic in later work.

Theorem V *For every $m,n \geq 1$, there exists a recursive function s_n^m of $m + 1$ variables such that for all $x, y_1, \ldots, y_m$,*

$$\lambda z_1 \cdots z_n[\varphi_x^{(m+n)}(y_1, \ldots, y_m, z_1, \ldots, z_n)] = \varphi^{(n)}_{s_n^m(x,y_1,\ldots,y_m)}.$$

Proof. Take the case $m = n = 1$. (Proof is analogous for the other cases.) Consider the family of all partial functions of one variable which are expressible as $\lambda z[\varphi_x^{(2)}(y,z)]$ for various x and y. Using our standard formal characterization for functions of two variables, we can view this as a new formal characterization for a class of partial recursive functions of one variable. By Part III of the Basic Result, there exists a uniform effective procedure for going from sets of instructions in this new characterization to sets of instructions in the old. Hence, by Church's Thesis, there must be a recursive function f of two variables such that

$$\lambda z[\varphi_x^{(2)}(y,z)] = \varphi_{f(x,y)}.$$

This f is our desired s_1^1.☒

The informal argument by appeal to Church's Thesis and Part III of the Basic Result can be replaced by a formal proof. (Indeed, the functions $s_n{}^m$ can be shown to be primitive recursive.) We refer the reader to Davis [1958] and Kleene [1952]. Theorem V is known as the *s-m-n theorem* and is due to Kleene. Theorem V (together with Church's Thesis) is a tool of great range and power. We give one illustrative application here and shall see a further application in §1.9. On many occasions in later chapters the *s-m-n* theorem, like Church's Thesis, will be tacitly used.

Theorem VI *There is a recursive function g of two variables such that for all x,y,*

$$\varphi_{g(x,y)} = \varphi_x\varphi_y \qquad (\text{i.e.}, = \lambda z[\varphi_x(\varphi_y(z))]).$$

Proof. By Church's Thesis it is immediate that, for any given x, y, $\eta = \varphi_x\varphi_y$ is a partial recursive function. It remains to get the recursive function g; i.e., we must show that a Gödel number for η can be found in a uniform effective way from x and y as x and y vary. The reader who has reflected on the Basic Result will regard this as quite likely. Theorem V gives a tool for proving it. Define

$$\theta(x,y,z) = \varphi_x(\varphi_y(z)) = \varphi_{x_1}(x,\varphi_{x_1}(y,z)),$$

where φ_{x_1} is the universal function of Theorem IV. By Church's Thesis, θ is partial recursive and has an index w_0. Applying Theorem V, we have

$$\varphi_x\varphi_y = \lambda z[\varphi_{w_0}(x,y,z)] = \varphi_{s_1{}^2(w_0,x,y)},$$

and $\lambda xy[s_1{}^2(w_0,x,y)]$ is our desired g. ⊠

§1.9 THE HALTING PROBLEM

Is there an effective procedure such that, given any x and y, we can determine whether or not $\varphi_x(y)$ is defined, i.e., whether or not P_x applied to input y yields an output? By Church's Thesis this question can be put in the following equivalent and precise form. Is there a recursive function g such that $g(x,y) = 1$ if $\varphi_x(y)$ is convergent and $g(x,y) = 0$ if $\varphi_x(y)$ is divergent? We answer the question in the following theorem.

Theorem VII *There is no recursive function g such that for all x, y,*

$$g(x,y) = \begin{cases} 1, & \text{if } \varphi_x(y) \text{ convergent;} \\ 0, & \text{if } \varphi_x(y) \text{ divergent.} \end{cases}$$

Informal proof. Assume there is such a recursive g. It can be used to define a new partial function ψ as follows:

$$\psi(x) = \begin{cases} 1, & \text{if } g(x,x) = 0; \\ \text{divergent}, & \text{if } g(x,x) = 1. \end{cases}$$

Since $g(x,x)$ must $= 0$ or 1, this gives an algorithm, and by Church's Thesis,

ψ will be a partial recursive function. (When $g(x,x) = 1$, the instructions for ψ can make use of some infinite cyclic condition—a pair of quadruples $q_n 1 1 q_n$, $q_n B B q_n$ will suffice—to guarantee that $\psi(x)$ be undefined.) Let y_0 be a Gödel number for ψ. Then, by the definition of ψ, $\varphi_{y_0}(y_0)$ convergent $\Leftrightarrow g(y_0,y_0) = 0$; but, by our initial assumption about g, $g(y_0,y_0) = 0 \Leftrightarrow \varphi_{y_0}(y_0)$ divergent. This is a contradiction, and hence there can be no such recursive g. ☒

The proof just given uses Church's Thesis in a proof by contradiction. The reader may well ask what formal counterpart can exist to the use of Church's Thesis in such "hypothetical" context. It is instructive to trace the path that a more formal proof takes.

More formal proof. Define ψ by

$$\psi(z,x) = \begin{cases} 1, & \text{if } \varphi_z^{(2)}(x,x) = 0; \\ \text{divergent}, & \text{if } \varphi_z^{(2)}(x,x) \neq 0 \text{ or divergent.} \end{cases}$$

By Church's Thesis, ψ is partial recursive. Applying Theorem V, a Gödel number for $\lambda x[\psi(z,x)]$ can be found uniformly effectively from z; i.e., there is a recursive function h such that for all z, $\varphi_{h(z)} = \lambda x[\psi(z,x)]$. ($\lambda z[s_1{}^1(w_1,z)]$ is such an h, where w_1 is a Gödel number for ψ.)

Now assume $g = \varphi_{z_0}$ for some z_0. By the definition of h, $\varphi_{h(z_0)}(x)$ convergent $\Leftrightarrow \varphi_{z_0}(x,x) = 0$. Substituting $h(z_0)$ for x, we have

$$\varphi_{h(z_0)}(h(z_0)) \text{ convergent} \Leftrightarrow \varphi_{z_0}(h(z_0),h(z_0)) = 0.$$

But then φ_{z_0} cannot be g; for, with input $\langle h(z_0),h(z_0)\rangle$, φ_{z_0} either is undefined or gives erroneous information. ☒

We see that no partial recursive function can satisfy the requirement for g, and our proof is constructive in the sense that for any z, $h(z)$ provides a specific example where φ_z fails. The following corollary to Theorem VII is immediate.

Corollary VII *There is no recursive function f such that*

$$f(x) = \begin{cases} 1, & \textit{if } \varphi_x(x) \textit{ convergent;} \\ 0, & \textit{if } \varphi_x(x) \textit{ divergent.} \end{cases}$$

Proof. By proof of the theorem. ☒

Our original question (in the first paragraph of this section) has been called the *halting problem*, where the word "halting" means "having an output." The fact stated as Theorem VII is known as the *recursive unsolvability of the halting problem.* It has precise content in terms of our formal characterization.† The Basic Result of §1.5 gives this fact fundamental

† "There is no Turing machine which can solve the halting problem (for Turing machines). Indeed, given any candidate 'solver' Turing machine, we can exhibit an instance of the halting problem upon which it fails; in fact, there *is* a single Turing machine which will compute, from the description of any given candidate, an appropriate counterinstance."

significance. The halting problem was one of the first "natural" combinatorial problems to be shown recursively unsolvable. Demonstration of the existence of easily described recursively unsolvable problems is one of the more striking achievements of twentieth-century mathematics. Prior to such demonstration (in the 1930's) many mathematicians were unwilling to concede that there could exist easily described combinatorial problems (such as the halting problem) which had no algorithmic solution. We give a more general discussion of such unsolvable problems in Chapter 2.

The results of the early formal work on recursive function theory (prior to 1940) can be summarized, by and large, under (1) the Basic Result (§1.6); (2) existence of universal functions (§1.8); and (3) unsolvability of the halting problem (§1.9). Turing's basic paper [1936] is chiefly concerned with these matters.

We conclude this section with another unsolvability result. It was first obtained by Kleene [1936], and it is a natural consequence of our discussion of diagonalization in §1.4.

Theorem VIII *There is no effective procedure for deciding, given any x, whether or not φ_x is a total function. That is to say, there is no recursive function f such that*

$$f(x) = \begin{cases} 1, & \text{if } \varphi_x \text{ is total;} \\ 0, & \text{if } \varphi_x \text{ is not total.} \end{cases}$$

Proof. We give an informal proof. The transition to a more formal proof is similar to that in Theorem VII (see Exercise 2-7).

Assume such a recursive f exists. Then define g by

$$g(0) = \mu y[f(y) = 1],$$
$$g(x+1) = \mu y[y > g(x) \text{ and } f(y) = 1].$$

Since $f(y) = 1$ for infinitely many y (Theorem I), g is a total function. By Church's Thesis, g is recursive. Now define h by

$$h = \lambda x[\varphi_{g(x)}(x) + 1].$$

By our assumption on f, h is total. By Church's Thesis, h is recursive. Let $h = \varphi_{z_0}$. By definition of g, $g^{-1}(z_0)$ is uniquely defined; call it y_0. Then $h(y_0) = \varphi_{g(y_0)}(y_0) + 1$ by our definition of h; but $h(y_0) = \varphi_{g(y_0)}(y_0)$ by our definition of y_0. Since h is total, this a contradiction.⊠

§1.10 RECURSIVENESS

In §1.5, we formally characterized a class of partial functions, known as the *partial recursive functions.* Henceforth, when we say that a partial function is "effective," "computable," "effectively computable," "recursively computable," "mechanically computable," or "algorithmic," we shall mean that it falls within this class. The property of being a member of this class is called *recursiveness.* (Some mathematicians refer to recursiveness as "general recursiveness"; others reserve the phrase "general recursiveness" for total functions and refer to the recursiveness of partial functions as "partial recursiveness." In either case, "general" serves to emphasize that functions from the broad class of §1.5 are in question rather than functions from a narrower subclass (such as the primitive recursive functions).)

In the present section we discuss further aspects of recursiveness. In particular, we consider (1) extension of the concept to nonnumerical inputs and outputs (*codings*); (2) certain structural features of the partial recursive functions and the relation of these features to question *10 in §1.1 (*the mu operator*); and (3) the nature and possible usefulness of our future theory (*final comments*). Goals and applications of the theory will be discussed in greater detail in Chapter 3.

Codings

The partial recursive functions are mappings from integers to integers, and their algorithms carry us from notations for integers to notations for integers. The original, unrestricted, informal notion of algorithm concerns procedures with more general kinds of symbolic input and output; e.g., the algorithm for differentiating polynomials. We have already used such unrestricted algorithms as intermediate stages in algorithms defining numerical partial functions (and have then used Church's Thesis to conclude that the numerical partial functions in question were partial recursive functions); e.g., in our definition of a universal function, we used an algorithm (with nonnumerical outputs) that carried us from an integer x to the set of instructions P_x. We now ask: is there any way to include such broader nonnumerical algorithms within our formal theory? Two approaches can be made to this problem.

The first approach is as follows. Given a class of nonnumerical inputs or outputs, choose some fixed one-one mapping from this class into the integers. Henceforth, for theoretical purposes, *identify* each symbolic entity in the nonnumerical class with its corresponding integer "label." Such a standard mapping is called a *coding*, and the integers used as labels are called *code numbers.* The coding is chosen so that (*a*) it is itself given by an informal algorithm in the unrestricted sense; and (*b*) it is *reversible;* i.e., there exists an informal algorithm (in the unrestricted sense) for recognizing code numbers and carrying out the reverse "decoding" mappings from code numbers to nonnumerical entities. Furthermore, it is stipulated that a coding shall be used only when (*c*) an informal algorithm exists for recognizing the expressions that constitute the uncoded, nonnumerical class.

By thus *identifying* expressions (of a nonnumerical class) with integers, we can bring the discussion of algorithms (for such a class) within our formal theory of algorithms for integers.† We have already seen an example of this in §1.8. Our fixed Gödel numbering for the partial recursive functions is a coding from sets of instructions onto the integers. (Codings are frequently referred to as *Gödel numberings*, and code numbers are then called *Gödel numbers*.‡)

The use of codings raises an immediate question of invariance. Once a coding is chosen, will the formal concept *partial recursive function on code numbers* correspond to the informal notion *algorithmic mapping on the uncoded expressions?* As the latter notion is informal, the answer must be, in part, empirical. Church's Thesis provides an affirmative answer. Let C be the uncoded class (see the diagram below). Let γ be the coding map from C into N, and let us assume, further, that γ takes C *onto* N (modification of the argument, for the case where γ does not map *onto* N, is straightforward.) Let γ^{-1} be the decoding map from N onto C. γ and γ^{-1} are (informally) algorithmic, by our definition of coding. Let φ be any partial recursive function. φ is (both formally and informally) algorithmic. Hence the mapping $\delta = \gamma^{-1}\varphi\gamma$ is an (informally) algorithmic mapping from C into C. Conversely, let δ be any (informally) algorithmic mapping from C into C; then $\varphi = \gamma\delta\gamma^{-1}$ is an (informally) algorithmic partial function, and by Church's Thesis, φ is a partial recursive function. Thus every formal φ has a corresponding informal δ, and every informal δ has a corresponding formal φ.

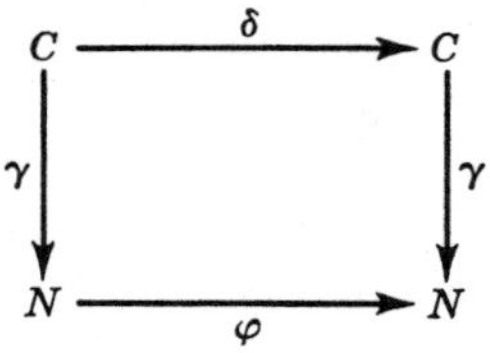

The second approach to a formal treatment of nonnumerical algorithms is as follows. The *formal characterization* of §1.5 is broadened to include directly, as inputs and outputs, expressions from wider "nonnumerical" classes. The Turing-machine characterization is especially convenient for

† The philosophically minded reader may ask why we take codings to be mappings into *integers* (mathematical objects) rather than into *numerals* for integers (symbolic objects), since algorithms for partial recursive functions must themselves operate on some form of numeral. It is true that use of numerals would be closer to our motivation; however, the distinction is unimportant for our purposes. The use of integers as labels has greater theoretical convenience and is the general practice.

‡ The first use of a coding was made by Gödel, who chose a fixed coding from the formulas of number theory into the integers; he was thus able to study the formulas and proof logic of number theory within number theory itself. An attempt to find self-referential paradoxes via this coding yields the *first incompleteness theorem* of Gödel. Tarski made a similar early construction of this kind.

this purpose.† It requires only that the expressions of the wider classes be expressible as finite strings in a fixed finite alphabet of basic symbols (other than B and 1). The basic operations of Turing machines are extended to include printing and erasure of symbols from this new alphabet. A nonnumerical mapping is then defined to be *recursive* (or *partial recursive*) if a Turing machine exists for carrying it out. After the formal characterization is so broadened, Parts I, II, and III of the Basic Result (§1.6) can themselves be modified and broadened to apply to this broader concept of recursiveness.

The second approach is evidently more direct. We use the first approach, in order to limit our subject matter and to emphasize it as a formal discipline about mappings on integers. For results obtained in this book, it makes no difference which approach is used.

The Mu Operator

The operator μ was defined in the Introduction.

Theorem IX *Let f be a recursive function of $k + 1$ variables; then $\lambda x_1 \cdots x_k[\mu y[f(x_1, \ldots, x_k, y) = 1]]$ is a partial recursive function of k variables.*

Proof. Immediate by Church's Thesis; for let

$$\psi = \lambda x_1 \cdots x_k[\mu y[f(x_1, \ldots, x_k, y) = 1]],$$

then to compute $\psi(x_1, \ldots, x_k)$ we need only compute, in succession,

$$f(x_1, \ldots, x_k, 0), f(x_1, \ldots, x_k, 1), f(x_1, \ldots, x_k, 2), \cdots.$$

If and when we find a y such that $f(x_1, \ldots, x_k, y) = 1$, we take it (the first such y) as value. The subcomputations for f always terminate, since f is a recursive function.☒

Theorem IX is known as the *mu theorem.* In Exercise 2-13 we shall see that Theorem IX does not hold in general when f is replaced by a *partial* recursive function of $k + 1$ variables.

There is a sense in which each partial recursive function can be obtained from (total) recursive functions by a single application of μ. Theorem X gives this.

Theorem X *There exist fixed recursive functions p and t of one and three variables, respectively, such that for all z,*

$$\varphi_z = \lambda x[p(\mu y[t(z,x,y) = 1])].$$

† Other extended formal characterizations for this purpose have been studied by Post [1943] and Smullyan [1961] (see also Asser [1960], Curry [1963], and Turing [1936]).

Proof. Define the function s as follows:

$$s(z,x,y,w) = \begin{cases} 1, & \text{if } P_z \text{ with input } x \text{ yields output } y, \text{ in fewer than } w \text{ steps;} \\ 0, & \text{otherwise.} \end{cases}$$

By Church's Thesis, s is recursive. Define p and q by

$$p = \lambda x[\textit{exponent of } 3 \textit{ in prime decomposition of } x + 1],$$
$$q = \lambda x[\textit{exponent of } 2 \textit{ in prime decomposition of } x + 1].$$

Define t by

$$t = \lambda zxy[s(z,x,p(y),q(y))].$$

By Church's Thesis, p, q, and hence t, are recursive. The theorem is now immediate from the definitions of s, p, q, and t.☒

Theorem X is known as the *Kleene normal-form theorem.* Once Theorem X is established, Theorem IV (the enumeration theorem) follows as a direct consequence. It is possible to show that both p and t are *primitive recursive.* Theorem X can be stated and proved for partial functions of k variables, $k > 1$, by introducing appropriate recursive functions s_k and t_k of $k + 3$ and $k + 2$ variables, respectively.

Corollary X *There exist recursive functions p and t_k such that for all z,*

$$\varphi_z^{(k)} = \lambda x_1 \cdots x_k[p(\mu y[t_k(z,x_1, \ldots ,x_k,y) = 1])].$$

Proof. As just indicated.☒

A proof of Theorem X can, of course, be based on the Kleene formal characterization. After a somewhat different sequence of definitions, a function $\hat{t}$ can be obtained analogous to our t. The relation

$$T = \{\langle x,y,z\rangle | \hat{t}(x,y,z) = 1\}$$

is called the *Kleene T-predicate,* and the assertion that $\langle x,y,z\rangle \in T$ is commonly abbreviated $T(x,y,z)$. The Kleene T-predicate occurs frequently in the literature.

The question of a stronger form for Theorem X, with p eliminated, naturally arises. Is there a recursive function t^* such that for all z,

$$\varphi_z = \lambda x[\mu y[t^*(z,x,y) = 1]]?$$

In Theorem 2-III, we shall show that no such t^* can exist. In fact, we shall find a partial recursive ψ of one variable such that for no recursive f is it true that $\psi = \lambda x[\mu y[f(x,y) = 1]]$.

The mu theorem emphasizes that our formal characterization permits a computation to take (in effect) the form of an unbounded search for an integer satisfying some given effective condition. The mu theorem is correlated with our failure to give an affirmative answer to question *10 in §1.1. It is a priori possible, of course, that some meaningful positive answer to question *10 might be deducible from the formal characterization. Unfortunately, this does not turn out to be true. Theorem XI and its proof show that any reasonable affirmative answer would permit a diagonalization leading to contradiction. The proof of Theorem XI is given as Exercise 2-8.

Theorem XI *There is no recursive function f of two variables such that for all x and z: P_z applied to x yields an output $\Leftrightarrow$ P_z applied to x yields an output in fewer than f(z,x) steps.*†

The nonexistence result in Theorem XI is similar to the unsolvability results of Theorems VII and VIII. Like Theorems VII and VIII, it is a consequence of the breadth of the formal notion of recursiveness. Because of such results, some mathematicians have argued that the formal notion of recursiveness is too broad to be a counterpart to their private informal notions of algorithmic computability. Be this as it may, recursiveness does express *one* notion of algorithmic computability and is sufficiently natural to merit further investigation in its own right.

Final Comments

The concept of recursiveness has virtues (breadth and clarity) and defects (unsolvabilities) that are characteristic of the theory now to be developed. Diagonal methods like those in Theorems VII and VIII will play an important role in the theory. It is not inaccurate to say that our theory is, in large part, a "theory of diagonalization."

Our theory is of limited practical usefulness at present. It is concerned with questions of existence or nonexistence of computational methods rather than with questions of efficiency and good design. Questions of the latter kind appear, not in our theory, but in more complex theories based on narrower concepts than recursiveness. Our theory can be viewed as a limiting, asymptotic version of these narrower, more difficult theories. As such, it has some practical value. The inclusive concept of Turing machine, the concept of universal machine (§1.8), and other combinatorial results and methods of our theory have been found useful and suggestive in work on computer programming. Perhaps the most direct practical applications have come from nonexistence (i.e., unsolvability) results, since these results carry over, a fortiori, to narrower theories.

It must be stated, however, that at the present time our theory derives its principal significance from its relevance to pure mathematics. It provides structures which possess considerable intrinsic beauty and naturalness. It gives new, and often deep, insights into other areas. These insights have been especially helpful in mathematical logic, and they have been increasingly useful in more classical areas as well.

† If we use the function s from the proof of Theorem X, Theorem XI can be restated: *there is no recursive f such that for all x, y, z,* $[(\exists w)[s(z,x,y,w) = 1] \Leftrightarrow s(z,x,y,f(z,x)) = 1]$.

2 *Unsolvable Problems*

§2.1 FURTHER EXAMPLES OF RECURSIVE UNSOLVABILITY

Theorem 1-VII, Corollary 1-VII, and Theorem 1-VIII gave examples of recursive unsolvability. Each concerned a "problem" which could be stated as the problem of effectively recognizing the members of a certain set or relation. For example, in Corollary 1-VII, this set was $\{x|\varphi_x(x)$ *convergent*$\}$. Each of the three results showed that the set or relation in question failed to have a recursive characteristic function. We summarized each result by saying that the problem was *recursively unsolvable.*

We now give several further examples of recursive unsolvability. We show, in each case, that a certain set (or relation) fails to have a recursive characteristic function. The "problems" are interesting and natural ones from within recursive function theory itself. All have to do with the general question: how much can be effectively determined about the behavior of a partial recursive function from a set of instructions for it; i.e., how much can be found out about the behavior of φ_x from the index x?

There are two different methods for demonstrating recursive unsolvability. The first, *direct* method is to make an argument, usually diagonal in flavor, showing that solvability would lead to contradiction. The second, *indirect* method is to take another problem already known to be unsolvable and then show that solvability of the problem under study would imply solvability of the problem known to be unsolvable. The latter method is called the method of *reduction;* in it, one shows that the known problem can be *reduced* to the problem under study (and hence would be solvable if the problem under study were solvable). The reduction method is often more convenient than the direct method. Theorems 1-VII and 1-VIII were obtained by the direct method. The examples now to be given use the indirect method. In each case, the problem of Corollary 1-VII will be used as the problem already known to be unsolvable.

We consider the following problems:

(*a*) The problem of deciding, for any x, whether or not φ_x is a constant function.

(*b*) The problem of deciding, for any x and y, whether y is in the range of φ_x.

(*c*) The problem of deciding, for any x, y and z, whether $\varphi_x(y) = z$.

(*d*) The problem of deciding, for any x and y, whether $\varphi_x = \varphi_y$.

(*e*) The problem of deciding, for any x, whether φ_x has infinite range.

(*f*) For each fixed y_0, the problem of deciding, for any given x, whether y_0 is in the range of φ_x.

(*g*) For each fixed x_0, the problem of deciding, for any given y, whether y is in the range of φ_{x_0}.

Theorem I *Problems* (*a*), (*b*), (*c*), (*d*), *and* (*e*) *are recursively unsolvable. For every choice of* y_0, *problem* (*f*) *is unsolvable. Problem* (*g*) *may or may not be recursively unsolvable, depending on the choice of* x_0.

Proof. (*a*) Let g be the characteristic function of $\{x|\varphi_x \textit{ is a constant function}\}$. We wish to show that g is not recursive.

Define ψ, a partial function of two variables, by the instructions: given input $\langle x,y \rangle$, find P_x and apply P_x to input x; if and when this converges, give output 0. By Church's Thesis, ψ is partial recursive. Clearly ψ satisfies the condition

$$\psi(x,y) = \begin{cases} 0, & \text{if } \varphi_x(x) \text{ convergent;} \\ \text{divergent,} & \text{if } \varphi_x(x) \text{ divergent.} \end{cases}$$

By Theorem 1-V (the *s-m-n* theorem), there is a recursive function h_1 such that $\lambda y[\psi(x,y)] = \varphi_{h_1(x)}$. ($h_1$ is $\lambda x[s_1{}^1(z_0,x)]$, where z_0 is an index for ψ.) Thus

$$\varphi_{h_1(x)}(y) = \begin{cases} 0, & \text{if } \varphi_x(x) \text{ convergent;} \\ \text{divergent,} & \text{if } \varphi_x(x) \text{ divergent.} \end{cases}$$

Therefore, $\varphi_{h_1(x)}$ is a constant function if and only if $\varphi_x(x)$ is convergent.

Now if g is recursive, then $gh_1(=\lambda x[g(h_1(x))])$ is a recursive function, and

$$gh_1(x) = \begin{cases} 1, & \text{if } \varphi_x(x) \text{ convergent;} \\ 0, & \text{if } \varphi_x(x) \text{ divergent.} \end{cases}$$

But gh_1 would be a recursive characteristic function for $\{x|\varphi_x(x) \textit{ convergent}\}$, contrary to Corollary 1-VII. Hence g cannot be recursive.

(*b*) Let f be the characteristic function of $\{<x,y>|\varphi_x \textit{ has } y \textit{ in its range}\}$. Let h_1 be the recursive function defined in the proof for (*a*). If f is recursive, then $\lambda x[f(h_1(x),0)]$ would be a recursive characteristic function for $\{x|\varphi_x(x) \textit{ convergent}\}$, contrary to Corollary 1-VII.

(*c*) Let f be a characteristic function for $\{<x,y,z>|\varphi_x(y) = z\}$. If f is recursive, then $\lambda x[f(h_1(x),0,0)]$ would be a recursive characteristic function for $\{x|\varphi_x(x) \textit{ convergent}\}$, contrary to Corollary 1-VII, where again h_1 is as in the proof for (*a*).

(*d*) Let f be the characteristic function for $\{<x,y>|\varphi_x = \varphi_y\}$. Take y_0 to be some fixed index for $\lambda x[0]$, then $\lambda x[f(h_1(x),y_0)]$ would be a recursive characteristic function for $\{x|\varphi_x(x) \textit{ convergent}\}$, contrary to Corollary 1-VII, where h_1 is as in the proof for (*a*).

(*e*) Let f be the characteristic function for $\{x|\varphi_x \textit{ has infinite range}\}$. By methods similar to those in the proof for (*a*), a recursive function h_2 can be defined such that

$$\varphi_{h_2(x)}(y) = \begin{cases} y, & \text{if } \varphi_x(x) \text{ convergent;} \\ \text{divergent}, & \text{if } \varphi_x(x) \text{ divergent.} \end{cases}$$

If f is recursive, then $\lambda x[f(h_2(x))]$ would be a recursive characteristic function for $\{x|\varphi_x(x) \textit{ convergent}\}$, contrary to Corollary 1-VII.

(*f*) Let y_0 be given, and let f be the characteristic function for $\{x|\varphi_x \textit{ has } y_0 \textit{ in its range}\}$. Then if f is recursive, $\lambda x[f(h_2(x))]$ would be a recursive characteristic function for $\{x|\varphi_x(x) \textit{ convergent}\}$, contrary to Corollary 1-VII, where h_2 is the recursive function defined in the proof for (*e*).

(*g*) Choose x_0 so that φ_{x_0} is the identity function, that is, $\varphi_{x_0}(y) = y$ for all y. Then $\{y|\varphi_{x_0} \textit{ has } y \textit{ in its range}\}$ has the recursive function $\lambda x[1]$ as its characteristic function. In Exercise 2-19 we shall see that values of x_0 also exist for which $\{y|\varphi_{x_0} \textit{ has } y \textit{ in its range}\}$ fails to have a recursive characteristic function.☒

In each of (*a*) to (*f*), we have demonstrated that the problem in question is unsolvable by showing that the halting problem of Corollary 1-VII can be *reduced* to it, i.e., by showing that if we could effectively solve the problem in question, we could use this to get an effective method for solving the halting problem of Corollary 1-VII.

The unsolvable problems in parts (*a*) to (*f*) of Theorem I (as well as Theorem 1-VII, Corollary 1-VII, and Theorem 1-VIII) are special cases of the following result, due to Rice [1953]. *Let $\mathcal{C}$ be a collection of partial recursive functions of one variable. Then $\{x|\varphi_x \in \mathcal{C}\}$ has a recursive characteristic function if and only if either $\mathcal{C}$ is empty or $\mathcal{C}$ consists of all partial recursive functions of one variable.* The proof of Rice's theorem is given as Exercise 2-39. Generalizations will be given and proved in a later chapter. Theorem I is given as above in order to provide specific introductory examples of reducibility and reducibility proofs.

§2.2 UNSOLVABLE PROBLEMS IN OTHER AREAS OF MATHEMATICS

A number of classical mathematical questions concern the existence of algorithms for solving certain "problems." By means of *codings*, as discussed in §1.10, these questions can be restated as questions about the existence of recursive functions. In this latter, more precise form, a number of these questions have been answered negatively.

Some of the first results of this kind occurred in logic, through work of Gödel, Church, and Turing. The results of Gödel and Church concerned the existence of algorithms ("decision procedures") for identifying the provable theorems in certain formal logical systems. We discuss these results briefly in §2.4 and more extensively in a later chapter.

Other unsolvability results have been obtained in number theory and algebra. Consider polynomials in any number of variables, with rational-integer coefficients. Consider the problem of deciding, for any such polynomial, whether or not that polynomial has a set of simultaneous real roots. ($<r_1, \ldots, r_k>$ is a *set of simultaneous roots* for the polynomial $p(x_1, \ldots, x_k)$, if $p(r_1, \ldots, r_k) = 0$.) This problem is solvable. Familiar methods of analysis (including, for example, Sturm's theorem) provide an algorithm for making this decision (see Tarski [1948]). Is there an algorithm for deciding, given any such polynomial, whether or not that polynomial has a set of simultaneous roots in the rational integers ("diophantine roots")? The "problem" concerned in this latter question is often referred to as *Hilbert's tenth problem*. At the present time it is not known whether this problem is recursively solvable. Davis, Putnam, and J. Robinson [1961] have shown that a closely related diophantine problem is recursively unsolvable, namely, the problem of deciding, for any equation between *exponential polynomials*, whether or not that equation has a solution in the nonnegative integers. An *exponential polynomial* is a polynomial-like form, with nonnegative coefficients, in which variables can occur as exponents.

The theory of groups provides an example of an unsolvability result in algebra. A *presentation* of a group is a finite list of *generators* and *relations* determining the group. (We do not define these terms here.) The problem of deciding, for any presentation and any string of generators, whether or not that string can be transformed to the identity by the relations of that presentation is called the *word problem for groups*. Novikov [1955] and Boone [1957] have shown this problem to be recursively unsolvable. In fact, they have each exhibited a single fixed presentation for which no algorithm for thus recognizing the identity exists. Their results, giving the recursive unsolvability of the word problem for groups, climaxed a sequence of results, by themselves and others, in which the word problems for various weaker kinds of algebraic structure were shown to be recursively unsolvable.† We remark in passing that the problem of deciding, for any presentation of a group, whether or not that presentation has a recursively solvable word problem, is itself known to be recursively unsolvable (see Exercise 2-29 for an analogous result).

Unsolvability results have also been obtained in topology. The problem of deciding, when any two triangulations for four-dimensional manifolds are explicitly given, whether or not the manifolds are homeomorphic is recursively unsolvable (see Markov [1958]). (The problem for two-dimensional manifolds is solvable by well-known methods.)

In the case of problems from outside recursive function theory, the method used to show recursive unsolvability is almost always that of reduction. The problem in question is related to an unsolvable problem of recursive function theory—usually some form of the halting problem. Thus, for their unsolvability result on exponential polynomials, Davis, Putnam, and J. Robinson [1961] exhibit an effective procedure whereby, given any x, an exponential polynomial equation can be obtained—call it E_x—such that E_x has a solution in nonnegative integers if and only if $\varphi_x(x)$ is convergent.

Description of such a reduction procedure, and demonstration that it has the desired reduction property, can be accomplished only by detailed treatment of the recursive function formalism and by study of deeper facts about the particular mathematical objects (e.g., exponential polynomial equations) occurring in the given problem. (One must show, in effect, that the objects of the problem are sufficiently flexible to "express" all instances of the halting problem.)

A number of unsolvability results from logic, number theory, and algebra are given in Davis [1958] together with detailed reduction proofs. Such results constitute an interesting and vital application of recursive function theory; however, we do not give them further detailed treatment in this book. For examples of recursive unsolvability, we confine ourselves to "problems" from recursive function theory itself—problems where Church's Thesis can be conveniently used to demonstrate unsolvability.

§2.3 EXISTENCE OF CERTAIN PARTIAL RECURSIVE FUNCTIONS

In §§2.1 and 2.2, we were concerned with the existence or nonexistence of certain recursive characteristic functions. We now turn to several questions that have to do with the existence or nonexistence of partial recursive functions.

† In weaker structures, which do not necessarily possess an identity, the *word problem* can be formulated as the problem of deciding, for any two strings, whether or not one string can be transformed to the other by relations of the presentation.

First, can every partial recursive function be extended to a total recursive function? Theorem II gives a negative answer to this question.

Theorem II *There exists a partial recursive function ψ such that for no recursive function f is it true that f is an extension of ψ, i.e., true that $(\forall x)[\psi(x)$ convergent $\Rightarrow \psi(x) = f(x)]$.*

Proof. A diagonal argument yields the result. Define ψ by the instructions: to compute $\psi(x)$, find P_x, apply P_x to x; and if and when an output occurs, take $\varphi_x(x) + 1$ as the value for $\psi(x)$. By Church's Thesis, ψ is partial recursive (ψ is $\lambda x[\varphi_{z_0}(x,x) + 1]$, if φ_{z_0} is the universal function of Theorem 1-IV). Thus

$$\psi(x) = \begin{cases} \varphi_x(x) + 1, & \text{if } \varphi_x(x) \text{ convergent;} \\ \text{divergent}, & \text{if } \varphi_x(x) \text{ divergent.} \end{cases}$$

Let f be any recursive function, and let y be an index of f; that is, $f = \varphi_y$. Since f is total, $f(y) = \varphi_y(y)$ is defined. Hence $\psi(y)$ is defined, and

$$\psi(y) = f(y) + 1.$$

Hence f cannot be an extension of ψ. ☒

Comment. Let P_{x_1} be the instructions for ψ in the above proof. It is informative to examine what occurs when P_{x_1} is applied to x_1, i.e., when an attempt is made to compute $\psi(x_1)$. P_{x_1} applied to x_1 instructs the computing agent first to find instructions with index x_1 and apply them to x_1. But this latter situation is again the application of P_{x_1} to x_1. An infinitely regressive sequence of similar situations is thus generated, each requiring the application of P_{x_1} to x_1; and the computation cannot terminate. $\psi(x_1)$ is for this reason undefined.

Theorem 1-X, the normal-form theorem, suggests several further questions about partial recursive functions. Can each partial recursive function be obtained from some recursive function by a single application of μ? In fact, can the function p be omitted from the statement of the normal-form theorem? Theorem III gives a negative answer to both these questions.

Theorem III *There exists a partial recursive function ψ such that for no recursive f is it true that $\psi = \lambda x[\mu y[f(x,y) = 1]]$.*

Proof. Define the partial function ψ as follows. To compute $\psi(x)$: find P_x, apply P_x to x; and if and when an output occurs, set $\psi(x) = x$. ψ is evidently partial recursive. Thus

$$\psi(x) = \begin{cases} x, & \text{if } \varphi_x(x) \text{ convergent;} \\ \text{divergent}, & \text{if } \varphi_x(x) \text{ divergent.} \end{cases}$$

Assume there is a recursive f such that for all x, $\psi(x) = \mu y[f(x,y) = 1]$. Then,

$$\varphi_x(x) \text{ convergent} \Rightarrow f(x,x) = 1;$$

and $$\varphi_x(x) \text{ divergent} \Rightarrow (\forall y)[f(x,y) \neq 1] \Rightarrow f(x,x) \neq 1.$$

Define g by

$$g(x) = \begin{cases} 1, & \text{if } f(x,x) = 1; \\ 0, & \text{if } f(x,x) \neq 1. \end{cases}$$

g is recursive, since $f(x,x)$ is defined for all x. But then g is a characteristic function for $\{x|\varphi_x(x)\ \mathit{convergent}\}$, contrary to Corollary 1-VII.⊠

Further results on partial recursive functions are given in Exercises 2-13 and 2-30 to 2-38.

§2.4 HISTORICAL REMARKS

The concept of recursive function has played a major role in recent developments in logic and foundations of mathematics. Some of the chief accomplishments of modern work in logic (in the nineteenth and twentieth centuries) can be informatively restated with the use of this concept. Perhaps the two most important results of modern logic are: (1) the discovery of a precise symbolic language within which all statements and proofs of mathematics can be made and in terms of which a combinatorial symbolic criterion of valid proof can be given and (2) the demonstration that there is no universal algorithmic procedure for determining whether or not a statement of this symbolic language is true.† The first result emerged early in this century through work of Frege, Russell, and Whitehead (which in turn built on work of Boole, Peirce, and others in the nineteenth century). In *Principia mathematica*, Whitehead and Russell [1910] presented a considerable body of mathematics within such a precise symbolic language. (The study of this language and of similar mathematical languages is a principal concern of *mathematical logic*.) Mathematicians then began to look for a universal algorithmic "decision procedure" for determining the true statements in such a language. During the 1920's a number of mathematicians were active in this search. Partial progress was made, and success was believed by some to be imminent. Then Gödel showed in his epochal paper [1931] that no algorithm out of a rather broad class could serve as such a decision procedure.‡ Gödel's work (the *Gödel incompleteness theorems*) preceded the

† Both these concerns were foreshadowed in work of the seventeenth-century mathematician and philosopher Leibnitz, who set forth two goals for science and philosophy: (1) the discovery of a universal precise symbolic language (*characteristica universalis*) within which all statements of science could be made and from which an especially clear insight into the meaning and validity of those statements could be obtained and (2) the discovery of a method (*calculus ratiocinator*) for manipulating the statements of this language in a way that directly elucidated their meaning and interrelationships.

‡ This conclusion is not explicit in Gödel's work. But it was evident that Gödel's methods could be applied to demonstrate the inadequacy of any algorithm from a class that included all algorithms of the type then being pursued (see Herbrand [1931, 1931*a*]; see also Gödel [1931, p. 197]). The argument by which this was evident can be reconstructed as follows. Gödel used a *coding* from formulas (i.e., statements in the formalism

general study of recursiveness. Subsequent work on the formal characterization of recursive function (§1.5) and on the Basic Result (§1.6) showed, in effect, that the broad class of algorithms to which the Gödel method applies includes all algorithms (Church [1936]), and hence that no universal decision procedure can exist. (As is indicated in work of Kleene (see Theorem 1-VIII), the existence of recursively unsolvable problems is a direct and almost trivial consequence of the formal characterization. The Basic Result, together with Church's Thesis, yields that any demonstration of recursive unsolvability is, in itself, an absolute demonstration that a universal decision procedure for mathematics cannot exist.)

As we shall see, recursive function theory provides valuable further insights into unsolvability and into the relation of unsolvability to mathematical logic. This book will be mainly concerned with concepts and structures that arise out of the unsolvability phenomenon.

§2.5 DISCUSSION

The results and examples of this chapter suggest that unsolvable problems can be classified according to the way and extent to which they are reducible to each other. How can the concept of reducibility be made precise? Several distinct reducibility concepts are defined, and a study of the resulting classification is begun, in Post [1944]. We shall present results of this and later work in Chapters 6 to 10.

Do all instances of unsolvability in some sense embody the unsolvability of the halting problem (Theorem and Corollary 1-VII)? That is to say, is the halting problem reducible to every unsolvable problem? To put it another way, is it true that every unsolvable problem can be demonstrated to be unsolvable by means of a reduction from the halting problem? This question (for an interesting class of problems known as the *recursively enu-*

---

of Whitehead and Russell) of number theory into integers. Let A be any set definable in number theory, i.e., expressible as $\{x| \cdots x \cdots\}$, where "$\cdots x \cdots$" is a statement of number theory involving the variable "x". Gödel presented a technique by which, given any such set A, a formula Θ with code number z_0 can be found whose clear and immediate meaning is $z_0 \notin A$. (This technique is described further in §11.6.) Now for every algorithm (of the kind being studied by mathematicians seeking a universal decision procedure), it was evident (*via coding*) that a set A could be found, definable in elementary number theory, such that for all x, $x \in A$ if and only if x is a code number of a formula asserted to be true by that algorithm. Applying Gödel's technique to get a corresponding Θ and z_0, we see that either (i) the formula Θ is true, in which case $z_0 \notin A$, and the algorithm fails to tell us that Θ is true, or (ii) Θ is false, in which case $z_0 \in A$, and the algorithm tells us that this false formula is true. Either way, the algorithm fails.

Gödel's original paper is explicitly concerned with matters of consistency and provability in certain formal theories. We discuss these specific results further in Chapters 7 and 11.

merable problems†) was one of Post's [1944] main concerns. He answered it for certain kinds of reducibility. The answer for one of the most general and important reducibility concepts, however, remained unknown until the work of Friedberg [1957] and Muchnik [1956] in 1956 (see Chapter 10).

§2.6 EXERCISES‡

§§*1.1* to *1.10*

△*2-1.* Show that the function g defined in §1.3 is primitive recursive. (*Hint:* Let $\hat{h} = \lambda x[max\ (0, x - 1)]$, $h = \lambda xy[max\ (0, x - y)]$, $f = \lambda x[min\ (x,1)]$. Show that $\hat{h}$ and f are primitive recursive. Use $\hat{h}$ to show h primitive recursive. Use h and f to get g.)

2-2. Define f by

$$f(x) = \begin{cases} 1, & \text{if } \varphi_x(x) = 1; \\ 0, & \text{otherwise.} \end{cases}$$

Is f recursive?

2-3. Consider the list of primitive recursive derivations described in §1.4. Let f_x be the primitive recursive function determined by the $(x + 1)$st derivation in this list $x = 0, 1, 2, \ldots$. Define $g = \lambda xy[f_x(y)]$. Is g recursive? Is g primitive recursive?

2-4. For any $k \geq 1$, define the "halting problem for partial recursive functions of k variables." Show that it is recursively unsolvable. (*Hint:* Reduction from Theorem 1-VII can be used.)

2-5. An instantaneous description (see §1.5) will be called *finite* if all but a finite number of cells on the tape are blank. Each finite instantaneous description possesses a unique minimal expression describing it, namely, that expression (see §1.5) which has no superfluous B's.

(*a*) Using these minimal expressions, describe a coding from finite instantaneous descriptions into integers.

(*b*) Given any Turing machine and any finite instantaneous description, the machine can be started in that instantaneous description, and a corresponding computation will follow. (If the internal state of the instantaneous description does not appear in the quadruples of the machine, the machine does nothing.) If and when a computation halts, a final finite instantaneous description is obtained. Hence, *via the fixed coding of* (*a*), every machine yields a partial function of one variable. Let ψ_z be the partial function yielded by machine P_z in this way. (Machine P_z is the machine that ordinarily computes φ_z. To obtain $\psi_z(y)$ for a given y, find the description with code number y; start P_z in this description; and if and when it halts, take the code number of the final description to be $\psi_z(y)$.) Is ψ_z partial recursive for all z?

(*c*) Does the class $\{\psi_0, \psi_1, \ldots\}$ contain all partial recursive functions of one variable? (*Hint:* Show that for all z, *domain* ψ_z is either empty or infinite.)

† A problem is called *recursively enumerable* if it concerns a set or relation of integers that can be effectively listed, i.e., is empty or is the range of some recursive function, where, for a relation to be the range of a function, a coding from k-tuples to integers must be used. All problems specifically discussed in §2.2 are recursively enumerable. The problems of Theorem 1-VII and Corollary 1-VII, and of parts (*b*), (*c*), (*f*), and (*g*) of Theorem I are recursively enumerable. The problems of Theorem 1-VIII and of parts (*a*), (*d*), and (*e*) of Theorem I are not. Recursive enumerability will be discussed in Chapter 5.

‡ Exercises are grouped according to related sections of the main text. Exercises on Chapter 1 occasionally use methods and concepts from Chapter 2.

(*d*) Does a universal partial recursive function exist for the class $\{\psi_0, \psi_1, \ldots\}$? (Such a function might be called a *general* universal function, since it can be used to duplicate the result of any Turing machine started in any finite instantaneous description.)

(*e*) Show, by trivial reduction from Theorem 1-VII, that $\{\langle x,y\rangle | \psi_x(y)$ *convergent*$\}$ does not have a recursive characteristic function. (This gives the unsolvability of what might be called the *general* halting problem for Turing machines.)

(*f*) Show that $\{z | domain\ \psi_z = \emptyset\}$ has a recursive characteristic function (in contrast to the result of Exercise 2-17(*b*)).

(*Note:* In a personal communication, Marvin Minsky and Hilary Putnam have announced that $\{z | \psi_z$ *is total*$\}$ does not have a recursive characteristic function (see also Exercise 2-9).)

2-6. Let $\{S_0, S_1, \ldots\}$ be an infinite alphabet of symbols available for use in Turing machines (in addition to B and 1). Modify the definition of Turing machine to allow use of any of these symbols (any machine, of course, consists of a finite number of quadruples). Modify the definition of finite instantaneous description to allow for appearance of these symbols on the tape.

(*a*) Describe a coding for these new finite instantaneous descriptions.

(*b*) Show that the results of Exercise 2-5 all carry through for these new machines and configurations, where the numbering indicated by the notation "P_z" is replaced by an appropriate numbering for the new machines.

2-7. Give a more formal proof for Theorem 1-VIII analogous to the more formal proof for Theorem 1-VII.

2-8. Prove Theorem 1-XI. (*Hint:* Use the unsolvability of the halting problem.)

2-9. Let $A = \{x | \varphi_x$ *is a total function*$\}$. Prove that there is no recursive function f such that *range* $f = A$. (*Hint:* See proof of Theorem 1-VIII.)

2-10. Let $\mathcal{P}$ be the class of all partial recursive functions of one variable. Let π be any map from N *onto* $\mathcal{P}$. π is called a *numbering* for the partial recursive functions of one variable. The standard Gödel numbers of §1.8 provide such a numbering; call it π_0. A numbering π is *acceptable* if it is possible to get back and forth effectively between π and π_0, i.e., if the following two conditions hold:

Condition 1. There exists a recursive function f (not necessarily one-one) such that $\pi_0 f = \pi$. (For every π number, we can find a π_0 number.)

Condition 2. There exists a recursive function g (not necessarily one-one) such that $\pi g = \pi_0$. (For every π_0 number, we can find a π number.)

(*a*) Show that any effective listing of all the Turing machines yields an acceptable numbering.

(*b*) Show that condition 1 is a necessary and sufficient condition that any π have a partial recursive universal function, i.e., satisfy an appropriate version of Theorem 1-IV, the enumeration theorem.

(*c*) Show that condition 2 is a necessary and sufficient condition that any π have an *s-m-n* theorem; i.e., satisfy an appropriate version of Theorem 1-V. (Assume that the indexing for partial recursive functions of more than one variable remains the standard one of §1.8. This assumption can be dropped if the definition of numbering and condition 2 are both appropriately generalized to apply to partial recursive functions of more than one variable.)

(*d*) Show that condition 2 implies that $\pi^{-1}(\psi)$ is infinite for every partial recursive function ψ of one variable, i.e., that π satisfies an appropriate version of Theorem 1-III. (*Hint:* Use result of Exercise 2-18(*b*).)

Comment. The above results give invariant significance to the notion of acceptable numbering and to Theorems 1-III, 1-IV, and 1-V (see discussion early in §1.8). In particular, any acceptable numbering has an enumeration theorem and an *s-m-n* theorem.

Any numbering π, by definition, includes all partial recursive functions of one variable.

Condition 1 may be viewed as requiring that the numbering be "algorithmic," i.e., that every number yield an algorithm. Condition 2 may be viewed as requiring that the numbering be "complete," i.e., "include all algorithms" (see comment after Part III of Basic Result in §1.6, and see Exercise 2-11).

△***2-11.*** Assume terminology of Exercise 2-10. Let g be a recursive function giving a one-one mapping from $N \times N$ onto N. (Any effective listing of $N \times N$ will supply such a g.) Define π_1 as follows. Given any y, find the x_1 and x_2 such that $g(x_1,x_2) = y$; then define $\psi = \pi_1(y)$ to be the partial recursive function satisfying the following conditions:

$$\psi(0) = \begin{cases} \text{divergent}, & \text{if } x_1 = 0; \\ x_1 - 1, & \text{if } x_1 \neq 0; \end{cases}$$

and

$$\psi(z) = \varphi_{x_2}(z), \qquad \text{all } z \neq 0.$$

(*a*) Show that π_1 is a numbering, i.e., includes all partial recursive functions of one variable.

(*b*) Show that, for π_1, condition 1 holds but condition 2 does not hold.

Comment. π_1 is hence a numbering for *all* the partial recursive functions of one variable but does not "include *all* algorithms." The enumeration theorem holds for it, but the *s-m-n* theorem does not.

2-12. Show that for every recursive f of one variable, there exists a recursive g such that $f = \lambda x[\mu y[g(x,y) = 1]]$.

△***2-13.*** Show that the partial recursive functions are not closed under μ; i.e., show that there exists a partial recursive ψ such that $\lambda x[\mu y[\psi(x,y) = 1]]$ is not partial recursive. (*Hint:* Define $\psi(x,y) = 1$ if $y = 1$ or if both $y = 0$ and $\varphi_x(x)$ convergent.)

2-14. Let a *finite-state machine* be a device with two tapes, each of which can move in only one direction. Input integers are presented to the device on one tape; outputs are printed on the other. The device has a finite number of internal states, and its operation is determined by a finite set of Turing-machine-like mechanical rules. Erasing operations are not used.

(*a*) Is the following problem recursively solvable: the problem of deciding, given any finite-state machine and any input, whether or not the resulting computation halts?

(*b*) Is the following problem recursively solvable: the problem of deciding, given any finite-state machine, whether or not that machine yields a total function?

(*c*) The class of functions computable on such finite-state machines will depend upon the symbolism allowed for expressing input and output integers. Let $\mathcal{A}_n$ be the class of functions resulting when input and output integers are both written to the base n, $n = 2, 3, \ldots$. Let $\mathcal{A}_1$ be the class of functions resulting when inputs and outputs are both written as sequences of 1's, with each integer x represented as a sequence of $x + 1$ 1's. Show that $\lambda x[2x]$ is in $\bigcap_{n=1}^{\infty} \mathcal{A}_n$.

△(*d*) Show that $\lambda x[x^2]$ is not in $\bigcup_{n=1}^{\infty} \mathcal{A}_n$.

(*e*) Give an example of $n,m \geq 2$ such that $\mathcal{A}_n \neq \mathcal{A}_m$. Show that $\mathcal{A}_1 \neq \mathcal{A}_n$ for all $n \geq 2$. Show that $\bigcap_{n=1}^{\infty} \mathcal{A}_n$ contains the nonnegative parts of all linear functions.

(*f*) Characterize $\mathcal{A}_1$. (*Answer:* A function is in $\mathcal{A}_1$ if and only if it is equal, at all but a finite number of places, to a function which can be expressed as the sum of a function of the form $m[x/n] + q$ (a *linear step function of nonnegative slope*) and a periodic function of period n.) Prove a unique-decomposition theorem.

2-15. Show that there exists a recursive h such that for all primitive recursive g, $(\exists x)(\forall y)[x \leq y \Rightarrow g(y) < h(y)]$.

▲*2-16.* Can we conclude from the following assumptions, and using no other facts about π_f, that f is primitive recursive: (i) f is the function mentioned, as f, in the first paragraph of §1.5; (ii) the function mentioned in (c) in §1.1 is primitive recursive; and (iii) f is total? (*Hint:* Show that there is a real number ρ such that $g = \lambda x$[*the integer* ≤ 9 *whose digit occurs at the* $(x + 1)$st *place in the decimal expansion of* ρ] is primitive recursive, and such that $\hat{f} = \lambda x$ [*the position of the left-hand digit of the first run of* x *or more* 5'*s in the expansion of* ρ] is total but not primitive recursive. Take t, p, q from §1.10 (they may be assumed primitive recursive), and use the function h constructed in Exercise 2-15 to get this ρ so that: the decimal of ρ consists entirely of 0's and 5's; g is primitive recursive (via use of p, q, and $\lambda xy[t(z_0,x,y)]$, where $\varphi_{z_0} = h$); and $(\forall x)[h(x) < \hat{f}(x)]$. Experience with primitive recursive functions is helpful but not necessary.)

§*2.1*

2-17. Show directly (i.e., without the use of Rice's theorem, stated at the end of §2.1) that none of the following has a recursive characteristic function:

(a) $\{x|\varphi_x$ *has infinite domain*$\}$.

(b) $\{x|\varphi_x$ *has empty domain*$\}$.

(c) $\{<x,y>|$*domain* $\varphi_x =$ *domain* $\varphi_y\}$.

2-18. Show directly that for any given z_0, neither of the following sets has a recursive characteristic function:

(a) $\{x|$*domain* $\varphi_x =$ *domain* $\varphi_{z_0}\}$.

(b) $\{x|\varphi_x = \varphi_{z_0}\}$.

(*Hint for* (a)*:* Treat cases *domain* $\varphi_{z_0} = \emptyset$ and *domain* $\varphi_{z_0} \neq \emptyset$ separately.)

2-19. Complete the proof of Theorem 2-I(g) by showing that there exists an x_0 such that *range* φ_{x_0} does not have a recursive characteristic function. (*Hint:* See proof of Theorem 2-III.)

2-20. Show that there exist x_0 and x_1 such that *domain* φ_{x_0} has a recursive characteristic function but *domain* φ_{x_1} does not.

2-21. Do parts (b) and (c) of Exercise 2-17 with "range" in place of "domain."

2-22. Do part (a) of Exercise 2-18 with "range" in place of "domain."

2-23. Show that the problem of Theorem 1-VII is directly reducible to the problem of Corollary 1-VII. (*Hint:* Use a construction similar to that used in the proof of Theorem 2-I(e).)

2-24. Show that problem (c) of Theorem 2-I is reducible to the halting problem.

△*2-25.* Show that problems (b) and (f) of Theorem 2-I are each reducible to the halting problem.

▲*2-26.* Show that none of problems (a), (d), or (e) of Theorem 2-I is reducible to the halting problem.

△*2-27.* Let f and g be recursive functions such that for some set A, $A =$ *range* f and $\bar{A} =$ *range* g. Does A have a recursive characteristic function?

△*2-28.* Modify the definition of Turing machine by allowing the special internal state q^* in addition to q_0, q_1, ... Modify the method of associating partial functions with machines by stipulating that an output is to be used only when a machine stops in state q^*. An input can hence fail to yield an output in either of two ways: (1) the computation does not terminate; (2) the computation terminates, but not in state q^*. We refer to the former situation as an *infinite singularity* and to the latter as a *block singularity.*

(a) Is every partial recursive function obtained in this way? If so, would an effective listing of these modified machines yield an *acceptable numbering* in the sense of Exercise 2-10?

(b) Is the following problem solvable: to decide, given any machine and input, whether or not an infinite singularity appears?

(*c*) Do part (*b*) with "infinite" replaced by "block."

(*d*) Can we "replace" block singularities by infinite singularities; i.e., is there an effective procedure for going from any machine M to a new machine M' such that M and M' represent the same partial function and such that M' has only infinite singularities?

(*e*) Can we "replace" infinite singularities by block singularities?

(*Hint:* Answers to (*d*) and (*e*) are "yes" and "no," respectively.)

§*2.2*

△***2-29.*** Consider the "metaproblem" of deciding, for any x_0, whether or not problem (*g*) of Theorem 2-I is recursively solvable. Show that this metaproblem is recursively unsolvable; i.e., show directly that $\{x|\textit{range } \varphi_x \textit{ has a recursive characteristic function}\}$ does not have a recursive characteristic function.

§*2.3*

2-30. Let two partial recursive functions have nonempty sets A and B as their respective domains, and assume that $A \cap B = \emptyset$.

(*a*) Does there necessarily exist a partial recursive function ψ such that $\psi(A) = \{0\}$ and $\psi(B) = \{1\}$?

△(*b*) Does there necessarily exist a recursive function f such that $f(A) = \{0\}$ and $f(B) = \{1\}$? (*Hint:* Prove Theorem 2-II (the extension theorem) for the case of functions and partial functions whose ranges are subsets of $\{0,1\}$.)

2-31. Is $\lambda x[\mu y[\varphi_x(y) \textit{ divergent}]]$ partial recursive?

△***2-32.*** Let two partial recursive functions have A and B as their respective domains. Give a necessary and sufficient condition that A and B satisfy the following: *there exists a partial recursive* ψ *such that for all* x, $[\textit{domain } \varphi_x = A \Rightarrow \psi(x) = 0]$ *and* $[\textit{domain } \varphi_x = B \Rightarrow \psi(x) = 1]$.

2-33. Consider the problem of deciding, for any x, whether or not φ_x is extendible to a recursive function. Is this problem recursively solvable?

△***2-34.*** Let A and B be sets. Consider the "problem" of deciding, for any given $x \in A$, whether or not $x \in B$. We say that this problem is *relatively solvable* if there exists a partial recursive function ψ such that $[x \in A \cap B \Rightarrow \varphi(x) = 1]$ and $[x \in A \cap \bar{B} \Rightarrow \varphi(x) = 0]$. Investigate relative solvability of the following problems:

(*a*) The problem of deciding, for any x such that φ_x is total, whether or not φ_x is a constant function. (In this case, $A = \{x|\varphi_x \textit{ is total}\}$ and $B = \{x|\varphi_x \textit{ is a constant function}\}$.)

(*b*) The problem of deciding, for any x such that φ_x is total, whether or not φ_x is *one-one.*

(*c*) The problem of deciding, for any x, y, and z such that φ_x is total, whether or not $\varphi_x(y) = z$.

(*d*) The problem of deciding, for any x such that φ_x is total, whether or not *range* φ_x is infinite.

(*Hint:* For unsolvability, make reductions from the halting problem (Corollary 1-VII) which proceed by associating with each z a total recursive function whose value for any argument y depends upon the result of performing y steps in the computation of P_z applied to z.)

Definition. If ψ is a partial function of two variables, define $\mu\psi = \lambda x[\mu y[\psi(x,y) = 0]]$.

2-35. What functions occur in $\{f|(\exists \textit{ recursive } g)[f = \mu g]\}$?

2-36. (Uspenskii [1957].) If ψ is partial recursive, must $\mu\psi$ be partial recursive?

△***2-37.*** Characterize $\{f|f \textit{ recursive } \& (\forall z)(\exists \textit{ recursive } g)[\varphi_z = f\mu g]\}$. (*Hint:* Consider functions which take each integer as value infinitely often. For related results, see Markov [1947] and Kuznecov [1950].)

△***2-38.*** Characterize $\{\psi | (\exists \textit{ recursive } g)[\psi = \mu g]\}$. (*Hint:* Consider those functions which, as relations, have a recursive characteristic function. By Theorem 2-III, this class does not include all partial recursive functions. For related results, see Skolem [1944] and Post [1946].) Exercises 2-35 to 2-38 complete the structural theory begun in Theorems 1-X and 2-III.

§*2.5*

2-39. ▲(*a*) Prove Rice's theorem, stated at the end of §2.1. (*Hint:* Show that a recursive characteristic function for any other case could be used to solve the halting problem.)

(*b*) Deduce, as an immediate consequence of this theorem, that $\{x | \textit{domain } \varphi_x \textit{ is infinite}\}$ cannot have a recursive characteristic function.

(*c*) Deduce the results of Exercises 2-17, 2-18, 2-29, and 2-33.

3 *Purposes; Summary*

§3.1 GOALS OF THEORY

The material presented in Chapters 1 and 2 concerned major features of recursive function theory as developed before 1940. Included were the Basic Result, certain more technical results, such as the enumeration theorem (Theorem 1-IV) and the *s-m-n* theorem (Theorem 1-V), and the recursive unsolvability of specific problems embodying the halting problem. Development since 1940 has been varied and extensive. For purposes of summary, the goals and results of this more recent research can be grouped into six areas: (1) *unsolvable problems;* (2) *unsolvability structures;* (3) *logic and foundations;* (4) *subrecursiveness;* (5) *recursive structure;* (6) *analogue structures.* In our brief description this grouping is somewhat artificial and ignores important interrelations.

1. Unsolvable Problems

This area was discussed in Chapter 2. It concerns results on the solvability or unsolvability of specific problems. Since 1940, increasingly refined techniques for demonstrating unsolvability have been developed, and, as a result, questions of unsolvability have been settled for increasingly broad classes of problems. A few typical results were stated in §2.2. Continuing effort has also gone into the study of solvable problems and into the identification of solvable subcases of more general problems known to be unsolvable.

2. Unsolvability Structures

This area concerns concepts and structures that arise in a more general analysis of the unsolvability phenomenon; i.e., it concerns concepts which prove useful in classifying unsolvable problems (e.g., various exact versions of the idea of *reducibility* mentioned in Chapter 2), and it concerns structures which appear as a result of such a classification. Work in this area has broad application to work in other areas. Its concepts, terminology, and results are widely used—especially in logic and foundations, where a number of major results find their most natural expression in these terms. More abstract and general kinds of classification can also be studied, and certain generalized kinds of computability can be defined and investigated. Area (2), sometimes called the area of *recursive invariance,*† includes the study of *hierarchies,* where classifications by reducibility are related to classifications by complexity of definition in certain formalized logical systems.

3. Logic and Foundations

Applications of recursive function theory to logic have been extensive and profound, yielding new results and new insights. As indicated in the historical discussion of §2.4, logic is, by its nature, concerned with algorithms. Many of the specific unsolvability results of area (1) have been obtained in logic (most often, such an unsolvability result will concern the problem of identifying the provable statements in some formalized system of deduction). A variety of syntactical and semantical notions (e.g. *axiomatizability* and *incompletableness*) can be usefully restated in recursive-function-theoretic form, and various more subtle distinctions can be drawn.‡ Results from area (2) can be used to study the relative power of various logical methods, systems, and means of expression. In modern work on foundations of mathematics, the notion of *constructivity* has also played a basic role; recursive function theory has proved useful in reformulating and analyzing various notions of constructive proof and in isolating the constructive content of known classical proofs.

4. Subrecursiveness

The study of generalized forms of recursiveness was included in area (2). Area (4) concerns concepts of computability more restrictive than recursiveness, and it concerns the way in which such concepts can be used to refine and stratify structures that arise in the study of ordinary recursiveness. Natural and significant results in this direction have been difficult to obtain. Considerable effort has been made in this area from the time of the earliest work on recursive function theory.§ (Primitive recursiveness is, of course, an example of such a more restrictive notion.)

5. Recursive Structure

This area concerns ways in which recursive structure can be imposed upon existing and familiar mathematical objects to yield a new and richer

† *Recursive invariance* will be defined in Chapter 4.

‡ *Syntax* is the study of formalized systems as pure formalisms, apart from intended meanings. *Semantics* (in logic) is the study of the relation between formalized systems and the mathematical objects (e.g., real numbers) about which the systems appear (or are intended) to speak. The distinction between syntax and semantics is intuitively useful but difficult to make fully precise.

§ A number of methods and results antedate the definition of recursiveness, and can be viewed as a part of the search for a satisfactory general definition of algorithm.

theory. The imposition of such structure is analogous to the imposition of topological structure on groups to yield the theory of topological groups.

6. Analogue Structures

Unlike area (5), where recursive function theory is used further to analyze existing mathematical objects and results, area (6) concerns the definition and study of new mathematical objects which are, in some sense, recursive analogues to more familiar (and nonrecursive) mathematical objects. The nonrecursive objects may use nondenumerable sets or have other nonconstructive features. In the recursive analogues, all basic sets are (at most) denumerable, all basic entities are given integer code numbers, and mappings are taken to be admissible only if they are partial recursive on code numbers. Most work done on analogue structures has been in the realm of set theory. Cardinal numbers, ordinal numbers, topological spaces, and real-variable analysis have all been studied in recursive-analogue form. Work on recursive analogues has been of continuing interest, both in the results and insights it affords and in the further questions it raises.

§3.2 EMPHASIS OF THIS BOOK

The chief concerns of this book will be area (2) (unsolvability structures) and (to a lesser extent) area (3) (logic and foundations). Many of the proofs will rely on the semiformal methods described and justified in Chapter 1. We shall omit area (1) (unsolvable problems) insofar as it concerns specific problems in other parts of mathematics. More formal methods necessary for treating such problems can be found in Davis [1958]. As was done in Chapter 2, we use Church's Thesis to give a semiformal treatment of unsolvable problems occurring within recursive function theory itself. Several illustrations from area (6) (analogue structures) will be presented. Some indication of work in areas (4) and (5) (subrecursiveness and recursive structure) will be given.

§3.3 SUMMARY

Chapters 1, 2, and 3 present the concept of recursiveness. The Basic Result is given, and particular instances of unsolvability and reducibility are described. The scope and usefulness of semiformal techniques are indicated.

In the remaining chapters (4 to 16), we define and consider certain general concepts that yield the examples and ideas of Chapter 2 as special cases, and we go on to treat further concepts that grow, in turn, out of these. Use and occurrence of these ideas in logic is also discussed. Chapters 4 to 16 thus fall in area (2) (unsolvability structures) with some material from area (3) (logic and foundations). Occasional examples from areas (4), (5), and (6) are given.

More specifically, Chapters 4 to 16 cover the following. Chapter 4 uses the concept of *recursive invariance* to characterize more accurately the subject matter of area (2). Chapter 5 introduces the basic concepts of recursive and recursively enumerable set. Chapters 6 to 9 define and investigate the basic reducibility concepts. Chapter 10 discusses a central problem about reducibilities left unsolved by Post (see §2.5). Chapter 11 presents the *recursion theorem.* In expositions of recursive function theory, this basic and useful tool is often presented at the beginning. The theorem is not necessary, however, for the material that precedes it below. We believe that its position in this book gives the student a better perspective and deeper understanding of its use. Chapter 11 includes a variety of applications of the recursion theorem and concludes with an introduction to ordinal notations (an analogue structure from area (6)). Chapters 12 and 13 give further results on the basic structures of Chapters 5 to 9; Chapter 12 considers the recursively enumerable sets as a lattice of sets, while Chapter 13 considers *degrees of unsolvability* (a *degree* is an equivalence class of sets under an equivalence relation of interreducibility) under partial ordering by reducibility. Chapter 14 introduces the *arithmetical hierarchy* (a classification of sets of integers according to minimum logical complexity of definition) and relates this structure to the degrees of unsolvability studied in Chapter 13. Chapter 15 extends the arithmetical hierarchy from a classification of sets of integers to a classification of sets of sets of integers. Chapter 15 also treats the notion of *recursive functional.* Chapter 16 introduces the *analytical hierarchy,* a classification similar to, but more inclusive than, the arithmetical hierarchy. Chapter 16 also considers generalizations of the notion of effective computability. Applications to logic occur throughout the book.

Bibliographical references are incomplete. We make acknowledgement when we know the correct attribution. References for more advanced study are occasionally suggested, although the choice of such references is somewhat arbitrary.

For examples of work in area (1), see Davis [1958]; Tarski, Mostowski, and Robinson [1953]; Rabin and Scott [1959]; Minsky [1961]; and Dreben [1962]; in area (4), see Grzegorczyk [1953]; Fischer [1962]; Ritchie [1963]; Axt [1959]; and Feferman [1961]; in area (5), see Rabin [1960]; in area (6), see Dekker and Myhill [1960]; Moschovakis [1963]; and Crossley [1965].

4 *Recursive Invariance*

§4.1 INVARIANCE UNDER A GROUP

The remaining chapters of this book will be primarily concerned with sets of integers and with functions and partial functions of integers. Certain properties of these objects will be of special interest. In the present chapter we isolate these properties.

To do this, we follow the approach of Felix Klein. He sought to define each "branch" of mathematics in terms of a *space* (i.e., set) and a *group of transformations* acting on that space. (A group of transformations on a space is a collection of mappings from the space onto the space such that the collection forms, algebraically, a *group* when composition of mappings is taken as rule of multiplication.) Let $\mathfrak{X}$ be a space and $\mathcal{G}$ a group. (For example $\mathfrak{X}$ might be (the set of points of) two-dimensional Euclidean space and $\mathcal{G}$ the group of all affine transformations of $\mathfrak{X}$ onto $\mathfrak{X}$.†) A property of subsets of $\mathfrak{X}$ is said to be $\mathcal{G}$-*invariant* if, whenever any subset possesses that property, so do all images of that subset under mappings in $\mathcal{G}$. (In the example just mentioned, the property of being a triangle is $\mathcal{G}$-invariant; the property of being an isosceles triangle is not.) In Klein's scheme, the branch of mathematics determined by $\mathfrak{X}$ and $\mathcal{G}$ is (by definition) the study of properties which are $\mathcal{G}$-invariant. We call this the $\mathcal{G}$-*theory*. (For $\mathfrak{X}$ the Euclidean plane and $\mathcal{G}$ the affine group, the $\mathcal{G}$-theory is known as *affine geometry of the Euclidean plane;* while for $\mathfrak{X}$ the Euclidean plane and $\mathcal{G}$ the group of Euclidean transformations,‡ the theory is ordinary Euclidean

† An *affine* transformation of the plane onto itself is a mapping which carries lines into lines, e.g., a uniform shearing of the plane. It does not, in general, preserve angles.

‡ A mapping of the plane onto itself is *Euclidean* if it can be described as a combination of translation, rotation, and reflection.

plane geometry. Concepts of affine geometry are also concepts of Euclidean geometry, since the Euclidean group is a subgroup of the affine group.)

Let $\mathfrak{X}$ and $\mathcal{G}$ be given; let A and B be subsets of $\mathfrak{X}$. Define $A \equiv_{\mathcal{G}} B$ if $B = g(A)$ for some transformation g in $\mathcal{G}$.† Since $\mathcal{G}$ is a group, it follows that $\equiv_{\mathcal{G}}$ is an equivalence relation (see Exercise 4-1). If $A \equiv_{\mathcal{G}} B$, we say that A and B are *$\mathcal{G}$-isomorphic*. The equivalence classes are called *$\mathcal{G}$-isomorphism types*. $\mathcal{G}$-invariant properties, then, are the properties (of subsets of $\mathfrak{X}$) which are well-defined with respect to $\mathcal{G}$-isomorphism; i.e., if any A has such a property, all sets $\mathcal{G}$-isomorphic to A have that property. In a sense, $\mathcal{G}$-isomorphism types are the basic objects of the $\mathcal{G}$-theory.

Let Ω be a collection of $\mathcal{G}$-invariant properties. Let $\mathfrak{I}$ be a $\mathcal{G}$-isomorphism type. If $\mathfrak{I}$ possesses all the properties in Ω, Ω is called a *set of invariants* for $\mathfrak{I}$. If, in addition, no other $\mathcal{G}$-isomorphism type possesses all the properties in Ω, Ω is called a *complete set of invariants* for $\mathfrak{I}$.

The notion of $\mathcal{G}$-invariance can be extended to relations on $\mathfrak{X}$ (i.e., subsets of $\mathfrak{X}^k$, $k > 1$) in an obvious way; we describe it below for the particular $\mathfrak{X}$ and $\mathcal{G}$ with which we work.

§4.2 RECURSIVE PERMUTATIONS

Consider the following collection $\mathcal{G}^*$ of transformations (i.e., functions) on N (the set of all nonnegative integers).

Definition $\mathcal{G}^* = \{f | f$ *is a one-one, onto recursive function*$\}$.

Theorem I *$\mathcal{G}^*$ is a transformation group on N.*

Proof. Let f and $g \in \mathcal{G}^*$. From the *onto* property of f and g, it is immediate that fg is *onto*. From the *one-one* property of f and g, it is immediate that fg is *one-one*. By Theorem 1-VI, fg is *recursive*. Hence $fg \in \mathcal{G}^*$.

Let $f \in \mathcal{G}^*$. Since f is *one-one*, f^{-1} is single-valued, i.e., a partial function. Since f is single-valued, f^{-1} is *one-one*. Since f is a (total) function, f^{-1} is *onto*. Since f is *onto*, f^{-1} is a total function. f^{-1} may be computed by the following instructions. To compute $f^{-1}(x)$: compute, in turn, $f(0)$, $f(1)$, $f(2)$, . . . until an n is found such that $f(n) = x$; give n as output. By Church's Thesis, f^{-1} is recursive. Hence $f^{-1} \in \mathcal{G}^*$.

Since $\mathcal{G}^*$ is closed under composition and inverses, it is a group.⊠

Definition f is a *recursive permutation* if $f \in \mathcal{G}^*$. We call $\mathcal{G}^*$ the *group of recursive permutations*.

† As indicated in the Introduction, $g(A)$ is an abbreviation for $\{y | (\exists x)[x \in A \;\&\; y = g(x)]\}$.

Algebraic structure of $\mathcal{G}^*$ has been studied by Kent [1962]. (Kent also considers various subrecursive subgroups of $\mathcal{G}^*$.) A few isolated facts about $\mathcal{G}^*$ are given in Exercises 4-6 and 4-7.

§4.3 RECURSIVE INVARIANCE

Definition *A recursively isomorphic to B* if $A \equiv_{\mathcal{G}^*} B$, that is,

$$(\exists f)[f \in \mathcal{G}^* \ \& \ f(A) = B].$$

We abbreviate A recursively isomorphic to B as $A \equiv B$.

We shall occasionally refer to recursive isomorphism simply as *isomorphism.*

Definition A collection of subsets of N is called a *recursive-isomorphism type* if it is a $\mathcal{G}^*$-isomorphism type.

Definition A property of subsets of N is *recursively invariant* if it is $\mathcal{G}^*$-invariant.

Almost all the concepts in this book will be recursively invariant. The notion of recursive invariance characterizes our theory and serves as a touchstone for determining possible usefulness of new concepts.†

Examples. Consider the following properties:

(i) *Contains* 7.

(ii) *Contains at least two members.*

(iii) *Possesses a recursive characteristic function.*

Property (i) is not recursively invariant. Properties (ii) and (iii) are recursively invariant (see Exercise 4-9).

The notion of recursive invariance is extended to relations, and hence to partial functions and functions, in the following way. Let $R \subset N^k$ for some k. Let $g \in \mathcal{G}^*$. Define $g(R) = \{<y_1, \ldots, y_k> | (\exists x_1) \cdots (\exists x_k) [<x_1, \ldots, x_k> \in R \text{ and } y_1 = g(x_1), y_2 = g(x_2), \ldots, y_k = g(x_k)]\}$.

Definition Let $R \subset N^k$ and $Q \subset N^k$ for some k. R is *recursively isomorphic to* Q if $(\exists g)[g \in \mathcal{G}^* \ \& \ Q = g(R)]$. We abbreviate this $R \equiv Q$.

Definition A property of k-ary relations on N is *recursively invariant* if, whenever a relation R possesses the property, so does $g(R)$ for all $g \in \mathcal{G}^*$.

The case of partial functions (and hence of functions) is especially interesting. Theorem II gives a special formulation of recursive isomorphism for this case.

† Concepts used in studies of *subrecursiveness* (area (4) of §3.1) may not be recursively invariant. Appropriate notions of invariance for such studies can be introduced via subgroups of $\mathcal{G}^*$.

Theorem II $\varphi \equiv \psi \Leftrightarrow (\exists f)[f \in \mathcal{G}^* \;\&\; \varphi = f^{-1}\psi f]$.

Proof. The proof is illustrated by the diagram:

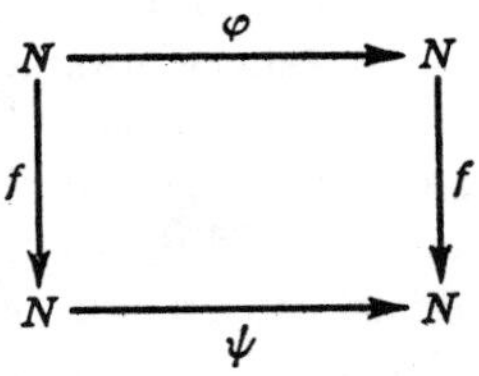

If $\varphi \equiv \psi$, then $\psi = f(\varphi)$† for some $f \in \mathcal{G}^*$, and

$$\begin{aligned} y = \varphi(x) &\Leftrightarrow f(y) = \psi f(x) \\ &\Leftrightarrow y = f^{-1}\psi f(x), \end{aligned}$$

and hence $\varphi = f^{-1}\psi f$. Conversely, if $\varphi = f^{-1}\psi f$, then

$$\begin{aligned} y = \varphi(x) &\Leftrightarrow y = f^{-1}\psi f(x) \\ &\Leftrightarrow f(y) = \psi f(x), \end{aligned}$$

and hence $\psi = f(\varphi)$. ☒

§4.4 RESEMBLANCE‡

Recursive isomorphism indicates similarity in structure. Indeed, for purposes of recursive function theory, it indicates identity of structure. In the case of partial functions, a somewhat weaker notion of recursive similarity can be defined.

Definition *φ resembles ψ* if there exist recursive permutations f and g such that $\psi = f^{-1}\varphi g$.

It is easy to see that resemblance is an equivalence relation (see Exercise 4-15).§

We refer to the equivalence classes as *resemblance types*. Examples of resemblance are given in §4.5 and in the exercises. Isomorphism clearly implies resemblance (see Exercises 4-14, 4-17, and 4-18).

§4.5 UNIVERSAL PARTIAL FUNCTIONS

To illustrate the above notions, we ask whether or not the concept of *universal partial function* is recursively invariant; §1.8 suggests the following definition of universal partial function.

† *Warning:* $f(\varphi)$ is in sense of $f(R)$; it does not indicate the composition $f\varphi$.

‡ Occasional exercises will refer to §§4.4 and 4.5, but no later results will depend on §§4.4 and 4.5.

§ For linear transformations on vector spaces, the words *equivalent* and *similar* have a traditional usage that is analogous to our use of *resemblance* and *isomorphism*, respectively.

Tentative Definition A universal partial function is a partial recursive function φ of two variables such that $(\forall x)[\varphi_x = \lambda y[\psi(x,y)]]$.

This concept, as a property of 3-ary relations, is obviously not recursively invariant (see Exercise 4-19). We therefore make a new definition in which, for simplicity, we replace the partial function of two variables by a partial function of one variable, and into which, hopefully, we put enough generality to yield recursive invariance. According to this new definition, ψ is universal if there is some way of coding ordered pairs into integers such that ψ applied to the code number for $\langle x,y \rangle$ yields the same result as φ_x applied to y. More formally, we have the following.

Definition ψ is *universal* if ψ is partial recursive and if there exists a recursive f of two variables such that $(\forall x)[\varphi_x = \lambda y[\psi f(x,y)]]$.

It is not immediate that this new concept is recursively invariant, since the definition involves our fixed indexing of the partial recursive functions. Theorem III, however, shows that it is not only recursively invariant but also invariant under resemblance.†

Theorem III *If ψ is universal and g and h are recursive permutations, then $\eta = g^{-1}\psi h$ is universal; i.e., the property of being universal is invariant under resemblance.*

Proof. By definition of *universal*, there is a recursive function f of two variables such that $\varphi_x = \lambda y[\psi f(x,y)]$.

For any x, $g\varphi_x$ is partial recursive. In fact, we have a recursive function k such that

$$\varphi_{k(x)} = g\varphi_x.$$

(Tacit use of *s-m-n* theorem.)

Define $\hat{f} = \lambda xy[h^{-1}f(k(x),y)]$. Then

$$\begin{aligned}
\lambda y[\eta\hat{f}(x,y)] &= \lambda y[g^{-1}\psi h h^{-1} f(k(x),y)], && \text{by definition of } \eta \text{ and } \hat{f},\\
&= \lambda y[g^{-1}\psi f(k(x),y)]\\
&= g^{-1}\varphi_{k(x)}, && \text{by definition of } f,\\
&= g^{-1}g\varphi_x, && \text{by definition of } k,\\
&= \varphi_x.
\end{aligned}$$

Hence η is universal.⊠

Corollary III(a) *If ψ is universal and g is a recursive permutation, then $g^{-1}\psi g$ is universal; i.e., the property of being universal is recursively invariant.*

† The definition of *universal partial function* is given, in terms of machines, by Davis [1957]. Davis had earlier [1956] defined (in effect) a "universal partial function" to be any partial recursive function to whose individual halting problem the usual halting problem (of Theorem 1-VII) can be reduced. The concept of Davis [1956] is different, from that of Davis [1957] (ours). The partial function of Exercise 4-20, for example, is universal in the former sense but not in the latter (see Exercise 2-23).

The theorem gives, incidentally, the result that under a recursive permutation of its range, a universal partial function remains universal. Thus we have Corollary III(*b*).

Corollary III(b) *If ψ is universal and g is a recursive permutation, then $g\psi$ is universal.*

Do the universal partial functions constitute a single resemblance type? In Chapter 11 we shall show, as the result of deeper work, that the answer to this question is "yes." In fact, we shall prove the stronger result, due to M. Blum, that the universal functions constitute a single isomorphism type. The property of being universal is thus, by itself, a complete set of recursive invariants.

What resemblance types are also isomorphism types? A final answer to this question has not been given. The only cases presently known are the type of the empty partial function, the type of all constant functions, and the type of all universal partial functions.†

§4.6 EXERCISES

§4.1

4-1. (*a*) Let $\mathcal{G}$ be a transformation group on $\mathfrak{X}$ and let e be its identity. Show that e, as a mapping on $\mathfrak{X}$, must be the identity mapping.

(*b*) Show that $\mathcal{G}$-isomorphism is an equivalence relation.

§4.2

4-2. Show that there exist *one-one, onto* functions ("permutations") which are not recursive.

△***4-3.*** If ψ is *one-one* and partial recursive, but not total, must ψ^{-1} be partial recursive?

4-4. Is it true that there exists a recursive f such that $(\forall x)[\varphi_x \in \mathcal{G}^* \Rightarrow [\varphi_{f(x)} \in \mathcal{G}^* \ \& \ \varphi_{f(x)} = \varphi_x{}^{-1}]]$? If true, give a proof similar to the proof of Theorem 1-VI.

4-5. Show that $\{x|\varphi_x \in \mathcal{G}^*\}$ does not have a recursive characteristic function. Give a direct reduction proof that does depend on Exercise 2-39.

4-6. (*a*) Show by a diagonal argument that there is no effective procedure by which we can list a set of Gödel numbers for recursive permutations such that every recursive permutation has at least one Gödel number in the set.

(*b*) Hence conclude that the group of recursive permutations is not finitely generated.

4-7. Show that there exists a recursive permutation g and a permutation h such that $h^{-1}gh$ is not recursive. Hence conclude that $\mathcal{G}^*$ is not a normal subgroup of the group of all permutations. (*Hint:* Take f a permutation that is not recursive. Define $h(x) = x$ if x odd; $h(x) = 2f(\frac{1}{2}x)$ if x even. Define $g(x) = x + 1$ if x even; $g(x) = x - 1$ if x odd.)

▲***4-8.*** Show that $\{f|f$ *is a primitive recursive permutation*$\}$ is not a group. (*Hint:* Use methods from the hint for Exercise 2-16.)

† A proof that these are the only cases possible could lead to an interesting algebraic axiomatization of the partial recursive functions.

§4.3

4-9. In the examples given in §4.3, show that properties (ii) and (iii) are recursively invariant, while property (i) is not.

4-10. Consider the following properties of sets:
- (i) Is empty;
- (ii) Is range of recursive function;
- (iii) Is domain of partial recursive function;
- (iv) Possesses recursive characteristic function;
- (v) Is infinite;
- (vi) Contains the even integers;
- (vii) Contains a set isomorphic to the even integers.

Which are recursively invariant?

$\triangle$***4-11.*** Do (v) and (vii) in 4-10 coincide as properties of sets? (*Hint:* Clearly (vii) $\Rightarrow$ (v). If (v) $\Rightarrow$ (vii), show how to get the appropriate isomorphism to the even integers. If (v) $\not\Rightarrow$ (vii), use a diagonal argument to get an appropriate counterexample.)

4-12. Consider the following properties of partial functions:
- (i) Has infinite domain;
- (ii) Has domain containing range;
- (iii) Is (total) recur ive;
- (iv) Is partial recursive;
- (v) Has domain containing the even integers;
- (vi) Has domain containing a set isomorphic to the even integers;
- (vii) Has domain with seven members.

Which are recursively invariant?

$\triangle$***4-13.*** Show that the intersection of (i) and (iv) in 4-12 coincides with the intersection of (vi) and (iv).

4-14. Do the recursive permutations form a single isomorphism type?

§4.4

4-15. Show that resemblance is an equivalence relation.

4-16. Which of the properties of 4-12 are invariant under resemblance?

4-17. Give examples of (*a*) a total function whose resemblance and isomorphism types differ; (*b*) a total function whose resemblance and isomorphism types coincide.

4-18. Do the recursive permutations form a single resemblance type?

§4.5

4-19. Show that the "tentative definition" in §4.5 is not recursively invariant. (*Hint:* Take $\varphi_{x_1} = \lambda x[x]$ and $\varphi_{x_2} = \lambda x[x + 1]$; use a recursive permutation f such that $f(x_1) = x_2$.)

4-20. Define the partial recursive function $\iota = \{<x,x>|\varphi_x(x)$ *convergent*$\}$. Show that ι is not universal. (*Hint:* Consider ranges.)

4-21. Show that no universal partial function can be extended to a recursive function. (*Hint:* Use Theorem 2-II.)

$\blacktriangle$***4-22.*** (*a*) Show that the universal partial functions form a single resemblance type.

(*b*) (Blum). Show that the universal partial functions form a single isomorphism type.

4-23. Let ψ be any universal function and let f be a recursive function of two variables mapping $N \times N$ onto N. Must the map which associates with each x the partial recursive function $\lambda y[\psi f(x,y)]$ be a *numbering* in the sense of Exercise 2-10?

5 Recursive and Recursively Enumerable Sets

§5.1 DEFINITIONS

We wish to study recursively invariant properties of sets of integers, in particular those properties which have to do with solvability and unsolvability. Most basic is the property of possessing a recursive characteristic function.

Definition A set is *recursive* if it possesses a recursive characteristic function. (That is to say, A is recursive if and only if there exists a recursive function f such that for all x, $x \in A \Rightarrow f(x) = 1$, and $x \in \bar{A} \Rightarrow f(x) = 0$.)

Intuitively, A is recursive if there exists an effective procedure for deciding, given any x, whether or not $x \in A$.

Examples. The following are all recursive sets:

(i) The set $\{0,2,4, \ldots\}$ of even integers;

(ii) N and $\emptyset$;

(iii) Any finite set;

(iv) Any set with finite complement.

Sets (iii) and (iv) are recursive since an explicit listing of the appropriate finite set can be used to give instructions for the characteristic function.

There are $\aleph_0$ recursive sets. By cardinality, nonrecursive sets must exist (since there are $2^{\aleph_0}$ subsets of N in all). Particular examples of nonrecursive sets were given in Chapters 1 and 2. Note, by trivial application of Church's Thesis, that A recursive $\Rightarrow \bar{A}$ recursive.

Closely related to recursiveness is a second and somewhat different property.

Definition A is *recursively enumerable* if either $A = \emptyset$ or there exists a recursive f such that $A = range\, f$.

Intuitively, a set is recursively enumerable if there exists an effective procedure for listing the members of the set (with repetition allowed). The properties of recursiveness and recursive enumerability are fundamental for our future work. We begin their study in the present chapter. We observe an immediate connection between recursiveness and recursive enumerability in the following theorems.

Theorem I *A recursive $\Rightarrow$ A recursively enumerable.*

Proof. Case (i): A is $\emptyset$. Then A is recursively enumerable by definition. Case (ii): A is finite and $\neq \emptyset$. Let $A = \{n_0, n_1, \ldots, n_k\}$. Define f by

$$f(x) = \begin{cases} n_x, & \text{for } x \leq k; \\ n_k, & \text{for } k < x. \end{cases}$$

Case (iii): A is infinite. Let g be its characteristic function. Define f by

$$f(0) = \mu y[g(y) = 1],$$
$$f(x+1) = \mu y[g(y) = 1 \text{ and } f(x) < y].$$

In cases (ii) and (iii), f is a recursive function by Church's Thesis, and $A = range\, f$.☒

This proof is nonconstructive in the sense that, given (an index of) a characteristic function for a set A, we may not know which case applies (compare Exercise 2-34(d)).

Theorem II *A recursive $\Leftrightarrow$ A and $\bar{A}$ are both recursively enumerable.*

Proof. $\Rightarrow$: Since A recursive $\Rightarrow$ $\bar{A}$ recursive, this is immediate from Theorem I.

$\Leftarrow$: (This appeared above as Exercise 2-27.) If either A or $\bar{A}$ is $\emptyset$, A is immediately recursive. If neither A nor $\bar{A}$ is $\emptyset$, then $A = range\, f$ and $\bar{A} = range\, g$ for certain recursive functions f and g. f and g give an effective procedure for testing membership in A; namely, to test a given x, examine, in turn, $f(0)$, $g(0)$, $f(1)$, $g(1)$, $f(2)$, ... If x appears as a value of f, then $x \in A$. If x appears as a value of g, then $x \in \bar{A}$. x must appear as a value of f or of g, since $A \cup \bar{A} = N$. This procedure can be more intuitively described as follows: generate lists for A and $\bar{A}$ simultaneously; at the same time carry out a *zigzag* hunt for x down the two lists. x must eventually appear. The list in which x appears determines whether x is in A or $\bar{A}$.☒

We shall see in §5.2 that the converse of Theorem I fails, and hence, by Theorem II, that there exist recursively enumerable sets whose complements are not recursively enumerable.

There are $\aleph_0$ recursively enumerable sets. By cardinality, sets that are not recursively enumerable must exist. $\{x \mid \varphi_x \textit{ total}\}$ is such a set. (Exercise

2-9, a corollary to Theorem 1-VIII, shows that $\{x|\varphi_x \textit{ total}\}$ is not recursively enumerable.)

The following special kinds of recursive enumerability yield partial converses to Theorem I.

Definition A is *recursively enumerable in nondecreasing order* if there exists a (total) recursive function f such that $A = \textit{range}\, f$ and f is a nondecreasing function $((\forall x)(\forall y)[x < y \Rightarrow f(x) \leq f(y)])$.

Definition A is *recursively enumerable in increasing order* if there exists a (total) recursive function f such that $A = \textit{range}\, f$ and f is an increasing function $((\forall x)(\forall y)[x < y \Rightarrow f(x) < f(y)])$.

Theorem III(a) *[A-recursive & $A \neq \emptyset$] $\Leftrightarrow$ A recursively enumerable in nondecreasing order.*

(***b***) *[A-recursive & A-infinite] $\Leftrightarrow$ A recursively enumerable in increasing order.*

Proof. $\Rightarrow$ for both (*a*) and (*b*): This is immediate by the functions constructed in cases (ii) and (iii) of the proof of Theorem I.

$\Leftarrow$ for (*a*): Consider the following two cases.

Case (i): A finite. Then A is recursive, since any finite set is recursive.

Case (ii): A infinite. Let f enumerate A in nondecreasing order. To test for x in A, generate the range of f until a number larger than x appears. Then $x \in A \Leftrightarrow x$ has already appeared in the list. Hence we have an effective test for membership in A, and A is recursive.

$\Leftarrow$ for (*b*): Use an argument similar to case (ii) in the proof for (*a*). (In this case, only $f(0), f(1), \ldots, f(x)$ need be generated.) ☒

Note that the proof of $\Leftarrow$ for (*a*) is nonconstructive in that we may not know whether case (i) or case (ii) applies.

Theorem III, though not profound, is often helpful. We give one application in Theorem IV and others in Exercises 5-6 and 5-7.

Theorem IV *Every infinite recursively enumerable set has an infinite recursive subset.*

Proof. Let A be infinite and recursively enumerable. Let f be recursive with $\textit{range}\, f = A$. Define a recursive function g by

$$g(0) = f(0),$$
$$g(x + 1) = f(\mu y[f(y) > g(x)]).$$

Let $B = \textit{range}\; g$. Then g enumerates B in increasing order. Hence, by Theorem III, B is infinite and recursive. Since $B \subset A$, the result is proved. ☒

The reader can easily verify that the concepts introduced in §5.1 are recursively invariant.

§5.2 BASIC THEOREM

The following theorem provides an alternative characterization of recursive enumerability. It is a simple but fundamental result.

***Theorem V* (*basic theorem on recursively enumerable sets*)** *A is recursively enumerable $\Leftrightarrow$ A is the domain of a partial recursive function (that is, $(\exists x)[A = domain\ \varphi_x]$).*

Proof. $\Rightarrow$: Case (i): $A = \emptyset$. Let ψ be the everywhere divergent partial recursive function. Then $A = domain\ \psi$.

Case (ii): $A \neq \emptyset$. Then $A = range\, f$. Define ψ by the instructions: to compute $\psi(x)$, generate *range* f; if and when x appears in *range* f, give output x. ψ is clearly partial recursive, and $A = domain\ \psi$.

$\Leftarrow$: Let $A = domain\ \psi$, with ψ partial recursive. We define an effective procedure that will list A if A is not $\emptyset$. The procedure is carried out in stages.

Stage 1. Compute 1 step in the computation of $\psi(0)$. If $\psi(0)$ converges in 1 step, place 0 in the list for A.

. . . .

Stage $n + 1$. Compute $n + 1$ steps in each of the $n + 1$ computations for $\psi(0), \psi(1), \ldots, \psi(n)$. For each of these which converges on or before its $(n + 1)$st step, add its *input* to the list for A.

. . . .

Now define η as follows:

$$\eta(0) = \text{first member of list;}$$
$$\eta(x+1) = \begin{cases} \mu y[y \textit{ has been added to the list in stage } x + 1 \textit{ and} \\ \quad y \notin \{\eta(0), \eta(1), \ldots, \eta(x)\}], & \text{if such a } y \text{ exists;} \\ \eta(0), & \text{otherwise.} \end{cases}$$

By Church's Thesis, η is a partial recursive function. If $A = \emptyset$, A is recursively enumerable by definition. If $A \neq \emptyset$, then, by the construction, η is a total function and $range\ \eta = domain\ \psi = A$; thus A is recursively enumerable. ☒

The above proof proceeded by "dovetailing" the computations for $\psi(0)$, $\psi(1), \ldots$. This is a common and useful construction in our theory. Later mention of this construction will sometimes be avoided by appeal to Theorem V.

We now introduce, as *names* for the recursively enumerable sets, the indices of corresponding partial recursive functions.

Definition $W_x = domain\ \varphi_x$.

x is called a *recursively enumerable index* or *Gödel number* for the recursively enumerable set W_x. Note, from Theorem 1-III, that each recursively enumerable set will have infinitely many recursively enumerable indices. We shall often refer to recursively enumerable indices as *r.e. indices*, or, more simply, as *indices*.

If we keep in mind the dovetailing procedure of Theorem V, an r.e. index for a set may be thought of, intuitively, as a name for a procedure which never terminates and which yields, intermittently (possibly only finitely often), a succession of output values—these output values constituting the set.

The methods of Theorem V yield the following corollary.

Corollary V(a) *A is recursively enumerable* $\Leftrightarrow$ *A is the range of a partial recursive function, (that is,* $(\exists x)[A = range\ \varphi_x]$).

Proof of corollary. $\Rightarrow$: Case (i). As before.

Case (ii). The ψ constructed in the proof of the theorem will serve here, for $A = domain\ \psi = range\ \psi$.

$\Leftarrow$: As before, except that, at the successive stages, *outputs* rather than *inputs* are added to the list being made.☒

It now follows that A is the domain of a partial recursive function if and only if A is the range of a partial recursive function. The following corollary expresses this fact in a stronger form.

Corollary V(b) *There exist recursive functions f and g such that for all x,*

$$range\ \varphi_{f(x)} = domain\ \varphi_x,$$

and

$$domain\ \varphi_{g(x)} = range\ \varphi_x.$$

Proof of corollary. For f. Given x, define

$$\psi(y) = \begin{cases} y, & \text{if } \varphi_x(y) \text{ converges;} \\ \text{divergent}, & \text{otherwise.} \end{cases}$$

Clearly $range\ \psi = domain\ \varphi_x$. Since instructions for ψ depend on x in a uniform effective way, $\psi = \varphi_{f(x)}$ for some recursive f. (Tacit use of *s-m-n* theorem.)

For g. Given x, define a listing procedure like that in the proof of Theorem V; except that *outputs* rather than *inputs* are listed, and φ_x is used in place of ψ. Define

$$\theta(y) = \begin{cases} y, & \text{if } y \text{ appears in list;} \\ \text{divergent}, & \text{otherwise.} \end{cases}$$

Clearly $domain\ \theta = range\ \varphi_x$. Since instructions for θ depend uniformly effectively on x, $\theta = \varphi_{g(x)}$ for some recursive g.☒

Still a third corollary is evident. It asserts that, for *nonempty* sets, one can go, in a uniform effective way, from domain or range instructions to instructions for listing by a total function.

Corollary V(c) *There exist recursive functions f' and g' such that:*
(i) *range* $\varphi_{f'(x)}$ = *domain* φ_x *and* [*domain* $\varphi_x \neq \emptyset \Rightarrow \varphi_{f'(x)}$ *is total*];
(ii) *range* $\varphi_{g'(x)}$ = *range* φ_x *and* [*range* $\varphi_x \neq \emptyset \Rightarrow \varphi_{g'(x)}$ *is total*].

Proof of corollary. For (i), take the partial recursive η constructed in Theorem V, with φ_x put in place of ψ. Then an index for η depends uniformly effectively on x; that is, $\eta = \varphi_{f'(x)}$ for some recursive f'.

For (ii), set $g' = f'g$, where g is as in Corollary V(b).☒

Finally, a fourth corollary asserts that one can go uniformly from domain or range instructions to instructions for listing without repetition.

Definition A is an *initial segment of* N if

$$(\forall x)(\forall y)[[y \in A \ \& \ x < y] \Rightarrow x \in A]$$

Corollary V(d) *There exist recursive functions f'' and g'' such that:*
(i) *range* $\varphi_{f''(x)}$ = *domain* φ_x, $\varphi_{f''(x)}$ *is one-one, and domain* $\varphi_{f''(x)}$ *is an initial segment of* N;
(ii) *range* $\varphi_{g''(x)}$ = *range* φ_x, $\varphi_{g''(x)}$ *is one-one, and domain* $\varphi_{g''(x)}$ *is an initial segment of* N.

Proof of corollary. The proof is similar to that for Corollary V(c) (see Exercise 5-11).☒

Theorem V is a powerful tool for showing recursive enumerability. By its use, many of the sets considered in Chapter 2 can be shown to be recursively enumerable. We give one such example here, $\{x|\varphi_x(x)$ *convergent*$\}$. We first introduce a special name for this set.

Definition $K = \{x|\varphi_x(x)$ *convergent*$\} = \{x|x \in W_x\}$. (We shall henceforth reserve the letter K to denote this set.)

Theorem VI *There exists a recursively enumerable but not recursive set, and K is such a set.*

Proof. Define

$$\psi(x) = \begin{cases} 1, & \text{if } \varphi_x(x) \text{ convergent;} \\ \text{divergent,} & \text{if } \varphi_x(x) \text{ divergent.} \end{cases}$$

ψ is evidently partial recursive, and K = *domain* ψ. Hence, by Theorem V, K is recursively enumerable.

We know that K is not recursive by Corollary 1-VII. However, we are now in a position to give a much shorter proof of this fact. Assume K

recursive. By Theorem II, $\bar{K} = W_m$ for some m. Then

$$m \in K \Leftrightarrow m \in W_m, \qquad \text{by the definition of } K;$$

but

$$m \in \bar{K} \Leftrightarrow m \in W_m, \qquad \text{by the choice of } m.$$

This is a contradiction, and hence K cannot be recursive. ☒†

Thus, while K is recursively enumerable, $\bar{K}$ is not recursively enumerable. Furthermore, by Theorem III, K itself is not recursively enumerable in increasing order.

The set K will be important in our later work. It embodies and abbreviates the halting problem. We shall use it in the construction of various counterexamples. We shall encounter a number of other, different sets that are recursively enumerable but not recursive. Various further basic properties of recursively enumerable sets and of recursive sets are set forth in theorems and exercises below.

We conclude with one further illustrative application of the basic theorem.

Theorem VII *If ψ is partial recursive and A is recursively enumerable then $\psi^{-1}(A)$ is recursively enumerable.*

Proof. To verify results of this kind in practice, it is often best to rely on intuition and to describe procedures in rough terms. For example, in this case: "List the members of A; at the same time list, by dovetailing, computations for ψ for all possible inputs; find those inputs which yield outputs in A." For a more formal proof, however, the basic theorem can be helpful. In this case, use the basic theorem to set A = *domain* φ_n for some n. Then $\psi^{-1}(A)$ = *domain* $\varphi_n\psi$, and, again by the basic theorem, $\psi^{-1}(A)$ is recursively enumerable. ☒

§5.3 RECURSIVE AND RECURSIVELY ENUMERABLE RELATIONS; CODING OF k-TUPLES

The concepts of recursiveness and recursive enumerability could be extended to k-ary relations by saying that a relation is recursive if it possesses a recursive characteristic function and by saying that a relation is recursively enumerable if it is the domain of a partial recursive function.‡ We do not extend the concepts in this way, however. We take a different (though equivalent) route which allows us to extend (to relations) any recursively invariant property of sets. In particular, it allows us to extend the "listing" definition of recursive enumerability used in §5.1.

We first introduce a particular coding from ordered pairs of integers (to

† The reader will note the similarity to the classical proof of Cantor's theorem (which asserts that the cardinality of any set is smaller than the cardinality of the collection of all its subsets).

‡ For a k-ary relation, the function or partial function would be of k variables.

integers). We then use this to define a particular coding from k-tuples of integers (to integers) for each k. These codings will remain fixed and will be used in later work.

Definition $\tau(x,y) = \frac{1}{2}(x^2 + 2xy + y^2 + 3x + y)$.

Lemma *τ is a recursive one-one mapping of $N \times N$ onto N.*

Proof. τ is evidently recursive. To the integers 0, 1, 2, 3, 4, 5, 6, 7, . . . let correspond the ordered pairs $<0,0>$, $<0,1>$, $<1,0>$, $<0,2>$, $<1,1>$, $<2,0>$, $<0,3>$, $<1,2>$, ... The reader can verify, with the help of the familiar identity $0 + 1 + 2 + \cdots + k = \frac{1}{2}k(k + 1)$, that this correspondence is given by τ.☒

Although a particular τ has been selected and defined, its specific form will not be important. The *only* properties that matter are that τ maps $N \times N$ *one-one onto* N and that τ is *recursive*. There is an infinite family of functions that can each serve in place of our particular τ. Such functions are called *pairing functions.*†

Definition π_1 and π_2 are the functions of one variable which yield the inverse mapping to τ. That is to say, for all z, $\tau(\pi_1(z),\pi_2(z)) = z$. π_1 and π_2 are evidently recursive.

Definition We shall use $<x,y>$ as an abbreviation for $\tau(x,y)$. We define $A \times B$ to be $\{<x,y> | x \in A \text{ and } y \in B\}$, that is, $A \times B = \tau(A \times B)$. $\times$, then, is an operation taking sets into sets.

For any $k > 0$, a coding τ^k mapping N^k *onto* N is defined from τ by the following inductive definition.

Definition

$$\tau^1 = \lambda x[x],$$
$$\tau^{k+1} = \lambda x_1 \cdots x_{k+1}[\tau(\tau^k(x_1, \cdots ,x_k),x_{k+1})].$$

(In particular, $\tau^2 = \tau$.)

$\pi_1{}^k, \ldots, \pi_k{}^k$ are the inverse functions for τ^k. That is to say, for all z, $\tau^k(\pi_1{}^k(z), \ldots ,\pi_k{}^k(z)) = z$. (In particular, $\pi_1{}^2 = \pi_1$ and $\pi_2{}^2 = \pi_2$.)

These functions are all, evidently, recursive. We abbreviate $\tau^k(x_1, \ldots ,x_k)$ as $<x_1, \ldots ,x_k>$.

Example 1. $<2,1,0> = <<2,1>,0> = <8,0> = 44$.

Example 2. $\tau^4(A^4) = ((A \times A) \times A) \times A$.

When the $\times$ notation is used without parentheses, we shall assume association to the left. For example, $A \times B \times C$ will be $(A \times B) \times C$.

By means of τ^k, we can go from any k-ary relation, and in particular, from any function or partial function, to a corresponding "encoding" set of integers. Any property of sets is extended to be a property of k-ary relations

† This phrase is also occasionally used for functions which map $N \times N$ one-one *into* N and have recursive range.

by stipulating that a relation has the given property if and only if its encoding, as a set, has that property.

Convention *Let P be a property of sets. A k-ary relation R will be said to have property P if* $\tau^k(R)$ *has property P.*

Accordingly, the notions of recursiveness and recursive enumerability, as defined in §5.1, are now extended to relations. Various basic results also extend. For example, from Theorem II, we have that R recursive $\Leftrightarrow$ R and $\bar{R}$ are recursively enumerable (here $\bar{R} = N^k - R$).

Are these extended definitions of recursiveness and recursive enumerability equivalent to the definitions suggested in the first sentence of §5.3? They are immediately equivalent by Theorem V (the basic theorem) together with the following theorem.

Theorem VIII(a) *Let* φ *be a partial recursive function of one variable, then* $\psi^{(k)} = \lambda x_1 \cdots x_k[\varphi(<x_1, \ldots, x_k>)]$ *is a partial recursive function of k variables.*

(*b*) *Let* $\psi^{(k)}$ *be a partial recursive function of k variables, then*

$$\varphi = \lambda z[\psi^{(k)}(\pi_1{}^k(z), \ldots, \pi_k{}^k(z))]$$

is a partial recursive function of one variable.

Proof. Immediate by Church's Thesis. Note that this associates total functions with total functions and that the association in (*a*) is inverse to the association in (*b*).☒†

To illustrate the extended definitions of recursiveness and recursive enumerability, we note the following. Proof is left for Exercise 5-17.

Theorem IX(a) *Let f be a function; then f is a recursive function* $\Leftrightarrow$ *f, as a binary relation, is recursive* $\Leftrightarrow$ *f, as a binary relation, is recursively enumerable.*

(*b*). *Let* ψ *be a partial function; then* ψ *is a partial recursive function* $\Leftrightarrow$ ψ*, as a binary relation, is recursively enumerable.*

Of course, in (*b*), ψ may be recursively enumerable without being recursive; $\psi = \{<x,x>|x \in K\}$ is one example. (This is a reason for calling the basic objects of our theory the "partial recursive functions" rather than the "recursive partial functions.") In other parts of the theory (as well as here), the phrase "partial recursive" is often associated with recursive enumerability (see, for example, §9.2).

Comment on recursive invariance. Via the τ^k codings, relations can (for, many purposes) be identified with sets. The particular mappings τ^k are not in themselves, recursively invariant objects, nor is the relationship $A = \tau^k(R)$ between set A and relation R a recursively invariant relationship. We can

† Theorem VIII shows that *dimension* will not play a role in the theory of recursive functions analogous to the role which it plays in the theory of continuous functions of real variables (where a counterpart to Theorem VIII fails).

show, however, that any recursively invariant *property* of sets, when extended according to the convention above, becomes a recursively invariant *property* of relations; i.e., we can show that $R \equiv Q \Rightarrow \tau^k(R) \equiv \tau^k(Q)$ for k-ary relations R and Q (see Exercise 5-19). But we must be warned: the converse does not hold. A property that is recursively invariant for k-ary relations does not always become a recursively invariant property for the corresponding sets; that is, $\tau^k(R) \equiv \tau^k(Q)$ does not in general imply that $R \equiv Q$ for k-ary R and Q. An example of this failure is suggested in Exercise 5-20.

§5.4 PROJECTION THEOREMS

In §5.1 we defined recursive enumerability. In §5.2 (the basic theorem) we characterized it in a second way. In the present section we give still a third characterization—in terms of projections of recursive relations.

We use the following abbreviation, which is common in metamathematical work: we abbreviate $<x_1, \ldots, x_k> \in R$ as $R(x_1, \ldots, x_k)$.

Definition The set

$$\{<x_1, \ldots, x_{j-1}, x_{j+1}, \ldots, x_k> | (\exists x_j) R(x_1, \ldots, x_k)\}$$

is called the *projection of R along the* jth *coordinate.* (The geometrical significance of the phrase is evident.)

Theorem X *If R is recursively enumerable, then there is a recursive relation S such that R is a projection of S. Specifically, if R is k-ary and recursively enumerable, then there exists a* $(k + 1)$*-ary recursive S such that* $R = \{<x_1, \ldots, x_k> | (\exists x_{k+1}) S(x_1, \ldots, x_{k+1})\}$.

Theorem XI *If R is recursively enumerable, then any projection of R is recursively enumerable. In particular, if R is k-ary and recursively enumerable, then* $\{<x_1, \ldots, x_{k-1}> | (\exists x_k) R(x_1, \ldots, x_k)\}$ *is recursively enumerable.*

The following corollary is immediate.

Corollary XI *R is recursively enumerable* $\Leftrightarrow$ *R is a projection of some recursive relation S.*

We refer to these theorems as the *projection theorems* or as the *existential-quantifier theorems* (in view of the role of the $\exists$ symbol in defining projection). The theorems are often useful in demonstrating recursive enumerability. Before proving the theorems, we give an illustrative application.

Example. Consider $A = \{x | 7 \in \mathit{range}\ \varphi_x\}$. To show that A is recursively enumerable, we express A as a projection.

$$A = \{x | (\exists y)[\varphi_x(y) = 7]\}$$
$$= \{x | (\exists y)(\exists z)[P_x \textit{ with input } y \textit{ yields output } 7 \textit{ in fewer than } z \textit{ steps}]\};$$

That is, $A = \{x | (\exists y)(\exists z) S(x,y,z)\}$,

where S is a recursive relation. By two applications of Theorem XI, A is recursively enumerable.

Proof of Theorem X. Let $\hat{R} = \tau^k(R)$. Then $\hat{R}$ is a recursively enumerable set. By Theorem V, there is a partial recursive function ψ such that $\hat{R} = domain\ \psi$. Let m be a fixed index for ψ and P_m a corresponding set of instructions for ψ. Then $R(x_1, \ldots, x_k) \Leftrightarrow <x_1, \ldots, x_k> \in \hat{R} \Leftrightarrow \psi(<x_1, \ldots, x_k>)$ *is convergent* $\Leftrightarrow (\exists x_{k+1})$ [P_m *with input* $<x_1, \ldots, x_k>$ *converges in fewer than* x_{k+1} *steps*]. By Church's Thesis, the final expression in brackets defines a recursive relation on N^{k+1}, and the theorem is proved.☒

Proof of Theorem XI. It is sufficient to prove this for projection of R along the kth coordinate. Let R be given recursively enumerable. By Theorem X, we can find a recursive S such that $R(x_1, \ldots, x_k) \Leftrightarrow (\exists x_{k+1})S(x_1, \ldots, x_{k+1})$. Let Q be the given projection of R. Then $Q(x_1, \ldots, x_{k-1}) \Leftrightarrow (\exists x_k)R(x_1, \ldots, x_k) \Leftrightarrow (\exists x_k)(\exists x_{k+1})S(x_1, \ldots, x_{k+1})$. Now define a partial recursive function ψ as follows. To compute $\psi(<x_1, \ldots, x_{k-1}>)$, test $S(x_1, \ldots, x_{k-1}, \pi_1(z), \pi_2(z))$ for $z = 0, 1, 2, \ldots$ in turn until an affirmative answer is obtained. If and when this occurs, give output 1. Then $Q(x_1, \ldots, x_{k-1}) \Leftrightarrow \psi(<x_1, \ldots, x_{k-1}>)$ *is convergent* $\Leftrightarrow <x_1, \ldots, x_{k-1}> \in domain\ \psi$. Hence by Theorem V, Q is recursively enumerable.☒

The projection theorems are an efficient, formal way of demonstrating recursive enumerability, though it is often easier to recognize recursive enumerability by means of the ideas of §§5.1 and 5.2. For example, consider $A = \{x | W_x \neq \emptyset\}$. (1) To show A recursively enumerable by the basic definition (§5.1), we must list the members of A. The practiced intuition will see that a double dovetailing construction (of "simultaneously" listing each W_0, W_1, W_2, . . . and then listing the indices of those lists which become nonempty) will suffice. (2) To show A recursively enumerable by the domain definition (Theorem V), define $\psi(x)$ to converge if $W_x \neq \emptyset$, and use a single dovetailing construction to list W_x. Then $A = domain\ \psi$ and is recursively enumerable. (3) To show A recursively enumerable by the projection theorems, redefine A as $\{x | (\exists y)(\exists z)$[*instructions* P_x *applied to input* y *yield an output in less than* z *steps*]$\}$ and recursive enumerability is immediate.

Definition If R is a k-ary relation, and n is fixed, the relation

$$\{<x_2, \ldots, x_k> | R(n, x_2, \ldots, x_k)\}$$

is called a *section of R at n.* Obviously, any section of a recursively enumerable or recursive relation is, respectively, a recursively enumerable or recursive relation. Does there exist a single recursive relation from which all recursively enumerable sets can be obtained by taking a section and a projection? The answer is implicit in the preceding proofs; we give it as Theorem XII.

***Theorem XII* (*enumeration theorem for recursively enumerable sets*)** *There exists a recursive 3-ary relation R such that for all z,*

$$W_z = \{x|(\exists w)R(z,x,w)\}.$$

Proof. Define R to be $\{\langle z,x,w\rangle | P_z$ *with input x yields an output in fewer then w steps*$\}$. R is recursive by Church's Thesis.☒

By the second projection theorem, we get the following corollary

Corollary XII $\{\langle z,x\rangle | x \in W_z\}$ *is a recursively enumerable set.*

§5.5 UNIFORMITY

Certain existence proofs in this chapter were nonconstructive in the sense that no effective procedure was given for deciding which of several cases held (see, for instance, the proof of Theorem III). It is important to distinguish those existence proofs for which effective procedures can be found. A theorem will be said to hold *uniformly*, or *uniformly effectively*, if such a procedure can be given. The exact significance of the phrase "uniformly effectively" will usually be clear from context. Several results already proved have given uniformity. For example, Corollary V(*b*) shows that we can "go uniformly" from domain to range, and vice versa. Corollary V(*c*) gives a similar explicit statement of uniformity. It is occasionally possible to show that uniformity cannot hold. The following theorems serve as illustrations.

Theorem XIII *The class of recursively enumerable sets is closed under the operations of* $\cup$, $\cap$, *and* $\times$ *uniformly effectively.*

Proof. Here, "uniformity" means that there exist recursive functions f, g, and h such that $W_{f(x,y)} = W_x \cup W_y$, $W_{g(x,y)} = W_x \cap W_y$, and $W_{h(x,y)} = W_x \times W_y$. We leave the proof for Exercise 5-23.☒

Theorem XIV *The class of recursive sets is closed under the operations of* $\cup$, $\cap$, $\times$, *and complementation. With respect to r.e. indices, the closure under* $\cup$, $\cap$, *and* $\times$ *is uniform, but the closure under complementation cannot be uniform.*

Proof. We leave proof of closure and of uniformity to Exercise 5-24. It remains to show that closure under complementation cannot be uniform, i.e., to show that there is no partial recursive ψ such that for all x, W_x recursive $\Rightarrow$ [$\psi(x)$ convergent and $W_{\psi(x)} = \bar{W}_x$].

Assume such ψ exists. We use the set K to derive a contradiction. Since K is recursively enumerable, and hence effectively listable, a recursive function g exists such that for any x, $\varphi_{g(x)}$ is convergent everywhere if $x \in K$, and divergent everywhere if $x \notin K$. Thus

$$W_{g(x)} = \begin{cases} N, & \text{if } x \in K; \\ \emptyset, & \text{if } x \notin K. \end{cases}$$

For all x, $W_{g(x)}$ is a recursive set. By assumption, $\psi g(x)$ is defined for all x. Let f be the total function ψg. We have

$$W_{f(x)} = \begin{cases} \emptyset, & \text{if } x \in K; \\ N, & \text{if } x \notin K. \end{cases}$$

Thus $\bar{K} = \{x|W_{f(x)} \neq \emptyset\}$. But $\{x|W_{f(x)} \neq \emptyset\} = \{x|(\exists y)[y = f(x) \ \& \ W_y \neq \emptyset]\} = \{x|(\exists y)(\exists z)(\exists w)[y = f(x)$ *and* P_y *with input* z *is convergent in fewer than* w *steps*$]\}$. Hence, by the second projection theorem, $\bar{K}$ is recursively enumerable. But $\bar{K}$ recursively enumerable $\Rightarrow K$ recursive, contrary to Theorem VI.☒

Remark. Closure with respect to complementation may be uniform if a different system of names for the recursive sets is used, e.g., indices for characteristic functions (see §5.6).

For further examples of uniformity and nonuniformity, we return to earlier theorems in this chapter. In Theorem III, if we name functions by ordinary Gödel numbers and recursive sets by indices for characteristic functions, then $\Rightarrow$ is uniform in both (a) and (*b*), and $\Leftarrow$ is uniform in (*b*); but $\Leftarrow$ cannot be uniform in (*a*) (see Exercise 5-26). Theorem IV is uniform, even if characteristic function indices are used for the recursive subset. In Theorem V, uniformity does not apply directly, because of the two separate clauses in the original definition of recursive enumerability. Corollaries V(*b*), V(*c*), and V(*d*) are, of course, explicit uniformity results. In Theorem VIII, uniformity (from indices to indices) holds in both parts. In Theorems X and XI, uniformity holds whether the recursive relations are named by r.e. indices or by characteristic function indices.

Final note. Formal proofs of uniformity almost always use the *s-m-n* Theorem. Since the s_n^m functions are primitive recursive, the resulting "uniformity functions" in such cases (like the f and g in Corollary V(*b*)) are usually primitive recursive.

§5.6 FINITE SETS

A *recursively enumerable* set can be named by giving an r.e. index for it. A *recursive* set can also be named by giving an index for its characteristic function. We call this a *characteristic index* for the recursive set. It is evident that we can go uniformly from characteristic indices to r.e. indices. We cannot go uniformly in the opposite direction. That is to say, there is no partial recursive ψ such that for all x, W_x recursive $\Rightarrow [\psi(x)$ convergent and $\varphi_{\psi(x)}$ is a characteristic function for $W_x]$. This is a consequence of Theorem XIV. If there were such a ψ, the last clause of Theorem XIV would be violated.

A *finite* set can be given still a third kind of name, a name that explicitly encodes the list of its members.

Definition Let A be the nonempty finite set $\{x_1, x_2, \ldots, x_k\}$, where $x_1 < x_2 < \cdots < x_k$. Then the integer $2^{x_1} + 2^{x_2} + \cdots + 2^{x_k}$ is called the *canonical index* of A. If A is empty, the *canonical index* assigned to A is 0.

Definition Let D_x be the finite set whose canonical index is x.

Evidently, every finite set has a unique canonical index, and every integer is the canonical index of some finite set. If x is written in binary notation, the members of D_x are given by the positions of the digit 1; specifically, for all z, z is a member of D_x if 1 appears as the $(z + 1)$st digit from the right in the binary expression for x.

Example. What is D_{13}? In binary notation 13 is 1101; 1's appear in the *first, third,* and *fourth* positions (from the right). Hence $D_{13} = \{0,2,3\}$.

It is evidently possible to go uniformly from canonical indices to characteristic indices, and hence to r.e. indices. The following theorem shows that it is not possible to go uniformly from characteristic indices to canonical indices.

Theorem XV(a) *There is a recursive f such that for all x, $f(x)$ is the size (i.e., cardinality) of D_x.*

(b) *There is no partial recursive ψ such that for all x, if φ_x is a characteristic function for a finite set A, then $\psi(x)$ is defined and $\psi(x)$ is the size of A.*

Proof. (*a*) This is immediate by Church's Thesis.

(*b*) Assume such a ψ exists. Given x, define η so that

$$\eta(z) = \begin{cases} 1, & \text{if } P_x \text{ applied to } x \text{ converges in } \textit{exactly } z \text{ steps;} \\ 0, & \text{otherwise.} \end{cases}$$

η is evidently recursive. A Gödel number for η can be found uniformly from x; call it $g(x)$. Then $\varphi_{g(x)}$ is a characteristic function for a set which has size 1 if $x \in K$ and size 0 if $x \notin K$. Hence ψg is a recursive function such that

$$\psi g(x) = \begin{cases} 1, & x \in K; \\ 0, & x \notin K. \end{cases}$$

This contradicts the nonrecursiveness of K.☒

Corollary XV *It is not possible to go uniformly from characteristic indices to canonical indices.*

Proof. If it were possible, then (*a*) would imply the existence of a ψ for (*b*) in the theorem.☒

Since we can go uniformly from characteristic indices to r.e. indices, it immediately follows from this corollary that it is not possible to go uniformly from r.e. indices to canonical indices. Exercise 5-29 gives further information on indices for finite sets and shows that it is not possible to go uniformly from r.e. indices for finite sets to characteristic indices for finite sets.

To summarize: three sorts of names for finite sets have been mentioned. One can go uniformly in the directions indicated in the following diagram and not in the reverse directions.

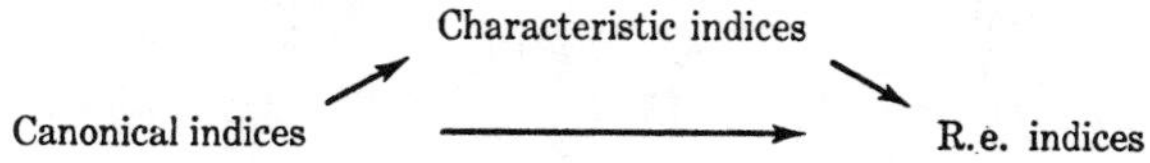

A Coding for All k-tuples as k Varies

Let $N^0 = \{\emptyset\}$. We can define a coding τ^* which maps $\bigcup_{k=0}^{\infty} N^k$ one-one onto N. ($\bigcup_{k=0}^{\infty} N^k = \{\emptyset\} \cup N \cup (N \times N) \cup (N \times N \times N) \cup \cdots$.)

Definition $\tau^*(\emptyset) = 0$.

$$\tau^*(<x_1, \ldots ,x_k>) = \tau(\tau^k(x_1, \ldots ,x_k), k - 1) + 1$$
$$= <x_1, \ldots , x_k, k - 1> + 1.$$

The mapping is evidently effective, one-one, and onto. This coding will be used in Chapter 16.

§5.7 SINGLE-VALUEDNESS THEOREM†

Definition A set A is *single-valued* if $\{<x,y>|<x,y> \in A\}$ is a single-valued relation. (*Warning:* As a property of sets, single-valuedness is not recursively invariant.)

Observe that A is single-valued if and only if $A = \tau(\psi)$ for some partial function ψ. It follows, by Theorem IX, that the single-valued members of the sequence $W_0, W_1, \ldots$ yield (under τ^{-1}) the family of all partial recursive functions. Unfortunately, $\{z|W_z \textit{ is single valued}\}$ is not recursive; indeed, it is not recursively enumerable. Nevertheless, we can enumerate a sequence of r.e. indices for single-valued sets that includes all the single-valued recursively enumerable sets (though not all indices for such sets) and hence yields all partial recursive functions. This is shown in the following theorem.

Definition *domain* $A = \{x|(\exists y)[<x,y> \in A]\}$.

***Theorem XVI* (single-valuedness theorem)** *There exists a recursive function f such that for all z,*

(*i*) $W_{f(z)}$ *is single-valued;*
(*ii*) $W_{f(z)} \subset W_z$;
(*iii*) *domain* $W_{f(z)} =$ *domain* W_z;
(*iv*) W_z *single-valued* $\Rightarrow W_{f(z)} = W_z$.

† Except for §16.5, no later results will depend on §5.7.

Proof. (iv) is immediate from (ii) and (iii). By Corollary V(d), a procedure for listing any W_z without repetition can be found effectively from z. Let any z be given. Define

$$A_z = \{<x,y> \mid <x,y> \in W_z \textit{ and } (\forall y')[[y' \neq y \;\&\; <x,y'> \in W_z] \Rightarrow <x,y> \textit{ precedes } <x,y'> \textit{ in the listing of } W_z \textit{ (given by Corollary } V(d))]\}.$$

It is evident, by either intuitive or formal argument, that A_z is recursively enumerable, and that an r.e. index for A_z can be found uniformly effectively from z. (For example, define a partial recursive function with A_z as domain.) Hence there is a recursive f such that for any z, $W_{f(z)} = A_z$. It follows from the definition of A_z that f has the desired properties for (i), (ii), and (iii). $\boxtimes$†

(Using the f of this theorem, we could define $\psi_z = \tau^{-1}(W_{f(z)})$. This would be a new numbering of the class of all partial recursive functions. It is not difficult to show that this numbering would be *acceptable* in the sense of Exercise 2-10.)

The single-valuedness theorem is important in two respects. The first and less important respect is that, although it yields no new insights beyond those of the basic theorem, it can serve as a formally elegant way of applying the basic theorem. The following theorem is an example.

***Theorem XVII* (reduction principle)** *Given any two recursively enumerable sets A and B, there exist recursively enumerable sets A' and B' such that $A' \subset A$, $B' \subset B$, $A \cup B = A' \cup B'$, and $A' \cap B' = \emptyset$.*

Proof. Define $C = (A \times \{0\}) \cup (B \times \{1\})$. By Theorem XIII, C is recursively enumerable. Let $C = W_n$, and then let $C' = W_{f(n)}$, where f is as in Theorem XVI (C' is the result of "single-valuizing" C). Then $\tau^{-1}(C')$ is a partial recursive function ψ. Set $A' = \psi^{-1}(0)$ and $B' = \psi^{-1}(1)$. $\boxtimes$

The second and more important respect in which the single-valuedness theorem is important concerns studies of generalized computability. These studies have to do with sets which are generalizations of the recursively enumerable sets, and partial functions which are generalizations of the partial recursive functions. Unlike the recursive case, it is more convenient in such studies *first* to define the generalization of recursively enumerable *set* (by a "projection" definition), *then* to prove a theorem analogous to the single-valuedness theorem, and *finally* to use this single-valuedness result to define the generalization of *partial recursive function*. (In order to go from a projection definition to a single-valuedness theorem, some counterpart to the basic theorem must be found. In such a counterpart, the ordinary sequential enumerations of our basic theorem are usually replaced by transfinitely well-ordered "enumerations.") Once the generalization of partial recursive function is obtained, further theory can then be developed along

† Theorem XVI can also be called a *uniformization theorem*. In point-set topology, a set A in the plane is said to *uniformize* a set B in the plane if (i) $(\forall x)(\exists$ *at most one* $y)[<x,y> \in A]$; (ii) $A \subset B$; and (iii) $\{x \mid (\exists y)[<x,y> \in A]\} = \{x \mid (\exists y)[<x,y> \in B]\}$ (see Kuratowski [1950]).

lines parallel to ordinary recursive function theory.† We develop such a generalized theory in §16.5.

We give one further application of the single-valuedness theorem. It tells us: given an effective enumeration of recursively enumerable sets, we can effectively find a nonempty set occurring in the enumeration if such a set exists. For recursively enumerable sets, this is a corollary to the basic theorem. The present proof, via the single-valuedness theorem, carries over to the generalized theory of §16.5.

***Theorem XVIII* (*selection theorem*)** *There is a partial recursive ψ such that for any z,*

(*i*) $\psi(z)$ *convergent* $\Leftrightarrow (\exists w)[w \in W_z \;\&\; W_w \neq \emptyset]$;

(*ii*) $\psi(z)$ *convergent* $\Rightarrow [\psi(z) \in W_z \;\&\; W_{\psi(z)} \neq \emptyset]$.

Proof. Let $A = \{<z,w>|(\exists u)[u \in W_w \;\&\; w \in W_z]\}$. Evidently, $[u \in W_w \;\&\; w \in W_z]$ describes a recursively enumerable relation; hence by projection, A is recursively enumerable. Single-valuizing A to A' and setting $\psi = \tau^{-1}(A')$, we obtain the desired ψ.☒

Exercise 5-36 shows that the selection theorem yields a more general form of the reduction principle.

Note that the selection theorem is just another restatement of the dovetailing phenomenon of the basic theorem. It tells us that, given any infinite effective sequence of computations, we can make an effective search that will locate a terminating computation if one exists.

§5.8 EXERCISES

§*5.1*

5-1. Show that the property *recursively enumerable* is recursively invariant.

5-2. A is *recursively enumerable without repetitions* if A = *range* f for some f that is recursive and one-one. Prove that A is infinite and recursively enumerable $\Leftrightarrow$ A is recursively enumerable without repetitions.

5-3. (*a*) Let A and B be infinite recursive sets with infinite complements. Show that $A \equiv B$. How many such sets are there?

(*b*) Describe the isomorphism types of recursive sets. How many such types are there?

5-4. Assume: f recursive; g recursive and one-one; *range* g recursive; and $(\forall x)[f(x) \geq g(x)]$. Deduce: *range* f recursive.

5-5. Let f be recursive. Let A be recursive and B be recursively enumerable. What can be concluded about recursiveness or recursive enumerability of the four sets $f(A)$, $f^{-1}(A)$, $f(B)$, and $f^{-1}(B)$?

△***5-6.*** A class $\mathcal{C}$ of recursively enumerable sets is a *recursively enumerable class* if there is a recursively enumerable set A such that $(\forall B[B \in \mathcal{C} \Leftrightarrow (\exists y)[y \in A \;\&\; W_y = B]]$, i.e., if an effective procedure exists which enumerates at least one r.e. index for every member of $\mathcal{C}$. (For definition of W_x notation, see §5.2.) Show that the class of all recursive sets is recursively enumerable. (*Hint:* Use Theorem III and the method of Theorem IV.) (This exercise requires Theorem 5 and its corollaries from §5.2.)

† Provided also that the family of generalized partial recursive functions has closure properties as "natural" as the closure properties (of the family of partial recursive functions) embodied in Church's Thesis (see, for example, Theorem 1-IX).

△*5-7.* Let R be a linear ordering of integers, which, as a relation, is recursively enumerable. Show that there exists a recursive relation which is a linear ordering of the same order type, i.e., which is *isomorphic* to R in the usual algebraic sense. (*Hint:* Use Theorem III(a).) (Recursiveness and recursive enumerability for relations are defined in §5.3.)

△*5-8.* Show that there exists an infinite set having no infinite recursively enumerable subset. (*Hint:* Use a nonconstructive diagonal argument.) Such sets are discussed further in Chapter 8.

§5.2

5-9. Where possible, classify the following as to (a) being recursive, (b) being recursively enumerable, (c) having recursively enumerable complement:

(i) $\{x|x \textit{ is prime}\}$;

(ii) $\{x|\textit{a run of at least } x \textit{ 7's occurs in } \pi\}$;

(iii) $\{x|\textit{a run of exactly } x \textit{ 7's occurs in } \pi\}$;

(iv) $\{x|W_x = \emptyset\}$;

△(v) $\{x|W_x \textit{ is infinite}\}$

(*Hint:* Show how recursive enumerability of this set would give recursive enumerability of $\{x|\varphi_x \textit{ total}\}$, contrary to Exercise 2-9. Show that recursive enumerability of $\{x|W_x \textit{ is finite}\}$ would solve the halting problem.)

(vi) $\{x|\varphi_x \textit{ total}\}$;

(vii) $\{x|W_x = W_n\}$ for a fixed n;

△(viii) $\{x|W_x \textit{ is recursive}\}$.

5-10. A is *enumerable by a partial recursive function in increasing order* if there is a partial recursive ψ such that $A = \textit{range } \psi$ and $(\forall x)(\forall y)[[x < y$ and $\psi(x)$ defined and $\psi(y)$ defined$] \Rightarrow \psi(x) < \psi(y)]$. Describe the class of all sets so enumerable.

5-11. (a) Prove Corollary V(d).

(b) Let $V_z = \{y|(\exists x)[\varphi_z(x) = y \;\&\; (\forall w)[w \leq x \Rightarrow \varphi_z(w) \textit{ defined}]]\}$. Show that A recursively enumerable $\Leftrightarrow (\exists z)[A = V_z]$, and prove an analogue to Corollary V(b).

5-12. Show that there are $\aleph_0$ sets which are recursively enumerable but not recursive.

▲*5-13.* Does every nonrecursive, recursively enumerable set have the property that it is isomorphic to K?

§5.3

5-14. (a) Is the operation $\times$ associative? Commutative? Distributive over union?

(b) Show that $\times$ is well-defined with respect to isomorphism types, i.e., that $A \equiv A' \;\&\; B \equiv B' \Rightarrow A \times B \equiv A' \times B'$. Show that the resulting operation on isomorphism types is commutative, associative, and possesses a zero. Show, however, that it does not possess a unity and that $A \times B \equiv A \times C$ does not imply $B \equiv C$.

5-15. Let τ' map $N \times N$ one-one into N, where τ' is recursive. (For example, $\tau'(x,y) = 2^x 3^y$. Kleene uses this pairing function.) Show that as z varies,

$$\psi_z = \lambda xy[\varphi_z(\tau'(x,y))]$$

yields all partial recursive functions of two variables.

5-16. Let B be recursive. How does the recursiveness or recursive enumerability of the following sets depend on the recursiveness or recursive enumerability of A: $A \times N$, $A \times \emptyset$, $A \times B$?

5-17. Prove Theorem IX.

5-18. Let R and S be binary relations. The *relative product* $R|S$ is defined to be $\{<x,z>|(\exists y)[R(x,y) \;\&\; S(y,z)]\}$. Show that R and S recursively enumerable $\Rightarrow R|S$ recursively enumerable.

5-19. Verify that $R \equiv Q \Rightarrow \tau^k(R) \equiv \tau^k(Q)$ for any k-ary relations R and Q.

5-20. Show that $\tau^k(R) \equiv \tau^k(Q)$ does not in general imply $R \equiv Q$. (*Hint:* Take $R = \{1\} \times N$, $Q = N \times \{1\}$.)

5-21. (*a*) Define $K_0 = \{<x,y> | x \in W_y\}$. Show that K_0 is recursively enumerable but not recursive.

▲(*b*) Show that $K_0 \equiv K$.

△***5-22.*** Consider the following alternative definition of universal function: ψ is "universal" if there exists a recursive f such that for all m, $\varphi_m = \lambda y[\psi(<f(m),y>)]$. Show that this definition is not recursively invariant. (*Hint:* Define $\psi = \lambda z[\varphi_{\pi_1(z)}(\pi_2(z))]$; that is, $\psi(<m,n>) = \varphi_m(n)$. Define $h(z) = \tau(\pi_2(z), \pi_1(z))$; that is, $h(<x,y>) = <y,x>$ for all x, y. Then define $\psi' = h^{-1}\psi h$. Show that ψ' is not "universal" by noting that for any θ, θ "universal" $\Rightarrow$ $[\lambda y[\theta(m,y)]$ is constant for some m but $\lambda x[\theta(x,n)]$ is constant for no n].)

§5.5

5-23. Verify Theorem XIII.

5-24. Verify the closure and uniformity parts of Theorem XIV.

5-25. Show that [A recursively enumerable and f recursive] $\Rightarrow f^{-1}(A)$ is recursively enumerable uniformly in indices for f and A.

5-26. Show that it is not possible effectively to decide, given an index for a nondecreasing function, whether or not the range of that function is finite.

5-27. (*a*) From cases (ii) and (iii) in the proof of Theorem I, it follows that [W_x recursive and $W_x \neq \emptyset$] $\Rightarrow W_x$ is the range of a nondecreasing recursive function which is either strictly increasing or else strictly increasing up to a point and constant beyond that point. Does this result hold uniformly?

△(*b*) Can Corollary V(*c*) be strengthened by adding "and [*domain* φ_x infinite $\Rightarrow$ $\varphi_{f'(x)}$ is one-one]" to part (i)?

§5.6

5-28. (Fischer, Luckham, Ritter). Let A be recursively enumerable, and let B be recursive.

(*a*) Show that $\bigcup_{x \in A} W_x$ is recursive enumerable. (The recursively enumerable sets are closed under recursively enumerable union.)

△(*b*) Show that $\bigcup_{x \in B} D_x$ need not be recursive. (*Hint:* Use the sequence of initial segments in an effective enumeration of K; select an appropriate subsequence.)

△(*c*) Show that $\bigcap_{x \in B} W_x$ need neither be recursively enumerable nor have recursively enumerable complement. (*Hint:* Show how to obtain $\{x | W_x \textit{ is infinite}\}$ as such an intersection; see Exercise 5-9.)

△(*d*) Let B be recursive and assume $(\forall x)[x \in B \Rightarrow W_x$ recursive]. Show that $\bigcap_{x \in B} W_x$ need neither be recursively enumerable nor have recursively enumerable complement. (*Hint:* Take a sequence of sets such that the nth set is $\{n, n+1, \ldots\} \cup \{x | W_x \textit{ has at least n members}\}$.)

△(*e*) If φ_x is a characteristic function, let C_x be the set determined by φ_x. Let B be recursive and assume $(\forall x)[x \in B \Rightarrow \varphi_x$ is a characteristic function]. Show that $\bigcap_{x \in B} C_x$ need not be recursively enumerable but must have recursively enumerable complement.

△***5-29.*** If φ_x is a characteristic function, let C_x be the set determined by φ_x.

(*a*) Show that it is not possible to go uniformly from recursively enumerable indices for finite sets to characteristic indices for finite sets, i.e., that there is no partial recursive ψ such that $(\forall x)[W_x$ finite $\Rightarrow [\psi(x)$ convergent & $W_x = C_{\psi(x)}]]$.

(*b*) (Rice). A class $\mathcal{C}$ of finite sets is *canonically enumerable* if there exists a recursively enumerable A such that $(\forall B)[B \in \mathcal{C} \Leftrightarrow (\exists y)[y \in A \;\&\; D_y = B]]$.

A class $\mathcal{C}$ of finite sets is *characteristically enumerable* if there exists a recursively enumerable A such that $(\forall x)[x \in A \Rightarrow \varphi_x$ is a characteristic function] and $(\forall B)[B \in \mathcal{C} \Leftrightarrow (\exists y)[y \in A \;\&\; C_y = B]]$.

A class $\mathcal{C}$ of finite sets is *recursively enumerable* if there exists a recursively enumerable A such that $(\forall B)[B \in \mathcal{C} \Leftrightarrow (\exists y)[y \in A \;\&\; W_y = B]]$.

(i) Show that a class of finite sets can be recursively enumerable without being characteristically enumerable. (*Hint:* Take $\mathcal{C} = \{D|(\exists x)[x \notin K \;\&\; D = \{x\}]$ *or* $(\exists x)[x \in K \;\&\; D = \{x, x+1\}]\}$.)

(ii) Show that a class of finite sets can be characteristically enumerable without being canonically enumerable. (*Hint:* Take $\mathcal{C} = \{D|(\exists x)[D = \{x\} \;\&\; x \notin K]$ *or* $(\exists x)(\exists y)[D = \{x, x+y+1\} \;\&\; P_x$ *with input* x *converges in exactly* y *steps*$]\}$.)

(iii) The second and third definitions above obviously can be extended, without change, to classes of infinite as well as finite sets. Exercise 5-6 showed that the class of all recursive sets is recursively enumerable. Show, by a diagonal argument, that this class is not characteristically enumerable.

5-30. Give an explicit formula for the inverse of τ^* in terms of the π_i^k functions.

§*5.7*

5-31. Show that $\{x|W_x$ *single-valued*$\}$ is not recursively enumerable. (*Hint:* Show that a recursive characteristic function for K could be obtained if this were false.) Is the class of all single-valued recursively enumerable sets recursively enumerable in the sense of Exercise 5-6?

5-32. Let f be the function whose existence is asserted in the single-valuedness theorem (Theorem XVI). Define $\psi_x = \tau^{-1}(W_{f(x)})$. Show that $\psi_0, \psi_1, \ldots$ gives a numbering of the partial recursive functions that is *acceptable* in the sense of Exercise 2-10.

5-33. A class $\mathcal{C}$ of sets satisfies the *separation principle* if $(\forall A)(\forall B)[[A$ and B in $\mathcal{C}$ and $A \cap B = \emptyset] \Rightarrow (\exists C)[A \subset C \;\&\; B \subset \bar{C} \;\&\; C \in \mathcal{C}$ and $\bar{C} \in \mathcal{C}]$. Show from Theorem XVII that the separation principle holds for $\mathcal{C}$ the class of all sets whose complements are recursively enumerable.

△***5-34.*** Show that the separation principle of 5-33 fails for $\mathcal{C}$ the class of all recursively enumerable sets. (*Hint:* See Exercise 2-30.)

5-35. A class $\mathcal{C}$ of sets satisfies the reduction principle if $(\forall A)(\forall B)[[A \in \mathcal{C} \;\&\; B \in \mathcal{C}] \Rightarrow (\exists A')(\exists B')[A' \in \mathcal{C} \;\&\; B' \in \mathcal{C} \;\&\; A' \cap B' = \emptyset \;\&\; A' \subset A \;\&\; B' \subset B \;\&\; A' \cup B' = A \cup B]]$. Show from 5-34 that the reduction principle fails for $\mathcal{C}$ the class of all sets whose complements are recursively enumerable.

△***5-36.*** Generalize Theorem XVII (the reduction principle) from two recursively enumerable sets A and B to an infinite family of recursively enumerable sets $W_{f(0)}$, $W_{f(1)}, \ldots$ given by a recursive function f. (*Hint:* For proof, use the selection theorem (Theorem XVIII).)

▲***5-37.*** (*a*) Prove the following theorem, which yields results from Exercises 5-9 and 5-31 as special cases.

Definition Let $\mathcal{C}$ be a class of recursively enumerable sets. Let $P_{\mathcal{C}} = \{x|W_x \in \mathcal{C}\}$. $\mathcal{C}$ is *completely recursively enumerable* if $P_{\mathcal{C}}$ is a recursively enumerable set.

Theorem (Rice, Shapiro, McNaughton, Myhill; see Rice [1956]). $\mathcal{C}$ *is completely recursively enumerable if and only if there exists a canonically enumerable class* $\mathcal{D}$ *of finite sets* (see Exercise 5-29 for definition) *such that* $(\forall A)[A \in \mathcal{C} \Leftrightarrow [A$ *is recursively enumerable* $\&\; (\exists D)[D \in \mathcal{D} \;\&\; D \subset A]]]$.†

(*b*) Deduce, as an immediate consequence of this theorem, that $\{x|W_x$ *is infinite*$\}$ neither is recursively enumerable nor has recursively enumerable complement.

† Such a class $\mathcal{D}$ is sometimes called a *key array* for $\mathcal{C}$.

6 Reducibilities

§6.1 GENERAL INTRODUCTION

In Chapter 2, we considered solvable and unsolvable problems. For purposes of our theory, any such problem is represented (via a coding) as a set of integers. The problem is *solvable* if the corresponding set of integers is recursive. In Chapter 2, we also considered, briefly and informally, the notion of *reducibility,* and used it to establish certain unsolvability results. According to our informal definition, one problem is *reducible* to another if a method for solving the second problem yields a method for solving the first. The precise content of this informal definition is not apparent, although the particular applications given in Chapter 2 were clear. In Chapters 6 to 9 we focus on reducibility. In view of the use of sets to represent problems, we shall formulate and study reducibility as a relation between sets of integers. We begin with two examples.

Example 1. Consider the two sets: $\{x|W_x \textit{ infinite}\}$ and $\{x|\varphi_x \textit{ total}\}$. In a reasonable sense, each of these sets is reducible to the other. For, first, assume we *could* test φ_x *total* for any x. Then, to test W_{x_0} *infinite* for a given x_0: use x_0 to get an x_1 such that φ_{x_1} lists the members of W_{x_0} without repetition and such that φ_{x_1} is total if W_{x_0} is infinite (see Corollary 5-V(d)); and see whether φ_{x_1} is total. Similarly, second, assume we *could* test W_x *infinite* for any x. Then, to test φ_{y_0} *total* for a given y_0: use y_0 to get a y_1 such that

$$\varphi_{y_1}(z) = \begin{cases} 1, & \text{if } \varphi_{y_0}(w) \text{ converges for all } w \leq z; \\ \text{divergent}, & \text{otherwise}; \end{cases}$$

and see whether W_{y_1} is infinite.

Example 2. Consider the two sets: K $(= \{x|\varphi_x(x) \textit{ converges}\})$ and $\{x|W_x \textit{ finite}\}$. Here, the first set is reducible to the second. For, to test x_0 in K: use x_0 to get x_1 such that

$$\varphi_{x_1}(z) = \begin{cases} 1, & \text{if } P_{x_0} \text{ with input } x_0 \text{ fails to converge in fewer than } z \text{ steps}; \\ \text{divergent}, & \text{otherwise}; \end{cases}$$

and see whether W_{x_1} is finite. Both Example 1 and our intuition might suggest that the second set is also reducible to the first. As we shall eventually show, this is not the case. (Exercise 6-5 below shows this failure for a restricted form of reducibility.)

Several different ways of formulating the concept of reducibility have been studied. These will be presented in Chapters 7 to 9. In the present chapter, we consider some of the common features that these formulations will possess and some of the related terminology that will be used.

Abbreviate *A reducible to B* as $A \leq_r B$. In all cases to be studied, this relation will be both reflexive and transitive; that is, $A \leq_r A$, for all A, and $[A \leq_r B \;\&\; B \leq_r C] \Rightarrow A \leq_r C$, for all A, B, C. Define $A \equiv_r B$ to mean $A \leq_r B$ and $B \leq_r A$. As an immediate consequence of the reflexiveness and transitivity of $\leq_r$, $\equiv_r$ is an equivalence relation. Furthermore, $\leq_r$ gives a well-defined partial ordering of the equivalence classes of $\equiv_r$. We say that one equivalence class is *below* another if members of the first are reducible to members of the second, but not conversely.

In all cases, the relation $\leq_r$ will be recursively invariant.† Hence, any such $\leq_r$ and $\equiv_r$ can be used to classify sets as to recursive structure, and to measure (in some sense) their similarity or dissimilarity. The equivalence classes of $\equiv_r$ are called *degrees of unsolvability with respect to* $\leq_r$. The partial ordering of these equivalence classes is called the *reducibility ordering of* $\leq_r$. (It might seem natural to call those classes which involve recursive sets "degrees of *solvability*"; it is customary, however, to speak, without exception, of "degrees of *unsolvability*.") Thus in Example 2, $\{x \mid W_x \text{ finite}\}$ is of *higher* degree of unsolvability than K. (And K, in turn, is evidently of higher degree than any recursive set.)

In our subsequent study of reducibilities, we shall pay especial attention to the structure of certain reducibility orderings over the *recursively enumerable* sets. We do this (1) because recursively enumerable sets possess, by definition, a considerable amount of constructiveness, and (2) because our applications to logic will largely concern formalized logical systems in which the set of provable statements is (under a simple coding) recursively enumerable. Even when limited to recursively enumerable sets, the reducibility orderings have, as we shall see, a structure that is complex and not yet fully understood.

We conclude with a preliminary and simple illustration of the kind of concept that will be useful in the study of reducibilities over recursively enumerable sets. The notion is due to Post. We define: A is *complete with respect to* $\leq_r$ if (i) A is recursively enumerable and (ii) $(\forall B)[B$ recursively enumerable $\Rightarrow B \leq_r A]$.‡ Thus, a complete set (if any exists) would have maximum degree of unsolvability among all recursively enumerable sets.

† This is equivalent to saying $A \equiv B \Rightarrow A \equiv_r B$, for all A, B (see Exercise 6-8).

‡ A is said to be *universal with respect to* $\leq_r$ if there is a B such that $B \leq_r A$ and B is complete with respect to $\leq_r$.

It is not difficult to see that complete sets exist (for our as yet informal notion of reducibility). For example, let $K_0 = \{<x,y> | x \in W_y\}$. K_0 is recursively enumerable (see Exercise 5-21(*a*)). Let any recursively enumerable B be given. Then $B = W_{y_0}$ for some y_0. Hence to test whether $x \in B$, we need only test whether $<x,y_0> \in K_0$. Thus B is reducible to K_0. Hence K_0 is complete.

The set K is also complete (see Exercise 6-3). Thus K and K_0 are of the same degree of unsolvability. We shall see in Chapter 7 that, in fact, $K \equiv K_0$. The set K (or K_0) represents the halting problem. So far, virtually all our demonstrations of unsolvability have been accomplished in the following way: we show the reducibility of the halting problem to the problem in question. We can now conclude, for example, that any problem shown unsolvable *in this way* must have degree of unsolvability at least as high as the maximum degree of unsolvability for recursively enumerable sets. (These statements have been made with respect to our informal notion of reducibility. We shall see that they hold for all the reducibilities to be precisely formulated.)

§6.2 EXERCISES

For the following exercises, define: $A \leq_r B$ if there exists a recursive function f such that $(\forall x)[x \in A \Leftrightarrow f(x) \in B]$.

6-1. Verify that $\{x | W_x \textit{ infinite}\} \equiv_r \{x | \varphi_x \textit{ total}\}$.

6-2. Verify that $K \leq_r \{x | W_x \textit{ finite}\}$.

6-3. Show that $K_0 \leq_r K$; hence conclude that K is complete with respect to $\leq_r$.

6-4. Show that $\{x | W_x \neq \emptyset\}$ is complete with respect to $\leq_r$.

6-5. (*a*) Show that $\bar{K} \leq_r \{x | W_x \textit{ finite}\}$.

$\triangle$(*b*) Show that $\{x | W_x \textit{ finite}\}$ is *not* $\leq_r K$. (*Hint:* Consider recursive enumerability of the sets in question.) (This is, essentially, Exercise 2-26.)

$\triangle$***6-6.*** Define A *rel* B if $A \leq_r \bar{B}$. Show that *rel* would not be acceptable as a reducibility relation in the sense of the general discussion of §6.1.

$\triangle$***6-7.*** Suggest several other, possibly different, formulations of the informal notion of reducibility introduced in Chapter 2.

6-8. Let $\leq_{r'}$ be a reflexive and transitive reducibility relation. Define: $\leq_{r'}$ is *recursively invariant* if $[[A_1 \equiv B_1 \ \& \ A_2 \equiv B_2 \ \& \ A_1 \leq_{r'} A_2] \Rightarrow B_1 \leq_{r'} B_2]$. Show that $\leq_{r'}$ is recursively invariant if and only if $[A \equiv B \Rightarrow A \equiv_{r'} B]$.

7 One-one Reducibility; Many-one Reducibility; Creative Sets

§7.1 ONE-ONE REDUCIBILITY AND MANY-ONE REDUCIBILITY

We now study two formulations of reducibility that are directly suggested by the informal examples of reducibility given in preceding chapters.

Definition A is *one-one reducible* to B (notation: $A \leq_1 B$) if there exists a one-one recursive function f such that $(\forall x)[x \in A \Leftrightarrow f(x) \in B]$.

We occasionally abbreviate "one-one reducibility" as "1-reducibility."

Definition A is *many-one reducible* to B (notation: $A \leq_{\mathrm{m}} B$) if there exists a recursive function f such that $(\forall x)[x \in A \Leftrightarrow f(x) \in B]$.

We occasionally abbreviate "many-one reducibility" as "m-reducibility."

In either of the above definitions, the condition $(\forall x)[x \in A \Leftrightarrow f(x) \in B]$ can be restated in any one of the following equivalent forms: $A = f^{-1}(B)$; $f(A) \subset B$ & $f(\bar{A}) \subset \bar{B}$; $c_A = c_B f$, where c_A and c_B are characteristic functions for A and B. If $A = f^{-1}(B)$, we say that A is reducible to B *via* f.

The following theorem gives some simple and basic facts about $\leq_1$ and $\leq_{\mathrm{m}}$.

Theorem I(a) *$\leq_1$ and $\leq_{\mathrm{m}}$ are reflexive and transitive.*

(b) $A \leq_1 B \Rightarrow A \leq_{\mathrm{m}} B$.

(c) $A \leq_1 B \Rightarrow \bar{A} \leq_1 \bar{B}$.

(d) $A \leq_{\mathrm{m}} B \Rightarrow \bar{A} \leq_{\mathrm{m}} \bar{B}$.

(e) $[A \leq_{\mathrm{m}} B$ & B *recursive*$] \Rightarrow A$ *recursive.*

(f) $[A \leq_{\mathrm{m}} B$ & B *recursively enumerable*$] \Rightarrow A$ *recursively enumerable.*

Proof. Parts (*a*), (*b*), (*c*), and (*d*) are immediate.

(*e*) Assume $A \leq_m B$ via f, then $c_A = c_B f$; hence if c_B is recursive, so is c_A.

(*f*) Assume $A \leq_m B$ via f, then $A = f^{-1}(B)$; hence by Theorem 5-VII, if B is recursively enumerable, so is A.☒

Corollary I *$[A \leq_1 B$ & B recursive$] \Rightarrow A$ recursive; $[A \leq_1 B$ & B recursively enumerable$] \Rightarrow A$ recursively enumerable.*

Proof. Immediate from (*b*), (*e*), and (*f*).☒

Definition

$$A \equiv_1 B, \quad \text{if } A \leq_1 B \;\&\; B \leq_1 A;$$
$$A \equiv_m B, \quad \text{if } A \leq_m B \;\&\; B \leq_m A.$$

By Theorem I(*a*), these are equivalence relations. Their equivalence classes are called *degrees of unsolvability with respect to one-one reducibility* and *degrees of unsolvability with respect to many-one reducibility*, respectively. We abbreviate the former to "one-one degrees" or "1-degrees," and the latter to "many-one degrees" or "m-degrees."

From Theorem I(*e*) and Corollary I, it follows that any degree containing a recursive set consists entirely of recursive sets. From Theorem I(*f*) and Corollary I, it follows that any degree containing a recursively enumerable set consists entirely of recursively enumerable sets. We can thus speak of *recursive degrees* and *recursively enumerable degrees* in each of these reducibility orderings. Theorem I(*f*) shows, further, that the property of being recursively enumerable is hereditary downward in both reducibility orderings; i.e., any degree below a recursively enumerable degree is a recursively enumerable degree. Solutions to Exercises 2-26 and 6-5 follow from Theorem I(*f*). Since there are recursively enumerable sets which are not recursive, there must be nonrecursive as well as recursive degrees among the recursively enumerable degrees in each ordering.

The following theorem gives information concerning the existence and properties of incomparable degrees in these reducibility orderings.

Definition $A \text{ join } B = \{y \mid [y = 2x \;\&\; x \in A] \text{ or } [y = 2x + 1 \;\&\; x \in B]\}$.

Theorem II(a) *There exist two nonrecursive sets which are incomparable with respect to $\leq_m$ (and hence with respect to $\leq_1$).*

(*b*) *The* m-*reducibility ordering is an upper semilattice; i.e., any two degrees have a unique least upper bound. Furthermore, the least upper bound of two recursively enumerable degrees is recursively enumerable.*

Proof. (*a*) (N and $\emptyset$ are incomparable sets; however, they are recursive.) Consider the sets K and $\bar{K}$. K is recursively enumerable but $\bar{K}$ is not (Theorem 5-VI). By Theorem I(*f*), $\bar{K}$ is not $\leq_m K$. By Theorem I(*d*), it follows that K is not $\leq_m \bar{K}$.

(*b*) Let $d(X)$ be the m-degree of X, for any set X. Let A and B be given. Then $d(A \text{ join } B)$ is the desired least upper bound to $d(A)$ and $d(B)$ in the m-reducibility ordering. For, $A \leq_m A \text{ join } B$ via $\lambda x[2x]$; $B \leq_m A \text{ join } B$ via $\lambda x[2x+1]$; and for any C, if $A \leq_m C$ via f and $B \leq_m C$ via g, then $A \text{ join } B \leq_m C$ via h, where h is defined by

$$h(2x) = f(x),$$
$$h(2x+1) = g(x). \boxtimes$$

From part (*a*) of the proof, we note a corollary.

Corollary II *$A \leq_m B$ does not in general imply $A \leq_m \bar{B}$.*

Of course, incomparable sets generate incomparable degrees. Do there exist nonrecursive, recursively enumerable m-degrees which are incomparable? We answer this question affirmatively in Chapter 10. Is the m-reducibility ordering a lattice (i.e., are there greatest lower bounds)? In Exercise 13-55, we see that this is not true. Is the 1-reducibility ordering an upper semilattice? In Chapter 10 we see that this is not true.

Frequently, in cases of special interest, $\leq_m$ and $\leq_1$ coincide. It can usually be shown that $\leq_1$ holds, once $\leq_m$ has been shown to hold. For this reason, $\leq_1$ and $\leq_m$ are sometimes grouped together as the *strong reducibilities.*

Various facts about $\leq_1$ and $\leq_m$ will be given in the remainder of this chapter and in exercises in §7.9 (see in particular, at this point, Exercises 7-6 to 7-8).

A general structure theorem relating $\leq_1$ and $\leq_m$ is given in §7.6.

§7.2 COMPLETE SETS

Does there exist a maximum degree among the recursively enumerable degrees for $\leq_m$ or for $\leq_1$? We repeat the definition given in Chapter 6.

Definitions A is *complete with respect to* $\leq_1$ (A is "1-complete") if
(i) A is recursively enumerable, and
(ii) $(\forall B)[B \text{ recursively enumerable} \Rightarrow B \leq_1 A]$.
A is *complete with respect to* $\leq_m$ (A is "m-complete") if
(i) A is recursively enumerable, and
(ii) $(\forall B)[B \text{ recursively enumerable} \Rightarrow B \leq_m A]$.

Let $K_0 = \{\langle x,y\rangle \mid x \in W_y\}$. The following theorem, foreshadowed in Chapter 6, shows that K_0 is 1-complete and hence m-complete.

Theorem III *K_0 is 1-complete.*

Proof. $K_0 = \{\langle x,y\rangle \mid (\exists z)[P_y \textit{ with input } x \textit{ converges in fewer than } z \textit{ steps}]\}$ and is hence recursively enumerable by the second projection theorem.

Let B be any recursively enumerable set. Then $B = W_{y_0}$ for some y_0. Thus $(\forall x)[x \in B \Leftrightarrow \langle x,y_0\rangle \in K_0]$; and $B \leq_1 K_0$ via $\lambda x[\langle x,y_0\rangle]$. $\boxtimes$

Theorem IV *K is 1-complete.*

Proof. It is enough to show $K_0 \leq_1 K$.

We first show $K_0 \leq_m K$. By familiar technique,† we find an f such that

$$\varphi_{f(x)}(z) = \begin{cases} 1, & \text{if } \varphi_{\pi_2(x)}(\pi_1(x)) \text{ convergent;} \\ \text{divergent,} & \text{otherwise.} \end{cases}$$

(The behavior of $\varphi_{f(x)}$ thus does not depend on input z.) Then $x \in K_0 \Leftrightarrow f(x) \in K$. Hence $K_0 \leq_m K$.

We next show that if f is not one-one, it can be altered to an f^* such that $K_0 \leq_1 K$ via f^*. We do this by a technique of "padding" whereby successively larger Gödel numbers for the same partial recursive function are constructed. Define, by the methods of Theorem 1-III, a recursive function t' such that

(i) $\varphi_{t'(x,y)} = \varphi_x$; and
(ii) $y_1 \neq y_2 \Rightarrow t'(x,y_1) \neq t'(x,y_2)$.

Then define a recursive function t by the following inductive definition: $t(0,0) = t'(0,0)$; if $t(x',y')$ is defined for all $<x',y'>$ such that $<x',y'> < <x,y>$, then set $t(x,y) = t'(x,z)$ where $z = \mu w[t'(x,w) \neq t(x',y')$ *for all* $<x',y'>$ *such that* $<x',y'> < <x,y>]$.

Thus,

(i) $\varphi_{t(x,y)} = \varphi_x$; and
(ii) $[x_1 \neq x_2$ or $y_1 \neq y_2] \Rightarrow t(x_1,y_1) \neq t(x_2,y_2)$.

(In fact, for any k, $\varphi^{(k)}_{t(x,y)} = \varphi^{(k)}_x$.)

Now f^* is defined by

$$f^* = \lambda x[t(f(x),x)].$$

Evidently f^* is one-one, and $K_0 \leq_1 K$ via f^*. ⊠

The construction above is due to Davis. We shall use the function t on several later occasions to go from $\leq_m$ to $\leq_1$. The structure theorem in §7.6 will give more information on the relation between $\leq_m$ and $\leq_1$.

Corollary IV. $K \equiv_1 K_0$.

Proof. Immediate from Theorems III and IV. ⊠

Obviously, 1-completeness implies m-completeness. If we return to earlier specific examples of nonrecursive, recursively enumerable sets, we can use the technique of Theorem IV to show, in every case, that K is not only m-reducible but also 1-reducible to the set being considered. This suggests several questions.

1. Do m-completeness and 1-completeness coincide? We answer this question affirmatively in §7.5.
2. Is every nonrecursive, recursively enumerable set m-complete?
3. Do $\leq_m$ and $\leq_1$ coincide on nonrecursive, recursively enumerable sets? (Exercises 7-5 and 7-6 show that they differ on recursive sets.) We give negative answers to questions 2 and 3 in Chapter 8.

† Which would involve, in a more formal proof, the *s-m-n* theorem.

Remark. We shall, on various occasions, make use of the particular sets K and K_0 for proofs of unsolvability and for the construction of counterexamples.

A productive & B⊂A r.e. ⇒ A-B productive cf. p.9

§7.3 CREATIVE SETS

The failure of the set $\bar{K}$ to be recursively enumerable can be asserted in a rather strong and constructive sense; namely, from the index of any recursively enumerable subset of $\bar{K}$ we can find an integer which is in $\bar{K}$ but not in that subset. More specifically, we have (by definition of K) that $W_x \subset \bar{K} \Rightarrow x \in \bar{K} - W_x$. This property of $\bar{K}$ is given a recursively invariant formulation in the following definition.

Definition A is *productive* if there exists a partial recursive ψ such that $(\forall x)[W_x \subset A \Rightarrow [\psi(x) \text{ convergent } \& \, \psi(x) \in A - W_x]]$. ψ is called a *productive partial function* for A. (The term *productive* is due to Dekker.)

We are especially interested in recursively enumerable sets whose complements are productive. Such sets are called, by Post, *creative* (see the remark following the proof of Theorem XI below).

Definition A is *creative* if

(i) A is recursively enumerable, and

(ii) $\bar{A}$ is productive.

Example 1. K is a *creative* set, since K is recursively enumerable and $\bar{K}$ is productive (with the identity function $\lambda x[x]$ as productive partial function).

Example 2. $\{x|\varphi_x \textit{ total}\}$ is *productive*, where a productive partial function can be based on the diagonalization construction of §1.4. More specifically, let W_x be given. Take $\varphi_{f'(x)}$ as in Corollary 5-V(*c*); i.e., *range* $\varphi_{f'(x)}$ is W_x, and $\varphi_{f'(x)}$ is total if W_x is nonempty. Define

$$\varphi_{g(x)}(z) = \begin{cases} 0, & \text{if } P_{f'(x)} \text{ with input 0 does not converge in } z \text{ steps or fewer;} \\ \varphi_{\varphi_{f'(x)}(z-z_0)}(z) + 1, & \text{otherwise, where } z_0 \text{ is the exact number of steps required for } P_{f'(x)} \text{ with input 0 to converge.} \end{cases}$$

Then g is the desired productive partial function (see Exercise 7-17).

The following theorem gives several basic properties of productiveness.

Theorem V(a) *A productive $\Rightarrow$ A not recursively enumerable.*

(*b*) *$[A$ productive & $A \le_m B] \Rightarrow B$ productive.*

Proof. (*a*) Immediate by definition.

(*b*) Let A have ψ as productive partial function. Let $A \le_m B$ via f. There is a recursive g such that $W_{g(x)} = f^{-1}(W_x)$ (see Exercise 5-25). Then

$f\psi g$ is a productive partial function for B. For,

$$W_x \subset B \Rightarrow W_{g(x)} \subset A \Rightarrow [\psi g(x) \text{ convergent } \& \ \psi g(x) \in A - W_{g(x)}] \Rightarrow [f\psi g(x) \text{ convergent } \& \ f\psi g(x) \in B - W_x]. \boxtimes$$

Theorem V(b) shows that the property of productiveness is hereditary upwards in the strong reducibility orderings. Theorem V(b) also gives a way of demonstrating the productiveness of $\{x|\varphi_x \textit{ total}\}$ that is more convenient than the direct diagonalization used in Example 2 above. Namely, show $\bar{K} \leq_m \{x|\varphi_x \textit{ total}\}$; from this, by Theorem V(b), the productiveness of $\{x|\varphi_x \textit{ total}\}$ immediately follows.

Corollary V(a) *A creative $\Rightarrow$ A not recursive.*

(*b*) *[A creative & $A \leq_m B$] $\Rightarrow$ $\bar{B}$ productive.*

(*c*) *A m-complete $\Rightarrow$ A creative.*

Proof. Immediate. $\boxtimes$

Is every nonrecursive, recursively enumerable set creative? This question is answered negatively in Chapter 8 (this implies a negative answer to question 2 at the end of §7.2). Does the converse of Corollary V(c) hold; i.e., is every creative set m-complete? This question is answered affirmatively in Chapter 11.

We discuss productive sets further in §7.7. Various aspects of productiveness are considered in Dekker [1955].

§7.4 ONE-ONE EQUIVALENCE AND RECURSIVE ISOMORPHISM

The concepts of one-one equivalence and recursive isomorphism coincide. This result is due to Myhill [1955]. In content and proof, the result is closely related to the Cantor-Schröder-Bernstein theorem in the theory of cardinal numbers.

Theorem VI (Myhill) $A \equiv B \Leftrightarrow A \equiv_1 B$.

Proof. $\Rightarrow$: Immediate.

$\Leftarrow$: We introduce the following definition. A finite sequence of ordered pairs $\langle\langle x_1,y_1\rangle, \ldots, \langle x_n,y_n\rangle\rangle$ is called a *finite correspondence between A and B* if (i) $i \neq j \Rightarrow [x_i \neq x_j \ \& \ y_i \neq y_j]$, $1 \leq i \leq n$, $i \leq j \leq n$; and (ii) $x_i \in A \Leftrightarrow y_i \in B$, $1 \leq i \leq n$. We prove a lemma.

Lemma *Assume $C \leq_1 D$. Then there is an effective procedure such that given any finite correspondence $\langle\langle x_1,y_1\rangle, \ldots, \langle x_n,y_n\rangle\rangle$ between C and D, and given any $x' \notin \{x_1, \ldots, x_n\}$, we can find a y' such that*

$$\langle\langle x_1,y_1\rangle, \ldots, \langle x_n,y_n\rangle, \langle x',y'\rangle\rangle$$

is a finite correspondence between C and D.

Proof of lemma. Assume $C \leq_1 D$ via f. Calculate $f(x')$. See whether $f(x') \neq y_i$ for all i, $1 \leq i \leq n$. If so, set $y' = f(x')$. If not, and $f(x) = y_{i_1}$, compute $f(x_{i_1})$ and see whether $f(x_{i_1}) \neq y_i$ for all i, $1 \leq i \leq n$. If so, set $y' = f(x_{i_1})$. If not, and $f(x_{i_1}) = y_{i_2}$, compute $f(x_{i_2})$ and see whether . . . . It is immediate from the one-one property of f and from the one-one property of the given finite correspondence that this procedure must terminate with a y' such that $y' \neq y_i$, $1 \leq i \leq n$. Furthermore, we see, from the reducibility condition on both f and the given finite correspondence, that $x' \in C \Leftrightarrow y' \in D$. This proves the lemma.

We return to the proof of the theorem. Since $A \equiv_1 B$, the lemma can be applied *either* with A and B in place of C and D *or* with B and A in place of C and D. We use this to construct an isomorphism between A and B as follows. We give an effective procedure for listing pairs of integers. The construction is arranged so that at each stage in the listing the ordered pairs already listed constitute a finite correspondence and, furthermore, so that every integer occurs eventually as the first member of some ordered pair in the list *and* so that every integer occurs eventually as the second member of some ordered pair in the list. The entire collection of ordered pairs listed thus constitutes a recursive permutation under which A is recursively isomorphic to B.

The procedure is as follows. Let $A \leq_1 B$ via g, and let $B \leq_1 A$ via h.

Stage 0. Take $\langle 0, g(0) \rangle$ as the first ordered pair.

Stage 1. See whether $g(0) = 0$. If so, proceed to stage 2. If not, use the lemma (with A for D, B for C, and h for f) to list a new ordered pair whose second member is 0.

. . . .

Stage $2k$. See whether k occurs as the first member of some pair already listed. If so, go to stage $2k + 1$. If not, use the lemma (with A for C, B for D, and g for f) to list a new ordered pair whose first member is k.

Stage $2k + 1$. See whether k occurs as the second member of some pair already listed. If so, go to stage $2k + 2$. If not, use the lemma (with A for D, B for C, and h for f) to list a new ordered pair whose second member is k.

. . . .

This concludes the proof. ☒

The theorem shows that 1-degrees and isomorphism types coincide. The 1-reducibility ordering is thus an ordering of the isomorphism types. (As we saw in Chapter 4, the isomorphism types are, in effect, the basic objects of our theory.)

Theorem VI gives a quick and convenient way to demonstrate recursive isomorphism. For example, from Corollary IV we can immediately conclude the following corollary.

Corollary VI $K \equiv K_0$.

For another example, take $\{x|W_x \textit{ infinite}\}$ and $\{x|\varphi_x \textit{ total}\}$. Using the t function from the proof of Theorem IV, we can show that $\{x|W_x \textit{ infinite}\} \equiv_1 \{x|\varphi_x \textit{ total}\}$ (see Exercise 7-13). Then, by Theorem VI, we have that $\{x|W_x \textit{ infinite}\} \equiv \{x|\varphi_x \textit{ total}\}$ (and, for the purposes of recursive function theory, the two sets are identical).

§7.5 ONE-ONE COMPLETENESS AND MANY-ONE COMPLETENESS

K is both m-complete and 1-complete. Is *every* m-complete set also 1-complete? We answer this question in the following theorem.

Theorem VII *A is m-complete $\Leftrightarrow$ A is 1-complete.*

Proof. (A shorter and more elegant proof will be indicated in Chapter 11; the present proof is of independent interest, as we shall see in §7.6.)

$\Leftarrow$: Immediate.

$\Rightarrow$: Assume A is m-complete. Then A is recursively enumerable and $K \leq_m A$ via some recursive function f. We show that $K \leq_1 A$. The 1-completeness of A will then follow from the 1-completeness of K.

Recall that D_x is the finite set with canonical index x (see §5.6). Assume that the following lemma has been proved.

Lemma *If $K \leq_m A$, then there exists a recursive g such that for all x,*

$$[D_x \neq \emptyset \;\&\; D_x \subset A] \Rightarrow g(x) \in A - D_x.$$

and

$$[D_x \neq \emptyset \;\&\; D_x \subset \bar{A}] \Rightarrow g(x) \in \bar{A} - D_x.$$

We can then go on to prove that $K \leq_1 A$ as follows. Define f' by the following instructions.

To compute $f'(0)$. Set $f'(0) = f(0)$. (Hence $0 \in K \Leftrightarrow f'(0) \in A$.)

To compute $f'(n+1)$. See whether $f(n+1) \in \{f'(0), \ldots, f'(n)\}$. If not, set $f'(n+1) = f(n+1)$. If so, and $f(n+1) = f'(m_0)$, take the finite set $D_{x_0} = \{f'(m_0)\}$ and, using g from the lemma, compute $g(x_0)$. (x_0 is of course obtainable from $f'(m_0)$; it is, in fact $2^{f'(m_0)}$.) See whether $g(x_0) \in \{f'(0), \ldots, f'(n)\}$. If not, set $f'(n+1) = g(x_0)$. If so, and $g(x_0) = f'(m_1)$, take the finite set $D_{x_1} = \{f'(m_0), f'(m_1)\}$ and compute $g(x_1)$. (Here, $x_1 = 2^{f'(m_0)} + 2^{f'(m_1)}$.) See whether $g(x_1) \in \{f'(0), \ldots, f'(n)\}$ .... By the lemma, this procedure eventually yields $f'(n+1) \notin \{f'(0), \ldots, f'(n)\}$.

Furthermore, by the lemma, $n+1 \in K \Leftrightarrow f'(n+1) \in A$. Hence $K \leq_1 A$ via f'.

It remains to prove the lemma. For this we shall need a sublemma concerning f.

Sublemma *If $K \leq_m A$ via f, then*
(a) B recursive $\Rightarrow f^{-1}(B)$ recursive;
(b) $f(K)$ is infinite.

Proof of sublemma. (*a*) For any B, $f^{-1}(B) \leq_m B$ via f. Hence, by Theorem I(*e*), B recursive $\Rightarrow f^{-1}(B)$ recursive.

(*b*) Since $K \leq_m A$ via f, $f^{-1}(f(K)) = K$. Hence, by part (*a*), if $f(K)$ were finite (and therefore recursive), K would be recursive. But K is not recursive.

Proof of lemma. Choose an effective procedure for listing K. This yields an effective procedure for listing the set $f(K)$, which, by the sublemma, is infinite.

Since A is recursively enumerable, a recursive h can be found such that

$$W_{h(x)} = \begin{cases} f^{-1}(D_x), & \text{if } D_x \cap A = \emptyset; \\ N, & \text{if } D_x \cap A \neq \emptyset. \end{cases}$$

Let any D_x be given. We compute $g(x)$ according to the following instructions.

If $fh(x) \notin D_x$, set $g(x) = fh(x)$. If $fh(x) \in D_x$, set $g(x) =$ the first member in the list for $f(K)$ which is not a member of D_x.

Since $f(K)$ is infinite, $g(x)$ is convergent for all x. It remains to show that g has the desired properties. By definition, $g(x) \notin D_x$. Furthermore,

$$[D_x \neq \emptyset \;\&\; D_x \subset A] \Rightarrow W_{h(x)} = N \Rightarrow h(x) \in W_{h(x)} \Rightarrow h(x) \in K \Rightarrow g(x) \in A;$$

and

$[D_x \neq \emptyset \;\&\; D_x \subset \bar{A}] \Rightarrow W_{h(x)} = f^{-1}(D_x) \Rightarrow h(x) \notin f^{-1}(D_x)$ (else $h(x) \in K \;\&\; fh(x) \in A \;\&\; fh(x) \in D_x \;\&\; D_x \cap A \neq \emptyset$)

$$\Rightarrow [h(x) \notin W_{h(x)} \;\&\; fh(x) \notin D_x]$$
$$\Rightarrow [h(x) \in \bar{K} \;\&\; fh(x) \notin D_x] \Rightarrow g(x) \in \bar{A}.$$

This completes the proof of the lemma and hence of the theorem. ☒

Theorems VI and VII together show that the m-complete sets constitute a single isomorphism type. (The property of m-completeness is thus, by itself, a complete set of recursive invariants.) As we have noted, K is m-reducible to each of the nonrecursive, recursively enumerable sets exhibited so far. We can hence conclude, by Theorems VI and VII, that every nonrecursive, recursively enumerable set exhibited so far is *isomorphic* to K. (This is also true for the recursively enumerable sets indirectly mentioned in §2.2.)

We summarize our position at this point. The properties of *isomorphism to K*, 1-*completeness*, and m-*completeness* all coincide. Each of our examples of a nonrecursive, recursively enumerable set has had these properties. We have yet to determine whether or not creativeness is a more inclusive property and whether or not being recursively enumerable but not recursive is a still more inclusive property.

§7.6 CYLINDERS

For all A and B, $A \equiv_1 B \Rightarrow A \equiv_m B$. Each m-degree may thus be viewed as consisting of one or more 1-degrees, i.e., isomorphism types. Can we specify more exactly the structure of a single m-degree in terms of its component 1-degrees?

Definition A is a *cylinder* if $A \equiv B \times N$ for some B. (The geometrical basis for this term is evident. Note that $\equiv$ is used instead of $=$ in order that the concept of *cylinder* be recursively invariant.)

The following elementary but pleasing theorem gives us some of the information we desire.

Theorem VIII(a) $A \leq_1 A \times N$.

(*b*) $A \times N \leq_m A$ *(and therefore* $A \equiv_m A \times N$*)*.

(*c*) A *is a cylinder* $\Leftrightarrow (\forall B)[B \leq_m A \Rightarrow B \leq_1 A]$.

(*d*) $A \leq_m B \Leftrightarrow A \times N \leq_1 B \times N$.

Proof. (*a*) $A \leq_1 A \times N$ is immediate via $\lambda x[<x,0>]$.

(*b*) $A \times N \leq_m A$ is immediate via π_1.

(*c*) $\Rightarrow$: Assume $A \equiv C \times N$ and assume $B \leq_m A$. Then $B \leq_m C$ via some f. Let $f' = \lambda x[<f(x),x>]$. Then $B \leq_1 C \times N$ via f'. Hence $B \leq_1 A$.

$\Leftarrow$: Assume $(\forall B)[B \leq_m A \Rightarrow B \leq_1 A]$. Consider $A \times N$. By (*b*), $A \times N \leq_m A$; hence, by our assumption, $A \times N \leq_1 A$. By (*a*), $A \leq_1 A \times N$. Therefore $A \equiv_1 A \times N$. Hence, by Theorem VII, $A \equiv A \times N$, and A is a cylinder.

(*d*) $\Rightarrow$: Assume $A \leq_m B$. Then $A \times N \leq_m A \leq_m B \leq_1 B \times N$. Hence $A \times N \leq_m B \times N$. Hence, by (*c*), $A \times N \leq_1 B \times N$.

$\Leftarrow$: Assume $A \times N \leq_1 B \times N$. Then $A \leq_1 A \times N \leq_1 B \times N \leq_m B$, and hence $A \leq_m B$.☒

Corollary VIII *A is a cylinder* $\Leftrightarrow A \times N \leq_1 A \Leftrightarrow A \equiv A \times N$.

Proof. Immediate.☒

Call $A \times N$ the *cylindrification* of A. From (*a*), (*b*), and (*c*) in the foregoing theorem, every m-degree contains a maximum 1-degree; and this 1-degree is obtainable from any set in the m-degree by cylindrification. Furthermore, from (*d*) in the foregoing theorem, the structure of the m-reducibility ordering is mirrored in the 1-ordering in a natural way; i.e., there is a canonical homomorphism from the m-ordering into the 1-ordering.

The proof of Theorem VII in §7.5 showed, in effect, that $[[K \leq_m A$ & A recursively enumerable$] \Rightarrow A$ is a cylinder$]$. The lemma used in that proof suggests still another characterization of cylinder.

Theorem IX *A is a cylinder $\Leftrightarrow$ there exists a recursive g such that for all x,*

$$[D_x \neq \emptyset \,\&\, D_x \subset A] \Rightarrow g(x) \in A - D_x,$$

and

$$[D_x \neq \emptyset \,\&\, D_x \subset \bar{A}] \Rightarrow g(x) \in \bar{A} - D_x.$$

(Thus a set is a cylinder if and only if it has a certain "doubly productive" property with respect to finite sets.) The proof of Theorem IX is given as Exercise 7-35. It is a straightforward modification of the proof of Theorem VII.

A special characterization of recursively enumerable cylinders is given in Exercise 7-37.

§7.7 PRODUCTIVENESS

The concept of *productiveness* is basic in applications of recursive function theory to logic. We study it further here. Some of the applications to logic will be indicated in §7.8; others will be given later (see also Dekker [1955]).

It follows from the definition of productiveness that if a set A is productive, then there is an effective procedure by which, given any recursively enumerable subset of A, we can get a larger recursively enumerable subset of A. Let us consider ways of iterating this procedure. We are led first to the following theorem.

Theorem X *A productive $\Rightarrow$ A has an infinite recursively enumerable subset.*

Proof. We first give an informal proof. Let ψ be a productive partial function for A. We get a recursive g whose range is an infinite subset of A. The instructions for g are inductive. Let z_0 be an r.e. index for the empty set.

To compute $g(0)$. Set $g(0) = \psi(z_0)$. Since $W_{z_0} \subset A$, $\psi(z_0)$ is defined and in A.

To compute $g(n+1)$. Let z_{n+1} be an r.e. index for the finite set $\{g(0), \ldots, g(n)\}$. Set $g(n+1) = \psi(z_{n+1})$. Since $W_{z_n} \subset A$, $\psi(z_{n+1})$ is defined and in A.

Evidently g is a one-one function. This concludes the informal proof.

The recursiveness of g becomes clearer in the following, more formal version of this proof. From Theorem 5-XIII, there is a recursive function f of two variables such that

$$W_{f(x,y)} = W_x \cup W_y.$$

By Church's Thesis, there is a recursive function h such that

$$W_{h(x)} = \{x\}.$$

As before, let z_0 be an r.e. index for $\emptyset$. Define the function k as follows:

$$k(0) = z_0,$$
$$k(n+1) = f(h\psi k(n), k(n)).$$

Then $k(n)$ corresponds to z_n in the preceding informal proof. We define

$$g = \psi k. \boxtimes$$

Corollary X *A productive $\Rightarrow$ A has an infinite recursive subset.*

Proof. Immediate by Theorem 5-IV. $\boxtimes$

This proof suggests that more extensive iterations of the productive procedure be tried. Consider the following *informal* instructions for listing a subset of A. The instructions involve transfinite induction over ordinal numbers. (We shall assume, for the moment, that the reader is familiar with this latter notion. Exercises 11-43 to 11-53 give a review of ordinal numbers.)

"Take the function g of the proof of Theorem X. Let x_0, x_1, x_2, . . . be the integers $g(0)$, $g(1)$, $g(2)$, ... Let z_ω be an r.e. index (obtained from the instructions for g) for the set $\{x_0,x_1, \ldots\}$ = *range* g. Then $\psi(z_\omega)$ is a new member of A. Let x_ω be $\psi(z_\omega)$. Then $\{x_\omega\} \cup \{x_0,x_1, \ldots\}$ is itself a recursively enumerable set. Let $z_{\omega+1}$ be an r.e. index for *this* set. Then $x_{\omega+1} = \psi(z_{\omega+1})$ is a new member of A. In this way we can list the set $\{x_0,x_1, \ldots\} \cup \{x_\omega,x_{\omega+1}, \ldots\}$. Let $z_{\omega+\omega}$ be an r.e. index (obtained from *these* instructions) for this latter set. Then $x_{\omega+\omega} = \psi(z_{\omega+\omega})$ is a new member of A. In this way, obtain $x_{\omega+\omega+\omega}$, x_{ω^ω}, etc. Now make a final listing (of a subset of A) by dovetailing all the preceding listings."

These instructions would appear to enumerate a subset of A that is, in some sense, closed under the productive procedure. It might be hoped, in fact, that they would yield all of A. But, by Theorem V(c), A is not recursively enumerable, and, by the definition of productiveness, no recursively enumerable subset of A *can* be closed under the productive procedure (any such subset must itself have an r.e. index w, and $\psi(w)$ will be in A but outside the subset). We thus reach an apparent paradox.†

The solution to this apparent paradox is that our informal instructions *cannot* be made fully precise. Any precise version put forward will be less inclusive than the loose phrasing of the informal instructions might seem to indicate. No matter what precise counterpart is given, a more inclusive precise version can be found. Precise versions involve systems of notation for ordinal numbers. We are dealing with an essential limitation on our

† It is similar in form to the Burali-Forti paradox in the theory of ordinal numbers. (The Burali-Forti paradox arises from the treatment of the ordered collection of all ordinals as a new ordinal.) The present apparent paradox, however, involves only countable ordinals.

ability to set up comprehensive systems of notation for ordinals. The subject of ordinal notations is fundamental in more advanced parts of recursive function theory. It is considered further in Chapter 11. As we shall see, the subject is closely related to logical incompleteness theorems.

The above "paradox" warns us of dangers in being too loose and informal in recursive-function-theoretic arguments. Exercise 7-46 shows that a precise, nonconstructive version of the informal instructions *can* be found, if we do not require that the resulting set be recursively enumerable.

We now turn to other characterizations of productiveness and to other concepts related to productiveness. Several results are given here. Further material will be found in exercises in §7.9.

Given a productive set, can a *total* productive partial function be found for it? In Chapter 11, a short and elegant proof will be given that this is so. We now give a longer proof; it gives valuable insight into the later proof, which it preceded historically.

Theorem XI *A is productive $\Leftrightarrow$ there is a (total) recursive function f such that A is productive with f as its productive partial function.*

Proof. $\Leftarrow$: Immediate.

$\Rightarrow$: Let A be productive with productive partial function ψ. We give instructions for computing f.

By methods that are now familiar to the reader, a recursive function g is found such that

$$W_{g(z,x)} = \begin{cases} W_x, & \text{if } z \in K; \\ \emptyset, & \text{otherwise.} \end{cases}$$

Let $k_0, k_1, \ldots$ be an effective listing of K.

To compute $f(x)$. Begin computations for $\psi(x), \psi g(k_0,x), \psi g(k_1,x), \ldots$. Take whichever converges first as the value for $f(x)$.

f is a productive partial function for A; for $x, g(k_0,x), g(k_1,x), \ldots$ are all indices for W_x, and $f(x)$ must converge if $\psi(x)$ converges.

f is total; for assume $f(x_0)$ were divergent for some x_0. Then $\psi g(y,x_0)$ would be divergent for all $y \in K$. However, for $y \in \bar{K}$, $W_{g(y,x_0)} = \emptyset$; and hence, for $y \in \bar{K}$, $\psi g(y,x_0)$ is convergent. Thus we would have $\bar{K} = \{y|\psi g(y,x_0)\ \textit{convergent}\}$, and, by the second projection theorem, $\bar{K}$ would be recursively enumerable, contrary to the nonrecursiveness of K. ☒

When a productive partial function is total, it is called a *productive function*. Stronger versions of Theorem XI are given in Exercises 7-50 and 7-51. (Post originally defined a creative set to be a recursively enumerable set whose complement is productive with a (total) productive function.)

The assertion that A is not recursively enumerable can be expressed: $(\forall x)[W_x \neq A]$, or, in more detail, $(\forall x)(\exists y)[y \in W_x - A \text{ or } y \in A - W_x]$. We might say that this assertion holds "constructively" (or "uniformly") if there is a recursive function f such that $(\forall x)[f(x) \in W_x - A$ or $f(x) \in A - W_x]$. This suggests the following definition.

Definition A is *completely productive* if there exists a recursive function f such that for all x, either $f(x) \in W_x - A$ or $f(x) \in A - W_x$. f is called a *completely productive function* for A.

Example. $\bar{K}$ is completely productive with $\lambda x[x]$ as completely productive function. For $x \in W_x \Rightarrow x \notin \bar{K}$, and $x \notin W_x \Rightarrow x \in \bar{K}$, by the definition of K.

Clearly a completely productive set must be productive. Does the converse hold? In Chapter 11, we shall see that it does and that the concepts of productiveness and complete productiveness are therefore equivalent. Still another characterization of productiveness is given in Exercise 11-15, where we see that A is productive $\Leftrightarrow \bar{K} \leq_1 A \Leftrightarrow \bar{K} \leq_m A$.

In Exercise 7 13, we see that there are $2^{\aleph_0}$ productive sets. Is it true that every non-recursively-enumerable set is productive? Theorem X, together with the result of Exercise 5-8, shows that this is not true; there exist nonproductive, non-recursively-enumerable sets. We study such sets further in Chapter 8.†

The following definition asserts that a productive procedure exists for going from recursively enumerable subsets to larger recursively enumerable subsets.

Definition A is *semiproductive* if there exists a partial recursive ψ such that $(\forall x)[W_x \subset A \Rightarrow [\psi(x)$ convergent & $W_x \subset W_{\psi(x)}$ & $W_x \neq W_{\psi(x)}$ & $W_{\psi(x)} \subset A]]$.

Clearly, A productive $\Rightarrow A$ semiproductive (Exercise 7-53). Furthermore, an analogue to Theorem XI holds for semiproductive sets (Exercise 7-54). Must every semiproductive set be productive? Exercise 8-42 shows that this is not true; a set can be semiproductive without being productive. The result is due to Shoenfield (see Exercise 8-42).

Disjoint Pairs of Recursively Enumerable Sets

It is possible to develop a theory of *disjoint pairs of recursively enumerable sets* that is analogous to our theory of recursively enumerable sets. This theory is significant in certain applications to logic (where, for instance, we may be interested in both the set of provable statements and the set of disprovable statements of some formal system).

In this theory, the analogue to recursiveness is as follows.

Definition A and B are *recursively separable* if there exists a recursive set C such that $A \subset C$ and $B \subset \bar{C}$.

† For such a set A, the assertion $(\forall x)(\exists y)[y \in W_x - A$ or $y \in A - W_x]$ must hold "nonconstructively" (or "nonuniformly"). The assertion is true, but y cannot be a recursive function of x.

The analogue to *creativeness* is as follows.

Definition A and B are *effectively inseparable* if there exists a partial recursive function ψ of two variables such that for any u and v,

$$[A \subset W_u \;\&\; B \subset W_v \;\&\; W_u \cap W_v = \emptyset] \Rightarrow [\psi(u,v) \text{ convergent } \& \; \psi(u,v) \in \overline{W_u \cup W_v}].$$

The following theorem is straightforward.

Theorem XII(a) *For any sets A and B, if A and B are effectively inseparable, then A and B are not recursively separable.*

(**b**) *For any disjoint pair of recursively enumerable sets A and B, if A and B are effectively inseparable, then A and B are each creative.*

(**c**) *There exists an effectively inseparable, disjoint pair of recursively enumerable sets.*

Proof. For (*a*) and (*b*); immediate (see Exercise 7-56).

(*c*) Let $A_0 = \{x|\varphi_x(x) = 0\}$ and $A_1 = \{x|\varphi_x(x) = 1\}$. A_0 and A_1 are evidently disjoint recursively enumerable sets.

Let u and v be given. Define η as follows.

To compute $\eta(y)$, simultaneously list W_u and W_v; if y appears first in W_u, set $\eta(y) = 1$; if y appears first in W_v, set $\eta(y) = 0$.

Then $\eta = \varphi_{f(u,v)}$ for some f. f is now the desired "productive function." For if $A_0 \subset W_u$, $A_1 \subset W_v$, and $W_u \cap W_v = \emptyset$, then, for example,

$$f(u,v) \in W_u \Rightarrow \eta(f(u,v)) = 1 \Rightarrow \varphi_{f(u,v)}(f(u,v)) = 1 \Rightarrow f(u,v) \in A_1 \Rightarrow f(u,v) \in W_v, \quad \text{a contradiction.} \boxtimes$$

A slightly different formulation of several of the above ideas was given in Exercise 2-30. From time to time, we shall give other results from the theory of pairs of recursively enumerable sets. A treatment of this theory, and of its applications to logic, is given in Smullyan [1961]. An extension from pairs of sets to sequences of sets is considered in Cleave [1961].

§7.8 LOGIC

Some Terminology

The *well-formed formulas* (we abbreviate: wffs) of a formalized logical system are a certain infinite class of finite strings of symbols from a certain basic finite alphabet. Invariably, the wffs are specified in such a way that an effective procedure exists for deciding which strings are wffs and which strings are not. In the investigation of a logical system, the wffs are the basic objects of study. In terms of motivation, the wffs are the class of "meaningful" strings. In most of our examples from quantificational logic, the wffs will coincide with what are often called, in quantificational logic, the *sentences* of a system.†

In order to apply the concepts of recursive function theory, one must use a coding from wffs to integers. We shall limit our discussion to codings which map *onto* N.‡ The integer that is associated with a wff under a coding is called the *Gödel number* of that wff (under that coding). (We assume, as usual, that the operations of encoding and decoding are effective.) A general discussion of codings was given in §1.10.

Let a logical system be given. Then a coding associates with each set of wffs a set of integers. Let a set of wffs be given. It is clear that any recursively invariant property that holds for the set of Gödel numbers obtained under one coding will also hold for the set of Gödel numbers obtained under any other coding (see §1.10). Recursively invariant properties may therefore be directly associated with sets of wffs. We may speak, for example, of a recursive set of wffs, or of a recursively enumerable set of wffs, or of a productive set of wffs.

In the study of a particular logical system, certain sets of wffs are usually distinguished as being of special interest. For example, one set of wffs may be distinguished as "provable" (under certain specified syntactical rules of proof), or another set of wffs may be distinguished as "true" (under some—usually nonconstructive—definition of *truth*). A set of wffs so distinguished is often called a *theory*. (Ordinarily, one requires that a "theory" be closed with respect to certain underlying rules of deducibility. Often, these are the rules of basic quantificational logic. This restriction is unimportant for our purposes; for the moment, we simply define a theory to be any *set* of wffs.)

The wffs of a theory can often be effectively listed. This is usually the case when the theory consists of the "provable" wffs under certain formal rules of proof.§ It has become customary to say that a theory, i.e., set of wffs, is *axiomatizable* if it can be effectively listed, i.e., if it is recursively enumerable. Similarly, a theory is said to be *decidable* if it is recursive.

Existence of recursively enumerable but nonrecursive sets (like K) suggests the possibility that certain well-known axiomatizable theories may not be decidable. This is indeed true. *Church's theorem* (Church [1936*a*])

† *Sentences* are expressions in which every variable is acted on by some quantifier. We note in passing that, in quantificational logic, the term "wff" is usually applied to expressions containing variables not acted on by a quantifier as well as to sentences. We do not wish, in our present general discussion, to distinguish more than a single class of "meaningful" strings in a given system, and we have chosen to call this class the "wffs" even though this conflicts, in some examples, with the conventional more inclusive usage of the term "wff."

‡ In practice, it is quite common to use codings that map onto a recursive, proper subset of N. In order to consider such cases we would need to make several minor modifications in our discussion.

§ One proceeds by making an exhaustive list of all proofs (much as the derivations for all primitive recursive functions were listed in §1.4).

shows that the provable wffs of basic quantificational logic form an undecidable theory. We shall see below that the provable wffs of elementary arithmetic (under any of various standard axiomatizations) also form an undecidable theory.

The Gödel Incompleteness Theorem

The "incompletableness phenomenon" of Gödel [1931] (see also Tarski [1932, 1936]) can be formulated in a number of different ways. Perhaps the simplest is to say that in any formal system which has a certain minimal complexity and for which the notion of "true" wff can be defined in a certain natural way, the set of "true" wffs is productive (and hence not recursively enumerable). Thus the "true" wffs do not form an axiomatizable theory.

We are already in a position to see, informally, that this must be so. Consider any formal mathematical system flexible and inclusive enough to make assertions about Turing machines and their indices. Such statements as "17 *is the index for a total recursive function*" will then be expressible within the theory. Consider statements of this form, i.e., of the form, "**n** *is the index for a total recursive function*," where **n** is the numeral for an integer. We cannot hope for a procedure that will list all true statements of this form and no false statements of this form, for $\{x|\varphi_x \textit{ total}\}$ is, as we know, a productive set. Indeed, from the instructions for any procedure which lists only true statements of this form, we can effectively obtain (by productiveness of $\{x|\varphi_x \textit{ total}\}$) a *new* true statement of this form which is not listed. Similarly, consider statements of the form "$P_{\mathbf{n}}$ *with input* **n** *diverges*," i.e., of the form "$\mathbf{n} \notin K$." We cannot hope to list all true and no false statements of this latter form, since $\bar{K}$ is productive.

Elementary Arithmetic

We now consider a specific logical system. The wffs of *elementary arithmetic* are those wffs which can be built up from $+$, $\times$, $=$, 0, 1, 2, . . . variable symbols for nonnegative integers, quantifiers over nonnegative integers, and the sentential connectives $\neg$, &, $\vee$, $\Rightarrow$, $\Leftrightarrow$: subject to the requirement that every variable in a wff be acted on by some quantifier. For example,

$$(\forall a)[\neg\, a = 2 \Rightarrow (\exists b)a = b \times b]$$

is a wff. (It makes the false assertion that every integer different from 2 is a square.)

It is possible to define the set of "true" wffs of elementary arithmetic in a way that is straightforward and entirely in accord with our intuition.†

† This theory (the set of true wffs) is sometimes called *elementary number theory* or *true elementary number theory*. The phrase "elementary number theory" is also sometimes applied to the set of all wffs of elementary arithmetic.

The definition of *true* wff of elementary arithmetic can itself be formalized. As Tarski has shown, however, such formalization requires a means of expression that is richer than the means of expression of elementary arithmetic (see Exercise 11-45).

We can thus speak of *true* wffs and of *false* wffs. We assume that this definition of *truth* has been made.

An unexpectedly wide variety of combinatorial statements can be expressed within elementary arithmetic. In particular (see Exercise 7-64) an expression F with one unquantified variable can be found such that, when the numeral for any integer x is substituted in place of the unquantified variable, the resulting wff (which we henceforth call F_x) asserts, intuitively, that $x \in K$. More precisely, it can be shown that

$$x \in K \Leftrightarrow [F_x \text{ is true}],$$

and

$$x \notin K \Leftrightarrow [F_x \text{ is false}] \Leftrightarrow [(\neg F_x) \text{ is true}],$$

where $(\neg F_x)$ is the negation of F_x.

The following lemma, though trivial, is basic for our further discussion.

Basic lemma *[B recursive & $A \cap B$ productive] $\Rightarrow$ A productive.*

Proof. See Exercise 7-60.☒

Since the set of wffs $\{F_x | x \in N\}$ is recursive, and $\{F_x | x \notin K\}$ (= $\{F_x | F_x$ *is false*$\}$) is productive (see Exercise 7-61), we can apply the basic lemma. We get the following.

(*a*) *The true wffs of elementary arithmetic form a productive set* (*and therefore a nonrecursive set*).

(*b*) *The false wffs of elementary arithmetic form a productive set.*

Thus the true wffs constitute an undecidable and unaxiomatizable theory.

There are various standard ways of specifying rules of proof in elementary arithmetic. One such is based on *Peano's axioms* (including those instances of the Peano induction axiom that can be expressed in elementary arithmetic); the resulting theory, i.e., set of provable wffs, is often called *Peano arithmetic.*† A different theory (of "provable" wffs) can be obtained by taking as "provable" all those wffs of elementary arithmetic which are provable in a standard set theory (whose own class of wffs may go beyond the wffs of elementary arithmetic). We call this a *set-theory arithmetic.* For either of these choices of axiomatizable theory, it is not difficult to show that

$$x \in K \Rightarrow [F_x \text{ is provable}].$$

(Indeed, the "proof" of F_x amounts to exhibiting the computation that puts x in K.)

† More precisely, *Peano arithmetic* consists of those wffs of elementary arithmetic that are deducible within elementary quantifier logic from the following axioms: $(\forall a)(\forall b)[a + 1 = b + 1 \Rightarrow a = b]$; $(\forall a)[\neg 0 = a + 1]$; $(\forall a)[a + 0 = a]$; $(\forall a)(\forall b)[a + (b + 1) = (a + b) + 1]$; $(\forall a)[a \times 0 = 0]$; $(\forall a)(\forall b)[a \times (b + 1) = (a \times b) + a]$; $0 + 1 = 1$; all wffs of the form $(\cdot ((1 + 1) + 1) \cdots + 1) = \mathbf{x}$, where $\mathbf{x}$ is the numeral of the integer x, and the x is the number of 1's appearing on the left; and all wffs of the form

$$[\cdots 0 \cdots \;\&\; (\forall a)[\cdots a \cdots \Rightarrow \cdots (a + 1) \cdots]] \Rightarrow (\forall a) \cdots a \cdots.$$

Consider either Peano arithmetic or a set-theory arithmetic. Let us make the *special assumption* that no false wffs of elementary arithmetic are provable. Then

$$x \in K \Leftrightarrow [F_x \text{ is provable}],$$

and therefore $$x \notin K \Leftrightarrow [F_x \text{ is not provable}].$$

Applying our basic lemma, we have the following.

(*c*) *The unprovable wffs form a productive set.*

(*d*) *The provable wffs form a creative (and therefore not recursive) set.*

Hence the provable wffs constitute an undecidable theory.

Next, consider $\{F_x | (\neg F_x)$ *is provable*$\}$. This set is a recursively enumerable subset of the unprovable wffs. Using the productiveness of $\bar{K}$, we have the following.

(*e*) *There is an* x_0 *such that neither* F_{x_0} *nor* $(\neg F_{x_0})$ *is provable.*

Such a formula is sometimes called an *undecidable* wff. Note that of these two wffs, we immediately have that $\neg F_{x_0}$ is true and F_{x_0} is false.

The *special assumption* that no false wff of elementary arithmetic is provable is, of course, essential.† For the case of Peano arithmetic, this assumption can itself be demonstrated within more general set-theoretic mathematics. For the case of a set-theory arithmetic, this assumption must rest either on empirical grounds or on intuitive insight that goes beyond the set theory used.

To summarize informally: no axiomatization of mathematics can exactly capture all true statements of elementary arithmetic; and from any axiomatization which yields only true statements of elementary arithmetic, a new true statement can be found not provable in that axiomatization.

These facts lend special significance to our study of productiveness. Post believed that such facts manifest an essentially creative quality of mathematics; hence the name *creative set.*

Remark. Results (*c*), (*d*), and (*e*) do not mention truth. For these results, it is possible to replace the assumption that no false wff is provable by a weaker syntactical assumption (i.e., combinatorial assumption about the symbolism) concerning *consistency* of the set of provable wffs. In his original proof, Gödel does this. In effect, he shows that $[x \in K \Leftrightarrow [F_x$ is provable]] holds provided the set of provable wffs has a property known to logicians as *ω-consistency.* Rosser later strengthened this result by (in effect) using a somewhat more complex expression G in place of Gödel's F and showing that $[x \in K \Leftrightarrow [G_x$ is provable]] holds provided the set of provable wffs is *consistent.* For definitions and an outline of methods, see Exercises 7-64 and 7-65. Further results about set-theory arithmetic are given in §14.7 and in Exercises 14-21 to 14-23.

† If no false wff occurs in a set of wffs, that set of wffs is said to be *sound.* Our *special assumption* asserts that the set of provable wffs is sound.

§7.9 EXERCISES

§7.1

7-1. Show that $[A \leq_m \bar{A}$ & A recursively enumerable$] \Rightarrow A$ recursive.

7-2. Show that it is possible to have $A \leq_m \bar{A}$ but A not recursive. (*Hint:* Use K *join* $\bar{K}$.)

7-3. (Dekker). (*a*) Assume: A and B recursively enumerable. Show:

$$[A \cup B = N \& A \cap B \neq \emptyset] \Rightarrow A \leq_m A \cap B.$$

$\triangle$(*b*) Assume: A and B recursively enumerable. Show:

$$[A \cup B = N \ \& A \cap B \text{ infinite}] \Rightarrow A \leq_1 A \cap B.$$

7-4. Assume that A and B are infinite sets such that A differs from B by a finite set; that is, $(A - B) \cup (B - A)$ is finite. Show that $A \equiv_m B$. (It need not be true that $A \equiv_1 B$; see Chapter 8.)

7-5. Investigate the relationship, under $\leq_m$, of the following sets.

(i) $\{x|x \text{ } prime\}$,
(ii) $\{x|x \text{ } even\}$,
(iii) $\{x|W_x = \emptyset\}$,
(iv) $\{x|W_x \text{ } infinite\}$,
(v) $\{x|\varphi_x \text{ } total\}$,
$\triangle$(vi) $\{x|W_x = W_n\}$ for a fixed n. (The relation of (vi) to the other sets will depend upon n. Three cases must be considered.)

7-6. (*a*) Analyze the structure of the recursive sets under $\leq_m$.
(*b*) Show that $[A$ recursive and B nonrecursive$] \Rightarrow A \leq_m B$.

7-7. (*a*) Analyze the structure of the recursive sets under $\leq_1$.
$\triangle$(*b*) Show that there exist A and B such that A is recursive, and B is nonrecursive, but $A \not\leq_1 B$. (*Hint:* Use result of Exercise 5-8.)

7-8. (*a*) Show that every m-degree contains at most $\aleph_0$ sets. Do any contain fewer than $\aleph_0$ sets?
(*b*) Show that every m-degree has at most $\aleph_0$ m-degrees below it in the m-reducibility ordering.
(*c*) Show that parts (*a*) and (*b*) hold for 1-degrees and the 1-reducibility ordering.

7-9. Show that the construction of Theorem II(*b*) cannot be used to prove that the 1-reducibility ordering is a semilattice. Show, however, that the recursive 1-degrees, by themselves, constitute an upper semilattice under $\leq_1$.

7-10. Define: $A \leq_{cf} B$ if there exists a recursive g such that $c_A = gc_B$. Is $\leq_{cf}$ reflexive and transitive? If so, give a complete discussion of the reducibility ordering it generates.

$\triangle$***7-11.*** Show that $\{x|\varphi_x \text{ } total\}$ and $\{x|\varphi_x \text{ } not \text{ } total\}$ are incomparable under $\leq_m$. (*Hint:* Show that, otherwise, pairs of indices of partial recursive functions could be listed so that, in each pair, one and only one partial function would be total; and show that this listing could be used to get a diagonal construction over the class of all (total) recursive functions yielding a contradiction. The diagonal construction is complex, and a given pair of indices may be diagonalized more than once.)

§7.2

7-12. Locate $\{x|W_x = \emptyset\}$ in the 1-reducibility ordering. (*Answer:* $\equiv_1 \bar{K}$.)

7-13. (*a*) Repeat Exercise 7-5 for relationship under $\leq_1$.

(*b*) Let D be a finite set. Investigate the relationship, under $\leq_1$, of $K \times \bar{K}$ and $\{x | W_x = D\}$.

▲*7-14.* What is the relation between $\{x | W_x \textit{ recursive}\}$ and $\{x | W_x \textit{ infinite}\}$ under $\leq_1$?

7-15. (*a*) Show that [φ partial recursive & φ one-one & $\varphi(N)$ recursive & A recursive] $\Rightarrow \varphi(A)$ recursive.

(*b*) Show that [φ partial recursive & $\varphi(N) = N$ & A recursive & $\varphi(A)$ nonrecursive] is possible.

7-16. In each of the following cases, show that the relation defined between A and B is reflexive and transitive. Analyze the structure of the corresponding "degrees of unsolvability" with special regard to recursively enumerable sets.

(i) ($\exists$ partial recursive φ)[$A \subset \varphi^{-1}(B)$].

(ii) ($\exists$ partial recursive φ)[$A = \varphi^{-1}(B)$].

(iii) ($\exists$ partial recursive φ)[φ is one-one & $A = \varphi^{-1}(B)$].

(*Partial answer:* (i) All sets fall into two degrees; (ii) recursively enumerable sets fall into two degrees; (iii) recursively enumerable sets fall into infinitely many degrees.)

§*7.3*

7-17. In example 2 of §7.3, verify that g is the desired productive partial function.

7-18. Show that [$A \neq \emptyset$ & ($\exists$ partial recursive ψ)$(\forall x)$[[$W_x \subset A$ & $W_x \neq \emptyset$] $\Rightarrow$ [$\psi(x)$ convergent & $\psi(x) \in A - W_x$]]] $\Rightarrow A$ productive.

7-19. Show by direct construction, i.e., not by Theorem V(*b*), that the following sets are productive: $\{x | \varphi_x \textit{ is a permutation}\}$ and $\{x | \varphi_x \textit{ is onto}\}$.

7-20. Show that $\{x | \varphi_x \textit{ not total}\}$ is productive. (*Hint:* Show $K \leq_m \{x | \varphi_x \textit{ total}\}$.) This is a set such that both it and its complement are productive.

7-21. Show that there are $\aleph_0$ creative sets.

7-22. Show that $\{x | (\exists y)[x \in W_y \ \& \ y \in W_x]\}$ is creative, with $\lambda x[x]$ as productive partial function.

7-23. Which sets in Exercise 7-5 are creative? Which have creative complements?

7-24. Discuss $f(A)$ and $f^{-1}(A)$ when f is a one-one recursive function and A is creative; when f is a one-one recursive function and A is productive.

7-25. (*a*) Show: [A productive & B recursively enumerable & $B \subset A$] $\Rightarrow A - B$ productive.

(*b*) Show: [A productive & B recursively enumerable & $A \subset B$] $\Rightarrow A \cup \bar{B}$ productive.

△*7-26.* Show: [A creative & B recursively enumerable & $A \leq_m B \times \bar{B}$] $\Rightarrow B$ creative.

△*7-27.* Show: A creative $\Rightarrow A \times A$ creative. (The converse of this has been shown by Lachlan.)

§*7.4*

7-28. Between what pairs of sets in Exercise 7-5 does recursive isomorphism hold?

7-29. Show: $A \textit{ join } \bar{A} \equiv \overline{A \textit{ join } \bar{A}}$.

△*7-30.* Show that [$f(A) = B$ & $f(\bar{A}) = \bar{B}$ & $g(B) = A$ & $g(\bar{B}) = \bar{A}$ & f recursive & g recursive] $\Rightarrow A \equiv B$.

△*7-31.* Let $R_x = \textit{range } \varphi_x$. Let V_x be defined as in Exercise 5-11(*b*). Then $\{R_x\}_{x=0}^{x=\infty}$, $\{V_x\}_{x=0}^{x=\infty}$, and $\{W_x\}_{x=0}^{x=\infty}$ each give an indexing for the recursively enumerable sets. Show that, given any two of these three indexings, one of them is, in an appropriately strong sense, a recursive permutation of the other. (*Hint:* Use methods from proof of Theorem VI.)

§*7.5*

▲*7-32.* (Myhill). Show that any two creative sets are recursively isomorphic.

§7.6

7-33. Show how the structure of the recursive sets under $\leq_1$ and under $\leq_m$ accords with Theorem VIII.

7-34. Define: $A \leq_1^r B$ if ($\exists$ recursive f)[f is one-one & $A = f^{-1}(B)$ & *range* f is recursive].

(*a*) Show that $\leq_1^r$ is reflexive and transitive and that the "degrees" under $\leq_1^r$ are just the recursive-isomorphism types.

(*b*) Show that $A \leq_m B \Leftrightarrow A \times N \leq_1^r B \times N$.

(*c*) Show that, when either A or B is a cylinder, $A \leq_1 B \Rightarrow A \leq_1^r B$. (R. W. Robinson has exhibited recursively enumerable sets A and B such that $A \leq_1 B$ but $A \not\leq_1^r B$).

7-35. Prove Theorem IX.

7-36. (*a*) Show that if either A or B is a cylinder, then $A \times B$ is a cylinder.

(*b*) Show that if both A and B are cylinders, then A *join* B is a cylinder. (*Hint:* Use Theorem IX.)

In Exercise 8-29, we shall see that A *join* B and $A \times B$ need not be of the same m-degree, even when A and B are recursively enumerable.

(*c*) Which of $K \times \bar{K}$ and K *join* $\bar{K}$ are cylinders?

(*d*) Show that if A *join* B is a cylinder, it need not follow that either A or B is a cylinder.

$\triangle$(*e*) What is the relation of $K \times \bar{K}$ and K *join* $\bar{K}$ in the m-ordering? (*Partial answer:* They are of distinct degrees.)

$\triangle$*7-37.* Define: A is a *one-one splinter* if ($\exists$ recursive, one-one f)($\exists x$)[$A = \{x, f(x), ff(x), fff(x), \ldots\}$]. (Splinters were studied by Ullian and later by Myhill; see Ullian [1960] and Myhill [1959*a*].)

(*a*) Show that the concept is recursively invariant.

(*b*) Show that, for A and $\bar{A}$ infinite, A is a one-one splinter $\Rightarrow A$ is a recursively enumerable cylinder. (*Hint:* Show: $(\forall B)[B \leq_m A \Rightarrow B \leq_1 A]$. Note that "cycles" may occur in $\bar{A}$.)

(*c*) Show that [$A \neq \emptyset$ & A is a recursively enumerable cylinder] $\Rightarrow A$ is a one-one splinter. (*Hint:* As $A \equiv A \times N$, it is enough to get $A \times N$ as a one-one splinter. Let $n \in A$. Define f so that $f(\langle x,y\rangle) = \langle x, y+1\rangle$, except that, as members of A appear in an effective listing of A, this rule is modified, by dovetailing, in such a way that the successors of $\langle n,0\rangle$, under f, gradually exhaust $A \times N$.) (*Note:* The definition of *splinter* is obtained from the definition of one-one splinter by omitting the condition that f be one-one. Young has shown the existence of an infinite splinter which is not a cylinder.)

(*d*) (Young). Define: A recursive permutation f is *cyclefree* if for every nonempty finite D, $f(D) \neq D$.

Prove: A is a cylinder $\Leftrightarrow$ there exists a cyclefree recursive permutation f such that $f(A) = A$.

7-38. Use Theorem IX to show that all sets in Exercise 7-5 are cylinders.

§7.7

7-39. f is a given recursive function. Let $\mathcal{A} = \{B | B$ *is productive with f as productive function*$\}$.

(*a*) Give examples of f such that $\mathcal{A} = \emptyset$ and such that $\mathcal{A} \neq \emptyset$.

(*b*) Show: [$A \in \mathcal{A}$ & $B \in \mathcal{A}$] $\Rightarrow A \cap B \in \mathcal{A}$.

(*c*) Show: [$A \in \mathcal{A}$ & $B \in \mathcal{A}$] does not always imply $A \cup B \in \mathcal{A}$.

$\blacktriangle$(*d*) Show: [$A \in \mathcal{A}$ & $B \in \mathcal{A}$] $\Rightarrow A \cup B$ productive.

$\blacktriangle$(*e*) Show that $\mathcal{A}$ generates a lattice of productive sets under the operations of $\cap$ and $\cup$.

7-40. (Dekker). Let *Dom A* be $\{x|W_x \subset A\}$. Show: A productive $\Rightarrow$ *Dom A* productive.

7-41. Show: $\bar{K}$ − *Dom* $\bar{K}$ is infinite.

7-42. (See 7-40.) Let A be productive with ψ as productive partial function. Let $center_\psi\ A = \psi\ (Dom\ A)$. Show that $center_\psi\ A \subset B \subset A \Rightarrow B$ productive.

7-43. Show that there are $2^{\aleph_0}$ productive sets. (*Hint:* Use 7-41 and 7-42.)

7-44. Show that every productive set has $2^{\aleph_0}$ productive subsets. (*Hint:* See 7-43. Find a center for the productive set such that infinitely many integers are in the set but not in its center.)

7-45. Let A be productive with ψ as productive partial function. Show that there is a productive set B with ψ as productive partial function such that $B = center_\psi\ B$. (*Hint:* Take intersection of all sets productive by ψ.) We call such a productive set *full* with respect to ψ.

△***7-46.*** (Dekker). (This exercise assumes some knowledge of countable ordinal numbers. Exercises 11-46 through 11-53 provide a review of ordinal numbers.) Let ψ be the productive partial function for a productive set. Define, by transfinite induction over the countable ordinals,

$$S_0 = \emptyset,$$
$$S_{\alpha+1} = S_\alpha \cup \{\psi(x)|W_x \subset S_\alpha\},$$
$$S_\beta = \bigcup_n S_{\alpha_n}, \qquad \text{for } \beta = \lim_n \alpha_n.$$

Let $B = \bigcup_\alpha S_\alpha$. Show that B is full (see Exercise 7-45) and coincides with the intersection of all sets productive with respect to ψ.

(We shall see later that any two such sets, for differing *one-one* ψ, are recursively isomorphic and of a high degree of unsolvability.)

7-47. Show that every infinite recursively enumerable set is the disjoint union of a creative set and a productive set.

7-48. Show that every productive set is the disjoint union of a creative and a productive set. (*Hint:* Use Theorem X and Exercise 7-47.)

7-49. Show that every one-one recursive function is the productive function for some creative set. (*Hint:* Try $\{f(x)|f(x) \in W_x\}$. A construction like that in 7-46 can also be used.)

△***7-50.*** Show that A productive $\Rightarrow A$ productive with a one-one productive function. (*Hint:* Let f be a given productive function. Define: $g(0) = f(0)$; $g(n+1) = f(n+1)$ if $f(n+1) \notin \{g(0), \ldots, g(n)\}$, otherwise use productive property. Warning: $\{g(0), \ldots, g(n)\}$ may not $\subset A$. Further hint: Take $W_{h(x)} = W_x \cup \{f(x)\}$. Consider $f(n+1)$, $fh(n+1)$, $fhh(n+1)$, . . . ; and consider the alternatives that a repetition does or does not occur.

7-51. Show that A productive $\Rightarrow A$ productive with a recursive permutation as productive function. (*Hint:* Clearly the method of 7-50 will make an *onto* function into a *one-one onto* function. It remains to show that an *into* function f can be made into an *onto* function h. Let D be an infinite recursive set of r.e. indices for N. Let g enumerate D without repetitions. Define: $h(x) = g^{-1}(x)$ on D, and $h(x) = f(x)$ on $\bar{D}$.)

▲***7-52.*** (Myhill). Show that A productive $\Leftrightarrow A$ completely productive.

7-53. Show that A productive $\Rightarrow A$ semiproductive.

7-54. (*a*) Prove Theorem X for semiproductive sets.
(*b*) Prove Theorem XI for semiproductive sets.

7-55. (*a*) Define a notion of "strong reducibility" appropriate to the theory of disjoint pairs of recursively enumerable sets.

(*b*) Show that the pair of sets constructed in Theorem XII(c) is, in an appropriate sense, "complete" with respect to this reducibility.

(c) Show that any complete disjoint pair of recursively enumerable sets is effectively inseparable.

$\triangle$(d) (Büchi). Use reducibility from the sets in Theorem XII(c) to show that the following two sets are not recursively separable and are, in fact, effectively inseparable: $\{<x,y>|\varphi_x(y)$ *convergent*$\}$ and $\{<x,y>|P_x(y)$ *cyclic*$\}$, where $P_x(y)$ *cyclic* means that some machine-tape configuration occurs more than once in the Turing-machine computation of P_x with input y.

7-56. Prove parts (a) and (b) of Theorem XII.

7-57. Adapt the proof of Theorem XI to show that any effectively inseparable disjoint pair of recursively enumerable sets has a total "productive function."

7-58. Modify the proof of Theorem XII(c) to get a collection of more than two sets any two of which are disjoint and effectively inseparable.

7-59. Show (without using 7-32) that there exists a creative set K' such that K and K' are disjoint and effectively inseparable. (*Hint:* Show $K \equiv A_0$, where A_0 is as in proof of Theorem XII.)

§7.8

7-60. Prove: [B recursively enumerable & $A \cap B$ productive] $\Rightarrow A$ productive (This implies the basic lemma of §7.8.)

7-61. Using Church's Thesis, explain why $\{F_x|x \in N\}$ (as defined in §7.8) must be recursive and why $\{F_x|x \notin K\}$ must be productive.

7-62. Assume that we have a logical system formulated in the symbolism of ordinary quantificational logic. The wffs may thus involve the symbol "&" (among others). We identify wffs with their Gödel numbers. A relation of *immediate derivability* is given as a relation between finite sets of wffs and wffs. This relation is recursive; i.e., $\{<x,y>|$ *wff* y *is immediately derivable from finite set* $D_x\}$ is recursive. Define a relation of *derivability* as follows. y is *derivable* from A if there exists a finite sequence $<y_0,y_1, \ldots ,y_n>$ such that $y = y_n$ and such that for all j, $0 \leq j \leq n$, either $y_j \in A$ or y_j is immediately derivable from some subset of $\{y_0, \ldots ,y_{j-1}\}$. Let $\widetilde{A} = \{x|x$ *is derivable from* $A\}$. If $B = \widetilde{A}$, we call A a *set of axioms* for B.

(a) Show that A recursively enumerable $\Rightarrow \widetilde{A}$ recursively enumerable.

(b) Show that $A \subset \widetilde{A}$, and $\widetilde{\widetilde{A}} = \widetilde{A}$.

Let $(x \,\&\, y)$ denote the wff obtained by placing "&" between wffs x and y. Assume that for all x and y, $(x \,\&\, y) \in \widetilde{\{x,y\}}$, $x \in \widetilde{\{(x \,\&\, y)\}}$, and $y \in \widetilde{\{(x \,\&\, y)\}}$.

(c) Suggest how $\widetilde{A \cup \widetilde{B}} = \widetilde{A} \cup \widetilde{B}$ may fail.

$\triangle$(d) (Craig). Show that B recursively enumerable $\Rightarrow (\exists A)$ [A recursive & $\widetilde{A} = \widetilde{B}$]. (*Hint:* Use Theorem 5-III. Note that for any x, members of the sequence x, $(x \,\&\, x)$, $((x \,\&\, x) \,\&\, x)$, ... are interderivable.) This is *Craig's theorem: every recursively enumerable theory (in ordinary logic) has a recursive set of axioms* (Craig [1953]).

7-63. (This exercise assumes some knowledge of logic and logical terminology.) Consider the *pure lower predicate calculus*, i.e., general quantificational logic. Is there a recursive set of valid biconditionals, i.e., equivalences, such that all valid biconditionals may be generated from them by "substituting equals for equals," i.e., by the operation of substituting one equivalent for another (as part of a possibly larger wff in some possibly different biconditional)? (*Hint:* Use method similar to that in 7-62.)

7-64. (Gödel's first incompleteness theorem, Gödel [1931]). Consider Peano arithmetic or a set-theory arithmetic. Assume there is an expression of elementary arithmetic with four unquantified variables, call it *Mabcd* (with a, b, c, d as the variables), such that (i) if P_z with input x yields output y in fewer than w steps, then **Mzxyw** is provable;† (ii) if P_z with input x does not yield output y in fewer than w steps, then $(\neg\, M\mathbf{zxyw})$ is provable (see §§14.4 and 14.7). Our theory is said to be *ω-consistent* if, for any expression Ga in a single unquantified variable a, whenever $(\exists a)Ga$ is provable, then, for some x, $(\neg\, G\mathbf{x})$ is not provable.

Assume that our theory incorporates the usual rules of elementary logic. Show: *if our theory is ω-consistent, then* (i) *it forms a creative set, and* (ii) *an undecidable wff must exist.* (*Hint:* Take F_x to be $(\exists c)(\exists d)M\mathbf{xx}cd$, and show $[[F_x \text{ provable}] \Leftrightarrow x \in K]$.)

△***7-65.*** (Gödel-Rosser incompleteness theorem, Rosser [1936]). Assume a theory and an expression $Mabcd$ as in 7-64. Let z_0 and z_1 be r.e. indices for A_0 and A_1 in Theorem XII. Assume that z_0 and z_1 can be chosen so that the following wff is provable: $(\forall b)\, \neg\, [(\exists c)(\exists d)M\mathbf{z}_0bcd \,\&\, (\exists c)(\exists d)M\mathbf{z}_1bcd]$. Our theory is said to be *consistent* if, for any wff H, whenever H is provable, then $(\neg H)$ is not provable.

Assume that our theory incorporates the usual rules of elementary logic. Show: *if our theory is consistent, then* (i) *it forms a creative set, and* (ii) *an undecidable wff must exist.* (*Hint:* Consider $W_u = \{x \mid (\exists c)(\exists d)M\mathbf{z}_0\mathbf{x}cd \textit{ is provable}\}$ and $W_v = \{x \mid \neg\,(\exists c)(\exists d)M\mathbf{z}_0\mathbf{x}cd \textit{ is provable}\}$, and use effective inseparability of A_0 and A_1. Note that *any* recursively enumerable B such that $A_0 \subset B \subset \bar{A}_1$ must be creative.)

† We adopt here the notational convention that if $\cdots a \cdots$ is an expression with an unquantified variable a, and x is an integer, then $\cdots \mathbf{x} \cdots$ is the result of substituting the numeral expression of x for a in $\cdots a \cdots$. Thus $M\mathbf{zxyw}$ denotes the wff obtained by substituting the numerals of the integers z, x, y, and w for the variables a, b, c, and d in the expression $Mabcd$; similarly below for $G\mathbf{x}$ and $(\exists c)(\exists d)M\mathbf{xx}cd$.

8 Truth-table Reducibilities; Simple Sets

§8.1 SIMPLE SETS

Among the questions left open in the last chapter were:

1. Is every nonrecursive, recursively enumerable set creative?
2. Is every nonrecursive, recursively enumerable set m-complete?
3. Do $\leq_1$ and $\leq_m$ coincide on the nonrecursive, recursively enumerable sets?

The intimate relation between "productive" diagonalization methods and unsolvability might suggest an affirmative answer to question 1. The examples of nonrecursive, recursively enumerable sets so far exhibited might suggest an affirmative answer to question 2. Theorem 7-VII might suggest an affirmative answer to question 3. All three questions, however, have negative answers, as we now prove.

Exercise 5-8 showed the existence of an infinite set having no infinite recursively enumerable subset. If a set with this property has a recursively enumerable complement, then the recursively enumerable complement is called *simple*. The term is due to Post [1944].

Definition A is *simple* if

(i) A is recursively enumerable,
(ii) $\bar{A}$ is infinite,
(iii) $(\forall B)[[B \text{ infinite } \& \ B \text{ recursively enumerable}] \Rightarrow B \cap A \neq \emptyset]$.

The concept is recursively invariant, since it is defined in terms of recursively invariant concepts. A few basic properties of simple sets are listed in Theorem I.

Theorem I(a) *A simple* $\Rightarrow$ *A not recursive.*

(b) *A simple* $\Rightarrow$ *A not creative.*

(c) *A simple* $\Rightarrow$ *A not* m-*complete.*

(d) *A simple* $\Rightarrow$ *A not a cylinder.*

Proof. (*a*) If A is recursive, and (ii) in the definition holds, then (iii) fails with $B = \bar{A}$.

(*b*) Since every productive set has an infinite recursively enumerable subset (Theorem 7-X), (iii) fails when A is creative.

(*c*) Immediate from (*b*) (Corollary 7-V).

(*d*) Assume $A \equiv C \times N$, and A simple. Then $\bar{A} \neq \emptyset$; hence

$$\overline{C \times N} = \bar{C} \times N \neq \emptyset$$

and $\bar{C} \neq \emptyset$. Let $m \in \bar{C}$. Then $\{m\} \times N$ is an infinite recursively enumerable subset of $\bar{C} \times N$. By recursive invariance, (iii) is violated for A, and A cannot be simple; this is a contradiction.⊠

Next, Theorem II shows that simple sets exist.

Theorem II (Post) *There exists a simple set.*

Proof. Let $C = \{<x,y>|y \in W_x \text{ and } y > 2x\}$. By the second projection theorem, C is recursively enumerable. Choose a particular effective method for listing C. Form the set $C' = \{<x,y>|<x,y> \in C \;\&\; (\forall z)[[z \neq y \;\&\; <x,z> \in C] \Rightarrow <x,z> \textit{ comes after } <x,y> \textit{ in the listing of } C]\}$. (In effect, we are applying Theorem 5-XVI, the single-valuedness theorem.) C' is evidently recursively enumerable, and $\{<x,y>|<x,y> \in C'\}$ is a partial recursive function (Theorem 5-IX). Let S be the range of this partial recursive function; that is, $S = \{y|(\exists x)[<x,y> \in C']\}$. We show that S is simple.

(i) S is recursively enumerable, since it is the range of a partial recursive function.

(ii) By the construction, at most k integers out of $\{0,1, \ldots ,2k\}$ can occur in S. This holds for any k; hence $\bar{S}$ is infinite.

(iii) Assume B is recursively enumerable and infinite. Let $B = W_{x_0}$. Then there are integers greater than $2x_0$ in B. By the construction, $<x_0,z>$ must be in C' for some z in B. Hence $z \in B \cap S$, and $B \cap S \neq \emptyset$.⊠

Several corollaries follow from Theorems I and II.

Corollary II(a) *There exist nonrecursive, recursively enumerable sets which are neither* m-*complete, creative, nor cylinders.*

Proof. Immediate.⊠

Corollary II(b) $\equiv_1$ *and* $\equiv_{\mathrm{m}}$ *do not coincide on the nonrecursive, recursively enumerable sets, and hence* $\leq_1$ *and* $\leq_{\mathrm{m}}$ *do not coincide on the nonrecursive, recursively enumerable sets.*

Proof. $S \equiv_m S \times N$ by Theorem 7-VIII, but $S \not\equiv_1 S \times N$ by Theorem I. Hence, by Theorem 7-VIII, $S \times N \leq_m S$ but $S \times N \not\leq_1 S$.☒ This answers questions 1, 2, and 3 at the beginning of this chapter.

In §8.6 we shall show that a single recursively enumerable m-degree can contain infinitely many 1-degrees. The "pathology" exhibited by a simple set is of interest in its own right, and will be studied further. In particular, it is useful in the investigation of a new and more general reducibility that will be formulated in §8.3. A number of other sorts of pathology can occur in recursively enumerable sets; we discuss some of these in §8.7 and in the exercises. In the absence of satisfactory representation theorems, a major part of our study of recursively enumerable sets must be the somewhat arbitrary cataloguing of various pathologies known to occur.

§8.2 IMMUNE SETS

Sets with the special property exhibited in Exercise 5-8, i.e., with the property attributed to the complement of a simple set, have been given a special name by Dekker.

Definition A is *immune* if
(i) A is infinite,
(ii) $(\forall B)[[B \text{ infinite } \& \ B \text{ recursively enumerable}] \Rightarrow B \cap \bar{A} \neq \emptyset]$.

The family of immune sets is similar in some ways to the family of finite sets. The family of immune or finite sets is, in certain respects, like the family of sets of measure zero in a measure space (see Exercise 8-11).

Definition A is *isolated* if A is finite or immune.

It will be helpful to have a standard way of indicating that the complement of a set has a specified property. For this purpose, we adopt the following convention.

Convention Let $(\cdot\ \cdot\ \cdot)$ be a property of sets. We say that a set is *co*-$(\cdot\ \cdot\ \cdot)$ if its complement is $(\cdot\ \cdot\ \cdot)$.

For example, a *co-immune* set is a set whose complement is immune, and a *co-finite* set is a set whose complement is finite. A set is simple if and only if it is both recursively enumerable and co-immune.

The sets which are co-finite or simple, i.e., the recursively enumerable co-isolated sets, bear some similarity to the sets of measure one in a measure space of total measure one. In particular, they form a dual ideal in the lattice of all recursively enumerable sets (under $\cap$ and $\cup$, see Exercise 8-11). This might suggest that the isolated sets form an ideal in the lattice of all sets. The following theorem, however, shows that the family of isolated sets is not closed under finite union, and, incidentally, that uncountably many isolated sets exist.

Theorem III *There exist* $2^{\aleph_0}$ *sets* A *such that both* A *and* $\bar{A}$ *are immune.*

Proof. Let $x_0, x_1, \ldots$ be the members of $\{x|W_x \text{ infinite}\}$ in increasing order. Define the sequence of pairs $\{y_0,z_0\}, \{y_1,z_1\}, \ldots$ as follows:

$$\{y_0,z_0\} = \text{the two smallest members of } W_{x_0},$$

where $y_0 < z_0$.

$$\{y_{k+1},z_{k+1}\} = \text{the two smallest of those members of } W_{x_{k+1}} \text{ that are greater than both } y_k \text{ and } z_k,$$

where $y_{k+1} < z_{k+1}$. We form a set A by choosing one integer from each member of this sequence of pairs. There are $2^{\aleph_0}$ distinct ways of doing this. Both A and $\bar{A}$ must intersect every infinite recursively enumerable set. Hence both A and $\bar{A}$ are immune.☒

Must every nonrecursive, recursively enumerable set be either creative or simple? The following theorem gives a negative answer to this question.

Theorem IV *There is a recursively enumerable set which is neither recursive, simple, nor creative.*

Proof. Let A be simple, and consider $A \times N$. $A \times N$ is recursively enumerable since A is. $A \times N$ is not recursive since $A \leq_1 A \times N$. $A \times N$ is not simple since it is a cylinder. It is easy to show that $A \times N$ creative $\Rightarrow$ A creative (Exercise 8-10); the theorem follows.☒

Corollary IV *There exist sets which are neither recursively enumerable, productive, nor immune.*

Proof. Take $\overline{A \times N}$ in the proof of the theorem.☒

The next theorem shows that, although isolatedness is not hereditary downward in the m-ordering (since $A \times N \leq_m A$ in the preceding proof), it is hereditary downward in the 1-ordering.

Theorem V $[A \leq_1 B$ *& B isolated*$] \Rightarrow A$ *isolated.*

Proof. Immediate.☒

Isolated sets have been used to build a recursive-function-theoretic analogue to the theory of cardinal numbers, and, in particular, to the theory of cardinals which are *mediate* in the sense of Whitehead and Russell. (A *mediate cardinal* is the cardinal of a set which is larger in size than any finite set but cannot be put into a one-one correspondence with a proper subset of itself. In ordinary set theory, the axiom of choice entails the nonexistence of mediate cardinals.) We indicate this analogue in Exercise 8-32 (see also Exercise 10-4).

Since the simple and co-finite sets form a dual ideal in the lattice of all recursively enumerable sets (see §12.1 for this concept), a quotient lattice can be formed in which two recursively enumerable sets are identified if and only if they differ by a finite or co-simple set; that is, A and B are identified $\Leftrightarrow$ $(A - B) \cup (B - A)$ is finite or co-simple. For example, in this quotient lattice, a set A is identified with N if and only if A is simple or co-finite. Thus, for simple A, A is identified with N, but $A \times N$ is not. Since $A \equiv_m A \times N$, the equivalence classes forming elements of the quotient lattice are not well-defined with respect to m-degrees. As our chief present concern is reducibility, we postpone general treatment of these lattices to Chapter 12 (see also Exercise 8-12).

§8.3 TRUTH-TABLE REDUCIBILITY

The sets K, $\bar{K}$, K *join* $\bar{K}$, and $K \times \bar{K}$ fall into four distinct m-degrees (see Exercise 7-36), with K and $\bar{K}$ incomparable. Nevertheless, these four sets are, in a natural and intuitive sense, interreducible. For example, if we could decide, for any integer, whether it is in K, we could, by making two such decisions, decide whether any $<x,y>$ is in $K \times \bar{K}$. More trivially, if we had a method for deciding whether any integer is in K, we obviously would have a method for deciding whether any integer is in $\bar{K}$. In this sense, the concept of m-reducibility is too narrow.

We formulate a new and more general reducibility, describing it first in rough form. We say that A is "reducible" to B if there is an effective procedure such that for any x, we can find (i) a finite set of integers $\{y_1,y_2, \ldots ,y_k\}$ and (ii) a specification of how answers to the questions "is y_1 in B?", "is y_2 in B?", . . . , "is y_k in B?" determine whether or not $x \in A$. For example, $A = K \times \bar{K}$ is "reducible" to $B = K$: for given any x, we take the set of integers $\{\pi_1(x),\pi_2(x)\}$; if $\pi_1(x) \in B$ and $\pi_2(x) \notin B$, then $x \in A$; if $\pi_1(x) \notin B$ or $\pi_2(x) \in B$, then $x \notin A$. Similarly $A = \bar{K}$ is "reducible" to $B = K$: for given any x, we take the set $\{x\}$; if $x \notin B$ then $x \in A$; while if $x \in B$, then $x \notin A$.

We make our concept precise in the following definitions.

Definition

$$I = \{0,1\},$$
$$I^k = I \times I \times \cdots \times I \qquad (k \text{ factors}).$$

Definition α is a k-ary *Boolean function* if α maps I^k into I. A Boolean function is a finite object. In logic, a Boolean function is sometimes called a *truth table.* There are, evidently, 2^{2^k} distinct k-ary Boolean functions. (If I is given the structure of the field of two elements, then any polynomial in k variables over this field will determine a k-ary Boolean function. Conversely, by the Lagrange interpolation theorem, every k-ary function is determined by a polynomial. We need only consider "reduced" polynomials in which each variable occurs to at most the first power. There are 2^{2^k} such reduced polynomials. Hence there must be a one-one correspondence between reduced polynomials and the Boolean functions. We call such polynomials *Boolean polynomials.*)

Definition. The ordered pair $<<x_1, \ldots, x_k>, \alpha>$, where

$$<x_1, \ldots, x_k>$$

is a k-tuple of integers and α is a k-ary Boolean function ($k > 0$) is called a *truth-table condition* (or *tt-condition*) of *norm* k. The set $\{x_1, \ldots, x_k\}$ is called the *associated set* of the tt-condition.

Definition The tt-condition $<<x_1, \ldots, x_k>, \alpha>$ is *satisfied* by A if $\alpha(c_A(x_1), \ldots, c_A(x_k)) = 1$, where c_A is the characteristic function for A.

Each tt-condition is a finite object; clearly an effective coding can be chosen which maps all the tt-conditions (of varying norm) onto N. Assume henceforth that a particular such coding has been chosen. When we speak of "tt-condition x," we shall mean the tt-condition with code number x.

Our reducibility concept can now be given.

Definition A is *truth-table reducible* to B (notation: $A \leq_{tt} B$) if there is a recursive f such that for all x, $[x \in A \Leftrightarrow$ tt-condition $f(x)$ is satisfied by $B]$. We also abbreviate "truth-table reducibility" as "tt-reducibility"

The following theorem is evident.

Theorem VI *$\leq_{tt}$ is reflexive and transitive.*

Proof. Reflexivity. Given x, let $f(x)$ be the tt-condition $<<x>, \alpha>$ where α is the identity 1-ary Boolean function.

Transitivity. Assume $A \leq_{tt} B$ by f, and $B \leq_{tt} C$ by g. We show how to compute a function h such that $A \leq_{tt} C$ by h. Let x be given. Find the tt-condition $f(x)$; let

$$f(x) \text{ be } <<y_1, \ldots, y_m>, \alpha>.$$

Find the tt-conditions $g(y_1), \ldots, g(y_m)$; let

$$g(y_1) \text{ be } <<z_{11}, \ldots, z_{1n_1}>, \beta_1>,$$
$$\cdots\cdots$$
$$g(y_m) \text{ be } <<z_{m1}, \ldots, z_{mn_m}>, \beta_m>.$$

Then the desired tt-condition $h(x)$ is

$$<<z_{11}, \ldots, z_{1n_1}, z_{21}, \ldots, z_{m1}, \ldots, z_{mn_m}>, \lambda w_{11} \cdots w_{mn_m} [\alpha(\beta_1(w_{11}, \ldots, w_{1n}), \beta_2(w_{21}, \ldots), \ldots, \beta_m(w_{m1}, \ldots, w_{mn_m}))]>.$$

The Boolean functions $\beta_1, \ldots, \beta_m$ are n_1-ary, $\ldots$, n_m-ary, and the Boolean function in $h(x)$ is q-ary, where $q = n_1 + \cdots + n_m$. ☒

Definition $A \equiv_{tt} B$ if $A \leq_{tt} B$ and $B \leq_{tt} A$. The equivalence classes of $\equiv_{tt}$ are called *truth-table degrees* or tt-*degrees.*

Definition A is *truth-table complete* ("tt-complete") if
(i) A is recursively enumerable,
(ii) $(\forall B)[B$ recursively enumerable $\Rightarrow B \leq_{tt} A]$.
The following theorem gives elementary facts about tt-reducibility.

Theorem VII(a) $A \leq_m B \Rightarrow A \leq_{tt} B.$
(*b*) $A \leq_{tt} \bar{A}$ *(and therefore* $A \equiv_{tt} \bar{A}$*).*
(*c*) *The* tt-*reducibility ordering is an upper semilattice.*
(*d*) $[B$ *recursive and* $A \leq_{tt} B] \Rightarrow A$ *recursive.*
(*e*) A *recursive* $\Rightarrow (\forall B)[A \leq_{tt} B]$.
Proof. (*a*) Immediate.
(*b*) Immediate; see example of K and $\bar{K}$ discussed above.
(*c*) A *join* B again serves to generate a least upper bound, since, if $A \leq_{tt} C$ and $B \leq_{tt} C$, it is immediate that A *join* $B \leq_{tt} C$.
(*d*) Immediate.
(*e*) Let A be recursive; choose f so that $f(x)$ is a tt-condition with Boolean function identically 1 if $x \in A$ and identically 0 if $x \notin A$. ☒

The recursive invariance of $\leq_{tt}$ is immediate from (*a*). It follows from (*d*) and (*e*) that the tt-reducibility ordering has a single minimum degree consisting of the recursive sets and no others. From (*b*) we see that a degree may contain both recursively enumerable and nonrecursively enumerable sets. We call a tt-degree *recursively enumerable* if it contains a recursively enumerable set. Thus the tt-degree of $\bar{K}$ is a recursively enumerable degree since it contains K. (In Chapter 11 we shall see that incomparable recursively enumerable tt-degrees exist.)

From (*a*) we have that $A \equiv_m B \Rightarrow A \equiv_{tt} B$. We can therefore view a tt-degree as composed of m-degrees. The relation between tt-degrees and m-degrees is discussed further in §8.4.

Remark. The definition of $\leq_{tt}$ can be modified in various ways (see Exercise 9-45, for example). One modification would be to require that, in the reduction of one set to another, the same Boolean function appear in all the tt-conditions for that pair of sets. An apparently less severe modification would be to require that, in the reduction of one set to another, there be a uniform bound to the norms of the tt-conditions. That is to say, A is reducible to B if $(\exists$ recursive $f)(\exists m)(\forall x)[$tt-condition $f(x)$ has norm $\leq m$, and $[x \in A \Leftrightarrow f(x)$ is satisfied by $B]]$. These two modifications prove to be equivalent (Exercise 8-28) apart from a trivial exceptional case. The resulting reducibility is studied in §8.5 below.

Is tt-reducibility the most general reducibility consistent with our intuitive ideas? A surprising negative answer to this question is supplied in Chapter 9.

§8.4 TRUTH-TABLE REDUCIBILITY AND MANY-ONE REDUCIBILITY

$\leq_{\text{m}}$ and $\leq_{\text{tt}}$ differ on sets in general. Do they differ on the recursively enumerable sets? In particular, are there tt-complete sets which are not m-complete? The following theorem and corollary give an affirmative answer to both questions. The construction in the theorem is due to Post [1944].

***Theorem VIII* (*Post*)** *There exists a set which is both simple and* tt-*complete.*

Proof. Let S be the simple set constructed in the proof of Theorem II. We noted that for any k, at most k integers out of $\{0,1, \ldots ,2k\}$ can occur in S. Hence any subset of $k + 1$ integers from $\{0,1, \ldots ,2k\}$ must intersect $\bar{S}$. In particular, each of the sets $\{0\}$, $\{1,2\}$, $\{3,4,5,6\}$, $\{7, \ldots ,14\}$, $\ldots$, $\{2^n - 1, \ldots , 2^{n+1} - 2\}$, $\ldots$ must intersect $\bar{S}$. Let

$$S_x = \{2^x - 1, \ldots , 2^{x+1} - 2\}.$$

Then, for all x, $S_x \cap \bar{S} \neq \emptyset$.

Define $S^* = S \cup (\bigcup_{x \in K} S_x)$. S^* is evidently recursively enumerable. Furthermore, $\overline{S^*}$ is infinite, since $\bar{K}$ is infinite. $\overline{S^*}$ is immune since it is infinite and is a subset of the immune set $\bar{S}$. Hence S^* is a simple set.

By our construction, $x \in K \Leftrightarrow S_x \subset S^*$. This is an instance of tt-reducibility. Given any x, the tt-condition corresponding to x is $\langle\langle 2^x - 1, \ldots , 2^{x+1} - 2\rangle,\alpha\rangle$, where

$$\alpha(w_1, \ldots ,w_m) = 1 \Leftrightarrow w_1 = w_2 = \cdots = w_m = 1;$$

this tt-condition has norm $m = 2^x$. Thus $K \leq_{\text{tt}} S^*$. Let A be any recursively enumerable set. Then $A \leq_{\text{m}} K$. Hence $A \leq_{\text{tt}} S^*$. Thus S^* is tt-complete.☒

Corollary VIII *There exists a set which is* tt-*complete but not* m-*complete; hence* $\equiv_{\text{m}}$ *and* $\equiv_{\text{tt}}$ *differ on the recursively enumerable sets, and* $\leq_{\text{m}}$ *and* $\leq_{\text{tt}}$ *differ on the recursively enumerable sets.*

Proof. $K \equiv_{\text{tt}} S^*$, but, by Theorem I(*c*), $K \not\leq_{\text{m}} S^*$.☒

What can be said about the structure of a tt-degree in terms of its component m-degrees and 1-degrees? The following structure theorem is a counterpart to Theorem 7-VIII on cylinders.

Definition $B^{\text{tt}} = \{x|x$ *is a* tt-*condition satisfied by* $B\}$.

Definition A is a tt-*cylinder* if $A \equiv B^{\text{tt}}$ for some B.

***Theorem IX*(*a*)** $A \leq_1 A^{\text{tt}}$.

(*b*) $A^{\text{tt}} \leq_{\text{tt}} A$ (*and therefore* $A \equiv_{\text{tt}} A^{\text{tt}}$).

(*c*) *A is a* tt-*cylinder* $\Rightarrow$ *A is a cylinder.*

(*d*) *A is a* tt-*cylinder* $\Leftrightarrow (\forall B)[B \leq_{tt} A \Rightarrow B \leq_1 A]$.

(*e*) $A \leq_{tt} B \Leftrightarrow A^{tt} \leq_1 B^{tt}$.

Proof. (*a*) Take f so that $f(x)$ is the tt-condition $<<x>,\alpha>$ with α the identity 1-ary Boolean function. Then $x \in A \Leftrightarrow f(x)$ is satisfied by $A \Leftrightarrow f(x) \in A^{tt}$.

(*b*) Take $g = \lambda x[x]$. Then $x \in A^{tt} \Leftrightarrow x$ is satisfied by $A \Leftrightarrow g(x)$ is satisfied by A. Hence $A^{tt} \leq_{tt} A$ by g.

(*c*) Let the tt-condition $<<x_1, \ldots ,x_n>,\alpha>$ be given. Form $<<x_1, \ldots ,x_n,x_{n+1}, \ldots ,x_m>,\beta>$, where $x_{n+1}, \ldots , x_m$ are chosen arbitrarily and where β is defined by $\beta(w_1, \ldots ,w_m) = \alpha(w_1, \ldots ,w_n)$. Obviously, the two tt-conditions are satisfied by exactly the same sets. Given any tt-condition, we can uniformly effectively list an infinite number of such "larger" conditions that are satisfied by exactly the same sets. It follows directly that the necessary and sufficient condition of Theorem 7-IX applies to any set of the form A^{tt}. Hence any tt-cylinder is a cylinder.

(*d*) $\Rightarrow$: Let $A \equiv C^{tt}$, and assume $B \leq_{tt} A$. Then, by (b), $B \leq_{tt} C$; and there is a recursive f such that $x \in B \Leftrightarrow f(x)$ is satisfied by $C \Leftrightarrow f(x) \in C^{tt}$. Thus $B \leq_m C^{tt}$ by f. By (*c*), C^{tt} is a cylinder, and by Theorem 7-VIII, $B \leq_1 C^{tt}$. Therefore $B \leq_1 A$.

$\Leftarrow$: Assume $(\forall B)[B \leq_{tt} A \Rightarrow B \leq_1 A]$. Consider A^{tt}. By (*b*), $A^{tt} \leq_{tt} A$; hence, by assumption, $A^{tt} \leq_1 A$. By (*a*), $A \leq_1 A^{tt}$. Therefore $A \equiv_1 A^{tt}$, and hence $A \equiv A^{tt}$. Thus A is a tt-cylinder.

(*e*) $\Rightarrow$: Assume $A \leq_{tt} B$. Then $A^{tt} \leq_{tt} A \leq_{tt} B \leq_1 B^{tt}$ by (*a*) and (*b*). Hence $A^{tt} \leq_{tt} B^{tt}$, and, by (*d*), $A^{tt} \leq_1 B^{tt}$.

$\Leftarrow$: Assume $A^{tt} \leq_1 B^{tt}$. Then $A \leq_1 A^{tt} \leq_1 B^{tt} \leq_{tt} B$, by (*a*) and (*b*), and hence $A \leq_{tt} B$.☒

Corollary IX *A is a* tt-*cylinder* $\Leftrightarrow A^{tt} \leq_1 A \Leftrightarrow A \equiv A^{tt}$.

Proof. Immediate ☒

Call A^{tt} the tt-*cylindrification* of A. From the foregoing theorem, every tt-degree contains a maximal 1-degree (and hence also a maximal m-degree); and this 1-degree is obtainable from any set in the tt-degree by tt-cylindrification.

Furthermore, from (*e*) in the foregoing theorem, there is a canonical homomorphism from the tt-reducibility ordering into the 1-reducibility ordering. Note that if A is not recursive, then A^{tt} is not recursively enumerable (since both A and $\bar{A}$ are $\leq_1 A^{tt}$, by Theorems VII(*b*) and IX(*d*)). Thus the nonrecursive 1-degrees occurring as images in this homomorphism are not recursively enumerable.

In §8.6 we shall show that a single recursively enumerable tt-degree can contain infinitely many recursively enumerable m-degrees. In Chapter 9 we shall show that there are recursively enumerable sets which are neither recursive nor tt-complete.

§8.5 BOUNDED TRUTH-TABLE REDUCIBILITY

Definition A is *bounded truth-table reducible* to B (notation: $A \leq_{\text{btt}} B$) if $(\exists$ recursive $f)(\exists m)(\forall x)[$tt-condition $f(x)$ has norm $\leq m$, and $[x \in A \Leftrightarrow f(x)$ is satisfied by $B]]$.

We abbreviate "bounded truth-table reducibility" as "btt-reducibility."

Examples. $K \leq_{\text{btt}} \bar{K}$, where 1 is a bound.

$K \times \bar{K} \leq_{\text{btt}} K$, where 2 is a bound.

The tt-conditions used in Theorem VIII to show $K \leq_{\text{tt}} S^*$ did not have bounded norm. (The tt-condition obtained for x had norm 2^x.)

Theorem X *$\leq_{\text{btt}}$ is reflexive and transitive.*

Proof. From the proof of Theorem VI we see that $A \leq_{\text{btt}} A$ with norm bounded by 1.

If $A \leq_{\text{btt}} B$ and $B \leq_{\text{btt}} C$, with norms bounded by m_1, m_2, respectively, then, from the construction of Theorem VI, $A \leq_{\text{btt}} C$ with norm bounded by $m_1 m_2$. ☒

Definition $A \equiv_{\text{btt}} B$ if $A \leq_{\text{btt}} B$ and $B \leq_{\text{btt}} A$. The equivalence classes of $\equiv_{\text{btt}}$ are called btt-*degrees*.

Definition A is btt-*complete* if

(i) A is recursively enumerable,

(ii) $(\forall B)[B$ recursively enumerable $\Rightarrow B \leq_{\text{btt}} A]$.

Theorem XI *Theorem VII holds with $\leq_{\text{btt}}$ in place of $\leq_{\text{tt}}$.*

Proof. The proof of Theorem VII carries over directly. ☒

Bounded truth-table reducibility has not been widely studied. Several elegant proofs exist concerning it. We give them below. Evidently $A \leq_{\text{btt}} B \Rightarrow A \leq_{\text{tt}} B$. Does $\leq_{\text{btt}}$ differ from $\leq_{\text{tt}}$ (i.e., can $A \leq_{\text{tt}} B$ without $A \leq_{\text{btt}} B$)? Do the btt-complete sets differ from the tt-complete sets? The following theorem of Post [1944], in conjunction with Theorem VIII, gives an affirmative answer to both questions.

Theorem XII (Post) *A btt-complete $\Rightarrow A$ not simple.*

Proof. Assume A is btt-complete. Then $K \leq_{\text{btt}} A$. Let m be a bound on the norm in the btt-reduction of K to A. We first prove a lemma.

Lemma *Let $K \leq_{\text{btt}} A$ with norm bounded by m. Then there exists an infinite recursively enumerable set of tt-conditions, $t_0, t_1, t_2, \ldots$, each of norm $\leq m$, with associated sets $T_0, T_1, T_2, \ldots$, such that*

(i) *for every i, t_i is satisfied by A;*

(ii) *for every i, $T_i \cap \bar{A} \neq \emptyset$;*

(iii) *for every i, $\{j \mid T_j \cap \bar{A} = T_i \cap \bar{A}\}$ does not contain more than 2^{2^m+1} members.*

Proof of lemma. Any tt-condition can, in an obvious way, be represented by a formula of sentential logic built up from basic parts of the form "$\mathbf{n} \in X$" ($\mathbf{n}$ a numeral for an integer), where the formula is true of any set X if and only if the tt-condition is satisfied by that set; and conversely, any such formula yields a tt-condition. The formula "$4 \notin X \vee 17 \in X$" can be associated, for example, with the tt-condition $\langle\langle 4,17\rangle,\alpha\rangle$, where α is determined by the Boolean polynomial $w_1w_2 + w_1 + 1$. Note that with each formula can be associated any of an infinite number of tt-conditions (corresponding to different repetitions and permutations in the k-tuple and to different values of k) and that with each tt-condition can be associated any of an infinite number of logically equivalent formulas. We say that a formula is *deducible* from a set of formulas if it is deducible by elementary sentential logic. For example, "$9 \notin X \vee 17 \in X$" is deducible from the three formulas "$4 \in X$," "$9 \in X$," and "$4 \notin X \vee 17 \in X$." We say that a tt-condition is deducible from a set of tt-conditions if deducibility holds for corresponding formulas. It follows that this notion of deducibility for tt-conditions is well-defined with respect to the (many-many) correspondence between formulas and tt-conditions and, from the soundness of sentential logic, that if x (a tt-condition) is deducible from A (a set of tt-conditions), then any B which satisfies all the tt-conditions in A must also satisfy x.†

Since $\bar{K} \leq_{\text{btt}} K$ with bound 1, and $K \leq_{\text{btt}} A$ with bound m, we have that $\bar{K} \leq_{\text{btt}} A$ with bound m. Let $\bar{K} \leq_{\text{btt}} A$ by f with bound m. We construct the effective sequence $t_0, t_1, \ldots$ as follows.

To find t_0. Consider the set C of all tt-conditions corresponding to formulas "$\mathbf{x} \in X$" with $x \in A$. Since A is recursively enumerable, this set is recursively enumerable. Let C_0 be the set of all tt-conditions deducible from tt-conditions in C. C_0 is evidently recursively enumerable. Let $B_0 = f^{-1}(C_0) = \{x | f(x) \in C_0\}$. Let x_0 be a recursively enumerable index for B_0; take $t_0 = f(x_0)$.

To find t_{k+1}. Let C_{k+1} be the set of all tt-conditions deducible from $C_k \cup \{t_k\}$. C_{k+1} is recursively enumerable since C_k is recursively enumerable. Let $B_{k+1} = f^{-1}(C_{k+1})$. Let x_{k+1} be an r.e. index for B_{k+1}; take $t_{k+1} = f(x_{k+1})$.

Evidently $C_0 \subset C_1 \subset C_2 \subset \cdots$. We show, by induction, that (i) and (ii) hold.

t_0. All conditions in C_0 are satisfied by A. Hence, by reduction of $\bar{K}$ to A, $B_0 \subset \bar{K}$. Then $x_0 \in \bar{K} - B_0$ by the productiveness of $\bar{K}$ with $\lambda x[x]$ as productive function. So, by reduction of $\bar{K}$ to A, $t_0 = f(x_0)$ is satisfied by A. This proves (i). $T_0 \cap \bar{A} \neq \emptyset$ for $[T_0 \cap \bar{A} = \emptyset$ & t_0 satisfied by $A] \Rightarrow$

† Deducibility could have been defined directly for tt-conditions; namely, x is *deducible* from A if $(\forall B)[(\forall y)[y \in A \Rightarrow y$ satisfied by $B] \Rightarrow x$ satisfied by $B]$. The use, in the text, of formulas serves to indicate (without further proof) that the set of all tt-conditions deducible from a recursively enumerable set of tt-conditions is itself recursively enumerable.

$t_0 \in C_0 \Rightarrow x_0 \in B_0$ contrary to the fact that $x_0 \in \bar{K} - B_0$. This proves (ii). Note that $t_0 \notin C_0$.

t_{k+1}. Assume (i) and (ii) hold for all $i \leq k$, and assume that all tt-conditions in C_k are satisfied by A. Then all conditions in C_{k+1} are satisfied by A. Hence $B_{k+1} \subset \bar{K}$, and $x_{k+1} \in \bar{K} - B_{k+1}$ by the productiveness of $\bar{K}$. Thus, by the reduction of $\bar{K}$ to A, $t_{k+1} = f(x_{k+1})$ is satisfied by A. This proves (i). $T_{k+1} \cap \bar{A} \neq \emptyset$, for $[T_{k+1} \cap \bar{A} = \emptyset \;\&\; t_{k+1}$ satisfied by $A] \Rightarrow t_{k+1} \in C_0 \Rightarrow t_{k+1} \in C_{k+1} \Rightarrow x_{k+1} \in B_{k+1}$, contrary to the fact that $x_{k+1} \notin B_{k+1}$. This proves (ii). Note that $t_{k+1} \notin C_{k+1}$.

Let n_i be the norm of t_i. For each i, consider the tt-condition t_i' obtained from t_i as follows. t_i' has the same associated set T_i. The n_i-tuple for t_i' is obtained by reordering the n_i-tuple for t_i so that the members of $T_i \cap \bar{A}$ come first in nondecreasing order and so that the members of $T_i \cap A$ follow (in any order). The Boolean function for t_i' is obtained by reordering the arguments of the Boolean function for t_i in a corresponding way. Thus for each i, t_i and t_i' are trivially equivalent (and interdeducible). $t_0', t_1', \ldots$ all have norm $\leq m$, since, by assumption, all tt-conditions in *range* f have norm $\leq m$. (The sequence $t_0', t_1', \ldots$ may not be recursively enumerable.)

To prove (iii), observe that there are $2^{2^0} + 2^{2^1} + \cdots + 2^{2^m} < 2^{2^{m+1}}$ distinct Boolean functions of m or fewer arguments. If

$$\{j \mid T_j \cap \bar{A} = T_i \cap \bar{A}\}$$

had more than $2^{2^{m+1}}$ members for any i, then, for some j_1 and j_2 such that $j_1 < j_2$, t_{j_1}' and t_{j_2}' would have the same Boolean function and

$$T_{j_1} \cap \bar{A} = T_{j_2} \cap \bar{A}$$

This is a contradiction, for t_{j_2}' would then be derivable from $C \cup \{t_{j_1}'\}$ (see Exercise 8-25), hence t_{j_2} would be derivable from $C \cup \{t_{j_1}\}$, and hence t_{j_2} would be in $C_{j_1+1} \subset C_{j_2}$, contrary to our conclusion that $t_k \notin C_k$ for all k.

This concludes the proof of the lemma.† We now complete the proof of the theorem.

Define $s(i)$ = the size of $T_i \cap \bar{A}$ (s need not be a recursive function). Let $q = \mu z[s(i) = z$ *for infinitely many* $i]$. Since $1 \leq s(i) \leq m$, such a q exists.

Then $s(i) < q$ for only finitely many i. Choose i_0 so that $s(i) \geq q$ for all $i \geq i_0$. Enumerate A. As soon as all but q members of any T_i ($i \geq i_0$) have been listed in A, we know that the remaining q members must be in $\bar{A}$. Hence we can list $j_0, j_1, \ldots$ such that $i_0 \leq j_0 < j_1 < \cdots$ and

$$s(j_0) = s(j_1) = \cdots = q$$

and for each j_i ($i = 0, 1, 2, \ldots$), we can identify the members of $T_{j_i} \cap \bar{A}$.

† This lemma can also be obtained as a consequence of the proof of a more general lemma to be proved in Theorem 9-XVIII in a slightly different way.

Thus $W = \bigcup_{i=0}^{\infty} (T_{j_i} \cap \bar{A})$ is recursively enumerable. Evidently, $W \subset \bar{A}$. By (iii) of the lemma, W must be infinite. Therefore, $\bar{A}$ contains an infinite recursively enumerable subset, and A cannot be simple.⊠

Corollary XII *There exists a set which is* tt-*complete but not* btt-*complete; hence* $\equiv_{\rm btt}$ *and* $\equiv_{\rm tt}$ *differ on the recursively enumerable sets, and* $\leq_{\rm btt}$ *and* $\leq_{\rm tt}$ *differ on the recursively enumerable sets.*

Proof. S^* from Theorem VIII is such a set. $K \equiv_{\rm tt} S^*$ but $K \not\leq_{\rm btt} S^*$.⊠

It is easily seen that $\leq_{\rm m}$ and $\leq_{\rm btt}$ differ on the recursive sets. The following result of Fischer [1963] shows that $\leq_{\rm m}$ and $\leq_{\rm btt}$ differ on the nonrecursive, recursively enumerable sets.

Theorem XIII (Fischer) *There exist nonrecursive, recursively enumerable sets A and B such that $A \leq_{\rm btt} B$ but $A \not\leq_{\rm m} B$.*

Proof. Take S^* from the proof of Theorem VIII. Evidently, $S^* \times S^* \leq_{\rm btt} S^*$. We assume $S^* \times S^* \leq_{\rm m} S^*$ and obtain a contradiction.

If $S^* \times S^* \leq_{\rm m} S^*$, then there is a recursive g such that $[x \in S^* \;\&\; y \in S^*] \Leftrightarrow \langle x,y \rangle \in S^* \times S^* \Leftrightarrow g(x,y) \in S^*$. Now by the proof of Theorem VIII, $x \in K \Leftrightarrow S_x \subset S^*$, where $S_x = \{2^x - 1, \ldots, 2^{x+1} - 2\}$. Define the recursive function f as follows:

$$
\begin{aligned}
f(0) &= 0,\\
f(1) &= g(1,2),\\
f(2) &= g(g(3,4),g(5,6)),\\
f(3) &= g(g(g(7,8),g(9,10)),g(g(11,12),g(13,14))),\\
&\ldots\ldots\ldots\ldots\ldots\ldots\ldots\ldots\ldots\ldots.
\end{aligned}
$$

Then $f(x) \in S^* \Leftrightarrow S_x \subset S^*$. Hence $K \leq_{\rm m} S^*$ by f, and S^* is m-complete. But S^* is simple, and this contradicts Theorem I(*c*).⊠

Corollary XIII $\equiv_{\rm btt}$ *and* $\equiv_{\rm m}$ *differ on the nonrecursive, recursively enumerable sets.*

Proof. In the above proof, $S^* \times S^* \equiv_{\rm btt} S^*$, but $S^* \times S^* \not\leq_{\rm m} S^*$.⊠

Is every btt-complete set also m-complete? This question was answered by Young [1963], who exhibited a btt-complete set which is not m-complete. His proof is briefly indicated in Exercise 10-18. The question is related to a question settled by Lachlan [1966] who showed that $A \times A$ creative $\Rightarrow A$ creative.

What analogues to Theorem IX hold for btt-reducibility? In particular, is there a maximum m-degree within each btt-degree? Jockusch has shown that the answer to this question is negative.

In Theorem 14-IX below, we use Exercise 8-28 to obtain the following characterization of btt-reducibility. For any B, let $\mathcal{B}_{\rm m}(B)$ be the Boolean algebra of sets generated by the collection of all sets m-reducible to B. Then if B is neither N nor $\emptyset$, $A \leq_{\rm btt} B \Leftrightarrow A \in \mathcal{B}_{\rm m}(B)$.

§8.6 STRUCTURE OF DEGREES

How many recursively enumerable degrees are there in the various reducibility orderings? How many 1-degrees can there be in a given m-degree; how many m-degrees in a given tt-degree? We now consider these questions.

***Theorem XIV* (Dekker)** *The* m-*degree of a simple set includes an infinite collection of* 1-*degrees linearly ordered under* $\leq_1$ *with order type of the rational integers* (. . . , −2, −1, 0, 1, 2, . . .) *and consisting entirely of simple sets.*

Proof. We begin with a lemma.

Lemma *Let A and B be recursively enumerable sets such that*

$$B = A \cup \{m\}$$

for some $m \in \bar{A}$. *Then*

(i) *A is simple* $\Leftrightarrow$ *B is simple;*

(ii) *A simple* $\Rightarrow [B \leq_1 A \;\&\; A \leq_m B \;\&\; A \not\leq_1 B]$.

Proof of lemma. (i) Immediate.

(ii) Let $n \in \bar{A}$, $n \neq m$. Then $A \leq_m B$ by the following f:

$$f(x) = \begin{cases} x, & \text{if } x \neq m; \\ n, & \text{if } x = m. \end{cases}$$

Let C be an infinite recursive subset of A. Let p be a recursive permutation which takes $C \cup \{m\}$ onto C. Then $B \leq_1 A$ by the following g:

$$g(x) = \begin{cases} x, & \text{if } x \notin C \cup \{m\}; \\ p(x), & \text{if } x \in C \cup \{m\}. \end{cases}$$

Assume $A \leq_1 B$. Then $A \equiv B$ by some recursive permutation h (Theorem 7-VI). Hence $m, h(m), hh(m), \ldots$ will be distinct and therefore form an infinite recursively enumerable subset of $\bar{A}$, contradicting the simpleness of A. Therefore $A \not\leq_1 B$. This proves the lemma.

The theorem is now immediate. Let A be a simple set. Let $a_0, a_1, \ldots$ be an infinite subset of A and let $b_0, b_1, \ldots$ be an infinite subset of $\bar{A}$ Then, by the lemma,

$$\ldots, A \cup \{b_0,b_1\}, A \cup \{b_0\}, A, A - \{a_0\}, A - \{a_0,a_1\}, \ldots$$

yields a linear ordering of 1-degrees of simple sets, as desired.☒

Corollary XIV *The* m-*degree of an immune set contains infinitely many* 1-*degrees.*

Proof. Make a construction similar to the final construction in the theorem. If any two sets were of the same 1-degree, we would have a contradiction similar to the contradiction used to prove $A \not\leq_1 B$ in the lemma.☒

***Theorem XV* (*Fischer*)** *The complete* tt-*degree includes a linear ordering of recursively enumerable* m-*degrees with order type of the integers* (0, 1, 2, . . .).

Proof. Take S^* from the proof of Theorem VIII. Consider the sets

$$\begin{aligned} A_0 &= S^*, \\ A_1 &= S^* \times S^*, \\ &\cdots\cdots\cdots, \\ A_k &= S^* \times S^* \times \cdots \times S^* \qquad \text{(with } 2^k \text{ factors)}, \\ &\cdots\cdots\cdots\cdots\cdots\cdots\cdots\cdots\cdots\cdots. \end{aligned}$$

Evidently, $A_0 \leq_1 A_1 \leq_1 A_2 \leq_1 \cdots$, and $A_0 \equiv_{tt} A_1 \equiv_{tt} A_2 \equiv_{tt} \cdots$. However, $j > i \Rightarrow A_j \not\leq_m A_i$. For otherwise, let u, v be such that $u < v$ and $A_v \leq_m A_u$. Returning to Theorem XIII, we see that a recursive function f can be found such that, for all x, $S_x \subset S^* \Leftrightarrow f(x) \in A_u$. We suggest the definition of f via an example. Assume $A_4 \leq_m A_2$ by h. (By definition, $A_4 = (S^*)^{16}$, and $A_2 = (S^*)^4$.) Then

$$\begin{aligned} f(0) &= \langle 0,0,0,0 \rangle, \\ f(1) &= \langle 1,2,1,2 \rangle, \\ f(2) &= \langle 3,4,5,6 \rangle, \\ f(3) &= h(\langle 7,8, \ldots ,14,7,8, \ldots ,14 \rangle), \\ f(4) &= h(\langle 15, \ldots ,30 \rangle), \\ f(5) &= h(\langle w_1, \ldots ,w_{16} \rangle), \end{aligned}$$

where

$$\langle w_1, \ldots ,w_4 \rangle = h(\langle 31, \ldots ,46 \rangle),$$

and

$$\langle w_5, \ldots ,w_8 \rangle = h(\langle 47, \ldots ,62 \rangle),$$

and

$$\langle w_9, \ldots ,w_{16} \rangle = \langle w_1, \ldots ,w_8 \rangle, \qquad \text{etc.}$$

We omit details of a general definition for f.

Since $x \in K \Leftrightarrow S_x \subset S^*$, we have $K \leq_m A_u$. But $A_u \leq_{btt} S^*$. Hence $K \leq_{btt} S^*$. Since S^* is simple, this contradicts Theorem XII.⊠

Corollary XV *There exists a* btt-*degree containing a linear ordering of recursively enumerable* m-*degrees with order type of the integers* (0, 1, 2, . . .).

Proof. In the proof of the theorem, $A_i \equiv_{btt} A_j$ for all i, j.⊠

Theorem XIV has recently been strengthened by Young [1966], who has shown that every nonrecursive m-degree either consists of a single 1-degree or else includes a linear ordering of 1-degrees with order type of the rationals. Jockusch has shown that there are recursively enumerable m-degrees, other than the m-degrees of $\emptyset$, N, and K, which consist of a single 1-degree. Jockusch has also shown that every nonrecursive tt-degree contains infinitely many m-degrees. See Jockusch [1966].

§8.7 OTHER RECURSIVELY ENUMERABLE SETS

Three different kinds of recursively enumerable sets have been emphasized: recursive, creative, and simple. Theorem IV showed that other kinds of recursively enumerable sets exist: if A is simple, then $A \times N$, the cylindrification of A, is neither recursive, creative, nor simple. Indeed, if every creative set is m-complete (as will be proved in Chapter 11), it follows from the structure theory developed so far that the cylindrification of any nonrecursive, non-m-complete set is neither recursive, creative, nor simple. Recursively enumerable sets which are neither recursive, creative, nor simple have been called *mezoic* by Dekker [1953].

No satisfactory representation theory exists for recursively enumerable sets. A number of elementary questions remain unsettled. A classification of recursively enumerable sets has been defined and studied by Dekker and Myhill, and by Uspenskii [1957]. It is a useful catalogue of the kinds of behavior discovered so far, but it remains, from a theoretical point of view, somewhat arbitrary. As a first step, all sets are grouped in the following five categories:

$\mathcal{B}_0 = \{A|A$ *is recursively enumerable*$\}$.

$\mathcal{B}_1 = \{A|A$ *immune*$\}$.

$\mathcal{B}_2 = \{A|A$ *is not recursively enumerable and A is the union of an infinite recursively enumerable set and an immune set*$\}$.

$\mathcal{B}_3 = \{A|(\forall$ *recursively enumerable* $B)[B \subset A \Rightarrow (\exists$ *recursively enumerable* $C)[C$ *infinite* & $C \subset A$ & $C \cap B = \emptyset]]$ *where no uniform effective method exists for finding an r.e. index for such a C from an r.e. index for B*$\}$.

$\mathcal{B}_4 = \{A|A$ *productive*$\}$.

It is trivial to show (Exercise 8-34) that this classification is mutually exclusive and exhaustive. Classes $\mathcal{B}_1$ to $\mathcal{B}_4$ may be viewed as regions along a spectrum of increasing "richness" in the possession of infinite recursively enumerable subsets.

A classification of recursively enumerable sets according to their complements can then be obtained from $\mathcal{B}_0, \ldots, \mathcal{B}_4$. Let

$$\mathcal{C}_i = \{A|A \textit{ recursively enumerable and } \bar{A} \in \mathcal{B}_i\} \qquad i = 0, 1, 2, 3, 4.$$

Then $\mathcal{C}_0$ is just the class of recursive sets, $\mathcal{C}_1$ the class of simple sets, and $\mathcal{C}_4$ the class of creative sets. Sets in $\mathcal{C}_2$ are sometimes called *pseudosimple,* and sets in $\mathcal{C}_3$ are sometimes called *pseudocreative* (these terms are occasionally used in a slightly different way, see below). $\mathcal{C}_2$ and $\mathcal{C}_3$ are both nonempty; for let A be simple, then (i) $\{2x|x \in A\}$ falls in $\mathcal{C}_2$, and (ii) $A \times N$ falls in $\mathcal{C}_3$ (see Exercises 8-35 and 8-36). Indeed any nonrecursive, noncreative, recursively enumerable cylinder falls in $\mathcal{C}_3$. Young has shown that not every set in $\mathcal{C}_3$ is a cylinder (see Exercise 8-49).

Sets in $\mathcal{C}_2$ have been further classified as follows.

$\mathcal{C}_{21} = \{A|A \in \mathcal{C}_2$ *and* $(\exists$ *recursive* $C)[C$ *infinite* & $C \subset \bar{A}$ & $(\forall$ *recursively enumerable* $B)[B \subset \bar{A} \Rightarrow B$ *has all but a finite number of its members in* $C]]\}$.

$\mathcal{C}_{22} = \{A|A \in \mathcal{C}_2$ *and* $A \notin \mathcal{C}_{21}$ *and* $(\exists$ *recursively enumerable* $C)[C$ *infinite* & $C \subset \bar{A}$ & $(\forall$ *recursively enumerable* $B)[B \subset \bar{A} \Rightarrow B$ *has all but a finite number of its members in* $C]]\}$.

$\mathcal{C}_{23} = \{A|A \in \mathcal{C}_2$ and $A \notin \mathcal{C}_{21}$ and $A \notin \mathcal{C}_{22}\}$.

If a recursively enumerable set C has the property that $[C$ infinite & $C \subset \bar{A}$ & $(\forall$ recursively enumerable $B)[B \subset \bar{A} \Rightarrow B$ has all but a finite number of its members in $C]]$, then C is called a *center* of $\bar{A}$. If $\bar{A} = C \cup D$, where C is infinite recursively enumerable and D is immune, C is said to be *simple in* $\bar{A}$. Thus, for example, $\mathcal{C}_2$ is the class of those nonrecursive, recursively enumerable sets which have sets simple in their complements; and $\mathcal{C}_{21}$ is the class of nonrecursive, recursively enumerable sets whose complements have recursive centers. Occasionally, in the literature, the term pseudosimple has been limited to sets in $\mathcal{C}_{21} \cup \mathcal{C}_{22}$, and the term pseudocreative has been applied to sets in $\mathcal{C}_{23} \cup \mathcal{C}_3$.

$\mathcal{C}_{21}$ is nonempty, since for any simple A, $\{2x|x \in A\} \in \mathcal{C}_{21}$. We shall see in Exercise 8-39 that $\mathcal{C}_{22}$ is nonempty, and in Exercise 8-42 that $\mathcal{C}_{23}$ is nonempty.

$\mathcal{C}_1$, $\mathcal{C}_{21}$, $\mathcal{C}_{22}$, $\mathcal{C}_{23}$, $\mathcal{C}_3$, $\mathcal{C}_4$ mark regions along a spectrum of nonrecursive, recursively enumerable sets according to increasing richness of their complements (see Exercise 8-46). Several further regions have been identified within $\mathcal{C}_1$ (the simple sets). We study these in Chapters 9 and 12.

§8.8 EXERCISES

§8.1

8-1. Let A and B be simple.

(*a*) Show A *join* B is simple.

(*b*) Show $A \times B$ is not simple.

(Part (*a*) shows that the join of two noncylinders can be a noncylinder.)

8-2. What is the cardinality of the family of simple sets?

8-3. Exhibit two simple sets whose union is N.

$\triangle$***8-4.*** (Dekker). (*a*) Show that the intersection of a productive and a simple set must be productive.

(*b*) Show that the intersection of a creative and a simple set must be creative.

8-5. (Rabin). Let A be an infinite set with infinite complement. Consider the following game between players I and II. I picks any integer x, then II, knowing x, picks y. If $x + y \in A$, I wins; otherwise II wins. In case A is not recursive, we assume the existence of an "oracle" that tells I and II who has won. Obviously a pure, i.e., deterministic, strategy for II takes the form of a total function f such that, when I plays an x, II plays $y = f(x)$. Since $\bar{A}$ is infinite, it is clear that for any x, a y exists such that $x + y \notin A$. Thus a winning strategy for II must exist. (This is a special and trivial instance of a general theorem about games.)

Show that the following occurs if A is taken to be a simple set:
(i) II has no recursive winning strategy.
(ii) If II selects any recursive strategy and sticks to it through repeated plays of the game, then there is an effective procedure by which I can discover an x that wins against II's strategy.

8-6. Assume $A \leq_m B$, B simple, C recursively enumerable and $C \cap A = \emptyset$. Show that A and C are recursively separable.

8-7. (Fischer). (*a*) Show: C creative $\Rightarrow C \times B$ creative, for any recursively enumerable nonempty B.

(*b*) Show: A simple $\Rightarrow A \times A$ not creative.

$\triangle$(*c*) Let A be simple. Show: $A \times B$ creative $\Rightarrow B$ creative, for any recursively enumerable B. (*Hint:* A proof can be made from first principles. This is one of the more difficult $\triangle$ exercises.)

§*8.2*

8-8. Let A be the set of even numbers. Show that a recursively enumerable 1-degree exists which is incomparable with A and with all finite degrees but which lies above all the co-finite 1-degrees.

8-9. Let f be a one-one recursive function. Discuss $f(A)$ and $f^{-1}(A)$ for (i) A simple; (ii) A immune.

8-10. Show that $A \times N$ creative $\Rightarrow A$ creative.

8-11. (Dekker). Let A and B be simple; let C be any recursively enumerable set. Show:
(i) $A \cap B$ is simple,
(ii) $A \cup C$ is simple or co-finite.

Hence conclude that the family of simple and co-finite sets forms a dual ideal in the lattice of recursively enumerable sets (for the concept of *dual ideal*, see §12.1).

8-12. Consider the quotient lattice obtained by identifying the recursively enumerable sets modulo the co-simple and finite sets (for the concept of *quotient lattice*, see §12.1).

(*a*) Show that any set identified with a creative set must be creative.

(*b*) Show that there are two creative sets which are not identified with each other.

§*8.3*

8-13. Show that A *join* $B \equiv_{tt} A \times B$, if neither A nor $B = \emptyset$.

8-14. Show that the family $\{A | A \leq_{tt} B\}$ forms a Boolean algebra, i.e., is closed under union, intersection, and complementation.

8-15. Show that every tt-degree contains exactly $\aleph_0$ sets.

8-16. Show that every nonrecursive tt-degree contains a set which is not recursively enumerable and does not have recursively enumerable complement.

▲***8-17.*** Show that K, $\{x | W_x$ *infinite*$\}$, and $\{x | W_x$ *recursive*$\}$ fall into distinct but comparable tt-degrees.

§*8.4*

8-18. Give an example of a cylinder which is not recursively enumerable and not a tt-cylinder.

8-19. Show how the structure of the recursive sets under the tt-ordering accords with Theorem IX.

8-20. (*a*) Show that A is a tt-cylinder $\Rightarrow A \equiv \bar{A}$.

$\triangle$(*b*) Show that $A \equiv \bar{A}$ does not in general imply that A is a tt-cylinder. (*Hint:* Take $A = K$ *join* $\bar{K}$, and see 7-36(*e*).)

§8.5

8-21. Show that the results of 8-13 to 8-16 hold with btt in place of tt.

8-22. (*a*) Show that $A \times A \leq_{\text{btt}} A$ (although, from Theorem XIII, there is a recursively enumerable set A such that $A \times A \nleq_{\text{m}} A$).

(*b*) Show that there are nonrecursive, recursively enumerable sets A and B such that $A \times B \not\equiv_{\text{m}} A$ *join* B.

8-23. Show that $(\forall A)(\exists B)[B \leq_{\text{btt}} A \;\&\; A \leq_1 B \;\&\; B \equiv \bar{B}]$. (*Hint:* Use an appropriate *join.*)

8-24. Consider btt-reducibility of norm 1. Show that it is transitive and reflexive. Is its ordering a semilattice? Show that its complete sets are creative. (It then follows, from the result that A creative $\Rightarrow A$ m-complete (to be proved in Chapter 11), that the complete degree in this ordering coincides with the complete m-degree (and therefore the complete 1-degree.))

8-25. In the proof of Theorem XII, show that if $T_{j_1} \cap \bar{A} = T_{j_2} \cap \bar{A}$, then t'_{j_2} is derivable from $C \cup \{t'_{j_1}\}$.

$\triangle$***8-26.*** Define a tt-condition to be *disjunctive* if it has an associated formula of the form $x_1 \in X \vee x_2 \in X \vee \cdots \vee x_k \in X$ (or, equivalently, if the Boolean polynomial for its Boolean function can be written as a sum of two terms, the first term being 1 and the second term a product of factors of the form $(w_i + 1)$). tt-reducibility sometimes occurs with only disjunctive tt-conditions. (Reduction of $\overline{\bar{A} \times \bar{A}}$ to A is one example.) Let us call this q-*reducibility.* We therefore have the following.

Definition. $A \leq_{\text{q}} B$ if $(\exists$ recursive $f)(\forall x)[x \in A \Leftrightarrow D_{f(x)} \cap B \neq \emptyset]$.

The question naturally arises, does q-reducibility coincide with tt-reducibility?

(i) Show: $\leq_{\text{q}}$ is reflexive and transitive.

(ii) Show: $A \leq_{\text{m}} B \Rightarrow A \leq_{\text{q}} B \Rightarrow A \leq_{\text{tt}} B$.

(iii) Describe the structure of the recursive sets under q-reducibility.

(iv) Define q-cylinder as follows: $A^{\text{q}} = \{x | D_x \cap A \neq \emptyset\} \times N$. Prove an analogue to Theorem IX. Show that this analogue fails if A^{q} is defined to be $\{x | D_x \cap A \neq \emptyset\}$. (*Hint:* Use A such that $\bar{A}$ is immune.)

(v) Show that $[A \leq_{\text{q}} B \;\&\; B$ recursively enumerable$] \Rightarrow A$ recursively enumerable.

(vi) Conclude from (v) that $\leq_{\text{q}}$ is implied by neither $\leq_{\text{tt}}$ nor $\leq_{\text{btt}}$.

(vii) Show that $\leq_{\text{q}}$ does not imply $\leq_{\text{btt}}$. (*Hint:* Show $\bar{K} \leq_{\text{q}} \overline{S^*}$, for S^* as in Theorem VIII.)

$\triangle$***8-27.*** (See Exercise 8-26.)

Definition A is q-*creative* if A is recursively enumerable and $(\exists$ recursive $f)(\forall x)[W_x \subset \bar{A} \Rightarrow [D_{f(x)} \subset \bar{A} \;\&\; D_{f(x)} \not\subset W_x]]$. (This definition is due to Shoenfield [1957], who called such sets *quasicreative.*)

(i) Show: A q-complete $\Rightarrow A$ q-creative. (The converse also holds; see Exercise 11-18.)

(ii) Show: A q-creative $\Rightarrow A$ not simple.

(iii) Conclude that the q-complete sets do not coincide with the tt-complete sets; hence conclude that $\leq_{\text{q}}$ and $\leq_{\text{tt}}$ differ on the nonrecursive, recursively enumerable sets.

(For more on q-reducibility, see Exercise 8-44.) (*Note:* If, in the discussion at the beginning of Exercise 8-26, we replace *disjunctive* by *conjunctive,* we are led to c-*reducibility,* where $A \leq_{\text{c}} B$ if $(\exists$ recursive $f)(\forall x)$ $[x \in A \Leftrightarrow D_{f(x)} \subset B]$. If we replace *disjunctive* by *positive,* we are led to p-*reducibility,* where $A \leq_{\text{p}} B$ if $(\exists$ recursive $f)$ $(\forall x)[x \in A \Leftrightarrow (\exists y)[y \in D_{f(x)} \;\&\; D_y \subset B]]$. Both c-reducibility and p-reducibility are studied in Jockusch [1966].)

△***8-28.*** (Fischer). Show that the more and less severe modifications on tt-reducibility mentioned in the remark at the end of §8.3 are equivalent. (The exceptional case is B or $\bar{B} = \emptyset$.)

§8.6

8-29. (Jockusch) (*a*) Let A and B be recursively enumerable cylinders. Show that A *join* $B \equiv_{\mathrm{m}} A \times B$ need not hold.

(*b*) Show that A *join* $A \leq_1 A \Leftrightarrow A$ is a cylinder, and conclude that there are nonrecursive recursively enumerable sets A and B such that A *join* B is not a least upper bound for A and B in the 1-ordering.

(*c*) Show that if either A or B is a cylinder, then A *join* B is a least upper bound for A and B in the 1-ordering.

(*d*) Show that every m-degree either consists of a single 1-degree or contains an infinite linearly ordered collection of 1-degrees.

8-30. If A is tt-complete, must it follow that every recursively enumerable set is m-reducible to $A \times \bar{A}$? (*Hint:* Take $A = S^*$.)

△***8-31.*** Define

$$B_1 = S^*,$$
$$B_2 = S^* \times S^*,$$
$$\ldots,$$
$$B_k = S^* \times S^* \times \cdots \times S^* \qquad (k \text{ factors}).$$
$$\ldots.$$

Show that $B_1, B_2, \ldots$ fall into distinct m-degrees.

△***8-32.*** Define *isolic reducibility* as follows.

Definition $A \leq_{\mathrm{i}} B$ if ($\exists$ partial recursive one-one φ)$[A = \varphi^{-1}(B)]$.

(i) Show that $\leq_{\mathrm{i}}$ is reflexive and transitive.

△(ii) Define $\equiv_{\mathrm{i}}$ in the obvious way. Prove that $A \equiv_{\mathrm{i}} B \Leftrightarrow$ ($\exists$ partial recursive one-one φ)$[A \subset$ *domain* φ & $B = \varphi(A)]$. (*Hint:* (Manaster.) Use a technique similar to that for Theorem 7-VI. Added difficulties appear.)

The i-degrees are known as *recursive equivalence types.* They have been extensively studied by Dekker, Myhill, Nerode, and others (see Dekker and Myhill [1960]). The i-degrees of isolated sets are called *isols.* (It is trivial that $[A$ isolated & $A \equiv_{\mathrm{i}} B] \Rightarrow B$ isolated.)

(iii) Show that if A and B are isolated, then A *join* B and $A \times B$ are isolated.

If the "sum" and "product" of the isols of A and B are defined as the isols of A *join* B and $A \times B$, respectively, a theory of *isolic arithmetic* can be developed in close analogy to the theory of mediate cardinals (see comment in §8.2 above). In particular, various desirable cancellation laws can be shown to hold.†

(iv) Show that there are $2^{\aleph_0}$ isols.

(v) Show that the co-simple sets fall into $\aleph_0$ distinct isols (use Theorem XIV).

(vi) Show that the infinite recursively enumerable sets form a single i-degree.

(vii) Show that there are infinitely many i-degrees which are not isols. (*Hint:* Use a cardinality argument on the number of nonisolated sets and the number of possible sets in each i-degree.)

(viii) Show: A immune $\Rightarrow A$ and N incomparable in the i-reducibility ordering.

(ix) Show that for sets A and B which are infinite and recursively enumerable, $A \equiv B \Leftrightarrow \bar{A} \equiv_{\mathrm{i}} \bar{B}$.

† From the point of view of homological algebra, the theory of recursive equivalence types arises when we consider the *category* of sets of integers and partial recursive mappings. The operations of *join* and $\times$ then appear as direct sum and direct product.

8-33. Define $\leq_{i'}$ as follows. $A \leq_{i'} B$ if ($\exists$ partial recursive one-one φ)$[A \subset \varphi^{-1}(B)]$.

(i) Show: $\leq_{i'}$ is reflexive and transitive.

(ii) Show that for isolated sets, the i′-degrees and the i-degrees (see Exercise 8-32) coincide.

(iii) Show that all nonisolated sets fall into a single i′-degree.

§8.7

8-34. Show that the classification $\mathcal{B}_0, \ldots, \mathcal{B}_4$ is exhaustive and mutually exclusive.

8-35. Show that A simple $\Rightarrow \{2x|x \in A\} \in \mathcal{C}_2$.

8-36. Show that A simple $\Rightarrow A \times N \in \mathcal{C}_3$.

8-37. (*a*) Show that all nonrecursive, recursively enumerable cylinders are in $\mathcal{C}_3 \cup \mathcal{C}_4$.

(*b*) Show that all *semicreative* sets (i.e., recursively enumerable sets with semiproductive complements) are in $\mathcal{C}_{23} \cup \mathcal{C}_3 \cup \mathcal{C}_4$.

8-38. Show that if C is a center of $\bar{A}$, then C is simple in $\bar{A}$.

△***8-39.*** This exercise concerns the theory of disjoint pairs of recursively enumerable sets (see §7.7). We present here an analogue to *simpleness* and use it to show the existence of a disjoint pair of recursively enumerable sets which is neither recursively separable nor effectively inseparable and also to show that the class $\mathcal{C}_{22}$ in §8.7 is not empty. The construction is similar to Post's construction of S (Theorem II) and is a modification of a construction of Muchnik [1956*a*] and of Tennenbaum.

Let $D = \{<x,y>|y \in W_x \;\&\; y > 3x\}$. D is recursively enumerable and can be effectively enumerated in a sequence $<x_0,y_0>, <x_1,y_1>, \ldots$. We enumerate D and, at the same time, form two lists A and B according to a procedure (presently to be defined) in which, as each $<x_i,y_i>$ from D is enumerated, y_i may or may not be added to one of the two lists. At any such stage, if y_i *is* added to one list or the other, we say that x_i *contributes* y_i to that list. The procedure for making the lists is defined as follows.

Stage i. See whether y_i has been contributed to either list at a previous stage. If so, go to stage $i + 1$. If not, see whether x_i has made any contribution to A. If not, put y_i in A. If so, see whether x_i has made any contribution to B. If not, put y_i in B. If so, go to stage $i + 1$.

Evidently, A and B form a disjoint pair of recursively enumerable sets. Prove the following:

(i) $\overline{A \cup B}$ is infinite.

(ii) $[C$ recursively enumerable $\& \; C \cap (\overline{A \cup B})$ infinite$] \Rightarrow [C \cap A \neq \emptyset \;\&\; C \cap B \neq \emptyset]$.

(iii) The pair A, B is not recursively separable.

(iv) The pair A, B is not effectively inseparable.

(v) B is a center of $\bar{A}$ which is recursively enumerable but not recursive.

(vi) $\bar{A}$ has no recursive center. (From (v) and (vi) we conclude that $A \in \mathcal{C}_{22}$ (see §8.7).)

A pair of recursively enumerable sets with properties (i) and (ii) is said to be *strongly inseparable*.

8-40. A is called a *maximal set* if A is recursively enumerable and $\bar{A}$ is infinite and $(\forall B)[[B$ recursively enumerable $\& \; B \supset A] \Rightarrow [B - A$ finite or $N - B$ finite$]]$.

(i) Show A maximal $\Rightarrow A$ simple.

(ii) Exhibit a simple set which is not maximal.

▲(iii) Show that a maximal set exists. (We prove this in Chapter 12; no easy proof is known.)

△***8-41.*** Let $S_0, S_1, \ldots$ be as in the proof of Theorem VIII. Let $\mathcal{S} = \{S_x|S_x \subset W_x\}$. Let $A = \{z|(\exists y)[S_y \in \mathcal{S} \;\&\; z$ *is the largest member of* $S_y]\}$.

(i) Show that A is semicreative (i.e., recursively enumerable with semiproductive complement), with a semiproductive function f such that $W_{f(x)} = W_x \cup S_x$.

(ii) Show that A is, in fact, creative. (*Hint:* Prove directly by getting productive function or indirectly by showing A to be m-complete.)

△***8-42.*** (Shoenfield). Now modify the construction of 8-41 as follows. Take S from Theorem II. Take $\mathcal{S}$ as in 8-41. Let $E = \bigcup_{S_x \in \mathcal{S}} S_x$. Form $B = S \cup E$. From Theorem VIII, $(\forall x)[\bar{S} \cap S_x \neq \emptyset]$. Hence $S_x \subset B \Leftrightarrow S_x \in \mathcal{S}$. Choose a single representative from each S_x in $\mathcal{S}$ by the following process. Enumerate B (it is obviously recursively enumerable); and, for each S_x in $\mathcal{S}$, take, as its representative, the last of its members to be given in the enumeration of B. Let A = the set of these representatives. Prove the following:

(i) A is recursively enumerable and infinite.

(ii) $B - A$ is recursively enumerable.

(iii) A is semicreative.

(iv) A is not recursive.

(v) B is simple.

(vi) A is not creative. (*Hint:* If it were, we could apply productiveness to $B - A$ and get an infinite recursively enumerable subset of $\bar{B}$.)

(vii) A falls in the class $\mathcal{C}_{23}$ of §8.7.

8-43. (*a*) Show that A simple $\Rightarrow A \times N$ not semicreative (and hence that $\mathcal{C}_3$ contains sets which are not semicreative).

(*b*) Show that A semicreative $\Rightarrow A \times N$ semicreative.

(*c*) Using 8-42, conclude that a non-m-complete, recursively enumerable cylinder need not be isomorphic to the cylindrification of a simple set.

△***8-44.*** (Shoenfield). (See Exercises 8-26 and 8-27.)

(*a*) Show that the set A in Exercise 8-42 is q-complete (and hence q-creative).

(*b*) Show that the set A in Exercise 8-42 is not btt-complete. (*Hint:* Assume A is btt-complete; use the recursive enumerability of $B - A$ and methods of Theorem XII to show that B cannot be simple, which is a contradiction.)

(*c*) Hence conclude that $\leq_q$ differs from $\leq_{btt}$ on the recursively enumerable sets.

△***8-45.*** (Shoenfield). Obtain the following improved version of Theorem XII: A btt-*complete* $\Rightarrow A \in \mathcal{C}_3 \cup \mathcal{C}_4$. (*Hint:* Use method of 8-44(*b*).)

8-46. (Young). Show that the classes $\mathcal{C}_1$, $\mathcal{C}_{21}$, $\mathcal{C}_{22}$, $\mathcal{C}_{23}$, $\mathcal{C}_3$, $\mathcal{C}_4$ lie on a linear spectrum in the following sense: no recursively enumerable set can be 1-reducible to a recursively enumerable set lying to its left in the spectrum.

8-47. (C. D. Wyman). Show that the complement of a set in $\mathcal{C}_{22}$ cannot be the disjoint union of a recursive and an immune set.

8-48. If A $\leq_1 B$ and A is a cylinder, must B be a cylinder? (*Hint:* Use $A = N \times N$, $B \in \mathcal{C}_2$.)

△***8-49.*** (Young). Show that $\mathcal{C}_3$ contains a noncylinder. (*Hint:* Establish the lemma [S *simple* & $S \times S$ *a cylinder*] $\Rightarrow S \times S \leq_m S$, then apply to S^* of Theorem XIII.) (Jockusch has shown that it is possible to have S simple and $S \times S$ a cylinder.)

△***8-50.*** (Young). Exhibit sets A and B in $\mathcal{C}_3$ such that $A \leq_1 B$, A is a cylinder, but B is not a cylinder. (*Hint:* Use $S^* \times N$ and $S^* \times S^*$ from 8-49.)

8-51. (Sacks). Define: A is effectively simple if A is recursively enumerable and $(\exists$ recursive $f)(\forall x)[W_x \subset \bar{A} \Rightarrow W_x$ has fewer than $f(x)$ members].

(*a*) Show that the set S of Theorem II is effectively simple.

▲(*b*) Show that there exists a simple set which is not effectively simple.

9 Turing Reducibility; Hypersimple Sets

§9.1 AN EXAMPLE

Intuitively, A is *reducible* to B if, given any method for calculating c_B, we could then obtain a method for calculating c_A. Does truth-table reducibility satisfactorily represent this intuitive notion, or are there instances where intuitive reducibility holds but truth-table reducibility does not?

Consider the following example. Let

$$\tilde{K} = \{x|(\exists y)[\varphi_x(x) = y \ \&\ \text{tt-}\textit{condition } y \textit{ is satisfied by } K]\}.$$

Theorem I $\tilde{K} \not\leq_{\text{tt}} K$.

Proof. Assume $\tilde{K} \leq_{\text{tt}} K$. Then $\bar{\tilde{K}} \leq_{\text{tt}} K$. Let $\bar{\tilde{K}} \leq_{\text{tt}} K$ via φ_{x_0}. We have $x_0 \in \bar{\tilde{K}} \Leftrightarrow \varphi_{x_0}(x_0)$ is satisfied by K (by reducibility of $\bar{\tilde{K}}$ to K via φ_{x_0}) $\Leftrightarrow$ $x_0 \in \tilde{K}$ (by definition of $\tilde{K}$). This is a contradiction.☒

We now argue that the following proposition holds.

Proposition *$\tilde{K}$ is reducible to K, intuitively.*

Argument. Assume that we have a means for testing membership in K. Let z be given. We wish to test whether $z \in \tilde{K}$. Our procedure is as follows.

We see whether $z \in K$. If $z \notin K$, then $\varphi_z(z)$ is divergent, and we can conclude $z \notin \tilde{K}$. If $z \in K$, then $\varphi_z(z)$ is convergent, and we can compute $\varphi_z(z)$ and, by asking further questions about K, see whether $\varphi_z(z)$ as a tt-condition is satisfied by K. If so, we conclude $z \in \tilde{K}$; if not, we conclude $z \notin \tilde{K}$.☒

Comment. $\bar{K}$ is intuitively reducible to K since, for any individual-membership question about $\bar{K}$, we need ask only a finite number of individual membership questions about K. Note, however, that we do not determine *ahead of time,* i.e., before we have obtained any answers about K, what the questions are we need to ask about K. To test $z \in \bar{K}$, we *first* test $z \in K$ and *then* use the answer to this question to determine what further questions (if any) are necessary.

This is in contrast to tt-reducibility. As we have previously noted, if A is tt-reducible to B, then, for any question about A, we can effectively and explicitly determine *ahead of time,* i.e., before we have obtained any answers about B, (i) a finite set of questions to ask about B and (ii) a specification which tells us how each possible combination of answers to those questions about B determines an answer to our question about A.†

In this chapter we define a reducibility which we call *Turing reducibility.* It is commonly accepted as formalizing the most general intuitive notion of reducibility. It is, perhaps, the most useful and significant of the reducibility concepts. We also introduce other ideas and methods, related to Turing reducibility, that are fundamental for further work.

§9.2 RELATIVE RECURSIVENESS

Let a set X be given. Consider a procedure that is determined by a finite set of instructions and by an input in the following way. Given the input, a computation is begun. This computation proceeds algorithmically except that (*a*) from time to time, the computing agent may be required to obtain an answer to a question of the form "*is* **n** *in* X?" (in general this question itself, i.e., the value of n, is the result of preceding calculation); (*b*) no means for answering such questions about X are given by the instructions; (*c*) obtaining an answer to such a question counts as a single step in the overall procedure; and (*d*) subsequent steps in the procedure depend, in general, upon that answer. If such answers are correctly and automatically supplied by some external agency, we have a well-defined and otherwise effective computation.‡ We call such a procedure an *algorithm relative to X.* In §9.1, for example, the procedure for calculating $c_{\bar{K}}$, the characteristic function for $\bar{K}$, is *algorithmic relative to K.*

How can this notion of *relative algorithm* be made precise? Various distinct but equivalent characterizations have been given, usually in the form of a straightforward extension of some one of the formal characterizations of

† These remarks suggest a (possibly) intermediate reducibility in which the set of questions (i) is given ahead of time but the specification (ii) is not. We call this *weak truth-table reducibility.* It is considered in Exercise 9-45.

‡ If the set X is recursive, we can make the entire procedure recursive by adding instructions for c_X.

§1.5. We adopt a different course and define relative algorithm directly in terms of our basic concept of partial recursive function. Before doing this, we summarize several of the better-known characterizations. We omit details.

Turing Machines With Oracles

An external agency which supplies correct answers to questions about X is often called an *oracle*. It is occasionally pictured as an otherwise unspecified "black box" associated with the computing agent. The notion of relative algorithm can be formally characterized in terms of objects which are, in effect, Turing machines with oracles. This is the approach of Davis. Turing machines are defined as before, except that (i) a machine can write the symbol "S" as well as the symbol "1",† and (ii) in addition to quadruples of the form <*state, cell condition, operation, state*>, a machine may contain quadruples of the form <*state, cell condition, state, state*>. If, for instance, $q_2 1 q_3 q_4$ is in a given machine, then, whenever (in any computation relative to X) the machine finds itself in state q_2 and examining a cell containing 1, the following occurs. Let n be the total number of 1's appearing on the tape (at that moment). If $n \in X$, the machine goes to state q_3; if $n \notin X$, the machine goes to state q_4. This counts as a single step, and no operation is performed on the tape in this step.

We call such a modified form of Turing machine an *oracle-machine.* Note that definition of an oracle-machine via a set of quadruples does not involve a particular choice of X. The same oracle-machine and same input can be used with various different X, and may yield correspondingly different computations.

The following oracle-machine, for example, will compute c_X relative to any given set X. (We make the same conventions regarding numerical inputs and outputs as before; see Chapter 1.)

$$\begin{gathered} q_0 1 B q_1 \\ q_1 B R q_2 \\ q_2 1 q_3 q_4 \\ q_2 B q_3 q_4 \\ q_3 B 1 q_7 \\ q_3 1 B q_5 \\ q_5 B R q_3 \\ q_4 1 B q_6 \\ q_6 B R q_4. \end{gathered}$$

This particular machine does not use S. In more general cases, S is used in order that the machine may record and preserve results of preceding computation while asking questions of the oracle.

† Davis allows more than one additional tape symbol.

Turing Machines with Auxiliary Tapes

A different modification of the Turing machine characterization is obtained by supplying each machine with an auxiliary infinite tape (instead of an oracle) upon which are listed, in increasing order (as sequences of 1's) the members of X. Operations L' and R', for moving the auxiliary tape left and right, are used; and quadruples are replaced by *quintuples* of the form <*state, cell condition, cell condition on auxiliary tape, operation, state*>. Under appropriate instructions, the machine searches through the auxiliary tape in order to get information about X. Answering a single question about X may take many steps. This characterization is equivalent to the preceding one in the sense that, for any X, the same class of partial functions is characterized. We do not go into further detail.

Extended Kleene Characterization

A third, but still equivalent, characterization is obtained from the Kleene characterization of partial recursive function (§1.5) as follows. A *relative function symbol* c is introduced (in addition to the usual main and auxiliary function symbols). c may be thought of as denoting the characteristic function for X. An equation of the form $c(\sigma) = \tau$ may be introduced into a computation provided *either* that it is deducible from preceding equations according to the usual rules *or* that σ is a numeral for some integer, and τ is 1 or 0 according as that integer is or is not a member of X. We do not discuss this characterization further here.

The equivalence of these three characterizations can be proved by methods similar to those used to prove equivalence in the Basic Result of §1.6. Ample and convincing evidence can be supplied to show that these characterizations capture the full generality of the informal notion of *algorithm relative to X*. A discussion closely parallel to the discussion in Chapter 1 can be carried out and, in the obvious way, a *relativized Church's Thesis* can be formulated and adopted.

Our own characterization of relative algorithm takes a different route. In order to motivate it, we first show, informally, how the oracle-machine characterization leads to it.

Make an effective list of all oracle-machines. Use this list to assign to each oracle-machine an index. Let P'_z be the oracle-machine with index z. To each oracle-machine and to each choice of X there corresponds, in the usual way, a partial function of one variable (numerical inputs and outputs being determined as in Chapter 1). Let ψ_z^X be the partial function computed by P'_z relative to the set X. If z is fixed and if some input x is given, we can effectively generate a (possibly infinite) *diagram* that describes all possible computations for $\psi_z^X(x)$ (as X varies). This is done as follows. Take the input x. Begin the computation determined by P'_z. Compute until the first question is asked (this might be, for example, "*is* 7 *in X?*"). At this point, begin two computations, one corresponding to an affirmative answer ($7 \in X$) and one corresponding to a negative answer ($7 \notin X$). Each of these branches is then subdivided, etc., as further questions occur. Of course, a question asked on one branch need not be asked on another branch. If on one branch a question is asked that has been previously answered on that branch, we use the previous answer and do not subdivide. This ramifying computation constitutes our desired diagram. In outline, such a diagram might appear, in part, as follows.

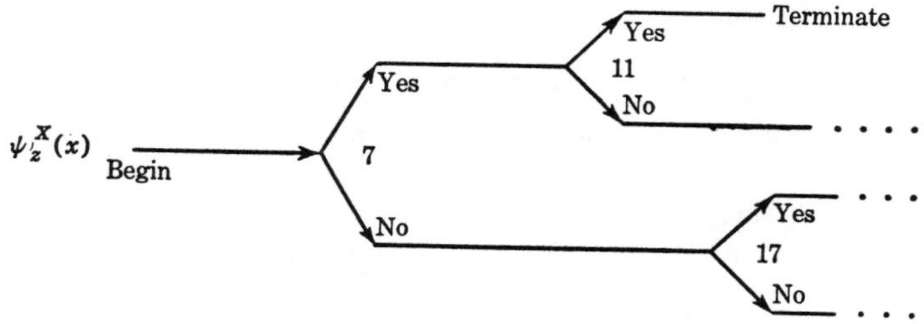

Any branch of the diagram determines two sets D' and D'', where $D' = \{w|$*an affirmative answer to "is* **w** *in X?" is used on the branch*$\}$, and $D'' = \{w|$*a negative answer to "is* **w** *in X?" is used on the branch*$\}$. (Thus, for the terminating branch shown in the above diagram, $D' = \{7,11\}$, $D'' = \emptyset$.) Clearly $D' \cap D'' = \emptyset$. *For any X such that* $D' \subset X$ *and* $D'' \subset \bar{X}$, *the oracle-machine computation of* $\psi_z{}^X(x)$ *must coincide with the computation along that branch.* Certain branches of the diagram may terminate (and therefore yield an output); certain others may not. On any terminating branch, D' and D'' are finite.

In the process of constructing the diagram, we can effectively generate the following (recursively enumerable) set: $\{<y,u,v>|$*a terminating branch exists for which* $D' = D_u$ *and* $D'' = D_v$, *and for which the output is* $y\}$. An r.e. index for this set will depend upon the input x and the machine index z. Therefore (by the *s-m-n* and projection theorems), there is a recursive function h such that

$W_{h(z)} = \{<x,y,u,v>|$*a terminating branch exists in the diagram for* P_z' *with input* x, *for which* $D' = D_u$ *and* $D'' = D_v$, *and for which the output is* $y\}$.

This suggests that we can characterize relative algorithms in terms of appropriate recursively enumerable sets. Such a characterization is now given.. We take it as basic in further work.

Definitions $<x,y,u,v>$ is *consistent* if $D_u \cap D_v = \emptyset$. $<x_1,y_1,u_1,v_1>$ and $<x_2,y_2,u_2,v_2>$ are *compatible* if

$$D_{u_1} \cap D_{v_2} = D_{u_2} \cap D_{v_1} = \emptyset.$$

Definition W_z is *regular* if

(i) $<x,y,u,v> \in W_z \Rightarrow <x,y,u,v>$ consistent;

(ii) $[<x,y_1,u_1,v_1> \in W_z \ \& \ <x,y_2,u_2,v_2> \in W_z \ \& \ <x,y_1,u_1,v_1> \neq <x,y_2,u_2,v_2>] \Rightarrow <x,y_1,u_1,v_1>$ and $<x,y_2,u_2,v_2>$ are not compatible. Evidently, $W_{h(z)}$ defined above is regular. Conversely, Blum has shown that for every regular recursively enumerable set A, $(\exists z)[A = W_{h(z)}]$ (see Exercises 9-4 and 9-5).

Theorem II *There exists a recursive function* ρ *such that for all* z,

(i) $W_{\rho(z)}$ *is regular;*

(ii) W_z *regular* $\Rightarrow W_{\rho(z)} = W_z$.

Proof. Take a fixed f' as in Corollary 5-V(c). Corollary 5-V(c) then yields, for each z, a standard listing of W_z. A list of $W_{\rho(z)}$ is obtained from the list of W_z by omitting any $<x,y,u,v>$ which is inconsistent and by omitting any $<x,y_1,u_1,v_1>$ which is compatible with a $<x,y_2,u_2,v_2>$ already placed in $W_{\rho(z)}$. (This proof is similar to the proof of Theorem 5-XVI, the single-valuedness theorem.) ☒

We now make our characterization. Let ρ be the recursive function of Theorem II.

Definition

$$\varphi_z^X = \{<x,y> | (\exists u,v)[<x,y,u,v> \in W_{\rho(z)} \ \& \ D_u \subset X \ \& \ D_v \subset \bar{X}]\}.$$
$$\varphi_z^{(k)X} = \lambda x_1 \cdots x_k[\varphi_z^X(<x_1, \ldots ,x_k>)].$$

The regularity of $W_{\rho(z)}$ ensures that, for any z and X, φ_z^X is single-valued.

Let A be fixed.

Definition η is a *partial A-recursive function* if $\eta = \varphi_z^A$ for some z.

f is an *A-recursive function* if, for some z, $f = \varphi_z^A$ and φ_z^A is total.

(Of course, for fixed z and varying X, the totality of φ_z^X may depend on X.)

The following theorem is immediate.

Theorem III *A is recursive* $\Rightarrow \varphi_z^A$ *is a partial recursive function.*

Proof. Recursiveness of A makes the entire computation recursive. ☒

From the above discussion, it is evident that our characterization is at least as general as the oracle-machine characterization. For example (for partial functions of one variable), $(\forall z)(\forall X)[\varphi_{h(z)}^X = \psi_z^X]$, where h is as above.

Conversely, for given X, z, and x, we can compute $\varphi_z^X(x)$ as follows. Enumerate $W_{\rho(z)}$. As each $<x',y',u',v'>$ appears, see whether $x' = x$, $D_u \subset X$, and $D_v \subset \bar{X}$. If so, terminate and yield y' as output. If we choose to use a relativized Church's Thesis (with respect to oracle-machines), we have ($\exists$ recursive f)$(\forall z)(\forall X)[\psi_{f(z)}^X = \varphi_z^X]$ (since φ_z^X is evidently intuitively algorithmic relative to X (with instructions obtainable from z)).†,‡

† Of course, with more detailed treatment, an explicit construction for f can be given.

‡ A similar argument, for partial functions of k variables ($k > 1$), can be made by way of Theorem 5-VIII.

Although the oracle-machine characterization is, in certain respects, more intuitive than ours, our characterization has the virtue that it depends only on concepts already defined and can be used informally without the necessity of a new, relativized Church's Thesis. The reader is urged, however, to keep the intuitive notion of relative algorithm well in mind. The proofs that follow can be expressed in any of the characterizations, but they are best explained in terms of that intuitive notion.

The following definitions and results are natural and immediate.

Definition *A is recursive in B* if c_A is B-recursive.

A is *recursively enumerable in B* if either $A = \emptyset$ or $A = \textit{range } f$ for some f which is B-recursive.

Theorem IV *A is recursive in B* $\Leftrightarrow$ *A and* $\bar{A}$ *are both recursively enumerable in B.*

Proof. As in Theorem 5-II.☒

The foregoing terminology can be generalized to k-ary relations (according to the convention of §5.3) if we associate with each k-ary relation R the set $\tau^k(R)$.

Definition Let R be a k_1-ary relation and S a k_2-ary relation.

$$R \text{ is } \begin{Bmatrix} \textit{recursive} \\ \textit{recursively enumerable} \end{Bmatrix} \text{ in } S \text{ if}$$

$$\tau^{k_1}(R) \text{ is } \begin{Bmatrix} \textit{recursive} \\ \textit{recursively enumerable} \end{Bmatrix} \text{ in } \tau^{k_2}(S).$$

ψ is *partial R-recursive* if ψ is partial $\tau^{k_1}(R)$-recursive.

f is *R-recursive* if f is $\tau^{k_1}(R)$-recursive.

Of course, a function is itself a relation. We hence have the following theorem.

Theorem V *f is A-recursive* $\Leftrightarrow$ *f is recursive in A.*

Proof. As for Theorem 5-IX.☒

The following theorem completes the analogue to Theorem 5-IX.

Theorem VI ψ *is partial A-recursive* $\Leftrightarrow$ ψ *is recursively enumerable in A.*

Proof. As for Theorem 5-IX.☒

Corollary VI *For any f, f is recursively enumerable in A* $\Leftrightarrow$ *f is recursive in A.*

Proof. Since f is total, f is recursively enumerable in $A \Leftrightarrow f$ is partial A-recursive (by Theorem VI) $\Leftrightarrow f$ is A-recursive (by definition) $\Leftrightarrow f$ is recursive in A (by Theorem V).☒

In much of the foregoing, a set and its characteristic function are interchangeable. We have, for example, the following theorem.

Theorem VII *A is recursive in B* $\Leftrightarrow c_A$ *is recursive in B* $\Leftrightarrow c_A$ *is recursive in* $c_B \Leftrightarrow$ *A is recursive in* c_B.

Proof. The first and third equivalences follow by definition and Theorem V. The second equivalence is immediate if we note that any question of the form "*is* **z** *in B*?" can be restated as "*is* $<\mathbf{z},1>$ *in* c_B?" and that any question of the form "*is* $<\mathbf{x},\mathbf{y}>$ *in* c_B?" can be restated as "*is it true that* $\mathbf{x} \in B$ *and* $\mathbf{y} = 1$ *or that* $\mathbf{x} \notin B$ *and* $\mathbf{y} = 0$?".⊠

As we shall point out in §9.7, "ψ *is partial recursive in* φ" is commonly defined in a sense which (i) coincides with "ψ is partial $\tau(\varphi)$-recursive" for φ total but (ii) is narrower when φ is not total.

Recursiveness in More than One Set

Procedures can be defined which make use of more than one oracle. We would say that a procedure is *algorithmic relative to* $X_1, \ldots, X_m$ if it could ask, at any point, questions of the form "*is* **n** *in* X_i?", $i = 1, \ldots, m$. We characterize this by the following definition.

Definition ψ is *partial* $(X_1, X_2, \ldots, X_m)$*-recursive* if ψ is partial (X_1 *join* X_2 *join* $\cdots X_m$)-recursive, where X_1 *join* X_2 *join* $\cdots X_m$

$$= (\cdot((X_1 \text{ join } X_2) \text{ join } X_3) \cdots X_m).$$

Notation We occasionally write $\varphi_z^{X_1 \text{join} \cdots X_m}$ as $\varphi_z^{X_1, \ldots, X_m}$.

The reader can easily verify that the various concepts defined above are recursively invariant.

§9.3 RELATIVIZED THEORY

In the recursive function theory developed in the preceding chapters, the partial recursive functions have been the basic objects. If a fixed set A is chosen and if, in this theory, partial recursive functions are replaced by partial A-recursive functions (in all definitions and theorems), we obtain a *relativized theory*. Under such *full relativization*, all of the theory obtained so far remains valid. Proofs involving length-of-computation arguments carry over directly, since a terminating A-computation requires that only finitely many questions be asked about A (and each question may be counted as a single step), or, to put it more rigorously, since a terminating computation for φ_z^A corresponds to a terminating search through $W_{\rho(z)}$ (and steps can be measured as steps in the enumeration of $W_{\rho(z)}$). Concepts obtained in such a full relativization are *recursively invariant relative to* A in the sense that they are invariant under the group $\{f|f$ *is one-one, onto, and A-recursive*$\}$.

Of more interest and significance are concepts and results obtained in a *partially relativized theory*. In many cases, a given concept can be partially relativized in several different ways. For instance, let the concept *immune* be (fully) relativized as follows.

Definition B is *A-immune* if B is infinite and B contains no infinite set recursively enumerable in A.

Three different ways are then evident for partially or fully relativizing the concept *simple*. We may define B is *A-simple* by

(*a*) B is recursively enumerable in A, and $\bar{B}$ is A-immune (full relativization);

(*b*) B is recursively enumerable, and $\bar{B}$ is A-immune (partial relativization);

(*c*) B is recursively enumerable in A, and $\bar{B}$ is immune (partial relativization).

In treating relativized concepts, we shall assume the full relativization unless we indicate otherwise.

As noted, the fully relativized results carry over directly. We mention several of the more fundamental of these results here.

Definition $W_z^X = domain\ \varphi_z^X$.

Theorem VIII **(*relativized basic theorem on recursively enumerable sets* (*Theorem 5-V*))** *A is recursively enumerable in $B \Leftrightarrow (\exists z)$ $[A = W_z^B]$.*

Proof. The proof of Theorem 5-V carries over directly.☒

Theorem IX **(*relativized projection theorems*)** *If R is recursively enumerable in A, then there is a relation S, recursive in A, such that R is a projection of S.*

If R is recursively enumerable in A, then any projection of R is recursively enumerable in A.

Proof. The proofs of Theorems 5-X and 5-XI, which were from first principles, carry over directly with all computations replaced by computations relative to A. More rigorous proofs, in terms of the formal characterization of §9.2, proceed by straightforward application of the unrelativized projection theorems.☒

Definition $K^A = \{x | \varphi_x^A(x)\ \ convergent\} = \{x | x \in W_x^A\}$.

Theorem X *$(\forall A)(\exists B)$ [B is recursively enumerable in A, and B is not recursive in A].*

Proof. Let A be given. Take $B = K^A$. K^A is recursively enumerable in A by Theorem IX (see Exercise 9-8). K^A cannot be recursive in A; for if so, then by Theorems IV and VIII, $\overline{K^A} = W_{x_0}^A$ for some x_0, and we would have $x_0 \in K^A \Leftrightarrow x_0 \in W_{x_0}^A$ (by definition of K^A) $\Leftrightarrow x_0 \in \overline{K^A}$ (by choice of x_0); a contradiction.☒

Partially relativized versions of earlier theorems often hold in forms that are stronger than the fully relativized versions. In particular, a strong partially relativized form of the *s-m-n* theorem holds. This in turn yields stronger partially relativized versions of various other results for which the original *s-m-n* theorem was used. For example, it will be a consequence of our relativized *s-m-n* theorem that, in relativized versions of the corollaries to Theorem 5-V, the functions f, g, f', g', f'', and g'' can be kept recursive.

***Theorem XI* (*relativized* s-m-n *theorem*)** *For every* $m,n \geq 1$, *there exists a recursive function* $\hat{s}_n{}^m$ *of* $m+1$ *variables such that for all* X *and for all* $z, x_1, \ldots, x_m$,

$$\lambda x_{m+1} \cdots x_{m+n}[\varphi_z^{(m+n)X}(x_1, \ldots, x_m, x_{m+1}, \ldots, x_{m+n})] = \varphi_{\hat{s}_n{}^m(z,x_1,\ldots,x_m)}^{(n)X}.$$

Proof. We take $\hat{s}_1{}^1$. Proof for other values of m and n is similar.

Apply the unrelativized *s-m-n* theorem to obtain a recursive f such that for all x_1 and z,

$$W_{f(z,x_1)} = \{<x_2,y,u,v> \mid <<x_1,x_2>,y,u,v> \in W_{\rho(z)}\}.$$

By this construction, $W_{f(z,x_1)}$ is regular for all z and x_1. It follows that

$$<x_2,y> \in \varphi_{f(z,x_1)}^X \Leftrightarrow <<x_1,x_2>,y> \in \varphi_z^{(2)X}.$$

Hence f is the desired $\hat{s}_1{}^1$.

More formally, $\hat{s}_1{}^1$ may be defined to be $\lambda zx_1[h_2s_4{}^1(h_1\rho(z),x_1)]$, where h_1 and h_2 are recursive functions yielding uniformity in the following specific cases (see Theorem 5-VIII and remark on uniformity at end of §5.5):

$$\varphi_{h_1(z)}^{(5)} = \lambda x_1 \cdots x_5[\varphi_z(<x_1, \ldots, x_5>)],$$

and

$$\varphi_{h_2(z)} = \lambda x[\varphi_z^{(4)}(\pi_1{}^4(x), \ldots, \pi_4{}^4(x))].$$

Then, noting that $<<x_1,x_2>,y,u,v> = <x_1,x_2,y,u,v>$ (see §5.3), we have

$$\textit{domain}\ \varphi_{h_1\rho(z)}^{(5)} = \{<x_1,x_2,y,u,v> \mid <<x_1,x_2>,y,u,v> \in W_{\rho(z)}\},$$

hence $\textit{domain}\ \varphi_{s_4{}^1(h_1\rho(z),x_1)}^{(4)} = \{<x_2,y,u,v> \mid <<x_1,x_2>,y,u,v> \in W_{\rho(z)}\}$,

and therefore,

$$W_{h_2s_4{}^1(h_1\rho(z),x_1)} = \{<x_2,y,u,v> \mid <<x_1,x_2>,y,u,v> \in W_{\rho(z)}\} = W_{\hat{s}_1{}^1(z,x_1)}. \boxtimes\dagger$$

For an example of a strong partial relativization (of an earlier result) that can now be obtained, consider Theorem 5-XIII. It asserted

$$(\exists \text{ recursive } h)(\forall x)(\forall y)[W_{h(x,y)} = W_x \cup W_y].$$

The full relativization of this result is

$$(\forall X)(\exists \text{ an } X\text{-recursive } h)(\forall x)(\forall y)[W_{h(x,y)}^X = W_x{}^X \cup W_y{}^X].$$

† The $\hat{s}_n{}^m$ can, in fact, be obtained primitive recursive.

If Theorem XI is used in a relativized proof, this can be strengthened not only to

$$(\forall X)(\exists \text{ recursive } h)(\forall x)(\forall y)[W^X_{h(x,y)} = W_x{}^X \cup W_y{}^X]$$

but, in fact, to

$$(\exists \text{ recursive } h)(\forall X)(\forall x)(\forall y)[W^X_{h(x,y)} = W_x{}^X \cup W_y{}^X].$$

Such strengthening often occurs in relativization of uniformity results.

Concepts and results already relativized to a set A can themselves, in turn, be relativized to another set B. With full relativizations, the new concepts and results are equivalent to those obtained by an original single relativization with respect to A *join* B. If, however, iterated partial relativizations are made, questions of interest can arise. We consider these occasionally in later parts of our theory.

§9.4 TURING REDUCIBILITY

Definition A is *Turing reducible* to B (notation: $A \leq_T B$) if A is recursive in B; i.e., if c_A is a B-recursive function. We also abbreviate "Turing reducibility" as "T-reducibility." The assertions "$A \leq_T B$" and "A is recursive in B" are equivalent. Some authors use the terminology "A is recursive in B" exclusively.

Example. From §9.1 we have that $\bar{K} \leq_T K$, although $\bar{K} \not\leq_{tt} K$.

Theorem XII *$\leq_T$ is reflexive and transitive.*

Proof. Reflexivity. Immediate, since c_A is trivially an A-recursive function (see Exercise 9-5).

Transitivity. Assume $A \leq_T B$ and $B \leq_T C$. Then $c_A = \varphi_{z_1}{}^B$, and $c_B = \varphi_{z_2}{}^C$. Proof proceeds by an application of the second projection theorem to a set constructed from $W_{\rho(z_1)}$ and $W_{\rho(z_2)}$ (see Exercise 9-14). This yields a z_3 such that $c_A = \varphi_{z_3}{}^C$. A more intuitive argument is as follows. We can evaluate c_A for any given input if we can evaluate c_B for a finite number of inputs. We can evaluate c_B for any given input if we can, in turn, evaluate c_C for a finite number of inputs. Hence we can evaluate c_A for a given input if we can evaluate c_C for a finite number of inputs.⊠

Definition $A \equiv_T B$ if $A \leq_T B$ and $B \leq_T A$.

The equivalence classes of $\equiv_T$ are called *Turing degrees* or T-*degrees*.

In the literature, Turing reducibility is commonly taken to be the most fundamental reducibility (although usually, in applications, when Turing reducibility holds, some stronger form of reducibility also holds). When the word "reducible" is used without qualification, *Turing reducibility* is usually intended. When the phrase "degree of unsolvability" is used without qualification, T-*degree* is usually intended. T-reducibility and tt-reducibility are often grouped together as the *weak* reducibilities (in contrast to m-reducibility and 1-reducibility, the *strong* reducibilities).

Definition A is *Turing complete* ("T-complete" or "complete") if
(i) A is recursively enumerable,
(ii) $(\forall B)[B$ recursively enumerable $\Rightarrow B \leq_{\mathrm{T}} A]$.
The set K is a T-complete set.
The following theorem gives some elementary facts about T-reducibility.

Theorem XIII(a) $A \leq_{\mathrm{tt}} B \Rightarrow A \leq_{\mathrm{T}} B$.
(*b*) *The* T-*reducibility ordering is an upper semilattice.*
(*c*) $[B$ *recursive* & $A \leq_{\mathrm{T}} B] \Rightarrow A$ *recursive.*
Proof. (*a*) Immediate.
(*b*) The degree of A *join* B is a least upper bound to the degrees of A and of B in the T-reducibility ordering.
(*c*) By Theorem III. ☒
From (*a*) we have that $\leq_{\mathrm{T}}$ is recursively invariant. It follows from (*c*) and Theorem 8-VII(*e*) that the T-reducibility ordering has a single minimum degree consisting of the recursive sets and no others. We say that a T-degree is *recursively enumerable* if it contains a recursively enumerable set. From (*a*) we have that $A \equiv_{\mathrm{tt}} B \Rightarrow A \equiv_{\mathrm{T}} B$. We can therefore view a T-degree as composed of tt-degrees. By the example of §9.1, the T-degree of K, i.e., the complete T-degree, consists of more than one tt-degree. In §9.6 we shall see that it contains more than one recursively enumerable tt-degree. The relationship between T-degrees and tt-degrees is considered further in §9.6.

§9.5 HYPERSIMPLE SETS; DEKKER'S THEOREM

Is there a nonrecursive, recursively enumerable set which is not tt-complete? More particularly, is there a recursively enumerable set which is T-complete but not tt-complete? Post obtained an affirmative answer to these questions through the study of *hypersimple* sets.

Definition Let A be an infinite set. f *majorizes* A if $(\forall n)[f(n) \geq z_n]$, where $z_0, z_1, \ldots$ are the members of A in strictly increasing order.

It is easy to show, by diagonalization, that sets exist which are majorized by no recursive function. Let $f_0, f_1, \ldots$ be a sequence of functions which includes all the recursive functions. Define

$$g(0) = f_0(0) + 1,$$
$$g(n+1) = \mu z[z > g(n) \;\&\; z > f_{n+1}(n+1)].$$

Then *range* g is majorized by no recursive function. We call such sets *hyperimmune.*

Definition A is *hyperimmune* if A is infinite and ($\forall$ recursive f) [f does not majorize A].

Theorem XIV *A hyperimmune $\Rightarrow$ A immune.*
Proof. See Exercise 9-23.⊠

Definition A is *hypersimple* if A is recursively enumerable and $\bar{A}$ is hyperimmune.

The set S of Theorem 8-II is an example of a set which is simple but not hypersimple, since $\bar{S}$ is majorized by $\lambda x[2x]$. We shall show below that hypersimple sets exist.

A useful and not entirely obvious characterization of hyperimmune sets is given in the following theorem (Medvedev [1955]). (Post defined hypersimple sets by this characterization.)

Theorem XV (Kuznecov, Medvedev, Uspenskii) *A hyperimmune $\Leftrightarrow$ A is infinite and $\neg(\exists$ recursive $f)[(\forall u)[D_{f(u)} \cap A \neq \emptyset]$ & $(\forall u)(\forall v)[u \neq v \Rightarrow D_{f(u)} \cap D_{f(v)} = \emptyset]]$*; that is, A is infinite and there is no effectively enumerable infinite sequence (by canonical indices) of disjoint finite sets each of which intersects A.

Proof. $\Rightarrow$: Assume such an f exists. Then $(\forall u)D_{f(u)} \neq \emptyset$. Define $g = \lambda y[\textit{maximum member of} \bigcup_{i=0}^{y} D_{f(i)}]$. Evidently g is recursive and g majorizes A.

$\Leftarrow$: Assume that a recursive g exists which majorizes A. Define a recursive function h as follows:

$$h(0) = g(0),$$
$$h(n+1) = g(h(n)+1).$$

Consider the sequence of finite sets

$$D^{(0)} = \{0,1, \ldots, h(0)\},$$
$$D^{(1)} = \{h(0)+1, \ldots, h(1)\},$$
$$\ldots\ldots\ldots$$
$$D^{(n+1)} = \{h(n)+1, \ldots, h(n+1)\},$$
$$\ldots\ldots\ldots.$$

This sequence is evidently recursively enumerable by canonical indices. Furthermore, for any n, $D^{(n+1)}$ has $h(n+1) = g(h(n)+1)$ as its greatest member. Hence, by the assumed majorizing property of g, at least $h(n)+2$ members of A are $\leq h(n+1)$. But $D^{(n+1)}$ contains all but $h(n)+1$ of the integers $\leq h(n+1)$. Hence $D^{(n+1)} \cap A \neq \emptyset$.⊠

The first proof of the existence of hypersimple sets was obtained by Post. Subsequently, a more general and elegant existence theorem was found by

Dekker [1954], who showed that every nonrecursive, recursively enumerable T-degree contains a hypersimple set.

Theorem XVI (Dekker) *[A not recursive & A recursively enumerable]* $\Rightarrow (\exists B)[B \equiv_T A$ *& B hypersimple.]*

Proof. Let A be given recursively enumerable but not recursive. Then $A = range\ f$ for some recursive one-one f. Define x is *minimal with respect to* f if $(\forall y)[x < y \Rightarrow f(x) < f(y)]$.

Let

$$B = \{x \mid x \textit{ is not minimal with respect to } f\} = \{x \mid (\exists y)[x < y \ \& \ f(y) \leq f(x)]\}.$$

We show that B is hypersimple and that $B \equiv_T A$.

(i) B is recursively enumerable. This is immediate by the second projection theorem.

(ii) $\bar{B}$ is hyperimmune. For if not, let g majorize $\bar{B}$. Then $x \in A \Leftrightarrow x \in \{f(0), f(1), \ldots, f(g(x))\}$. This gives an effective test for A, and A is recursive, contrary to assumption.

(iii) $B \leq_{tt} A$. For $x \in B \Leftrightarrow f(y) < f(x)$ for some $y > x \Leftrightarrow C \cap A \neq \emptyset$, where $C = (\{0,1, \ldots, f(x)\} - \{f(0), f(1), \ldots, f(x)\})$, and this last can be expressed as a tt-condition.

(iv) $A \leq_T B$. For we can test $z \in A$ as follows. Test 0, 1, 2, . . . for membership in B until $z + 1$ nonmembers of B have been obtained. Let the largest of these be n_z. See whether z occurs in $\{f(0), f(1), \ldots, f(n_z)\}$. If so, $z \in A$; if not, $z \notin A$. This test is satisfactory since n_z is minimal and $z \leq f(n_z)$.☒

Corollary XVI(a) *Every nonrecursive, recursively enumerable* T-*degree contains a simple set.*

(b) *For every nonrecursive, recursively enumerable set A, there is a hypersimple set B such that* $B \leq_{tt} A$.

Proof. Immediate.☒

Maximal sets were defined in Exercise 8-40. Their existence will be shown in Chapter 12. In Exercise 9-26, we show that every maximal set is hypersimple. The class $\mathcal{C}_1$ of simple sets (see discussion in §8.7) can be divided into the following subclasses:

$$\mathcal{C}_{11} = \{A \mid A \textit{ is maximal}\}.$$
$$\mathcal{C}_{12} = \{A \mid A \textit{ is hypersimple } \& \ A \notin \mathcal{C}_{11}\}.$$
$$\mathcal{C}_{13} = \{A \mid A \textit{ is simple } \& \ A \notin \mathcal{C}_{11} \cup \mathcal{C}_{12}\}.$$

Sets in these classes have complements which are progressively less "thin." $\mathcal{C}_{13}$ is nonempty by the example of S from Theorem 8-II (see remark above after the definition of hypersimple set). In Exercise 9-36, we show that $\mathcal{C}_{12}$ is not empty. In Chapter 12 we consider a further subdivision of these classes.

§9.6 TURING REDUCIBILITY AND TRUTH-TABLE REDUCIBILITY; POST'S PROBLEM

The construction of §9.1 can be used to show that there is no way of obtaining a T-reducibility analogue to Theorems 7-VIII and 8-IX.

Theorem XVII *The* T-*degree of* K *contains, among its members, no set maximal with respect to* tt-*reducibility.*

Proof. Define, for any A, $\tilde{A} = \{x | \varphi_x(x) \in A^{\text{tt}}\}$. Then for any A such that $K \leq_{\text{T}} A$: $A \leq_1 \tilde{A}$, $\tilde{A} \not\leq_{\text{tt}} A$, and $\tilde{A} \leq_{\text{T}} A$. We prove this as follows.

(i) $A \leq_1 \tilde{A}$. For any given x, let ψ be the constant function whose value is the tt-condition $\langle\langle x\rangle,\alpha\rangle$ where α is the 1-ary Boolean identity function. Since instructions for ψ can be found uniformly from x, we have a recursive f such that $\psi = \varphi_{f(x)}$. f is evidently one-one, and $x \in A \Leftrightarrow \varphi_{f(x)}(f(x))$ is satisfied by $A \Leftrightarrow f(x) \in \tilde{A}$.

(ii) $\tilde{A} \not\leq_{\text{tt}} A$. If $\tilde{A} \leq_{\text{tt}} A$, then $\bar{\tilde{A}} \leq_{\text{tt}} A$ via φ_{x_0} for some x_0. Hence, $x_0 \in \bar{\tilde{A}} \Leftrightarrow \varphi_{x_0}(x_0)$ is satisfied by $A \Leftrightarrow x_0 \in \tilde{A}$, a contradiction.

(iii) $\tilde{A} \leq_{\text{T}} A$. To test whether $z \in \tilde{A}$, first use the T-reducibility of K to A to test whether $z \in K$. If $z \notin K$, then $z \notin \tilde{A}$. If $z \in A$, compute $\varphi_z(z)$ and see whether $\varphi_z(z)$ is satisfied by A. If so, $z \in \tilde{A}$; if not, $z \notin \tilde{A}$.

As a result of (i), (ii), and (iii) there can be no set maximal with respect to tt-reducibility in the T-degree of K. ☒

(Theorem XVII can also be proved as a corollary to Exercises 9-18(a) and 9-20.) (Jockusch [1966] shows that every nonrecursive recursively enumerable T-degree contains infinitely many recursively enumerable m-degrees.)

Corollary XVII(a) *No* T-*degree above that of* K *(in the* T-*reducibility ordering) can have a maximal* tt-*degree.*

(***b***) *Every* T-*degree above that of* K *contains an infinite sequence of* tt-*degrees, linearly ordered with respect to* $\leq_{\text{tt}}$ *and having order type of the integers* (0, 1, 2, . . .).

Proof. Immediate. ☒

Does every recursively enumerable T-degree contain a recursively enumerable tt-degree that is maximal with respect to $\leq_{\text{tt}}$ among the recursively enumerable tt-degrees contained in that T-degree? This question is unsettled. For the T-degree of K, the tt-degree of K is maximal.

Do the T-complete sets coincide with the tt-complete sets? Post used the following theorem to answer this question negatively.

Theorem XVIII (Post) *A* tt-*complete* $\Rightarrow$ *A not hypersimple.*

Proof. The proof is closely related to the proof of Theorem 8-XII, which showed that no btt-complete set is simple.

Assume A is tt-complete. Then $K \leq_{tt} A$. We prove a lemma similar to the lemma used in the proof of Theorem 8-XII.

Lemma *Let* $K \leq_{tt} A$. *Then there exists an infinite recursively enumerable set of* tt-*conditions* $t_0, t_1, \ldots$, *having associated sets* $T_0, T_1, \ldots$, *such that*

(i) *for all* i, t_i *is satisfied by* A;

(ii) *for all* i, $T_i \cap \bar{A} \neq \emptyset$;

(iii) *for any finite set* D, $\{i | T_i \cap \bar{A} \subset D\}$ *has no more than* $2^{2^{d+1}}$ *members, where* d *is the cardinality of* D.

Proof of lemma. $t_0, t_1, \ldots$ are constructed and (i) and (ii) are proved, exactly as for Theorem 8-XII. We prove (iii) as follows.

Definition Let $t = \langle\langle x_1, \ldots, x_k\rangle, \alpha\rangle$. Let T be the associated set of t, that is, $T = \{x_1, \ldots, x_k\}$. Assume that n members of $\langle x_1, \ldots, x_k\rangle$ are in $T \cap \bar{A}, n > 0$. Define $\alpha' = \lambda w_1 w_2 \cdots w_n[\alpha(a_1, a_2, \ldots, a_k)]$ where $\langle a_1, \ldots, a_k\rangle$ is obtained from $\langle x_1, \ldots, x_k\rangle$ by substituting 1 for each x_i which is in A and by substituting $w_1, w_2, \ldots, w_n$ in order for the remaining x_i. α' is called the *derived truth table* for t. For example, if

$$t = \langle\langle x_1, x_2, x_3, x_4\rangle, \alpha\rangle,$$

and $x_1, x_3 \in A$, and $x_2, x_4 \in \bar{A}$, then $\alpha' = \lambda w_1 w_2[\alpha(1, w_1, 1, w_2)]$ is the derived truth table for t.

Let C be as in the proof of Theorem 8-XII. Form $t'_0, t'_1, \ldots$ from $t_0, t_1, \ldots$ as in the proof of Theorem 8-XII. Let t_i and t_j be such that $T_i \cap \bar{A} = T_j \cap \bar{A}$. If t'_i and t'_j have the same derived truth tables, then t'_j must be *deducible* (in the sense of Theorem 8-XII) from t'_i and C, and t'_i must be *deducible* from t'_j and C. Hence, by the construction of $t'_0, t'_1, \ldots$, if $T_i \cap \bar{A} = T_j \cap \bar{A}$, then t'_i and t'_j must have distinct derived truth tables (see Exercise 9-43).

Now take a finite set D of cardinality d. D has 2^d subsets. A fortiori, at most 2^d subsets of D can be contained in $\bar{A}$. Each of these subsets has cardinality $\leq d$. Hence each subset has, corresponding to it,† at most 2^{2^d} distinct derived truth tables. $2^{2^d} \cdot 2^d = 2^{2^d + d} < 2^{2^{d+1}}$. Hence, if there were more than $2^{2^{d+1}}$ members of $\{t'_i | T_i \cap \bar{A} \subset D\}$, at least two members would yield a common intersection with $\bar{A}$ and have the same derived truth tables, which is a contradiction. This proves the lemma.

To prove the theorem, we show that $\bar{A}$ is not hyperimmune. Define a recursive function f as follows:

$$f(0) = \text{maximum member of } T_0.$$

$$f(n+1) = \text{maximum member of } \bigcup_{i=0}^{2^{2^{f(n)+1}}} T_i.$$

† A derived truth table α' *corresponds* to a finite set D' if, for some i, $T_i \cap \bar{A} = D'$ and α' is the derived truth table for t'_i.

By (iii) of the lemma, at most $2^{2^{f(n)+2}}$ of the T_i can have

$$T_i \cap \bar{A} \subset \{0,1,2, \ldots ,f(n)\}.$$

Hence, for each n, $\{f(n) + 1, \ldots ,f(n + 1)\} \cap \bar{A} \neq \emptyset$. Hence $\{0, \ldots ,f(0)\}$, $\{f(0) + 1, \ldots ,f(1)\}$, $\{f(1) + 1, \ldots\}$, $\ldots$ is a sequence of disjoint finite sets each of which intersects $\bar{A}$. By Theorem XV, $\bar{A}$ is not hyperimmune. (Indeed, f is a majorizing function for $\bar{A}$.) ☒

Corollary XVIII *There exists a set which is* T*-complete but not* tt*-complete; hence* $\leq_{\text{T}}$ *and* $\leq_{\text{tt}}$ *differ on the recursively enumerable sets, and* $\equiv_{\text{T}}$ *and* $\equiv_{\text{tt}}$ *differ on the recursively enumerable sets.*

Proof. Let $\hat{K}$ be a hypersimple set associated with K by Dekker's theorem. Then $K \equiv_{\text{T}} \hat{K}$; but, by the above theorem, $K \not\leq_{\text{tt}} \hat{K}$. ☒

A somewhat deeper insight into the relation between $\leq_{\text{T}}$ and $\leq_{\text{tt}}$ is given by the following theorem of Nerode [1957]. Recall that, by definition,

$$A \leq_{\text{T}} B \Leftrightarrow (\exists z)[c_A = \varphi_z^B].$$

Theorem XIX **(*Nerode*)**

$$A \leq_{\text{tt}} B \Leftrightarrow (\exists z)[c_A = \varphi_z^B \;\&\; (\forall X)[\varphi_z^X \textit{ is a characteristic function}]]$$
$$\Leftrightarrow (\exists z)[c_A = \varphi_z^B \;\&\; (\forall X)[\varphi_z^X \textit{ is total}]].$$

Proof. Let 1*, 2*, and 3* be the three statements to be shown equivalent, in the order given.

(i) 2* $\Rightarrow$ 3*. This is immediate.

(ii) 1* $\Rightarrow$ 2*. Let $A \leq_{\text{tt}} B$ via f. Take z_0 so that for any x,

$$\varphi_{z_0}^X(x) = \begin{Bmatrix} 1 \\ 0 \end{Bmatrix} \text{ if tt-condition } f(x) \begin{Bmatrix} \text{is} \\ \text{is not} \end{Bmatrix} \text{ satisfied by } X.$$

$\varphi_{z_0}^X$ is evidently a characteristic function for any X, and $c_A = \varphi_{z_0}^B$.

(iii) 3* $\Rightarrow$ 1*. Let z_1 be given, where $\varphi_{z_1}^X$ is total for all X, and $c_A = \varphi_{z_1}^B$. We wish to show $A \leq_{\text{tt}} B$; we therefore seek a recursive function f such that $x \in A \Leftrightarrow f(x)$ is a tt-condition satisfied by B. Let x be given. We compute $f(x)$ by the following instructions. Take $\varphi_{z_1}^X(x)$ and generate the computation diagram described and illustrated in §9.2. By assumption, every branch must terminate. By the basic compactness theorem for trees (stated, proved, and discussed in Exercise 9-40 below), the entire diagram must be finite. In this diagram, label each fork with the integer about which information is requested at that fork. Let D be the set of all such integers (occurring as labels in our finite diagram). Let $x_1, \ldots, x_k$ be the members of D in increasing order. $\langle x_1, \ldots ,x_k\rangle$ will be the k-tuple for our desired tt-condition $f(x)$. A Boolean function α is obtained as follows. For any $\langle w_1, \ldots ,w_k\rangle \in I^k$, let $Y = \{x_i | w_i = 1\}$. Compute $\varphi_{z_1}^Y(x)$. If the value is 0, set $\alpha(w_1, \ldots, w_k) = 0$; if the output is different from 0, set $\alpha(w_1, \ldots, w_k) = 1$. Then

$$x \notin A \Rightarrow [\varphi_{z_1}{}^B(x) = 0 \;\&\; \alpha(c_B(x_1), \ldots, c_B(x_k)) = 0],$$

and $x \in A \Rightarrow [\varphi_{z_1}{}^B(x) = 1 \;\&\; \alpha(c_B(x_1), \ldots, c_B(x_k)) = 1]$. Thus $A \leq_{tt} B$ via f.⊠

This theorem suggests that reducibilities be viewed as "effective" mappings from 2^N (the family of all sets) into 2^N. We consider this further in §§9.7, 9.8, and 13.7.

Post's Problem

We have left unsettled the following major question concerning the reducibility orderings of the recursively enumerable sets. Is there a nonrecursive, non-T-complete set? Equivalently: are there more than two recursively enumerable T-degrees? Equivalently: must any two nonrecursive, recursively enumerable sets be of the same T-degree? Post [1944] raised this question. It became known as *Post's problem* and remained unsolved until 1956, when Friedberg and Muchnik independently and almost simultaneously found solutions. We give a solution in the next chapter.

Post's problem was significant in two respects: (1) it concerned the variety of structures possible among the nonrecursive, recursively enumerable sets; (2) it therefore concerned the variety possible among axiomatizable theories and among other sorts of recursively enumerable problems. All known theories (prior to 1956) were $\equiv_T K$ (and, in fact, $\equiv K$). If all theories were $\equiv_T K$, then reducibility from K (at least T-reducibility from K) would be a general method for demonstrating the undecidability of axiomatizable theories.

Post himself devoted considerable effort to the problem. Just as simple sets cannot be m-complete, and hypersimple sets cannot be tt-complete, so might there be, Post conjectured, a special form of hypersimple set that could not be T-complete. Post suggested the concept of *hyperhypersimple* set as one possibility.

Definition A is *hyperhyperimmune* if A is infinite and $\neg(\exists$ recursive $f)[(\forall u)[W_{f(u)}$ is finite &

$$W_{f(u)} \cap A \neq \emptyset] \;\&\; (\forall u)(\forall v)[u \neq v \Rightarrow W_{f(u)} \cap W_{f(v)} = \emptyset]].$$

(This parallels the characterization of hyperimmune sets in Theorem XV, but with r.e. indices appearing in place of canonical indices.)

Definition A is *hyperhypersimple* if A is recursively enumerable and $\bar{A}$ is hyperhyperimmune (see Exercise 9-48(*b*) for a simplification in the definition of hyperhypersimple set due to Yates).

Post did not prove either the existence of hyperhyperimmune sets or the existence of hypersimple sets which are not hyperhypersimple. Nor did he demonstrate that a hyperhypersimple set cannot be T-complete. (We now know that this is false; see below.) We consider certain features of hyperhypersimple sets and hyperhyperimmune sets in Exercises 9-46 to 9-48. We discuss these sets further in Chapter 12. The existence of hyperhypersimple sets is a consequence of a theorem of Friedberg to be given in Chapter 12, (see Exercise 9-47). The existence of hypersimple sets which are not hyperhypersimple will be shown in Exercise 10-17. A characterization of hyperhypersimple sets in terms of retraceable sets (see Exercise 9-44) is given in Exercise 10-19. Yates has exhibited a hyperhypersimple set which is T-complete.

§9.7 ENUMERATION REDUCIBILITY

We now present a technical concept that is closely related both to the various reducibilities studied above and to other parts of recursive function theory. For purposes of exposition we call it a "reducibility" although it is, in certain respects, more inclusive than our basic intuitive notion of reducibility. Our study of it will not parallel our study of the other reducibilities. Calling it a reducibility serves to emphasize a naturalness that is sometimes obscured in treatments of it and related concepts.

Let sets A and B be given. Consider a procedure that is determined by a finite set of instructions in the following way. A computation is begun. The computation proceeds algorithmically except that, from time to time, the computing agent may be requested to obtain an "input" integer, and, from time to time, the procedure yields an "output" integer. When an input is requested, any integer, or no integer, may be supplied. Assume that when the members of B are supplied, *in any order whatever*, as inputs, then the computation always eventually yields the set A, in *some order*, as outputs. The order in which the members of A appear may vary as the order of inputs varies. (We permit repetitions in the listing of B and in the listing of A.) If such a procedure exists, we say that A is *enumeration reducible* to B. To put it as briefly as possible: A is *enumeration reducible* to B if there is an effective procedure for getting an enumeration of A from *any* enumeration of B.

Example 1. Consider a procedure which asks for an input, gives twice that input integer as output; asks for another input, etc. Under this procedure, $\{2,4\}$ is enumeration reducible to $\{1,2\}$, and E, the set of even integers, is enumeration reducible to N.

Example 2. Consider a procedure which asks for no inputs and lists K as outputs. Under this procedure, K is enumeration reducible to any set.

How can the notion of enumeration reducibility be made precise? One way of characterizing it is to define an appropriately modified version of Turing machine. For example, we might take machines that write "S" as well as "1" and that have two additional operations I and O (for "input" and "output"). When operation I is called for, then, in a single step, an external agency may or may not print an input number as a string of 1's on a portion of the tape that lies to the right of both the nonblank cells and the cell being examined by the machine. When operation O is called for, then, in a single step, an external agency records, on a separate list of outputs, the total number of 1's appearing on the tape. Of course, it will not be clear ahead of time, for a given machine and set of inputs, whether or not the set of outputs varies with the order of inputs.† Furthermore, it is not evident, without more detailed investigation, that this Turing-machine characterization is sufficiently general to include all instances of enumeration reducibility that are intuitively acceptable.

We avoid both difficulties by defining enumeration reducibility in terms of the basic concept of partial recursive function. Our approach to enumeration reducibility is similar, in this respect, to our approach to T-reducibility.

In order to motivate our definition, we show how the intuitive notion leads to it. Let a procedure be given by which, intuitively, A is *enumeration reducible* to B. Consider all finite sequences of distinct integers. Let us say that an integer x is *caused* by a sequence $<y_1, \ldots, y_k>$ if the procedure lists x as an output after $y_1, \ldots, y_k$ are given in order as inputs but before another input is requested.‡ By examining, in turn, all finite sequences, we can recursively enumerate the set $\{<x,u>|D_u$ *consists of the members of a finite sequence of distinct integers that causes* $x\}$. Let this set be W_z. It is easily verified, from our assumption that the set A is listed regardless of the order in which B is presented, that

$$x \in A \Leftrightarrow (\exists u)[<x,u> \in W_z \mathbin{\&} D_u \subset B].$$

This suggests our definition.

Definition A is *enumeration reducible* to B (notation: $A \leq_e B$) if

$$(\exists z)(\forall x)[x \in A \Leftrightarrow (\exists u)[<x,u> \in W_z \mathbin{\&} D_u \subset B]].$$

A is *enumeration reducible* to B *via* z if

$$(\forall x)[x \in A \Leftrightarrow (\exists u)[<x,u> \in W_z \mathbin{\&} D_u \subset B]].$$

Note, conversely, that any z yields a procedure of the intuitive kind first described. This procedure is as follows. Simultaneously begin an enumeration of W_z and start requesting inputs. Whenever it is found that an $<x,u>$ has already appeared in W_z for which D_u is contained in the inputs already obtained and for which x has not yet been given as output, add x to the

† As we shall see from our ultimate definition, this difficulty can be solved by placing further, rather complex (but uniform) restrictions on the machines.

‡ It might be that after $y_1, \ldots y_{k-1}$ are given as inputs, no further input is requested. In this case *no* integer is *caused* by $y_1, \ldots, y_k$.

output list. For any given set of inputs, this procedure yields the same set of outputs, regardless of the order in which the set of inputs is presented.

Thus any z and any B determine a unique corresponding A such that $A \leq_e B$ via z, namely, $\{x|(\exists u)[<x,u> \in W_z \,\&\, D_u \subset B]\}$. Hence each z determines a total mapping from 2^N into 2^N. We call such mappings *enumeration operators*, and denote the operator corresponding to z as Φ_z.

Definition $\Phi_z(X) = Y$ if $Y \leq_e X$ via z.

The enumeration operators are an analogue, at the level of sets, to the recursive functions. They are simpler than recursive functions in that an enumeration theorem holds; i.e., they have a Gödel numbering onto N.

Theorem XX *If Φ and Ψ are enumeration operators, then the composition $\Phi\Psi$ is an enumeration operator.*

Proof. Immediate by Church's Thesis. A more rigorous proof uses the second projection theorem. We omit details.☒

The following theorem lists several basic structural properties of enumeration operators.

Theorem XXI *Let Ψ be an enumeration operator.*

(*a*) $A \subset B \Rightarrow \Psi(A) \subset \Psi(B)$ ("*Ψ is monotone*").

(*b*) $x \in \Psi(A) \Rightarrow (\exists D)[D$ *finite* $\&\ D \subset A \ \&\ x \in \Psi(D)]$ ("*Ψ is continuous*") (see Exercise 11-35).

Proof. Immediate from the definition of enumeration reducibility.☒

Application of enumeration operators to single-valued sets (and hence, by the convention of §5.3, to partial functions) raises a number of interesting questions which we consider further in §9.8.

Definition $A \equiv_e B$ if $A \leq_e B$ and $B \leq_e A$.

Notation Let ψ and φ be partial functions. We abbreviate $\tau(\psi) \leq_e \tau(\varphi)$ as $\psi \leq_e \varphi$, and we abbreviate $\tau(\psi) \equiv_e \tau(\varphi)$ as $\psi \equiv_e \varphi$.

In the literature, the phrase *ψ is partial recursive in φ* is usually defined to coincide with $\psi \leq_e \varphi$. The degrees in the e-reducibility ordering, i.e., the equivalence classes of $\equiv_e$, are sometimes called *partial degrees.* These degrees are considered further in §13.6.

Among the three relations, A *recursive in* B, $A \leq_e B$, and A *recursively enumerable in* B, the following implications hold; and no other implications hold in general:

A recursive in $B \Rightarrow A$ recursively enumerable in B; $A \leq_e B$
$\Rightarrow A$ recursively enumerable in B.

(See Exercise 9-57.) This remains the case when A and B are limited to being single-valued. For example, ψ recursively enumerable in φ does not in general imply that $\psi \leq_e \varphi$.

§9.8 RECURSIVE OPERATORS

A number of interesting questions arise concerning the relationship between enumeration operators and single-valued sets. We consider some of these here, and later, in §15.3, give a fuller treatment.

Let $\mathbf{\mathcal{P}}$ be the class of all partial functions of one variable. A mapping from a subset of $\mathbf{\mathcal{P}}$ into $\mathbf{\mathcal{P}}$ will be called a *functional operator*. Every partial function φ has a corresponding single-valued set $\tau(\varphi)$, and every single-valued set A determines a partial function $\tau^{-1}(A)$. Let $\mathcal{P}$ be the class of all single-valued sets. Then τ yields a one-one standard map of $\mathbf{\mathcal{P}}$ onto $\mathcal{P}$. Every functional operator $\mathbf{\Psi}$ determines a mapping Γ from a subset of $\mathcal{P}$ into $\mathcal{P}$ where $\Gamma = \tau\mathbf{\Psi}\tau^{-1}$, and every mapping Γ from a subset of $\mathcal{P}$ into $\mathcal{P}$ determines a functional operator $\mathbf{\Psi}$ where $\mathbf{\Psi} = \tau^{-1}\Gamma\tau$.

Definition Let Φ be any (total) mapping from 2^N into 2^N. Φ determines a mapping $\Phi_{\mathcal{P}}$ from a subset of $\mathcal{P}$ into $\mathcal{P}$ as follows:

(i) domain $\Phi_{\mathcal{P}} = \Phi^{-1}(\mathcal{P}) \cap \mathcal{P}$, and

(ii) $\Phi_{\mathcal{P}} = \Phi$ on domain $\Phi_{\mathcal{P}}$.

(Thus $\Phi_{\mathcal{P}}$ is Φ restricted to those single-valued sets A for which $\Phi(A)$ is also single-valued.) Then $\Phi_{\mathcal{P}}$, in turn, determines a functional operator $\mathbf{\Psi} = \tau^{-1}\Phi_{\mathcal{P}}\tau$. We say that the mapping Φ *defines* the functional operator $\mathbf{\Psi}$.

Definition $\mathbf{\Psi}$ is a *partial recursive operator* if:

(i) $\mathbf{\Psi}$ is a functional operator, and

(ii) for some z, Φ_z defines $\mathbf{\Psi}$.

(That is to say, a functional operator is a partial recursive operator if it is defined by an enumeration operator.)

Definition $\mathbf{\Psi}$ is a *recursive operator* if

(i) $\mathbf{\Psi}$ is a partial recursive operator, and

(ii) domain $\mathbf{\Psi} = \mathbf{\mathcal{P}}$.

(That is to say, a partial recursive operator is a recursive operator if it is defined by an enumeration operator Φ_z such that $\Phi_z(\mathcal{P}) \subset \mathcal{P}$.)

Example 1. $\lambda A[A]$ defines a recursive operator, the identity mapping from $\mathbf{\mathcal{P}}$ onto $\mathbf{\mathcal{P}}$.†

Example 2. $\lambda A[N]$ defines the partial recursive operator whose domain is empty.

Example 3. Let η be a fixed partial recursive function. The "composition" mapping from $\mathbf{\mathcal{P}}$ into $\mathbf{\mathcal{P}}$ given by $\lambda\varphi[\eta\varphi]$ is a recursive operator, and so is $\lambda\varphi[\varphi\eta]$.

Let $\mathfrak{F}$ be the class of all total functions of one variable.

† We here extend use of the λ notation to set-valued and function-valued expressions in an obvious way.

Definition Ψ is a *general recursive operator* if

(i) Ψ is a partial recursive operator,

(ii) $\mathfrak{F} \subset$ domain Ψ, and

(iii) $\Psi(\mathfrak{F}) \subset \mathfrak{F}$.

(Thus a partial recursive operator is a general recursive operator if it is defined on all total functions and takes total functions into total functions.)

It is easy to show that not every partial recursive operator can be extended to a recursive operator (Exercise 9-64). On the other hand, the following theorem shows that every general recursive operator is a recursive operator.

Theorem XXII *Let Ψ be a partial recursive operator. If $\mathfrak{F} \subset$ domain Ψ, then Ψ is a recursive operator.*

Proof. Let Φ_z define Ψ. If Ψ is not a recursive operator, then, by Theorem XXI, there must exist A and D such that $A = \Phi_z(D)$, D is single-valued and finite, and A is not single-valued. But D can be extended to a B (that is, $D \subset B$) such that $B = \tau(f)$ for some f. By Theorem XXI, $A \subset \Phi_z(B)$, and hence $\Phi_z(B)$ is not single-valued, contrary to our assumption that $\mathfrak{F} \subset$ domain Ψ.☒

Corollary XXII *Every general recursive operator is a recursive operator.*

Proof. Immediate.☒

We now turn to the main result of the present section, the *fundamental operator theorem.* This theorem shows that if we limit our attention to inputs from $\mathfrak{F}$ (the total functions), then every partial recursive operator can be extended to a recursive operator. As a theorem about partial recursive operators, this result stands in contrast to Theorem 2-II on the nonextendability of all partial recursive functions. The theorem holds uniformly, and hence provides us with an effective enumeration of the recursive operators on $\mathfrak{F}$. Thus it also stands in contrast to Exercise 2-9 on the nonenumerability of the recursive functions.

Theorem XXIII (***the fundamental operator theorem***) *There exists a recursive function σ such that for every z, if Φ_z defines the partial recursive operator Ψ, then $\Phi_{\sigma(z)}$ defines a recursive operator Ψ' such that for every (total) f, $f \in$ domain $\Psi \Rightarrow \Psi'(f) = \Psi(f)$.*

Proof. We first give an intuitive argument, then give a more detailed construction. For reasons of technical simplicity, the detailed construction follows a different path from the intuitive argument.

Intuitive argument. Given an enumeration, in any order, of a total function, we can test for membership in that function (see Corollary VI). Hence from any enumeration of $\tau(f)$, we can get an enumeration of $\tau(f)$ in the standard order $<0,f(0)>$, $<1,f(1)>$, ... Putting this standard enumeration through Φ_z, we get $\Phi_z(\tau(f))$ in a fixed order. The output $\Phi_z(\tau(f))$ can be single-valuized with respect to this order (as in the proof of Theorem 5-XVI). To obtain $\Phi_{\sigma(z)}$, we first form the composition of the three operations: (i) convert input to standard order, (ii) apply Φ_z, (iii) single-valuize output (this determines a well-defined operator from $\mathcal{P}$ into $\mathcal{P}$); and then we add the condition that for any non-single-valued input, the output is to be N (the output is thus independent of the order in which the inputs are given, and we have a well-defined operator from 2^N into 2^N with the desired properties).

Detailed construction. Let f'' be a fixed recursive function as in Corollary 5-V(d). Let any z be given. Consider W_z enumerated in the "standard" order determined by $\varphi_{f''(z)}$. If $v \in W_z$, we say that *u precedes v in W_z* if $u \in W_z$ and u occurs before v in this standard order; i.e., if $\varphi^{-1}_{f''(z)}(u)$ exists and $\varphi^{-1}_{f''(z)}(u) < \varphi^{-1}_{f''(z)}(v)$.

Consider

$$(*) \quad \{<<x,y>,t>|D_t \textit{ is single-valued } \& \ (\exists s)[D_s \subset D_t \ \& \ <<x,y>,s> \in W_z \ \& \ (\forall s')(\forall y')[<<x,y'>,s'> \textit{ precedes } <<x,y>,s> \textit{ in } W_z \Rightarrow D_{s'} \cup D_t \textit{ is not single-valued}]]\}.$$

This set is evidently recursively enumerable with r.e. index depending uniformly on z. Take σ to be a recursive function such that $W_{\sigma(z)}$ is this set. It remains to show that $\Phi_{\sigma(z)}$ has the desired properties. We first prove a lemma.

Lemma $[<<x,y>,t> \in W_{\sigma(z)} \ \& \ D_t \subset D_{t_1} \ \& \ D_{t_1} \textit{ single-valued}] \Rightarrow <<x,y>,t_1> \in W_{\sigma(z)}$.

Proof. Assume D_{t_1} single-valued, $D_t \subset D_{t_1}$, $<<x,y>, t> \in W_{\sigma(z)}$, and $<<x,y>,t_1> \notin W_{\sigma(z)}$. Take D_s as given from D_t by (*), and $D_{s'}$ as given from D_s and D_{t_1} by (*) (with t_1 in place of t in (*)). Then, by (*), $D_{s'} \cup D_{t_1}$ is single-valued but $D_{s'} \cup D_t$ is not single-valued. This contradicts the assumption that $D_t \subset D_{t_1}$, and the lemma is proved.

We next show that $\Phi_{\sigma(z)}$ defines a recursive operator. Assume otherwise. Then for some single-valued A: $<<x,y_1>,t_1> \in W_{\sigma(z)} \ \& \ D_{t_1} \subset A$; $<<x,y_2>,t_2> \in W_{\sigma(z)} \ \& \ D_{t_2} \subset A$; and $y_1 \neq y_2$. Let D_{s_1} be obtained from D_{t_1} by (*) (with s_1 for s and t_1 for t), and let D_{s_2} be obtained from D_{t_2} by (*) (with s_2 for s and t_2 for t). Without loss of generality, we may assume that $<<x,y_1>,t_1>$ precedes $<<x,y_2>,t_2>$ in W_z. Then, by (*), $D_{s_1} \cup D_{t_2}$ is not single-valued. Since $D_{s_1} \subset D_{t_1}$, $D_{t_1} \cup D_{t_2}$ is not single-valued. But $D_{t_1} \cup D_{t_2} \subset A$, and hence A is not single-valued; this is a contradiction.

Finally, we show that if $\Phi_z(\tau(f))$ is single-valued, then

$$\Phi_{\sigma(z)}(\tau(f)) = \Phi_z(\tau(f)).$$

Assume f given and $\Phi_z(\tau(f))$ single-valued.

Then $<x,y> \in \Phi_{\sigma(z)}(\tau(f)) \Rightarrow (\exists t)[<<x,y>,t> \in W_{\sigma(z)} \ \& \ D_t \subset \tau(f)] \Rightarrow (\exists s)[<<x,y>,s> \in W_z \ \& \ D_s \subset \tau(f)]$ (by (*)) $\Rightarrow <x,y> \in \Phi_z(\tau(f))$.

Conversely, let $<x,y> \in \Phi_z(\tau(f))$. Then there must be an s such that

$<<x,y>,s> \in W_z$ and $D_s \subset \tau(f)$. Take the first such s to appear under the "standard" order of W_z. Let
$S = \{s_1, \ldots, s_k\} = \{s' | (\exists y')[<<x,y'>,s'> \textit{ precedes } <<x,y>,s> \textit{ in } W_z]\}$. Since $\Phi_z(\tau(f))$ is single-valued, $D_{s_i} \not\subset \tau(f)$ $(1 \leq i \leq k)$. Define D_t so that $D_t \subset \tau(f)$ and $\textit{domain } D_t = \textit{domain } D_s \cup (\bigcup_{i=1}^{k} \textit{domain } D_{s_i})$. Since $D_{s_i} \not\subset \tau(f)$ $(1 \leq i \leq k)$, we have that $D_{s_i} \cup D_t$ is not single-valued $(1 \leq i \leq k)$. Hence, by (*), $<<x,y>,t> \in W_{\sigma(z)}$. (If $S = \emptyset$, define $D_t = D_s$, and again $<<x,y>,t> \in W_{\sigma(z)}$.) Thus $<x,y> \in \Phi_{\sigma(z)}(\tau(f))$.

This completes the proof of the theorem. ☒

The following question is left open by the fundamental theorem. Do there exist partial functions ψ and φ such that φ is taken to ψ by a partial recursive operator but not by a recursive operator? We answer this question affirmatively in §13.6 (see Exercise 9-69).

In some parts of the literature, functional operators are studied only as mappings from $\mathfrak{F}$ to $\mathcal{P}$ (rather than from $\mathcal{P}$ to $\mathcal{P}$). Under this limitation, every partial recursive operator can be extended to a recursive operator (by Theorem XXIII—indeed, the regularity provided by Theorem XXIII is perhaps the chief reason for adopting this limitation), and the phrase "partial recursive operator" is sometimes used (where we have used "recursive operator") to mean a mapping from $\mathfrak{F}$ to $\mathcal{P}$ defined on all of $\mathfrak{F}$.

We next use the fundamental theorem to relate functional operators to the various reducibilities previously considered.

Theorem XXIV *A recursive in B* $\Leftrightarrow c_A <_e c_B$.

Proof. As with Theorem XXIII, we give first an intuitive argument and then a detailed construction.

Intuitive argument. $\Rightarrow$: Assume A recursive in B. Any enumeration of c_B can be used to answer individual-membership questions about B, hence to answer individual-membership questions about A, and hence to enumerate c_A.

$\Leftarrow$: Assume $c_A \leq_e c_B$, and assume that we have an oracle for B. Using the oracle, we can enumerate c_B; from this we can enumerate c_A, and using this enumeration we can answer individual-membership questions about A.

Detailed construction. $\Rightarrow$: Assume A recursive in B. Then $c_A = \varphi_z^B$ for some z. That is,

$$c_A = \{<x,y> | (\exists u)(\exists v)[<x,y,u,v> \in W_{\rho(z)} \;\&\; D_u \subset B \;\&\; D_v \subset \bar{B}]\}.$$

Take h to be a recursive function of two variables such that

$$D_{h(u,v)} = (D_u \times \{1\}) \cup (D_v \times \{0\}).$$

Take g_1 to be a recursive function such that

$$W_{g_1(z)} = \{<w,t> | (\exists x)(\exists y)(\exists u)(\exists v)[w = <x,y> \;\&\; t = h(u,v) \;\&\; <x,y,u,v> \in W_{\rho(z)}]\}.$$

Then $\langle x,y\rangle \in c_A \Leftrightarrow (\exists t)[\langle\langle x,y\rangle,t\rangle \in W_{g_1(z)} \;\&\; D_t \subset \tau(c_B)]$.

Hence $c_A \leq_e c_B$ via $g_1(z)$.

$\Leftarrow$: Assume $c_A \leq_e c_B$ via z. Then, by Theorem XXIII,

$$\tau(c_A) = \Phi_{\sigma(z)}(\tau(c_B)),$$

where $\Phi_{\sigma(z)}$ defines a recursive operator.

Let Θ be a recursive function such that

$$W_{\Theta(z)} = \{\langle\langle x,y\rangle,t\rangle \mid \langle\langle x,y\rangle,t\rangle \in W_{\sigma(z)} \;\&\; D_t \subset N \times \{0,1\}\}:$$

$W_{\Theta(z)}$ must be infinite. Let $\langle\langle x_1,y_1\rangle,t_1\rangle$, $\langle\langle x_2,y_2\rangle,t_2\rangle, \ldots$ be an effective listing of the members of $W_{\Theta(z)}$. Recall from the lemma in the proof of Theorem XXIII that if $\langle\langle x,y\rangle,t\rangle \in W_{\sigma(z)}$, then $\langle\langle x,y\rangle,t'\rangle \in W_{\sigma(z)}$ for all t' such that $D_t \subset D_{t'}$ and $D_{t'}$ is single-valued. We give instructions for a new effective listing (with repetition) of $W_{\Theta(z)}$ as follows.

Stage 1. List $\langle\langle x_1,y_1\rangle,t_1\rangle$. Go to stage 2.

. . . . . .

Stage n. List $\langle\langle x_n,y_n\rangle,t_n\rangle$, and then list $\langle\langle x_n,y_n\rangle,t'\rangle$ for all t' such that $D_{t_n} \subset D_{t'}$, $D_{t'} \subset N \times \{0,1\}$, *domain* $D_{t'} = \bigcup_{k=1}^{n}$ *domain* D_{t_k} and $D_{t'}$ is single-valued. Then go to stage $n + 1$.

. . . . . .

We denote the successive members of this new listing of $W_{\Theta(z)}$ as $\langle\langle x_1',y_1'\rangle,t_1'\rangle$, $\langle\langle x_2',y_2'\rangle,t_2'\rangle, \ldots$. Define h_1 and h_2 to be recursive functions such that

$$D_{h_1(t)} = \{x \mid \langle x,1\rangle \in D_t\},$$

and

$$D_{h_2(t)} = \{x \mid \langle x,0\rangle \in D_t\}.$$

Let $C = \{\langle x,y,u,v\rangle \mid (\exists t)[\langle\langle x,y\rangle,t\rangle \in W_{\Theta(z)} \;\&\; u = h_1(t) \;\&\; v = h_2(t)]\}$. Consider the effective listing of C: $\langle x_1',y_1',u_1',v_1'\rangle$, $\langle x_2',y_2',u_2',v_2'\rangle, \ldots$, where $u_i' = h_1(t_i')$ and $v_i' = h_2(t_i')$ $(i = 1, 2, \ldots)$. Note that all quadruples in this listing are consistent (that is, $D_{u'_i} \cap D_{v'_i} = \emptyset$). We now use this listing to make an effective listing of a certain subset C^* of C as follows.

Stage 1. Put $\langle x_1',y_1',u_1',v_1'\rangle$ in C^*. Go to stage 2.

. . . . . .

Stage n. See whether $(\exists y)(\exists u)(\exists v)[\langle x_n',y,u,v\rangle$ has already been put in C^* and $\langle x_n',y_n',u_n',v_n'\rangle$ is compatible with $\langle x_n',y,u,v\rangle$ (that is,

$$D_{u_n} \cap D_v = D_{v_n} \cap D_u = \emptyset)].)$$

If so, go directly to stage $n + 1$. If not, put $\langle x_n',y_n',u_n',v_n'\rangle$ in C^* and then go to stage $n + 1$.

. . . . . .

Evidently C^* is recursively enumerable with index uniform in z. Let g_2 be a recursive function such that $W_{g_2(z)} = C^*$. It remains to show that $c_A = \varphi^B_{g_2(z)}$.

First observe, from the construction of C^* from C, that C^* is regular. Hence by Theorem II, $W_{g_2(z)} = W_{\rho g_2(z)}$. If $<x,y> \in \varphi^B_{g_2(z)}$, then for some u and v, $<x,y,u,v> \in W_{g_2(z)}$ & $D_u \subset B$ & $D_v \subset \bar{B}$. Hence, by the construction, $<<x,y>,t> \in W_{\sigma(z)}$ & $D_t \subset \tau(c_B)$, where $u = h_1(t)$ and $v = h_2(t)$. Hence $<x,y> \in \Phi_{\sigma(z)}(\tau(c_B))$, and $<x,y> \in c_A$.

Conversely, assume $<x,y> \in c_A$. Then $<<x,y>,t> \in W_{\sigma(z)}$ & $D_t \subset \tau(c_B)$, for some t. Hence there is a stage n in the listing procedure for $W_{\Theta(z)}$ used above, such that $<<x,y>,t'>$ is listed in $W_{\Theta(z)}$, where $x = x_n$, $y = y_n$, $D_t \subset D_{t'}$, $D_{t'} \subset \tau(c_B)$, and $\textit{domain } D_{t'} = \bigcup_{k=1}^{n} \textit{domain } D_{t_k}$. Let this $<<x,y>,t'>$ be the mth member of the listing of $W_{\Theta(z)}$. Then

$$<<x'_m,y'_m>,t'_m> = <<x,y>,t'>,$$

and $\textit{domain } D_{t'} = \bigcup_{k=1}^{m} \textit{domain } D_{t'_k}$. If $<x'_m,y'_m,u'_m,v'_m>$ is listed in C^* at stage m, then $<x,y,u'_m,v'_m> \in W_{\rho g_2(z)}$ & $D_{u'_m} \subset B$ & $D_{v'_m} \subset \bar{B}$, and hence $<x,y> \in \varphi^B_{g_2(z)}$. If $<x'_m,y'_m,u'_m,v'_m>$ is not listed in C^* at stage m, then there has to be a $q < m$ such that $<x'_q,y'_q,u'_q,v'_q>$ is listed in C^* at stage q, $x'_m = x'_q$, and $<x'_m,y'_m,u'_m,v'_m>$ is compatible with $<x'_q,y'_q,u'_q,v'_q>$. But compatibility, together with the condition that $\textit{domain } D_{t'} = \bigcup_{k=1}^{m} \textit{domain } D_{t'_k}$, implies that $D_{u'_q} \subset D_{u'_m}$ and $D_{v'_q} \subset D_{v'_m}$. Hence $<x'_q,y'_q> = <x,y'_q> \in \varphi^B_{g_2(z)}$. Hence, by the previous paragraph, $<x,y'_q> \in c_A$. Hence $y'_q = y$, and we have $<x,y> \in \varphi^B_{g_2(z)}$, as desired.☒

Corollary XXIV *For any total functions f and g, f is recursive in $g \Leftrightarrow f \leq_e g$.*

Proof. See Exercise 9-65.☒

Theorems XX and XXIV together give another proof of transitivity for $\leq_T$.

Enumeration reducibility generates a reducibility ordering in the family of all sets (see Exercise 9-56). By Theorem XXIV, the T-reducibility ordering is isomorphic to a subordering of the e-reducibility ordering; the isomorphism is obtained by identifying the T-degree of A with the e-degree of c_A. Is this subordering the entire e-reducibility ordering? Equivalently (see Exercise 9-61), is there, for each partial function ψ, a total function f such that $f \equiv_e \psi$? We answer this question negatively in §13.6 (see Exercise 9-70).

The following characterization of Turing reducibility is immediate.

Theorem XXV $A \leq_T B \Leftrightarrow$ *there is a recursive operator carrying* c_B *to* c_A.

Proof. By Theorems XXIII and XXIV.☒

If one characteristic function is carried to another by a recursive operator, must it be carried to the other by a general recursive operator? The following theorem answers this question negatively by showing that existence of a general recursive operator coincides with tt-reducibility. The theorem is a reformulation of Theorem XIX.

Theorem XXVI (Nerode) $A \leq_{tt} B \Leftrightarrow$ *there is a general recursive operator carrying* c_B *to* c_A.

Proof. $\Rightarrow$: By Theorem XIX, there is a z such that $c_A = \varphi_z^B$ & $(\forall X)[\varphi_z^X$ is a characteristic function]. Take $\Phi_{g_1(z)}$ as in the proof of Theorem XXIV. Let Φ be an enumeration operator which defines the recursive operator Ψ, where

$$[\Psi(\varphi)](x) = \begin{cases} 1, & \text{if } \varphi(x) > 0; \\ 0, & \text{if } \varphi(x) = 0; \\ \text{undefined}, & \text{if } \varphi(x) \text{ undefined}. \end{cases}$$

Then $\Phi_{g_1(z)}\Phi$ defines the desired general recursive operator.

$\Leftarrow$: Let Φ_z define the given general recursive operator. Take g_2 as in the proof of Theorem XXIV; then $c_A = \varphi^B_{g_2(z)}$ and for all X, $\varphi^X_{g_2(z)}$ is total. Hence, by Theorem XIX, $A \leq_{tt} B$.☒

Corollary XXVI(a) $A \leq_{tt} B \Leftrightarrow$ *there is a general recursive operator which carries characteristic functions to characteristic functions and which carries* c_B *to* c_A.

Proof. By the construction for $\Rightarrow$.☒

Corollary XXVI(b) $(\exists f)(\exists g)[f$ *is carried to* g *by a recursive operator, but* f *is not carried to* g *by a general recursive operator*].

Proof. By Theorem I.☒

The other reducibilities studied so far correspond to special kinds of recursive operator. $A \leq_1 B \Leftrightarrow [c_A = c_B f$ for some one-one recursive $f]$. $A \leq_m B \Leftrightarrow [c_A = c_B f$ for some recursive $f]$.

In §§13.6 and 15.3 we shall consider certain topological aspects of operators and some of the extension questions that arise in connection with operators.

§9.9 EXERCISES

§9.1

9-1. Let $\tilde{K}$ be as in §9.1. Is $\tilde{K}$ recursively enumerable?

9-2. Show that $K \leq_1 \tilde{K}$.

9-3. Let $K^* = \{x | (\exists y)[\varphi_x(x) = y$ & tt-*condition* y *is not satisfied by* $\bar{K}]\}$. What reducibility relations hold between K and K^*? Is $K^* \leq_{tt} K$?

§9.2

9-4. Show that $W_{h(z)}$, as defined in §9.2, is regular.

9-5. (Blum). Show that A recursively enumerable and regular $\Rightarrow (\exists z)[A = W_{h(z)}]$. (For purposes of this exercise, each P'_z may be identified simply as a consistent effective way of generating diagrams.)

9-6. Let A be given. Show how to find a z such that $\varphi_z{}^A = c_A$.

9-7. Assume that for every A, ψ is partial A-recursive. Must ψ be partial recursive?

§9.3

9-8. Show that K^A is recursively enumerable in A.

$\triangle$***9-9.*** (*a*) Show that the relation *A recursive in B* is transitive. (This exercise should be attempted before a reading of §9.4. It occurs again as 9-14.)

(*b*) Show that the relation *A recursively enumerable in B* is not transitive. (*Hint:* Use $\emptyset$, K, and $\bar{K}$.)

$\triangle$***9-10.*** Disprove the following conjecture: $[A$ not recursive & $(\forall B)[A$ recursive in $B \Leftrightarrow A$ recursively enumerable in $B]] \Rightarrow (\exists f)[A \equiv \tau(f)]$. (*Hint:* Find (nonconstructively) a set A such that $\bar{A}$ is immune and $(\forall x)[2x \in A$ or $2x + 1 \in A$ but not both$]$. Demonstrate and use the fact that $(\exists f)[C \equiv \tau(f)] \Rightarrow \bar{C}$ not immune.)

9-11. Show that K^A is maximum with respect to 1-reducibility among all sets recursively enumerable in A.

9-12. State and verify relativized versions of Theorems 5-III and 5-IV.

9-13. Let A be fixed; discuss the existence of the three kinds of *A-simple* sets mentioned in §9.3.

§9.4

9-14. Supply details for the rigorous proof of transitivity suggested in Theorem XII. Show that transitivity is uniform (in an appropriate sense).

9-15. (*a*) Show that every T-degree contains $\aleph_0$ sets.

(*b*) Show that every T-degree has at most $\aleph_0$ T-degrees below it in the T-reducibility ordering.

9-16. Show that the family $\{A|A \leq_{\mathrm{T}} B\}$ forms (for any B) a Boolean algebra.

9-17. Show that every T-degree has an infinite sequence of T-degrees above it, linearly ordered with respect to $\leq_{\mathrm{T}}$ and of order type of the integers (0, 1, 2, . . .). (*Hint:* Use Theorem X and Exercise 9-11.)

9-18. (*a*) Let **d** be a T-degree. Is the following true: **d** has a maximal 1-degree $\Leftrightarrow$ **d** has a maximal m-degree $\Leftrightarrow$ **d** has a maximal tt-degree?

(*b*) Show that K has the following property: for any recursively enumerable degree **d** there exists a recursive A such that $K \cap A$ is in **d**. (*Hint:* Use K_0.)

$\triangle$***9-19.*** Show: B recursively enumerable in $A \Leftrightarrow [(\exists f)[\tau(f) \equiv_{\mathrm{T}} A$ & $B = \mathit{range}\ f]$ or B has fewer than two members$]$.

$\triangle$***9-20.*** Let A be given such that $K \leq_{\mathrm{T}} A$. Let $B = A\ \mathit{join}\ \bar{A}$ and let $C = \{x|\varphi_x(x)$ *defined* & $\varphi_x(x) \notin B\}$. Show that $A \leq_1 C$, $C \leq_{\mathrm{T}} A$, but $C \not\leq_1 A$.

$\triangle$***9-21.*** (*a*) Show: $A \leq_{\mathrm{tt}} B \Leftrightarrow (\exists$ recursive $f)(\exists$ recursive $g)(\forall x)[[x \in A \Leftrightarrow (\exists u)(\exists v)[<u,v> \in D_{f(x)}$ & $D_u \subset B$ & $D_v \subset \bar{B}]]$ & $[x \notin A \Leftrightarrow (\exists u)(\exists v)[<u,v> \in D_{g(x)}$ & $D_u \subset B$ & $D_v \subset \bar{B}]]]$.

(*b*) Show: $A \leq_{\mathrm{T}} B \Leftrightarrow (\exists$ recursive $f)(\exists$ recursive $g)(\forall x)[[x \in A \Leftrightarrow (\exists u)(\exists v)[<u,v> \in W_{f(x)}$ & $D_u \subset B$ & $D_v \subset \bar{B}]]$ & $[x \notin A \Leftrightarrow (\exists u)(\exists v)[<u,v> \in W_{g(x)}$ & $D_u \subset B$ & $D_v \subset \bar{B}]]]$.

$\triangle$***9-22.*** Let A and B be recursively enumerable. Show that $A \leq_{\mathrm{T}} B \Leftrightarrow (\exists x)[\bar{A} = W_x{}^B] \Leftrightarrow (\exists$ recursive $f)(\forall x)[x \notin A \Leftrightarrow (\exists y)[y \in W_{f(x)}$ & $D_y \subset \bar{B}]]$.

§9.5

9-23. Show that A hyperimmune $\Rightarrow A$ immune.

9-24. Determine whether or not the set S^* of Theorem 8-VIII is hypersimple.

9-25. Verify that the construction in Dekker's theorem is uniform in the following sense: $(\exists$ recursive $g)(\forall x)[W_x$ not recursive $\Rightarrow [W_{g(x)}$ hypersimple & $W_{g(x)} \leq_{tt} W_x$ & $W_x \leq_T W_{g(x)}]]$. What is $W_{g(x)}$ if W_x is recursive?

△*9-26.* (*a*) Define maximal set as in Exercise 8-40. Show that A maximal $\Rightarrow A$ hypersimple.

(*b*) Show that $\mathcal{C}_{11}$, $\mathcal{C}_{12}$, and $\mathcal{C}_{13}$ lie along the "spectrum" of §8.7 (see also Exercise 8-46) by showing that if $A \leq_1 B$, then (i) B simple $\Rightarrow A$ simple, (ii) B hypersimple $\Rightarrow A$ hypersimple, and (iii) B maximal $\Rightarrow A$ maximal.

9-27. Let f be recursive and one-one. Discuss $f(A)$ and $f^{-1}(A)$ for A hypersimple and for A hyperimmune.

9-28. (Post). Prove the following characterization of hyperimmune set.

Theorem *A is hyperimmune $\Leftrightarrow A$ is infinite and* $\neg$ $(\exists$ *recursive* $f)[(\forall u)[D_{f(u)} \cap A \neq \emptyset]$ & $(\forall z)(\exists u)$[*every member of* $D_{f(u)}$ *is greater than* z]].

9-29. Show by a nonconstructive diagonal construction that there exists an A such that A and $\bar{A}$ are hyperimmune.

9-30. Show that there exists an A such that A and $\bar{A}$ are immune and such that $A \leq_T K$. (*Hint:* Show that the construction of Theorem 8-III is effective provided we have an oracle that can solve any instance of the halting problem, i.e., an oracle for K.)

△*9-31.* (*a*) Show: $(\exists A)[A$ and $\bar{A}$ are immune, and $A \equiv_T K]$. (*Hint:* Modify the construction of 9-30 so that A encodes (e.g., by the size of its gaps in increasing order) knowledge of the membership of K.)

(*b*) Show: $(\forall B)[K \leq_T B \Rightarrow (\exists A)[A$ and $\bar{A}$ immune, and $A \equiv_T B]]$.

9-32. Do Exercise 9-30 with "hyperimmune" for "immune." (*Hint:* Use construction of 9-29; apply hint from 9-30.)

△*9-33.* Do Exercise 9-31 with "hyperimmune" for "immune." (*Hint:* As in 9-31.)

9-34. (Dekker). Let A and B be hypersimple. Let C be any recursively enumerable set. Show:

(i) $A \cap B$ is hypersimple;

(ii) $A \cup C$ is hypersimple or co-finite.

Hence conclude that the family of hypersimple or co-finite sets forms a dual ideal in the lattice of recursively enumerable sets (for the concept of *dual ideal* see §12.1).

9-35. Let A be any set. Show that there is a set which is A-immune but not hyperimmune. (*Hint:* Relativize Theorem 8-II.)

△*9-36.* Show the existence of a hypersimple set which is not maximal. (*Hint:* Let A be hypersimple. Define f by $f(2x) = 2x + 1$ and $f(2x + 1) = 2x$. Show that $f(A) \cap A$ is hypersimple but not maximal; see Exercise 9-34.)

9-37. Define, for any set A: $n_A(x) =$ the cardinality of $[A \cap \{0,1, \ldots ,x\}]$. Define: $A \leq_d B$ ("*B is as dense as A*") if $(\exists$ recursive $f)(\forall x)[n_A(x) \leq n_B(f(x))]$.

(i) Show that $\leq_d$ gives a partial ordering.

(ii) There is a maximal "degree" in this ordering. What sets are in it?

(iii) Let A be hyperimmune. Let $p \in A$ and $q \notin A$. Show that $A \cup \{q\} \not\leq_d A$ and $A \not\leq_d A - \{p\}$, although $A - \{p\} \leq_d A \leq_d A \cup \{q\}$.

△*9-38.* (Tennenbaum). Define $d(A,f)$ (the *density* of A in f) to be *lim inf*$_{n\to\infty}\left(\frac{\alpha_n}{n}\right)$, where α_n is the size of the set $[\{0,1, \ldots ,n\} \cap f^{-1}(A)]$.

(i) Show A hyperimmune $\Rightarrow (\forall$ recursive one-one $f)$ $[d(A,f) = 0]$.

(ii) Show that $[f$ recursive & A hyperimmune & $d(A,f) > 0] \Rightarrow (\exists x)[x \in A$ & $d(\{x\},f) > 0]$.

§*9.6*

9-39. Must every nonrecursive, recursively enumerable set which is not tt-complete be hypersimple?

9-40. Let $\mathfrak{T}$ be the set of all finite sequences of 0's and 1's. Let a and b denote members of $\mathfrak{T}$. Define $a \leq b$ if a is an initial segment of b. This gives a partial ordering of $\mathfrak{T}$. Under this ordering, $\mathfrak{T}$ is sometimes called the *binary tree.* It can be diagrammed as follows:

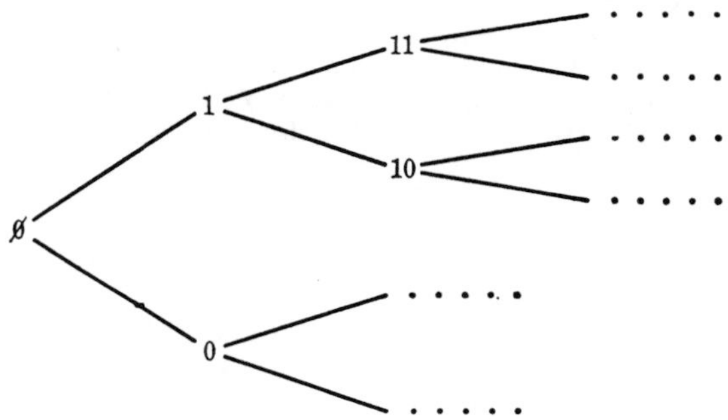

$\mathfrak{S}$ is called a *subtree* of $\mathfrak{T}$ if $\mathfrak{S} \subset \mathfrak{T}$ and $(\forall a)[a \in \mathfrak{S} \Rightarrow \{b|b < a\} \subset \mathfrak{S}]$. Let $\mathfrak{S}$ be any subtree, a maximal linearly ordered subset of $\mathfrak{S}$ is called a *branch* of $\mathfrak{S}$.

(a) Prove the following *basic compactness theorem for trees.* (It is sometimes called the *fan theorem* or the *Endlichkeitslemma;* it can be viewed as a graphic-theoretic restatement of the Bolzano-Weierstrass theorem for closed intervals on the real line.)

Theorem *If a subtree of* $\mathfrak{T}$ *has no infinite branch, then that subtree is finite.*

(b) Prove or disprove the following: if every member of a subtree is contained in a finite branch of that subtree, then the subtree is finite.

9-41. P_x is the Turing machine with index x (see §1.8). P_x can be applied to input tapes other than those of the special form $\cdots BB11 \cdots 1B \cdots$ (which we have used (§1.5), in our basic characterization of partial recursive function, to represent input integers). An input tape will be called *finite* if it is blank at all but a finite number of cells; a tape will be called *infinite* otherwise. (Thus a tape is finite or infinite according as it contains a finite or infinite number of 1's.)

△(i) Assume that Turing machines are always started in state q_0. Show that $\{x|$*for all input tapes (finite or infinite) and for all starting positions,* P_x *halts*$\}$ is recursively enumerable. (*Hint:* Use the fan theorem, Exercise 9-40.)

▲(ii) Assume that Turing machines are always started in state q_0. Show that $\{x|$*for all finite input tapes and for all starting positions,* P_x *halts*$\}$ is neither recursively enumerable nor has recursively enumerable complement. (*Hint:* Show that both K and $\bar{K}$ are m-reducible to this set.)

(iii) Conclude from (i) that $\{x|$*for all input tapes, all starting positions, and all initial states,* P_x *halts*$\}$ is recursively enumerable.†

9-42. See Exercise 9-40. Let $\mathfrak{T}^*$ be the set of all finite sequences of integers. Let a and b denote members of $\mathfrak{T}^*$. Define $a \leq b$ if a is an initial segment of b. Under this ordering, $\mathfrak{T}^*$ is sometimes called the *function tree.* Define *subtree* and *branch* as before.

(i) Does the Endlichkeitslemma hold for $\mathfrak{T}^*$?

△(ii) Show that (as a partial ordering) $\mathfrak{T}^*$ can be embedded in $\mathfrak{T}$.

† Hooper [1966] shows that this set is not recursive.

9-43. Explain why, in the proof of Theorem XVIII, t'_i and t'_j must have distinct derived truth tables when $i \neq j$ and $T_i \cap \bar{A} = T_j \cap \bar{A}$.

9-44. (Dekker, Myhill, Tennenbaum). Define: A is *retraceable* if ($\exists$ partial recursive ψ)$(\forall x)[x \in A \Rightarrow [\psi(x)$ is defined; $\psi(x) = x$ if x is the smallest member of A; and $\psi(x) =$ the next smaller member of A if x is not the smallest member of A]]. Show the following:

(i) [A retraceable & $B \subset A$ & B infinite] $\Rightarrow A \leq_T B$.

(ii) A retraceable $\Rightarrow A$ recursive or immune.

$\triangle$(iii) [A retraceable & $\bar{A}$ recursively enumerable] $\Rightarrow A$ recursive or hyperimmune. (*Hint:* Undertake, for each segment $\{0, \ldots, n\}$, computations which (a) generate a single retraced chain, and also (b) put all nonmembers of the chain in $\bar{A}$.)

$\triangle$(iv) There exists a set which is retraceable but neither recursive nor hyperimmune. (*Hint:* Consider the *binary tree* (Exercise 9-40) relabeled as follows:

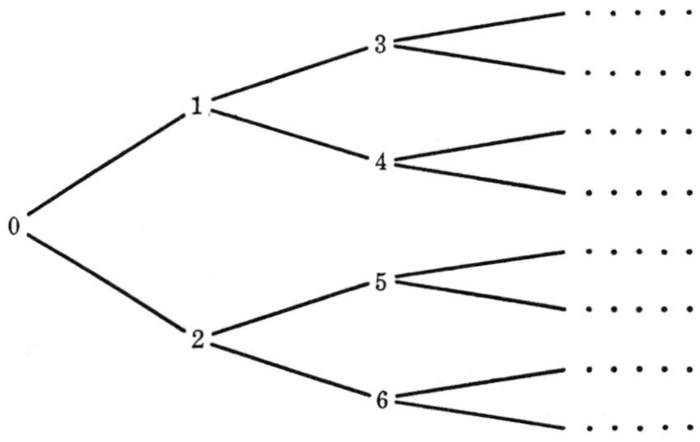

Each branch gives a retraceable set. No branch is hyperimmune.)

$\triangle$(v) There exist sets which are hyperimmune and retraceable. (*Hint:* Use the function tree of Exercise 9-42.)

$\triangle$(vi) The set B constructed in Theorem XVI (Dekker's theorem) has a retraceable complement.

$\triangle$(vii) (Yates). If $\bar{A}$ is recursively enumerable, then [A retraceable $\Leftrightarrow$ ($\exists$ recursive f)$[\bar{A} = \{x | (\exists y)[y > x \ \& \ f(y) \leq f(x)]\}]$]. (*Hint:* See hint for (iii) above.) This representation theorem is a partial converse to (vi). It will follow from Exercise 10-19 that not every hypersimple set has a retraceable complement.

(viii) Show that the property *contains an infinite retraceable subset* is recursively invariant. (McLaughlin has shown that the property *retraceable* is not recursively invariant.)

(ix) Define: A is *regressive* if there is some enumeration (not necessarily effective) without repetitions of A such that ($\exists$ partial recursive ψ)$(\forall x)[x \in A \Rightarrow [\psi(x)$ defined; $\psi(x) = x$ if x is the first member of the enumeration; and $\psi(x)$ immediately precedes x in the enumeration if x is not the first member of the enumeration]]. Show that there is a regressive set which is not retraceable.

(x) Show that A contains an infinite regressive subset $\Leftrightarrow A$ contains an infinite retraceable subset. (For more on retraceable and regressive sets, see Dekker [1962] and Appel and McLaughlin [1965].)

$\triangle$***9-45.*** Define *weak truth-table reducibility* ($\leq_w$) by the following definition, which is easily made more precise. $A \leq_w B$ if $(\exists z)[c_A = \varphi_z^B$ and ($\exists$ recursive f)$(\forall x)[D_{f(x)}$ contains all integers whose membership or nonmembership in B is used in the computation of $\varphi_z^B(x)$] (see discussion in §9.1).

(i) Show that $K \leq_{tt} B \;\&\; A \leq_{w} B \Rightarrow A \leq_{tt} B$. (*Hint:* For a tt-reduction of A to B, use K to determine which combinations of answers about B make the w-reduction procedure terminate.)

(ii) Conclude, via Theorem I, that $\leq_w$ is distinct from $\leq_T$.

(iii) Show that w-*complete* is distinct from T-*complete*. (*Hint:* Show w-complete $\Rightarrow$ not hypersimple.) Lachlan [1965] shows that $\leq_w$ and $\leq_{tt}$ are distinct for recursively enumerable sets.

9-46. Show, by a nonconstructive diagonal argument, that hyperhyperimmune sets exist.

9-47. (Friedberg). Let A be a maximal recursively enumerable set (Exercise 8-40). Show that A is hyperhypersimple.

$\triangle$***9-48.*** (*a*) (Post). Show that no set A exists with the following property: A is recursively enumerable and $\bar{A}$ is infinite and $\neg(\exists$ recursive $f)[(\forall u)[W_{f(u)}$ finite & $W_{f(u)} \cap \bar{A} \neq \emptyset]$ & $(\forall n)(\exists v)$[every member of $W_{f(v)}$ is greater than n]]. Thus an analogue to Exercise 9-28 fails for hyperhypersimple sets.

(*b*) (Yates). Show: A is hyperhypersimple $\Leftrightarrow$ [A is recursively enumerable and $\bar{A}$ is infinite and $\neg(\exists$ recursive $f)[(\forall u)[W_{f(u)} \cap \bar{A} \neq \emptyset]$ & $(\forall u)(\forall v)[u \neq v \Rightarrow W_{f(u)} \cap W_{f(v)} = \emptyset]]]$.

9-49. For any A, let $A^* = \{x|(\exists u)(\exists v)[\langle u,v\rangle \in D_x \;\&\; D_u \subset A \;\&\; D_v \subset \bar{A}]$. Show that $A^* \equiv A^{tt}$.

$\triangle$***9-50.*** Exercises 9-21 and 9-49 suggest approaching the question of T-cylinders as follows.

Define: $A^{\mathrm{T}} = \{x|(\exists u)(\exists v)[\langle u,v\rangle \in W_x \;\&\; D_u \subset A \;\&\; D_v \subset \bar{A}]\}$.

Show:

(i) A^{T} is recursively enumerable in A.

(ii) $A \leq_1 A^{\mathrm{T}}$.

(iii) $A^{\mathrm{T}} \not\leq_{\mathrm{T}} A$.

(*Hint:* Show: $A^{\mathrm{T}} \equiv K^A$, where K^A is as in §9.3.)

Thus this attempt to parallel Theorems 7-VIII and 8-IX fails. We do not get $A \leq_1 B^{\mathrm{T}} \Rightarrow A \leq_{\mathrm{T}} B$. (Take $A = B^{\mathrm{T}}$.)

▲***9-51.*** See Exercise 9-50. Show that $A \leq_{\mathrm{T}} B \Leftrightarrow A^{\mathrm{T}} \leq_1 B^{\mathrm{T}}$. This gives an isomorphism from the T-ordering into the 1-ordering (although the image of a T-degree does not lie within that T-degree.) This result will be proved in Chapter 13.

9-52. Show that there is no maximal T-degree.

▲***9-53.*** Define a set A to be an $\forall\exists$ set if $A = \{x|(\forall y)(\exists z)R(x,y,z)\}$ for some recursive relation R. $\exists\forall$ sets are defined similarly. Show: A is both $\exists\forall$ and $\forall\exists \Leftrightarrow A \leq_{\mathrm{T}} K$.

▲***9-54.*** Show that $\{x|W_x$ *finite*$\}$ is an $\exists\forall$ set and that $\{x|W_x$ *finite*$\} \equiv K^K$. This solves a problem left open by Post [1944]. In Chapter 14 we shall study general techniques for treating such problems.

▲***9-55.*** (Tennenbaum). Define $A \leq_{\mathrm{Q}} B$ if $(\exists$ recursive $f)(\forall x)[x \notin A \Leftrightarrow W_{f(x)} \cap \bar{B} \neq \emptyset]$. Show that $\leq_{\mathrm{Q}}$ does not coincide with $\leq_{\mathrm{T}}$ on the recursively enumerable sets (see Exercise 9-22). (*Hint:* Take A and B as in Exercise 8-39. Let C be a simple set such that $C \leq_{\mathrm{T}} A$ (by Dekker's theorem). Assume $C \leq_{\mathrm{Q}} A$ via f and obtain a contradiction by considering $\{x|W_{f(x)} \cap B \neq \emptyset\}$ and $\{x|W_{f(x)} \subset \bar{B} \;\&\; W_{f(x)} \cap \bar{A} \neq \emptyset\}$).

§9.7

9-56. (*a*) Show that $\leq_e$ is reflexive and transitive.

(*b*) Show that the e-reducibility ordering is a semilattice.

(*c*) Is there a minimum degree? If so what is it?

9-57. Consider the three relations: A *recursive in* B, $A \leq_e B$, and A *recursively enumerable in* B.

(i) Prove the implications that necessarily hold.

△(ii) Give counterexamples to the implications that do not hold. (*Hint:* Use K and $\emptyset$; $\bar{K}$ and K; and K^K and K.)

9-58. Let Ψ be an enumeration operator. What containment relations must hold for the following pairs:

$$\Psi(A \cup B), \quad \Psi(A) \cup \Psi(B);$$
$$\Psi(A \cap B), \quad \Psi(A) \cap \Psi(B)?$$

9-59. Show: A recursively enumerable in $B \Leftrightarrow A \leq_e \tau(c_B)$.

9-60. Let A and B be recursively enumerable. Show that $A \leq_T B \Leftrightarrow \bar{A} \leq_e \bar{B}$.

9-61. (*a*) Show: $(\forall A)(\exists \varphi)[A \equiv_e \tau(\varphi)]$.

(*b*) Show: $(\forall A)[(\exists f)[A \equiv_e \tau(f)] \Rightarrow (\exists B)[A \equiv_e \tau(c_B)]]$. (See also Exercise 9-70.)

9-62. Define: $A \leq_s B$ if ($\exists$ recursive f)$(\forall x)[x \in A \Leftrightarrow W_{f(x)} \cap B \neq \emptyset]$.

(i) Describe the s-degrees of recursively enumerable sets.

(ii) Show that s-reducibility $\Rightarrow$ e-reducibility but not conversely.

(Exercises 9-55 and 9-61 show that s-reducibility differs from e-reducibility in a nontrivial way.)

§*9.8*

9-63. (*a*) Give an example of a recursive operator that is not a general recursive operator.

(*b*) Can two distinct enumeration operators define the same partial recursive operator?

9-64. Define the mapping Ψ by

$$\Psi(\{<0,0>\}) = \{<0,0>\};$$
$$\Psi(\{<1,0>\}) = \{<0,1>\}.$$

Show that Ψ can be extended to a partial recursive operator but not to a recursive operator.

9-65. Prove Corollary XXIV.

9-66. Discuss recursiveness of the following sets.

(i) $\{z|\Phi_z$ *defines a recursive operator*$\}$.

(ii) $\{z|\Phi_z$ *defines a general recursive operator*$\}$.

(iii) $\{z|$*the partial recursive operator defined by* Φ_z *can be extended to a general recursive operator*$\}$.

△***9-67.*** Let $\mathcal{C}$ be a class of enumeration operators. Show that $\{x|\Phi_x \in \mathcal{C}\}$ is recursive if and only if either $\mathcal{C}$ is empty or $\mathcal{C}$ consists of all enumeration operators (compare Exercise 2-39(*a*)).

▲***9-68.*** (*a*) Let $\mathcal{C}$ be a class of partial recursive functions. Formulate and prove an analogue to the Rice-Shapiro theorem given in Exercise 5-37.

(*b*) Let $\mathcal{C}$ be a class of enumeration operators. Formulate and prove an analogue to the Rice-Shapiro theorem given in Exercise 5-37.

▲***9-69.*** Show that if $\psi \leq_e \varphi$, then there need not be a recursive operator carrying φ to ψ.

▲***9-70.*** Use the result of 9-69 to show that there exists a φ such that for all f, $\varphi \not\equiv_e f$. (Hence, by 9-61, the image of the T-reducibility ordering in the e-reducibility ordering (given by Theorem XXIV) is a *proper* subordering of the e-reducibility ordering.)

10 Post's Problem; Incomplete Sets

§10.1 CONSTRUCTIVE APPROACHES

For any set A, let $\mathbf{d}_T(A)$ be the T-degree of A. *Post's problem* asks the question: are there any recursively enumerable T-degrees other than $\mathbf{d}_T(\emptyset)$ and $\mathbf{d}_T(K)$; i.e., are there recursively enumerable sets which are neither recursive nor complete? Since $\mathbf{d}_T(\emptyset)$ and $\mathbf{d}_T(K)$ are, respectively, a minimum and a maximum among all recursively enumerable T-degrees, this question can be equivalently formulated: is there a recursively enumerable T-degree between $\mathbf{d}_T(\emptyset)$ and $\mathbf{d}_T(K)$? Some aspects of the history and significance of Post's problem were discussed in §9.6.

In this chapter we solve Post's problem by showing that there are *incomparable* (with respect to $\leq_T$) recursively enumerable sets. We thus, incidentally, show that there are incomparable recursively enumerable degrees in each of the four chief reducibility orderings, i.e., the 1-, m-, tt-, and T-orderings.

In the present section we present, by way of introduction, a result (Theorem II, below) that explains some of the difficulties encountered in reaching a solution to Post's problem.

Let A be a recursively enumerable set. Consider the assertion that A is not recursive. By Theorem 5-II, this can be expressed

$$(\forall x)[\bar{A} \neq W_x];$$

or, equivalently,

$$(\forall x)(\exists y)[y \in A \Leftrightarrow y \in W_x].$$

If the existential quantifier in this assertion can be given by a recursive function, i.e., if

$$(\exists \text{ recursive } f)(\forall x)[f(x) \in A \Leftrightarrow f(x) \in W_x],$$

we say that A is *constructively nonrecursive.* For a recursively enumerable A, the assertion that A is constructively nonrecursive is equivalent to the assertion that $\bar{A}$ is completely productive (see §7.7). We hence have the following result.

Theorem I *If A is recursively enumerable and constructively nonrecursive, then A is creative.*

Proof. Immediate by definition.☒†

Theorem I shows that the construction of noncreative, nonrecursive, recursively enumerable sets will require special nonconstructive methods. If creative sets happened to be the only known nonrecursive, recursively enumerable sets, Theorem I might lead us to suppose that there were no other nonrecursive, recursively enumerable sets. Of course, we know that other such sets do exist, namely, all sets in the classes $\mathcal{C}_1$, $\mathcal{C}_2$, and $\mathcal{C}_3$ of §8.7. In accord with Theorem I, the construction of such sets did use indirect methods of a rather subtle kind. The first such indirect construction was made by Post and was given in Theorem 8-II. (An example of an assertion that is true but not "constructively true" is given in Kleene [1943].)

A similar situation arises in connection with Post's problem. In order to prove that a recursively enumerable set exists which is neither recursive nor complete, we must find a set C such that C is recursively enumerable and such that both $C \not\leq_T \emptyset$ and $K \not\leq_T C$. By Theorem 9-IV, the assertion that $A \not\leq_T B$, for recursively enumerable A and B, can be expressed

$$(\forall x)[\bar{A} \neq W_x^B],$$

or, equivalently, $$(\forall x)(\exists y)[y \in A \Leftrightarrow y \in W_x^B].$$

If

$$(\exists \text{ recursive } f)(\forall x)[f(x) \in A \Leftrightarrow f(x) \in W_x^B],$$

we say that A is *constructively nonrecursive in B.*

Theorem II *If A and B are recursively enumerable, and if A is constructively nonrecursive in B, then B is recursive.*

Proof. Let any z be given. Define a partial B-recursive function ψ as follows:

$$\psi(x) = \begin{cases} x, & \text{if } z \notin B; \\ \text{divergent}, & \text{if } z \in B. \end{cases}$$

A Gödel number for ψ can be obtained uniformly from z. Hence there is a recursive function g such that for all z,

$$W^B_{g(z)} = \begin{cases} N, & \text{if } z \notin B; \\ \emptyset, & \text{if } z \in B. \end{cases}$$

† The converse will follow from Chapter 11. We then have: for a recursively enumerable A, A is creative $\Leftrightarrow$ A is constructively nonrecursive.

By assumption there is a recursive function f such that $(\forall x)\,[f(x) \in A \Leftrightarrow f(x) \in W_x^B]$. Consider $fg(z)$, for any z. We have

$$fg(z) \in A \Leftrightarrow fg(z) \in W^B_{g(z)} \text{ (by choice of } f) \Leftrightarrow W^B_{g(z)} = N \text{ (by choice of } g) \Leftrightarrow z \notin B.$$

Thus
$$\bar{B} = g^{-1}f^{-1}(A).$$

By assumption, A and B are recursively enumerable. Therefore (by Theorem 5-VII) $\bar{B}$ is recursively enumerable, and hence B must be recursive.⊠

It is natural, in a first attempt on Post's problem, to try to find a recursively enumerable C such that C is constructively nonrecursive in $\emptyset$ and K is constructively nonrecursive in C. By Theorem II, any such *constructive approach* must fail. Prior to the final solution of Post's problem, a number of investigators conjectured that there were only two recursively enumerable T-degrees. Manifestations of Theorem II were partly responsible for these mistaken conjectures.

Theorem II yields a corollary showing that it is possible to have A nonrecursive in B even though A is not constructively nonrecursive in B. This corollary thus restores some plausibility to the assertion (to be proved in §10.2) that there are more than two recursively enumerable T-degrees.

Corollary II(a) *If A and B are recursively enumerable, and if A is constructively nonrecursive in B, then B is recursive and A is creative.*

(***b***) *There exist recursively enumerable sets A and B such that A is nonrecursive in B although A is not constructively nonrecursive in B.*

Proof. (*a*) The assertion that A is constructively nonrecursive in B is a partial relativization of the assertion that $\bar{A}$ is completely productive, and it implies the complete productiveness of $\bar{A}$ (see Exercise 10-2).

(*b*) Let A be any simple set, and let $B = \emptyset$. Then A is nonrecursive in B, but, by (*a*), A is not constructively nonrecursive in B.⊠†

§10.2 FRIEDBERG'S SOLUTION

Post's problem was solved independently and almost simultaneously by Friedberg and Muchnik in 1956 (see Friedberg [1957] and Muchnik [1956]). In key steps, the methods used by Friedberg and Muchnik are similar.‡ We present a modified form of Friedberg's proof.

Theorem III (the Friedberg-Muchnik theorem) *There exist recursively enumerable sets A and B such that A and B are incomparable with respect to $\leq_T$.*

† The converse of Corollary II(*a*) also holds. See Exercise 10-2 and the preceding footnote.

‡ In their initial presentations, both Friedberg and Muchnik built on earlier ideas and results of Kleene and Post. We do not describe or use this dependence here. Some of the Kleene-Post results in question will be presented in Chapter 13.

Proof. We must (1) give instructions for the enumerations of A and B and (2) show the existence of two functions f and g such that $(\forall x)[f(x) \in A \Leftrightarrow f(x) \in W_x{}^B]$ and $(\forall x)[g(x) \in B \Leftrightarrow g(x) \in W_x{}^A]$. (From Theorem II, we know that the functions f and g cannot be recursive.)

For expository purposes, we give the instructions in somewhat anthropomorphic terms; the instructions are, nonetheless, entirely precise. We begin with two identical, infinite, vertical lists, which we call the *A-list* and the *B-list*. Each list consists of all the integers arranged in increasing order downward. At any point in the computation, we shall have used only a finite portion of each list.† In the course of the computation, we write a *plus* (+) beside certain integers in each list. The set of integers eventually receiving a *plus* in the *A-list* constitutes the set A. The set of integers eventually receiving a *plus* in the *B-list* constitutes the set B. The computation takes place in *stages*.

Definition $A_0 = B_0 = \emptyset$.
$A_n = \{x|x \textit{ receives a plus in the A-list by the end of stage } n\}, n = 1, 2, \ldots.$
$B_n = \{x|x \textit{ receives a plus in the B-list by the end of stage } n\}, n = 1, 2, \ldots.$

Then $A_0 \subset A_1 \subset A_2 \subset \cdots$; $B_0 \subset B_1 \subset B_2 \subset \cdots$; $A = \bigcup_{n=0}^{\infty} A_n$; and $B = \bigcup_{n=0}^{\infty} B_n$.

Two infinite sets of *movable markers* are used for bookkeeping purposes during the computation. We denote the markers in the first collection as $\boxed{0}_1, \boxed{1}_1, \boxed{2}_1, \ldots$ and in the second collection as $\boxed{0}_2, \boxed{1}_2, \boxed{2}_2, \ldots$. Markers of the first collection will be associated with integers in the *A-list*, and markers in the second collection will be associated with integers in the *B-list*. These markers are movable in the following sense: when a marker is introduced into the computation, it is associated with some integer in one of the lists; at a later time in the computation, its association with that integer may be terminated, and it will then be associated with some integer further down in the list; at a still later time, it may be moved downward again; etc. A given marker may be moved several times in this way.

Our motivation is as follows: we wish ultimately to associate each $\boxed{j}_1$ with an integer x_j such that $x_j \in A \Leftrightarrow x_j$ has a *plus* $\Leftrightarrow x_j \in W_j{}^B$. If we can successfully arrange this for each $\boxed{j}_1$, we shall have $(\forall x)(\exists y)[y \in A \Leftrightarrow y \in W_x{}^B]$ and hence $A \not\leq_{\mathrm{T}} B$. Similarly for each $\boxed{j}_2$ and $B \not\leq_{\mathrm{T}} A$.

Our construction will make use of the following definition and lemma. Let z and a finite set D be given. Then

$$W_z{}^D = \{x|(\exists y)(\exists u)(\exists v)[<x,y,u,v> \in W_{\rho(z)} \;\&\; D_u \subset D \;\&\; D_v \subset \bar{D}]\},$$

by §§9.2 and 9.3. The following is a (uniform) algorithm for listing $W_z{}^D$: list the members of $W_{\rho(z)}$; as each member $<x,y,u,v>$ appears, test whether

† Each infinite list is analogous, in this respect, to the infinite tape of a Turing machine.

$D_u \subset D$ and $D_v \subset \bar{D}$; whenever this test is successful, add x to the list for $W_z{}^D$.

Definition *Let* z and a finite D be given. By n *steps in the listing of* $W_z{}^D$, we shall mean the above algorithm carried as far as n machine steps in the listing of $W_{\rho(z)}$.

Lemma *For any sequence of finite sets* $A_0, A_1, \ldots$ *such that* $A_0 \subset A_1 \subset \cdots$, *if* $A = \bigcup_{n=0}^{\infty} A_n$ *and if* $a \in W_z{}^A$, *then* $(\exists m)(\forall n)[m \leq n \Rightarrow a$ *appears within* n *steps in the listing of* $W_z{}^{A_n}]$.

Proof of lemma. The proof is immediate (see Exercise 10-5).

Continuing the proof of the theorem, we give instructions for the main computation.

Definition At a given time, we say that an integer is *free* in a given list if neither that integer nor any integer below it has any mark or marker associated with it. At a given time, we say that an integer is *vacant* in a given list if it does not have a *plus* beside it.

Stage 1. Associate $\boxed{0}_1$ with 0 in the *A-list.*

Stage 2. Associate $\boxed{0}_2$ with 0 in the *B-list.*

. . . .

Stage $2n + 1$. Associate $\boxed{n}_1$ with the first free integer in the *A-list.* Let $a_0^{(n)}, \ldots, a_n^{(n)}$ be the current positions of $\boxed{0}_1, \boxed{1}_1, \ldots, \boxed{n}_1$. Compute n *steps in the listing* for each of $W_0^{B_{2n}}, \ldots, W_n^{B_{2n}}$. Let $a_j^{(n)}$ be the least member of $\{a_0^{(n)}, \ldots, a_n^{(n)}\}$ such that $a_j^{(n)}$ is vacant and $a_j^{(n)}$ occurs in the partial listing of $W_j^{B_{2n}}$ just made. (If there is no such $a_j^{(n)}$, go on to stage $2n + 2$.) Put a *plus* beside $a_j^{(n)}$ in the *A-list,* and a *minus* $(-)$ beside each of those vacant integers in the B-list whose membership in $\bar{B}_{2n}$ was used to show that $a_j^{(n)} \in W_j^{B_{2n}}$.† Then, if $j < n$, go to the *B-list* and *move all the markers* $\boxed{i}_2$, $j \leq i < n$, *down to free integers in the B-list.*

Stage $2n + 2$. Associate $\boxed{n}_2$ with the first free integer in the *B-list.* Let $b_0^{(n)}, \ldots, b_n^{(n)}$ be the current positions of $\boxed{0}_2, \ldots, \boxed{n}_2$. Compute n steps in a listing for each of $W_0^{A_{2n+1}}, \ldots, W_n^{A_{2n+1}}$. Let $b_j^{(n)}$ be the least member of $\{b_0^{(n)}, \ldots, b_n^{(n)}\}$ such that $b_j^{(n)}$ is vacant and $b_j^{(n)}$ occurs in the partial listing of $W_j^{A_{2n+1}}$ just made. (If there is no such $b_j^{(n)}$, go on to stage $2n+3$.) Put a *plus* beside $b_j^{(n)}$ in the *B-list,* and a *minus* beside each of those vacant integers in the *A-list* whose membership in $\bar{A}_{2n+1}$ was used to show that $b_j^{(n)} \in W_j^{A_{2n+1}}$. Then, if $j < n$, go to the *A-list* and *move all the markers* $\boxed{i}_1$, $j < i \leq n$, *down to free integers in the A-list.*

. . . .

† That is to say, if $a_j^{(n)} \in W_j^{B_{2n}}$ as a result of $<a_j^{(n)},y,u,v> \in W_{\rho(j)}$ (in n steps), where $D_u \in B_{2n}$ and $D_v \in \bar{B}_{2n}$, then put a *minus* beside each member of D_v in the *B-list.*

(Note that it is possible, in either list, for an integer to have first a *minus* and then a *plus* beside it.)

Observe, by a simple inductive argument, that each $\boxed{n}_1$ and each $\boxed{n}_2$ can be moved only finitely often: for $\boxed{0}_1$ is never moved; $\boxed{0}_2$ can be moved at most once; $\boxed{1}_1$ can be moved at most once for each position of $\boxed{0}_2$; $\boxed{1}_2$ can be moved at most once by $\boxed{0}_1$ and at most once for each position of $\boxed{1}_1$; etc.†

Let $f(x)$ be the final position of $\boxed{x}_1$, $g(x)$ the final position of $\boxed{x}_2$. Assume $f(x) \in A$. Then $f(x)$ receives a *plus* at some stage $2n + 1$. No *minus* written down (in the *B-list*) during stage $2n + 1$ can ever be changed to a *plus:* because all $\boxed{k}_2$, $x \leq k$, are placed below all such *minuses;* and if $\boxed{k}_2$, $k < x$ changed one of these *minuses* to a *plus*, then $\boxed{x}_1$ would be moved, contrary to our assumption that $f(x)$ is the final position of $\boxed{x}_1$. Hence $f(x) \in W_x^B$. Conversely, if $f(x) \in W_x^B$, then, by the *lemma*, there is an m such that for all n greater than m, $W_x^{B_{2n}}$ yields $f(x)$ in fewer than n steps; since $\boxed{x}_1$ eventually reaches $f(x)$, $f(x)$ must eventually receive a *plus;* and $f(x) \in A$. Thus $f(x) \in A \Leftrightarrow f(x) \in W_x^B$. Similarly, $g(x) \in B \Leftrightarrow g(x) \in W_x^A$. Hence $A \not\leq_{\text{T}} B$ and $B \not\leq_{\text{T}} A$, and our proof is complete. ☒

An argument like the above is sometimes called a *priority argument*, and an ordering like $\boxed{0}_1$, $\boxed{0}_2$, $\boxed{1}_1$, . . . is called a *priority listing.* In the above proof, whenever a marker gets a *plus* beside it, all markers of *lower priority* must "yield," i.e., they must be moved down so that they cannot subsequently affect any *minuses* that have just been introduced. The basic feature of such a priority argument is, of course, the fact that each marker is moved only a finite number of times. Several other priority arguments are given in exercises below, and further examples are given in Chapter 12.

Two corollaries are immediate.

Corollary III(a) **(solution to Post's problem).** *There are more than two recursively enumerable* T-*degrees.*

Corollary III(b) *There exist two recursively enumerable sets which are incomparable with respect to* $\leq_1$, $\leq_{\text{m}}$, *and* $\leq_{\text{tt}}$.

A further corollary is obtained by generalizing the construction in the theorem.

Corollary III(c) *There exists a denumerably infinite family of recursively enumerable* T-*degrees such that any two members of this family are incomparable with respect to* $\leq_{\text{T}}$.

Proof. See Exercise 10-7. ☒

Corollary III(d) *There exist* $\aleph_0$ *recursively enumerable* T-*degrees.*
There exists $\aleph_0$ *recursively enumerable* tt-*degrees.*
There exist $\aleph_0$ *recursively enumerable* m-*degrees.*

† Bounds on the number of positions that can be occupied by $\boxed{0}_1$, $\boxed{0}_2$, $\boxed{1}_1$, $\boxed{1}_2$, $\boxed{2}_1$, . . . are given by successive values in the Fibonacci sequence 1, 2, 3, 5, 8, ...

There exist $\aleph_0$ *recursively enumerable* 1-*degrees.*

Proof. Immediate from Corollary III(c).⊠

The last two statements in Corollary III(d) are already known to us from §8.6.

§10.3 FURTHER RESULTS AND PROBLEMS

Theorem III and the lemma of Theorem 8-XIV can be used to show that the 1-reducibility ordering is not an upper semilattice.

***Theorem IV* (Young)** *There exist two recursively enumerable sets having no least upper bound in the* 1-*reducibility ordering.*

Proof. By Theorem III and Theorem 9-XVI (Dekker's theorem), there exist two simple sets which are incomparable with respect to 1-reducibility. Call these sets A and B. Our proof proceeds as follows. First, we show that A and B have an upper bound which is simple. Second, we conclude that any least upper bound must be simple. Finally, we use the lemma of Theorem 8-XIV to show that no least upper bound can be simple.

Let $C = A$ *join* B. By Exercise 8-1, C is simple. $A \leq_1 C$ via $\lambda x[2x]$, and $B \leq_1 C$ via $\lambda x[2x + 1]$.

Assume that D is a least upper bound to A and B. Then $D \leq_1 C$, and hence D must be simple (Exercise 8-9). Let $A \leq_1 D$ via f and $B \leq_1 D$ via g. f and g may be chosen so that $f(A) = D$ and $g(B) = D$ (Exercise 10-10). Then $\bar{D} - f(N) \neq \emptyset$ and $\bar{D} - g(N) \neq \emptyset$ (otherwise $D \equiv A$ or $D \equiv B$, contrary to our choice of A and B as incomparable).

Take $m \in \bar{D} - f(N)$ and $n \in \bar{D} - g(N)$. Then $A \leq_1 D \cup \{m\}$ via f and $B \leq_1 D \cup \{m\}$ via h, where

$$h(x) = \begin{cases} g(x), & \text{if } g(x) \neq m; \\ n, & \text{if } g(x) = m. \end{cases}$$

Hence $D \cup \{m\}$ is an upper bound to A and B. By the lemma of Theorem 8-XIV, $D \not\leq_1 D \cup \{m\}$. This contradicts our choice of D as least upper bound. Therefore A and B are two recursively enumerable sets having no least upper bound in the 1-reducibility ordering.⊠

More information about the T-reducibility ordering of the recursively enumerable T-degrees is given in the following result announced by Sacks. We state it here without proof. It is a strong version of our solution to Post's problem. The proof uses a priority argument.

Theorem (Sacks) *Let* Π *be any denumerable partial ordering and let* A *be any recursively enumerable, nonrecursive set. Then there exists a collection of recursively enumerable sets, all of which are* T-*reducible to* A, *whose partial ordering under* T-*reducibility is isomorphic to* Π.

The theorem immediately yields such particular results as the following: there exists a collection of recursively enumerable degrees which is linearly ordered under T-reducibility with order type of the rationals.

In Exercise 10-11 we obtain a special case of Sacks's theorem. As a corollary, we obtain a result first announced by Muchnik: there exists no nonrecursive, recursively enumerable set which is minimal (with respect to $\leq_T$) among the nonrecursive, recursively enumerable sets.

The following theorem of Sacks [1964*a*] shows that there is no recursively enumerable set which is maximal (with respect to $\leq_T$) among the incomplete recursively enumerable sets.

Definition $A <_T B$ if $A \leq_T B$ and $B \not\leq_T A$.

***Theorem* (*Sacks*)** *If A and B are recursively enumerable, and* $A <_T B$, *then there is a recursively enumerable C such that* $A <_T C <_T B$.

An interesting question concerning the T-reducibility ordering of the recursively enumerable T-degrees was a conjecture of Shoenfield, which we state after making several preliminary definitions.

Definition $\mathbf{d}_T(A)$ *join* $\mathbf{d}_T(B) = \mathbf{d}_T(A$ *join* $B)$.

By Theorem 9-XIII, this *join* operation is well defined on T-degrees, and the join of two T-degrees is their least upper bound in the T-reducibility ordering. Note that the join of two recursively enumerable degrees is recursively enumerable.

The T-reducibility ordering of T-degrees can be viewed as an algebraic structure involving (i) a certain binary relation (the relation $\leq_T$); (ii) a certain binary operation (the operation *join*); and (iii) two distinguished elements (the recursive and complete degrees). Consider *all* algebraic structures in (i) a binary relation; (ii) a binary operation; and (iii) two distinguished elements. We write the binary relation as $\leq$, the binary operation as $\cup$, and the two distinguished elements as $\mathbf{0}$ and $\mathbf{1}$. We say that two such structures, $\mathbf{L}$ and $\mathbf{M}$, are *algebraically isomorphic* if there exists a one-one mapping f from $\mathbf{L}$ onto $\mathbf{M}$ such that $f(\mathbf{0}) = \mathbf{0}$, $f(\mathbf{1}) = \mathbf{1}$,

$$f(a \cup b) = f(a) \cup f(b),$$

and $a \leq b \Leftrightarrow f(a) \leq f(b)$. If $a_1, \ldots, a_n$ are elements in a structure $\mathbf{L}$, $\mathbf{L}(a_1, \ldots, a_n)$ will denote the substructure of $\mathbf{L}$ *generated* by $a_1, \ldots, a_n$ (the intersection of all substructures containing $\mathbf{0}$, $\mathbf{1}$, $a_1, \ldots, a_n$ and closed under $\cup$). If $a_1, \ldots, a_n$ are elements of $\mathbf{L}$ and $b_1, \ldots, b_n$ are elements of $\mathbf{M}$, we say that $\langle a_1, \ldots, a_n \rangle$ is *strongly equivalent* to $\langle b_1, \ldots, b_n \rangle$ if there exists an algebraic isomorphism f from $\mathbf{L}(a_1, \ldots, a_n)$ onto $\mathbf{M}(b_1, \ldots, b_n)$ such that $f(a_i) = b_i$, $1 \leq i \leq n$.

Definition Let $\mathfrak{A}$ be any collection of structures. $\mathbf{L}$ is a *dense structure* (with respect to $\mathfrak{A}$) if $\mathbf{L}$ is in $\mathfrak{A}$ and for any $\mathbf{M}$ in $\mathfrak{A}$, any n, any $a_1, \ldots, a_n$

in **L**, any $b_1, \ldots, b_n, b_{n+1}$ in **M**: $<a_1, \ldots, a_n>$ strongly equivalent to $<b_1, \ldots, b_n> \Rightarrow$ there exists a_{n+1} in **L** such that $<a_1, \ldots, a_{n+1}>$ is strongly equivalent to $<b_1, \ldots, b_{n+1}>$.

Example. Let $\mathcal{A}$ be the collection of all structures such that $\leq$ is a dense (in the usual sense) linear ordering, $\cup$ is the least-upper-bound operation in this ordering, $\mathbf{0} \leq \mathbf{1}$ and $\mathbf{1} \nleq \mathbf{0}$. Then the rational numbers, under their usual ordering, form a *dense* (in the sense defined above) *structure* with respect to this collection.

It is not difficult to show that any two countable dense structures with respect to the same collection $\mathcal{A}$ are algebraically isomorphic (Exercise 10-12). (This generalizes a theorem of Cantor for the special case given as an example above.) Dense structures need not exist (see Exercise 10-13).

We can now state Shoenfield's conjecture. Let $\mathcal{A}_T$ be the collection of all structures satisfying the following conditions: (i) $\leq$ is a partial ordering; (ii) $\cup$ gives least upper bound with respect to $\leq$; (iii) $\mathbf{0}$ is a minimum element with respect to $\leq$; and (iv) $\mathbf{1}$ is a maximum element with respect to $\leq$. It is not difficult to show that $\mathcal{A}_T$ possesses a countable dense structure (see Exercise 10-14). By Exercise 10-12, this countable dense structure is unique up to isomorphism.

Shoenfield's conjecture *The* T-*reducibility ordering of the recursively enumerable* T-*degrees is a countable dense structure with respect to* $\mathcal{A}_T$ (with $\leq_T$ as $\leq$, *join* as $\cup$, the recursive degree as $\mathbf{0}$, and the complete degree as $\mathbf{1}$).

Two immediate consequences of Shoenfield's conjecture should be noted (as Shoenfield pointed out).

Consequence 1. For any two nonrecursive, recursively enumerable sets A and B such that $A <_T B$, there exists a recursively enumerable set C such that $A \leq_T C, C \leq_T A$, and $\mathbf{d}_T()B = \mathbf{d}_T(A)$ *join* $\mathbf{d}_T(C)$.

Consequence 2. If A and B are recursively enumerable sets, $A \leq_T B$, and $B \leq_T A$, then $\mathbf{d}_T(A)$ and $\mathbf{d}_T(B)$ do not have a greatest lower bound in the T-reducibility ordering.

Both theorems of Sacks quoted above are also direct consequences of Shoenfield's conjecture (see Exercises 10-15 and 10-16). Shoenfield's conjecture, if true, would have been an ultimate strongest form of the solution to Post's problem.† In 1965 Lachlan disproved consequence 1 above and Yates disproved consequence 2 (by constructing two recursively enumerable T-incomparable sets A and B such that $[C \leq_T A \;\&\; C \leq_T B] \Rightarrow C$ recursive). Shoenfield's conjecture therefore fails.

† It might also have led to a solution of the following still open problems: (1) Is the first-order theory of $\leq_T$ over the recursively enumerable T-degrees axiomatizable, i.e., recursively enumerable? (2) Is this theory decidable, i.e., recursive? (The theory consists of the wffs built up from "$\leq_T$" within quantificational logic (with no unquantified variables) which are true of the T-reducibility ordering.)

§10.4 INSEPARABLE SETS OF ANY RECURSIVELY ENUMERABLE DEGREE

We now prove that every nonrecursive, recursively enumerable T-degree contains a recursively inseparable pair of disjoint recursively enumerable sets. (Inseparable pairs of sets were introduced and discussed in §7.7.) The construction used in the proof is of special interest. It is related to the construction suggested in Exercise 10-11 below. The theorem is due to Shoenfield [1958*a*].

Theorem V (***Shoenfield***) *Let C be recursively enumerable but not recursive. Then there exists a recursively inseparable pair of disjoint recursively enumerable sets A and B such that* $A \equiv_T B \equiv_T C$.

Proof. Let f be a recursive function which enumerates C without repetitions. Define a partial function ψ as follows:

$$\psi(<x,y>) = \begin{Bmatrix}1\\0\end{Bmatrix} \text{ if } (\exists z)(\exists w)[f(z) = x \;\&\; w < z \;\&\; P_y \text{ with input } <x,y> \text{ yields } \begin{Bmatrix}0\\1\end{Bmatrix} \text{ as output in } w \text{ steps}].$$

By the second projection theorem, ψ is partial recursive. Let

$$A = \{x|\psi(x) = 1\},$$
$$B = \{x|\psi(x) = 0\}.$$

A and B are disjoint and, by the projection theorems, recursively enumerable. It remains to show that $A \equiv_T B \equiv_T C$ and that A and B are recursively inseparable.

1. $A \leq_T C$. We can test $<x,y> \in A$ as follows. See whether $x \in C$. If not, $<x,y> \notin A$. If so, find $z = f^{-1}(x)$; then see whether P_y with input $<x,y>$ yields 0 in fewer than z steps. If so, $<x,y> \in A$. If not, $<x,y> \notin A$.

2. $B \leq_T C$. Similarly.

3. $C \leq_T A$. We can test $x \in C$ as follows. Choose y_0 so that φ_{y_0} is identically zero, that is, $\varphi_{y_0} = \lambda z[0]$. See whether $<x,y_0> \in A$. If so, $x \in C$. If not, let u be the number of steps necessary to compute $\varphi_{y_0}(<x,y_0>) = 0$ from instructions P_{y_0}, and see whether $f(z) = x$ for some z such that $z \leq u$. If so, $x \in C$; if not, $x \notin C$.

4. $C \leq_T B$. Similarly.

5. A and B are recursively inseparable. Assume otherwise. Then we can find a recursive h such that *range* $h = \{0,1\}$, $h(A) = 1$, and $h(B) = 0$. Let $h = \varphi_{y_1}$. For any $x \in C$, the computation of P_{y_1} with input $<x,y_1>$ must take at least $z = f^{-1}(x)$ steps; otherwise we would have (by the original construction) $\varphi_{y_1}(<x,y_1>) = 0 \Leftrightarrow \varphi_{y_1}(<x,y_1>) = 1$, which is a contradiction. Since h is total, there must be a recursive g such that for any x,

$g(x)$ = the length of the computation of P_{y_1} with input $<x,y_1>$. Hence we have

$$x \in C \Leftrightarrow (\exists z)[z < g(x) \;\&\; f(z) = x].$$

But this gives an effective test for C, contrary to our assumption that C is not recursive.☒

We shall see, in Theorem 11-V, that A and B cannot be *effectively* inseparable unless C is complete.

§10.5 THEORIES OF ANY RECURSIVELY ENUMERABLE DEGREE†

In §7.8 we defined a *theory* to be a collection of wffs of some given logical system. For purposes a recursive function theory, we identified a theory with a set of corresponding code numbers under some chosen coding from wffs to integers. We now consider theories of a special kind.

Definition Let $\mathcal{K}$ be a collection of predicate symbols (e.g., "$\leq$"), of operator symbols (e.g., "$\cup$"), and of individual constant symbols (e.g., "1"). Let $\mathcal{L}_{\mathcal{K}}$ be the collection of all expressions built up within elementary quantificational logic from symbols in $\mathcal{K}$ together with the logical symbols of quantificational logic ("=", "$\forall$", "$\exists$", "$\neg$", "$\vee$", "&", "$\Rightarrow$", "$\Leftrightarrow$", and the individual variables "a", "a_1", "a_2", . . . , "b", "c", . . .), where we require that every variable in an expression be acted on by some quantifier. The expressions in $\mathcal{L}_{\mathcal{K}}$ are called the *first-order sentences in* $\mathcal{K}$.

Example. Let $\mathcal{K}$ consist of the binary operator symbols "+" and "×", and the individual constant symbols "0", "1", "2", ... Then $\mathcal{L}_{\mathcal{K}}$ consists of the wffs of *elementary arithmetic* as defined in §7.8.

Definition T is a *first-order theory in* $\mathcal{K}$ if (i) T is a subcollection of $\mathcal{L}_{\mathcal{K}}$, and (ii) T is closed in $\mathcal{L}_{\mathcal{K}}$ with respect to quantificational deducibility; i.e., any sentence in $\mathcal{L}_{\mathcal{K}}$ which is deducible from sentences in T is itself in T.‡

Definition T is a *first-order theory* if, for some $\mathcal{K}$, T is a first-order theory in $\mathcal{K}$.

Throughout §10.5, we limit ourselves to first-order sentences and first-order theories.

We ask two questions about first-order theories: (1) Is there (a set of code numbers for) a first-order theory in each degree of unsolvability? (2) Is there an axiomatizable, i.e., recursively enumerable, theory in each recursively enumerable degree of unsolvability?

† In §10.5 we assume knowledge of quantificational *deducibility* and of the *soundness* and *completeness* theorems for quantificational logic. Material in later chapters will not depend on §10.5.

‡ A definition of quantificational deducibility may be found in Quine [1959] or Suppes [1957].

For 1-degrees, the answer to each question is easily seen to be negative, since no first-order theory can have an immune complement. (If x is not in the theory, then neither are $x \& x$, $(x \& x) \& x$, ...)†

For m-degrees, the answer to each question is also negative. Consider the set S^* of Theorem 8-VIII, and assume that $T \equiv_m S^*$ for some first-order theory T. Let $S^* \leq_m T$ via f and $T \leq_m S^*$ via g. Then $<x,y> \in S^* \times S^* \Leftrightarrow <f(x),f(y)> \in T \times T \Leftrightarrow (f(x) \& f(y)) \in T \Leftrightarrow g(f(x) \& f(y)) \in S^*$; but this yields $S^* \times S^* \leq_m S^*$, contrary to the proof of Theorem 8-XIII.

For btt-degrees, both questions are open. (See also Exercise 10-24.)

For tt-degrees (and hence for T-degrees) both questions have an affirmative answer, as Feferman [1957] has shown.

Theorem VI (Feferman) *For every set A there exists a theory T_A such that $A \leq_1 T_A$ and $T_A \leq_{tt} A$; furthermore, if A is recursively enumerable, then T_A is axiomatizable.*

Proof. Take $\mathcal{K}$ to be empty and consider $\mathcal{L}_{\mathcal{K}}$, which we henceforth designate $\mathcal{L}_0$. ($\mathcal{L}_0$ is sometimes called "the pure calculus of identity.") Typical sentences in $\mathcal{L}_0$ are

(i) $(\forall a)(\exists b) \neg a = b$,

(ii) $(\exists a)(\exists b) \neg a = b$,

(iii) $\neg (\exists a)(\forall b)[(\forall c)[c = a \vee c = b] \Rightarrow b = a]$.

Given any set of objects, the quantifiers in a sentence in $\mathcal{L}_0$ can be interpreted as ranging over that set of objects. Therefore, for any such set of objects, a given sentence in $\mathcal{L}_0$ must be either true or false. (For example, (i) is true for any set of cardinality greater than 1.) Clearly, the truth or falsity of a given sentence for a set of objects depends only on the cardinality of that set. In what follows, we limit ourselves to nonempty sets of objects.‡ We define the *spectrum* of a sentence in $\mathcal{L}_0$ to be the collection of positive finite cardinalities for which that sentence is true. We abbreviate "spectrum of x" as *spectrum* (x). Both (i) and (ii) above have the spectrum $\{x|2 \leq x\}$; (iii) has the spectrum $\{2\}$. Certain fundamental properties of spectra (of sentences in $\mathcal{L}_0$) are established in the following lemma.

Lemma (a) *Every spectrum is either finite or co-finite.*

(b) *There is a uniform effective procedure for going from a sentence to its spectrum; more specifically, there are recursive functions f and g such that for any sentence x, if spectrum (x) is finite, then $g(x) = 0$ and $D_{f(x)} =$ spectrum (x), and if spectrum (x) is cofinite, then $g(x) = 1$ and $\bar{D}_{f(x)} =$ spectrum (x).*

(c) *For every sentence x, x is true for some infinite cardinality $\Leftrightarrow$ x is true for every infinite cardinality $\Leftrightarrow$ x has a co-finite spectrum.*

† We identify a sentence with its code number (as in Exercise 7-62) and use "$x \& y$" to denote (the code number of) the conjunction of x and y.

‡ This restriction is necessary in order to apply the soundness and completeness theorems later in the proof.

(*d*) *Every finite or co-finite set of positive integers occurs as a spectrum.*

(*e*) *There is a uniform effective procedure for going from a given finite or co-finite set of positive integers (given by canonical index) to a sentence having that set as spectrum.*

Proof of lemma. Exercise 10-21 below gives a uniform effective method for testing whether or not a given integer occurs in the spectrum of a given sentence. Exercise 10-22 shows that a sentence with exactly n quantifiers has n in its spectrum if and only if its spectrum contains $\{m|n \leq m\}$; (*a*) and (*b*) follow. Exercise 10-23 generalizes Exercise 10-22 to prove (*c*).

Let F_1 be the sentence "$(\exists a)a = a$"; and, for each $n > 1$, let F_n be the sentence

$$\text{“}(\exists a_1) \cdots (\exists a_n)[[\neg\, a_1 = a_2] \,\&\, [\neg\, a_1 = a_3] \,\&\, \cdots \,\&\, [\neg\, a_{n-1} = a_n]].\text{”}$$

For each $n > 0$, let E_n be the conjunction of F_n with the negation of F_{n+1}. For any finite set of positive integers $\{n_1, \ldots, n_k\}$, let $E_{n_1} \vee \cdots \vee E_{n_k}$ denote the disjunction of $E_{n_1}, E_{n_2}, \ldots, E_{n_k}$; and let $\neg\,(E_{n_1} \vee \cdots \vee E_{n_k})$ denote the negation of this disjunction. Then $E_{n_1} \vee \cdots \vee E_{n_k}$ has spectrum $\{n_1, \ldots, n_k\}$, and $\neg\,(E_{n_1} \vee \cdots \vee E_{n_k})$ has spectrum $\overline{\{0, n_1, \ldots, n_k\}}$. This proves (*d*) and (*e*). This concludes the proof of the lemma.

The soundness of quantificational logic implies that any sentence x deducible (by quantificational logic) from a collection of sentences B which are true in a given nonempty interpretation must be true in that interpretation. From this it follows that if x is deducible from a collection of sentences B, then *spectrum* $(x) \supset \bigcap_{y \in B}$ *spectrum* (y).

The completeness of quantificational logic implies that if x is true for all those nonempty interpretations for which every sentence in a collection B is true, then x is deducible from B. From this it follows that if *spectrum* $(x) \supset \bigcap_{y \in B}$ *spectrum* (y), then x is deducible from B.

Now let A be any given set of integers. We can assume, without loss of generality, that $0 \in A$. Let E_0 be "$(\forall a) \neg\, a = a$." For each $n > 0$, let E_n be as in the proof of the lemma. For all n, let $\neg\, E_n$ denote the negation of E_n. Then, for all n, $\neg\, E_n$ has $N - \{0,n\}$ as its spectrum. Let $B = \{\neg\, E_n | n \in A\}$. Define T_A to be the collection of all sentences deducible from B. Then, for any sentence x,

$$x \in T_A \Leftrightarrow x \text{ is deducible from } B \Leftrightarrow \textit{spectrum } (x) \supset \bigcap_{y \in B} \textit{spectrum } (y).$$

But $\bigcap_{y \in B}$ *spectrum* $(y) = \bar{A}$.

Therefore, $$x \in T_A \Leftrightarrow \textit{spectrum } (x) \supset \bar{A}.$$

Using the recursive functions f and g from part (*b*) of the lemma, we have

$$x \in T_A \Leftrightarrow [[g(x) = 0 \,\&\, D_{f(x)} \supset \bar{A}] \text{ or } [g(x) = 1 \,\&\, D_{f(x)} \subset A]].$$

This gives two consequences: (1) it gives that for all n, $n \in A \Leftrightarrow \neg E_n$ is in T_A; and hence that $A \leq_1 T_A$; (2) it provides, for each x, a tt-condition such that $x \in T_A$ if and only if that tt-condition is satisfied by A (consider the cases $\bar{A}$ finite and $\bar{A}$ infinite separately); hence it gives that $T_A \leq_{tt} A$.

Finally, by the definition of quantificational deducibility, if B is recursively enumerable, then the collection of all sentences deducible from B is recursively enumerable (enumerate all possible *proofs* from finite sets of sentences in B). Hence we have that T_A is axiomatizable if A is recursively enumerable. (Indeed, since $A \leq_1 T_A$, T_A is axiomatizable if and only if A is recursively enumerable.) This concludes the proof of the theorem. ☒

Hanf [1962] strengthens this result to show that there is a *finitely axiomatizable* theory in each recursively enumerable tt-degree. (A theory is finitely axiomatizable if there is a finite collection of sentences in the theory from which all sentences in the theory are deducible.) Shoenfield [1958*a*] uses Theorem V together with an elaboration of Theorem VI to show that there is an axiomatizable, essentially undecidable theory in each nonrecursive, recursively enumerable T-degree. (A first-order theory is *essentially undecidable* if there is no consistent decidable first-order theory containing it.)

§10.6 EXERCISES

§*10.1*

10-1. Let an axiomatizable theory T (i.e., a recursively enumerable set of wffs in a logical system) be given. Assume a Gödel numbering of wffs into integers of the usual kind (see §7.8). Assume that an effective procedure exists such that given any x, a wff F_x can be found such that $F_x \in T \Leftrightarrow x \in K$.

(i) Must T be of complete degree of unsolvability in each of the reducibility orderings?

(ii) Must any two such theories be recursively isomorphic?

10-2. Let A and B be recursively enumerable. Show that A is constructively nonrecursive in $B \Leftrightarrow$ [A is constructively nonrecursive & B is recursive].

10-3. (i) Define, in a reasonable way: A is *constructively non-*m*-reducible to* B.

(ii) State and prove analogues to Theorem II and Corollary II.

(iii) Show that A is constructively nonrecursive in $B \Leftrightarrow A$ is constructively non-m-reducible to B. (*Hint:* See Exercise 10-2.)

10-4. Define: A is *constructively infinite* if $(\exists \text{ recursive } f)(\forall x)[x \in A \Rightarrow [x < f(x)\ \&\ f(x) \in A]]$.

(*a*) Show that A is constructively infinite if and only if A is not isolated.

Define: A is *recursively equivalent* to B if $(\exists$ partial recursive one-one $\varphi)[A \subset$ *domain* φ and $B = \varphi(A)]$ (see Exercise 8-32). Recursive equivalence is a recursive-function-theoretic analogue to the relation of equal cardinality in set theory (see comment after Theorem 8-V).

(*b*) Show that A is constructively infinite if and only if A is recursively equivalent to a proper subset of itself.

§*10.2*

10-5. Show that the algorithm described before the lemma in §10.2 has the property asserted in the statement of the lemma.

10-6. State and prove a relativized version of Theorem III.

10-7. (*a*) Show that there exists a family of three recursively enumerable T-degrees which are pairwise incomparable with respect to $\leq_T$. (*Hint:* Use three lists and six collections of markers: $\boxed{n}_{12}$, $\boxed{n}_{13}$, $\boxed{n}_{21}$, $\boxed{n}_{23}$, $\boxed{n}_{31}$, $\boxed{n}_{32}$; $n = 0, 1, 2, \ldots$. Marker $\boxed{n}_{ij}$ is used in the ith list. At stage $3n + i$, a *plus* is put beside a marker $\boxed{k}_{ij}$ at a (least vacant) position x, if x is yielded after n steps in a listing of $W_k^{S_j(n)}$, where $S_j^{(n)}$ is the current set of plus positions in the jth list. Use an appropriate priority listing for the movement of markers.)

△(*b*) Generalize (*a*) to prove Corollary III(*c*).

10-8. Show that the sets A and B constructed in Theorem III are not simple. (*Hint:* To get an infinite recursively enumerable subset of $\bar{A}$, take a recursively enumerable increasing sequence $x_0, x_1, \ldots$ such that for all X and all i, $W_{x_i}^X = \emptyset$. Consider the initial positions of $\boxed{x_i}_1$ in the *A-list.*)

10-9. Show: $(\exists f)(\exists A)[\textit{range}\, f = A \;\&\; f \not\leq_T A \;\&\; A \not\leq_T f]$. (*Hint:* Take A and B as in Theorem III. Take $n =$ the least member of A. Define f so that $f(B) = \{n\}$, $f(\bar{B}) = A - \{n\}$, and $f \equiv_T B$.)

§*10.3*

10-10. (Tennenbaum). Show that if A is infinite, B is infinite and recursively enumerable, and $A \leq_1 B$, then there exists a recursive f such that $B = f(A)$ and $A \leq_1 B$ via f.

△***10-11.*** Prove the following special case of Sacks's first theorem. (It yields the corollary that there is no minimal recursively enumerable degree among the nonrecursive, recursively enumerable degrees.)

Theorem *Let C be any nonrecursive, recursively enumerable set. There exist recursively enumerable sets A and B such that $A \leq_T C$, $B \leq_T C$, $A \not\leq_T B$, and $B \not\leq_T A$.*

(*Hint:* Modify the procedure of Theorem III as follows. The *A-list* and the *B-list* each consist of all the integers, but each is arranged in the form of a doubly infinite array whose nth row consists of $\langle n,0 \rangle, \langle n,1 \rangle, \ldots$ in order. At any point in the procedure, only finitely many rows and columns from each list are in use.

As before, the movable markers $\boxed{0}_1$, $\boxed{1}_1$, . . . and $\boxed{0}_2$, $\boxed{1}_2$, . . . are used in the *A-list* and the *B-list*, respectively. As the computation progresses, certain *columns* in each list are marked at the top with a *star* (*). At a given time, an integer in a given list is said to be *specified* if it falls in a column that has already been *starred.* At a given time, a *row* in a given list is said to be *free* if there are no marks or markers in that row or in any row below it. In addition to the movable markers $\boxed{n}_1$ and $\boxed{n}_2$, $n = 0, 1, \ldots$, and to the *star*, the following marks will be used: *plus, minus,* and *circle* (○). In any row, these marks are written immediately below the integers with which they are associated. From time to time, certain *circles* are erased. The following combinations of marks can appear below an integer: *circle; minus; circle-minus; circle-plus;* and *circle-minus-plus* (a movable marker may appear as well). As before, the *plus* is never erased; those integers in the *A-list* receiving a *plus* constitute the set A, and those integers in the *B-list* receiving a *plus* constitute the set B. As the computation proceeds, movable markers are introduced and associated with integers. Under certain circumstances, a movable marker may be moved to the right in the same row; under certain other circumstances a movable marker may be moved down to a new row. If at a given time, in a given list, a row contains both a *plus* and a marker $\boxed{j}$, possibly in different positions in that row, we say that the integer j is *fulfilled.* If j is fulfilled, it may become unfulfilled at a later time, if $\boxed{j}$ is moved down to a new row.

Let f enumerate C without repetitions. The procedure is as follows. $A_0, A_1, \ldots$ and $B_0, B_1, \ldots$ are defined as in Theorem III. The markers are given the following priority listing: $\boxed{0}_1$, $\boxed{0}_2$, $\boxed{1}_1$, $\boxed{1}_2$, ...

Stage $2n + 1$. (Unless stated otherwise, all references are to the *A-list.*) (i) Associate $\boxed{n}_1$ with the first unspecified member of the first free row in the *A-list.* (ii) Compute $f(n)$. (iii) *Star* the $f(n)$th column, i.e., the column containing $<x,f(n)>$, $x = 0, 1, 2, \ldots$. (iv) Write a *plus* in the first *circle* occurring in that column, if any. (This *circle* may already have a *minus.*) If there is such a *circle*, let $\boxed{j}_1$ be the marker in the row of that *circle* (there will be one and only one such marker); move all markers $\boxed{k}_2$ ($k < n$) of lower priority than $\boxed{j}_1$ down to free rows in the *B-list* and associate them with the first unspecified members of these new rows; and erase all *circles* in the old rows (in the *B-list*) from which these lower-priority $\boxed{k}_2$ have been moved. (v) Move all unfulfilled markers remaining in the $f(n)$th column to the first unspecified integers to the right in their respective rows. (vi) Compute n steps in each of $W_0^{B_{2n}}, \ldots, W_n^{B_{2n}}$. Let the current positions of $\boxed{0}_1, \boxed{1}_1, \ldots, \boxed{n}_1$ be $a_0^{(n)}, \ldots, a_n^{(n)}$. Take the least unfulfilled j, if any, such that $a_j^{(n)}$ appears in this partial listing of $W_j^{B_{2n}}$. (If no such j exists, go on to stage $2n + 2$.) Put a *circle* under $a_j^{(n)}$ and move $\boxed{j}_1$ to the first unspecified integer to the right in the same row. Write a *minus* under each integer in the *B-list* whose membership in $\bar{B}_{2n}$ was used to show that $a_j^{(n)} \in W_j^{B_{2n}}$. (Some of these *minuses* may go into existing *circles.*) Move all $\boxed{k}_2$ of lower priority than $\boxed{j}_1$ down to free rows in the *B-list*, and associate them with the first unspecified members of those rows. Erase all *circles* in the rows from which these lower priority $\boxed{k}_2$ have been moved.

Stage $2n + 2$. Similarly (for the *B-list*).

It remains to verify that A and B have the desired properties.

1. Each marker can move only finitely often. Otherwise take the highest-priority marker that moves infinitely often. It must ultimately move infinitely often in a single row, and never be fulfilled in that row. Let this row be $<m,0>, <m,1>, \ldots$. It follows from the construction that $\{x| <m,x> \textit{ has a circle}\}$ must be $\bar{C}$. But this makes $\bar{C}$ recursively enumerable, contrary to the assumed nonrecursiveness of C.

2. $A \leq_T C$. To test $<x,y> \in A$, see whether $y \in C$. If $y \notin C$, $<x,y> \notin A$. If $y \in C$, compute $m = f^{-1}(y)$, and see whether $<x,y>$ receives a plus during the $(2m + 1)$st stage of the computation. If so, $<x,y> \in A$; if not, $<x,y> \notin A$.

3. $B \leq_T C$. Similarly.

4. $A \not\leq_T B$. Otherwise $\bar{A} = W_n{}^B$ for some n. Consider the final row in which $\boxed{n}_1$ appears. Either n is fulfilled in that row or it is not. Assume n is ultimately fulfilled in that row. Let the first *plus* occurring in that row be at a. Then by the construction, $a \in W_n{}^B$ (the priority arrangement guarantees this by protecting the *minuses* needed to put a in $W_n{}^B$). Hence $a \in A \cap W_n{}^B$, and $\bar{A} \neq W_n{}^B$. Alternatively, assume n is never fulfilled in that row. Let a' be the final position of $\boxed{n}_1$. Then, by the lemma of Theorem III, $a' \in \bar{A} \cap \bar{W}_n{}^B$; and again $\bar{A} \neq W_n{}^B$. Either way, we have a contradiction.

5. $B \not\leq_T A$. Similarly.)

(*Note:* Sacks has obtained a stronger form of the above theorem, in which $A \cup B = C$ and $A \cap B = \emptyset$.)

10-12. Given a collection of structures $\mathfrak{A}$, show that any two countable dense strucures (with respect to $\mathfrak{A}$) are algebraically isomorphic.

10-13. Give an example of a collection of structures $\mathfrak{A}$ for which no dense structure exists. (*Hint:* Take $\mathfrak{A}$ to consist of cyclic groups of orders 2 and 3, with $\mathbf{\cup}$ as the group operation.)

$\triangle$***10-14.*** Show that $\mathfrak{A}_T$ possesses a countable dense structure. (*Hint:* Observe that for any $\mathbf{L}$ in $\mathfrak{A}_T$ and any $a_1, a_2, \ldots, a_n$ in $\mathbf{L}$, $\mathbf{L}(a_1, \ldots, a_n)$ is finite. Observe further that if, $\mathbf{L}$, $\mathbf{N}_1$, and $\mathbf{N}_2$ in $\mathfrak{A}_T$ are finite, if $\mathbf{L}$ is algebraically isomorphic to a substructure of $\mathbf{N}_1$, and if $\mathbf{L}$ is algebraically isomorphic to a substructure of $\mathbf{N}_2$, then there is a structure $\mathbf{M}$ in $\mathfrak{A}_T$ such that $\mathbf{N}_1$ is algebraically isomorphic to a substructure of $\mathbf{M}$, $\mathbf{N}_2$ is algebraically isomorphic to a substructure of $\mathbf{M}$, and the image of $\mathbf{L}$ in $\mathbf{M}$ via $\mathbf{N}_1$ coincides with the image of $\mathbf{L}$ in $\mathbf{M}$ via $\mathbf{N}_2$. Then make an inductive construction.)

10-15. In §10.3, deduce consequences 1 and 2 and Sacks's second theorem from Shoenfield's conjecture.

△***10-16.*** Deduce Sacks's first theorem from Shoenfield's conjecture.

△***10-17.*** Let B be a hypersimple set obtained by the construction of Dekker's theorem (Theorem 9-XVI). Show that B is not hyperhypersimple. (*Hint* (suggested by Sacks): Let f, A, and B be as in Theorem 9-XVI. Use a priority argument as follows. Take an infinite list of the integers. We shall associate markers $\boxed{0}$, $\boxed{1}$, . . . with integers in the list. An integer is *free* at a given moment if it does not have and has not had any marker associated with it.

Stage $n + 1$. *Substep* 1. Let z be the least free integer $\geq n$. Calculate $f(z)$. Let $b_0, \ldots, b_n$ be the current positions of $\boxed{0}, \ldots, \boxed{n}$. Take the least $i \leq n$, if any, such that $b_i < z$ and $f(z) < f(b_i)$. If no such i exists, go to substep 2. If there is such an i, move marker $\boxed{i}$ down to z, and all markers $\boxed{j}$, $i < j \leq n$, down to free integers $c_{i+1} < \ldots < c_n$ such that $f(z) < f(c_{i+1}) < \ldots < f(c_n)$. Then go to substep 2.

Substep 2. Let c be the current position of $\boxed{n}$. Associate $\boxed{n+1}$ with the first free integer d below c such that $f(c) < f(d)$. Then go to stage $n + 2$.

Define g such that $W_{g(x)}$ = the set of all integers with which $\boxed{x}$ is associated during the procedure. Then, taking g for f in the definition of hyperhyperimmunity in §9.6, we can show that $\bar{B}$ is not hyperhyperimmune.) (*Note:* This exercise is an immediate consequence of Exercise 9-44 (vi) and Exercise 10-19 below.)

▲***10-18.*** (Young). Show that there exists a set which is btt-complete but not m-complete. (*Hint:* Construct an A such that $A \subset K$, $A \equiv K - A$, and A is not a cylinder. It immediately follows that $K \leq_{\text{btt}} A$ and hence that A is btt-complete; A is not m-complete since A is not a cylinder. A is constructed as follows. Define a recursive permutation p such that $p(K) = K$, and $(\forall x)[p(x) \neq x \;\&\; pp(x) = x]$. Define a recursive g such that *range* $g = K$ and $(\forall x)[g(2x + 1) = pg(2x)]$. Observe that for any cylinder C, there exists a recursive one-one f such that $(\forall x)[f(x) \neq x \;\&\; [x \in C \Leftrightarrow f(x) \in C]]$. Arrange a priority argument by which A is constructed so that no such f exists for A and so that $A \subset K$ and $p(A) = K - A$. It is convenient to treat (and generate) A and $K - A$ in symmetric fashion. Note incidentally, by Exercise 8-45, that A is a noncylinder in $\mathcal{C}_3$ having no sets from $\mathcal{C}_1$ or $\mathcal{C}_2$ in its btt-degree.)

▲***10-19.*** (Yates [1962]). Show that if A is recursively enumerable and $\bar{A}$ is infinite then [$\bar{A}$ hyperhyperimmune $\Leftrightarrow$ $\bar{A}$ has no infinite retraceable subset] (see Exercise 9-44). (*Hint:* $\Leftarrow$: Assume $\bar{A}$ not hyperhyperimmune. Construct a partial recursive ψ as follows. Begin simultaneously listing the disjoint finite sets $W_{f(0)}, W_{f(1)}, \ldots$ and, at the same time, enumerating A. (f is as in definition of hyperhyperimmunity.) More specifically, at stage n, list each $W_{f(x)}$, $x \leq n$ up through the first member which does not appear in A by n steps in the enumeration of A. Define ψ on all those final members of $W_{f(0)}, \ldots, W_{f(n)}$ which have been listed, have not yet appeared in A, and for which ψ has not previously been defined. Do so in such a way that a maximal chain $m > \psi(m) > \psi\psi(m) > \cdots$ of final members exists containing all final members for which ψ has not previously been defined. Then ψ will be a retracing function for an infinite subset of $\bar{A}$.

$\Rightarrow$: This generalizes the proof of 10-17 (see (vi) of Exercise 9-44). Consider the retracing "tree" of ψ (defined in the obvious way). Associate markers $\boxed{0}$, $\boxed{1}$, . . . with each level in the tree. List A and move the markers (each within its own level) according to a priority scheme such that each marker comes to final rest on a member of $\bar{A}$ and all markers come to rest on a single branch. This branch is then an infinite retraced subset of $\bar{A}$, and, for each n, the positions of $\boxed{n}$ constitute the recursively enumerable set $W_{f(n)}$. (This defines f.) This family of recursively enumerable sets shows that $\bar{A}$ is not hyperhyperimmune.) (*Note:* A number of "priority" arguments (including Theorem III) can be cast in this form; i.e., markers are moved, each through its own level, in a certain given tree, so that all come to rest on a single final branch of that tree.) (Exercise 12-50 gives a strengthened version of Exercise 10-19 (see Exercises 9-48(*b*), 12-48, and 12-49).)

§*10.4*

10-20. Modify the proof of Theorem V to show that every nonrecursive, recursively enumerable T-degree contains

(i) a collection of three recursively enumerable sets any two of which are disjoint and recursively inseparable;

△(ii) a collection of $\aleph_0$ recursively enumerable sets any two of which are disjoint and recursively inseparable. (*Hint:* Use N in place of $\{0,1\}$ in the construction of Theorem V.)

§*10.5*

10-21. Describe a uniform effective procedure such that given any sentence x in $\mathfrak{L}_0$ and any $n > 0$, we can test whether or not x is true for cardinality n. (*Hint:* Use the numerals 1, 2, . . . , **n**. Replace universal quantifiers by finite conjunctions and existential quantifiers by finite disjunctions, then compute truth or falsity—as illustrated in the following example for $n = 3$ and $x = (\exists a)(\forall b) \neg a = b$:

$$(\exists a)(\forall b) \neg a = b,$$
$$(\exists a)[\neg a = 1 \,\&\, \neg a = 2 \,\&\, \neg a = 3],$$
$$[1 \neq 1 \,\&\, 1 \neq 2 \,\&\, 1 \neq 3] \vee [2 \neq 1 \,\&\, 2 \neq 2 \,\&\, 2 \neq 3] \vee [3 \neq 1 \,\&\, 3 \neq 2 \,\&\, 3 \neq 3],$$
$$[f \,\&\, t \,\&\, t] \vee [t \,\&\, f \,\&\, t] \vee [t \,\&\, t \,\&\, f],$$
$$f \vee f \vee f,$$
$$f.)$$

△***10-22.*** Show that for any sentence x in $\mathfrak{L}_0$, if x has exactly n quantifiers and $m > n$, then [x is true for cardinality $n \Leftrightarrow x$ is true for cardinality m]. (*Hint:* Consider the procedure of 10-21 and show that the computation for m cannot reach a different conclusion from the computation for n. This is shown most conveniently if "$\Leftrightarrow$" and "$\Rightarrow$" are first eliminated in favor of "$\neg$", "&", and "$\vee$".)

△***10-23.*** Prove part (*c*) of the lemma in Theorem VI. (*Hint:* Generalize the proof of 10-22 to the case where m is replaced by an infinite cardinal. Arguments concerning finite conjunctions and disjunctions must be replaced by more general set-theoretical arguments.)

10-24. (Jockusch). Show that $A \times A \leq_m A \Leftrightarrow$ there is a first-order theory T such that $A \equiv_m T$.

11 *The Recursion Theorem*

§11.1 INTRODUCTION

Various kinds of recursively invariant structure have been studied in preceding chapters. Sets of integers with rather special pathological properties have been constructed. Our study of pathological structure culminated, in Chapter 10, in the construction of recursively enumerable sets which are neither recursive nor T-complete.

We now present a simple, general method for producing a variety of pathological structures in recursive function theory. Although the method does not suffice to yield Friedberg's theorem (Chapter 10), it does provide a means for answering a number of questions that have been left unsettled in earlier chapters. In §11.3, we use it to show that every creative set is m-complete and that every productive set is completely productive. In §§11.4 to 11.8, we use it to establish various other results.

We give the method in the form of a theorem. The theorem is usually called the *recursion theorem* or the *fixed-point theorem of recursive function theory*. As we have indicated, it is a basic research tool. It is a deep result in the sense that it provides a method for handling, with elegance and intellectual economy, constructions that would otherwise require extensive, complex treatment.

The theorem can be given in several forms and can be looked at intuitively from several points of view. In one sense, the theorem summarizes a certain class of diagonalization methods, including that used to construct a recursively enumerable but nonrecursive set (Theorem 5-VI). In another sense, the theorem embodies a fixed-point result and, like the fixed-point theorems of analysis, can be used to show the existence of various implicitly defined functions.

For a simple illustration of the content, power, and use of the recursion theorem, consider the following question. Does there exist an m such that $W_m = \{m\}$? At first glance, the existence of such an m might appear to be an arbitrary and accidental feature of the indexing of recursively enumerable sets. Indeed, it might seem likely that, in our indexing, no such m exists. In §11.2, however, we use the recursion theorem to show that such an m must exist for our indexing of recursively enumerable sets and for all other acceptable indexings of the recursively enumerable sets.

The recursion theorem is due to Kleene (as are many of its known applications).

§11.2 THE RECURSION THEOREM

The recursion theorem is given in its strongest and most general form in Theorem IV below. For expository purposes, three simpler versions of the theorem are given first. The results stated explicitly in Theorems II, III, and IV are already implicit in the proof for Theorem I.

Theorem I *Let f be any recursive function; then there exists an n such that*

$$\varphi_n = \varphi_{f(n)}.$$

(We call n a *fixed-point value* for f.)

Proof. Let any u be given. Define a recursive function ψ by the following instructions: to compute $\psi(x)$, first use P_u with input u; if and when this terminates and gives an output w, use P_w with input x; if and when this terminates, take its output as $\psi(x)$. We summarize this:

$$\psi(x) = \begin{cases} \varphi_{\varphi_u(u)}(x), & \text{if } \varphi_u(u) \text{ convergent;} \\ \text{divergent,} & \text{if } \varphi_u(u) \text{ divergent.} \end{cases}$$

The instructions for ψ depend uniformly on u. Take $\tilde{g}$ to be the recursive function which yields, from u, the Gödel number for these instructions for ψ. Thus

$$\varphi_{\tilde{g}(u)}(x) = \begin{cases} \varphi_{\varphi_u(u)}(x), & \text{if } \varphi_u(u) \text{ convergent;} \\ \text{divergent,} & \text{if } \varphi_u(u) \text{ divergent.} \end{cases}$$

Now let any recursive function f be given. Then $f\tilde{g}$ is a recursive function. Let v be a Gödel number for $f\tilde{g}$. Since $\varphi_v = f\tilde{g}$ is total, $\varphi_v(v)$ is convergent. Hence, putting v for u in the definition of $\tilde{g}$, we have

$$\varphi_{\tilde{g}(v)} = \varphi_{\varphi_v(v)} = \varphi_{f\tilde{g}(v)}.$$

Thus $n = \tilde{g}(v)$ is a fixed-point value, as desired.☒

Corollary I *Let f be any recursive function; then there exists an n such that $W_n = W_{f(n)}$.*

Proof. Immediate.☒

Example. Consider the illustrative question suggested in §11.1: *does there exist an m such that $W_m = \{m\}$?* By Church's Thesis (and the *s-m-n* theorem), there is a recursive function f such that for any x, $W_{f(x)} = \{x\}$. Applying Corollary I, we have an n such that $W_n = W_{f(n)}$. Hence $W_n = \{n\}$, and the question asked in §11.1 has an affirmative answer. (Note that this proof can be carried through for any Gödel numbering that is *acceptable* in the sense of Exercise 2-10; see also Exercise 11-12.)

Theorem I can be strengthened in three ways: (1) we can show that n depends uniformly on a Gödel number for f; (2) we can show that when f involves other parameters recursively, then n can be made to depend uniformly on those parameters; and (3) we can show that for any f, an infinite set of fixed-point values can be recursively enumerated. We give (1) and (2) separately in Theorems II and III. We combine (1), (2), and (3) in our most general formulation, Theorem IV. Theorems I, II, and III are all included in Theorem IV. Theorems I and III will be the forms most commonly used in applications.

Theorem II *There exists a recursive function n such that for any z, if φ_z is total, then*

$$\varphi_{n(z)} = \varphi_{\varphi_z(n(z))}.$$

Proof. Let $\varphi_z = f$ and consider the proof of Theorem I. By Theorem 1-VI, v, the Gödel number for $f\tilde{g}$, can be obtained uniformly from z. Let $\tilde{v}$ be a recursive function such that $\varphi_{\tilde{v}(z)} = \varphi_z\tilde{g}$. Then our desired recursive function n is obtained by defining $n(z) = \tilde{g}\tilde{v}(z)$.☒

Corollary II *There exists a recursive function n such that for any z, if φ_z is total, then*

$$W_{n(z)} = W_{\varphi_z(n(z))}.$$

Proof. Immediate.☒

Theorem III *Let f be a recursive function of $k + 1$ variables. Then there exists a recursive function n_f of k variables such that for all $x_1, \ldots, x_k$,*

$$\varphi_{n_f(x_1,\ldots,x_k)} = \varphi_{f(n_f(x_1,\ldots,x_k),x_1,\ldots,x_k)}.$$

Proof. The construction parallels that in Theorem I. Define $\tilde{g}$ to be a function of $k + 1$ variables such that

$$\varphi_{\tilde{g}(u,x_1,\ldots,x_k)}(y) = \begin{cases} \varphi_{\varphi_u^{(k+1)}(u,x_1,\ldots,x_k)}(y), & \text{if } \varphi_u^{(k+1)}(u,x_1,\ldots,x_k) \text{ is convergent;} \\ \text{divergent}, & \text{if } \varphi_u^{(k+1)}(u,x_1,\ldots,x_k) \text{ is divergent.} \end{cases}$$

Then let v be a Gödel number for

$$\lambda u x_1 \cdots x_k[f(g(u,x_1, \ldots ,x_k),x_1, \ldots ,x_k)],$$

a function of $k + 1$ variables. Since f and $\tilde{g}$ are total, $\varphi_v^{(k+1)}$ is total. Hence we have

$$\varphi_{\tilde{g}(v,x_1,\ldots,x_k)} = \varphi_{f(\tilde{g}(v,x_1,\ldots,x_k),x_1,\ldots,x_k)}.$$

Setting $n_f = \lambda x_1 \cdots x_k[\tilde{g}(v,x_1, \ldots ,x_k)]$, we have the desired result.☒

Corollary III *Let f be a recursive function of $k + 1$ variables. Then there is a recursive function n_f of k variables such that for all $x_1, \ldots ,x_k$,*

$$W_{n_f(x_1,\ldots,x_k)} = W_{f(n_f(x_1,\ldots,x_k),x_1,\ldots,x_k)}.$$

Proof. Immediate.☒

We now give our most general formulation.

Definition A function f of k variables is *one-one* if, for every y, there is at most one k-tuple $<x_1, \ldots ,x_k>$ such that $f(x_1, \ldots ,x_k) = y$.

***Theorem IV* (Kleene)** *For each k, there exists a recursive function n of $k + 2$ variables such that n is one-one and for any z, if $\varphi_z^{(k+1)}$ is total, then for all $x_1, \ldots , x_k, y$,*

$$\varphi_{n(z,x_1,\ldots,x_k,y)} = \varphi_{\varphi_z^{(k+1)}(n(z,x_1,\ldots,x_k,y),x_1,\ldots,x_k)}.$$

Proof. Let t be the recursive function of two variables defined in the proof of Theorem 7-IV. Then t is one-one and for all x and y, $\varphi_{t(x,y)} = \varphi_x$. Take $\tilde{g}$ as in the proof of Theorem III. Define $\tilde{\tilde{g}}$ by

$$\tilde{\tilde{g}}(u,x_1, \ldots ,x_k) = t(\tilde{g}(u,x_1, \ldots ,x_k),<u,x_1, \ldots ,x_k>).$$

$\tilde{\tilde{g}}$ is a one-one function with the property asserted of $\tilde{g}$ in the proof of Theorem III. Let $\tilde{\tilde{v}}$ be a recursive function such that $\tilde{\tilde{v}}(z)$ is a Gödel number for

$$\lambda u x_1 \ldots x_k[\varphi_z^{(k+1)}(\tilde{\tilde{g}}(u,x_1, \ldots ,x_k),x_1, \ldots ,x_k)].$$

Set

$$n = \lambda z x_1 \cdots x_k y[\tilde{\tilde{g}}(t(\tilde{\tilde{v}}(z),y),x_1, \ldots ,x_k)].$$

n evidently possesses the desired properties.☒

A corollary to Theorem IV, parallel to Corollaries I, II, and III, can be given. We omit it.

The basic objects in Theorems I to IV are partial recursive functions of one variable. For any $k > 1$, statements and proofs carry over directly for partial recursive functions of k variables.

Kleene's formulation and proof for Theorem I are given in Exercise 11-4. Kleene's formulation differs slightly, and inessentially, from ours. It allows his proofs to be formally simpler in certain respects.

§11.3 COMPLETENESS OF CREATIVE SETS; COMPLETELY PRODUCTIVE SETS

We now use the recursion theorem to answer two questions that were left unsettled in earlier chapters: (1) Is every creative set m-complete (and hence, by Theorem 7-VII, 1-complete)? (2) Is every productive set completely productive?

Theorem V *(**Myhill** [1955])* *A creative* $\Rightarrow$ *A* m-*complete.*

Proof. Let A be creative. Then $\bar{A}$ is productive, and by Theorem 7-XI, $\bar{A}$ possesses a productive function. Let h be such a productive function. Let B be any recursively enumerable set. For fixed x and y, define a partial recursive ψ by the following instructions: given input z, search for y in B; if and when y appears in B, compute $h(x)$; if $z = h(x)$, give 0 as output; if $z \neq h(x)$, give no output. These instructions depend uniformly on x and y. Hence there is a recursive f such that, for any x and y,

$$\psi = \varphi_{f(x,y)}.$$

Taking domains, we have

$$W_{f(x,y)} = \begin{cases} \{h(x)\}, & \text{if } y \in B; \\ \emptyset, & \text{if } y \notin B. \end{cases}$$

Applying Corollary III, we obtain a recursive function n such that

$$W_{n(y)} = W_{f(n(y),y)} = \begin{cases} \{hn(y)\}, & \text{if } y \in B; \\ \emptyset, & \text{if } y \notin B. \end{cases}$$

Now

$$y \in B \Rightarrow hn(y) \in W_{n(y)} \Rightarrow W_{n(y)} \not\subset \bar{A} \text{ (by productiveness)} \Rightarrow hn(y) \in A;$$

and $\quad y \notin B \Rightarrow W_{n(y)} = \emptyset \Rightarrow hn(y) \in \bar{A} \qquad$ (by productiveness).

Thus, for any y, $y \in B \Leftrightarrow hn(y) \in A$. Hence $B \leq_{\text{m}} A$. Since B is any recursively enumerable set, A is m-complete. ☒

Corollary V *A is creative* $\Leftrightarrow$ *A is* 1-*complete* $\Leftrightarrow$ *A is* m-*complete.*

Proof. By Theorem 7-VII, m-completeness $\Leftrightarrow$ 1-completeness. By Corollary 7-V, m-completeness $\Rightarrow$ creativeness. The corollary follows. ☒

The creative sets therefore constitute a single isomorphism type; from the point of view of recursive function theory, any two creative sets are identical.

Theorem VI *A productive* $\Rightarrow$ *A completely productive.*

Proof. Let A be productive and by Theorem 7-XI, let h be a productive function for A. By a similar construction to that in Theorem V, obtain a recursive f such that

$$W_{f(x,y)} = W_y \cap \{h(x)\}.$$

Applying Corollary III, obtain a recursive function n such that for all y,

$$W_{n(y)} = W_{f(n(y),y)} = W_y \cap \{hn(y)\}.$$

Now

$$\begin{aligned} hn(y) \in W_y \Rightarrow W_{n(y)} = \{hn(y)\} &\Rightarrow W_{n(y)} \not\subset A \qquad \text{(by productiveness)} \\ &\Rightarrow hn(y) \in \bar{A}; \end{aligned}$$

and

$$\begin{aligned} hn(y) \notin W_y \Rightarrow W_{n(y)} = \emptyset \Rightarrow W_{n(y)} \subset A \Rightarrow hn(y) & \\ \in A & \qquad \text{(by productiveness).} \end{aligned}$$

Hence $$hn(y) \in \bar{A} \Leftrightarrow hn(y) \in W_y.$$

Thus A is completely productive with hn as completely productive function.☒

Both the above proofs use Theorem 7-XI. Use of this theorem can be easily omitted in a proof for Theorem VI, and Theorem 7-XI then becomes a corollary to Theorem VI (see Exercise 11-13). Theorem 7-XI can also be given a simple direct proof by means of the recursion theorem as follows.

Alternative proof for Theorem 7-XI. Let A be a productive set with productive partial function ψ. Take a recursive f such that

$$W_{f(x,y)} = \begin{cases} W_y, & \text{if } \psi(x) \text{ convergent;} \\ \emptyset, & \text{if } \psi(x) \text{ divergent.} \end{cases}$$

Corollary III yields a fixed-point function n such that

$$W_{n(y)} = W_{f(n(y),y)} = \begin{cases} W_y, & \text{if } \psi n(y) \text{ convergent;} \\ \emptyset, & \text{if } \psi n(y) \text{ divergent.} \end{cases}$$

Then

$$\psi n(y) \text{ divergent} \Rightarrow W_{n(y)} = \emptyset \Rightarrow \psi n(y) \text{ convergent} \qquad \text{(by productiveness).}$$

Hence ψn is a total recursive function. Hence, for all y, $W_{n(y)} = W_y$. Therefore

$$\begin{aligned} W_y \subset A \Rightarrow W_{n(y)} \subset A &\Rightarrow \psi n(y) \in A - W_{n(y)} \qquad \text{(by productiveness)} \\ &\Rightarrow \psi n(y) \in A - W_y. \end{aligned}$$

Hence ψn is a (total) productive function for A.☒

Theorem 7-VII was also used in obtaining Corollary V. Use of Theorem 7-VII can be omitted if, in the proof of Theorem V, the result of Exercise 7-50 is used to make h a one-one productive function and Theorem IV is used to make n a one-one function. The modified proof then yields directly that A creative $\Rightarrow$ A 1-complete, and Corollary V follows from Corollary 7-V alone.

In this way, the results of this section together with Theorems 7-VII and 7-XI are all obtained by simple recursion-theorem constructions. It is instructive (in such cases as Theorem 7-XI) to compare proofs that use the recursion theorem with proofs that do not. If the reader has successfully proved Theorems V and VI prior to this chapter (as Exercises 7-32 and 7-52), he should compare his proofs with the recursion-theorem proofs above. In the development of recursive function theory, a number of results have been obtained first by complex direct argument and then, later, by shorter and more elegant recursion-theorem constructions.

Although the immediate effect of the recursion theorem is to produce pathologies, we note that, in Theorems V and VI, it serves to establish regularity among concepts previously defined.

§11.4 OTHER APPLICATIONS AND CONSTRUCTIONS

We now give five further illustrations of the use of the recursion theorem. Still other illustrations will be given in later sections of this chapter or indicated in exercises. The first illustration concerns set theory; the second concerns logic; the third concerns the theory of automata; and the fourth and fifth concern recursive function theory.

First Illustration: Recursively Enumerable Sets and Set Theory

Consider the relation $E = \{\langle m,n\rangle | m \in W_n\}$. If $\langle m,n\rangle \in E$, we write mEn. A rough analogy exists between the relation E and the relation $\in$ of set theory.† This analogy suggests several questions. (1) Can we have an m such that mEm? Clearly for any member m of K, mEm by definition. We sharpen question (1) as follows. (2) Define: m is a *unit set* if $W_m = \{n\}$ for some n. Can we have a unit set m such that mEm? This is the same as asking: can we have $W_m = \{m\}$? From §11.2, we know that this is possible. (3) Can we have unit sets m and n such that $m \neq n$, mEn, and nEm? In Exercise 11-24 we show that this is possible. (4) Can we have an infinite *ascending* chain of unit sets? That is, can we find distinct $m_0, m_1, \ldots$ such that $m_0Em_1, m_1Em_2, \ldots$? This can be done as follows. Define a recursive h such that for all y,

$$W_{h(y)} = \{y\}, \qquad \text{and} \qquad h(y) > y.$$

Set

$$m_0 = h(0),$$
$$m_{n+1} = h(m_n).$$

The sequence $m_0, m_1, \ldots$ is the desired sequence (and is, in fact, recursively enumerable). (5) Can we have an infinite *descending* chain of units sets?

† Among the usual basic axioms for set theory (in the Zermelo-Fraenkel formulation) the axioms of *pairing, sum, choice,* and *infinity* hold. The axioms of *power, extensionality,* and *foundation* fail. The axioms of *comprehension* and *replacement* fail in general but hold in a restricted form. The *continuum hypothesis* holds (vacuously, as all sets are finite or denumerable). In this analogy, the theorem which asserts the non-recursive-enumerability of $\bar{K}(= \{x | not\ xEx\})$ can be viewed as a counterpart to the *Russell paradox.* It is also related to *Cantor's theorem* in both statement and proof. (In set theory, *Cantor's theorem* shows that nondenumerable sets exist; the *Russell paradox* arises when the collection of all sets which are not members of themselves is itself considered as a set.) See Exercise 11-23.

This question is more difficult. To answer it, we make use of the recursion theorem.

Theorem VII *There exists an infinite sequence of distinct integers* $m_0, m_1, \ldots$ *such that for all* i, $W_{m_i} = \{m_{i+1}\}$.

Proof. Our previous construction of an *ascending* chain might suggest that we find a recursive function g such that for all y, $W_y = \{g(y)\}$. This is impossible since not all y are unit sets. Can we attain the more modest goal of finding a recursive g such that for all y,

$$W_{g(y)} = \{gg(y)\}?$$

We get such a g as follows. Define a recursive f so that

$$W_{f(x,y)} = \begin{cases} \{\varphi_x\varphi_x(y)\}, & \text{if } \varphi_x\varphi_x(y) \text{ convergent;} \\ \emptyset, & \text{if } \varphi_x\varphi_x(y) \text{ divergent.} \end{cases}$$

Then define a recursive f' so that for all x and y, $W_{f'(x,y)} = W_{f(x,y)}$ and $f'(x,y) > y$. (The function t from Theorem 7-IV may be used for this purpose; set $f'(x,y) = t(f(x,y),\mu z[t(f(x,y),z) > y])$.) Define a recursive h such that

$$\varphi_{h(x)}(y) = f'(x,y),$$

and note that, for all x, $\varphi_{h(x)}$ is a total function. Apply Theorem I and get an n such that

$$\varphi_n = \varphi_{h(n)}.$$

φ_n is total, and for any y,

$$W_{\varphi_n(y)} = W_{\varphi_{h(n)}(y)} = W_{f'(n,y)} = \{\varphi_n\varphi_n(y)\}.$$

Furthermore, for all y, $\varphi_n(y) > y$. Then φ_n is the desired g such that

$$W_{g(y)} = \{gg(y)\}.$$

Set

$$m_0 = g(0),$$
$$m_{n+1} = g(m_n),$$

and the sequence $m_0, m_1, \ldots$ constitutes the desired descending chain. ☒

In Exercise 11-25, Theorem VII is used to show the existence of *full* sets other than those of the special kind described in Exercises 7-45 and 7-46.

Second Illustration: Axioms Asserting Their Own Consistency

Let a logical system be given together with a coding from wffs onto N. Assume that some desired notion of "consistency" (for sets of wffs) has been defined and that there is a recursive function $\hat{c}$ such that for any x, the wff (with code number) $\hat{c}(x)$ is understood to assert the consistency of the set of wffs W_x. Let a recursively enumerable set W_p of wffs be given. Can we add to W_p a wff (with code number) q which will assert the consistency of the set $W_p \cup \{q\}$? This question is answered in the following theorem. The construction is similar to that for Theorem VI.

Theorem VIII *For any integer p and any recursive function $\hat{c}$, there exists an integer n such that*

$$W_n = W_p \cup \{\hat{c}(n)\}.$$

Proof. Take a recursive f such that for any x, $W_{f(x)} = W_p \cup \{\hat{c}(x)\}$. From Corollary I the result is immediate. Indeed, by Corollary II, n can be found uniformly from p and an index for $\hat{c}$. ☒

Thus, under the previously mentioned assumptions about $\hat{c}$, Theorem VIII shows that we can find an r.e. index n such that W_n consists of W_p together with a wff which asserts the consistency of W_n. We make three further comments.

Comment 1. The theorem can evidently be applied to properties of sets (of wffs) other than consistency, e.g., inconsistency, completeness, independence, provided that for such a property P, a recursive function c_P exists, where $c_P(x)$ is understood to assert that W_x has property P.

Comment 2. Among the various r.e. indices for a given recursively enumerable set, certain indices may be considered more "natural" (in some sense) than others. (Instructions for a fixed recursively enumerable set may vary greatly in structure; see Exercise 11-26.) Is there some "natural" index n such that $W_n = W_p \cup \{\hat{c}(n)\}$? For example, in the proof of Theorem VIII, if we stipulate that (for any x) $f(x)$ is the one and only "natural" index for $W_{f(x)}$, we can ask: is there an m such that

$$W_{f(m)} = W_p \cup \{\hat{c}f(m)\};$$

i.e., is there an m such that $W_{ff(m)} = W_{f(m)}$? In general, the answer to this question is negative; and such a natural index cannot be found (see Exercise 11-27). For particular logical systems and choices of W_p, however, this criterion of naturalness can be so weakened (in a reasonable way) that an affirmative answer becomes possible. Consider a logical system that includes the sentential connective "$\Leftrightarrow$" and the usual sentential rules associated with "$\Leftrightarrow$". For any fixed x and y, let "$[\hat{c}(x) \Leftrightarrow \hat{c}(y)]$" denote (in our discussion) the wff obtained by putting "$\Leftrightarrow$" between the wffs (with code numbers) $\hat{c}(x)$ and $\hat{c}(y)$. Take f to be the function constructed above such that $W_{f(x)} = W_p \cup \{\hat{c}(x)\}$. Define an r.e. index y of $W_{f(x)}$ to be *natural* if the wff $[\hat{c}(y) \Leftrightarrow \hat{c}(f(x))]$ is derivable (in the given logical system) from wffs in W_p. Consider in particular the case where the logical system is elementary arithmetic, W_p is the set of Peano's axioms, and the function $\hat{c}$ is obtained as outlined in §11.6 below. If n is the fixed-point value for f yielded by Corollary I, it can be shown that the wff $[\hat{c}(n) \Leftrightarrow \hat{c}f(n)]$ is derivable from Peano's axioms (by first showing that a wff expressing the assertion "n and $f(n)$ are r.e. indices for the same set" is derivable from Peano's axioms). Hence, in the proof of

Theorem VIII, n becomes a *natural* index for $W_p \cup \{\hat{c}(n)\}$; and $\hat{c}f(n)$ can be derived from $W_p \cup \{\hat{c}(n)\}$ even though $\hat{c}f(n)$ may not be a member of $W_p \cup \{\hat{c}(n)\}$.

Comment 3. Will W_n in fact *be* consistent? In the case of elementary arithmetic, with $\hat{c}$ formed as in §11.6, W_n may or may not be consistent depending upon the choice of W_p and of the index p for W_p. If W_p is chosen to be empty, W_n will be consistent; if W_p is chosen to be Peano's axioms, and p is an index for the usual enumeration of Peano's axioms, then W_n *cannot* be consistent. This is a consequence of Gödel's second incompleteness theorem (see §11.6).

Third Illustration: Self-reproducing Machines

Does there exist a machine with the ability to reproduce itself? Several ways of making this question precise, and of answering it, have appeared in the literature. (An early affirmative theorem of this kind is due to von Neumann.) The constructions made often bear a strong resemblance to proofs of the recursion theorem. We now use the recursion theorem to get a general theorem that yields various known results on self-reproducing machines as special cases.

Let $\mathfrak{M}$ be a countable class of objects. We call the members of $\mathfrak{M}$ *machines*. Let $\mathfrak{R}$ be a countable class of objects. We call the members of $\mathfrak{R}$ *representations*. Assume that each machine M in $\mathfrak{M}$ has associated with it a partial mapping from N into $\mathfrak{R}$, i.e., a single-valued subset of $N \times \mathfrak{R}$. Given machine M, we denote its partial map as M also, and we use the notation $M(x) = R$ to indicate that integer x is mapped to representation R by M. Assume further that for every partial recursive function ψ and every machine M, there is a machine M' such that the map M' coincides with the map $M\psi$. (We might say, "the class $\mathfrak{M}$ is closed under composition with Turing machines.") Finally, assume that a fixed coding from $\mathfrak{M}$ onto N is given; and that there is a recursive function h such that for all x_1 and x_2, $M_{h(x_1,x_2)} = M_{x_1}\varphi_{x_2}$ (where, for any z, M_z is the machine with code number z).

Theorem IX *Let D be any given machine. Then there is an integer m such that for all inputs, M_m yields $D(m)$ as output.*

(For intuitive motivation, think of each representation in $\mathfrak{R}$ as, in some sense, a *replica* of a machine in $\mathfrak{M}$, and think of D as a special machine in $\mathfrak{M}$ which, given any x, constructs a replica of the machine with code number x. The theorem then shows that there must be a machine which, regardless of input, constructs its own replica.)

Proof. Let d be a code number for D. For any given x, let ψ be the constant function whose output (for any input) is $h(d,x)$. Instructions for ψ will depend on x; hence we can find a recursive f such that

$$\varphi_{f(x)}(y) = h(d,x), \qquad \text{for all } y.$$

Applying the recursion theorem, we obtain an n such that

$$\varphi_n(y) = \varphi_{f(n)}(y) = h(d,n), \qquad \text{for all } y.$$

Hence

$$D\varphi_n(y) = Dh(d,n), \qquad \text{for all } y.$$

But $D\varphi_n = M_{h(d,n)}$ (by the definition of h). Setting $m = h(d,n)$, we have

$$M_m(y) = D(m) \qquad \text{for all inputs } y. \boxtimes$$

We briefly consider two examples.

Example 1. Take $\mathfrak{M}$ to be the class of Turing machines and let M_x denote the machine with Gödel number x under our original indexing of Turing machines. Let a *representation* of a given Turing machine be a sequence of 1's and B's that encodes the quadruples of that machine according to the following rules:

q_i	$111 \cdots 1B$	($i + 1$ 1's);
1	11	
B	BB	
L	$1B$	
R	$B1$.	

Thus the machine $\{q_1 1 R q_2, q_2 B 1 q_1\}$ has

$$11B11B1111B111BBB1111B$$

as a representation. Take $\mathfrak{R}$ to be the class of all such representations of Turing machines. It is possible (and not difficult) to show that there exists a Turing machine D which, given any input x, will terminate with a representation for M_x (and nothing else) on its tape. Theorem IX then shows that there is a particular Turing machine which, for any string of 1's as input, terminates with a representation for itself (and nothing else) on its tape.†

Example 2. Take $\mathfrak{M}$ to be a more general class of machines, some of which have the ability to manufacture new physical objects and absorb raw material from their environment. Take $\mathfrak{R}$ to be $\mathfrak{M}$ itself. Take D to be a special "universal machine" which, given any code number ("blueprint") for a machine, will then construct a copy of that machine. Theorem IX now shows that there is a particular machine that can construct a copy of itself.

Of course, the existence of the appropriate machine D must be proved in both these examples. In Example 2, this is a matter of some subtlety. A precise definition of machines and environments must be made, and a more detailed study of possible environmental configurations must be carried out. Nevertheless, our application of the recursion theorem displays the intellectually central features of a self-reproduction proof and is directly relevant

† This application was suggested by C. Y. Lee.

to fallacious arguments (by "infinite regress") that self-reproducing machines cannot exist.†

Fourth Illustration: A Double Recursion Theorem

Can we find a common fixed-point value for two given recursive functions? In general, this is not possible. The following theorem, however, shows that there is a slightly weakened sense in which this can be done. The theorem is due to Smullyan [1961]. The proof gives a fourth illustration of the use of the recursion theorem.

Theorem X(a) (Smullyan) *For any recursive functions g and h, there exist m and n such that*

$$\varphi_m = \varphi_{g(<m,n>)}, \qquad \text{and} \qquad \varphi_n = \varphi_{h(<m,n>)}.$$

Proof. Take a recursive f such that

$$\varphi_{f(x,y)} = \varphi_{g(<x,y>)}.$$

Applying Theorem III, we find a recursive function $\hat{m}$ such that

$$\varphi_{\hat{m}(y)} = \varphi_{f(\hat{m}(y),y)} = \varphi_{g(<\hat{m}(y),y>)}.$$

Take a recursive k such that

$$\varphi_{k(y)} = \varphi_{h(<\hat{m}(y),y>)}.$$

Applying Theorem I, we find an $\hat{n}$ such that

$$\varphi_{\hat{n}} = \varphi_{k(\hat{n})} = \varphi_{h(<\hat{m}(\hat{n}),\hat{n}>)}.$$

Setting $n = \hat{n}$ and $m = \hat{m}(\hat{n})$, we have

$$\varphi_m = \varphi_{g(<m,n>)}, \qquad \text{and} \qquad \varphi_n = \varphi_{h(<m,n>)}$$

as desired. ☒

An application of Theorem X will be presented in Exercise 11-29 below.

† In Example 2, the question of measurement tolerances and gradual introduction of error can be raised (for macroscopic machines). This is ultimately a question of how we define $\mathfrak{M}$ and how we design D. One solution is to incorporate absolute standards (atomic clocks, spectral-line wavelengths, etc.) into D.

The approach in Theorem IX differs somewhat from the self-reproducing-machine constructions most often given. In the latter, the self-reproducing machine can be pictured as $<D,C,E,<b,i>>$, where D is a "blueprint realizer" (that will build an object from any given "blueprint"), C is a "program copier," E is supplementary equipment for handling inputs and outputs for C and D, and $<b,i>$ is a "program" consisting of b, a blueprint for $<D,C,E>$, and i, a set of supplementary instructions. The machine takes its orders from i and operates as follows. b is placed in D, and replicas D', C', and E' are produced. Then $<b,i>$ is placed in C, and a copy of the program $<b',i'>$ is made. The "offspring" $<D',C',E',<b',i'>>$ is then assembled.

Our Theorem IX does not use such a final blueprint-copying step. Instead, as we saw, the self-reproducing machine takes the form $<D,\varphi_n>$, where φ_n first *computes* a "total blueprint" for $<D,\varphi_n>$, and D then realizes it.

Fifth Illustration: Isomorphism of Universal Partial Functions

Recall that a partial function ψ is *universal* if there exists a recursive f of two variables such that $(\forall x)[\varphi_x = \lambda y[\psi f(x,y)]]$. In Theorem 4-III, we showed that the property of being universal is recursively invariant. We now use the recursion theorem to obtain the converse result.

Theorem X(b) (Blum) *Any two universal partial functions are recursively isomorphic.*

Proof.

Definition f is an *encoder* for ψ if f is recursive and

$$(\forall x)[\varphi_x = \lambda y[\psi f(x,y)]].$$

Let $\psi = \lambda x[\varphi_{\pi_1(x)}(\pi_2(x))]$. Then ψ is universal with $\lambda xy[<x,y>]$ as encoder. Let θ be any given universal partial function with h as encoder. To prove the theorem we must show the existence of a recursive permutation g such that $g\theta = \psi g$.

Lemma *Let θ be universal. Then there is a uniform effective procedure by which, given any $v_1, v_2, \ldots, v_k$ such that $\theta(v_1) = \theta(v_2) = \cdots = \theta(v_k)$, we can find $v_{k+1} \notin \{v_1, \ldots, v_k\}$ such that $\theta(v_1) = \cdots = \theta(v_k) = \theta(v_{k+1})$.*

Proof of lemma. Define a recursive f' so that

$$\varphi_{f'(t)}(x) = \begin{cases} \theta(v_1), & \text{if } h(t,0) \notin \{v_1, \ldots, v_k\}; \\ \text{divergent}, & \text{if } h(t,0) \in \{v_1, \ldots, v_k\}. \end{cases}$$

Applying Theorem I, obtain n such that $\varphi_{f'(n)} = \varphi_n$. If $h(n,0) \notin \{v_1, \ldots, v_k\}$, set $v_{k+1} = h(n,0)$. If $h(n,0) \in \{v_1, \ldots, v_k\}$, then $\theta(v_1), \ldots, \theta(v_k)$ must be divergent. Define a recursive f'' so that

$$\varphi_{f''(t)}(x) = \begin{cases} 0, & \text{if } h(t,0) \in \{v_1, \ldots, v_k\}; \\ \text{divergent}, & \text{if } h(t,0) \notin \{v_1, \ldots, v_k\}. \end{cases}$$

Applying Theorem I, obtain n' such that $\varphi_{f''(n')} = \varphi_{n'}$. Since $\theta(v_1), \ldots, \theta(v_k)$ are divergent, $\varphi_{f''(n')}(0)$ must be divergent and $h(n',0) \notin \{v_1, \ldots, v_k\}$. Set $v_{k+1} = h(n',0)$. By Theorem II, the entire procedure is uniformly effective in $v_1, \ldots, v_k$. This concludes the proof of the lemma. (*Corollary to lemma.* Every universal partial function has a one-one encoder.)

Let any z be fixed. Take z' so that

$$\varphi_{z'} = \{<x,y> | <y,x> \in \varphi_z \;\&\; (\forall y')[[y' \neq y \;\&\; <y',x> \in \varphi_z] \Rightarrow <y,x> \textit{ precedes } <y',x> \textit{ in (a standard enumeration of) } \varphi_z]\}.$$

z' may be obtained uniformly from z, and $\varphi_{z'}$ is a single-valued "inverse" to φ_z.

We give instructions for computing a certain partial recursive η. More specifically, we give instructions for enumerating the ordered pairs of η.

Stage $2t$. See whether t has already appeared in *domain* η. If so, go to stage $2t + 1$. If not, obtain, by the method of Theorem 7-IV, a sequence of distinct indices for $\varphi_z\theta$. Let w be the first index in this sequence such that $<w,t>$ has not yet been placed in *range* η. Add $<t,<w,t>>$ to η, and go to stage $2t + 1$.

Stage $2t + 1$. See whether t has already appeared in *range* η. If so, go to stage $2t + 2$. If not, let s be an index for $\varphi_{z'}\psi$. Use the lemma to generate a sequence of distinct integers $v_1, v_2, \ldots$ such that

$$\theta(v_1) = \theta(v_2) = \cdots = \varphi_s(t).$$

Let v be the first member of this sequence such that v has not yet been placed in *domain* η. Add $<v,t>$ to η, and go to stage $2t + 2$.

. . . . . .

It is evident, by the construction, that η is single-valued and hence a partial recursive function.

Observe that for t added to *domain* η at stage $2t$,

$$\varphi_z\theta(t) = \varphi_w(t) = \psi(<w,t>) = \psi\eta(t);$$

and observe that for v added to *domain* η at stage $2t + 1$,

$$\varphi_z\theta(v) = \varphi_z\varphi_s(t) = \varphi_z\varphi_{z'}\psi(t) = \varphi_z\varphi_{z'}\psi\eta(v).$$

It is evident, by the construction, that for every z, η is total, one-one, and onto. Moreover, an index for η depends uniformly on z. Thus we have a recursive $\hat{f}$ such that for any z, $\eta = \varphi_{\hat{f}(z)}$ and η is a recursive permutation. Applying Theorem I, obtain m such that $\varphi_m = \varphi_{\hat{f}(m)}$. Let $g = \varphi_{\hat{f}(m)}$.

Then, in the above construction, taking $z = m$, we have $\eta = g$, $\varphi_z = g$, and $\varphi_{z'} = g^{-1}$. Hence for t added to *domain* η at stage $2t$,

$$g\theta(t) = \varphi_z\theta(t) = \psi\eta(t) = \psi g(t);$$

and for v added to *domain* η at stage $2t + 1$.

$$g\theta(v) = \varphi_z\theta(v) = \varphi_z\varphi_{z'}\psi\eta(v) = gg^{-1}\psi g(v) = \psi g(v).$$

Thus $g\theta = \psi g$, and the proof is complete.☒

Exercises 11-30 to 11-33 give several other applications of the recursion theorem. Exercise 11-33 is a simple prototype of an argument that will be used in §§11.7 and 11.8 and later theory.

§11.5 OTHER FORMS OF THE RECURSION THEOREM

Strictly speaking, the recursion theorem as formulated above (in Theorem IV) is not a fixed-point theorem, although we have used the phrase "fixed-point value" in connection with it. We now present a genuine fixed-point theorem that is closely related to Theorem IV. It is, as we shall see, somewhat weaker and somewhat less useful than Theorem IV.

Theorem XI *Let Φ be an enumeration operator.*† *There exists a set A such that*

(i) $\Phi(A) = A$,
(ii) $(\forall B)[\Phi(B) = B \Rightarrow A \subset B]$,
(iii) *A is recursively enumerable.*

Proof.‡ Define a sequence of sets $A_0, A_1, \ldots$ as follows:

$$A_0 = \emptyset;$$
$$A_{n+1} = \Phi(A_n).$$

Take

$$A = \bigcup_{n=0}^{\infty} A_n.$$

Recall from Theorem 9-XX that Φ has the properties

(*a*) $A \subset B \Rightarrow \Phi(A) \subset \Phi(B)$ (monotonicity),
(*b*) $x \in \Phi(A) \Rightarrow (\exists D)[D \text{ finite } \& \ D \subset A \ \& \ x \in \Phi(D)]$ (continuity).

To prove (i):

$$x \in A \Rightarrow (\exists n)[n > 0 \ \& \ x \in A_n] \text{ (by definition of } A)$$
$$\Rightarrow (\exists n)[n > 0 \ \& \ x \in \Phi(A_{n-1})] \text{ (by definition of } A_n)$$
$$\Rightarrow x \in \Phi(A) \text{ (by monotonicity);}$$

and

$$x \in \Phi(A) \Rightarrow (\exists D)[D \text{ finite } \& \ D \subset A \ \& \ x \in \Phi(D)] \text{ (by continuity)}$$
$$\Rightarrow (\exists D)(\exists n)[D \text{ finite } \& \ D \subset A_n \ \& \ x \in \Phi(D)] \text{ (by definition of } A)$$
$$\Rightarrow (\exists n)[x \in A_{n+1}] \text{ (by monotonicity)} \Rightarrow x \in A \text{ (by definition of } A).$$

Hence $x \in A \Leftrightarrow x \in \Phi(A)$, and we have (i).

To prove (ii), assume $\Phi(B) = B$. Then

$$A_0 = \emptyset \subset B.$$

If $A_n \subset B$, then, by monotonicity, $A_{n+1} = \Phi(A_n) \subset \Phi(B) = B$. Therefore, by induction, $(\forall n)[A_n \subset B]$. Hence $A \subset B$, and we have (ii).

To prove (iii), let $\Phi = \Phi_z$. Then for any y,

$$\Phi(W_y) = \{x | (\exists u)[\langle x,u \rangle \in W_z \ \& \ D_u \subset W_y]\} \qquad \text{(by the definition in §9.7).}$$

† Enumeration operators were defined in §9.7. An enumeration operator maps 2^N into 2^N.

‡ Parts (i) and (ii) of Theorem XI are a special case of the Knaster-Tarski theorem, which asserts that if $\mathcal{L}$ is a complete lattice, then any monotone and continuous mapping F from $\mathcal{L}$ into $\mathcal{L}$ has a fixed point; where *monotone* means that $x \leq y \Rightarrow F(x) \leq F(y)$, and *continuity* is taken with respect to the natural topology induced by closure under the least-upper-bound operation in the lattice. For Theorem XI, the lattice may be taken as the lattice of all sets under the ordering $\subset$. Monotonicity and continuity are given by Theorem 9-XX (see Exercise 11-35).

$\Phi(W_y)$ is a recursively enumerable set (by the second projection theorem), and we have, by familiar methods, a recursive f such that for all y

$$W_{f(y)} = \Phi_z(W_y).$$

Let q be any r.e. index for the empty set. Define a recursive function g as follows:

$$g(0) = q,$$
$$g(n+1) = fg(n).$$

Then for all i, $A_i = W_{g(i)}$; and therefore

$$x \in A \Leftrightarrow (\exists y)[x \in W_{g(y)}].$$

By the second projection theorem, A is recursively enumerable. This proves (iii).☒

In the proof for (iii) f may be made to depend uniformly on z. We thus state the following corollary.

Corollary XI(a) *There exists a recursive* $\tilde{f}$ *such that for all* z *and* y, $W_{\tilde{f}(y,z)} = \Phi_z(W_y)$.

Proof. Immediate.☒

It also follows from the proof for (iii) that an r.e. index for the fixed point A can be obtained uniformly from z. We thus have the following strengthened form of the theorem.

Corollary XI(b) *There exists a recursive function* h *such that for all* z,

(i) $\Phi_z(W_{h(z)}) = W_{h(z)}$;

(ii) $(\forall B)[\Phi_z(B) = B \Rightarrow W_{h(z)} \subset B]$.

Proof. An r.e. index for $\{x|(\exists y)[x \in W_{g(y)}]\}$ depends uniformly on an index for g, which in turn depends uniformly on an index for f, which in turn depends uniformly on z (in the proof of (iii) for the theorem).☒†

A fixed-point theorem for recursive operators follows from Theorem XI.

Theorem XII (Kleene) *Let* Ψ *be a recursive operator. Then there is a* φ *such that*

(i) $\Psi(\varphi) = \varphi$;

(ii) $(\forall\psi)[\Psi(\psi) = \psi \Rightarrow \varphi \subset \psi]$;

(iii) φ *is partial recursive.*

Proof. Let Φ_z be an enumeration operator defining Ψ. Returning to the proof of Theorem XI and putting Φ_z for Φ, we note that

$$A_0 = \emptyset \text{ is a single-valued set,}$$

and that A_n single-valued $\Rightarrow A_{n+1}$ single-valued (since Φ_z defines a recursive

† The question naturally arises, what happens if we apply Theorem III to Corollary XI(a)? This will yield a fixed point; but must it be minimal? We consider this in Theorems XIII and XIV below.

operator). It follows that the minimal fixed point $A = \bigcup_{n=0}^{\infty} A_n$ must be single-valued.

Take

$$\varphi = \{<x,y>|<x,y> \in A\}.$$

Parts (i), (ii), and (iii) now follow from (i), (ii), and (iii) of Theorem XI.☒

Two corollaries, parallel to Corollaries XI(a) and XI(b), can be given.

Corollary XII(a) *There is a recursive* $\tilde{\tilde{f}}$ *such that for any* z *and* y, *if* $\Phi_z(\tau(\varphi_y))$ *is single-valued, then*

$$\varphi_{\tilde{\tilde{f}}(y,z)} = \tau^{-1}\Phi_z(\tau(\varphi_y)).$$

In particular, if Φ_z *defines a recursive operator* Ψ, *then for all* y,

$$\varphi_{\tilde{\tilde{f}}(y,z)} = \Psi(\varphi_y).$$

Proof. Consider the following sequence of steps, for any y and z. Take φ_y, go to a recursively enumerable set $W_{y'} = \tau(\varphi_y)$ via Theorem 5-IX. Get $W_{\tilde{f}(y',z)}$ by Corollary XI(a); single-valuize $W_{\tilde{f}(y',z)}$ by Theorem 5-XVI, and then apply Theorem 5-IX to get a partial recursive function. This sequence of steps is uniform, and hence the desired $\tilde{\tilde{f}}$ exists.☒

Corollary XII(b) *There is a recursive* h *such that for any* z, *if* Φ_z *defines a recursive operator* Ψ, *then*

(i) $\Psi(\varphi_{h(z)}) = \varphi_{h(z)}$;

(ii) $(\forall\psi)[\Psi(\psi) = \psi \Rightarrow \varphi_{h(z)} \subset \psi]$.

Proof. From Corollary XII(a), as in proof for Corollary XI(b).☒

Theorem XII may fail for a *partial* recursive operator, since the fixed point of the defining enumeration operator need not be single-valued. In the case of a general recursive operator, the minimal fixed point may not be total; indeed, a general recursive operator may have no total fixed point (see Exercise 11-36). Theorem XI can be deduced from Theorem XII (see Exercise 11-39).

Corollaries XI(a) and XII(a) do not depend on any fixed-point argument. Given Corollary XI(a), part (i) of Corollary XI(b) and hence parts (i) and (iii) of Theorem XI follow directly by Corollary III (the recursion theorem). Similarly, given Corollary XII(a), part (i) of Corollary XII(b) and parts (i) and (iii) of Theorem XII follow directly by Theorem III (the recursion theorem). Part (ii), the minimality result, is not so immediate, however, in either Theorem XI or Theorem XII. For this reason, Theorems XI and XII are usually presented and proved (as above) independently of Theorem IV. Theorem XI and Theorem XII are themselves usually called *recursion theorems*. Indeed, as we shall see below, the name "recursion theorem" has its origins in special applications of Theorem XII. Theorems XI and XII together are sometimes called the "*first* recursion theorem," and Theorem IV is then called the "*second* recursion theorem."†

Henceforth, as previously, the phrase "recursion theorem," by itself, will refer to Theorem IV or to one of its special cases (Theorems I, II, and III). The special fixed-point theorems, Theorems XI and XII, will be referred to, together or separately, as the *weak recursion theorem.* In any context where confusion may arise, the recursion theorem (Theorem IV) will be referred to as the *strong recursion theorem.* The strong recursion theorem does not follow directly from the weak recursion theorem since, for example, the applications given in §§11.3 and 11.4 cannot be directly obtained by the weak theorem.

What more can be said about the relation between the strong and weak recursion theorems? In order to discuss this further, we state a theorem due to Myhill and Shepherdson [1955].

Definition f is *extensional* if for all x and y, $\varphi_x = \varphi_y \Rightarrow \varphi_{f(x)} = \varphi_{f(y)}$.

Theorem (Myhill and Shepherdson) (a) *Every recursive operator* Ψ *determines an extensional* f *such that* $\varphi_{f(x)} = \Psi(\varphi_x)$, *for all* x.

(***b***) *Every extensional* f *determines a unique recursive operator* Ψ *such that* $\varphi_{f(x)} = \Psi(\varphi_x)$ *for all* x.‡

Part (*a*) is immediate from Corollary XII(*a*). Part (*b*) will be proved in Theorem 15-XXIX. A proof will be indicated in Exercise 11-43.

Given an extensional f, must it be the case that the construction of Theorem II yields (for every index of f) a fixed-point value n such that φ_n is a minimal fixed point for the corresponding recursive operator Ψ? The following theorem shows that this is not the case. Ψ will be the identity operator, for which, of course, the minimal fixed point is the empty partial function. The proof uses the recursion theorem.

Theorem XIII *Let* n *be the recursive function constructed in Theorem II. There exists a recursive* f *with index* z *such that* $\varphi_{f(x)} = \varphi_x$ *for all* x, *but such that* $\varphi_{fn(z)} = \varphi_{n(z)} \neq \emptyset$.

Proof. Take t to be some fixed index for the constant function 0; that is, $\varphi_t = \lambda y[0]$. For any z, define the partial recursive function ψ as follows:

$$\psi(x) = \begin{cases} x, & \text{if } x \neq n(z); \\ t, & \text{if } x = n(z). \end{cases}$$

A Gödel number for ψ can be found uniformly from z; i.e., there is a recursive function h such that

$$\psi = \varphi_{h(z)}.$$

† The reverse of the present order of presentation.

‡ The mapping from partial recursive functions to partial recursive functions determined by an extensional f is called, by Myhill and Shepherdson, an *effective operation.*

By Theorem I, find a fixed-point value for h; call it m. We then have

$$\varphi_m(x) = \begin{cases} x, & \text{if } x \neq n(m); \\ t, & \text{if } x = n(m). \end{cases}$$

Take $f = \varphi_m$, then

$$f(x) = x, \qquad \text{for } x \neq n(m),$$

and, by Theorem II,

$$\varphi_{f(x)} = \varphi_x, \qquad \text{for } x = n(m).$$

Hence
$$\varphi_{f(x)} = \varphi_x, \qquad \text{for all } x.$$

Now the fixed-point value given by Theorem II is $n(m)$, and

$$\varphi_{n(m)} = \varphi_{fn(m)} = \varphi_t = \lambda y[0] \neq \emptyset. \boxtimes\dagger$$

Does this mean that the weak recursion theorem (Theorem XII) is not a direct consequence of the strong recursion theorem (Theorem IV)? The next theorem shows that, in spite of Theorem XIII, Theorem XII does follow from Theorem IV. There is a uniform procedure for going from an index for any extensional f to an index for an extensional h such that h determines the same recursive operator as f and such that Theorem II, applied to the index for h, yields the minimal fixed point for the recursive operator. The theorem is stated in a form that does not assume the Myhill-Shepherdson theorem.

Theorem XIV *Let n be the recursive function constructed in Theorem II. There is a recursive g such that for any recursive f and for any recursive operator* Ψ, *if* $(\forall x)[\varphi_{f(x)} = \Psi(\varphi_x)]$ *and if z is an index for f, then* $\varphi_{g(z)}$ *is a total function, call it h, such that*

(i) $(\forall x)[\varphi_{h(x)} = \varphi_{f(x)}]$; *and*

(ii) $\varphi_{ng(z)}$ *is a minimal fixed point for the recursive operator* Ψ.

Proof. A proof for Theorem XIV will be given in §11.6.

Theorems XI and XII thus become direct consequences of Theorem IV, and the strong recursion theorem is shown to be the more general and fundamental of the two recursion theorems.

Let Ψ be a recursive operator. Consider possible partial functions φ such that the implicit relationship

$$\varphi = \Psi(\varphi)$$

is satisfied. Theorem XII shows that a partial recursive φ must exist satisfying this implicit relationship. Such implicit relationships may be special cases of what are sometimes called *definitions by recursion.* Hence the name "recursion theorem."

† Note that the proof holds for any recursive function n satisfying the condition of Theorem II.

Example. The following set of equations gives a definition by recursion.

(*E*)
$$\psi(1) = 2,$$
$$\psi(2x) = 2\psi(x).$$

Consider the equations

(*E**)
$$\psi(1) = 2,$$
$$\psi(2x) = 2\varphi(x),$$

which define ψ from φ. These determine a recursive operator Ψ such that $\psi = \Psi(\varphi)$. We may apply Theorem XII directly to this operator, or we may, as in Theorem XIV, take h so that $\varphi_{h(x)} = \Psi(\varphi_x)$ and apply Theorem IV. Either way, we obtain a minimal partial recursive ψ satisfying (*E*). This minimal partial recursive function is $\hat{\psi}$, where

$$\hat{\psi}(x) = \begin{cases} 2x, & \text{if } x \text{ is a power of 2;} \\ \text{divergent,} & \text{if } x \text{ is not a power of 2.} \end{cases}$$

(In view of Theorem XIV, it makes no difference whether the strong or weak recursion theorem is used to solve this implicit relationship.)†‡

In §§11.7 and 11.8, as well as in later chapters, the recursion theorem will provide methods for defining partial recursive functions over well-ordered

† The process of finding an object to satisfy a given implicit relationship is often spoken of as *solving* that relationship. Thus $\hat{\psi}$ above is a *solution* to (*E*). The recursion theorem therefore serves as an *existence theorem* for partial recursive solutions to certain kinds of recursion equations. This use of the recursion theorem is similar to the use of fixed-point theorems in analysis to obtain existence of solutions to certain kinds of differential equations, or the use of "recursion theorems" in set theory to define functions on ordinals.

‡ Not every set of recursion equations of the form

$$\psi = \cdots \psi \cdots$$

can be formulated as

$$\psi = \Psi(\psi),$$

where Ψ is a recursive operator. In some cases, the left side may not be uniquely determined by the right side; in other cases, *no* solution may exist. In nonunique cases, it may be possible to obtain uniqueness by introducing additional conditions. For example, the nonunique equations

$$\psi(4x) = 2\psi(2x),$$
$$\psi(2) = 2$$

can be carried over to a recursive operator Ψ, where $\psi = \Psi(\varphi)$ is given by the equations

$$\psi(x) = \begin{cases} 2, & \text{if } x = 2; \\ 2\varphi(z), & \text{if } z \text{ is even and } x = 2z; \\ \text{divergent,} & \text{otherwise.} \end{cases}$$

The additional conditions may then rule out certain partial functions that satisfy the original recursion equations; e.g., in the above example, the function $\psi = \lambda x[x]$. It is possible for a set of recursion equations to have a unique total solution where that solution is not recursive (see Exercise 11-41).

sets. In this, as in most other applications, the strong (rather than the weak) recursion theorem will be used (see Exercise 11-33).

§11.6 DISCUSSION

In §11.5 we isolated and discussed (in Theorems XI, XII, and XIV) certain *fixed-point* aspects of the recursion theorem. In the present section we consider certain *self-referential* features of the recursion theorem. For example, if $W_m = \{m\}$, then the integer m in some sense *names* itself. Self-reference is closely related to diagonalization, and, as we indicated in §11.1, the recursion theorem incorporates and summarizes a wide class of diagonalization techniques. Let us first consider Theorem I and explore in somewhat more detail the mechanism by which the fixed-point value n is produced. Recall that $\tilde{g}$ was defined to make

$$\varphi_{\tilde{g}(u)} = \begin{cases} \varphi_{\varphi_u(u)}, & \text{if } \varphi_u(u) \text{ convergent;} \\ \text{everywhere divergent,} & \text{if } \varphi_u(u) \text{ divergent.} \end{cases}$$

More particularly, for any u, $\tilde{g}(u)$ is a *name* (Gödel number) for a set of instructions. To put the matter informally,

$\tilde{g}(u)$ = "*given any input, use* $P_{\mathbf{u}}$ *to compute* $\varphi_{\mathbf{u}}(\mathbf{u})$; *if and when that computation converges, apply instructions* $\varphi_{\mathbf{u}}(\mathbf{u})$ *to the given input*"

or, more briefly,

$\tilde{g}(u)$ = "*compute* $\varphi_{\mathbf{u}}(\mathbf{u})$; *if and when* $\varphi_{\mathbf{u}}(\mathbf{u})$ *converges, do* $\varphi_{\mathbf{u}}(\mathbf{u})$."

("$\mathbf{u}$" here represents the numeral for the integer u.) Note that $\tilde{g}(u)$ can be directly computed (as a *name*) regardless of whether or not $\varphi_u(u)$ converges. Next, using the f assumed in Theorem I, and taking Gödel numbers for f and $\tilde{g}$, we have, by Theorem 1-VI, a Gödel number for $f\tilde{g}$; call it v. Then,

$\tilde{g}(v)$ = "*compute* $\varphi_{\mathbf{v}}(\mathbf{v})$; *if and when* $\varphi_{\mathbf{v}}(\mathbf{v})$ *converges, do* $\varphi_{\mathbf{v}}(\mathbf{v})$."

To assist our discussion, we indicate this same name in a slightly different way.

$\tilde{g}(v)$ = "*compute* $f\tilde{g}(\mathbf{v})$; *if and when* $f\tilde{g}(\mathbf{v})$ *converges, do* $f\tilde{g}(\mathbf{v})$."†

Since f is total and $\tilde{g}(v)$ is the name given above, $f\tilde{g}(v)$ can be computed. $\tilde{g}(v)$ instructs us *first* to compute $f\tilde{g}(v)$ and *then* to apply $f\tilde{g}(v)$. Thus $f\tilde{g}(v)$ and $\tilde{g}(v)$ are instructions for the same partial recursive function, but the computation of that partial function by $\tilde{g}(v)$ will be somewhat longer than

† Of course $\tilde{g}(v)$ is a particular integer. We have used two different informal expressions to indicate that same integer.

the computation by $f\tilde{g}(v)$, since $\tilde{g}(v)$ requires the extra stage of computing $f\tilde{g}(v)$.

The name $\tilde{g}(v)$ *thus includes not itself* (which is manifestly impossible) *but a name for instructions for computing itself* (as a part of instructions for computing $f\tilde{g}(v)$). This is characteristic of self-referential constructions in logic and recursive function theory. For such constructions we must have not only names for certain objects but also names for instructions for computing names for those objects (see discussion below of the *Gödel substitution function*).

We are now in a position to give, in outline, a proof for Theorem XIV.

Theorem XIV *Let n be the recursive function constructed in Theorem II. There is a recursive g such that for any recursive f and for any recursive operator* Ψ, *if* $(\forall x)[\varphi_{f(x)} = \Psi(\varphi_x)]$ *and if z is an index for f, then* $\varphi_{g(z)}$ *is a total function, call it h, such that*

(i) $(\forall x)[\varphi_{h(x)} = \varphi_{f(x)}]$; *and*

(ii) $\varphi_{ng(z)}$ *is a minimal fixed point for the recursive operator* Ψ.

Proof. Let f and Ψ be given as in the assumptions and let z be an index for f. We describe how $g(z)$ is to be computed.

Any *finite* partial function φ is partial recursive. If D_u is single-valued, then instructions for computing $\{<x,y>|<x,y> \in D_u\}$ may be obtained uniformly from u. That is to say, there is a recursive function d such that $(\forall u)[D_u \text{ single-valued} \Rightarrow \tau(\varphi_{d(u)}) = D_u]$.

Define F as follows:

$$F = \{<<x,t>,u>|D_u \textit{ is single-valued } \& <x,t> \in \varphi_{fd(u)}\}.$$

F is evidently recursively enumerable with an index that depends uniformly on z. Let w be a recursive function such that $F = W_{w(z)}$. From basic properties of enumeration operators, we have that $\Phi_{w(z)}$ is an enumeration operator which defines Ψ. (This is proved in Exercise 11-42.) Now (see Corollary XII(a)), $w(z)$ determines a recursive function h such that $\varphi_{h(y)} = \Psi(\varphi_y)$. (Take $h = \lambda y[\tilde{\tilde{f}}(y,w(z))]$, where $\tilde{\tilde{f}}$ is the function for Corollary XII(a).) An index for h can be found uniformly from z. Let $g(z)$ be such an index for h.† Note that h has the following special property which f, in general, may not possess: for any y, the instructions $h(y)$ tell us to compute $\varphi_{h(y)}$ by putting $\tau(\varphi_y)$ through a certain enumeration operator, namely, $\Phi_{w(z)}$. More specifically, $h(y)$ $(= \tilde{\tilde{f}}(y,w(z)))$ instructs us as follows. "Given an input x, begin an enumeration of the ordered pairs of φ_y. Simultaneously, begin listing $W_{w(z)}$ and look for a member of the form $<<x,t>,u>$ such that $\tau^{-1}(D_u)$ is a subset of the ordered pairs of φ_y already listed. If and when this occurs, give t as output." We indicate this more briefly by: "given input x,

† The definition of $\tilde{\tilde{f}}$ ultimately traces back, via the second projection theorem, to the s_n^m functions. Similarly, a particular choice of g, tracing back to the s_n^m functions in a natural way, is assumed (but not described).

put the ordered pairs of φ_y through the enumeration operator $\Phi_{w(z)}$ and look for an output $\langle x,t\rangle$; if and when this occurs, take t as output."

It remains to show that $ng(z)$ is a minimal fixed point. From the proof of Theorem II, $ng(z) = \tilde{g}(v)$ where v is index for $h\tilde{g}$. From our preceding discussion, we have that

$$\tilde{g}(v) = \text{"}\textit{compute } h\tilde{g}(\mathbf{v})\textit{; if and when } h\tilde{g}(\mathbf{v}) \textit{ converges, do } h\tilde{g}(\mathbf{v})\text{."}$$

Now what are the instructions $h\tilde{g}(v)$? Informally stated, they are "given input x, enumerate the ordered pairs of $\varphi_{\tilde{g}(v)}$ and put them through $\Phi_{w(z)}$ and look for an output of the form $\langle x,t\rangle$. Take such a t as final output." Thus the *instructions* $\tilde{g}(v)$ tell us to find and follow the instructions $h\tilde{g}(v)$ and these latter in turn tell us to begin enumerating the ordered pairs of $\varphi_{\tilde{g}(v)}$ and simultaneously to put these ordered pairs through $\Phi_{w(z)}$. Instructions for enumerating $\varphi_{\tilde{g}(v)}$ thus require that $\varphi_{\tilde{g}(v)}$ itself be enumerated in a *subcomputation.* The situation is not contradictory since, in general, the subcomputation of $\varphi_{\tilde{g}(v)}$ will occur at a slower rate than the overall computation of $\varphi_{\tilde{g}(v)}$.†

Let $\langle x_0,t_0\rangle$, $\langle x_1,t_1\rangle$, . . . be the enumeration of $\varphi_{\tilde{g}(v)}$. By the instructions for $\varphi_{\tilde{g}(v)}$,

$$\langle x_0,t_0\rangle \in \Phi_{w(z)}(\emptyset),$$
$$\langle x_{n+1},t_{n+1}\rangle \in \Phi_{w(z)}(\{\langle x_0,t_0\rangle, \ldots, \langle x_n,t_n\rangle\}).$$

Hence, if

$$A_0 = \Phi_{w(z)}(\emptyset),$$

and if

$$A_{n+1} = \Phi_{w(z)}(A_n),$$

we have by monotonicity that $\langle x_n,t_n\rangle \in A_n$ for all n. Hence

$$\tau(\varphi_{\tilde{g}(v)}) \subset \bigcup_{n=0}^{\infty} A_n.$$

But by the construction of Theorem XI, $\bigcup_{n=0}^{\infty} A_n$ is the minimal fixed point for $\Phi_{w(z)}$; hence $\varphi_{\tilde{g}(v)}$ is contained in the minimal fixed point for Ψ. On the

† The computation for enumerating $\varphi_{\tilde{g}(v)}$ is analogous to a nontrivial sequence of integers which contains infinitely many proper subsequences identical with itself. For example, the sequence x_0, x_1, . . . determined by the recursion equations

$$x_{2n} = n^2,$$
$$x_{2n+1} = x_n$$

begins

0, 0, 1, 0, 4, 1, 9, 0, 16, 4, 25, 1, . . .

and has the subsequence structure

0	0	1	0	4	1	9	0	16	4	25	1
	0		0		1		0		4		1
			0				0				1
							0				.

other hand, by Theorem II, $\varphi_{\tilde{g}(v)}$ is a fixed point for $\mathbf{\Psi}$. Hence $\varphi_{\tilde{g}(v)}$ coincides with the minimal fixed point for $\mathbf{\Psi}$ and our proof is complete.☒

The following question remains open. Call the recursive function $\hat{n}$ a *fixed-point function* if, for all z,

$$\varphi_{\hat{n}(z)} = \begin{cases} \varphi_{\varphi_z(\hat{n}(z))}, & \text{if } \varphi_z(\hat{n}(z)) \text{ convergent;} \\ \text{everywhere divergent}, & \text{if } \varphi_z(\hat{n}(z)) \text{ divergent.} \end{cases}$$

Does Theorem XIV hold for every fixed-point function $\hat{n}$ (in place of n)? An affirmative answer might yield a more invariant proof for Theorem XIV.

The Gödel Substitution Function

Consider a logical system formulated within ordinary quantificational logic. Assume that numerals for all the nonnegative integers occur among the *terms* (i.e., expressions for individuals) of the system. In discussing substitutions, we denote the numeral of any integer x as $\mathbf{x}$. Thus if Fa is an expression with an unquantified variable a, $F\mathbf{x}$ will be the result of substituting the numeral of x for a in that expression. Assume a coding from all expressions of the system (including terms, and assertions with unquantified ("free") individual variables) onto N.

Define a function σ of two variables such that for any x and y,

$\sigma(x,y)$ = *the code number of the expression that results when the numeral for x is substituted at all occurrences of free variables (if any) in the expression whose code number is y.*

Since our coding of expressions is a coding in the sense of §1.10, σ is a recursive function. σ is called the *Gödel substitution function* for the given system and coding.

The Gödel substitution function is useful when the system is strong enough to allow assertions about σ within the system. The function can then be used to construct a variety of self-referential expressions. For example, assume that the system includes a symbol $\boldsymbol{\sigma}$ which is understood to denote σ. If a is an individual variable, and if x is the code number for the term $\boldsymbol{\sigma}(a,a)$, then $\boldsymbol{\sigma}(\mathbf{x},\mathbf{x})$ is an expression which denotes its own code number.†

† Examples of this phenomenon can be constructed in ordinary English. The phrase

"the result of substituting "25" in "x is a prime""

denotes, of course, the phrase

"25 *is a prime.*"

The phrase

"the result of substituting "the result of substituting "x" in "x"" in "the result of substituting "x" in "x"""

denotes itself, as can readily be computed.

More specifically, consider elementary arithmetic. Assume the usual concepts of *truth* and *falsity* for wffs (with no free variables), as described in §7.8. As is proved in Davis [1958], there is an expression $Mabcd$ in the free variables a, b, c, and d such that $M\mathbf{zxyw}$ is true if and only if P_z with input x yields output y in fewer than w steps (see §§14.4 and 14.7). Let s be an index for the recursive function $\lambda x[\sigma(x,x)]$. We give several examples of the construction of self-referential statements.

Example 1. Choose a fixed effective method for enumerating all logical consequences (under the rules of elementary logic) of the wffs occurring in any given recursively enumerable set W_x. That is to say, choose a recursive f such that for any x, $W_{f(x)}$ is the set of all logical consequences of wffs in W_x. Let k be an index for f. Let a set of wffs (call them *axioms*) W_p be given. To assert that a wff (with code number) y is provable from W_p is to assert that $y \in W_{f(p)}$, i.e., that $(\exists a)[(\exists b)M\mathbf{kp}ab \;\&\; (\exists c)(\exists d)Ma\mathbf{y}cd]$ is true. We abbreviate this last wff as $Pr_{\mathbf{p}}(\mathbf{y})$. If we had a symbol $\boldsymbol{\sigma}$ for the Gödel substitution function, we could proceed as follows. Let u be a code number for $\neg Pr_{\mathbf{p}}(\boldsymbol{\sigma}(b,b))$. Then $\neg Pr_{\mathbf{p}}(\boldsymbol{\sigma}(\mathbf{u},\mathbf{u}))$ asserts its own unprovability.

Although we have no symbol $\boldsymbol{\sigma}$ in our system, it is easy to carry out a parallel construction using s. Let u be a code number for

$$(\exists b)[(\exists d)M\mathbf{s}abd \;\&\; \neg Pr_{\mathbf{p}}(b)].$$

Then

$$(\exists b)[(\exists d)M\mathbf{su}bd \;\&\; \neg Pr_{\mathbf{p}}(b)]$$

is true if and only if it itself is unprovable from W_p. We have obtained a wff which asserts its own unprovability.

Example 2. Let $\neg Pr_{\mathbf{p}}(b)$ be as in Example 1. Let q be the code number of some fixed wff of the form $[F \;\&\; \neg F]$, for example, "$0 = 0 \;\&\; \neg 0 = 0$." Then $\neg Pr_{\mathbf{p}}(\mathbf{q})$ is true if and only if W_p is consistent. We abbreviate $\neg Pr_{\mathbf{p}}(\mathbf{q})$ as $Con(\mathbf{p})$. If we replace $\mathbf{p}$ by an individual variable c, we abbreviate the result as $Con(c)$. (If for all x, $\hat{c}(x)$ is a code number for $Con(\mathbf{x})$ then $\hat{c}$ is a recursive function of the kind considered in the second illustration in §11.4.) Can we get a wff which asserts its own consistency? By Theorem VIII, such a wff exists. It can be obtained by substitution-function methods as follows.

Let m be an index for a recursive function h such that $W_{h(x)} = \{x\}$ for all x. Let u be a code number for

$$(\exists b)[(\exists d)M\mathbf{s}abd \;\&\; (\exists c)[(\exists d)M\mathbf{m}bcd \;\&\; Con(c)]].$$

Then the wff

$$(\exists b)[(\exists d)M\mathbf{su}bd \;\&\; (\exists c)[(\exists d)M\mathbf{m}bcd \;\&\; Con(c)]]$$

is true if and only if it is itself, as a single axiom, consistent.†

† Our intuitive guide in this construction is: let u be a code number for "$\boldsymbol{\sigma}(a,a)$ *is consistent*"; then form "$\boldsymbol{\sigma}(\mathbf{u},\mathbf{u})$ *is consistent.*"

In this fashion, self-referential wffs of elementary arithmetic can be constructed in connection with a variety of syntactical properties.†‡ (For a general treatment, see Feferman [1960].)

The substitution function originates with Gödel, both as heuristic guide and as formal technique. When Peano's axioms are adopted as a fixed set of axioms (or when any set from which Peano's axioms are deducible is adopted), certain further results about the expression *Mabcd* can be obtained (see Exercise 7-64).§ From these results and from the wff in Example 1 above, Gödel's first incompleteness theorem in its usual form (with no reference to truth or falsity) can be derived. In Exercise 7-64, the substitution function was not explicitly used. It is used, however, in Gödel's original proof (see Exercise 11-44). In a similar way, the substitution function can be used to show that the notion *n is* (*a code number for*) *a true wff of elementary arithmetic* is not expressible within elementary arithmetic. This theorem is due to Tarski (see Exercise 11-45).

The recursion theorem is closely related to the Gödel substitution function. Define the recursive function $\hat{\sigma}$ by

$$\varphi_{\hat{\sigma}(x,y)} = \begin{cases} \varphi_{\varphi_x(y)}, & \text{if } \varphi_x(y) \text{ is defined;} \\ \text{everywhere undefined,} & \text{otherwise.} \end{cases}$$

Let n be an index for $\lambda x[f\hat{\sigma}(x,x)]$. Then

$$\varphi_{\hat{\sigma}(n,n)} = \varphi_{f\hat{\sigma}(n,n)}.$$

This puts the proof of the recursion theorem into a form closely analogous to applications of σ.

Indeed, the functions $s_n{}^m$ are a form of substitution function. $s_1{}^1(x,y)$, for example, is an index for the set of 1-ary instructions that result when y is inserted as first input in the 2-ary instructions with index x (see Exercises 11-4 and 11-5).

† Theorem VIII can also be approached in this way. The recursion-theorem approach taken in §11.4, however, makes fewer assumptions about the system.

‡ Examples parallel to the above can also be constructed in ordinary English. Consider the assertion

> *"the following is not provable: the result of substituting "the following is not provable: the result of substituting "x" in "x"" in "the following is not provable: the result of substituting "x" in "x"""*

This assertion asserts its own unprovability.

In effect, what Gödel does in proving his first incompleteness theorem is to take this example and show that the notions of substitution and provability (as well as names for expressions) can be expressed within elementary arithmetic. (He also, of course, eliminates assumptions concerning truth, as described in Exercise 7-64.)

§ The facts summarized in Exercise 7-64 can also be proved when a certain finite set of axioms (weaker than Peano's axioms) is adopted. This useful result is due to Tarski, Mostowski, and R. M. Robinson [1953].

§11.7 ORDINAL NOTATIONS†

Church and Kleene initiated the general theory of systems of notation for ordinal numbers (Church and Kleene [1937], Church [1938], and Kleene [1938]). In §§11.7 and 11.8 we give a few chief results of this theory. The recursion theorem is fundamental in the development of these results.

We assume the concept of ordinal number and the more familiar notions and notations associated with this concept. (These matters are reviewed in Exercises 11-46 to 11-53.) Much of the traditional theory of ordinals can be formulated as a theory of *notations* for ordinals. For example, consider ordinals expressible by exponential polynomials in ω. These form an important and inclusive class. In a natural way, results about this class can be formulated as results about corresponding exponential polynomial expressions.

A study of various familiar systems of notation suggests the following general definition.

Definition A *system of notation* S is a mapping ν_S from a set of integers D_S onto a segment of the ordinal numbers such that

(i) there exists a partial recursive function k_S such that‡

$$\begin{aligned} &\nu_S(x) = 0 \Rightarrow k_S(x) = 0,\\ &\nu_S(x) \text{ is a successor} \Rightarrow k_S(x) = 1,\\ &\nu_S(x) \text{ is a limit} \Rightarrow k_S(x) = 2; \end{aligned}$$

(ii) there exists a partial recursive function p_S such that

$$\nu_S(x) \text{ a successor} \Rightarrow [p_S(x) \text{ convergent } \& \ \nu_S(x) = \nu_S(p_S(x)) + 1];$$

(iii) there exists a partial recursive function q_S such that

$$\nu_S(x) \text{ a limit} \Rightarrow [q_S(x) \text{ convergent } \& \ \varphi_{q_S(x)} \text{ total } \& \ \{\nu_S(\varphi_{q_S(x)}(n))\}_{n=0}^{n=\infty} \text{ is}$$
$$\text{an increasing sequence with } \nu_S(x) \text{ as limit}].$$

This definition is due to Kleene.§ The members of D_S are called the *notations* of S. All the usual notational schemes for segments of the denumerable ordinals become (after coding into N) systems in this sense. Note (by induction) that if a system S gives a notation to an ordinal α, then S gives notations to all ordinals less than α.

† Except where otherwise noted, material in later chapters will not depend on §§11.7 and 11.8.

‡ We are using Greek letters to denote ordinal numbers. In order to avoid confusion, we depart from our usual convention that partial functions are denoted by Greek letters. The condition on k_S requires that k_S be defined in a certain way on D_S. The convergence or divergence of k_S on $\bar{D}_S$ is not specified. Similarly for the partial functions p_S and q_S.

§ Kleene calls such systems of notation *r-systems*.

Definitions A system of notation S is
(*a*) *univalent* if ν_S is one-one;
(*b*) *recursive* if D_S is recursive;
(*c*) *recursively related* if

$$R_S = \{<x,y> | x \in D_S \ \& \ y \in D_S \ \& \ \nu_S(x) \leq \nu_S(y)\}$$

is recursive.

Since $D_S = \{x| <x,x> \in R_S\}$, a recursively related system must be recursive. In Exercise 11-54 we see that a univalent recursive system must be recursively related. The system of exponential polynomials mentioned above is recursively related.

Definition α is a *constructive ordinal* if there is a system of notation which assigns at least one notation to α.

Constructive ordinals were first defined and studied by Church. (Church's definition is somewhat different from that given here.)

Every notation system covers at most a denumerable segment of the ordinals; hence every constructive ordinal is denumerable. We shall see that not every denumerable ordinal is constructive.

Definition A system of notation S is *maximal* if S gives a notation to every constructive ordinal.

An apparent obstacle to the study of maximal systems of notation is presented in the following theorem and corollary.

Theorem XV *Let S be a recursively related system of notation, and let α be the least ordinal not given a notation by S. Then there exists a recursively related system S' which gives a notation to α.*

Proof. We assume α is a limit. The proof is similar (and easier) if α is a successor. (If $\alpha = 0$, the proof is trivial.)

Define S' as follows:

$$D_{S'} = \{x | x = 1 \text{ or } (\exists y)[x = 2y \ \& \ y \in D_S]\}.$$

$$\nu_{S'}(x) = \begin{cases} \nu_S(y), & \text{if } x = 2y \ \& \ y \in D_S; \\ \alpha, & \text{if } x = 1. \end{cases}$$

$$k_{S'}(x) = \begin{cases} k_S(y), & \text{if } x = 2y; \\ 2, & \text{if } x = 1.\dagger \end{cases}$$

$$p_{S'}(x) = 2p_S(y), \quad \text{if } x = 2y.$$

$$q_{S'}(x) = \begin{cases} \text{an index for } 2\varphi_{q_{S'}(y)}, & \text{if } x = 2y; \\ m, & \text{if } x = 1, \text{ where } m \text{ is obtained as follows. Let } n_0 \text{ be a notation for 0 in } S. \text{ Define } \psi \text{ so that} \\ & \quad \psi(0) = n_0, \\ & \quad \psi(j+1) = \mu u[u \in D_S \ \& \ \nu_S(u) \not\leq \nu_S(\psi(j))]. \\ & \text{Take } m \text{ to be an index of } 2\psi. \end{cases}$$

† The clause "divergent otherwise" is tacitly intended in each of the definitions for $k_{S'}$, $p_{S'}$, and $q_{S'}$.

Since S is recursively related and α is a limit, ψ is a (total) recursive function, and $\{\nu_{S'}(2\psi(n))\}_{n=0}^{n=\infty}$ is an increasing sequence with limit α.

We thus have a notation system S' which gives the notation 1 to α. Now

$$R_{S'} = \{<x,y>|[x = 2x' \;\&\; y = 2y' \;\&\; <x',y'> \in R_S] \text{ or } [x = 2x' \;\& \\ y = 1 \;\&\; <x',x'> \in R_S] \text{ or } [x = 1 \;\&\; y = 1]\},$$

and this is evidently recursive. Hence S' is recursively related.☒

Corollary XV *There is no maximal recursively related system of notations.*
Proof. Immediate.☒

Definition A system of notation S is *universal* if for any system S', there is a partial recursive function φ, mapping $D_{S'}$ into D_S, such that $x \in D_{S'} \Rightarrow \nu_{S'}(x) \leq \nu_S(\varphi(x))$.

Obviously, a universal system must be maximal.

If we drop the assumption of recursive relatedness (made in Theorem XV), it is possible to show that maximal systems exist, and, indeed, that universal systems exist. This remarkable result is due to Kleene and makes use of the recursion theorem.

Definition The system S_1 is defined as follows:
0 receives the notation 1.
Assume all ordinals $< \gamma$ have received their notations, then
(i) If $\gamma = \beta + 1$, γ receives $\{2^x | x$ *is a notation for* $\beta\}$ as notations;
(ii) If γ is a limit, γ receives $\{3 \cdot 5^y | \{\varphi_y(n)\}_{n=0}^{n=\infty}$ *are notations for an increasing sequence of ordinals with limit* $\gamma\}$ as notations.

These conditions define ν_{S_1} and D_{S_1} by transfinite induction. Then k_{S_1}, p_{S_1}, and q_{S_1} are defined by

$$\begin{aligned} k_{S_1}(1) &= 0, \\ k_{S_1}(2^x) &= 1, \\ k_{S_1}(3 \cdot 5^y) &= 2,\dagger \\ p_{S_1}(2^x) &= x, \\ q_{S_1}(3 \cdot 5^y) &= y. \end{aligned}$$

This completes the definition and shows that S_1 is a system. Let α be the least ordinal not receiving a notation; α must be a limit and no y can exist such that $\{\nu_{S_1}(\varphi_y(n))\}_{n=0}^{n=\infty}$ is an increasing sequence with α as limit.‡

Theorem XVI (Kleene) *S_1 is a universal system of notation. Indeed, given any system S, there is a partial recursive φ, mapping D_S into D_{S_1}, such that $x \in D_S \Rightarrow \nu_S(x) = \nu_{S_1}(\varphi(x))$.*

† Again the clause "divergent otherwise" is tacit.

‡ Kleene [1938] defines S_1 in a slightly different way. He allows notations $3 \cdot 5^y$ so that φ_y defined on 1, 2, 2^2, 2^{2^2}, . . . yields an increasing sequence of notations. (This was to permit a projected incorporation of S_1 into generalized systems of notation intended as analogues to higher ordinal-number classes.) The factor 3 in $3 \cdot 5^y$ also occurs for this reason. We have kept the factor 3 for historical reasons, although 5^y alone would serve our purposes as well. There is an ordinal-preserving recursive permutation between Kleene's S_1 and the S_1 defined here.

Proof. Let a system S be given with associated partial recursive functions k_S, p_S, and q_S. For any z, define a partial recursive ψ as follows:

$$\psi(x) = \begin{cases} 1, & \text{if } k_S(x) = 0; \\ 2^{\varphi_z p_S(x)}, & \text{if } k_S(x) = 1; \\ 3 \cdot 5^{y'}, & \text{if } k_S(x) = 2 \text{ and } q_S(x) \text{ is convergent, where } y' \text{ is an} \\ & \quad \text{index for } \lambda n[\varphi_z \varphi_{q_S(x)}(n)]; \\ \text{divergent}, & \text{otherwise.} \end{cases}$$

An index for ψ depends uniformly on z; hence there is a recursive f such that $\varphi_{f(z)} = \psi$. Applying Theorem I, we get an m such that $\varphi_{f(m)} = \varphi_m$, i.e., such that

$$\varphi_m(x) = \begin{cases} 1, & \text{if } k_S(x) = 0; \\ 2^{\varphi_m p_S(x)}, & \text{if } k_S(x) = 1; \\ 3 \cdot 5^{y'}, & \text{if } k_S(x) = 2 \text{ and } q_S(x) \text{ is convergent, where } y' \text{ is an} \\ & \quad \text{index for } \lambda n[\varphi_m \varphi_{q_S(x)}(n)]; \\ \text{divergent}, & \text{otherwise.} \end{cases}$$

φ_m is evidently the desired partial recursive φ. For, φ_m is defined on all of D_S, and $x \in D_S \Rightarrow \nu_{S_1}(\varphi_m(x)) = \nu_S(x)$ (otherwise take the least ordinal having an S-notation where either φ_m is undefined or this equation fails; a contradiction is immediate).⊠

Corollary XVI *There exist denumerable ordinals which are not constructive.*

Proof. The ordinals given notations by S_1 form a denumerable segment. Hence the least ordinal not in this segment is denumerable.⊠

Next, we define a universal system with certain additional useful properties. Our definition will impose, incidentally, a certain partial ordering on the set of notations defined.

Definition The system O is defined as follows. We define both ν_O and a partial ordering $<_O$ on D_O.

0 receives notation 1.

Assume all ordinals $<\gamma$ have received their notations, and assume that $<_O$ has been defined on these notations.

(i) If $\gamma = \beta + 1$, then for each x such that β has x as a notation, γ receives 2^x as a notation; and the ordered pairs $<z,2^x>$ are added to the relation $<_O$ for all z for which either $z = x$ or $<z,x>$ is already in $<_O$.

(ii) If γ is a limit, then for each y such that $\{\varphi_y(n)\}_{n=0}^{n=\infty}$ are notations for an increasing sequence of ordinals with limit γ *and* such that $(\forall i)(\forall j)[i < j \Rightarrow <\varphi_y(i),\varphi_y(j)>$ is already in $<_O]$, γ receives $3 \cdot 5^y$ as a notation; and the ordered pairs $<z,3 \cdot 5^y>$ are added to the relation $<_O$ for all z for which $(\exists n)[<z,\varphi_y(n)>$ is already in $<_O]$.

The partial recursive functions k_O, p_O, and q_O are identical with the partial recursive functions k_{S_1}, p_{S_1}, and q_{S_1}. This completes the definition of O.†

The set D_O will also be denoted as O. The notation $|x|_O$ will be used for $\nu_O(x)$. If $<x,y> \in <_O$, we write $x <_O y$. Note that $|x|_O < |y|_O$ does not in general imply $x <_O y$ (see diagram below).

The system O is clearly a subsystem of S_1. It has further useful properties. Given any $y \in O$, $\{x|x <_O y\}$ constitutes a univalent system of notation. Furthermore $\{x|x <_O y\}$ is recursively enumerable uniformly in y; i.e., there is a recursive f such that $(\forall y)[y \in O \Rightarrow W_{f(y)} = \{x|x <_O y\}]$ (see Exercise 11-55). The partial ordering $<_O$ is an infinitely ramifying tree with the structure suggested in the following diagram:

$$1 <_O 2 <_O 2^2 <_O \cdots \begin{cases} 3 \cdot 5^{y_1} <_O 2^{3 \cdot 5^{y_1}} <_O 2^{2^{3 \cdot 5^{y_1}}} <_O \cdots \begin{cases} \cdots \\ \cdots \end{cases} \\ 3 \cdot 5^{y_2} <_O 2^{3 \cdot 5^{y_2}} <_O \cdots \qquad \begin{cases} \\ \end{cases} \\ \cdots \\ \cdots \end{cases},$$

where the successive ramifications occur at notations for successive limit numbers. In this diagram $3 \cdot 5^{y_1}$ and $3 \cdot 5^{y_2}$ are two (out of infinitely many) notations for ω. Note that if $3 \cdot 5^{y_1} <_O z$, then z and $3 \cdot 5^{y_2}$ are incomparable with respect to $<_O$. A definition of D_O and $<_O$ in terms of integers and sets of integers (without reference to ordinals) can be given without difficulty (see Exercise 16-6).

The following result gives another application of the recursion theorem. It concerns a recursive function of two variables. We denote the function as $+_O$, and we write $+_O(x,y)$ as $x +_O y$.

Theorem XVII **(Kleene)** *There exists a recursive function $+_O$ of two variables such that for all x and y in O,*

(i) $x +_O y \in O$;
(ii) $|x +_O y|_O = |x|_O + |y|_O$;
(iii) $y \neq 1 \Rightarrow x <_O x +_O y$.

Proof. Let x and z be given, define ψ by

$$\psi(y) = \begin{cases} x, & \text{if } y = 1; \\ 2^{\varphi_z(u)}, & \text{if } y = 2^u; \\ 3 \cdot 5^{v'} & \text{if } y = 3 \cdot 5^v, \text{ where } v' \text{ is an index for } \lambda n[\varphi_z\varphi_v(n)]; \\ 0, & \text{otherwise.} \end{cases}$$

Then ψ is partial recursive and an index for ψ can be obtained uniformly from x and z. That is to say, there is a recursive f such that

$$\psi = \varphi_{f(z,x)}.$$

† The preceding footnote applies to O as well as to S_1.

Applying Theorem III, we obtain a recursive function n such that

$$\varphi_{n(x)} = \varphi_{f(n(x),x)}.$$

We now take

$$+_O = \lambda xy[\varphi_{n(x)}(y)].$$

Then (i), (ii), and (iii) must hold for $+_O$. For, otherwise, take a least x (in ordinal position), and for that x a least y (in ordinal position), such that either (i), (ii), or (iii) fails; a contradiction is immediate. In Exercise 11-57, we note that $+_O$ is a total function.☒

The function $+_O$ can be used, along with another application of the recursion theorem, to show that the system O is universal (and hence maximal).

***Theorem XVIII* (Kleene)** *O is a universal system of notation.*

Proof. Let S be any system of notation. Given z, define ψ as follows:

$$\psi(x) = \begin{cases} 1, & \text{if } k_S(x) = 0; \\ 2^{\varphi_z p_S(x)}, & \text{if } k_S(x) = 1; \\ 3 \cdot 5^{y'}, & \text{if } k_S(x) = 2 \text{ and } q_S(x) \text{ is convergent, where } y' \text{ is an} \\ & \quad \text{index for } \eta, \text{ and } \eta \text{ is defined by} \\ & \qquad \eta(0) = \varphi_z\varphi_{q_S(x)}(0), \\ & \qquad \eta(n+1) = \eta(n) +_O \varphi_z\varphi_{q_S(x)}(n+1); \\ \text{divergent}, & \text{otherwise.} \end{cases}$$

Taking a recursive f such that $\psi = \varphi_{f(z)}$, and applying Theorem I, we obtain a partial recursive $\varphi_n = \varphi_{f(n)}$ which will serve as the φ required by the definition of universality. To show that φ_n has the desired properties, assume otherwise, take the least ordinal position at which difficulty occurs, and obtain an immediate contradiction.☒

Corollary XVIII(a) *The system O is maximal; i.e., it associates a notation with every constructive ordinal.*

Proof. Immediate.☒

The system O derives much of its usefulness in later applications from the following corollary, which shows that the theorem holds in a stronger form for univalent systems.

Corollary XVIII(b) *Given any univalent system S, there is a partial recursive φ mapping D_S into O such that*

(i) $x \in D_S \Rightarrow \nu_S(x) = |\varphi(x)|_O$;

(ii) $x,y \in D_S \Rightarrow [x < y \Leftrightarrow \varphi(x) <_O \varphi(y)]$.

Proof. Construction is as in the theorem, except that η is defined

$$\eta(0) = \varphi_z\varphi_{q_S(x)}(0);$$
$$\eta(n+1) = \varphi_z\varphi_{q_S(x)}(n+1).☒$$

§11.8 CONSTRUCTIVE ORDINALS

We present several further properties of the constructive ordinals. Although there can be no maximal, recursively related system of notation, the following theorem holds.

Theorem XIX *For every constructive ordinal, there is a recursively related, univalent system assigning a notation to that ordinal.*

Proof. Let α be a constructive ordinal. By Corollary XVIII, there is a $z \in O$ such that $|z|_O = \alpha$. Let $A_z = \{y|y <_O z \text{ or } y = z\}$. A_z is recursively enumerable (see Exercise 11-55); A_z is linearly ordered by $<_O$, and each ordinal $\leq \alpha$ has one and only one notation in A_z. If A_z is finite, α is finite and the result is immediate. If A_z is infinite, let f be a one-one recursive function having A_z as range. Define a system of notations S as follows:

$$\begin{aligned} D_S &= N, \\ \nu_S(x) &= |f(x)|_O, \\ k_S(x) &= k_O(f(x)), \\ p_S(x) &= f^{-1}p_Of(x), \\ q_S(x) &= y', \end{aligned}$$

where y' is an index for $\lambda n[f^{-1}\varphi_{q_Of(x)}(n)]$. S is recursively related since, for any x and y, either $f(x) = f(y)$, $f(x) <_O f(y)$, or $f(y) <_O f(x)$. As $\{u|u <_O v\}$ is recursively enumerable uniformly in v, we can effectively test which of these three cases occurs.

S is our desired recursively related, univalent system, and it assigns the notation $f^{-1}(z)$ to α.☒

Definition α is a *recursive ordinal* if there exists a relation R such that: (i) R is a well-ordering (of some set of integers); (ii) R is recursive; and (iii) the well-ordering given by R is order-isomorphic to α.

Corollary XIX *Every constructive ordinal is recursive.*

Proof. The univalent system of notations constructed in the proof of Theorem XIX provides the desired recursive well-ordering.☒

The converse of Corollary XIX also holds (Markwald [1954], Spector [1955]).

Theorem XX (Markwald, Spector) *Every recursive ordinal is constructive.*

Proof. Let α be a recursive ordinal. Let R be a recursive well-ordering isomorphic to α. Note that for every u in the ordering, $\{z|\langle z,u\rangle \in R\}$ is recursively enumerable uniformly in u. We define a new linear-ordering relation as follows:

$$\hat{R} = \{\langle\langle x_1,y_1\rangle,\langle x_2,y_2\rangle\rangle|\langle x_1,x_2\rangle \in R \;\&\; [x_1 = x_2 \Rightarrow y_1 \leq y_2]\}.$$

$\hat{R}$ is evidently recursive. Its ordering is evidently order-isomorphic to $\beta = \omega \cdot \alpha$. Let m be the first integer in the ordering of R. $\hat{R}$ provides a system of notations S as follows:

$$D_S = \{x | \langle x,x\rangle \in \hat{R}\},$$

$$k_S(x) = \begin{cases} 0, & \text{if } x = \langle m,0\rangle; \\ 1, & \text{if } x = \langle u,v\rangle \text{ and } v > 0; \\ 2, & \text{if } x = \langle u,0\rangle \text{ and } u \neq m. \end{cases}$$

$$p_S(x) = \langle u, v - 1\rangle, \quad \text{if } x = \langle u,v\rangle \ \&\ k_S(x) = 1.$$

$q_S(x) = y$, if $x = \langle u,0\rangle \ \&\ u \neq m$, where y is an index for a partial recursive h defined as follows: $h(0) = \langle m,0\rangle$; and if $h(n) = \langle s,t\rangle$, then

$$h(n+1) = \begin{cases} \langle w,0\rangle, & \text{where } w = \mu x[x \textit{ occurs by the } n\text{th } \textit{step in the enumeration of } \{z|\langle z,u\rangle \in R \ \&\ z \neq u\} \textit{ and } s \textit{ occurs by the } n\text{th } \textit{step in the enumeration of } \{z|\langle z,x\rangle \in R \ \&\ z \neq x\}], \\ \langle s, t+1\rangle, & \text{if no such } x \text{ exists.} \end{cases}$$

These definitions evidently determine a mapping ν_S uniquely, and we have a recursively related system assigning notations to all ordinals $< \omega \cdot \alpha$. By Theorem XV, a system exists assigning a notation to $\omega \cdot \alpha$. Hence $\omega \cdot \alpha$ is constructive. Since $\alpha \leq \omega \cdot \alpha$, α is constructive.☒

Corollary XX *An ordinal is recursive if and only if it is constructive.*

Proof. Immediate.☒

The constructive ordinals can be viewed as a recursive analogue to the second number class. Ordinals in the second number class may be characterized either by rules of generation or as representatives of denumerable well-orderings (see Exercise 11-46). The definitions of constructive and of recursive ordinals are, respectively, analogues to these two set-theoretic characterizations. Corollary XX is a recursive analogue to the result, in Exercise 11-46, that the two set-theoretic characterizations yield the same class of ordinals.

The constructive ordinals and the universal system O occur frequently in the literature of recursive function theory and logic. The recursion theorem is a useful, versatile, and fundamental tool in applications of the constructive ordinals and of O. For a final illustration, let ψ be a *one-one* productive partial function for a productive set, and consider the sets S_α and B defined as in Exercise 7-46.

Theorem XXI (Parikh) **(a)** *For every constructive* β, $B \neq \bigcup_{\alpha \leq \beta} S_\alpha$.

(b) $B = \bigcup_{\alpha \text{ constructive}} S_\alpha$.

Proof. Let q be an index for $\emptyset$. Let f be a recursive function such that $(\forall y)[y \in O \Rightarrow W_{f(y)} = \{x|x <_O y\}]$. Use the recursion theorem to obtain a partial recursive φ' mapping O into B, where

$$\varphi'(1) = \psi(q);$$
$$\varphi'(2^x) = \psi(y),$$

where y is an r.e. index for $\varphi'(W_{f(x)})$;

$$\varphi'(3 \cdot 5^x) = \psi(y),$$

where y is an r.e. index for $\varphi'(W_{f(3 \cdot 5^x)})$. It follows, by a simple inductive argument, that for every $x \in O$, $\varphi'(x) \in S_{|x|_O+1} - S_{|x|_O}$ and $\varphi'(x) \in B$. This proves (*a*).

Next we use the recursion theorem to define a partial recursive φ'' mapping B into O such that if $x \in S_{\alpha+1} - S_\alpha$, then $\alpha < |\varphi''(x)|_O$. This will prove (*b*). We define φ'' as follows:

$$\varphi''(x) = 3 \cdot 5^y,$$

where y is an index for a partial recursive η to be computed as follows. Compute $\psi^{-1}(x)$. If and when this converges, enumerate $W_{\psi^{-1}(x)}$. Set $\eta(0) = 1$,

$$\eta(n+1) = \begin{cases} \eta(n) +_O \varphi''(m), & \text{if } m \text{ is a new member of } W_{\psi^{-1}(x)} \text{ occurring on the } n\text{th step in the enumeration of } W_{\psi^{-1}(x)}; \\ \eta(n) +_O 2, & \text{if no such new member occurs on the } n\text{th step.} \end{cases}$$

By a simple inductive argument, φ'' has the desired properties. This completes the proof. ☒

In Exercise 11-63, we note that any such B must be recursively isomorphic to O.

§11.9 EXERCISES

§*11.2*

△*11-1.* (*a*) Show that there is a recursive function f whose set of fixed-point values is not recursively enumerable.

(*b*) Show that if the set of fixed-point values for f is recursive, then $\{\varphi_x|\varphi_x = \varphi_{f(x)}\}$ includes all partial recursive functions. (*Hint:* Show that otherwise a recursive function with no fixed points can be found.)

11-2. State and prove Theorem II in a form that applies to all φ_z, total or not.

11-3. Theorem I can be relativized by (i) taking f to be recursive in A, or (ii) putting φ^A for φ, or (iii) doing both. Which of these relativized versions hold? Prove or give counterexamples.

11-4. Let ψ be any partial recursive function of two variables.

(i) Show that there is an n such that $\varphi_n = \lambda y[\psi(n,y)]$.

(ii) Show that result (i) is equivalent to Theorem I.

(iii) Let v be a Gödel number for $\lambda xy[\psi(s_1^1(x,x),y)]$. Show that n in (i) can be taken to be $s_1^1(v,v)$.

Part (i) is Kleene's version of Theorem I, and (iii) is his proof.

11-5. Let f be a recursive function. Let v be a Gödel number for $\lambda xy[\varphi_{fs_1^1(x,x)}(y)]$. Set $n = s_1^1(v,v)$. Show that $\varphi_n = \varphi_{f(n)}$. (This presents our proof for Theorem I in a different way.)

11-6. Show that the following is directly equivalent to Corollary I. *For every recursively enumerable A, there exists an n such that* $W_n = \{x| <x,n> \in A\}$.

11-7. Use Theorems 5-IX and 5-XVI to show that Corollary I directly yields Theorem I.

11-8. Prove or disprove the following statement. Let f be a recursive function such that for all x, φ_x total $\Rightarrow \varphi_{f(x)}$ total; then there exists an n such that $\varphi_n = \varphi_{f(n)}$ and φ_n is total.

11-9. (*a*) In each of the following cases, prove or disprove the existence of an m with the asserted property.

(i) $W_m = \{m^2\}$.

(ii) $W_m = \{10^m\}$.

(iii) $W_m = N - \{m\}$.

(iv) $W_m = \{x|x$ *is a quadratic residue mod* $m\}$.

(v) $W_m = \{x|\varphi_m(x)$ *is divergent*$\}$.

(vi) $\emptyset \neq W_m = \{x|(\exists y)[x = 2y \;\&\; \varphi_m(y)$ *convergent*$]\}$.

(vii) $W_m = \{3\} \cup \{x|(\exists y)[x = 2y \;\&\; \varphi_m(y)$ *convergent*$]\}$.

(*b*) Prove or disprove the existence of a recursive function f such that

$$W_{f(x)} = \{f(x) + x\}.$$

11-10. (G. C. Wolpin). A short and simple proof of the theorem in Exercise 2-39(*a*) can be obtained from Theorem I.

(*a*) Let $\mathcal{C}$ be any class of partial recursive functions. Let $A = \{x|\varphi_x \in \mathcal{C}\}$. Deduce from Theorem I that $A \not\leq_m \bar{A}$.

(*b*) Deduce the theorem of Exercise 2-39(*a*).

11-11. Use Theorem I and Theorem 5-IV to prove the last part of Theorem 5-XIV (on closure under complementation). (*Hint:* Assume the contrary and find an n such that $W_n \supset \bar{W}_n$ and $W_n \neq N$.)

$\triangle$***11-12.*** Let $\psi_0, \psi_1, \ldots$ be a sequence of partial recursive functions. Assume that recursive functions h and g exist such that $\psi_{h(x)} = \varphi_x$ and $\varphi_{g(x)} = \psi_x$; i.e., that the sequence $\psi_0, \psi_1, \ldots$ yields an *acceptable* numbering in the sense of Exercise 2-10. Show that the recursion theorem holds for this numbering; i.e., show that for any recursive f there is an n such that $\psi_n = \psi_{f(n)}$. (*Hint:* Use Theorem I to get an m such that $\varphi_m = \varphi_{gfh(m)}$.)

§*11.3*

11-13. Carry through the proof of Theorem VI with a partial productive function ψ in place of the productive function h.

11-14. Let A and B be an effectively inseparable disjoint pair of recursively enumerable sets. Show that every recursively enumerable set C, such that $A \subset C \subset \bar{B}$, must be 1-complete.

11-15. Show, as a corollary to Theorem V, that for all A, A productive $\Leftrightarrow \bar{K} \leq_1 A$ $\Leftrightarrow \bar{K} \leq_m A$. (*Hint:* See Theorem 7-V(*b*).)

$\triangle$***11-16.*** (Kreider). (*a*) Define A to be *unit-productive* if there is a recursive f such that for all x, $[W_x \subset A \;\&\; W_x$ contains at most one member$] \Rightarrow f(x) \in A - W_x$. Show: A is unit productive $\Leftrightarrow A$ is productive.

(*b*) Do parts (*d*) and (*e*) of Exercise 7-39.

11-17. (*a*) In the proof of Theorem VI, show that $center_{hn}\ A \subset h(\{x|W_x = \emptyset\})$ (see Exercise 7-42 for definition of $center_{hn}\ A$).

(*b*) Use Theorem IV to show that A has infinitely many pairwise disjoint *centers* (as the productive function is varied) contained in $h(\{x|W_x = \emptyset\})$.

△*11-18.* (Shoenfield). See Exercises 8-26 and 8-27. Show: A is q-creative $\Rightarrow A$ is q-complete. (*Hint:* A proof paralleling the proof of Theorem V can be made; or we can show that a set is q-creative if and only if its q-cylinder is creative and then use Theoem V directly.) (q-creativeness was also considered in Exercise 8-44. Yates has shown the existence of semicreative sets not T-complete, and hence that there are semicreative sets which are not q-creative.)

11-19. Define: A is q-*productive* if there is a partial recursive ψ such that for all x, $W_x \subset A \Rightarrow [\psi(x)$ convergent & $D_{\psi(x)} \subset A$ & $D_{\psi(x)} \not\subset W_x]$. Show that the partial recursive function can be replaced by a recursive function in this definition.

△*11-20.* Define an appropriate version of *complete* q-*productiveness.* Prove an analogue to Theorem VI.

△*11-21.* Define: A is T-*productive* if there is a recursive g such that for all x,

$$(\exists y)[y \in W_{g(x)} \ \&\ D_y \subset A] \Leftrightarrow \neg (\exists y)[y \in W_{g(x)} \ \&\ D_y \subset W_x].$$

Define: A is T-*creative* if A is recursively enumerable and $\bar{A}$ is T-productive.

Show: A is T-creative $\Leftrightarrow$ A is T-complete. (*Hint:* To show $\Leftarrow$, use reducibility from K and see Exercise 9-22; to show $\Rightarrow$, use the recursion theorem to obtain, for any recursively enumerable B, a recursive function n such that

$$W_{n(y)} = \begin{cases} \{x|(\exists z)[z \in W_{gn(y)} \ \&\ x \in D_z]\}, & \text{if } y \in B; \\ \emptyset, & \text{if } y \notin B, \end{cases}$$

and use Exercise 9-22.)

△*11-22.* Show that for every *universal* partial function (in the sense of §4.5) ψ there is an m such that $\psi(m) = m$.

§*11.4*

11-23. Consider the following axioms (Zermelo-Fraenkel) for set theory (Fraenkel [1928]; see also Quine [1963] and Part I of Bernays and Fraenkel [1958]).

(i) $(\forall x)(\forall y)[(\forall z)[z \in x \Leftrightarrow z \in y] \Rightarrow x = y]$ (extensionality).

(ii) $(\forall x)(\forall y)(\exists z)(\forall u)[u \in z \Leftrightarrow [u = x \vee u = y]]$ (pairing).

(iii) $(\forall x)(\exists z)(\forall u)[u \in z \Leftrightarrow (\exists y)[u \in y \ \&\ y \in x]]$ (sum).

(iv) $(\forall x)(\exists z)(\forall u)[u \in z \Leftrightarrow (\forall y)[y \in u \Rightarrow y \in x]]$ (power).

(v) $(\forall x)(\exists z)(\forall u)[u \in z \Leftrightarrow [u \in x \ \&\ S(u)]]$ (comprehension).

(vi) $(\forall u)(\forall v)(\forall w)[[R(w,u) \ \&\ R(w,v)] \Rightarrow u = v] \Rightarrow (\forall x)(\exists y)(\forall z)[z \in y \Leftrightarrow (\exists t)[t \in x \ \&\ R(t,z)]]$ (replacement).

(vii) $(\forall x)(\exists z)[x \in z \ \&\ (\forall u)[u \in z \Rightarrow (\exists v)[v \in z \ \&\ (\forall y)[y \in v \Leftrightarrow y = u]]]]$ (infinity).

(viii) $(\forall x)[(\forall u)(\forall v)[[u \in x \ \&\ v \in x \ \&\ (\exists r)[r \in u \ \&\ r \in v]] \Rightarrow u = v] \Rightarrow (\exists y)(\forall z)[[z \in x \ \&\ (\exists w)[w \in z]] \Rightarrow (\exists t)(\forall s)[[s \in y \ \&\ s \in z] \Leftrightarrow s = t]]]$ (choice).

(ix) $(\exists x)S(x) \Rightarrow (\exists y)[S(y) \ \&\ (\forall z)[z \in y \Rightarrow \neg S(z)]]$ (foundation).

Show that for the relation E $(= \{\langle x,y \rangle | x \in W_y\})$, assertions (ii), (iii), (vii), and (viii) hold, but (i) and (iv) fail. Show that (v), (vi), and (ix) do not hold (in general) as S and R vary over all sets and relations. Show that if S and R are restricted to be recursively enumerable, then (v) and (vi) hold, but (ix) does not.

11-24. Use the recursion theorem to show that there exist m and n such that $m \neq n$, $m \in W_n$, and $n \in W_m$.

△*11-25.* (*a*) Given any recursive h, show that there exists an infinite sequence of distinct integers $m_0, m_1, \ldots$ such that for all i, $W_{m_i} = \{h(m_{i+1})\}$. (*Hint:* Modify the proof of Theorem VII.)

(*b*) Define *full* with respect to ψ as in Exercise 7-45. Show that there are nondenumerably many productive sets which are full with respect to the identity function. (*Hint:* Use Theorem VII and then a construction similar to Exercise 7-46.)

(*c*) Let h be any one-one recursive function. Show that there are nondenumerably many productive sets which are full with respect to h. (*Hint:* Use (*a*).)

11-26. Explain why no axiomatizable formal theory can hope to include all true and no false assertions of the form *x and y are r.e. indices for the same set.* (Assume that the Gödel number for this assertion depends uniformly on x and y.)

11-27. Take f and $\hat{c}$ as in the proof of Theorem VIII. Assume that $\hat{c}$ is one-one and that for all x, $f(x) \neq x$. Under what circumstances can there be an m such that $W_{ff(m)} = W_{f(m)}$? (*Comment:* In specific cases, $\hat{c}$ will almost invariably be one-one, and f (constructed by way of Theorem 5-XIII) will almost invariably have the property that $(\forall x)[f(x) \neq x]$.)

11-28. The machine described in Example 1 after Theorem IX has the property that given any string of 1's as input, it prints its own quadruples as output. Show that a nontrivial machine exists with the property that, given a blank tape as input, it prints its own quadruples as output.

$\triangle$***11-29.*** (Smullyan). (*a*) Let $\{A_1,A_2\}$ and $\{B_1,B_2\}$ be disjoint pairs of recursively enumerable sets. We say that $\langle A_1,A_2\rangle$ is *strongly reducible* to $\langle B_1,B_2\rangle$ if there exists a recursive f such that $f(A_1) \subset B_1$, $f(A_2) \subset B_2$, and $f(\overline{A_1 \cup A_2}) \subset \overline{B_1 \cup B_2}$. Let $\{C_1,C_2\}$ be an effectively inseparable disjoint pair of recursively enumerable sets. Show that for *any* disjoint pair of recursively enumerable sets $\{A_1,A_2\}$, $\langle A_1,A_2\rangle$ is strongly reducible to $\langle C_1,C_2\rangle$. (This is an analogue, for pairs of sets, to Theorem V.) (*Hint:* Define

$$W_{g(z)} = \begin{cases} C_1 \cup \{\psi(\pi_1(z),\pi_2(z))\}, & \text{if } x \in A_2; \\ C_1, & \text{if } x \notin A_2, \end{cases}$$

and define $W_{h(z)}$ similarly for C_2 and A_1; where ψ is the productive partial function for $\langle C_1,C_2\rangle$. Apply Theorem X.)

(*b*) Conclude as a corollary that the partial productive function for any effectively inseparable pair of sets can be replaced by a total function.

$\triangle$***11-30.*** (*a*) (Parikh). Let g be any recursive function. Show that there is an n such that

(i) W_n recursive;
(ii) $\mu y[W_y = \bar{W}_n] > g(n)$.

(*Hint:* Define $W_{f(x)}$ to intersect each nonempty W_i, $i \leq g(x)$, and to be finite. Apply the recursion theorem. Note that this yields W_n finite.)

(*b*) Do part (*a*) subject to the requirement that W_n be co-finite.

$\triangle$***11-31.*** Let h be any recursive function. Show that there exists an m and a u such that $W_m = D_u$ and $u > h(m)$. (*Hint:* Use (*a*) of the preceding exercise.) We thus have finite sets with r.e. indices "arbitrarily smaller" than their canonical indices.

$\triangle$***11-32.*** (Parikh). Let g be any recursive function. Show that there is an m such that

(i) $m = \mu y[W_y = W_m]$;
(ii) $m > g(\mu y[W_y{}^K = W_m])$.

(*Hint:* Take a recursive h such that $W^K_{h(x)} = \bar{W}_x$ for all x. Apply Exercise 11-30(*a*) with gh in place of g. Note that this yields W_m co-finite.)

$\triangle$***11-33.*** Let a set A be well-ordered (under some given ordering relation). For any y in A, let S_y be the set of members of A that precede y in the well-ordering. Assume that a partial function ψ is given, and assume that there exists a recursive f such that for any x and for any member y of A, if φ_x coincides with ψ on S_y, then $\varphi_{f(x,y)}$ coincides with ψ on $S_y \cup \{y\}$. Show that there is a partial *recursive* function which coincides with ψ on A. (*Hint:* Apply the recursion theorem to $\varphi_{g(x)} = \lambda y[\varphi_{f(x,y)}(y)]$.)

This exercise provides a recursive-function-theoretic analogue to the *principle of transfinite induction* by which functions over the ordinal numbers are defined. Note that no recursive structure is required directly of A itself or of the ordering of A. The result is generalized in the *recursion lemma* of §16.4.

§*11.5*

11-34. (*a*) Prove a recursion theorem for enumeration operators. Namely, show that for any recursive f there is an n such that

$$\Phi_n = \Phi_{f(n)}.$$

(*b*) Does there exist an m such that for all A,

$$\Phi_m(A) = \{m\}?$$

△*11-35.* Let $\mathfrak{N} = 2^N$. (*a*) For each finite set D, consider the set (of sets)

$$\mathfrak{D}_D = \{A | D \subset A\}.$$

(i) Show that the family of all such sets (of sets) is closed under finite intersection, and hence forms the basis for a topology on $\mathfrak{N}$.

(ii) Show that the closed subsets of $\mathfrak{N}$ in this topology are just the sets (of sets) closed under subset and increasing union.†

(iii) Let Φ be a mapping from $\mathfrak{N}$ into $\mathfrak{N}$. Show that Φ is continuous with respect to this topology if and only if Φ is both "monotone" and "continuous" in the sense of Theorem 9-XX. (*Hint:* Recall that a map f on a topological space $\mathfrak{X}$ is continuous if and only if for any $x \in \mathfrak{X}$ and any $\mathfrak{A} \subset \mathfrak{X}$, $x \in$ *closure* $(\mathfrak{A}) \Rightarrow f(x) \in$ *closure* $(f(\mathfrak{A}))$.)

(*b*) Let $\mathfrak{A} \subset \mathfrak{N}$. Call $\mathfrak{A}$ *closed* if $\mathfrak{A}$ is closed under increasing union.

(i) Show that this defines a topology on $\mathfrak{N}$.

(ii) Let Φ be a monotone mapping from $\mathfrak{N}$ into $\mathfrak{N}$. Show that Φ is continuous with respect to this topology if and only if Φ is "continuous" in the sense of Theorem 9-XX.

(*c*) For each pair $<D_1,D_2>$ of finite sets, consider the set (of sets)

$$\mathfrak{D}_{D_1,D_2} = \{A | D_1 \subset A \;\&\; D_2 \subset \bar{A}\}.$$

(i) Show that the family of all such sets (of sets) forms the basis for a topology on $\mathfrak{N}$. (This is the *Cantor set* topology often given to $\mathfrak{N}$.)

(ii) Show that the closed subsets of $\mathfrak{N}$ in this topology are closed under increasing union.

(iii) Let Φ be a mapping from $\mathfrak{N}$ into $\mathfrak{N}$. Show that if Φ is continuous with respect to this topology, then Φ is "continuous" in the sense of Theorem 9-XX.

(iv) Exhibit an enumeration operator which is not continuous in this topology. (*Hint:* Consider Φ such that

$$\Phi(A) = \begin{cases} N, & \text{if } A \neq \emptyset; \\ \emptyset, & \text{if } A = \emptyset.) \end{cases}$$

11-36. Give examples of
(i) A partial recursive operator with no fixed point;
(ii) A partial recursive operator with nonempty domain and no fixed point;
(iii) A recursive operator with no total fixed point;

† $\mathfrak{A}$ is *closed under subset* if $A \in \mathfrak{A} \;\&\; B \subset A \Rightarrow B \in \mathfrak{A}$. $\mathfrak{A}$ is *closed under increasing union* if $[(\forall i)[A_i \in \mathfrak{A}] \;\&\; (\forall i)(\forall j)[i \leq j \Rightarrow A_i \subset A_j]] \Rightarrow \bigcup_{i=0}^{\infty} A_i \in \mathfrak{A}$.

(iv) A recursive operator with total fixed points, but whose minimal fixed point is not total;

(v) A recursive operator with exactly three fixed points.

11-37. Show that if a recursive operator has only total fixed points, then it has exactly one fixed point.

11-38. Define

$$f(x) = \begin{cases} x - 1, & \text{if } x > 0; \\ 0, & \text{if } x = 0. \end{cases}$$

Define an enumeration operator Φ by

$$\Phi(A) = \{<f(x),f(y)>|<x,y> \in A\}.$$

Consider the partial recursive operator defined by Φ. Call it Ψ.

(i) Is Ψ a recursive operator?

(ii) Describe the fixed points of Ψ.

11-39. Let $\iota_A = \{<x,x>|x \in A\}$. With every enumeration operator Φ a recursive operator Ψ can be associated such that

$$\Psi(\iota_A) = \iota_{\Phi(A)}.$$

Use this construction to obtain Theorem XI directly from Theorem XII.

11-40. Describe all solutions to the set of equations

$$\psi(4x) = 2\psi(2x),$$
$$\psi(2) = 2.$$

$\triangle$***11-41.*** (*a*) (Kreisel). Consider the function g defined by

$$g(x,y,z) = \begin{cases} 1, & \text{if } P_x \text{ with input } x \text{ converges in fewer than } y \text{ steps;} \\ 2z, & \text{otherwise.} \end{cases}$$

g is recursive and, by §1.5, a set of recursion equations can be found which uniquely determine g. Call them (E). Take the set of recursion equations obtained by adjoining to (E) the equation

$$f(x,y) = g(x,y,f(x, y + 1)),$$

where f is a function symbol not occurring in (E).

Show that f is uniquely determined (as a total function) but that f is not a recursive function.

(Total functions which are uniquely determined by a set of recursion equations are called *Herbrand-recursive* functions. As we shall later note, they form a natural class that goes beyond the recursive functions.)

(*b*) Exhibit a general recursive operator Ψ such that

(i) Ψ possesses one and only one total fixed point,

(ii) the total fixed point is not recursive.

(*Hint:* See (*a*).) (Note that Ψ must, by Theorem XII, possess a nontotal fixed point that is partial recursive.)

§*11.6*

11-42. In the proof of Theorem XIV, show that the enumeration operator $\Phi_{w(z)}$ defines Ψ. (*Hint:* Use monotonicity and continuity.)

$\triangle$***11-43.*** Prove the theorem of Myhill and Shepherdson stated in §11.5. (*Hint:* Part (*a*) follows from Corollary XII(*a*). To prove part (*b*), show that the mapping on partial recursive functions determined by f is both monotone and continuous (otherwise the halting problem could be solved). Then construct $\Phi_{w(z)}$ as in the proof of Theorem XIV and use Exercise 11-42 to get uniqueness.)

△*11-44.* Consider Peano arithmetic and take the assumptions about $Mabcd$ made in Exercise 7-64. Let $(\exists b)[(\exists d)M\mathbf{subd}\ \&\ \neg\ Pr_{\mathbf{p}}(b)]$ be constructed to assert its own unprovability as in §11.6. Show that if Peano arithmetic is ω-consistent, then neither this wff nor its negation is provable. (This is closer to Gödel's original proof than is Exercise 7-64.)

△*11-45.* (Tarski). (*a*) Assume the properties of M described in §11.6. Show that there can be no expression Ta of elementary arithmetic with a single free variable a such that for all wffs (with code number) x,

$$T\mathbf{x} \text{ is true if and only if } x \text{ is true.}$$

(*Hint:* Use a substitution function argument to show that, otherwise, a wff could be constructed which asserted its own falsity, and hence would be true if and only if it were false.)†

Definition. A is *completely arithmetically productive* if there exists a recursive f such that for any expression Fa of elementary arithmetic having code number x and a single free variable a,

$$f(x) \in (A - \{y|F\mathbf{y} \textit{ is true}\}) \cup (\{y|F\mathbf{y} \textit{ is true}\} - A).$$

(*b*) Show that the set (of code numbers) of all true wffs of elementary arithmetic is completely arithmetically productive.

§*11.7*

Exercises 11-46 to 11-53 review some basic definitions and results concerning ordinal numbers. A fully satisfactory treatment requires an axiomatization of set theory, since the distinction between *sets* (which can always be members) and *properties* of sets (which sometimes cannot) must be made with care. If, for example, we consider the family of all ordinals (i.e., the *property* of being an ordinal) as a set, we reach a contradiction (the Burali-Forti paradox). In the following review, we proceed without an axiomatization; we try to avoid areas of possible paradox and naively assume principles that require more detailed justification in an axiomatic development. All our results can be obtained from axioms (i) to (ix) in Exercise 11-23.

Definition. A linearly ordered set is *well-ordered* if every nonempty subset has a least element.

By the axiom of choice, it is possible to show that every set can be well-ordered, and that an ordering is a well-ordering if and only if it has no infinite descending chains.

Two ordered sets are said to be of the same *order type* if they are order-isomorphic, i.e., isomorphic as ordered sets.

It is possible to show that certain special ordered sets, called ordinal numbers, exist such that every well-ordered set is order-isomorphic to one and only one ordinal number. The ordinal numbers are constructed in such a way that the well-ordered linear ordering of each is given by $\in$, the membership relation. For example, $\{\emptyset, \{\emptyset\}, \{\emptyset, \{\emptyset\}\}\}$ is the ordinal number representing any well-ordering of three elements. The general definition of ordinal is as follows.

Definition. z is an *ordinal* if

(i) $(\forall x)[x \in z \Rightarrow (\forall y)[y \in x \Rightarrow y \in z]]$;

(ii) $(\forall x)(\forall y)[[x \in z\ \&\ y \in z\ \&\ x \neq y] \Rightarrow [x \in y \vee y \in x]]$; and

(iii) $(\forall x)[[(\forall y)[y \in x \Rightarrow y \in z]\ \&\ x \neq \emptyset] \Rightarrow (\exists u)[u \in x\ \&\ u \cap x = \emptyset]]$.

† This construction can be paralleled in ordinary English by replacing "provable" with "true" in the example given in the footnote on p. 204. The existence of such assertions (which can be neither true nor false) is a compelling reason for rejecting ordinary English (with its apparent ordinary meanings) as an interpreted formal system. As such a system, ordinary English is inconsistent.

The ordinal numbers, as a family, are linearly ordered by $\in$, and each ordinal coincides with the set of all its predecessors in this ordering. (Of course, the family of all ordinals cannot be treated as a *set.*)

The ordinals $\emptyset$, $\{\emptyset\}$, $\{\emptyset,\{\emptyset\}\}$, . . . are abbreviated 0, 1, 2, . . . ; and the ordinal $\{0,1,2, \ldots\}$ is called ω. In general we use "α", "β", "γ", . . . as symbols for ordinals.

Every ordinal has a unique immediate successor. The successor of α is $\alpha \cup \{\alpha\}$ and is denoted $\alpha + 1$. If $\alpha_0, \alpha_1, \ldots$ is an increasing sequence of ordinals, $\lim_n \alpha_n$ denotes its least upper bound. The notation $\{\alpha_n\}\uparrow$ denotes an increasing (and denumerably infinite) sequence of ordinals. If $\gamma = \lim_n \alpha_n$, for some $\{\alpha_n\}\uparrow$, $\{\alpha_n\}\uparrow$ is called a *fundamental sequence* for γ.

Ordinal numbers can be classified into three *kinds:*

(i) The ordinal 0;

(ii) Ordinals having an immediate predecessor; such ordinals are called *successor* ordinals;

(iii) Ordinals (like ω) having no immediate predecessor; such ordinals are called *limit* ordinals.

$\triangle$***11-46.*** Consider: (i) the family of all ordinals that are finite or denumerable; (ii) the smallest family of ordinals that contains 0, is closed under the operation of taking successor, and is closed under limits of increasing, denumerably infinite sequences. Show that these two families coincide.

The family defined in 11-46 (it is, in fact, a set) is traditionally called the *second number class;* we denote it as C_{II}. (The *first number class* consists of the finite ordinals. As defined here, the second number class includes the first number class.) It has the remarkable property that it is a nondenumerable well-ordered set each proper initial segment of which is denumerable. *We henceforth confine our discussion to* (ordinals in) *the second number class.*

Consider functions which map C_{II} into itself. A function f is *monotone* if $(\forall\alpha)(\forall\beta)[\alpha \leq \beta \Rightarrow f(\alpha) \leq f(\beta)]$. A function is *continuous* if it is continuous in the natural least-upper-bound topology (see footnote on page 193); i.e., if $f(\lim_n \alpha_n) = \lim_n f(\alpha_n)$ for all $\{\alpha_n\}\uparrow$.

11-47. Is the function $f = \lambda\alpha[\alpha + 1]$ continuous? Is it monotone?

The *principle of definition by transfinite induction* asserts that given any monotone (but not necessarily continuous) f, and any ordinal ν, a unique continuous function g is defined by the conditions

(i) $g(0) = \nu$,

(ii) $g(\alpha + 1) = f(g(\alpha))$,

(iii) $g(\lim_n \alpha_n) = \lim_n g(\alpha_n)$, for any $\{\alpha_n\}\uparrow$.

11-48. Verify the preceding principle. (*Hint:* Consider, for each β, the class of all functions defined on the predecessors of β which satisfy (i), (ii), and (iii).)

We can now define various familiar functions over C_{II}.

Addition

$$\begin{aligned} \oplus(\alpha,0) &= \alpha, \\ \oplus(\alpha, \beta + 1) &= \oplus(\alpha,\beta) + 1, \\ \oplus(\alpha, \lim_n \beta_n) &= \lim_n \oplus(\alpha,\beta_n). \end{aligned}$$

We write $\oplus(\alpha,\beta)$ as $\alpha + \beta$.

Multiplication

$$\begin{aligned} \otimes(\alpha,0) &= 0, \\ \otimes(\alpha, \beta+1) &= \otimes(\alpha,\beta) + \alpha, \\ \otimes(\alpha, \lim_n \beta_n) &= \lim_n \otimes(\alpha,\beta_n). \end{aligned}$$

We write $\otimes(\alpha,\beta)$ as $\alpha \cdot \beta$.

11-49. Is $\lambda\beta[\beta \cdot 2]$ continuous?

Exponentiation

$$\begin{aligned} e(\alpha,0) &= 1, \\ e(\alpha, \beta+1) &= e(\alpha,\beta) \cdot \alpha, \\ e(\alpha, \lim_n \beta_n) &= \lim_n e(\alpha,\beta_n). \end{aligned}$$

We write $e(\alpha,\beta)$ as α^β.

11-50. Is $\lambda\beta[\beta^\beta]$ continuous?

△11-51. (*a*) Verify the following rules of ordinal arithmetic:

$$\begin{aligned} (\alpha+\beta)+\gamma &= \alpha+(\beta+\gamma), \\ \alpha(\beta+\gamma) &= \alpha\cdot\beta+\alpha\cdot\gamma, \\ \alpha^{\beta+\gamma} &= \alpha^\beta+\alpha^\gamma, \\ (\alpha^\beta)^\gamma &= \alpha^{\beta\cdot\gamma}. \end{aligned}$$

(*b*) Disprove the following as rules:

$$\begin{aligned} \alpha+\beta &= \beta+\alpha, \\ \alpha\cdot\beta &= \beta\cdot\alpha, \\ (\alpha+\beta)\gamma &= \alpha\cdot\gamma+\beta\cdot\gamma. \end{aligned}$$

(*c*) Show that there exists an α such that

$$\alpha = \omega \cdot \alpha.$$

Consider those ordinals which can be obtained from finite ordinals and ω by finitely many applications of addition, multiplication, and exponentiation. Call this class C_p ("the exponential polynomials in 0, 1, 2, . . . , ω").

11-52. (*a*) Show that the ordinals in C_p form an initial segment of C_{II}.

△(*b*) Show that for every ordinal α in C_p, there is a unique finite sequence of smaller ordinals $\beta_1, \beta_2, \ldots, \beta_k$ and a unique finite sequence of nonzero finite ordinals $n_1, n_2, \ldots, n_k$ such that $\beta_1 > \beta_2 > \cdots > \beta_k$ and $\alpha = \omega^{\beta_1}\cdot n_1 + \omega^{\beta_2}\cdot n_2 + \cdots + \omega^{\beta_k}\cdot n_k$ (*normal-form theorem*).

11-53. (*a*) (Veblen). Show that every monotone continuous function on C_{II} has a fixed point. (This is a special case of the Knaster-Tarski theorem; see footnote on page 193.)

(*b*) Find a normal-form expression (see 11-52(*b*)) for the smallest fixed point of

(i) $\lambda\beta[\omega+\beta]$,
(ii) $\lambda\beta[\omega\cdot\beta]$.

(*c*) Show that the least fixed point for $\lambda\beta[\omega^\beta]$ is the least ordinal not in C_p.

(*d*) Let ϵ_0 be the least ordinal not in C_p. ϵ_0 is the limit of the sequence $\omega, \omega^\omega, \omega^{\omega^\omega}, \ldots$; and by Exercise 11-53, $\omega^{\epsilon_0} = \epsilon_0$. (The successive fixed points of $\lambda\beta[\omega^\beta]$ are usually denoted $\epsilon_0, \epsilon_1, \ldots$.) Increasingly large segments of C_{II} can be provided with notations in a systematic way by generalization of the definition of C_p. For example, we can form $C_{p'}$ by taking all ordinals expressible as exponential polynomials in ϵ_0 and its predecessors. This yields all ordinals less than the least fixed point of $\lambda\beta[\epsilon_0{}^\beta]$. A much broader extension can be obtained as follows. Define

$$\gamma_0 = \epsilon_0,$$

$\gamma_{\alpha+1}$ = the least ordinal not expressible as an exponential polynomial in γ_α and its predecessors (or, equivalently, the least fixed point of $\lambda\beta[\gamma_\alpha{}^\beta]$),

$$\gamma_{\lim_n \beta_n} = \lim_n \gamma_{\beta_n}.$$

This generates notations for all ordinals up to the least fixed point of $\lambda\beta[\gamma_\beta]$. Are these ordinals constructive?

§*11.8*

11-54. (*a*) Show that a univalent and recursive system of notation must be recursively related.

(*b*) Show that a recursive system need not be recursively related.

11-55. Show that there is a recursive f such that

$$(\forall x)[x \in O \Rightarrow W_{f(x)} = \{y | y <_O x\}].$$

(*Hint* (Moschovakis): Define an enumeration operator Φ, such that $A \subset O \Rightarrow \Phi(A) = A \cup B$ where B consists of the *first-level* predecessors of members of A (in an appropriate sense). Apply the weak recursion theorem.)

11-56. Show that the normal forms of Exercise 11-52 provide a univalent system of notations in the sense of §11.7.

11-57. Show that $+_O$ in Theorem XVII is a (total) recursive function.

11-58. (*a*) Does there exist a partial recursive function m such that

$$[x \in O \text{ and } y \in O] \Rightarrow [m(x,y) \in O \text{ and } |m(x,y)|_O = |x|_O \cdot |y|_O]?$$

(*b*) Does there exist a partial recursive function e such that

$$[x \in O \text{ and } y \in O] \Rightarrow [e(x,y) \in O \text{ and } |e(x,y)|_O = |x|_O^{|y|_O}]?$$

Δ11-59. (Spector). Show that there is a maximal univalent system of notations. (*Hint:* Take $O = \{x_0, x_1, x_2, \ldots\}$ (in any order), and consider the *branch* of $<_O$ determined by the sequence: x_0, $x_0 +_O x_1$, $(x_0 +_O x_1) +_O x_2$, ...)

11-60. Show that there are branches of $<_O$ which do not extend through the position of ω^2. (*Hint:* Use a cardinality argument.)

11-61. Let $W = \{z | \varphi_z^{(2)}$ *is the characteristic function of a well-ordering* $(\leq)$ *of some set of integers*$\}$. (We can think of W as a collection of notations for the recursive ordinals.) Show that $W \equiv O$. (*Hint:* See proofs of Theorems XIX and XX.)

11-62. Define α to be a *recursively enumerable* ordinal if α is order-isomorphic to some recursively enumerable well-ordering of a set of integers. Show that every recursively enumerable ordinal is recursive.

Δ11-63. (Parikh). Let B be as in Theorem XXI. Show that $B \equiv O$.

12 Recursively Enumerable Sets as a Lattice

§12.1 LATTICES OF SETS

In preceding chapters we have used the concepts of *reducibility* and *degree of unsolvability* to study recursively invariant properties of sets of integers. For the recursively enumerable sets, *lattice* concepts provide another, somewhat different approach. We describe this approach below. It was first studied by Dekker and Myhill (see Myhill [1956]).

Definition A *lattice* is a partial ordering in which any two members have a least upper bound and any two members have a greatest lower bound.†

We refer to the members of a lattice as *elements*. We use $a, b, \ldots$ to denote elements of a lattice. If a is less than or equal to b in the lattice, we write "$a \leq b$". We define $a < b$ to mean $[a \leq b \;\&\; a \neq b]$. If a is an element of a lattice $\mathfrak{L}$, we write "$a \in \mathfrak{L}$".

Definition If a and b are elements of a lattice $\mathfrak{L}$, then $a \cup b$ denotes the least upper bound ("join" or "union") of a and b; and $a \cap b$ denotes the greatest lower bound ("meet" or "intersection") of a and b.

The collection of all subsets of any fixed set forms a lattice under set

† Here, *partial ordering* means a relation R (on some set) such that for all a, b, and c (in the set),

(i) $[\langle a,b\rangle \in R \;\&\; \langle b,c\rangle \in R] \Rightarrow \langle a,c\rangle \in R$;

(ii) $[\langle a,b\rangle \in R \;\&\; \langle b,a\rangle \in R] \Leftrightarrow a = b$.

We conventionally read "$\langle a,b\rangle \in R$" as "a is less than or equal to b."

c is a *least upper bound* for a and b if (i) $\langle a,c\rangle \in R$ and $\langle b,c\rangle \in R$, and (ii) for any d (in the set), $[\langle a,d\rangle \in R \;\&\; \langle b,d\rangle \in R] \Rightarrow \langle c,d\rangle \in R$. Similarly for *greatest lower bound*. For a more extensive treatment of lattices, see Birkhoff [1940].

inclusion ($\subset$), with join and meet given by ordinary union and intersection, respectively. In particular, $\mathfrak{N} = \{A|A \subset N\}$ ($= 2^N$) is a lattice under $\subset$. Note that in any lattice,

$$a \leq b \Leftrightarrow b = a \cup b \Leftrightarrow a = a \cap b.$$

The partial ordering $\leq$ can thus be defined in terms of $\cup$ or $\cap$. Indeed, one can formulate the concept of lattice entirely in terms of the operations $\cup$ and $\cap$ by placing appropriate axiomatic restrictions on $\cup$ and $\cap$ (see Exercise 12-1).

Definition Let $\mathfrak{L}$ be a lattice. $\mathfrak{M}$ is a *sublattice* of $\mathfrak{L}$ if
(i) $\mathfrak{M}$ is a lattice;
(ii) (the partial ordering of) $\mathfrak{M}$ is contained in (the partial ordering of) $\mathfrak{L}$;
(iii) $\mathfrak{M}$ is closed under the operations $\cup$ and $\cap$ in $\mathfrak{L}$. ((iii) does not follow from (i) and (ii); see Exercise 12-2.)

Examples. Let $\mathcal{E}$ be the partial ordering of recursively enumerable sets under $\subset$. Let $\mathfrak{R}$ be the partial ordering of recursive sets under $\subset$. Then by Theorems 5-XIII and 5-XIV, $\mathfrak{R}$ and $\mathcal{E}$ are sublattices of $\mathfrak{N}$, and $\mathfrak{R}$ is a sublattice of $\mathcal{E}$.

Definitions A lattice is *distributive* if the identities

$$a \cup (b \cap c) = (a \cup b) \cap (a \cup c)$$

and

$$a \cap (b \cup c) = (a \cap b) \cup (a \cap c)$$

are satisfied. Clearly every sublattice of a distributive lattice is distributive. Henceforth, we shall be concerned exclusively with distributive lattices.

If a lattice contains a minimum element, that (necessarily unique) element is called the *zero element.* If a lattice contains a maximum element, that (necessarily unique) element is called the *unit element.* We sometimes denote the zero element of a lattice as *0* and the unit element as *1*. (Note that when $\mathfrak{M}$ is a sublattice of $\mathfrak{L}$, zero and unit elements for $\mathfrak{M}$ may not coincide with zero and unit elements for $\mathfrak{L}$.)

Let a and b be elements of a lattice containing a zero element (0); a and b are *disjoint* if $a \cap b = 0$.

Let a and b be elements of a lattice containing a zero element (0) and a unit element (1). b is called a *complement* of a if $a \cap b = 0$ and $a \cup b = 1$.

If a lattice $\mathfrak{L}$ contains zero and unit elements, and if every element of $\mathfrak{L}$ has a complement, $\mathfrak{L}$ is called a *complemented* lattice.

A distributive, complemented lattice is called a *Boolean algebra.* In a Boolean algebra, every element a has one and only one complement which we denote $\bar{a}$ (see Exercise 12-3).

Examples. $\mathfrak{N}$ is a Boolean algebra.

$\mathcal{E}$ contains unit and zero elements, but is not complemented.

$\mathfrak{R}$ is a Boolean algebra.

The *Stone representation theorem* (Exercise 12-9) asserts that for every Boolean algebra $\mathcal{B}$ there is a set of objects Σ such that $\mathcal{B}$ is isomorphic to a sublattice of the lattice of all subsets of Σ. It follows from the Stone theorem that every Boolean algebra (and hence every sublattice of a Boolean algebra) must satisfy the familiar identities (in $\cup$, $\cap$, and $^{-}$) that hold for lattices of sets (see Exercise 12-5).

Definitions Let $\mathcal{L}$ be a lattice and let $\mathcal{I}$ be a nonempty set of elements of $\mathcal{L}$. We say that $\mathcal{I}$ is an *ideal in* $\mathcal{L}$ if

(i) $[c_1 \in \mathcal{I} \;\&\; c_2 \in \mathcal{I}] \Rightarrow c_1 \cup c_2 \in \mathcal{I}$;

(ii) $[a \in \mathcal{L} \;\&\; c \in \mathcal{I}] \Rightarrow a \cap c \in \mathcal{I}$.

Let $\mathcal{L}$ be a lattice and let $\mathcal{D}$ be a nonempty set of elements of $\mathcal{L}$. We say that $\mathcal{D}$ is a *dual ideal* (or "filter") if

(i) $[c_1 \in \mathcal{D} \;\&\; c_2 \in \mathcal{D}] \Rightarrow c_1 \cap c_2 \in \mathcal{D}$;

(ii) $[a \in \mathcal{L} \;\&\; c \in \mathcal{D}] \Rightarrow a \cup c \in \mathcal{D}$.

An ideal or dual ideal in $\mathcal{L}$ is, trivially, a sublattice of $\mathcal{L}$.

Examples. The family $\mathcal{F}$ of all finite sets is an ideal in $\mathcal{N}$.

The family $\mathcal{S}$ of all sets which are either simple or co-finite is a dual ideal in $\mathcal{E}$, by Exercise 8-11.

The following theorem is basic for subsequent discussion.

Theorem I(a) *Let $\mathcal{L}$ be a distributive lattice, and let $\mathcal{I}$ be an ideal in $\mathcal{L}$. Define $a \approx_{\mathcal{I}} b$ if $(\exists c_1)(\exists c_2)[c_1 \in \mathcal{I} \;\&\; c_2 \in \mathcal{I} \;\&\; [a \cup c_1 = b \cup c_2]]$. Then $\approx_{\mathcal{I}}$ is an equivalence relation on the elements of $\mathcal{L}$. The operations $\cup$ and $\cap$ are well-defined on its equivalence classes, and these equivalence classes form a distributive lattice under $\cup$ and $\cap$. We call this lattice of equivalence classes the* quotient lattice $\mathcal{L}/\mathcal{I}$. *$\mathcal{I}$ itself is an equivalence class and serves as the zero element of $\mathcal{L}/\mathcal{I}$.*

(b) *Let $\mathcal{L}$ be a distributive lattice, and let $\mathcal{D}$ be a dual ideal in $\mathcal{L}$. Define $a \approx_{\mathcal{D}} b$ if $(\exists c_1)(\exists c_2)[c_1 \in \mathcal{D} \;\&\; c_2 \in \mathcal{D} \;\&\; [a \cap c_1 = b \cap c_2]]$. Then $\approx_{\mathcal{D}}$ is an equivalence relation on the elements of $\mathcal{L}$. The operations $\cup$ and $\cap$ are well-defined on its equivalence classes, and these equivalence classes form a distributive lattice under $\cup$ and $\cap$. We call this lattice of equivalence classes the* quotient lattice $\mathcal{L}/^*\mathcal{D}$ (where * indicates that a *dual* ideal is being used). *$\mathcal{D}$ itself is an equivalence class and serves as the unit element of $\mathcal{L}/^*\mathcal{D}$.*

(c) *Let $\mathcal{L}$ be a Boolean algebra. If $\mathcal{I}$ is an ideal in $\mathcal{L}$, then $\mathcal{L}/\mathcal{I}$ is a Boolean algebra. If $\mathcal{D}$ is a dual ideal in $\mathcal{L}$, then $\mathcal{L}/^*\mathcal{D}$ is a Boolean algebra. Let $\mathcal{I}$ be a set of elements in $\mathcal{L}$ and let $\mathcal{D} = \{a \mid \bar{a} \in \mathcal{I}\}$. Then $\mathcal{I}$ is an ideal if and only if $\mathcal{D}$ is a dual ideal; and if $\mathcal{I}$ is an ideal, then $\mathcal{L}/\mathcal{I}$ and $\mathcal{L}/^*\mathcal{D}$ are identical.*

(d) *Let $\mathcal{L}$ be a distributive lattice, let $\mathcal{I}$ be an ideal in $\mathcal{L}$, and let $\mathcal{M}$ be a sublattice of $\mathcal{L}$. Then those equivalence classes of $\approx_{\mathcal{I}}$ which each contain at least one element of $\mathcal{M}$ form a sublattice of $\mathcal{L}/\mathcal{I}$. We call this sublattice $\mathcal{M}/\mathcal{I}$* in $\mathcal{L}$. *(Note that $\mathcal{I}$ need not be contained in $\mathcal{M}$.) Similarly for $\mathcal{M}/^*\mathcal{D}$ in $\mathcal{L}$, where $\mathcal{D}$ is a dual ideal in $\mathcal{L}$.*†

† Occasionally, in applications of (*d*), the lattice $\mathcal{L}$ will not be explicitly mentioned Its presence and nature will always be clear from the context or from preceding discussion.

Proof. (*a*) $\approx_{\mathcal{I}}$ is an equivalence relation, as is easily verified (see Exercise 12-6).

Assume $a_1 \approx_{\mathcal{I}} a_2$ and $b_1 \approx_{\mathcal{I}} b_2$. Then $a_1 \cup c_1 = a_2 \cup c_2$, and $b_1 \cup c_3 = b_2 \cup c_4$; where c_1, c_2, c_3, and c_4 are in $\mathcal{I}$. Combining these equations under $\cup$, we have

$$(a_1 \cup c_1) \cup (b_1 \cup c_3) = (a_2 \cup c_2) \cup (b_2 \cup c_4),$$

and hence, $$(a_1 \cup b_1) \cup (c_1 \cup c_3) = (a_2 \cup b_2) \cup (c_2 \cup c_4).$$

Therefore $a_1 \cup b_1 \approx_{\mathcal{I}} a_2 \cup b_2$, since $c_1 \cup c_3$ and $c_2 \cup c_4$ must be in $\mathcal{I}$. Similarly, combining the equations under $\cap$, we have

$$(a_1 \cup c_1) \cap (b_1 \cup c_3) = (a_2 \cup c_2) \cap (b_2 \cup c_4),$$

and hence, by distributivity,

$$(a_1 \cap b_1) \cup (a_1 \cap c_3) \cup (c_1 \cap b_1) \cup (c_1 \cap c_3) = (a_2 \cap b_2) \cup (a_2 \cap c_4) \cup (c_2 \cap b_2) \cup (c_2 \cap c_4).$$

Since $a_1 \cap c_3$, $c_1 \cap b_1$, $c_1 \cap c_3$, $a_2 \cap c_4$, $c_2 \cap b_2$, and $c_2 \cap c_4$ must all be in $\mathcal{I}$, and since $(a_1 \cap c_3) \cup (c_1 \cap b_1) \cup (c_1 \cap c_3)$ and $(a_2 \cap c_4) \cup (c_2 \cap b_2) \cup (c_2 \cap c_4)$ must hence both be in $\mathcal{I}$, we have that $a_1 \cap b_1 \approx_{\mathcal{I}} a_2 \cap b_2$.

Thus $\cup$ and $\cap$ yield well-defined operations on the equivalence classes. It follows that any identity that holds for the operations of $\mathcal{L}$ must hold for the operations on these equivalence classes. Thus (by Exercise 12-1), $\mathcal{L}/\mathcal{I}$ is a distributive lattice. In Exercise 12-8 we show that $\mathcal{I}$ forms a single equivalence class which is the zero element in $\mathcal{L}/\mathcal{I}$.

(*b*) Proof is similar to proof for (*a*).

(*c*) To show that $\mathcal{L}/\mathcal{I}$ is a Boolean algebra, we need only show that the complementation operation of $\mathcal{L}$ is well-defined on equivalence classes. Let $a \cup c_1 = b \cup c_2$, with c_1 and c_2 in $\mathcal{I}$. Then $\overline{a \cup c_1} = \overline{b \cup c_2}$. Hence (see Exercise 12-5), $\bar{a} \cap \bar{c}_1 = \bar{b} \cap \bar{c}_2$. Hence

$$(\bar{a} \cap \bar{c}_1) \cup (\bar{a} \cap c_1) \cup (\bar{b} \cap c_2) = (\bar{b} \cap \bar{c}_2) \cup (\bar{b} \cap c_2) \cup (\bar{a} \cap c_1).$$

Applying Exercise 12-5, we have

$$\bar{a} \cup (\bar{b} \cap c_2) = \bar{b} \cup (\bar{a} \cap c_1).$$

This yields $\bar{a} \approx_{\mathcal{I}} \bar{b}$, since $\bar{b} \cap c_2$ and $\bar{a} \cap c_1$ must be in $\mathcal{I}$.

Similarly for $\mathcal{L}/^*\mathcal{D}$ when $\mathcal{D}$ is a dual ideal.

Let $\mathcal{I}$ be a set of elements in $\mathcal{L}$, and let $\mathcal{D} = \{a \mid \bar{a} \in \mathcal{I}\}$. Then $c \in \mathcal{I} \Leftrightarrow \bar{c} \in \mathcal{D}$. Take c_1 and c_2 in $\mathcal{I}$; then $c_1 \cup c_2 \in \mathcal{I} \Leftrightarrow \overline{c_1 \cup c_2} \in \mathcal{D} \Leftrightarrow \bar{c}_1 \cap \bar{c}_2 \in \mathcal{D}$. Take c in $\mathcal{I}$ and $a \in \mathcal{L}$; then $c \cap a \in \mathcal{I} \Leftrightarrow \overline{c \cap a} \in \mathcal{D} \Leftrightarrow \bar{c} \cup \bar{a} \in \mathcal{D}$. Hence $\mathcal{I}$ is an ideal if $\mathcal{D}$ is a dual ideal. By similar argument, $\mathcal{D}$ is a dual ideal if $\mathcal{I}$ is an ideal. Now assume $\mathcal{I}$ is an ideal. Then $a \approx_{\mathcal{I}} b \Leftrightarrow \bar{a} \approx_{\mathcal{I}} \bar{b}$ (since complementation is well-defined on equivalence classes) $\Leftrightarrow \bar{a} \cup c_1 = \bar{b} \cup c_2 \Leftrightarrow \overline{a \cap \bar{c}_1} = \overline{b \cap \bar{c}_2} \Leftrightarrow a \cap \bar{c}_1 = b \cap \bar{c}_2$ (since complements are unique in

$\mathfrak{L}) \Leftrightarrow a \approx_{\mathfrak{D}} b$. Hence the relations $\approx_{\mathfrak{s}}$ and $\approx_{\mathfrak{D}}$ coincide, and $\mathfrak{L}/\mathfrak{s}$ and $\mathfrak{L}/^*\mathfrak{D}$ must be identical.

(*d*) This is immediate from (*a*), the definition of sublattice, and Exercise 12-1.☒

A converse to (*c*) and a partial converse to (*a*) are given in Exercise 12-12.

The following lattices, ideals, and dual ideals have been mentioned:

$\mathfrak{N}$, the lattice of all sets of integers.

$\mathcal{E}$, the lattice of all recursively enumerable sets.

$\mathcal{R}$, the lattice of all recursive sets.

$\mathfrak{F}$, the ideal (in $\mathfrak{N}$) of finite sets.

$\mathcal{S}$, the dual ideal (in $\mathcal{E}$) of co-finite or simple sets. By Exercise 9-34, we also have

$\mathcal{H}$, the dual ideal (in $\mathcal{E}$) of co-finite or hypersimple sets. (The co-finite or hyperhypersimple sets also form a dual ideal in $\mathcal{E}$; see Exercise 12-55.) Theorem I then enables us to form the various quotient lattices: $\mathfrak{N}/\mathfrak{F}$, $\mathcal{E}/\mathfrak{F}$, $\mathcal{R}/\mathfrak{F}$, $\mathcal{E}/^*\mathcal{S}$, $\mathcal{E}/^*\mathcal{H}$, $\mathcal{R}/^*\mathcal{S}$, and $\mathcal{R}/^*\mathcal{H}$. (Exercise 12-13 shows that $\mathcal{R}/^*\mathcal{H} = \mathcal{R}/^*\mathcal{S} = \mathcal{R}/\mathfrak{F}$.) In $\mathfrak{N}/\mathfrak{F}$, any element containing a recursively enumerable set is made up entirely of recursively enumerable sets. We call such elements *recursively enumerable*. Similarly for *recursive* elements of $\mathfrak{N}/\mathfrak{F}$. $\mathcal{R}/\mathfrak{F}$ is a sublattice of $\mathcal{E}/\mathfrak{F}$, and $\mathcal{E}/\mathfrak{F}$ is a sublattice of $\mathfrak{N}/\mathfrak{F}$. The present chapter is chiefly concerned with the quotient lattice $\mathcal{E}/\mathfrak{F}$.

In what sense is the lattice $\mathcal{E}/\mathfrak{F}$ invariant? Any two sets of integers, $\neq\emptyset$ and $\neq N$, belonging to the same element of $\mathcal{E}/\mathfrak{F}$ are m-equivalent. On the other hand, recursively isomorphic sets can fall into different elements of $\mathcal{E}/\mathfrak{F}$ (take the sets of even and odd integers, for example). Thus, particular elements of $\mathcal{E}/\mathfrak{F}$ (other than *0* and *1*), like particular elements of $\mathcal{E}$ (other than $\emptyset$ and N), are not, by themselves, recursively invariant objects. Nevertheless, in an important sense, the lattice $\mathcal{E}/\mathfrak{F}$ is recursively invariant. We first define the notion of *lattice-theoretic property*.

Definition A property, i.e., collection, of elements of a lattice $\mathfrak{L}$ is *lattice-theoretic in* $\mathfrak{L}$ if it is invariant under all automorphisms of $\mathfrak{L}$. (Motivation: A property is to be lattice-theoretic in $\mathfrak{L}$ if it can be "defined" in terms of the lattice structure of $\mathfrak{L}$.)†

Definition Let the lattice $\mathfrak{L}$ have sublattice $\mathfrak{M}$. A property of elements of $\mathfrak{L}$ is *lattice-theoretic in* $\mathfrak{L}$ *with respect to* $\mathfrak{M}$ if it is invariant under all automorphisms of $\mathfrak{L}$ that carry $\mathfrak{M}$ onto itself. (Motivation: A property is to be lattice-theoretic in $\mathfrak{L}$ with respect to $\mathfrak{M}$ if it can be "defined" in terms of the lattice structure of $\mathfrak{L}$ and membership in $\mathfrak{M}$.)

We abbreviate "lattice-theoretic in $\mathfrak{N}$ with respect to $\mathcal{E}$" to "lattice-theoretic with respect to $\mathcal{E}$." Since every automorphism of $\mathcal{E}$ extends to an

† f is an *automorphism* of a lattice $\mathfrak{L}$ if f is a one-one mapping from $\mathfrak{L}$ onto $\mathfrak{L}$ such that $f(a \cup b) = f(a) \cup f(b)$ and $f(a \cap b) = f(a) \cap f(b)$ for all $a,b \in \mathfrak{L}$.

automorphism of $\mathfrak{N}$ (both are determined by the induced automorphism on $\mathfrak{F}$), a property of recursively enumerable sets is lattice-theoretic with respect to $\mathcal{E}$ if and only if it is lattice-theoretic in $\mathcal{E}$.

Examples. The property of *finiteness* is lattice-theoretic in $\mathcal{E}$.

The property of *immunity* is lattice-theoretic with respect to $\mathcal{E}$.

Definition A property of recursively enumerable sets is *lattice-theoretic in* $\mathcal{E}/\mathfrak{F}$ if it is the union of a property (i.e., collection) of elements of $\mathcal{E}/\mathfrak{F}$ that is lattice-theoretic in $\mathcal{E}/\mathfrak{F}$.†

Examples. Recursiveness is lattice-theoretic in $\mathcal{E}/\mathfrak{F}$, for A is recursive if and only if A belongs to a complemented element of $\mathcal{E}/\mathfrak{F}$.

So is *simplicity,* for A is simple if and only if A belongs to a non-unit element of $\mathcal{E}/\mathfrak{F}$ that has no nonzero element disjoint from it.

Relations between sets can also be lattice-theoretic (in the obvious way: invariance in $\mathfrak{L}^n$ under automorphisms of $\mathfrak{L}$). For example, recursive inseparability and strong inseparability (Exercise 8-39) are lattice-theoretic in $\mathcal{E}/\mathfrak{F}$.

Theorem II *If a property is lattice-theoretic in* $\mathcal{E}/\mathfrak{F}$, *then it is recursively invariant.*

Proof. Every recursive permutation f induces an automorphism of $\mathcal{E}$, namely, the mapping that carries A to $f(A)$. This automorphism carries $\mathfrak{F}$ onto $\mathfrak{F}$, and hence induces an automorphism of $\mathcal{E}/\mathfrak{F}$. A property invariant under all automorphisms of $\mathcal{E}/\mathfrak{F}$ must therefore be invariant under all recursive permutations.⊠

If a recursively invariant property (of recursively enumerable sets) is well-defined on $\mathcal{E}/\mathfrak{F}$, must it be lattice-theoretic in $\mathcal{E}/\mathfrak{F}$? This converse to Theorem II is an open question. A negative answer seems probable. Creativeness, hypersimplicity, m-reducibility (for sets other than N and $\emptyset$), and the weak reducibilities are examples of recursively invariant properties which are well-defined on $\mathcal{E}/\mathfrak{F}$ but not known to be lattice-theoretic in $\mathcal{E}/\mathfrak{F}$. In the remainder of this chapter (and especially in §12.6) we shall study a

† A property of sets is *elementary lattice-theoretic in* $\mathcal{E}/\mathfrak{F}$ if there is an expression of quantificational logic in the single predicate symbol $\leq$ and with a single free individual variable, such that if quantifiers are interpreted as ranging over $\mathcal{E}/\mathfrak{F}$ and $\leq$ is interpreted as the lattice ordering of $\mathcal{E}/\mathfrak{F}$, then the sets with that property are precisely the sets belonging to elements of $\mathcal{E}/\mathfrak{F}$ which satisfy the expression. For example, the expression $(\forall b)[a \leq b]$ determines the elementary lattice-theoretic property of *finiteness.*

The *first-order theory* of $\mathcal{E}/\mathfrak{F}$ consists of those expressions in $\leq$ with no free variable which are true of $\mathcal{E}/\mathfrak{F}$. It is not known whether this theory is decidable, or even axiomatizable. A decision procedure for this theory would yield Theorems IV, XI, and XIV (below) directly and would settle several open problems. Let $\mathcal{K}$ be the dual ideal (in $\mathcal{E}$) of co-finite or hyperhypersimple sets. Lachlan has shown that the first-order theories of $\mathcal{K}$, $\mathcal{R}$, $\mathcal{K}/\mathfrak{F}$, and $\mathcal{R}/\mathfrak{F}$ are all decidable. He has also shown that the decision problem for the first-order theory of $\mathcal{E}$ is reducible to the decision problem for the first-order theory of $\mathcal{E}/\mathfrak{F}$.

number of interrelated recursively invariant properties some but not all of which are known to be lattice-theoretic in $\mathcal{E}/\mathfrak{F}$.†

Note that the lattice $\mathcal{E}/\mathfrak{F}$ possesses the following homogeneity property: for any nonzero element a, $\{b|b \leq a\}$ constitutes a sublattice which is isomorphic to $\mathcal{E}/\mathfrak{F}$ (Exercise 12-18).

We shall consider several questions about $\mathcal{E}/\mathfrak{F}$.

1. Is every noncomplemented element of $\mathcal{E}/\mathfrak{F}$ the join of two disjoint noncomplemented elements? This question can be put in more familiar terms. Is every recursively enumerable, nonrecursive set the disjoint union of two recursively enumerable, nonrecursive sets? The solution to this problem is given in §12.2. It is due to Friedberg.

2. Is $\mathcal{E}/\mathfrak{F}$ densely ordered? That is, if $a < b$, must there exist a c such that $a < c$ and $c < b$? The lattices $\mathfrak{N}$, $\mathcal{E}$, and $\mathcal{R}$ are, of course, not densely ordered. The lattices $\mathfrak{N}/\mathfrak{F}$ and $\mathcal{R}/\mathfrak{F}$ are densely ordered, as is easily shown. The following definition and theorem put the question of density for $\mathcal{E}/\mathfrak{F}$ in a somewhat different form.

Definition An element in $\mathcal{E}/\mathfrak{F}$ is *maximal* if it is maximal among the nonunit elements of $\mathcal{E}/\mathfrak{F}$. (It is easily shown that there are no minimal elements among the nonzero elements of $\mathcal{E}/\mathfrak{F}$.)

***Theorem III* (*Myhill*)** *$\mathcal{E}/\mathfrak{F}$ is densely ordered if and only if $\mathcal{E}/\mathfrak{F}$ has no maximal element.*

Proof. $\Rightarrow$: Immediate.

$\Leftarrow$: Assume $\mathcal{E}/\mathfrak{F}$ is not densely ordered. Then there exist elements a and b such that $a < b$ but for no c is $a < c$ and $c < b$. Taking representatives, we have sets A and B such that $A \in a$, $B \in b$ and $A \subset B \cup D$, where D is finite (see Exercise 12-7). Let f be a one-one recursive function with $B \cup D$ as range (this is possible since $b \neq 0$). Take $A' = f^{-1}(A)$. Then the equivalence class of A' is the desired maximal element, as can readily be verified.⊠

In §12.4 we shall solve the density problem by showing that $\mathcal{E}/\mathfrak{F}$ possesses a maximal nonunit element. The construction is due to Friedberg.

3. Is every nonunit element of $\mathcal{E}/\mathfrak{F}$ contained in a maximal element? In

† The following question is related to the converse to Theorem II. If a property is lattice-theoretic in $\mathcal{E}$ and well-defined on $\mathcal{E}/\mathfrak{F}$, must it then be lattice-theoretic in $\mathcal{E}/\mathfrak{F}$? This is in turn related to the question: is every automorphism of $\mathcal{E}/\mathfrak{F}$ induced by some recursive permutation? Both questions are open; an affirmative answer to the second implies an affirmative answer to the first.

The following implications are easily shown to hold (Exercise 12-17): lattice-theoretic in $\mathcal{E}/\mathfrak{F} \Rightarrow$ [well-defined on $\mathcal{E}/\mathfrak{F}$ and lattice-theoretic in $\mathcal{E}$] $\Rightarrow$ well-defined on $\mathcal{E}/\mathfrak{F}$ and recursively invariant. Both converse implications are open questions. They are related to the open question: is every automorphism of $\mathcal{E}/\mathfrak{F}$ induced by some automorphism of $\mathcal{E}$?

Exercise 12-32 shows that a recursively invariant property need not be lattice-theoretic in $\mathcal{E}$.

§12.5, we give a negative answer to this question. The construction is due to Martin.

In §12.3 we consider a particular ideal in $\mathfrak{N}$, the ideal generated by the *cohesive* sets, and consider some of the properties of its members. The cohesive sets are "almost finite" in the sense that they cannot be split into smaller infinite parts by recursively enumerable sets.

In §12.6, a variety of other *almost-finiteness* concepts, e.g., hyperhyper-immunity, are considered, and several open problems are mentioned.

§12.2 DECOMPOSITION

The following theorem answers question 1 in §12.1.

Theorem IV (**Friedberg** [**1958b**]) *Let A be any set which is recursively enumerable but not recursive. Then there exist sets B_1 and B_2 such that*

(i) $B_1 \cup B_2 = A$,
(ii) $B_1 \cap B_2 = \emptyset$,
(iii) *B_1 and B_2 are each recursively enumerable but not recursive.*

Proof. Let f be a one-one recursive function with range A. We give a procedure for enumerating B_1 and B_2 in successive stages. Let $B_1^{(n)}$ be the members of B_1 which have been enumerated by the end of stage n. Similarly for $B_2^{(n)}$.

Stage 0. Put $f(0)$ into B_1.

Stage $n + 1$. Perform n steps in the listing of each of $W_0, W_1, \ldots, W_n$. Call the resulting sets $W_0^{(n)}, W_1^{(n)}, \ldots, W_n^{(n)}$. See whether there is any x, $x \leq n$, such that $f(n + 1) \in W_x^{(n)}$ and either $B_1^{(n)} \cap W_x^{(n)} = \emptyset$ or $B_2^{(n)} \cap W_x^{(n)} = \emptyset$. If so, take the least such x and call it x_n. If

$$B_1^{(n)} \cap W_{x_n}^{(n)} = \emptyset,$$

put $f(n + 1)$ in B_1. If $B_1^{(n)} \cap W_{x_n}^{(n)} \neq \emptyset$, put $f(n + 1)$ in B_2. If there is no such x, put $f(n + 1)$ in B_1.

Clearly B_1 and B_2 are each recursively enumerable, $B_1 \cup B_2 = A$, and $B_1 \cap B_2 = \emptyset$. It remains to show that neither B_1 nor B_2 can be recursive.

Assume B_1 recursive. Then $\bar{B}_1$ is recursively enumerable. Let $\bar{B}_1 = B_2 \cup \bar{A} = W_m$. Then $W_m \cap B_1 = \emptyset$ and $\bar{A} \subset W_m$. Observe that as n varies, x_n takes each value at most twice. Hence there must be some n_0 such that for all $n > n_0$, $x_n > m$ (if x_n exists). Then $n > n_0 \Rightarrow f(n + 1) \notin W_m^{(n)}$ (otherwise stage $n + 1$ would put $f(n + 1)$ into B_1, and $W_m \cap B_1$ would be $\neq \emptyset$). Therefore, after stage n_0, any member of B_2 which appears as a value of f must do so *before* it appears in W_m. On the other hand, since $\bar{A} \subset W_m$, any member of $\bar{A}$ must appear in W_m *before* it appears as a value of f (since it never appears as a value of f). Thus $\{z | (\exists n)[z \in W_m^{(n)} \ \& \ z \notin \{f(0), \ldots, f(n + 1)\}]\}$ must differ from $\bar{A}$ by at most a finite set. Hence, by the second projection theorem, $\bar{A}$ must be recursively enumerable. Hence A must be recursive, contrary to assumption. Thus B_1 cannot be recursive.

By parallel argument, B_2 cannot be recursive.☒

Sacks has combined Theorem IV with Exercise 10-11 (as was mentioned at the end of Exercise 10-11). He has shown that any nonrecursive recursively enumerable set A is the union of two disjoint sets, where the disjoint sets are of incomparable T-degree of unsolvability, and where each has lower T-degree than A. We do not give Sacks's proof.

A straightforward generalization of the proof of Theorem IV shows that any nonrecursive, recursively enumerable set can be decomposed into a uniformly enumerable, infinite collection of pairwise disjoint, nonrecursive, recursively enumerable sets (see Exercise 12-22).

§12.3 COHESIVE SETS

The concepts of *immunity* and *hyperimmunity* can be strengthened to the following.

Definition A is *cohesive* if

(i) A is infinite; and

(ii) $(\forall B)[B$ recursively enumerable $\Rightarrow [A \cap B$ finite or $A \cap \bar{B}$ finite$]]$.†

Thus an infinite set is cohesive if and only if it cannot be divided into two infinite parts by a recursively enumerable set. To put it another way, a set is cohesive if and only if it is a member of an element a of $\mathfrak{N}/\mathfrak{F}$ such that $a \neq 0$ and a is disjoint from every recursively enumerable element which is not above it.

Theorem V *A cohesive $\Rightarrow$ A hyperimmune.*

Proof. Assume A not hyperimmune. Then, by Theorem 9-XV, there exists a recursive f such that

$$(\forall x)(\forall y)[x \neq y \Rightarrow D_{f(x)} \cap D_{f(y)} = \emptyset] \;\&\; (\forall x)[D_{f(x)} \cap A \neq \emptyset].$$

Consider $B = \bigcup_{x \text{ even}} D_{f(x)}$. Then $B \cap A$ must be infinite and $\bar{B} \cap A$ must be infinite. Hence A is not cohesive.☒

(A similar argument shows that A cohesive $\Rightarrow$ A hyperhyperimmune.)

The following theorem shows the existence of cohesive sets.

Theorem VI (Dekker and Myhill) *Every infinite set possesses a cohesive subset.*

† This notion is due to Rose and Ullian [1963]. It is related to the notion of *infinite indecomposable* set as defined by Dekker and Myhill [1960]. In §12.6, we show that the two notions are distinct. The term *almost-finite* has occasionally been used to mean cohesive.

Proof. Let A be a given infinite set. Define a sequence $A_0, A_1, \ldots$ as follows:

$$A_0 = A;$$
$$A_{n+1} = \begin{cases} A_n \cap W_n, & \text{if } A_n \cap W_n \text{ is infinite;} \\ A_n \cap \bar{W}_n, & \text{otherwise.} \end{cases}$$

Then $A_0 \supset A_1 \supset A_2 \supset \cdots$.

Assume A contains no cohesive set. Then for every n there exists a z such that $z \geq n$ and W_z divides A_n into two infinite parts. It follows that the sequence $A_0, A_1, A_2, \ldots$ contains a strictly decreasing subsequence $B_0, B_1, B_2, \ldots$. Define a sequence of integers $x_0, x_1, \ldots$ by taking

$$x_n = \mu y[y \in B_n - B_{n+1}].$$

Let C be the set of integers in this sequence, that is, $C = \{x_0, x_1, \ldots\}$. Let any z be given. By our construction, $A_{z+1} \subset W_z$ or $A_{z+1} \subset \bar{W}_z$. Hence $B_{z+1} \subset W_z$ or $B_{z+1} \subset \bar{W}_z$. Since all but a finite number of members of C must lie in B_{z+1}, either $C \cap W_z$ or $C \cap \bar{W}_z$ is finite. This holds for all z, and thus C is a cohesive subset of A contrary to our assumption that A has no cohesive subset. ⊠

(The above proof can be made somewhat more constructive. It is here arranged so that the only fact used about the recursively enumerable sets is that they form an at most denumerably infinite family. Like the above proof, a number of other proofs below will be seen to be purely set-theoretic (rather than recursive-function-theoretic) in nature—requiring only that the family of recursively enumerable sets be at most denumerable, or, in addition, requiring that it be closed under intersection or that it include the finite and co-finite sets. The proof of Theorem VI is generalized in Exercises 12-24 and 12-31.)

Since every infinite subset of a cohesive set is cohesive, there are clearly $2^{\aleph_0}$ cohesive sets. We say that a cohesive set C *belongs* to a recursively enumerable set B if $C \cap B$ is infinite, i.e., if the element of C lies below the element of B in $\mathfrak{N}/\mathfrak{F}$. We call two cohesive sets c-*equivalent* if they belong to the same recursively enumerable sets.

Definition Let A and B be cohesive. A is c-*equivalent* to B if $(\forall x)[A \cap W_x \text{ is infinite} \Leftrightarrow B \cap W_x \text{ is infinite}]$.

Theorem VII *Let A and B be cohesive. Then*
(i) *A is c-equivalent to $B \Leftrightarrow A \cup B$ is cohesive;*
(ii) *$A \cap B$ infinite $\Rightarrow A \cup B$ cohesive.*

Proof. Immediate from definition. See Exercise 12-23. ⊠

If any fixed c-equivalence class of cohesive sets is taken together with the finite sets, the result is an ideal in $\mathfrak{N}$, and the quotient of this ideal with $\mathfrak{F}$ is an ideal in $\mathfrak{N}/\mathfrak{F}$. Each c-equivalence class contains $2^{\aleph_0}$ elements. How many c-equivalence classes are there? The following theorem answers this question.

Theorem VIII *The cohesive sets fall into* $2^{\aleph_0}$ *c-equivalence classes.*

Proof. A proof can be made with simple concepts of recursive function theory and set theory. It is convenient and efficient, however, to abbreviate the proof by using familiar properties of real and rational numbers.

Observe that the integers can be effectively mapped onto the rationals in the following sense: there exists a one-one mapping ζ from N onto the rationals and a recursive function f such that for all n, $\zeta(n) = \pi_1 f(n)/\pi_2 f(n)$. (Any of the usual enumerations of the rationals can be used to obtain such a mapping.)

Assume that such a mapping has been fixed. Sets of rationals can now be identified with sets of integers. In particular, every closed interval of rationals with rational end points will constitute a recursive set.

For each real number, choose an infinite convergent sequence of rationals having that real as limit. Each of these chosen convergent sequences is, as a set of integers, infinite; and hence by Theorem VI, each contains a cohesive subset. Choose one cohesive subset from each convergent sequence. Cohesive sets coming from distinct sequences cannot be c-equivalent, since their union can be separated into two infinite parts by an appropriate interval with rational end points. As there are $2^{\aleph_0}$ sequences (since there are $2^{\aleph_0}$ reals), there exist at least $2^{\aleph_0}$ cohesive sets no two of which are c-equivalent. There are at most $2^{\aleph_0}$ c-equivalence classes, since there are $2^{\aleph_0}$ subsets of N.☒

The concept of cohesive set provides a means for proving the following theorem.

Theorem IX (Kent) *There exists a permutation f such that*

(i) *for all recursively enumerable* B, $f(B)$ *and* $f^{-1}(B)$ *are recursively enumerable (and hence for all recursive* A, $f(A)$ *and* $f^{-1}(A)$ *are recursive);*

(ii) f *is not recursive.*

Proof. Take D to be cohesive. Consider

$$\mathfrak{D} = \{f | f \textit{ is a permutation} \ \& \ (\forall x)[x \notin D \Rightarrow f(x) = x]\}.$$

Since each f in $\mathfrak{D}$ is arbitrary on D, $\mathfrak{D}$ contains $2^{\aleph_0}$ members. Since the class of recursive functions is denumerable, $\mathfrak{D}$ contains some member that is not recursive. It remains to verify that for any f in $\mathfrak{D}$, $f(A)$ and $f^{-1}(A)$ are recursive for any recursive A, and $f(B)$ and $f^{-1}(B)$ are recursively enumerable for any recursively enumerable B. This follows directly from the cohesiveness of D (see Exercise 12-23).☒

Corollary IX *The lattice* $\mathcal{E}$ *has automorphisms other than those induced by the recursive permutations.*

Proof. Immediate.☒

In Exercise 12-32, Corollary IX is strengthened to show that there are recursively enumerable sets A and B such that A is carried to B by an automorphism of $\mathcal{E}$, but A is not recursively isomorphic to B. It follows that a recursively invariant property need not be lattice-theoretic in $\mathcal{E}$.

Definition A cohesive set A is *completed* if $(\forall B)[[B$ cohesive & B c-equivalent to $A] \Rightarrow B \cap \bar{A}$ is finite$]$. Thus a c-equivalence class contains a completed set if and only if the ideal in $\mathfrak{N}/\mathfrak{F}$ determined by that c-equivalence class possesses a maximum element.†

In §12.4, we show that completed sets exist. In §12.6, we show that not every c-equivalence class contains a completed member. In Exercise 12-56 we show, further, that there are only countably many completed sets. In Exercise 12-60, we show that A completed cohesive $\Rightarrow \bar{A}$ recursively enumerable. The following concept is due to Rose and Ullian [1963].

Definition A is *quasicohesive* if A is the union of a finite (nonzero) number of cohesive sets.

If A is quasicohesive, then any decomposition of A into cohesive sets determines a set of corresponding c-equivalence classes. It follows from Theorem VII that this set of c-equivalence classes is uniquely determined, independent of the decomposition used (see Exercise 12-27).

The quasicohesive sets together with the finite sets form an ideal $\mathfrak{F}_1$ in $\mathfrak{N}$; it is the ideal generated by the cohesive sets. The quotient lattices $\mathfrak{N}/\mathfrak{F}_1$, $\mathcal{E}/\mathfrak{F}_1$, $\mathcal{R}/\mathfrak{F}_1$ can hence be formed.

The quasicohesive sets are hyperimmune (and in fact hyperhyperimmune; see Exercise 12-26). In Exercise 12-30, we see that a set can be hyperhyperimmune without being quasicohesive.

§12.4 MAXIMAL SETS

We next consider recursively enumerable sets with cohesive complements.

Definition A is a *maximal* set if
(i) A is recursively enumerable,
(ii) $\bar{A}$ is cohesive.

Theorem X *A is a maximal set $\Leftrightarrow$ A belongs to an element of $\mathcal{E}/\mathfrak{F}$ that is maximal among the nonunit elements of $\mathcal{E}/\mathfrak{F}$.*

Proof. $\Rightarrow$: If $A \in a$, and a is not maximal among the nonunit elements of $\mathcal{E}/\mathfrak{F}$, then there must be an element b such that $a < b < 1$. Take $B \in b$ such that $A \subset B$. $N - B$ must be infinite and $B - A$ must be infinite. Hence $\bar{A} \cap B$ is infinite and $\bar{A} \cap \bar{B}$ is infinite. Hence $\bar{A}$ is not cohesive, and A cannot be maximal.

$\Leftarrow$: If A is recursively enumerable but not maximal, then $\bar{A}$ is not cohesive. Take C to be a recursively enumerable set such that $\bar{A} \cap C$ is

† That is to say, is a *principal* ideal.

infinite and $\bar{A} \cap \bar{C}$ is infinite. Set $B = A \cup C$. Let b be the element of B in $\mathcal{E}/\mathfrak{F}$. Then $a < b < 1$, where $A \in a$.☒

The existence of maximal sets is given by the following theorem, due to Friedberg [1958*b*]. In the proof, we use an elegant simplification, due to Yates [1965], of Friedberg's construction. Existence of complete maximal sets (Yates) and of incomplete maximal sets (Sacks) can be shown by appropriate modifications of this construction; see Exercise 12-59.

Theorem XI *There exists a maximal set.*

Proof. We first describe a procedure for enumerating a set A. We then prove that A is a maximal set. The enumeration of A is carried out in stages. We begin with two definitions.

Definition $W_z^{(n)} = \{x|x$ *is enumerated within n steps in an enumeration without repetitions of* $W_z\}$.†

Definition For any z, any stage $n + 1$, and any integer x, we define the *z-state of x at stage* $n + 1$ to be $b_0b_1 \cdots b_z$, where $b_0b_1 \cdots b_z$ is a finite sequence of 0's and 1's such that

$$b_i = \begin{cases} 1, & \text{if } x \in W_i^{(n+1)} \text{ and } i \leq n; \\ 0, & \text{if } x \notin W_i^{(n+1)} \text{ or } n < i, \end{cases}$$

for $0 \leq i \leq z$.

There are clearly 2^{z+1} possible z-states. We take them to be ordered lexicographically (from "lower" to "higher"); for example, 10111 will be *below* 11000 (in the ordering of 4-states).

We note two fundamental properties of z-states and their ordering: (i) for fixed z and x, and for $n < m$, the z-state of x at stage m must be at least as high as the z-state of x at stage n; (ii) for fixed stage n, for given x and y, and for $z < z'$, if the z-state of y is higher than the z-state of x, then the z'-state of y must be higher than the z'-state of x.

The procedure for enumerating A is as follows. We begin with a *list* of all the integers and assume that we have associated with each integer x in this list the marker $\boxed{x}$. We give instructions for moving certain of these markers downward. At any stage, an integer in the list that appears without a marker will remain without a marker at all later stages. Integers of the list that (eventually) appear without markers constitute the set A. (Although, in our description, we begin with infinitely many markers, and although, at a given stage, we speak of moving infinitely many markers, it will be evident that we have an effective procedure for listing A. It is possible, with some increase in intuitive complexity, to give our procedure in a form which uses only finitely many markers at each stage.)

Stage $n + 1$. For each m, let $x_m{}^{(n)}$ be the position of marker $\boxed{m}$ at the

† By "n steps in an enumeration without repetitions of W_z," we mean n machine steps in the following procedure: compute $\varphi_{f''(z)}(0)$; if and when this converges, compute $\varphi_{f''(z)}(1)$; if and when . . . ; where f'' is as in Corollary 5-V(d).

end of stage n. Compute $W_z^{(n+1)}$ for all $z \leq n$. Find the least m, if any, such that for some $q' > m$, $x_{q'}^{(n)}$ is in higher m-state (at stage $n + 1$) than $x_m^{(n)}$. If there is no such m, go to stage $n + 2$. If there is such a least m, let q be the least corresponding q', move marker $\boxed{m}$ down to $x_q^{(n)}$, and for each $p > m$, move marker $\boxed{p}$ down to $x_{p+q-m}^{(n)}$. Place in A all members of $\{y | x_m^{(n)} \leq y < x_q^{(n)}\}$ that are not yet in A. (These are the only integers in the list to lose markers during stage $n + 1$.) Then go to stage $n + 2$.

The following two lemmas show that A must be maximal.

Lemma 1 *$\bar{A}$ is infinite.*

Proof of Lemma 1. It is clearly enough to show that each marker moves only finitely often. (The final position of each marker is then a member of $\bar{A}$.) Assume otherwise. Let m be the least m' such that $\boxed{m'}$ moves infinitely often. By fundamental property (i) of z-states, after $\boxed{m-1}$ reaches its final position, $\boxed{m}$ must move to positions in successively higher m-states. But there are only finitely many m-states. This is a contradiction, and the lemma is proved.

Lemma 2 *For every z, either $W_z \cap \bar{A}$ is finite or $\bar{W}_z \cap \bar{A}$ is finite.*

Proof of Lemma 2. Fix z. For each integer x in $\bar{A}$, x must reach a final z-state β as stage n increases. In this case we say that x *terminates* in β. Since $\bar{A}$ is infinite, at least one z-state must have infinitely many terminating members of $\bar{A}$. We show that at most one z-state can have infinitely many terminating members of $\bar{A}$. Assume otherwise. Let β be the lowest z-state with infinitely many terminating members of $\bar{A}$, and let β' be another z-state with infinitely many terminating members of $\bar{A}$. It follows that there must be integers m, n, x, and y, such that $z < m < n$, x is the final position of $\boxed{m}$, y is the final position of $\boxed{n}$, x ultimately terminates in z-state β, and y ultimately terminates in z-state β'. By fundamental property (ii) of z-states, y reaches and terminates in a higher m-state than x. But this means, by the construction, that marker $\boxed{m}$ must eventually be moved, contrary to our assumption that x is its final position. This proves Lemma 2.

Lemmas 1 and 2 show that $\bar{A}$ is cohesive. Hence A is a maximal set. ☒

Corollary XI(a) *The lattice $\mathcal{E}/\mathfrak{F}$ is not densely ordered.*

Proof. By Theorems III and X. ☒

Corollary XI(b) *There exists a completed cohesive set.*

Proof. Let A be maximal. Then $\bar{A}$ is cohesive. If $\bar{A}$ is not completed, we have a contradiction to Theorem VII with A as the dividing recursively enumerable set. ☒

The unusual properties of maximal sets are useful in the construction of various examples and counterexamples (see Exercises 12-35 and 12-37 and §12.6).

$\{A | A$ *is maximal or cofinite*$\}$ does not form a dual ideal in $\mathcal{E}$ (Exercise 12-36). In this respect, maximal sets differ from simple and hypersimple sets.

Definition A is *quasimaximal* if A is recursively enumerable and $\bar{A}$ is quasicohesive. In Exercise 12-36 below we show that recursively enumerable sets exist which are quasimaximal but not maximal and that $\{A|A$ *is quasimaximal or co-finite*$\}$ is a dual ideal in $\mathcal{E}$.

§12.5 SUBSETS OF MAXIMAL SETS

Is every nonunit element of $\mathcal{E}/\mathcal{F}$ contained in a maximal element? That is to say, is every non-co-finite recursively enumerable set contained in a maximal set? This question is answered in Theorem XIV below. We first give some related results in the following theorem.

Theorem XII(a) *If a recursively enumerable set is neither simple nor co-finite, then it is contained in a maximal set.*

(***b***) *If a recursively enumerable set is not co-finite, then it is contained in a hypersimple set.*

Proof. (*a*) Let A be recursively enumerable with $\bar{A}$ neither finite nor immune. Then $\bar{A}$ contains an infinite recursive set B. Let f be a recursive function mapping N one-one onto B, and let C be the image in B (under f) of some maximal set. Then $\bar{B} \cup C$ is the desired maximal set containing A.

(*b*) Let A be a recursively enumerable set with $\bar{A}$ infinite. Assume A is not hypersimple (otherwise the result is immediate). By Exercise 12-39, take a recursive f such that $D_{f(0)}, D_{f(1)}, \ldots$ is a sequence of disjoint finite sets, each intersecting $\bar{A}$, whose union is N. Let B be any hypersimple set. Let $C = A \cup \bigcup_{i \in B} D_{f(i)}$. Then C is the desired hypersimple set containing A (see Exercise 12-40).☒

Definitions $|A|$ = the cardinality of A.

$\overline{lim}\ (f,A) = \lim_{n\to\infty} (\sup_{m>n} |W_{f(m)} \cap A|)$. (Thus, to assert $\overline{lim}\ (f,A) \geq p$ is to assert that for infinitely many values of m, $W_{f(m)} \cap A$ has at least p members.)

f is a *disjoint recursively enumerable sequence* if f is recursive and

$$(\forall u)(\forall v)[u \neq v \Rightarrow W_{f(u)} \cap W_{f(v)} = \emptyset].$$

$\overline{lim}\ (A) = \sup_{f\in\mathfrak{D}} (\overline{lim}\ (f,A))$, where $\mathfrak{D}$ is the collection of all disjoin recursively enumerable sequences. (Thus, to assert $\overline{lim}\ (A) \geq p$ is to assert that $\overline{lim}\ (f,A) \geq p$ for some disjoint recursively enumerable sequence f.)

Thus we have, for example, that $\overline{lim}\ (A) = 0 \Leftrightarrow A$ is finite.

The following theorems due to Martin [1963] answer our initial question.

Theorem XIII (Martin) *A maximal $\Rightarrow \overline{lim}\ (\bar{A}) = 1$.*

Proof. By definition, $\bar{A}$ is cohesive. Assume $\overline{lim}\ (\bar{A}) \geq 2$. Then for some disjoint f, $\overline{lim}\ (f,A) \geq 2$. The following instructions define a recursively enumerable set B.

Enumerate A; at the same time begin enumerations of $W_{f(0)}$, $W_{f(1)}$, ... At each stage, and for each i, place members of $W_{f(i)}$ into B according to the following rule. Each integer that has appeared in both A and $W_{f(i)}$ goes into B, and the first integer (if any) that has appeared in $W_{f(i)}$ but not yet in A goes into B. Since, for infinitely many i, $W_{f(i)}$ contains at least two members of $\bar{A}$, infinitely many members of $\bar{A}$ go into B and infinitely many members of $\bar{A}$ never go into B. But then B splits $\bar{A}$ into two infinite parts, contrary to the cohesiveness of $\bar{A}$. Hence $\overline{lim}\ (\bar{A}) \leq 1$. Since $\bar{A}$ is infinite, $\overline{lim}\ (\bar{A}) > 0$. Hence $\overline{lim}\ (\bar{A}) = 1$.⊠

***Theorem XIV* (*Martin*)** *There exists a recursively enumerable set A such that $\bar{A}$ is infinite and A is contained in no maximal set.*

Proof. We exhibit a recursively enumerable A such that $\bar{A}$ is infinite and for all recursively enumerable B, $B \supset A \Rightarrow \overline{lim}\ (\bar{B}) \neq 1$. By Theorem XIII, the result follows. We give instructions for enumerating A.

Define the sequence of disjoint finite sets

$$S_0 = \{0\},$$
$$S_1 = \{1,2\},$$
$$S_2 = \{3,4,5\},$$
$$\cdots\cdots\cdots;$$

where, for each n, S_n contains $n+1$ members. Begin enumerations of W_0, W_1, W_2 .... At each stage, and for each i, see whether, for some $j < i$, exactly i of the $i+1$ members of S_i have been already listed in W_j; for each such j, put into A the remaining member of S_i that is not yet in W_j.

Since S_i contains $i+1$ elements, and at most i of these can be put in A (one for each $j < i$), $\bar{A}$ must be infinite. (Indeed, by Theorem 9-XV, A is not hypersimple.) Let B be any recursively enumerable set such that $B \supset A$. Then $B = W_j$ for some j, and by the construction, for all $i > j$, either $S_i \subset B$ or $S_i \cap \bar{B}$ has at least two members. Hence either $\bar{B}$ is finite (and $\overline{lim}\ (\bar{B}) = 0$), or, for infinitely many i, $S_i \cap \bar{B}$ has at least two members (and $\overline{lim}\ (\bar{B}) \geq 2$). In either case, $\overline{lim}\ (\bar{B}) \neq 1$, and the theorem is proved.⊠

With a minor modification of the construction, the result can be strengthened as follows.

***Corollary XIV* (*Martin*) (*a*)** *If a recursively enumerable set is neither hypersimple nor co-finite, then it is contained in a simple, nonhypersimple set C such that C is contained in no maximal set.*

(*b*) *If a recursively enumerable set is neither hypersimple nor co-finite, then it is contained in a hypersimple set D such that D is contained in no maximal set.*

Proof. (*a*) Assume B is recursively enumerable but neither co-finite nor hypersimple. Take (by Theorem 9-XV) an effective enumeration by canonical indices of disjoint finite sets each intersecting $\bar{B}$. Taking unions of these finite sets, construct an effective sequence of disjoint finite sets $S'_0, S'_1, \ldots$ such that for each n, $S'_n \cap \bar{B}$ has at least $n + 1$ members and such that $(\forall x)(\exists i)[x \in S'_i]$. Apply the method of Theorem XIV to get a set A from $S'_0, S'_1, \ldots$. Set $C = A \cup B$. By the construction from Theorem XIV, C is not co-finite and not contained in a maximal set. Hence by Theorem XII(a), C is simple. The sequence $S'_0, S'_1, \ldots$ shows (by Theorem 9-XV) that C is not hypersimple.

(b) This is immediate from (a) and Theorem XII(b).⊠

The converse to Theorem XIII (for recursively enumerable A) is known to fail. Is every non-co-finite recursively enumerable set contained in a hyperhypersimple set? Martin [1963] gives a strengthening of Corollary XIV in which "maximal" is replaced by "hyperhypersimple."

For any element a in $\mathcal{E}/\mathfrak{F}$, consider the maximal chains of nonunit elements lying between a and the unit element. (The existence of such maximal chains is easy to show.) Theorem XIV shows that nonunit elements exist none of whose maximal chains can have a greatest element. Are there nonunit elements in $\mathcal{E}/\mathfrak{F}$ with maximal chains that are dense? Are there nonunit elements all of whose maximal chains are dense? These questions are open.†

Can we make a conjecture about the structure of $\mathcal{E}/\mathfrak{F}$ analogous to Shoenfield's conjecture on the structure of the T-reducibility ordering of the T-degrees (§10.3)? The irregularities indicated by the existence of maximal elements and by the existence of elements not contained in maximal elements appear to make the formulation of such a conjecture difficult.

How widely distributed are the maximal sets among the degrees of unsolvability? In this connection we give Theorem XV below. (We state a more general result, due to Martin, in Chapter 13.) Are there maximal sets of complete T-degree (see §9.6)? Are there maximal sets of incomplete T-degree? Are there nonrecursive, recursively enumerable T-degrees containing no maximal set? Affirmative answers to these three questions have been obtained by Yates [1965], Sacks [1964*b*], and Martin [1965].

***Theorem XV* (*Young*)** *If A is recursively enumerable but not recursive, B is maximal, and $A \leq_m B$, then $A \equiv_m B$.*

Proof. Assume $A \leq_m B$ via f. $f(\bar{A})$ must be infinite (otherwise $\bar{A}$ would be recursively enumerable). Hence $C = \bar{B} - f(\bar{A})$ must be finite (otherwise the recursively enumerable set $f(N)$ would split the cohesive set $\bar{B}$). Let m be a fixed member of A and n be a fixed member of $\bar{A}$. Define a recursive g as follows. To compute $g(y)$, see whether $y \in C$; if so set $g(y) = n$; if not begin an enumeration of B and, at the same time, begin a

† A *chain* in a lattice is a linearly ordered subset. A chain between two elements is maximal if no chain between those elements can be formed by adding a new element to the given chain. A chain is *dense* if between any two members of the chain there is a third member of the chain.

search for an x such that $f(x) = y$; if y appears in B before such an x is found, set $g(y) = m$; otherwise such an x must occur (since $(\bar{B} - C) \subset f(N)$), set $g(y) =$ the first such x. Then $B \leq_m A$ via g.☒

§12.6 ALMOST-FINITENESS PROPERTIES

The lattices, ideals, and quotient lattices mentioned in this chapter (and generalizations thereof) have not been extensively studied. It is not clear how much further insight into the structure of recursively enumerable sets (in connection with the ultimate goal of a representation theorem for recursively enumerable sets) such study will provide. A number of rather easily formulated questions remain unanswered.

In this section we consider certain properties that determine subclasses of the class of immune sets. In view of the similarity between immune sets and finite sets, we refer to these as *almost-finiteness properties.* We first discuss properties known to be lattice-theoretic in $\mathfrak{N}$ with respect to $\mathcal{E}$; we then discuss recursively invariant properties not known to be lattice-theoretic with respect to $\mathcal{E}$; and finally we discuss the restrictions of these properties to sets with recursively enumerable complements. We consider properties only as they apply to infinite sets.

Lattice-theoretic Properties

We make the following definitions.

Definition A is *indecomposable* if there do not exist two recursively enumerable sets B_1 and B_2 such that $B_1 \cap B_2 = \emptyset$, $A \subset B_1 \cup B_2$, $B_1 \cap A$ is infinite, and $B_2 \cap A$ is infinite.

Definition A is *recursively indecomposable* if there is no recursive set B such that $B \cap A$ is infinite and $\bar{B} \cap A$ is infinite.

We list certain properties, all of which are evidently lattice-theoretic with respect to $\mathcal{E}$:

(1) *Cohesive and* c-*equivalent to a completed set;*
(2) *Cohesive;*
(3) *Quasicohesive;*
(4) *Quasicohesive or generalized cohesive* (see Exercise 12-30);
(5) *Infinite and indecomposable;*
(6) *Infinite and recursively indecomposable;*
(7) *Immune.*

The following implications are immediate: (1) $\Rightarrow$ (2); (2) $\Rightarrow$ (3); (3) $\Rightarrow$ (4); (4) $\Rightarrow$ (7); (2) $\Rightarrow$ (5); (5) $\Rightarrow$ (6); and (6) $\Rightarrow$ (7). None of the converse implications holds. By Theorem VIII, (3) $\not\Rightarrow$ (2). Exercise 12-30 shows that (4) $\not\Rightarrow$ (3). The existence of simple but not hypersimple sets shows that (7) $\not\Rightarrow$ (4). Exercise 12-46 shows that (7) $\not\Rightarrow$ (6). Theorems XVI, XVII, and XVIII below show, respectively, that (2) $\not\Rightarrow$ (1), (5) $\not\Rightarrow$ (2), and (6) $\not\Rightarrow$ (5). In addition, Exercise 12-46 shows (3) $\not\Rightarrow$ (6), and Exercise 12-57 implies that (5) $\not\Rightarrow$ (4) (see discussion below of recursively invariant properties.) We summarize this information in the following diagram (in which all other implications fail) and then prove the relevant theorems.

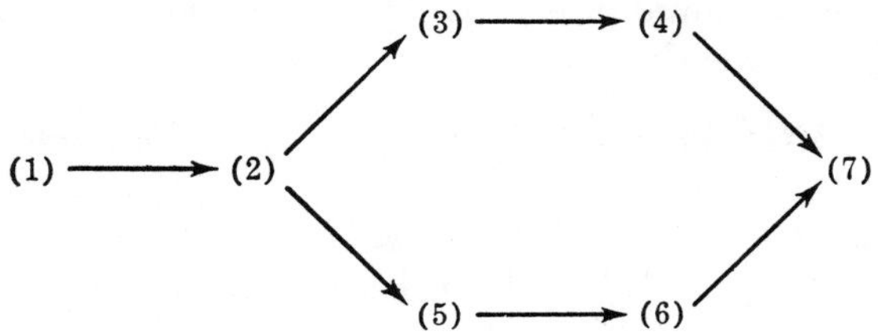

***Theorem XVI* (*Young*)** *There exists a cohesive set which has no completed set in its* c*-equivalence class.*

Proof. The construction of Theorem VI is modified as follows. Define a sequence of sets

$$A_0 = N;$$
$$A_{n+1} = \begin{cases} A_n \cap W_n, & \text{if } both\ A_n \cap W_n \text{ and } A_n \cap \bar{W}_n \text{ are infinite;} \\ A_n, & \text{otherwise.} \end{cases}$$

Note that $A_0 \supset A_1 \supset \cdots$ and that for each n, A_n is recursively enumerable. Since every infinite recursively enumerable set is divided into two infinite parts by some other recursively enumerable set, we can conclude that $A_{m+1} \neq A_m$ for infinitely many m. Hence we can define a properly decreasing subsequence by

$$B_0 = A_0,$$
$$B_{n+1} = A_m,$$

where $m = \mu y[A_y$ *is properly contained in* $B_n]$. By construction, $B_n - B_{n+1}$ is infinite for each n.

Let $x_n = \mu y[y \in B_n - B_{n+1}]$; and let $C = \{x_0, x_1, \ldots\}$. Let any W_z be given. By our construction, all but a finite number of members of C lie in A_{z+1}. If $A_{z+1} = A_z$, then either $A_z \cap W_z$ or $A_z \cap \bar{W}_z$ is finite, and hence either $C \cap W_z$ or $C \cap \bar{W}_z$ is finite. If $A_{z+1} \neq A_z$, then $A_{z+1} \cap \bar{W}_z = \emptyset$ and $C \cap \bar{W}_z$ is finite. In either case, $C \cap W_z$ or $C \cap \bar{W}_z$ is finite. Since C is infinite, we conclude that C is cohesive.

Assume there is a completed set D such that D is c-equivalent to C. Then by Theorem VII, there can be no recursively enumerable set which divides $D \cup C$ into two infinite parts. It follows that for each n, $D \cap (B_n - B_{n+1})$ must be finite. Hence $(B_n - B_{n+1}) - D \neq \emptyset$. Define

$$y_n = \mu y[y \in (B_n - B_{n+1}) - D].$$

Define $D' = \{y_0, y_1, \ldots\}$. Then for every z, either $(C \cup D') \cap W_z$ is finite or $(C \cup D') \cap \bar{W}_z$ is finite. Hence $C \cup D'$ is cohesive. By Theorem VII, C is c-equivalent to D'. Hence D is c-equivalent to D'. But $D \cap D' = \emptyset$, contrary to our assumption that D is completed.☒

Corollary XVI *Every infinite recursively enumerable set contains a cohesive set which has no completed set in its c-equivalence class.*

Proof. Given an infinite recursively enumerable set B, take $A_0 = B$ in the construction of the theorem.☒

Theorem XVII *There exists an infinite set which is indecomposable but not cohesive.*

Proof. Let D be a completed cohesive set. Let $x_0, x_1, \ldots$ be the sequence, in increasing order, of all r.e. indices x such that $D \subset W_x$. Define

$$A_0 = W_{x_0},$$
$$A_{n+1} = A_n \cap W_{x_n}.$$

The sequence $A_0, A_1, \ldots$ must contain a strictly decreasing subsequence; otherwise D would be recursively enumerable. Define

$$B_0 = A_0,$$
$$B_{n+1} = A_m,$$

where $m = \mu y[A_y$ *is properly contained in* $B_n]$.

Let $y_n = \mu y[y \in B_n - B_{n+1}]$. Form $C = \{y_0, y_1, \ldots\}$. Let D' be a cohesive subset of C (by Theorem VI). Then $D' \cap D = \emptyset$ and $D' \cap \bar{D}$ is infinite. Since D is completed, D' cannot be c-equivalent to D. Hence, by Theorem VII, $D' \cup D$ is not cohesive.

On the other hand, $D' \cup D$ is indecomposable. Assume otherwise. Then, since D' and D are each cohesive, we must have $D' \cap B_1$ infinite and $D \cap B_2$ infinite for two disjoint recursively enumerable sets B_1 and B_2. Furthermore, $D \cap \bar{B}_2$ must be finite. Hence $B_2' = B_2 \cup (D \cap \bar{B}_2)$ is recursively enumerable and $D \subset B_2'$. By the construction of D', $D' \cap B_2'$ must be infinite. But then D' is divided into two infinite parts by B_2', contrary to the cohesiveness of D'. Thus $D' \cup D$ is an infinite indecomposable set which is not cohesive.☒

Theorem XVIII **(*McLaughlin*)** *There exists a set which is recursively indecomposable but not indecomposable.*

Proof. Let A be a maximal set. By Theorem IV, decompose A into two recursively enumerable, nonrecursive sets A_1 and A_2. Let $\mathcal{B}$ be the family of all recursively enumerable sets B such that $B \supset \bar{A}$. The members of $\mathcal{B}$ are clearly closed under $\cap$. Furthermore, for each $B \in \mathcal{B}$, both $B \cap A_1$ and $B \cap A_2$ must be infinite; for if, say, $B \cap A_1$ were finite, then $B \cup A_2$ would be a complement to A_1 in $\mathcal{E}/\mathfrak{F}$, contrary to the nonrecursiveness of A_1.

Let $B_0, B_1, \ldots$ be the members of $\mathcal{B}$. Define

$$C_0 = B_0;$$
$$C_{n+1} = C_n \cap B_{n+1}.$$

Since all co-finite sets containing $\bar{A}$ occur in $\mathcal{B}$, both the sequence $C_0 \cap A_1$, $C_1 \cap A_1$, $C_2 \cap A_1, \ldots$ and the sequence $C_0 \cap A_2$, $C_1 \cap A_2$, $C_2 \cap A_2, \ldots$ each contain a strictly decreasing subsequence. Hence the sequence $C_0, C_1, \ldots$ contains a subsequence $C'_0, C'_1, \ldots$ such that for all n, $(C'_{n+1} - C'_n) \cap A_1 \neq \emptyset$ and $(C'_{n+1} - C'_n) \cap A_2 \neq \emptyset$. Choose $\{x_0, x_1, \ldots\}$ so that $x_n \in (C'_{n+1} - C'_n) \cap A_1$, and choose $\{y_0, y_1, \ldots\}$ so that $y_n \in (C'_{n+1} - C'_n) \cap A_2$. Take E_1 to be a cohesive subset of $\{x_0, x_1, \ldots\}$ and E_2 to be a cohesive subset of $\{y_0, y_1, \ldots\}$. Consider the set $E_1 \cup E_2$. This set is evidently decomposable (by A_1 and A_2). It is, however, not recursively decomposable. For, otherwise, let B be a recursive set which decomposes it. Let E_1 *belong* to B (in the sense of §12.3). Then by the decomposition, E_2 belongs to $\bar{B}$. But $\bar{A}$ belongs to either B or $\bar{B}$, and, by the construction, if $\bar{A}$ belongs to a set, both E_1 and E_2 belong to that set. This is a contradiction. ☒

This proof is generalized in Exercise 12-47.

Recursively Invariant Properties

The following properties are not known to be lattice-theoretic with respect to $\mathcal{E}$:

(8) *Infinite with no infinite retraceable subset* (see Exercise 9-44);
(9) *Hyperhyperimmune;*
(10) *Hyperimmune.*

The following implications are immediate: (9) $\Rightarrow$ (10), and (10) $\Rightarrow$ (7). In Exercises 12-48 and 12-49, we see that (4) $\Rightarrow$ (8) and (8) $\Rightarrow$ (9). Using the set S of Theorem 8-II, we see that (7) $\not\Rightarrow$ (10). Exercise 10-17 showed that (10) $\not\Rightarrow$ (9). Exercises 12-51 and 12-52 show that (9) $\not\Rightarrow$ (8) and (8) $\not\Rightarrow$ (4). Exercise 12-53 shows that (6) $\Rightarrow$ (10). Corollary XIX below and Exercises 12-45 and 12-62 show that (5) $\not\Rightarrow$ (9). We summarize this information in the following diagram (in which all other implications fail).

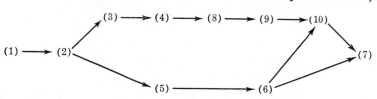

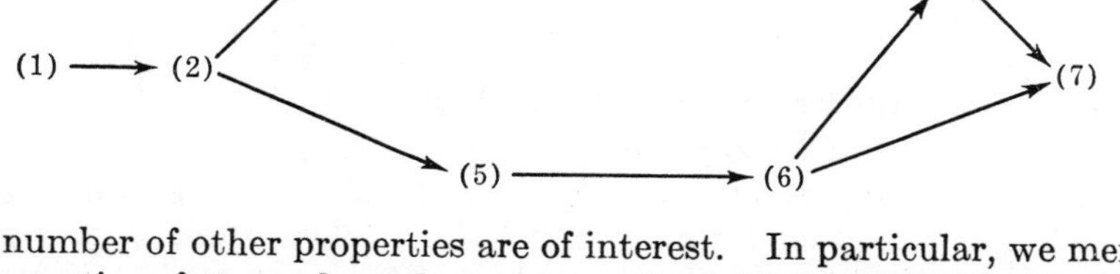

A number of other properties are of interest. In particular, we mention the properties of strong hyperhyperimmunity (which coincides with (8), see Exercises 12-48 and 12-50) and strong hyperimmunity (see Exercise 12-54) and the properties, for a set A, that $\overline{lim}\ (A) = 1$, that $\overline{lim}\ (A)$ be finite, and that $\overline{lim}\ (f,A)$ be finite for all recursively enumerable disjoint sequences f.

(Corresponding $\overline{lim}$ properties could also be defined using sequences of finite sets and using sequences of finite sets given by canonical indices.)

Recursively Enumerable Complements

A number of interesting questions arise when properties (1) to (10) are restricted to sets whose complements are recursively enumerable. We denote these restricted properties as (1)*, (2)*, . . . , (10)*.

All previous implications must continue to hold. Several converse implications that previously failed now hold. The proof of Corollary XI(b) showed that (2)* $\Rightarrow$ (1)*. Exercise 10-19 showed that (9)* $\Rightarrow$ (8)*. In Exercise 12-45 we show that (6)* $\Rightarrow$ (5)*. The counterexamples previously used show that (as before) (7)* $\not\Rightarrow$ (10)* (and hence (7)* $\not\Rightarrow$ (6)*), (10)* $\not\Rightarrow$ (9)*, (3)* $\not\Rightarrow$ (2)*, and (3)* $\not\Rightarrow$ (6)*. In Corollary XIX below, we have the result, due to Lachlan, that no set not in (2)* can be in both (6)* and (9)*. In Exercise 12-62 we obtain a set in (6)* but not in (2)*. Hence (6)* $\not\Rightarrow$ (9)*. This result was first obtained by R. W. Robinson. Hence (6)* $\not\Rightarrow$ (2)*, (6)* $\not\Rightarrow$ (3)*, and (6)* $\not\Rightarrow$ (4)*. Robinson has also shown that (4)* $\not\Rightarrow$ (3)*, see Exercise 12-63. The implication (9)* $\Rightarrow$ (4)* remains an open question. Sets in (6)* are known as *r-maximal* sets*. In Theorem XX below, we show that hyperhypersimplicity is lattice-theoretic in $\mathcal{E}/\mathcal{F}$. The remarkable characterization there given is due to Lachlan. It is the first result showing a property to be lattice-theoretic whose definition is not ostensibly so. A proof that (9)* $\Rightarrow$ (4)* would give an alternative lattice-theoretic characterization of sets in (9)*. We conjecture, however, that the implication (9)* $\Rightarrow$ (4)* fails.

Theorem XIX **(*Lachlan*)** *Let A and B be recursively enumerable sets such that $A \subset B$. If $B - A$ is hyperhyperimmune, then there exists a recursive C such that $B - A \subset C \subset B$.*

Proof. Construct a family of recursively enumerable sets $S_0, S_1, S_2, \ldots$ as follows. $S_i^{(n)}$ will be the members of the set S_i enumerated by stage n. $A^{(n)}$ and $B^{(n)}$ will be the sets obtained by computing n steps in the respective enumerations of A and B.

Stage $n + 1$. Find the least integer y, if any, such that $y \in B^{(n+1)} - A^{(n+1)}$, such that $(\forall i)[y \notin S_i^{(n)}]$, and such that for some j', $S_{j'}^{(n)} \subset A^{(n+1)}$ and $y > \max_{i<j'} |S_i^{(n)}|$. If no such y exists, go to stage $n + 2$. If such a least y exists, take j to be the least such corresponding j'. Define $S_j^{(n+1)} = S_j \cup \{y\}$ and $S_i^{(n+1)} = S_i^{(n)}$ for all $i \neq j$. Go to stage $n + 2$.

Note that, at any stage, for any i, S_i contains at most one member of $B - A$. Assume that for all i, S_i is finite. Then, since $B - A$ is infinite, the construction must give a member of $B - A$ to each S_i. But the sequence $\{S_i\}_{i=0}^{\infty}$ $(= \{W_{f(i)}\}_{i=0}^{\infty}$ for some recursive $f)$ violates our assumption that $B - A$ is hyperhyperimmune. Hence, for some least k, S_k is infinite. By the construction, $S_k \subset A$. Let $m = \max_{i<k} |S_i|$. Once the finite sets

S_i, $i < k$, are completely enumerated, every new integer y such that both $y > m$ and $y \in B - A$ must be put in S_j for some $j > k$. (Otherwise, y would be put in S_k.) Let $C' = \bigcup_{i \neq k} S_i$. Then $C' \subset B$ and $(B - A) - C'$ is finite. We show that C' is recursive. To test whether $z \in C'$ for any given z, proceed as follows. See whether or not z is in the finite set $\bigcup_{i<k} S_i$. If so, then $z \in C'$. If not, carry out the procedure for listing the S_i until at least z members have been placed in S_k. If, by this time, z has appeared in some S_j, $j > k$, we have $z \in C'$. If not, by the construction, we have $z \notin C'$. Let $C = (B - A) \cup {}'C$. This completes the proof. ☒

Corollary XIX *If A is recursively enumerable but not maximal, then $\bar{A}$ cannot be both hyperhyperimmune and recursively indecomposable.*

Proof. Assume A is recursively enumerable but not maximal, and assume $\bar{A}$ is hyperhyperimmune. Since A is not maximal, there is, by definition, a recursively enumerable set B such that $A \subset B$, $B - A$ is infinite, and $N - B$ is infinite. Since $B - A \subset \bar{A}$, $B - A$ is hyperhyperimmune. By the theorem, there is a recursive C such that $B - A \subset C \subset B$. Evidently, C gives a recursive decomposition of $\bar{A}$ into two infinite sets. ☒

In Exercise 12-60 below, we use Theorem XIX to show that every completed cohesive set is co-maximal. The proof given above for Theorem XIX is due to Martin. Note that ((6)*, (10)*) and ((3)*, (4)*, (9)*, (10)*) give two different (and, by Corollary XIX, incomparable) ways of subdividing the class $\mathcal{C}_{12}$ (of §9.5) along the spectrum of §§8.7 and 9.5. In Exercise 12-55, we see that the sets which are hyperhypersimple or co-finite form a dual ideal in $\mathcal{E}$.

Theorem XX now gives a lattice-theoretic (in both $\mathcal{E}$ and $\mathcal{E}/\mathcal{F}$) characterization of hyperhypersimple sets.

Theorem XX (Lachlan) *A is hyperhypersimple $\Leftrightarrow$ A is recursively enumerable, $\bar{A}$ is infinite, and for every recursively enumerable B such that $A \subset B$, the set $A \cup (N - B)$ is recursively enumerable.*

Proof. $\Rightarrow$: Assume A is hyperhypersimple, $A \subset B$, and B is recursively enumerable. We wish to show that $A \cup (N - B)$ is recursively enumerable. If $B - A$ is finite, then $A \cup (N - B)$ is co-finite and hence recursively enumerable. If $B - A$ is infinite, then $B - A$ is hyperhyperimmune since it is an infinite subset of a hyperhyperimmune set. By Theorem XIX, there is a recursive C such that $B - A \subset C \subset B$. Then $A \cup (N - B) = A \cup \bar{C}$ and is hence recursively enumerable.

$\Leftarrow$: Assume that A is recursively enumerable with $\bar{A}$ infinite, but that $\bar{A}$ is not hyperhyperimmune. We wish to find a recursively enumerable B such that $A \subset B$ and $A \cup (N - B)$ is not recursively enumerable. By the definition of hyperhyperimmunity, there is a recursive f such that $(\forall x)[W_{f(x)} \cap \bar{A} \neq \emptyset]$ and $(\forall x)(\forall y)[x \neq y \Rightarrow W_{f(x)} \cap W_{f(y)} = \emptyset]$. Define

$B = A \cup (\bigcup_{x \in N} (W_x \cap W_{f(x)}))$. Assume that $A \cup (N - B) = W_y$. By choice of f, $W_{f(y)} \cap \bar{A} \neq \emptyset$. Take $z \in W_{f(y)} \cap \bar{A}$. Then $z \in N - B \Rightarrow$ (by choice of y) $z \in W_y \cap W_{f(y)} \Rightarrow$ (by definition of B) $z \in B - A$. Conversely $z \in B - A \Rightarrow$ (since $x \neq y \Rightarrow W_{f(x)} \cap W_{f(y)} = \emptyset$) $z \in W_y \cap W_{f(y)} \Rightarrow$ (by choice of y) $z \in N - B$. Since $z \in \bar{A}$, this is a contradiction.☒

Note that the proof of the theorem yields Exercise 9-48(b) as an immediate corollary. We also have the following.

***Corollary XX* (*Lachlan*)** *A is co-finite or hyperhypersimple* $\Leftrightarrow$ *the collection of recursively enumerable sets containing A forms a Boolean algebra under the ordering* $\subset$.

Proof. Immediate.☒

§12.7 EXERCISES

§12.1

△***12-1.*** (a) Let $\mathfrak{L}$ be a set with binary (algebraic) operations $\cup$ and $\cap$ such that:

(i) $\cup$ and $\cap$ are both commutative;

(ii) $\cup$ and $\cap$ are both associative;

(iii) $\cup$ and $\cap$ are both idempotent (that is, $a \cup a = a$ and $a \cap a = a$);

(iv) For all a and b, $(a \cup b) \cap a = a$ and $(a \cap b) \cup a = a$.

Define: $a \leq b$ if $a \cup b = b$, and show that $\mathfrak{L}$ forms a lattice under $\leq$, with $\cup$ and $\cap$ as the join and meet operations of that lattice.

(b) Show that the operations of join and meet in any lattice satisfy (i) to (iv).

12-2. Give an example of two lattices $\mathfrak{M}$ and $\mathfrak{L}$ where $\mathfrak{M}$ is a subordering of $\mathfrak{L}$ but not a sublattice of $\mathfrak{L}$. (*Hint:* Take $\mathfrak{L} = \mathfrak{N}$ and $\mathfrak{M}$ = the lattice formed by $\{A | A \in \mathfrak{N}$ & $[[1 \in A \;\&\; 2 \in A] \Rightarrow 3 \in A]\}$ under $\subset$.)

12-3. Show that every element in a Boolean algebra has at most one complement. (*Hint:* Let b and c be complements for a. Consider $(a \cup b) \cap c$ and $(a \cup c) \cap b$.)

12-4. In the definition of distributive lattice, show that either identity implies the other.

12-5. Prove (independently of the Stone representation theorem) that the following dentities hold in any Boolean algebra:

$$\bar{\bar{a}} = a,$$
$$a \cup 1 = 1, \qquad a \cap 1 = a,$$
$$a \cup 0 = a, \qquad a \cap 0 = 0,$$
$$\overline{a \cup b} = \bar{a} \cap \bar{b}, \qquad \overline{a \cap b} = \bar{a} \cup \bar{b},$$
$$(a \cap b) \cup (a \cap \bar{b}) = a.$$

12-6. In part (a) of Theorem I, show that $\approx_{\mathfrak{g}}$ is an equivalence relation.

12-7. In part (a) of Theorem I, let a and b be elements of $\mathfrak{L}$. Let $\hat{a}$ and $\hat{b}$ be the equivalence classes determined by a and b respectively. Show that $\hat{a} \leq \hat{b}$ in $\mathfrak{L}/\mathfrak{g}$ if and only if, for some c in $\mathfrak{g}$, $a \leq b \cup c$ in $\mathfrak{L}$.

12-8. In part (a) of Theorem I, show that $\mathfrak{g}$ forms an equivalence class which is the zero element of $\mathfrak{L}/\mathfrak{g}$.

△***12-9.*** (The Stone Representation Theorem). Define: a dual ideal $\mathfrak{D}$ in a lattice $\mathfrak{L}$ is *proper* if $\mathfrak{D} \neq \mathfrak{L}$. A dual ideal $\mathfrak{D}$ in a lattice $\mathfrak{L}$ is *maximal* if $\mathfrak{D}$ is proper and is contained in no proper dual ideal other than itself.

Assume: in any Boolean algebra, every proper dual ideal is contained in a maximal dual ideal.

Prove: the Stone representation theorem.

(*Hint:* Let $\mathcal{B}$ be the given Boolean algebra. Define Σ to be the collection of all maximal dual ideals in $\mathcal{B}$. Associate with each $b \in \mathcal{B}$ the following subset of Σ: $\{\mathfrak{D}|\mathfrak{D}$ *is a maximal dual ideal containing* $b\}$.)

12-10. See Exercise 12-9. Define: a dual ideal $\mathfrak{D}$ in a lattice $\mathcal{L}$ is *prime* if $\mathfrak{D} \neq \mathcal{L}$ and for all a and b in $\mathcal{L}$, $a \cup b \in \mathfrak{D} \Rightarrow [a \in \mathfrak{D}$ or $b \in \mathfrak{D}]$.

Prove: (i) In a distributive lattice, every maximal dual ideal is prime.

(ii) In a Boolean algebra, every prime dual ideal is maximal.

12-11. *Zorn's lemma* asserts that if $\mathcal{P}$ is any partial ordering with the property that every linearly ordered subset of $\mathcal{P}$ has an upper bound in $\mathcal{P}$, then $\mathcal{P}$ contains a maximal element.

Assume: Zorn's lemma.

Prove: in any Boolean algebra, every proper dual ideal is contained in a maximal dual ideal. (This is the special assumption made in 12-9.) (*Hint:* Let $\mathfrak{D}$ be a given proper dual ideal. Consider the collection of all proper dual ideals containing $\mathfrak{D}$.) (*Note:* In the presence of the axioms of set theory (see Exercise 11-23) other than the axiom of choice, Zorn's lemma may be deduced from (and is, in fact, equivalent to) the axiom of choice.)

12-12. Let $\mathcal{L}$ be a lattice and let $\mathcal{M}$ be a lattice with zero element. Assume that a homomorphism (with respect to $\cup$ and $\cap$) is given from $\mathcal{L}$ onto $\mathcal{M}$. Let $\mathcal{I}$ be the set of elements in $\mathcal{L}$ which are mapped onto the zero element of $\mathcal{M}$ by the homomorphism.

(i) Show that $\mathcal{I}$ is an ideal in $\mathcal{L}$.

(ii) Show that if $\mathcal{L}$ is a Boolean algebra, then $\mathcal{L}/\mathcal{I}$ is isomorphic to $\mathcal{M}$.

(iii) Show that if $\mathcal{L}$ is distributive, $\mathcal{L}/\mathcal{I}$ need not be isomorphic to $\mathcal{M}$. (*Hint:* Take $\mathcal{L}$ to be a linear ordering of three elements and $\mathcal{M}$ a linear ordering of two elements.)

12-13. Let $\mathcal{L}$ be a distributive lattice, $\mathcal{M}$ a sublattice of $\mathcal{L}$, and $\mathcal{I}$ an ideal in $\mathcal{L}$ such that $\mathcal{M} \cap \mathcal{I} \neq \emptyset$.

(*a*) Show that $\mathcal{M} \cap \mathcal{I}$ is an ideal in $\mathcal{M}$.

(*b*) Show that $\mathcal{M}/\mathcal{I} = \mathcal{M}/\mathcal{M} \cap \mathcal{I}$ need not hold.

(*c*) Assume that $\mathcal{M}$ is a Boolean algebra and that $\mathcal{L}$ and $\mathcal{M}$ have the same zero and unit elements. Show that $\mathcal{M}/\mathcal{I} = \mathcal{M}/\mathcal{M} \cap \mathcal{I}$.

(*d*) Conclude that $\mathcal{R}/^*\mathcal{S} = \mathcal{R}/^*\mathcal{H} = \mathcal{R}/\mathcal{F}$.

12-14. (*a*) Show that $\mathcal{S}$ and $\mathcal{H}$ are not dual ideals in $\mathcal{N}$.

(*b*) Show that neither the collection of all finite or immune sets nor the collection of all finite or hyperimmune sets forms an ideal in $\mathcal{N}$.

12-15. (Myhill). See §8.7. Show that the following concepts are lattice-theoretic in $\mathcal{E}/\mathcal{F}$: $A \in \mathcal{C}_1$, $A \in \mathcal{C}_2$, $A \in \mathcal{C}_3 \cup \mathcal{C}_4$. (It is not known whether $A \in \mathcal{C}_3$ and $A \in \mathcal{C}_4$ are lattice-theoretic.)

12-16. Show that $\mathcal{R}/\mathcal{F}$ is densely ordered.

12-17. Show that every automorphism of $\mathcal{E}$ induces an automorphism of $\mathcal{E}/\mathcal{F}$.

12-18. Prove that for any nonzero a in $\mathcal{E}/\mathcal{F}$, $\{b|b \leq a\}$ constitutes a sublattice isomorphic to $\mathcal{E}/\mathcal{F}$. (*Hint:* Let $A \in a$; use a one-one recursive f with range $= A$.)

12-19. (*a*) Complete the proof of Theorem III.

(*b*) Show that $\mathcal{E}/^*\mathcal{S}$ contains no maximal nonunit elements. (This does not settle the question of density for $\mathcal{E}/^*\mathcal{S}$.)

12-20. (*a*) (Myhill). Show that for any $a \in \mathcal{E}/\mathcal{F}$, if $a \neq 0$, then there exist b, $c \in \mathcal{E}/\mathcal{F}$ such that $a = b \cup c$, where b consists of recursive sets and c consists of creative sets. (*Hint:* Use Theorem 5-IV.)

(*b*) (McLaughlin). Let $\mathcal{C}_0$, $\mathcal{C}_1$, $\mathcal{C}_2$ be as in §8.7. Show that $\mathcal{C}_0 \cup \mathcal{C}_1 \cup \mathcal{C}_2$ forms a sublattice of $\mathcal{E}$.

§*12.2*

△***12-21.*** (*a*) (K. Ohaŝhi). Show that the sets B_1 and B_2 constructed in Theorem IV are not recursively separable. (*Hint:* Show, by similar argument to final part of proof of Theorem IV, that a recursive separating set and its complement would make $\bar{A}$ recursively enumerable.)

(*b*) (Yates). Show that the sets B_1 and B_2 have the following property. $(\forall z)[[W_z \cap B_1 = \emptyset$ or $W_z \cap B_2 = \emptyset] \Rightarrow W_z - A$ is recursively enumerable] (where A is as in Theorem IV).

△***12-22.*** (McLaughlin [1962]). Let A be recursively enumerable but not recursive. Find a recursive f such that $(\forall u)[W_{f(u)}$ not recursive], $(\forall u)(\forall v)[u \neq v \Rightarrow W_{f(u)} \cap W_{f(v)} = \emptyset]$, and $A = \bigcup_{u \in N} W_{f(u)}$. (*Hint:* In place of B_1 and B_2, list an infinite collection $B_0, B_1, B_2, \ldots$.)

§*12.3*

12-23. (*a*) Prove Theorem VII.

(*b*) Complete the proof of Theorem IX.

△***12-24.*** (Dekker). Let $\mathfrak{A}$ be any denumerable collection of nonzero elements in $\mathfrak{N}/\mathfrak{F}$ which is closed under $\cap$. Show that there is a nonzero element c in $\mathfrak{N}/\mathfrak{F}$ such that $(\forall a)[a \in \mathfrak{A} \Rightarrow c \leq a]$. (*Hint:* Apply general method of Theorem VI.)

12-25. Show that N is the disjoint union of a denumerable collection of cohesive sets all from the same c-equivalence class. (*Hint:* Take any cohesive set A. Decompose A into an infinite collection of infinite sets. Adjoin members of $\bar{A}$ to these sets in an appropriate way.)

12-26. Show that A quasicohesive $\Rightarrow A$ hyperimmune (and in fact A hyperhyperimmune). (*Hint:* See proof of Theorem V.)

12-27. Show that decomposition of a quasicohesive set into cohesive sets is unique up to c-equivalence.

12-28. Is $\mathfrak{N}/\mathfrak{F}_1$ densely ordered?

12-29. Show that the finite sets together with all finite unions of those cohesive sets which are c-equivalent to completed sets form an ideal in $\mathfrak{N}$. (Note that any collection of cohesive sets generates (in this way) an ideal contained in $\mathfrak{F}_1$.)

△***12-30.*** Define: A is *generalized cohesive* if $A \notin \mathfrak{F}_1$, but for every x, either $W_x \cap A$ or $\bar{W}_x \cap A$ is in $\mathfrak{F}_1$. Show that generalized cohesive sets exist and that they are hyperimmune (and in fact hyperhyperimmune). (This generalization can be iterated; see Exercise 12-31.)

△***12-31.*** Let $\mathfrak{C}$ be a finite or denumerably infinite subcollection of $\mathfrak{N}$.

Let $\mathfrak{I}_0$ be an ideal in $\mathfrak{N}$.

Define

$$\mathfrak{G}_0 = \{A \mid A \notin \mathfrak{I}_0 \,\&\, (\forall B)[B \in \mathfrak{C} \Rightarrow [A \cap B \in \mathfrak{I}_0 \text{ or } A \cap \bar{B} \in \mathfrak{I}_0]]\}$$

(the "almost-$\mathfrak{I}_0$ sets"),

$\mathfrak{I}_1$ = the ideal generated by $\mathfrak{G}_0$.

Assume: (i) $\mathfrak{I}_0 \subset \mathfrak{I}_1$.

(ii) Every set not in $\mathfrak{I}_0$ has a subset which is in $\mathfrak{G}_0$.

Define

$$\mathfrak{G}_1 = \{A \mid A \notin \mathfrak{I}_1 \,\&\, (\forall B)[B \in \mathfrak{C} \Rightarrow [A \cap B \in \mathfrak{I}_1 \text{ or } A \cap \bar{B} \in \mathfrak{I}_1]]\}$$

(the "almost-$\mathfrak{I}_1$ sets").

Prove: every set not in $\mathfrak{I}_1$ has a subset which is in $\mathfrak{G}_1$.

(*Hint:* Generalize the proof of Theorem VI.) (For the special case where $\mathfrak{C}$ is the recursively enumerable sets and $\mathfrak{I}_0 = \{\emptyset\}$, this exercise yields Theorem VI with $\mathfrak{I}_1 = \mathfrak{F}$ and $\mathfrak{G}_1$ = the cohesive sets. Applied inductively, this exercise enables us to construct a chain of ideals as suggested in Exercise 12-30. The construction can be extended to the transfinite.)

△*12-32.* Show that there exist recursively enumerable sets A and B and an automorphism f of $\mathcal{E}$ such that $f(A) = B$ but A is not recursively isomorphic to B. (The property of being recursively isomorphic to A is then a recursively invariant property which is not lattice-theoretic in $\mathcal{E}$.) (*Hint:* Take A to be a simple set, $m \in \bar{A}$, and $B = A \cup \{m\}$. By the lemma in Theorem 8-XIV, $B \not\equiv A$. To get the automorphism f, apply the general technique of Theorem IX using two cohesive sets, one from A and one from $\bar{A}$.)

▲*12-33.* (Rose and Ullian). Given functions f and g and a partial function φ, define $error_{f,g,\varphi}(n)$ to be the number of integers m, $m \leq n$, such that $\varphi(g(m)) \neq f(g(m))$.

Definition. f is *recursively approximable* if, for every one-one recursive g, there exists a partial recursive φ such that

$$\lim_{n\to\infty}\left(\frac{1}{n}\cdot error_{f,g,\varphi}(n)\right) = 0.$$

(*a*) Show that recursively approximable functions exist which are not recursive. (*Hint:* Let A be any cohesive set. There are $2^{\aleph_0}$ functions f such that $f(\bar{A}) = 0$. Show that these are all recursively approximable. To do this take $\varphi = \lambda x[0]$ and observe that for any p, $\{m|g(m) \in A\}$ must fall (except for a finite set) into at most one residue class mod p.)

Definition. f is *constructively nonrecursive* if there is a recursive h such that $(\forall z)[f(h(z)) \neq \varphi_z(h(z))]$.

(*b*) Show that constructively nonrecursive functions exist. (*Hint:* Consider a simple diagonalization.)

(*c*) Show that a constructively nonrecursive function is not recursively approximable. (*Hint:* (i) Show that there is a uniform effective method such that, given any z and m, we can find a w such that $w > m$ and $f(w) \neq \varphi_z(w)$ (consider all possible partial functions that agree with φ_z beyond m and apply h to indices for these.) Then (ii) use this to construct a recursively enumerable sequence (g) with longer and longer segments on which f disagrees, in turn, with $\varphi_0, \varphi_1, \ldots$. Do this so that, simultaneously for all partial recursive φ,

$$\lim_{n\to\infty}\left(\frac{1}{n}\cdot error_{f,g,\varphi}(n)\right) = 1.)$$

▲*12-34.* Let $I = \{0,1\}$. Consider functions f with range in I. For any such f, define: $s_f(0) = 0$, $s_f(n+1) = \tau^*(<f(0), f(1), \ldots, f(n)>)$, where τ^* is as in §5.6. Let ψ be a partial function such that *range* $\psi \subset I \times I$. Define: ψ is *satisfactory for f* if $\lambda x[\psi s_f(x)]$ is a total function and $\pi_1\psi s_f(x) = 1$ for infinitely many values of x. Given an f and ψ such that ψ is satisfactory for f, let g be defined by $g(0) = \mu u[\pi_1\psi s_f(u) = 1]$ and $g(n+1) = \mu u[u > g(n)\ \&\ \pi_1\psi s_f(u) = 1]$, and let $\varphi = \lambda x[\pi_2\psi s_f(x)]$. For given f and ψ, define: $e_{f,\psi}(n) = error_{f,g,\varphi}(n)$ (see Exercise 12-33).

Definition (Church). f is *random* if for every partial recursive ψ that is satisfactory for f,

$$\lim_{n\to\infty}\left(\frac{1}{n}\cdot e_{f,\psi}(n)\right) = \tfrac{1}{2}.$$

(The partial function ψ may be thought of as a "gambling system"; $\pi_1\psi$ tells whether to bet, and $\pi_2\psi$ tells which way to bet. f is "random" if there is no effective system that wins (more than half the time on the average) against f.)

Put the measure $m(\{0\}) = m(\{1\}) = \frac{1}{2}$ on I and consider the product measure space

I^N. (Any function f with range in I is an element of I^N.) Prove the following theorem on the impossibility of a gambling system.

Theorem. *The measure of* $\{f|f$ *is random*$\}$ *is* 1. (*Hint:* The problem does not concern cohesive sets and can be handled by purely measure-theoretic methods. See methods for proving the strong law of large numbers in Loève [1955].)

§12.4

12-35. Show that the existence of strongly inseparable pairs of sets (see Exercise 8-39) follows directly from Theorems IV and XI.

12-36. (Yates). (i) Show that the intersection of two maximal sets need not be a maximal set.

(ii) Show that the intersection of two maximal sets is quasimaximal.

(iii) Show that $\{A|A$ *quasimaximal or co-finite*$\}$ is a dual ideal in $\mathcal{E}$.

$\triangle$***12-37.*** Show that N is the disjoint union of a collection of completed cohesive sets.

12-38. Show that there exist two maximal sets which are not recursively isomorphic. (*Hint:* Use construction from Theorem 8-XIV.)

§12.5

12-39. Prove the following stronger version of Theorem 9-XV. A is hyperimmune $\Leftrightarrow A$ is infinite and there exists no recursive f such that $(\forall u)[D_{f(u)} \cap A \neq \emptyset]$, $(\forall u)(\forall v)[u \neq v \Rightarrow D_{f(u)} \cap D_{f(v)} = \emptyset]$, and $(\forall x)(\exists u)[x \in D_{f(u)}]$.

12-40. Complete the proof of (*b*) in Theorem XII.

12-41. Let A be quasimaximal and let q be the number of c-equivalence classes represented in $\bar{A}$. Show that $\overline{lim}\ (\bar{A}) \leq q$. (*Hint:* The proof is similar to that of Theorem XIII.)

$\triangle$***12-42.*** Let A be maximal and f be a one-one recursive function such that A = *range* f. Define $f^0(A) = N$, $f^{n+1}(A) = f(f^n(A))$. Show that (i) for any $q > 0$, $f^q(A)$ is quasimaximal with exactly q c-equivalence classes represented in its complement; and that (ii) for any q, $\overline{lim}\ (\overline{f^q(A)}) = q$. (*Hint:* Use splinters of f, see Exercise 7-37.)

$\triangle$***12-43.*** (Martin). Generalize the proof of Theorem XIV to show that there is a recursively enumerable A with $\bar{A}$ infinite such that for all recursively enumerable B, if $B \supset A$ then either $\overline{lim}(\bar{B}) = 0$ or, for some recursive f, $\overline{lim}(f,\bar{B})$ is infinite.

12-44. Consider the spectrum $\mathcal{C}_{11}, \ldots, \mathcal{C}_4$ of §§8.7 and 9.5. By Exercises 8-46 and 9-26(*b*), this spectrum is linear with respect to $\leq_1$. Is it linear with respect to $\supset$? That is to say, if A and B are nonrecursive, recursively enumerable sets, and $A \supset B$, is it impossible for B to lie to the left of A in this classification?

§12.6

12-45. (Young). Show that if A is recursively enumerable and $\bar{A}$ is recursively indecomposable, then $\bar{A}$ is indecomposable.

12-46. (*a*) Give an example of an immune but recursively decomposable set. (*Hint:* Consider A *join* A for an immune A.)

$\triangle$(*b*) Give an example of a set that is quasicohesive but recursively decomposable. (*Hint:* Take $f^2(A)$ as in 12-42. Use splinters of f to get a decomposition of $\overline{f^2(A)}$. By Exercise 12-45, this gives a recursive decomposition.)

$\blacktriangle$***12-47.*** (McLaughlin). Give an example of a set that is recursively indecomposable but neither finite nor quasicohesive. (*Hint:* Generalize the proof of Theorem XVIII using Exercise 12-22.)

12-48. Define: A is ***strongly hyperhyperimmune*** if A is infinite and there is no recursive f such that $(\forall u)[W_{f(u)} \cap A \neq \emptyset]$ & $(\forall u)(\forall v)[u \neq v \Rightarrow W_{f(u)} \cap W_{f(v)} = \emptyset]$.

(*a*) Show that if A is strongly hyperhyperimmune then A has no infinite retraceable subset. (*Hint:* Let A have an infinite retraceable subset with ψ as retracing function; define $W_{f(n)}$ to be the nth level of the retracing tree of ψ, that is, $W_{f(n)} = \{x|\psi^n(x) = m\}$, where m is the least member of the retraceable subset.)

(*b*) Show that if A is generalized cohesive then A is strongly hyperhyperimmune. (*Hint:* Generalize the proof of Theorem V.)

▲***12-49.*** (Yates). Show that if A is infinite and not hyperhyperimmune, then A has an infinite retraceable subset. (*Hint:* Let $W_{f(0)}, W_{f(1)}, \ldots$ be a recursively enumerable sequence of disjoint finite sets each intersecting A. Construct a recursively enumerable tree from $W_{f(0)}, W_{f(1)}, \ldots$ and define a retracing function ψ from this tree. More specifically, carry out the following procedure. In the course of this procedure certain integers will be *covered* (in a sense to be defined); at a later time they may cease to be covered; however, all integers involved eventually become, and remain, covered. Let $x_0, x_1, \ldots$ be a recursive enumeration of $B = W_{f(0)} \cup W_{f(1)} \cup \cdots$. As we proceed, we list ordered pairs for ψ. Initially, no member of B is covered.

Stage 1. Assign $W_{f(0)}$ to x_0; x_0 is now, by definition, *covered.* As members of $W_{f(0)}$ occur in the enumeration of B, ordered pairs mapping these members to x_0 are put in ψ.

Stage $n + 1$. Assign $W_{f(n)}$ to the first uncovered member of $x_0, x_1, \ldots$. This member is now, by definition, covered. As members of $W_{f(n)}$ occur .... If, at any time, an ordered pair $<x,y>$ is put in ψ (defining $\psi(x) = y$) such that $x \leq y$, then y ceases to be covered.)

▲***12-50.*** (Yates). Show that if A is infinite and not strongly hyperhyperimmune, then A has an infinite retraceable subset. Thus conclude that for any set A, A is strongly hyperhyperimmune $\Leftrightarrow$ A is infinite and has no infinite retraceable subset. (*Hint:* The method suggested for Exercise 12-49 generalizes directly.)

△***12-51.*** Give an example of a set that is hyperhyperimmune but has an infinite retraceable subset. (*Hint:* Take an effective decomposition of N into infinitely many infinite recursive sets, for example, $\{0\} \times N$, $\{1\} \times N$, ... Then define (by nonconstructive diagonalization) a set A which contains a member of each of these recursive sets but is disjoint from at least one set in every recursively enumerable sequence of disjoint finite sets. Then A is hyperhyperimmune but not strongly hyperhyperimmune. By Exercise 12-50, it contains an infinite retraceable subset.)

△***12-52.*** Give an example of an infinite set that contains no infinite retraceable subset but that is not generalized cohesive. (*Hint:* Generalize Theorem V to iterations of the construction in Exercise 12-31.)

12-53. Show that if A is infinite and not hyperimmune, then A is recursively decomposable. (*Hint:* The construction of Theorem V gives a recursive decomposition.)

△***12-54.*** (Young). Define: A is *strongly hyperimmune* if A is infinite and there is no recursive f such that $(\forall u)[W_{f(u)} \cap A \neq \emptyset]$, $(\forall u)(\forall v)[u \neq v \Rightarrow W_{f(u)} \cap W_{f(v)} = \emptyset]$, and $(\forall x)[x \in A \Rightarrow (\exists u)[x \in W_{f(u)}]]$. Define: A is strongly hypersimple if A is recursively enumerable and $\bar{A}$ is strongly hyperimmune.

(*a*) Show that A hyperhypersimple $\Rightarrow$ A strongly hypersimple. (*Hint:* See Exercise 9-48(*b*).)

(*b*) Show that A strongly hypersimple $\Rightarrow$ A hypersimple. (*Hint:* See Exercise 12-39.)

(*c*) Show that no set with retraceable complement can be strongly hypersimple. (*Hint:* See hint for Exercise 12-48(*a*).)

(*d*) Conclude that there are hypersimple sets which are not strongly hypersimple. (*Hint:* See (vi) of Exercise 9-44.) (The existence of sets strongly hypersimple but not hyperhypersimple is shown in Exercise 12-64.)

△***12-55.*** (McLaughlin, Martin). Show that the sets which are co-finite or hyperhypersimple form a dual ideal in $\mathcal{E}$. (*Hint:* Show that [A and B recursively enumerable & $\overline{A \cap B}$ has infinite retraceable subset] $\Rightarrow$ $\bar{A}$ or $\bar{B}$ has infinite retraceable subset. The result follows from $(8)^* \Leftrightarrow (9)^*$, i.e., from Exercise 10-19.)

△***12-56.*** (McLaughlin). Show that there are at most $\aleph_0$ completed cohesive sets. (*Hint:* It suffices to show that if C is completed cohesive, then, for some z, $C \subset W_z$, and $W_z - C$ is recursively enumerable. Assume otherwise. Consider $\mathcal{B} = \{B|(\exists z)(\exists y)[B = W_z - (W_y \cup C) \,\&\, C \subset W_z \,\&\, W_y \subset W_z - C]\}$. $\mathcal{B}$ is closed under intersection and all sets in $\mathcal{B}$ are infinite. Applying the general method of Theorem XVII (see Exercise 12-24), find a cohesive C' such that $C' - B$ is finite for all $B \in \mathcal{B}$. Conclude that $C \cup C'$ is cohesive, contrary to the completedness of C.)

△***12-57.*** (McLaughlin). Show that there exists an indecomposable set with an infinite retraceable subset. (*Hint:* Let A be some recursively enumerable, nonrecursive set. (i) As in Exercise 12-22, there is a recursive f such that $W_{f(x)}$, $x = 0, 1, 2, \ldots$, are a sequence of pairwise disjoint sets whose union is A. (ii) Observe that the property of 12-21(b) holds for this sequence. (iii) Find a cohesive set C such that $C \subset \bar{A}$ and $(\forall x)[W_x \subset \bar{A} \Rightarrow W_x \cap C$ is finite] (see Exercise 12-24). (iv) Consider $\mathcal{B} = \{B|B$ *is recursively enumerable and* $B \supset C\}$. Observe from (ii) that for any $B \in \mathcal{B}$, $(\forall x)[W_{f(x)} \cap B$ is infinite]. (v) Apply the method of Theorem XVIII, as generalized in Exercise 12-47, to find cohesive sets $C_0, C_1, \ldots$ such that $(\forall x)[C_x \subset W_{f(x)}]$ and such that for any B in $\mathcal{B}$, $(\forall x)[C_x \cap B$ is infinite]. (vi) Consider the set $C \cup C_0 \cup C_1 \cup \cdots$. By (v) this set is indecomposable. By Exercise 12-50 this set has an infinite retraceable subset.)

12-58. (Tennenbaum). Let f be a function such that *range* $f = B$ and $(\forall z)(\exists n)(\forall m)[[n < m \,\&\, \varphi_z(m)$ convergent$] \Rightarrow \varphi_z(m) < f(m)]$. Show that $K \leq_T B$.

△***12-59.*** (a) (Yates). Show that there exist complete maximal sets. (*Hint:* Use Exercise 12-58 to modify the construction of Theorem XI.) (The significance of this result for Post [1944] is discussed in §9.6.)

(b) (Sacks). Show that there exist incomplete maximal sets. (*Hint:* Modify the construction of Theorem XI by using a second set of markers on the given list and by using an additional list with markers so as to simultaneously obtain a construction similar to that in Theorem 10-III.)

12-60. Show that every completed cohesive set is the complement of a maximal set. (*Hint:* Use Theorem XIX together with the construction of Exercise 12-56.)

12-61. (a) Show that if A is quasimaximal and $\bar{A}$ is the union of cohesive sets from n distinct c-equivalence classes, then A is the intersection of maximal sets from n distinct elements of $\mathcal{E}/\mathcal{F}$. (*Hint:* Use Theorem XX.)

(b) Show that the complement of any quasimaximal set is the union of completed cohesive sets. (*Hint:* Use Exercise 12-60.)

▲***12-62.*** (R. W. Robinson). (a) Show that there exists a set which is r-maximal but not maximal. (*Hint:* Construct recursively enumerable sets A and B, $A \subset B$, such that B is maximal, $B - A$ is infinite, and $\bar{A}$ is recursively indecomposable. Modify the Yates construction of Theorem XI as follows. Odd and even markers are treated differently. The *z-state of x at stage* $n + 1$ is defined as before. The *modified z-state of x with respect to y at stage* $n + 1$ is defined by changing the definition of z-state so that

$$b_i = \begin{cases} 1, & \text{if } x \in W_i^{(n+1)} \text{ and } y \in W_i^{(n+1)} \text{ and } i \leq n; \\ 0, & \text{otherwise,} \end{cases}$$

for $0 \leq i \leq z$. Markers are moved as in Theorem XI except that an even marker $\boxed{2m}$ must be moved to $x_{2q}{}^{(n)}$, $q > m$, such that $x_{2q}{}^{(n)}$ is in higher m-state than $x_{2m}{}^{(n)}$; and an odd marker $\boxed{2m+1}$ must be moved to $x_p{}^{(n)}$, $p > 2m + 1$, such that $x_p{}^{(n)}$ is in higher modified m-state with respect to $x_{2m}{}^{(n)}$ than is $x_{2m+1}^{(n)}$. Define $A = \{x|x$ *eventually has no marker*$\}$, $B = A \cup \{x|x$ *eventually has an odd marker*$\}$. Prove (1) that all markers come to rest, (2) that B is maximal, and (3) that for any z, $W_z \cap \bar{B}$ infinite $\Rightarrow \bar{A} - W_z$ finite. Conclude from this last (4) that $\bar{A}$ is recursively indecomposable. Proof (1) is almost identical with Lemma 1 of Theorem XI. Proof (2) concerns the even markers and is almost identical with Lemma 2 of Theorem XI. Proof (3) is analogous to Lemma 2 of

Theorem XI; one proves, for given z, that for sufficiently large m, there is exactly one modified z-state with respect to x_{2m} having infinitely many members of $\bar{A}$ terminating in it (where x_{2m} is the final position of marker $\boxed{2m}$.) Conclusion (4) is immediate.)

(*b*) Generalize the above construction to show that for any recursively enumerable set B there exists a recursively enumerable set A such that $B - A$ is infinite and for all z, $\bar{B} - W_z$ finite $\Rightarrow \bar{A} - W_z$ finite.

(Note that this exercise together with Exercise 12-50 implies the results of Exercises 12-47 and 12-57.)

△***12-63.*** (R. W. Robinson). Show that there is a recursively enumerable set A such that $\bar{A}$ is generalized cohesive but not quasicohesive. (*Hint:* Construct recursively enumerable sets H_i, $i = 0, 1, 2, \ldots$, such that for all i, $H_i \subset H_{i+1}$, $H_{i+1} - H_i$ is cohesive, and $\bar{H}_0$ is generalized cohesive. The construction is identical with the Yates construction of Theorem XI except that the condition "$x_q^{(n)}$ is in higher m-state (at stage $n + 1$) than $x_m^{(n)}$" is augmented by the condition "*and* $\pi_1(q) \geq \pi_1(m)$." Set $H_0 = \{x|x$ *eventually has no marker*$\}$ and, for $i > 0$, $H_i = \{x|x$ *eventually has a marker* $\boxed{m}$ *such that* $\pi_1(m) < i\}$. It follows directly, from proofs analogous to those for Lemmas 1 and 2 in Theorem XI, that the sets H_i, $i = 0, 1, 2, \ldots$ have the desired properties.)

12-64. (R. W. Robinson). Show that there is a set which is strongly hypersimple but not hyperhypersimple. (*Hint:* Use the result of Exercise 12-62.)

△***12-65.*** (Young). Show that there exist A and B such that $A \leq_1 B$ but $A \not\leq_1^r B$, where $\leq_1^r$ is as defined in Exercise 7-34. (*Hint:* Take A and B as in the construction suggested for Exercise 12-62. Let f be a one-one recursive function enumerating A. Then $A \leq_1 f(A)$ via f. Show that for no recursive g is it the case that $A \leq_1^r f(A)$ via g.)

12-66. (Martin). If φ_x is a characteristic function, define C_x to be $\{y|\varphi_x(y) = 1\}$. Define A is c-*strongly hyperimmune* if A is infinite and there is no recursive f such that $(\forall u)[\varphi_{f(u)}$ is a characteristic function$]$, $(\forall u)[C_{f(u)} \cap A \neq \emptyset]$, and $(\forall u)(\forall v)[u \neq v \Rightarrow C_{f(u)} \cap C_{f(v)} = \emptyset]$. Define A is c-*strongly hypersimple* if A is recursively enumerable and $\bar{A}$ is c-strongly hyperimmune.

(*a*) Show that A is c-strongly hypersimple $\Leftrightarrow A$ is strongly hypersimple.

△(*b*) Show that A is c-strongly hyperimmune $\Leftrightarrow A$ contains no infinite retraceable subset possessing a total retracing function. (*Hint:* See Exercises 12-48 to 12-50.)

12-67. (*a*) Show that every quasimaximal set is contained in a maximal set.

▲(*b*) (Lachlan). Show that there is a hyperhypersimple set which is contained in no maximal set.

▲(*c*) (R. W. Robinson). Show that there is an r-maximal set which is contained in no maximal set.

13 Degrees of Unsolvability

§13.1 THE JUMP OPERATION

This chapter concerns the reducibility ordering of T-degrees of unsolvability. We consider this structure over all T-degrees, recursively enumerable or not. There has been considerable interest in this topic, and a variety of results have been obtained. A number of the results to be presented here preceded the results presented in Chapter 10.

A basic difference between the ordering of recursively enumerable degrees and the ordering of all degrees is that the latter is not denumerable. As we shall see, a number of other structural properties of the ordering of recursively enumerable degrees (e.g., density) fail in the general case. Techniques of proof in the general case can be highly nonconstructive, since recursive enumerability does not usually need to be demonstrated. Nevertheless, proofs often have combinatorial content that is both simple and elegant. Our primary goal below will be to illustrate these techniques and to give the reader some feeling for what can and cannot be done, rather than to prove results in strongest form. A number of results not proved in the text will be given in exercises or indicated in §13.8. As we shall note in §13.8, some of the strongest results have combined the techniques now to be described with the priority methods described in Chapter 10.

The *jump operation* is a fundamental concept in the study of degrees. We define it first on sets, then on degrees.

Definition For any A, let

$$A' = \{x|\varphi_x{}^A(x) \textit{ is convergent}\} = \{x|x \in W_x{}^A\}.$$

A' is called the *jump* or *completion* of A. (A' is identical with the set K^A defined in §9.3.)

A number of basic facts about the jump operation on sets are given in Theorem I. The jump operation appears to depend on a choice of Gödel numbering for the partial recursive functions. It will follow from Theorem I, however, that if A_1 and A_2 are jumps of A based on two different acceptable Gödel numberings (in the sense of Exercise 2-10), then $A_1 \equiv A_2$ (see Exercise 13-1.†)

Theorem I(a) *$B \leq_T A \Leftrightarrow$ both B and $\bar{B}$ are recursively enumerable in A.*
(*b*) *A' is recursively enumerable in A.*
(*c*) *$A' \not\leq_T A$.*
(*d*) *B is recursively enumerable in $A \Leftrightarrow B \leq_1 A'$.*
(*e*) *$A \leq_T B \Leftrightarrow A' \leq_1 B'$.*

Proof. Part (*a*) is Theorem 9-IV. Parts (*b*) and (*c*) were obtained in the proof of Theorem 9-X.

(*d*) $\Rightarrow$: An informal proof would parallel the proofs of Theorems 7-III and 7-IV. More formally, the proof is as follows. Let $B = W_z^A$. Let any w be given. Consider $\{<x,y,u,v> | <w,y,u,v> \in W_{\rho(z)}\}$. By the second projection theorem, this set is recursively enumerable and an index for it can be obtained uniformly effectively from w and z. Let g be a recursive function such that $W_{g(w,z)} = \{<x,y,u,v> | <w,y,u,v> \in W_{\rho(z)}\}$. This set is immediately seen to be regular (in the sense of §9.2). Hence

$$W_{\rho g(w,z)} = W_{g(w,z)}.$$

Then $g(w,z) \in A' \Leftrightarrow g(w,z) \in W^A_{g(w,z)} \Leftrightarrow (\exists y)(\exists u)(\exists v)[<w,y,u,v> \in W_{\rho(z)}$ & $D_u \subset A$ & $D_v \subset \bar{A}] \Leftrightarrow w \in W_z^A \Leftrightarrow w \in B$. Hence $B \leq_1 A'$ via $\lambda w[g(w,z)]$.

$\Leftarrow$: This follows from (*b*) by Theorem 9-IX, the relativized projection theorem.

(*e*) $\Rightarrow$: Assume $A \leq_T B$. By (*b*), A' is recursively enumerable in A. By (*c*), $A' \neq \emptyset$. Hence $A' =$ *range* f for some f recursive in A. But f recursive in $A \Rightarrow f$ recursive in B, since $A \leq_T B$. Therefore A' is recursively enumerable in B, and, by (*d*), $A' \leq_1 B'$.

$\Leftarrow$: Assume $A' \leq_1 B'$. By elementary constructions, A is recursively enumerable in A and $\bar{A}$ is recursively enumerable in A. Therefore, $A \leq_1 A'$ and $\bar{A} \leq_1 A'$, by (*d*). Therefore, $A \leq_1 B'$ and $\bar{A} \leq_1 B'$. Therefore, by (*d*), both A and $\bar{A}$ are recursively enumerable in B. Hence, by (*a*), $A \leq_T B$. ☒

Corollary I(a) *$A \equiv_T B \Leftrightarrow A' \equiv B'$.*
(*b*) *$B \leq_T A \Leftrightarrow [B \leq_1 A'$ & $\bar{B} \leq_1 A']$.*
(*c*) *Parts (b), (d), and (e) of Theorem I hold uniformly in the following senses. Part (b): $(\exists z)(\forall A)[A' = W_z^A]$. Part (d): there exist recursive f and h such that $B = W_z^A \Rightarrow B \leq_1 A'$ via $\varphi_{f(z)}$, and $B \leq_1 A'$ via $\varphi_z \Rightarrow B = W^A_{h(z)}$.*

† Note that the indexing of partial A-recursive functions is determined, via Theorem 9-II, by the indexing of partial recursive functions.

Part (e): there exist recursive f and h such that $c_A = \varphi_z^B \Rightarrow A' \leq_1 B'$ *via* $\varphi_{f(z)}$, *and* $A' \leq_1 B'$ *via* $\varphi_z \Rightarrow c_A = \varphi_{h(z)}^B$.

Proof. Parts (*a*) and (*b*) are immediate. Part (*c*) follows immediately from examination of the formal constructions for Theorem I as follows. For part (*b*) of Theorem I, take z so that $W_z = \{<x,y,u,v> | <x,y,u,v> \in W_{\rho(x)}\}$. It follows that $W_{\rho(z)} = W_z$ and that for all A, $A' = W_z^A$. For part (*d*) of Theorem I, take f so that $\varphi_{f(z)} = \lambda w[g(w,z)]$, and obtain h by similar formal construction. For part (*e*) of Theorem I, proceed similarly. ☒

We have shown, as a result of Theorem I and Corollary I, that there is an isomorphism from the ordering of T-degrees into the ordering of 1-degrees. (Unlike the isomorphisms associated with m-cylinders and tt-cylinders, however, the image of a T-degree in this isomorphism does not fall within that T-degree.) It follows from Corollary I(*a*) that the jump operation is well-defined on T-degrees, and that it carries each T-degree to a unique higher T-degree. The following notation will be used (and is common in discussions of degrees.)

Notation **a**, **b**, **c**, . . . will denote degrees. **0** will denote the recursive degree. For any degree **a**, **a′** will be the (uniquely determined) degree of A', where A is any set in **a**. **a′** is called the *jump* or *completion* of **a**. If **b** = **a′** for some **a**, **b** is called a *complete* degree.

For any degrees **a** and **b**, **a** ≤ **b** will mean that **a** is T-reducible to **b**; **a** < **b** will mean that **a** ≤ **b** and **b** ≰ **a**; **a**|**b** will mean that **a** ≰ **b** and **b** ≰ **a** (we then say that **a** and **b** are *incomparable*); and **a** ∪ **b** will denote the *join* of **a** and **b**.

a *recursively enumerable* in **b** will mean that there is an A in **a** such that for some B in **b**, A is recursively enumerable in B (see Exercise 13-6). **a** *recursive in* **b**, as before, means that **a** ≤ **b**. If we say that A is recursive in **b** (or that f is recursive in **b**), we mean that A is recursive in B (or that f is recursive in B) for some $B \in$ **b**.

Examples. $\emptyset \in$ **0**. $K \in$ **0′**. **a** ≤ **a′**. **a′** ≰ **a**.

Theorem I yields the following *basic relation:* **a** ≤ **b** ⇒ **a′** ≤ **b′**. (In §13.3, the converse of this relation will be shown to fail.)

We sometimes indicate iterations of the jump as follows. For any A,

$$A^{(0)} = A;$$
$$A^{(n+1)} = (A^{(n)})'.$$

($A^{(n)}$ is called the nth *jump* of A.)

Correspondingly, for any **a**,

$$\mathbf{a}^{(0)} = \mathbf{a};$$
$$\mathbf{a}^{(n+1)} = (\mathbf{a}^{(n)})'.$$

Example. $(\forall n)[\emptyset^{(n)} \in \mathbf{0}^{(n)}]$.

Corollary I(d) *There exists a recursive f such that for any A, x, and y,* $[x \leq y \Rightarrow A^{(x)} \leq_1 A^{(y)}$ *via* $\varphi_{f(x,y)}]$. *There exists a recursive g such that for any A, B, and x,* $A \leq_1 B$ *via* $\varphi_z \Rightarrow A^{(x)} \leq_1 B^{(x)}$ *via* $\varphi_{g(z,x)}]$.

Proof. This follows from Corollary I(*c*) and the existence of a recursive k such that $A \leq_1 B$ via $\varphi_z \Rightarrow c_A = \varphi^B_{k(z)}$.☒

The members of the increasing sequence **0**, **0′**, **0″**, . . . , $\mathbf{0}^{(n)}$, . . . are especially useful as reference points in the ordering of degrees. We study **0′** and **0″** in §13.2. In Chapter 14, we find certain further natural properties of this sequence and see that it has significance for logic as well as for recursive function theory.

Certain facts about the ordering of degrees have been previously noted. Each degree contains $\aleph_0$ sets; hence there are $2^{\aleph_0}$ degrees in all. Each set has $\aleph_0$ sets reducible to it; hence each degree has at most $\aleph_0$ degrees below it.

Results obtained about recursively enumerable degrees in Chapter 10 give further information. (Note that the recursively enumerable degrees are all ≤**0′**.) The Friedberg-Muchnik theorem (Theorem 10-III) tells us that there exist incomparable degrees between **0** and **0′**. Indeed, Corollary 10-III tells us that $\aleph_0$ degrees fall between **0** and **0′**. Other results, stated but not proved in Chapter 10, give still further information. Sacks's result on density of the recursively enumerable degrees (see §10.3) tells us that there exist dense chains of degrees between **0** and **0′**. (The Friedberg-Muchnik result can be directly relativized to show that for any A, there exist incomparable sets recursively enumerable in A. Hence, for any **a**, there exist incomparable degrees between **a** and **a′**. Corollary 10-III and Sacks's result can be similarly relativized.) In what follows, we shall assume only the results that have been proved in previous chapters. Thus, in §13.4, we shall give a proof that dense chains exist between **0** and **0′**. Our proof (due to Kleene and Post) will be simpler than an argument by way of Sacks's theorem, since we shall not require that the degrees constructed be recursively enumerable.

The jump operation can be iterated into the transfinite.

Definition $A^{(\omega)} = \{\langle x,y\rangle \mid x \in A^{(y)}\}$. $A^{(\omega)}$ is called the *ω-jump* or *ω-completion* of A.

Obviously, for any n, $A^{(n)} \leq_1 A^{(\omega)}$ and $A^{(\omega)} \not\leq_T A^{(n)}$. The following theorem shows that the ω-jump is well-defined on degrees.

Theorem II $A \leq_T B \Rightarrow A^{(\omega)} \leq_1 B^{(\omega)}$.

Proof. Let $A \leq_T B$. From Corollary I, we can show that there is a recursive g such that $A^{(y)} \leq_1 B^{(y+1)}$ via $\varphi_{g(y)}$. We therefore have

$$\langle x,y\rangle \in A^{(\omega)} \Leftrightarrow x \in A^{(y)} \Leftrightarrow \varphi_{g(y)}(x) \in B^{(y+1)}$$
$$\Leftrightarrow \langle \varphi_{g(y)}(x),\, y+1\rangle \in B^{(\omega)}.$$

Defining $f(\langle x,y\rangle) = \langle \varphi_{g(y)}(x),\, y+1\rangle$, we have $A^{(\omega)} \leq_1 B^{(\omega)}$ via f.☒

The converse to Theorem II fails, for let A be given and $B = A^{(n)}$ for some $n > 0$. By Theorem II, $A^{(\omega)} \leq_1 B^{(\omega)}$. However, $<x,y> \in B^{(\omega)} \Leftrightarrow x \in B^{(y)} \Leftrightarrow x \in A^{(n+y)} \Leftrightarrow <x, n + y> \in A^{(\omega)}$. Hence $B^{(\omega)} \leq_1 A^{(\omega)}$. Thus $B^{(\omega)} \equiv A^{(\omega)}$, although $B \not\equiv_T A$.

Definition $\mathbf{a}^{(\omega)}$ is the (uniquely determined) degree of $A^{(\omega)}$, where A is any set in $\mathbf{a}$.

(For any $\mathbf{a}$, $\mathbf{a}^{(\omega)}$ is an example of a degree having an infinite chain of degrees below itself.)

***Theorem III* (*Shoenfield, Kleene, Post*)** *There exist nondenumerable chains of degrees.*

Proof. By Zorn's lemma (Exercise 12-11), there exists a maximal chain of degrees. This chain cannot have a greatest element, otherwise the jump of that element could be used to form a longer chain. Assume the chain is countable; then there is a sequence of degrees $\mathbf{a}_0, \mathbf{a}_1, \mathbf{a}_2, \ldots$ co-final in the chain. Let $A_0, A_1, \ldots$ be a sequence of sets such that $(\forall i)[A_i \in \mathbf{a}_i]$. Then $B = \{<x,y> | x \in A_y\}$ is above all members of the sequence and hence the degree of $\mathbf{b}$ is above all members of the chain, contrary to the maximality of the chain. ☒

Corollary III *There is a nondenumerable chain above any degree* $\mathbf{a}$. *Every maximal chain of degrees is nondenumerable.*

Proof. Immediate. ☒

The following concept will be studied more extensively in the next chapter.

Definition A is *arithmetical in* B if $A \leq_T B^{(n)}$ for some n. A is *arithmetical* if A is arithmetical in $\emptyset$.

These concepts are well-defined on degrees, since the jump is well-defined on degrees. Thus, for example, we say that a degree $\mathbf{a}$ is *arithmetical* if $\mathbf{a} \leq \mathbf{0}^{(n)}$ for some n, and that $\mathbf{a}$ is *arithmetical* in $\mathbf{b}$ if $\mathbf{a} \leq \mathbf{b}^{(n)}$ for some n. We shall see (in Chapter 14) that the arithmetical sets are the sets definable in elementary arithmetic, and that $\emptyset^{(\omega)}$ is recursively isomorphic to the set of true sentences of elementary arithmetic. Evidently A arithmetical in $B \Rightarrow A^{(\omega)} \leq_1 B^{(\omega)}$ (see Exercise 13-9). (Using methods of Cohen, Sacks has shown that the converse implication fails.) Define $\mathbf{a}$ is *arithmetically equivalent* to $\mathbf{b}$ if $\mathbf{a}$ is arithmetical in $\mathbf{b}$ and $\mathbf{b}$ is arithmetical in $\mathbf{a}$. The relation *arithmetical in* is easily seen to be reflexive and transitive (Exercise 13-10), hence arithmetical equivalence is an equivalence relation (and we can speak of the equivalence classes as "degrees"). If $\mathbf{a}$ is arithmetically equivalent to $\mathbf{b}$, must there exist m and n such that $\mathbf{a}^{(m)} = \mathbf{b}^{(n)}$? Martin and Lachlan have shown that this is not the case (see §13.8).

How far can the jump be iterated into the transfinite? We could attempt to generalize the construction of $A^{(\omega)}$ as follows. Let $\beta = \alpha + 1$ and assume $A^{(\alpha)}$ defined; set $A^{(\beta)} = (A^{(\alpha)})'$. Let γ be a countable limit ordinal, let

$\{\beta_n\}\uparrow$ be such that $\lim_n \beta_n = \gamma$, and assume $A^{(\beta_n)}$ defined for all n. Set $A^{(\gamma)} = \{<x,y>|x \in A^{(\beta_y)}\}$. If we use this definition, the degree of $A^{(\gamma)}$ is not well-defined (see Exercise 13-8). In order to avoid this difficulty, the definition for $A^{(\gamma)}$ must be restricted to sequences of ordinals that are, in some sense, effective. The study of iterations of the jump is hence intimately connected with the study of constructive ordinals. We consider this further in Chapter 16.

The main concern of the present chapter is the structure of degrees under the relation $\leq$ and the operations $'$ and $\cup$ (and, to a lesser extent, under the relation *recursively enumerable in* and the operation $^{(\omega)}$). Certain technical concepts and notations will be useful.

1. We shall find it convenient to work with characteristic functions rather than with sets. Recall (from Theorem 9-VII) that for any A and B, A recursive in $B \Leftrightarrow A$ recursive in $c_B \Leftrightarrow A$ recursive in $\tau(c_B)$, and that the first of these equivalences holds uniformly in each direction (the second equivalence holds by definition). We use the notation φ_z^f to abbreviate $\varphi_z^{\tau(f)}$.

2. A partial function with recursive domain will be called a *segment.* (In particular, every total function is a segment.) In contrast to earlier usage, we use the symbols $f, g, h, \ldots$ for segments (which may not be total). If f is a segment and either *domain* $f = \emptyset$ or *domain* $f = \{0,1, \ldots ,n\}$ for some n, f will be called an *initial segment.* The least integer not in the domain of an initial segment is called the *length* of that initial segment. If f and g are segments and $f \supset g$, we say that f is an *extension* of g. *In proofs below, unless otherwise stated, all segments will be assumed to have range contained in* $\{0,1\}$. If f is a segment, φ_z^f abbreviates $\varphi_z^{\tau(f)}$. An initial segment is a finite object. If we say that we know (or can compute) an initial segment f, we mean that we know (or can compute) the canonical index for $\tau(f)$.

3. We introduce the following special notation, where f is a segment

$$\varphi_z^{[f]}(x) = \begin{cases} \varphi_z^f(x), & \text{if the only questions asked about } f \text{ during the computation (from index } z\text{) involve arguments for which } f \text{ is defined;} \\ \text{divergent}, & \text{otherwise.} \end{cases}$$

More formally,

$$\varphi_z^{[f]}(x) = y \Leftrightarrow (\exists u)(\exists v)[<x,y,u,v> \in W_{\rho(z)} \mathbin{\&} D_u \subset \tau(f) \mathbin{\&} D_v \subset \overline{\tau(f)} \mathbin{\&} (\forall x)(\forall y)[<x,y> \in D_v \Rightarrow (\exists y')[<x,y'> \in \tau(f)]]].$$

If $\varphi_z^{[f]}(x) = y$, it follows that for all extensions g of f, $\varphi_z^{[g]}(x) = y$. Note that for initial segments f, $\varphi_z^{[f]}(x) = y$ is recursively enumerable as a relation in f, x, y, and z (where f is represented by the canonical index for $\tau(f)$); that is to say, $\{<u,x,y,z>|(\exists f)[f \textit{ is an initial segment and } D_u = \tau(f) \mathbin{\&} \varphi_z^{[f]}(x) = y\}$

is recursively enumerable (by the second projection theorem). Let $W_z^{[f]}$ denote *domain* $\varphi_z^{[f]}$.

Many of our proofs will involve the construction of a total function as the union of a chain of segments. (This general approach to degree problems was initiated by Kleene and Post [1954].) Friedberg's proof in Theorem 10-III can be viewed as the construction of two total functions by initial segments. (In 10-III, we chose not to use the segment concept, but rather to emphasize recursive enumerability of the sets in question by more anthropomorphic terminology.) If the requirement of recursive enumerability is dropped, the Friedberg proof becomes much simpler (since nonconstructive steps can be used). We give this simpler proof as an illustration of segment techniques.

***Theorem IV* (Friedberg, Kleene, Post)** *There exist sets A and B such that A is not recursively enumerable in B and B is not recursively enumerable in A.*

Proof. We define two chains of initial segments: $f_0, f_1, \ldots$ and $g_0, g_1, \ldots$. $f = \bigcup_i f_i$ will be c_A and $g = \bigcup_i g_i$ will be c_B.† We proceed in stages.

Stage 0. Let $f_0 = g_0 = \emptyset$.

Stage $2n + 1$. Let $m = 2n$. Let x_m be the length of f_m. *See whether* there is an initial segment extension $\tilde{g}$ of g_m such that $x_m \in W_n^{[\tilde{g}]}$. (This step is nonconstructive.) *If so,* set $f_{m+1} = f_m \cup \{<x_m,0>\}$ and $g_{m+1} = \tilde{g}$. *If not,* set $f_{m+1} = f_m \cup \{<x_m,1>\}$ and $g_{m+1} = g_m$. (Stage $2n + 1$ thus assures that for the eventually defined g, $x_m \in A \Leftrightarrow x_m \notin W_n{}^g$.)

Stage $2n + 2$. Let $m = 2n + 1$. The definition for stage $2n + 2$ is as for stage $2n + 1$ with the symbols "f" and "g" (with or without subscripts) interchanged throughout.

Let $f = \bigcup_i f_i$, $g = \bigcup_i g_i$, $A = \{x|f(x) = 1\}$, and $B = \{x|g(x) = 1\}$. It follows from the construction that for all n, $A \neq W_n{}^B$ and $B \neq W_n{}^A$.☒

(In §13.2, we shall see that both A and B in the above proof are recursive in $\mathbf{0}'$.)

The following theorem of Shoenfield [1960] gives further basic information on the degree ordering and, at the same time, another illustration of segment techniques.

***Theorem V* (Shoenfield, Kleene, Post)** *There exists a nondenumerable set of pairwise incomparable degrees.*

Proof. It follows from Zorn's lemma that there exists a maximal set of pairwise incomparable nonrecursive degrees. It is enough to show that such a set cannot be finite or countably infinite. This is implied by the following lemma.

† We abbreviate $\bigcup_{i=0}^{\infty} f_i$ as $\bigcup_i f_i$.

Lemma *Let A_0, A_1, . . . , A_n, . . . be any sequence of nonrecursive sets. Then there exists a set B such that $(\forall n)$[B is incomparable with A_n].*

Proof of lemma. We define f, a characteristic function for B, by a chain of initial segments. We proceed in stages.

Stage 0. Let $f_0 = \emptyset$.

Stage $2n + 1$. Let $m = 2n$. Take y, z so that $n = <y,z>$. Let x_m be the length of f_m. *See whether* $\varphi_z^{A_y}(x_m) = 0$. *If so,* set

$$f_{m+1} = f_m \cup \{<x_m,1>\}.$$

If not, set $f_{m+1} = f_m \cup \{<x_m,0>\}$. (This stage assures, for the eventually defined f, that $f \neq \varphi_z^{A_y}$.)

Stage $2n + 2$. Let $m = 2n + 1$. Let $n = <y,z>$. *See whether* there exists an x and an initial segment extension $\tilde{f}$ of f_m such that $\varphi_z^{[\tilde{f}]}(x)$ is defined and unequal to $c_{A_y}(x)$. *If so,* set $f_{m+1} = \tilde{f}$. *If not,* set $f_{m+1} = f_m$. (This stage assures, for the eventually defined f, that $c_{A_y} \neq \varphi_z^f$. In the case that $\tilde{f}$ exists, this is clearly so. In the case that such $\tilde{f}$ does not exist, then for all extensions $\tilde{f}$, $\varphi_z^{[\tilde{f}]} \subset c_{A_y}$. Hence the recursively enumerable relation $\{<u,v>|(\exists$ *initial segment* $\tilde{f})[f_m \subset \tilde{f}$ & $\varphi_z^{[\tilde{f}]}(u) = v\} \subset c_{A_y}$. If $c_{A_y} = \varphi_z^f$ for some total extension f of f_m, then the relation must equal c_{A_y}; but then c_{A_y} is a recursive function, contrary to the assumed nonrecursiveness of A_y.)

Define $f = \cup_i f_i$, and let $B = \{x|f(x) = 1\}$. We have shown that for all y, B is not recursive in A_y and A_y is not recursive in B. This concludes the proof of the lemma and hence of the theorem.☒

Corollary V *For any* $\mathbf{a} > \mathbf{0}$, *there are uncountably many degrees incomparable with* $\mathbf{a}$.

Proof. Immediate.☒

Final Comment

A property of degrees or an operation on degrees is *order-theoretic* if it is invariant under all automorphisms of the degree ordering (see discussion in §12.1). The degree $\mathbf{0}$ and the operation $\cup$ are clearly order-theoretic. It is not known whether the operation *jump* and the relation *recursively enumerable in* are order-theoretic. Indeed, it is not known whether the degree ordering possesses nontrivial automorphisms. (The recursive permutations induce only the identity automorphism.) The following *homogeneity problems* are open questions: (1) Is there, for every $\mathbf{a}$, an isomorphism between the degree ordering and the degree ordering restricted to $\{\mathbf{b}|\mathbf{a} \leq \mathbf{b}\}$? (2) Is there, for every $\mathbf{a}$, an isomorphism of the kind described in (1) which preserves the operation *jump* and the relation *recursively enumerable in*? As the reader will note in subsequent theorems and exercises, an affirmative answer to (2) would effect considerable simplification in existing proofs. Various "relativized" forms of theorems would become immediate corollaries to their simpler unrelativized versions.

§13.2 SPECIAL SETS AND DEGREES

We next consider the significance of the degrees **0**, **0′**, and **0″** (and of the degrees **a′** and **a″** for any **a**), and we look at some typical sets belonging to these degrees.

The Degree 0

This is, by definition, the degree of all recursive sets. It includes $\emptyset$ and N as members, and is a minimum in the degree ordering. We often use $\emptyset$ as a representative member of this degree.

The Degree 0′

This is, by definition, the degree of $\emptyset'$. It includes K as a member. (In fact $\emptyset' \equiv K$, since, by Theorem I, $\emptyset'$ must be complete with respect to $\leq_1$.) It also includes $\bar{K}$, $\{x | W_x = \emptyset\}$, and $\{x | W_x \neq \emptyset\}$. The recursive isomorphism of $\{x | W_x = \emptyset\}$ to $\bar{K}$ (and hence of $\{x | W_x \neq \emptyset\}$ to K) follows by Theorem 7-VI from Exercise 7-12.

0′ is the degree of the halting problem. A procedure is recursive in **0′** (i.e., recursive in $\emptyset'$) if it is effective except for requiring the solution to certain individual and explicit instances of the halting problem, i.e., to questions that ask whether a certain specified (and effectively recognizable) kind of event occurs at least once in the course of a certain effective computation. Consider, for example, the construction, in Theorem IV above, of sets A and B neither of which is recursively enumerable in the other. In stage $2n + 1$, the question was asked: is there an initial segment extension $\tilde{g}$ of g_m such that $x_m \in W_n^{[\tilde{g}]}$? Given g_m, the collection of all such extensions is recursively enumerable with an r.e. index that depends uniformly on g_m, x_m, and n. Hence this question asks for the solution to an explicitly given instance of the halting problem. A similar question is asked in stage $2n + 2$. Apart from these questions, the procedure is entirely effective. (For example, in stage $2n + 1$, if the answer to the question is known to be affirmative, an appropriate extension $\tilde{g}$ can be found by effective search.) It follows that both f and g are recursive in **0′**. We therefore have the following.

Corollary IV(a) *There exist sets A and B such that $A \leq_T K$, $B \leq_T K$, A is not recursively enumerable in B, and B is not recursively enumerable in A.*

(***b***) *There exist degrees* **a** *and* **b** *such that* $\mathbf{0} < \mathbf{a} < \mathbf{0}'$, $\mathbf{0} < \mathbf{b} < \mathbf{0}'$, *and* $\mathbf{a} | \mathbf{b}$.

Proof. Part (*a*) was proved above. For (*b*), take **a** and **b** to be the respective degrees of A and B in (*a*). Since, for any A and B, $A \leq_T B \Rightarrow A$ recursively enumerable in B, **a** and **b** must be incomparable and distinct from **0** and **0′**.⊠

(Of course, Theorem IV and Corollary IV can be obtained as corollaries to Theorem 10-III; the converse derivation, however, cannot be made directly, for, as we shall see in §13.3, not every degree below $\mathbf{0}'$ contains a recursively enumerable set.)

The Degree $\mathbf{a}'$

The intuitive significance of $\mathbf{a}'$ is parallel to that of $\mathbf{0}'$. A procedure is recursive in $\mathbf{a}'$ if it is effective except for requiring the solution to individual instances of the (relativized) halting problem for procedures recursive in A, where A is some given set in $\mathbf{a}$. The proof of the following theorem illustrates this.

Theorem VI **(*Kleene, Post*)** *For any* $\mathbf{a}$, $\mathbf{0} < \mathbf{a} \Rightarrow (\exists \mathbf{b})[\mathbf{b}|\mathbf{a} \;\&\; \mathbf{b} < \mathbf{a}']$.

Proof. The construction is similar to that for Theorem V. Let $A \in \mathbf{a}$. We construct f by initial segments $f_0, f_1, \ldots$. $\mathbf{b}$ will be the degree of the set whose characteristic function is f.

Stage 0. Let $f_0 = \emptyset$.

Stage $2n + 1$. Let $m = 2n$. Let x_m be the length of f_m. *See whether* $\varphi_n{}^A(x_m) = 0$. *If so*, set $f_{m+1} = f_m \cup \{<x_m,1>\}$. *If not*, set

$$f_{m+1} = f_m \cup \{<x_m,0>\}.$$

Stage $2n + 2$. Let $m = 2n + 1$. *See whether* there exist an x and an initial segment extension $\tilde{f}$ of f_m such that $\varphi_n^{[\tilde{f}]}(x)$ is defined and unequal to $c_A(x)$. *If so*, set $f_{m+1} = \tilde{f}$. *If not*, set $f_{m+1} = f_m$.

Let $f = \bigcup_i f_i$, and let $B = \{x|f(x) = 1\}$. Exactly as in Theorem V, the construction assures that B is not recursive in A and that A is not recursive in B. Hence $\mathbf{b}|\mathbf{a}$, where $\mathbf{b}$ is the degree of B. The question asked in stage $2n + 1$ is an instance of the halting problem for procedures recursive in A. Similarly, in stage $2n + 2$, the collection of desired $<x,\tilde{f}>$ is recursively enumerable in A; hence the question asked is an instance of the halting problem for procedures recursive in A. We therefore have $B \leq_T A'$, and hence $\mathbf{b} \leq \mathbf{a}'$. Since $\mathbf{b}|\mathbf{a}$, we have $\mathbf{b} < \mathbf{a}'$. ☒

Corollary VI *There exist degrees between* $\mathbf{0}$ *and* $\mathbf{0}''$ *which are incomparable with* $\mathbf{0}'$.

Proof. Take $\mathbf{0}'$ for $\mathbf{a}$ in the theorem. ☒

The Degree $\mathbf{a}' \cup \mathbf{b}'$

If a computation is effective except for requiring answers to questions which are either instances of the halting problem for procedures recursive in some fixed $A \in \mathbf{a}$ or instances of the halting problem for procedures recursive in some fixed $B \in \mathbf{b}$, then that computation is clearly recursive in $\mathbf{a}' \cup \mathbf{b}'$. We shall see, however, that a computation involving the halting problem for procedures recursive in some fixed $C \in \mathbf{a} \cup \mathbf{b}$ is not always recursive in $\mathbf{a}' \cup \mathbf{b}'$. (Theorem X will give $\mathbf{a}$ and $\mathbf{b}$ such that $\mathbf{a}' \cup \mathbf{b}' < (\mathbf{a} \cup \mathbf{b})'$.)

Theorem VI can be put in the following relativized form. The proof illustrates a frequently useful technique for giving a preassigned lower bound to a degree being constructed.

Theorem VII **(*Kleene, Post*)** *For* **a** *and* **d**,

$$\mathbf{d} < \mathbf{a} \Rightarrow (\exists \mathbf{b})[\mathbf{b}|\mathbf{a} \,\&\, \mathbf{d} < \mathbf{b} < \mathbf{a}'].$$

Proof. Let $A \in \mathbf{a}$ and $D \in \mathbf{d}$. The construction is as in Theorem VI except that $f_0, f_1, \ldots$ are not initial segments, but are taken from the class of all *special* segments $\tilde{f}$ such that either $domain\, \tilde{f} = \{2x|x \in N\}$ or

$$domain\, \tilde{f} = \{2x|x \in N\} \cup \{0,1, \ldots ,n\}.$$

Stage 0. f_0 is defined by

$$f_0(2x) = c_D(x);$$
$$f_0(2x+1) = \text{undefined}.$$

Stage $2n + 1$. As in Theorem VI, except that x_m is the least integer not in $domain\, f_m$.

Stage $2n + 2$. As in Theorem VI, except that $\tilde{f}$ is a special segment rather than an initial segment.

Let $f = \bigcup_i f_i$, $B = \{x|f(x) = 1\}$, and $\mathbf{b}$ = the degree of B. As before, $\mathbf{b}|\mathbf{a}$. Stage 0 is recursive in $\mathbf{d}$. Stages $2n + 1$ and $2n + 2$ are recursive in $\mathbf{d}$ except for requiring the solution to instances of the halting problem for procedures recursive in $\mathbf{a} \cup \mathbf{d}$. (In stage $2n + 2$, we seek, in effect, a finite segment that can be added to f_{2n+1} to form $\tilde{f}$.) Therefore $\mathbf{b} \leq \mathbf{d} \cup (\mathbf{a} \cup \mathbf{d})'$. Since $\mathbf{d} < \mathbf{a}$, $\mathbf{d} \cup (\mathbf{a} \cup \mathbf{d})' = \mathbf{d} \cup \mathbf{a}' = \mathbf{a}'$. Hence $\mathbf{b} \leq \mathbf{a}'$. Since $\mathbf{b}|\mathbf{a}$, $\mathbf{b} < \mathbf{a}'$.☒

The Degree 0″

This is, by definition, the degree of $\emptyset''$ and hence of K'. It includes, as we now show, the sets $\{x|\varphi_x \textit{ is total}\}$ and $\{x|W_x \textit{ is infinite}\}$.

Theorem VIII $K' \equiv \mathbf{0}'' \equiv \overline{\{x|\varphi_x \textit{ is total}\}} \equiv \overline{\{x|W_x \textit{ is infinite}\}}$.

Proof. Since $K \in \mathbf{0}'$ and $\emptyset' \in \mathbf{0}'$, it follows from Theorem I that $K' \equiv \emptyset''$.

That $\{x|\varphi_x \textit{ is total}\} \equiv \{x|W_x \textit{ is infinite}\}$ was indicated at the end of §7.4.

It remains to show that $\overline{\{x|\varphi_x \textit{ is total}\}} \equiv K'$. First observe that $\{x|\varphi_x \textit{ is not total}\} = \{z|(\exists x)[\varphi_z(x) \textit{ is divergent}]\}$. By the relativized projection theorem, this set is recursively enumerable in K (since $\{\langle x,z\rangle|\varphi_z(x) \textit{ is divergent}\}$ is recursive in K); and hence, by Theorem I, $\{x|\varphi_x \textit{ is not total}\} \leq_1 K'$.

Since K' is recursively enumerable in K, we have $K' = W_{z_0}{}^K$ for some z_0. Hence $K' = \{x|(\exists y)(\exists u)(\exists v)[\langle x,y,u,v\rangle \in W_{\rho(z_0)} \,\&\, D_u \subset K \,\&\, D_v \subset \bar{K}]\}$. Let x be given. Define ψ as follows. To compute $\psi(w)$, *see whether* $(\exists y)(\exists u)(\exists v)[\langle x,y,u,v\rangle$ occurs in $W_{\rho(z_0)}$ by the wth step in the enumeration of $W_{\rho(z_0)}$]; *if not*, set $\psi(w) = w$; *if so*, consider those $<x,y,u,v>$ which do occur by the wth step and for which all members of D_u occur in K by the wth step in the enumeration of K; if there are none such, set $\psi(w) = w$; if there are such, begin enumerating K and if and when it is shown, for all such, that $D_v \cap K \neq \emptyset$, set $\psi(w) = w$; otherwise $\psi(w)$ is undefined.

ψ is partial recursive, and a one-one recursive f exists such that $\varphi_{f(x)} = \psi$. It follows from the definition of ψ (as the reader can verify, see Exercise 13-15) that $x \in K' \Leftrightarrow (\exists w)[\psi(w)$ is undefined]. Hence $x \in K' \Leftrightarrow f(x) \in \{x|\varphi_x \textit{ is not total}\}$, and we have $K' \leq_1 \{x|\varphi_x \textit{ is not total}\}$.

Thus $K' \equiv \overline{\{x|\varphi_x \textit{ is total}\}}$, and the proof is complete. ☒

Theorem VIII suggests the intuitive significance of the degree $\mathbf{0}''$. A procedure is recursive in $\mathbf{0}''$ if and only if it is effective except for questions that ask whether or not a certain specified (and effectively recognizable) kind of event occurs infinitely often in the course of a certain effective computation. For example, W_x is infinite if and only if the event *an output appears* occurs infinitely often in the enumeration without repetitions of W_x; and φ_x is total if and only if the event *φ_x is now defined for a larger initial segment* occurs infinitely often in the simultaneous computation of $\varphi_x(0)$, $\varphi_x(1)$, ...

The Degree a″

Similarly, a procedure is recursive in $\mathbf{a}''$ if it is effective except for questions that ask whether or not a certain specified (and effectively recognizable) kind of event occurs infinitely often in the course of a certain computation recursive in some fixed $A \in \mathbf{a}$.

Chapter 14 will generalize the above results, and will develop more economical methods for handling such proofs as that of Theorem VIII.

§13.3 COMPLETE DEGREES; CATEGORY AND MEASURE

How are the complete degrees distributed in the ordering of degrees? If $\mathbf{a}' = \mathbf{b}'$, must $\mathbf{a} = \mathbf{b}$? Must $(\mathbf{a} \cup \mathbf{b})' = \mathbf{a}' \cup \mathbf{b}'$? These and other questions are answered in the theorems below.

Since, for any $\mathbf{a}$, $\mathbf{0} \leq \mathbf{a}$, it follows from Theorem I that every complete degree is greater than or equal to $\mathbf{0}'$. Does the converse hold, i.e., is every member of the nondenumerable family $\{\mathbf{a}|\mathbf{0}' \leq \mathbf{a}\}$ a complete degree? The following theorem of Friedberg [1957*a*] shows that this is so.

Theorem IX (Friedberg) $(\forall \mathbf{a})[\mathbf{0}' \leq \mathbf{a} \Rightarrow (\exists \mathbf{b})[\mathbf{b}' = \mathbf{a}]]$.

Proof. In terms of sets, the theorem asserts $(\forall A)[K \leq_{\mathrm{T}} A \Rightarrow (\exists B)[B' \equiv_{\mathrm{T}} A]]$. Let A be given such that $K \leq_{\mathrm{T}} A$, and let $g = c_A$. Construct a function f by initial segments $f_0, f_1, \ldots$ as follows.

Stage 0. Let $f_0 = \emptyset$.

Stage $n + 1$. *See whether* $\varphi_n^{[\tilde{f}]}(n)$ is convergent for some initial segment extension $\tilde{f}$ of f_n. *If so,* let f^* be the *first* such extension (in the effective enumeration of all such extensions that is obtained from $W_{p(n)}$ (see paragraph 3 on page 259). *If not,* let $f^* = f_n$. In either case, let x_n be the length of f^* and define $f_{n+1} = f^* \cup \{<x_n, g(n)>\}$.

Let $f = \bigcup_i f_i$, and let $B = \{x|f(x) = 1\}$. It remains to show that $B' \equiv_T A$.

(i) Given the set A, one can compute g, since $g = c_A$. Given A, one can also answer questions about the halting problem, since $K \leq_T A$. Hence, given A, one can test, at each stage, for the existence of the desired extension $\tilde{f}$. Hence one can reconstruct the segments $f_0, f_1, \ldots$ and, in the process, test for membership in $\{n|$*the extension* $\tilde{f}$ *is found to exist at stage* $n + 1\}$. But, by the construction, this latter set is precisely

$$\{x|\varphi_x{}^f(x) \textit{ is convergent}\} = (\tau(f))'.$$

Hence $(\tau(f))' \leq_T A$. As $B' \equiv (\tau(f))'$ (since $B \equiv_T f$), we have $B' \leq_T A$.

(ii) Conversely, given the set B', we can compute $f = c_B$ (since $B \leq_1 B'$), and we can solve instances of the halting problem (since $K \leq_1 B'$ by Theorem I). Hence we can reconstruct the individual segments using our ability to solve the halting problem to test for existence of the desired extension $\tilde{f}$ at each stage. But $g(n)$ = the last value in f_{n+1}. Hence, given B', we can compute g. Since $g = c_A$, we have $A \leq_T B'$.

Thus, by (i) and (ii), $B' \equiv_T A$.☒

The proof yields additional information which we give in the following corollary.

Corollary IX(*a*) $(\forall \mathbf{a})(\exists \mathbf{b})[\mathbf{b}' = \mathbf{b} \cup \mathbf{0}' = \mathbf{a} \cup \mathbf{0}']$.

Proof. Let A be given and $g = c_A$. Construct f and B as in the proof of the theorem. Argument (i) shows that B' is recursive in A and K; hence $\mathbf{b}' \leq \mathbf{a} \cup \mathbf{0}'$. Argument (ii) shows that A is recursive in B and K; hence $\mathbf{a} \leq \mathbf{b} \cup \mathbf{0}'$; and hence $\mathbf{a} \cup \mathbf{0}' \leq \mathbf{b} \cup \mathbf{0}'$. Since $\mathbf{b} \leq \mathbf{b}'$ and $\mathbf{0}' \leq \mathbf{b}'$, we have $\mathbf{b} \cup \mathbf{0}' \leq \mathbf{b}'$. Thus $\mathbf{b}' \leq \mathbf{a} \cup \mathbf{0}' \leq \mathbf{b} \cup \mathbf{0}' \leq \mathbf{b}'$, and the degrees must be equal.☒

Corollary IX(*b*) **(*Spector, Shoenfield*)**

(i) $(\exists \mathbf{b})(\exists \mathbf{c})[\mathbf{b}|\mathbf{c}\ \&\ \mathbf{c}' = \mathbf{b}']$.

(ii) $(\exists \mathbf{b})(\exists \mathbf{c})[\mathbf{b}|\mathbf{c}\ \&\ \mathbf{c}' < \mathbf{b}']$.

Proof. (i) Apply Corollary IX(*a*) with $\mathbf{a} = \mathbf{0}''$. Then

$$\mathbf{b}' = \mathbf{0}'' = \mathbf{b} \cup \mathbf{0}'.$$

Hence $\mathbf{b} \cup \mathbf{0}' \neq \mathbf{b}$, and $\mathbf{b} \cup \mathbf{0}' \neq \mathbf{0}'$. Thus $\mathbf{b}|\mathbf{0}'$, $\mathbf{b}' = (\mathbf{0}')'$, and (i) holds with $\mathbf{c} = \mathbf{0}'$.

(ii) Apply Corollary IX(*a*) with $\mathbf{a} = \mathbf{0}'''$. Then $\mathbf{b}' = \mathbf{0}'''$, but, as in (i), $\mathbf{b}|\mathbf{0}'$. Thus (ii) holds with $\mathbf{c} = \mathbf{0}'$.☒

Part (i) shows that the jump operation on degrees is not one-one (see also Exercise 13-19). Note, incidentally, that Corollary VI follows from the proof of Corollary IX(*b*), and that Corollary IX(*b*) implies the failure of the assertion that $\mathbf{a}' \leq \mathbf{b}' \Rightarrow \mathbf{a} \leq \mathbf{b}$ (the converse to the *basic relation* derived from Theorem I and asserted after Corollary I).

We now ask four questions.

1. Given that $\mathbf{a}' = \mathbf{b}'$, which of the possibilities $\mathbf{a} < \mathbf{b}$, $\mathbf{a} = \mathbf{b}$, $\mathbf{b} < \mathbf{a}$, $\mathbf{a}|\mathbf{b}$ can hold?
2. Given that $\mathbf{a}' < \mathbf{b}'$, which of the same four possibilities can hold?
3. Given that $\mathbf{a} < \mathbf{b} < \mathbf{a}'$, which of the three possibilities $\mathbf{a}' = \mathbf{b}' < \mathbf{a}''$, $\mathbf{a}' < \mathbf{b}' < \mathbf{a}''$, and $\mathbf{a}' < \mathbf{b}' = \mathbf{a}''$ can hold?
4. Must $\mathbf{a}' \cup \mathbf{b}' = (\mathbf{a} \cup \mathbf{b})'$?

For question 1, $\mathbf{a} = \mathbf{b}$ can hold trivially, and, by Corollary IX, $\mathbf{a}|\mathbf{b}$ can hold.

For question 2, neither $\mathbf{a} = \mathbf{b}$ nor $\mathbf{b} < \mathbf{a}$ can hold (by Theorem I). $\mathbf{a} < \mathbf{b}$ can hold, by the example of $\mathbf{b} = \mathbf{a}'$. $\mathbf{a}|\mathbf{b}$ can hold by Corollary IX. Question 2 is thus completely answered.

The following theorem of Spector [1956] completes the answer to question 1, answers question 4, and answers part of question 3. (Part (i) of Corollary IX(*b*) was first obtained as a corollary to this theorem.)

***Theorem X* (*Spector*)** $(\exists \mathbf{a})(\exists \mathbf{b})[\mathbf{a}' \leq \mathbf{a} \cup \mathbf{b}$ & $\mathbf{b}' \leq \mathbf{a} \cup \mathbf{b}$ & $\mathbf{a} \leq \mathbf{0}'$ *and* $\mathbf{b} \leq \mathbf{0}']$.

Proof. We construct two functions f and g by initial segments. f will be c_A, g will be c_B, and A and B will determine the degrees $\mathbf{a}$ and $\mathbf{b}$, respectively. For each n, f_n and g_n will have the same length.

Stage 0. Let $f_0 = g_0 = \emptyset$.

Stage $2n + 1$. Let $m = 2n$. By inductive assumption, f_m and g_m have the same length. *See whether* there exists an initial segment extension $\tilde{f}$ of f_m such that $\varphi_n^{[\tilde{f}]}(n)$ converges. *If so*, let $\tilde{f}$ be the first such extension (in the enumeration by $W_{\rho(n)}$), and let p be the length of $\tilde{f}$. Define

$$f_{m+1} = \tilde{f} \cup \{<p,1>\},$$

and define $g_{m+1} = g_m \cup (\tilde{f} - f_m) \cup \{<p,0>\}$. Then f_{m+1} and g_{m+1} have the same length. *If not*, let p be the length of f_m. Define $f_{m+1} = f_m \cup \{<p,0>\}$ and $g_{m+1} = g_m \cup \{<p,1>\}$.

Stage $2n + 2$. Let $m = 2n + 1$. The definition for stage $2n + 2$ is as for stage $2n + 1$ with the symbols f and g (with or without subscripts) interchanged throughout.

Let $f = \bigcup_i f_i$, $g = \bigcup_i g_i$, $A = \{x|f(x) = 1\}$, and $B = \{x|g(x) = 1\}$. Let $\mathbf{a}$ and $\mathbf{b}$ be the degrees of A and B, respectively.

In the construction of f we only need answers to instances of the halting problem in order to test for extensions and hence compute f and g. Therefore $\mathbf{a} \leq \mathbf{0}'$ and $\mathbf{b} \leq \mathbf{0}'$. Let $x_0, x_1, \ldots$ be the members of $\{x|f(x) \neq g(x)\}$

in increasing order. Then $\{x|\varphi_x^f(x)$ *is convergent*$\} = \{n|f(x_{2n}) = 1\}$. Hence, given both f and g, we can test for membership in $(\tau(f))'$. Thus $\mathbf{a}' \leq \mathbf{a} \cup \mathbf{b}$. By parallel argument, $\mathbf{b}' \leq \mathbf{a} \cup \mathbf{b}$.☒

Corollary X(a) $(\exists \mathbf{a})(\exists \mathbf{b})[\mathbf{a}' \cup \mathbf{b}' = \mathbf{a} \cup \mathbf{b} = \mathbf{a}' = \mathbf{b}' = \mathbf{0}' \;\&\; \mathbf{a} < \mathbf{0}' \;\&\; \mathbf{b} < \mathbf{0}' \;\&\; \mathbf{a}|\mathbf{b}]$.

Proof. Any **a** and **b** satisfying the theorem must have the additional properties asserted in the Corollary.☒

Corollary X(a) gives a negative answer to question 4, since

$$\mathbf{a}' \cup \mathbf{b}' = \mathbf{a} \cup \mathbf{b} < (\mathbf{a} \cup \mathbf{b})'.$$

A negative answer can also be deduced from Corollary IX(a); it is also possible to have $\mathbf{a}|\mathbf{b}$ and $\mathbf{a}' \cup \mathbf{b}' = (\mathbf{a} \cup \mathbf{b})'$ (see Exercise 13-20). Corollary X(a) completes the answer to question 1 by showing that the remaining possibilities, $\mathbf{a} < \mathbf{b}$ and $\mathbf{b} < \mathbf{a}$, can occur (since $\mathbf{0} < \mathbf{a}$ and $\mathbf{0}' = \mathbf{a}'$ in the corollary). It also shows that the first possibility in question 3 can occur (since $\mathbf{0} < \mathbf{a} < \mathbf{0}'$ and $\mathbf{0}' = \mathbf{a}' < \mathbf{0}''$). Shoenfield has shown (see §13.8) that the other two possibilities can also occur. Corollary X(a) also gives another and more direct proof for (i) in Corollary IX(b).

Theorem X and Corollary X can be put into the following relativized form.

Corollary X(b) $(\forall \mathbf{d})(\exists \mathbf{a})(\exists \mathbf{b})[\mathbf{a}' \cup \mathbf{b}' = \mathbf{a} \cup \mathbf{b} = \mathbf{a}' = \mathbf{b}' = \mathbf{d}' \;\&\; \mathbf{d} < \mathbf{a} < \mathbf{d}' \;\&\; \mathbf{d} < \mathbf{b} < \mathbf{d}']$.

Proof. See Exercise 13-22.☒

Corollary X(b) has the following consequence.

Corollary X(c) $\mathbf{0}' < \mathbf{c} \Rightarrow \mathbf{c}$ *is a least upper bound to the set of all degrees below* **c**.

Proof. If $\mathbf{0}' < \mathbf{c}$, then by Theorem IX, $\mathbf{c} = \mathbf{d}'$ for some **d**. Take **a** and **b** as in Corollary X(b). Then $\mathbf{a} < \mathbf{d}'$, $\mathbf{b} < \mathbf{d}'$, and $\mathbf{a} \cup \mathbf{b} = \mathbf{d}'$.☒

Consider the denumerable family of degrees: $\{\mathbf{a}|\mathbf{a} \leq \mathbf{0}'\}$. Is every such degree recursively enumerable? The following theorem of Shoenfield [1959] shows that this is not so.

Theorem XI (Shoenfield) $(\exists \mathbf{a})[\mathbf{a} \leq \mathbf{0}' \;\&\; (\forall x)[W_x \notin \mathbf{a}]]$.

Proof. We make the following definition. Let f, g, and h be initial segments and y and z be integers. We say that $R(f,y,g,z,h)$ holds if $g \subset \varphi_y^{[f]}$ and $h \subset \varphi_z^{[g]}$. Note that this relation is recursively enumerable.

Let c_x denote (for the remainder of this proof) the characteristic function of W_x. We define a function f by initial segments $f_0, f_1, \ldots$. We do so in such a way that for any pair $\langle y,z \rangle$ and for any x, either $c_x \neq \varphi_y^f$ or $f \neq \varphi_z^{c_x}$. We then observe that f is recursive in $\mathbf{0}'$ and take **a** to be the degree of f.

Stage 0. Let $f_0 = \emptyset$.

Stage $n + 1$. Let $n = <x,y,z>$. We shall construct f_{n+1} so that for all extensions $\tilde{f}$ of f_{n+1}, either $\varphi_y^{[\tilde{f}]} \not\subset c_x$ or $\tilde{f} \not\subset \varphi_z^{c_x}$. Let m be the length of f_n. Let $f^0 = f_n \cup \{<m,0>\}$ and $f^1 = f_n \cup \{<m,1>\}$.

Substage (*a*). *See whether* there exist initial segments $\tilde{f}$ and g such that $f_n \subset \tilde{f}$ and $R(\tilde{f},y,g,z,f^0)$. *If not,* set $f_{n+1} = f^0$ and go to stage $n + 2$. *If so,* go to substage (*b*).

Substage (*b*). *See whether* there exist initial segments $\tilde{f}$ and g such that $f_n \subset \tilde{f}$ and $R(\tilde{f},y,g,z,f^1)$. *If not,* set $f_{n+1} = f^1$ and go to stage $n + 2$. *If so,* go to substage (*c*).

Substage (*c*). Take $\tilde{f}^0$, g^0, $\tilde{f}^1$, and g^1 such that $f_n \subset \tilde{f}^0$, $f_n \subset \tilde{f}^1$, $R(\tilde{f}^0,y,g^0,z,f^0)$, and $R(\tilde{f}^1,y,g^1,z,f^1)$. Then, since f^0 and f^1 disagree for argument m, g^0 and g^1 must disagree for some argument. Hence either g^0 or g^1 disagrees with c_x for that argument. If g^0 disagrees, set $f_{n+1} = \tilde{f}^0$; otherwise set $f_{n+1} = \tilde{f}^1$.

Let $f = \bigcup_i f_i$. This completes the construction.

For any x, y, and z, let $n = <x,y,z>$. Then at stage $n + 1$, if f_{n+1} is constructed at substage (*c*), we have $\varphi_y^{[f_{n+1}]} \not\subset c_x$. If f_{n+1} is constructed at substages (*a*) or (*b*), and if $\varphi_y^f = c_x$, then $f_{n+1} \not\subset \varphi_z^{c_x}$. Hence either $c_x \neq \varphi_y^f$ or $f \neq \varphi_z^{c_x}$.

Observe that at each stage the construction is effective if we can solve instances of the halting problem (since the tests in (*a*) and (*b*), and the computation of $c_x(w)$ for any w, are questions of individual membership in a recursively enumerable set). Hence f is recursive in $\mathbf{0'}$. Let

$$A = \{x|f(x) = 1\}.$$

Then $(\forall x)[W_x \not\equiv_T A]$. Let $\mathbf{a}$ be the degree of A. Then $\mathbf{a} \leq \mathbf{0'}$ and $(\forall x)[W_x \notin \mathbf{a}]$.☒

A relativized version of Theorem XI is given in Exercise 13-23.

Category

We have shown that there are uncountably many complete degrees and uncountably many incomplete degrees. Is there any reasonable sense in which we can say that there are *more* degrees of one kind than of the other? Two concepts from analysis, *category* and *measure,* can be used to give distinct but consistent answers to this question. The usefulness of these concepts (in treating this and other questions) has been pointed out by Myhill [1961*a*] (for the case of category) and by Sacks (for the case of measure; see Exercise 13-29). Initial applications of the measure concept were made by Spector.

Definition Let $\mathfrak{C}$ be the collection of all characteristic functions. Define a distance function δ on $\mathfrak{C}$ as follows:

$$\delta(f,g) = \begin{cases} 0, & \text{if } f = g; \\ \dfrac{1}{\mu x[f(x) \neq g(x)] + 1}, & \text{otherwise} \end{cases}$$

It is easy to verify that δ is a metric on $\mathcal{C}$ (see Exercise 13-24). δ therefore yields a metric space topology on $\mathcal{C}$. The *spherical neighborhoods* are sets of the form $\{f|f_0 \subset f\}$, where f_0 is an initial segment.

An alternative way of putting a topology on $\mathcal{C}$ is as follows. Let $I = \{0,1\}$. Then $\mathcal{C}$ is the cartesian product I^N. Take the discrete topology on I and form the product topology on $\mathcal{C} = I^N$. The *basis neighborhoods* are then sets of the form $\{f|f_0 \subset f\}$ for any finite segment f_0 (initial or not).

It is easy to show that these topologies are the same (Exercise 13-24). We henceforth assume this topology in our discussion. (This topology is sometimes called the *Cantor-set* topology, since $\mathcal{C}$ can be canonically mapped onto the Cantor ternary set, and this topology then coincides with the usual relative topology on the Cantor set. Other reasonably natural topologies can be placed on $\mathcal{C}$ (see Exercises 13-24 and 11-35). The space $\mathcal{C}$ (under our chosen topology) is sometimes called *Cantor space.* We sometimes refer to members of $\mathcal{C}$ as *points* in $\mathcal{C}$. As a topological space, $\mathcal{C}$ is *complete* (see Exercise 13-25 for definition and proof).

Definition A set $\mathcal{A}$ in a topological space is *nondense* if the closure of $\mathcal{A}$ contains no nonempty open set.

A set $\mathcal{A}$ is of *first category* if it is a finite or denumerable union of nondense sets.

A set $\mathcal{A}$ is of *second category* if it is not of first category.

The following consequences of these definitions are immediate: a denumerable union of sets of first category is of first category; and the difference between a set of second category and a set of first category is of second category. The following result is commonly known as Baire's theorem.

***Theory XII* (*Baire*)** *Every complete metric space is (in its own topology) a set of second category.*

Proof. See Exercise 13-26.☒

Every collection of degrees determines a subset of $\mathcal{C}$, namely, the collection of all characteristic functions of sets in those degrees. The concept of category can therefore be used to measure the abundance of degrees of any particular kind. For any point f in $\mathcal{C}$, the unit set $\{f\}$ is of first category. Hence any individual degree is (as a denumerable collection of points) of first category, and every denumerable collection of degrees is of first category. In particular, $\{\mathbf{a}|\mathbf{a} \leq \mathbf{0}'\}$ is of first category. What can be said about $\{\mathbf{a}|\mathbf{0}' \leq \mathbf{a}\}$ and $\{\mathbf{a}|\mathbf{0}'|\mathbf{a}\}$?

Theorem XIII $\{\mathbf{a}|\mathbf{0}' \leq \mathbf{a}\}$ *is of first category in Cantor space.*

Proof. $\mathbf{0}' \leq \mathbf{a} \Leftrightarrow (\exists A)(\exists z)[A \in \mathbf{a} \,\&\, c_K = \varphi_z{}^A]$. We wish to show that $\mathcal{A} = \{f|(\exists z)[c_K = \varphi_z{}^f]\}$ is of first category. (We restrict our attention, throughout the proof, to f in $\mathcal{C}$.) It is enough to show that for any z, $\{f|c_K = \varphi_z{}^f\}$ is nondense (since $\mathcal{A}$ is the denumerable union of such sets). Let $\mathcal{B} = \{f|c_K = \varphi_z{}^f\}$ be given. Assume $\mathcal{B}$ is not nondense. Then there is a spherical neighborhood $\mathcal{S} = \{f|f_0 \subset f\}$ contained in the closure of $\mathcal{B}$. Every neighborhood contained in $\mathcal{S}$ must intersect $\mathcal{B}$. Thus $c_K = \varphi_z^{f^*}$ for *some* extension f^* of segment f_0, and $\varphi_z^{[\tilde{f}]} \subset c_K$ for every initial segment extension $\tilde{f}$ of f_0 (otherwise, $(\exists w)[\varphi_z^{[\tilde{f}]}(w) \neq c_K(w)]$ and the neighborhood $\{f|\tilde{f} \subset f\}$ is contained in $\mathcal{S}$ but disjoint from $\mathcal{A}$). Hence

$$c_K = \{<x,y>|(\exists \tilde{f})[f_0 \subset \tilde{f} \;\&\; \varphi_z^{[\tilde{f}]}(x) = y]\};$$

but the latter is a recursively enumerable relation and therefore c_K is a recursive function, contrary to the nonrecursiveness of K. So $\mathcal{B}$ is nondense, and $\mathcal{A}$ is of first category. ☒

A further comment on the above proof is given in Exercise 13-28.

Corollary XIII(a) $\{\mathbf{a}|\mathbf{0}'|\mathbf{a}\}$ *is of second category.*

(***b***) *For any* $\mathbf{c} \neq \mathbf{0}$, $\{\mathbf{a}|\mathbf{a} \leq \mathbf{c}$ *or* $\mathbf{c} \leq \mathbf{a}\}$ *is of first category and* $\{\mathbf{a}|\mathbf{c}|\mathbf{a}\}$ *is of second category.*

Proof. (*a*) Immediate.

(*b*) We can use any nonrecursive set in place of K in the proof of the theorem. ☒

Therefore, in the sense of category, there are *more* incomplete degrees than complete degrees. Corollary XIII(*b*) gives us a simple and direct proof for the lemma in Theorem V, as we now show.

Corollary XIII(c) *For any countable collection* $\mathcal{A}$ *of nonrecursive degrees, there exists a degree* $\mathbf{b}$ *such that* $(\forall \mathbf{a})[\mathbf{a} \in \mathcal{A} \Rightarrow \mathbf{b}|\mathbf{a}]$.

Proof. Let $\mathbf{a}_0, \mathbf{a}_1, \ldots$ be the members of $\mathcal{A}$. Define

$$\mathcal{A}_x = \{\mathbf{c}|\mathbf{c} < \mathbf{a}_x \text{ or } \mathbf{a}_x < \mathbf{c}\}.$$

By Corollary XIII(*b*), $\mathcal{A}_x$ is of first category. Hence $\mathcal{A}^* = \cup_x \mathcal{A}_x$ is of first category. Hence there must be a degree not in $\mathcal{A}^*$. Let $\mathbf{b}$ be any such degree. ☒

Measure

A different approach to questions of abundance of degrees is provided by the concept of *measure.* This approach allows us to use terminology from probability theory.

We make Cantor space into a *measure space* (see Halmos [1950] for definition and properties of measure) by associating the *equiprobable measure* μ $(\mu(\{0\}) = \mu(\{1\}) = \frac{1}{2})$ with I, and then taking the *product measure* (which we also denote by μ) on $\mathcal{C}$.

Every denumerable subset of $\mathcal{C}$ has measure zero, and hence every degree and every denumerable collection of degrees can be assigned measure zero.

The concepts of *zero measure* and *first category* are independent. It is possible for a set of first category to have nonzero measure and for a set of second category to have zero measure (see Exercise 13-27).

The assertion that a subset $\mathcal{A}$ of $\mathcal{C}$ has measure zero can be given the following probabilistic interpretation: if an infinite sequence of equiprobable independent trials ("coin tosses") is used to determine a member of $\mathcal{C}$ (i.e., an infinite sequence of 0's and 1's), the probability that this member of $\mathcal{C}$ falls into $\mathcal{A}$ is zero. We give the following theorem as an example.

Theorem XIV $\{\mathbf{a}|\mathbf{0}' \leq \mathbf{a}\}$ *has measure zero in Cantor space.*

Proof. Probabilistic terminology will be used to abbreviate certain elementary arguments about measure. Let $\mathcal{A}$ and $\mathcal{B}$ be measurable subsets of $\mathcal{C}$. We define $\Pr[f \in \mathcal{A}] = \mu(\mathcal{A})$, and call this value the *probability of* $\mathcal{A}$. If $\mu(\mathcal{B}) \neq 0$, we define $\Pr[f \in \mathcal{A}|f \in \mathcal{B}] = \mu(\mathcal{A} \cap \mathcal{B})/\mu(\mathcal{B})$, and call this value the *conditional probability of* $\mathcal{A}$ *given* $\mathcal{B}$.

Assume the theorem false. Then $\Pr[(\exists z)[c_K = \varphi_z^f]] > 0$. Hence for some z, $\Pr[c_K = \varphi_z^f] = \epsilon > 0$. Taking this z, consider the identity (for any n)

$$\Pr[c_K = \varphi_z^f] = \Pr[(\forall x)[x \leq n \Rightarrow c_K(x) = \varphi_z^f(x)]] \cdot \Pr[c_K = \varphi_z^f|(\forall x)[x \leq n \Rightarrow c_K(x) = \varphi_z^f(x)]] = \epsilon.$$

Since

$$\bigcap_{n=0}^{\infty} \{f|(\forall x)[x \leq n \Rightarrow c_K(x) = \varphi_z^f(x)]\} = \{f|c_K = \varphi_z^f\},$$

we have $\lim_{n\to\infty} \Pr[(\forall x)[x \leq n \Rightarrow c_K(x) = \varphi_z^f(x)]] = \epsilon$, and hence

$$\lim_{n\to\infty} \Pr[c_K = \varphi_z^f|(\forall x)[x \leq n \Rightarrow c_K(x) = \varphi_z^f(x)]] = 1.$$

Take n_0 so that $n \geq n_0 \Rightarrow \Pr[c_K = \varphi_z^f|(\forall x)[x \leq n \Rightarrow c_K(x) = \varphi_z^f(x)]] \geq \frac{3}{4}$. Consider the class $\mathcal{S}$ of all initial segments f' such that (i)

$$(\forall x)[x \leq n_0 \Rightarrow c_K(x) = \varphi_z^{[f']}(x)];$$

and (ii) for no shorter initial segment $f'' \subset f'$ is it true that

$$(\forall x)[x \leq n_0 \Rightarrow c_K(x) = \varphi_z^{[f'']}(x)].$$

For any initial segment f', define $P_1(f') = \Pr[f' \subset f]$ and

$$P_2(f') = \Pr[c_K = \varphi_z^f|f' \subset f].$$

Then $\Pr[(\forall x)[x \leq n_0 \Rightarrow c_K(x) = \varphi_z^f(x)]] = \Sigma_{f' \in \mathcal{S}} P_1(f')$;

$$\Pr[c_K = \varphi_z^f] = \Sigma_{f' \in \mathcal{S}} P_1(f') \cdot P_2(f');$$

and

$$\Pr[c_K = \varphi_z^f|(\forall x)[x \leq n_0 \Rightarrow c_K(x) = \varphi_z^f(x)]] = [\Sigma_{\mathcal{S}} P_1(f') \cdot P_2(f')]/\Sigma_{\mathcal{S}} P_1(f') \geq \tfrac{3}{4}.$$

Since the last expression is a weighted average, there must be some $f^* \in \mathcal{S}$ such that $P_2(f^*) \geq \frac{3}{4}$. Consider the following instructions for computing a function g.

Given input x, enumerate $W_{\rho(z)}$; for each member of the form $\langle x,y,u,v\rangle$, compute $\Pr[D_u \subset \tau(f)\ \&\ D_v \subset \overline{\tau(f)} | f^* \subset f]$ (this is an elementary calculation); call each such value a *probability for potential output y;* when the total probabilities accumulated for some potential output y exceed $\frac{1}{2}$, give that y as the value of $g(x)$.

Since $\Pr[c_K = \varphi_z^f | f^* \subset f] \geq \frac{3}{4}$, g must be total and equal to c_K. Hence c_K is recursive, contrary to the nonrecursiveness of K.☒

Corollary XIV(a) *$\{\mathbf{a}|\mathbf{a}|\mathbf{0}'\}$ has measure one.*

(b) *For any $\mathbf{c} \neq 0$, $\{\mathbf{a}|\mathbf{a} \leq \mathbf{c}$ or $\mathbf{c} \leq \mathbf{a}\}$ has measure zero, and $\{\mathbf{a}|\mathbf{a}|\mathbf{c}\}$ has measure one.*

Proof. Immediate.☒

Therefore, in the sense of measure (as well as in the sense of category), there are *more* incomplete degrees than complete degrees.

§13.4 ORDERING OF DEGREES

In this section we present two results concerning the ordering of degrees. They give further evidence of the wide variety of relationships that can exist in the degree ordering. The first theorem, due to Kleene and Post, shows that a chain of degrees of order type of the rationals exists between $\mathbf{0}$ and $\mathbf{0}'$. The second theorem, also due to Kleene and Post, shows that the degree ordering is not a lattice and, incidentally, that the degree $\mathbf{0}^{(\omega)}$ is not minimal among the upper bounds to the arithmetical degrees.

Theorem XV (Kleene, Post) *There exists a denumerably infinite collection of degrees between $\mathbf{0}$ and $\mathbf{0}'$ which is linearly ordered (in the degree ordering) with order type of the rationals.*

Proof. We call f a *segment of two variables* if f is a partial function of two variables with recursive domain. f is an *initial segment of two variables* if, for some m, *domain* $f = \{\langle x,y\rangle | x < m$ *and* $y < f^*(x)\}$, where f^* is an initial segment (of one variable) of length m. If f and g are segments of two variables, and $f \supset g$, we say that f is an extension of g. If f is a (possibly total) segment of two variables, we define $f^x = \lambda y[f(x,y)]$; and

$$f^{(x)} = \lambda zy[f(s_x(z),y)],$$

where $s_x(z) = z$ if $z < x$, and $s_x(z) = z + 1$ if $x \leq z$. (Thus f^x is the xth *section* of f, and $f^{(x)}$ is f with its xth section deleted.) If f is an initial segment of two variables, then f^x and $f^{(x)}$ are initial segments of one variable and two variables, respectively. For f a segment of two variables, the definition of $\varphi_z^{[f]}$ is parallel to the (previously given) definition of $\varphi_z^{[f]}$ for f a segment of one variable (see Exercise 13-31).

The proof is in four parts.

1. We define f, a function of two variables, such that for all x, $f^x \not\leq_{\mathrm{T}} f^{(x)}$, i.e., for all x and y, $f^x \neq \varphi_y^{f^{(x)}} (= \varphi_y^{\tau^2(f^{(x)})})$. We construct f as the union of a chain of initial segments $f_0, f_1, \ldots$.

Stage 0. Set $f_0 = \emptyset$.

Stage $n + 1$. Let $n = \langle x,y \rangle$. Let x_n be the length of f_n^x. We shall construct f_{n+1} so that for all extensions $\tilde{f}$ of f_{n+1}, $\tilde{f}^x(x_n) \neq \varphi_y^{[\tilde{f}^{(x)}]}(x_n)$. This will guarantee that $f^x \neq \varphi_y^{f^{(x)}}$. We proceed as follows. *See whether* $\varphi_y^{[\tilde{f}]}(x_n)$ is defined for some initial-segment extension $\tilde{f}$ of $f^{(x)}$. *If not*, set

$$f_{n+1} = f_n \cup \{\langle x,x_n,1 \rangle\}.$$

If so, take such an $\tilde{f}$, and let $w = \varphi_y^{[\tilde{f}]}(x_n)$. Define f_{n+1} so that $f_{n+1}^{(x)} = \tilde{f}$ and $f_{n+1}^x = f_n^x \cup \{\langle x_n, w + 1 \rangle\}$.

Let $f = \bigcup_i f_i$. Evidently, for all x and y, $f^x \neq \varphi_y^{f^{(x)}}$.

2. Given f constructed as above, we show that if g and h are one-one recursive functions, then [*range* $g \subset$ *range* $h \Leftrightarrow \lambda xy[f(g(x),y)] \leq_{\mathrm{T}} \lambda xy[f(h(x),y)]$].

$\Rightarrow$: Let x, y be given, and assume *range* $g \subset$ *range* h. To compute $f(g(x),y)$, find z such that $h(z) = g(x)$, and compute $f(h(z),y)$.

$\Leftarrow$: Assume *range* $g \not\subset$ *range* h. Let x_0 be in *range* g but not in *range* h. Assume $\lambda xy[f(g(x),y)] \leq_{\mathrm{T}} \lambda xy[f(h(x),y)]$. Then, in particular, $f^{x_0} \leq_{\mathrm{T}} \lambda xy[f(h(x),y)]$. But $\lambda xy[f(h(x),y)] \leq_{\mathrm{T}} f^{(x_0)}$ since $f^{(x_0)} = \lambda xy[f(s_{x_0}(x),y)]$ and *range* $h \subset$ *range* s_{x_0} (see proof for $\Rightarrow$). But then $f^{x_0} \leq_{\mathrm{T}} f^{(x_0)}$, contrary to part 1.

3. We associate with each rational r a one-one recursive function g_r such that for any rationals r and s: $r \leq s$ (in the usual ordering of the rationals) $\Leftrightarrow$ *range* $g_r \subset$ *range* g_s. To do this, we take a coding from the rationals (expressed as fractions in lowest terms) onto the integers and let r_n be the rational with code number n. For any rational r, $\{n | r_n \leq r\}$ is a recursive set. Take g_r to be a recursive function with range $= \{n | r_n < r\}$. Now for each rational r, take $\mathbf{a}_r$ to be the degree of $\lambda xy[f(g_r(x),y)]$. It is immediate from part 2 that for any rationals r and s, $r \leq s \Leftrightarrow \mathbf{a}_r \leq \mathbf{a}_s$.

4. It remains to show that for all rationals r, $\mathbf{a}_r \leq \mathbf{0}'$. This will be proved if we can show that $f \leq_{\mathrm{T}} K$. But calculation of the successive extensions $f_0, f_1, \ldots$ is effective if we can solve instances of the halting problem. Hence $f \leq_{\mathrm{T}} K$. This completes the proof. ☒

***Corollary XV* (*Kleene, Post*)** *For any* $\mathbf{d}$, *there exists a denumerably infinite collection of degrees between* $\mathbf{d}$ *and* $\mathbf{d}'$ *which is linearly ordered with order type of the rationals.*

Proof. See Exercise 13-31. ☒

***Theorem XVI* (*Kleene, Post*)** *There exist degrees* $\mathbf{b}$ *and* $\mathbf{c}$ *such that*

(i) $\mathbf{b} \leq \mathbf{0}^{(\omega)}$ *and* $\mathbf{c} \leq \mathbf{0}^{(\omega)}$;

(ii) $(\forall n)[\mathbf{0}^{(n)} \leq \mathbf{b}$ *and* $\mathbf{0}^{(n)} \leq \mathbf{c}]$;

(iii) $(\forall \mathbf{d})[[\mathbf{d} \leq \mathbf{b} \;\&\; \mathbf{d} \leq \mathbf{c}] \Rightarrow (\exists n)[\mathbf{d} \leq \mathbf{0}^{(n)}]]$.

Proof. We define two functions f and g by means of segments $f_0, f_1, \ldots$ and $g_0, g_1, \ldots$. Except for f_0 and g_0, these segments will have infinite domains. **b** will be the degree of B and **c** will be the degree of C, where $f = c_B$ and $g = c_C$.

Let c_n denote (in this proof) the characteristic function of $\emptyset^{(n)}$. We call $\tilde{f}$ a *finite extension* of f if $\tilde{f} \supset f$ and if (*domain* $\tilde{f}$ − *domain* f) is finite.

Stage 0. Let $f_0 = g_0 = \emptyset$.

Stage $n + 1$. *Substage* (a). Let $n = \langle x,y \rangle$. *See whether* there are finite extensions $\tilde{f}$ and $\tilde{g}$ of f_n and g_n, respectively, such that for some z, $\varphi_x^{[\tilde{f}]}(z)$ and $\varphi_y^{[\tilde{g}]}(z)$ are defined and unequal. *If so*, define $\hat{f}_n = \tilde{f}$ and $\hat{g}_n = \tilde{g}$. *If not*, define $\hat{f}_n = f_n$ and $\hat{g}_n = g_n$.

Substage (b). Extend $\hat{f}_n$ to form f_{n+1} by setting $f_{n+1}(\langle n,x \rangle) = c_n(x)$ for all $\langle n,x \rangle$ not in *domain* $\hat{f}_n$. (It is easily verified, inductively, that only finitely many $\langle n,x \rangle$ will be in *domain* $\hat{f}_n$.) Similarly, extend $\hat{g}_n$ to g_{n+1} be setting $g_{n+1}(\langle n,x \rangle) = c_n(x)$ for all $\langle n,x \rangle$ not in *domain* $\hat{g}_n$.

Let $f = \bigcup_i f_i$, $g = \bigcup_i g_i$, $B = \{x|f(x) = 1\}$, and $C = \{x|g(x) = 1\}$. Let **b** and **c** be the degrees of B and C respectively. It remains to prove assertions (i), (ii), and (iii) of the theorem.

(i) By the uniformity results in Corollary I, for $m < n$, $\emptyset^{(m)} \leq_1 \emptyset^{(n)}$ uniformly; and for any n, $\emptyset^{(n)} \leq_1 \emptyset^{(n+1)}$ uniformly. It follows that stage $n + 1$ of the construction of f and g is uniformly recursive in $\emptyset^{(n+1)}$ in the following sense: f_{n+1} and g_{n+1} are recursive in $\emptyset^{(n+1)}$ uniformly (i.e., with indices effectively computable from n) and *domain* f_{n+1} and *domain* g_{n+1} are recursive in $\emptyset^{(n+1)}$ uniformly (i.e., with indices for their characteristic functions (as $\emptyset^{(n+1)}$-recursive functions) effectively computable from n). For, f_0 and g_0 are evidently recursive in $\emptyset^{(0)}$ with domains recursive in $\emptyset^{(0)}$, and if, by inductive assumption, f_n and g_n can be computed recursively in $\emptyset^{(n)}$ with domains recursive in $\emptyset^{(n)}$, then the existence of the desired extensions $\tilde{f}$ and $\tilde{g}$ in substage (a) is an instance of the halting problem for procedures recursive in $\emptyset^{(n)}$, and $\hat{f}_n$ and $\hat{g}_n$ can be computed recursively in $\emptyset^{(n+1)}$ with domains recursive in $\emptyset^{(n+1)}$; but then in substage (b), f_{n+1} and g_{n+1} can be computed recursively in $\emptyset^{(n+1)}$ with domains recursive in $\emptyset^{(n+1)}$. Since $\emptyset^{(n+1)}$ is recursive in $\emptyset^{(\omega)}$ uniformly in n, we have that f and g are each recursive in $\emptyset^{(\omega)}$. Thus $\mathbf{b} \leq \mathbf{0}^{(\omega)}$ and $\mathbf{c} \leq \mathbf{0}^{(\omega)}$.

(ii) For any n, $[\langle n,x \rangle \in B \Leftrightarrow c_n(x) = 1 \Leftrightarrow x \in \emptyset^{(n)}]$ holds for all but a finite number of x. Hence $\mathbf{0}^{(n)} \leq \mathbf{b}$. Similarly, for any n, $\mathbf{0}^{(n)} \leq \mathbf{c}$.

(iii) Assume $D \leq_T B$ and $D \leq_T C$. Then $c_D = \varphi_x{}^f$ and $c_D = \varphi_y{}^g$ for some x and y. Let $n = \langle x,y \rangle$. At stage $n + 1$ it must have been the case, in substage (a), that there were no finite extensions $\tilde{f}$ and $\tilde{g}$ of f_n and g_n, respectively, for which either $\varphi_x^{[\tilde{f}]}$ or $\varphi_y^{[\tilde{g}]}$ took on values different from c_D. (Otherwise, let $\tilde{f}$, for example, be a finite extension of f_n such that $\varphi_x^{[\tilde{f}]}(w) \neq c_D(w)$ for some w; by assumption, there is a finite extension $\tilde{g}$ of g_n such that $\varphi_y^{[\tilde{g}]}(w) = c_D(w)$; but then substage ($a$) would have assured $\varphi_x{}^f \neq \varphi_y{}^g$.) Hence if we can compute f_n and g_n, we can list all finite extensions $\tilde{f}$ and $\tilde{g}$ and thereby compute c_D. Since f_n and g_n are recursive in $\emptyset^{(n)}$ (by proof for (i)), c_D is recursive in $\emptyset^{(n)}$. This proves (iii).☒

***Corollary XVI* (*Kleene, Post*) (*a*)** *The ordering of degrees is not a lattice.*

(***b***) $\mathbf{0}^{(\omega)}$ *is not a minimal upper bound to the degrees* $\mathbf{0}, \mathbf{0}', \mathbf{0}'', \ldots$.

Proof. (*a*) By (iii) of the theorem, $\mathbf{b}$ and $\mathbf{c}$ can have no greatest lower bound.

(*b*) By (*a*), $\mathbf{b}$ and $\mathbf{c}$ must be incomparable. Hence, by (i), $\mathbf{b} < \mathbf{0}^{(\omega)}$ and $\mathbf{c} < \mathbf{0}^{(\omega)}$. But, by (ii), both $\mathbf{b}$ and $\mathbf{c}$ are upper bounds to $\mathbf{0}, \mathbf{0}', \mathbf{0}'', \ldots$.☒

A relativized form of Theorem XVI is given in Exercise 13-33.

§13.5 MINIMAL DEGREES

Is the degree ordering dense? We now give a theorem of Spector which shows that there exist minimal nonrecursive degrees and, hence, that (in contrast to the ordering of recursively enumerable degrees) the ordering of degrees is not dense.

In preceding proofs, each construction by successive extensions has been "simple" in the sense that, at the end of each stage, the only restriction placed on the function f (being constructed) is that it contain a certain segment, i.e., fall in a certain closed set in Baire space (see Exercise 13-24). The construction in the following proof is not of this kind, and no "simple" construction by extensions is known. The reader should also note Exercise 13-29, which shows that the collection of minimal nonrecursive degrees must have measure zero.

***Theorem XVII* (*Spector*)** $(\exists \mathbf{a})[\mathbf{0} < \mathbf{a} \;\&\; (\forall \mathbf{b})[\mathbf{b} < \mathbf{a} \Rightarrow \mathbf{b} = \mathbf{0}]]$.

Proof. It suffices to show that $(\exists A)[A$ not recursive and $(\forall B)[B \leq_T A \Rightarrow [B$ recursive or $A \leq_T B]]$. We construct a function g which will be c_A. The construction of g will proceed in stages. For each n, we shall have, at the end of stage n, three functions h^n, $f_1{}^n$, and $f_2{}^n$, such that the following conditions hold.

Conditions. h^n, $f_1{}^n$, and $f_2{}^n$ are recursive functions; *range* $f_1{}^n \subset \{0,1\}$; and *range* $f_2{}^n \subset \{0,1\}$. h^n is monotone; that is, $x < y \Rightarrow h^n(x) < h^n(y)$. $f_1{}^n$ is identical with $f_2{}^n$ on the interval $\{z \mid z < h^n(0)\}$; and for every y, $f_1{}^n$ and $f_2{}^n$ are not identical on the interval $\{z \mid h^n(y) \leq z < h^n(y+1)\}$. We speak of $\{z \mid z < h^n(0)\}$ and $\{z \mid h^n(y) \leq z < h^n(y+1)\}$, $y = 0, 1, 2, \ldots$, as the *intervals determined* by h^n. For all $m < n$: *range* $h^m \supset$ *range* h^n (hence the intervals determined by h^n are finite unions of the intervals determined by h^m); and on each (sub)interval determined by h^m, $f_1{}^n$ is identical either with $f_1{}^m$ or with $f_2{}^m$, and $f_2{}^n$ is identical either with $f_1{}^m$ or with $f_2{}^m$ (in particular, $f_1^{(n)} = f_2^{(n)} = f_1^{(m)} = f_2^{(m)}$ on $\{x \mid x < h^m(0)\}$). Finally, for all $m < n$, $h^m(0) < h^n(0)$.

Assume that h^n, $f_1{}^n$, and $f_2{}^n$ have been defined for all n. Define g_n to be the initial segment of length $h^n(0)$ such that $g_n = f_1{}^n = f_2{}^n$ on $\{z|z < h^n(0)\}$. From the Conditions above, it follows that $g_0, g_1, \ldots$ must be a chain of initial segments whose union is a total function. Define $g = \bigcup_i g_i$. Note from the Conditions that given any n and any interval determined by h^n, either g is identical with $f_1{}^n$ or g is identical with $f_2{}^n$ on that interval.

It now remains to define h^n, $f_1{}^n$, and $f_2{}^n$ and to prove that g has the properties ultimately desired. We proceed as follows. (Note that the procedure for finding h^{n+1}, f_1^{n+1}, and f_2^{n+1} from h^n, $f_1{}^n$, and $f_2{}^n$ is not *uniformly* effective (and that g will not be a recursive function) but that, nevertheless, at each stage n, h^n, $f_1{}^n$, and $f_2{}^n$ are recursive.)

Stage 0. Let $h^0 = \lambda x[x], f_1{}^0 = \lambda x[0]$, and $f_2{}^0 = \lambda x[1]$. (Stage 0 thus places no constraints whatever on the eventual g.) Note that the Conditions hold trivially.

Stage $2n + 1$. Let $m = 2n$. By inductive assumption, let the Conditions hold for h^m, $f_1{}^m$, and $f_2{}^m$. Then $f_1{}^m$ and $f_2{}^m$ must differ on the interval $\{z|h^m(0) \leq z < h^m(1)\}$. *See whether*, on that interval, $\varphi_n \neq f_1{}^m$. *If so*, define $h^{m+1} = \lambda x[h^m(x + 1)]$; define $f_1^{m+1} = f_2^{m+1} = f_1{}^m$ on

$$\{z|z < h^{m+1}(0) = h^m(1)\};$$

and let f_1^{m+1} and f_2^{m+1} be equal, respectively, to $f_1{}^m$ and $f_2{}^m$ on $\{z|h^{m+1}(0) \leq z\}$. *If not*, then $\varphi_n \neq f_2{}^m$ on $\{z|h^m(0) \leq z < h^m(1)\}$, and we define h^{m+1}, f_1^{m+1}, and f_2^{m+1} as before except that $f_1^{m+1} = f_2^{m+1} = f_2{}^m$ on $\{z|z < h^{m+1}(0)\}$. It follows immediately that h^{m+1}, f_1^{m+1}, and f_2^{m+1} satisfy the Conditions. Stage $2n + 1$ assures that $g \neq \varphi_n$; hence g will not be a recursive function.

Stage $2n + 2$. Let $m = 2n + 1$. By inductive assumption, let the Conditions hold for h^m, $f_1{}^m$, and $f_2{}^m$. We call $\tilde{g}$ an *available segment at* m if *domain* $\tilde{g} = \{z|z < h^m(y)\}$ for some y, and if, on each interval determined by h^m in this domain, $\tilde{g}$ is identical either with $f_1{}^m$ or with $f_2{}^m$.† If $\tilde{g}$ is an available segment at m, we call g^* an *available extension at* m *of* $\tilde{g}$ if g^* is an available segment at m and is an extension of $\tilde{g}$.

Substage (*a*). *See whether* there exist an integer w and an available segment (at m) $\tilde{g}$ such that for all available extensions (at m) g^* of $\tilde{g}$, $\varphi_n^{[g^*]}(w)$ is undefined. *If not*, go to substage (*b*). *If so*, define $h^{m+1} = \lambda x[h^m(x + u)]$ where $\{z|z < h^m(u)\}$ is the domain of $\tilde{g}$; define $f_1^{m+1} = f_2^{m+1} = \tilde{g}$ on *domain* $\tilde{g}$; let f_1^{m+1} and f_2^{m+1} be equal, respectively, to $f_1{}^m$ and $f_2{}^m$ on

$$\{z|h^{m+1}(0) = h^m(u) \leq z\};$$

and go on to stage $2n + 3$. (Then the Conditions hold, and we have assured that $\varphi_n{}^g$ cannot be a total function.)

Substage (*b*). *See whether* there is some available segment (at m) $\tilde{g}$ such that for all w, $\varphi_n^{[g^*]}(w)$ takes on at most one value as g^* varies over all

† That is to say, $(\forall \text{ interval})(\exists i)[i = 1 \text{ or } i = 2$, and $\tilde{g}$ is identical with f_i on that interval]. We are not asserting that $(\exists i)(\forall \text{ interval})[\cdot\ \cdot\ \cdot]$.

available extensions (at m) of $\tilde{g}$.† *If not,* go to substage (c). *If so,* define h^{m+1}, f_1^{m+1}, and f_2^{m+1} from $\tilde{g}$ as in the last part of substage (a), and go on to stage $2n + 3$. (The Conditions hold, and, since the available extensions at m of $\tilde{g}$ are recursively enumerable, we have assured that $\varphi_n{}^g$ must be recursive if it is total.)

Substage (c). If substage (c) is reached, we know that for every available segment $\tilde{g}$, (i) for every w, there exists an available extension g^* such that $\varphi_n^{[g^*]}(w)$ is defined (see substage (a)); and (ii) there exist w and available extensions g_1^* and g_2^* such that $\varphi_n^{[g_1^*]}(w)$ and $\varphi_n^{[g_2^*]}(w)$ are defined and unequal (see substage (b)). We now define h^{m+1}, f_1^{m+1}, and f_2^{m+1} inductively as follows:

$$h^{m+1}(0) = h^m(1),$$
$$f_1^{m+1} = f_2^{m+1} = f_1{}^m \qquad \text{on } \{z|z < h^m(1) = h^{m+1}(0)\}.$$

(This choice is arbitrary; $f_2{}^m$ could have been used in place of $f_1{}^m$.)

Assume that h^{m+1} has been defined for all $x \leq p$, and that f_1^{m+1} and f_2^{m+1} have been defined on $\{z|z < h^{m+1}(p)\}$. h^{m+1} determines $p + 1$ intervals on $\{z|z < h^{m+1}(p)\}$. Hence there are 2^p segments of length $h^{m+1}(p)$ that will be available at $m + 1$. We denote these $\tilde{g}_0, \tilde{g}_1, \ldots, \tilde{g}_{2^p-1}$ (in, say, lexicographical order). We proceed through 2^p substeps in order to reach our definition of $h^{m+1}(p + 1)$ (and of f_1^{m+1} and f_2^{m+1} on the interval $\{z|h^{m+1}(p) \leq z < h^{m+1}(p + 1)\}$).

Substep 0. By remark (ii) above, there exist available (at m) extensions s_0 and t_0 of segment $\tilde{g}_0$ and there exists a z_0, such that $\varphi_n^{[s_0]}(z_0)$ and $\varphi_n^{[t_0]}(z_0)$ are defined and unequal; furthermore s_0 and t_0 can be taken to be of equal length. s_0, t_0, and z_0 can be found effectively, since we can recursively enumerate all such $\langle s_0,t_0,z_0 \rangle$. Define $u_0 = s_0 - \tilde{g}_0$ and $v_0 = t_0 - \tilde{g}_0$.

Substep $q + 1$. By remark (ii), there exist available (at m) extensions s' and s'' of $\tilde{g}_{q+1} \cup u_q$ and there exists a z_{q+1} such that $\varphi_n^{[s']}(z_{q+1})$ and $\varphi_n^{[s'']}(z_{q+1})$ are defined and unequal. By remark (i), there exists an available extension t of $\tilde{g}_{q+1} \cup v_q$ such that $\varphi_n^{[t]}(z_{q+1})$ is defined. Hence we can get s_{q+1} and t_{q+1} so that s_{q+1} is an available extension of $\tilde{g}_{q+1} \cup u_q$ and t_{q+1} is an available extension of $\tilde{g}_{q+1} \cup v_q$ and $\varphi_n^{[s_{q+1}]}(z_{q+1})$ and $\varphi_n^{[t_{q+1}]}(z_{q+1})$ are defined and unequal; furthermore s_{q+1} and t_{q+1} can be taken to be of equal length. As at substep 0, z_{q+1}, s_{q+1}, and t_{q+1} can be found effectively, since the relevant extensions and computations can be recursively enumerated. Define $u_{q+1} = s_{q+1} - \tilde{g}_{q+1}$ and $v_{q+1} = t_{q+1} - \tilde{g}_{q+1}$.

Observe that $u_0 \subset u_1 \subset \cdots \subset u_{2^p-1}$ and $v_0 \subset v_1 \subset \cdots \subset v_{2^p-1}$. Let w be the least integer greater than all members of *domain* u_{2^p-1} (it is easily seen that *domain* u_{2^p-1} (= *domain* v_{2^p-1}) is not empty). We define

$$h^{m+1}(p + 1) = w;$$
$$f_1^{m+1} = u_{2^p-1} \qquad \text{on } \{z|h^{m+1}(p) \leq z < h^{m+1}(p + 1)\};$$
$$f_2^{m+1} = v_{2^p-1} \qquad \text{on } \{z|h^{m+1}(p) \leq z < h^{m+1}(p + 1)\}.$$

† By substage (a), it takes at least one value.

This completes the inductive definition of h^{m+1}, f_1^{m+1}, and f_2^{m+1}. (It is immediate that the Conditions hold. Furthermore, if substage (*c*) is used, then g is recursive in $\varphi_n{}^g$. For we can compute g if we can decide, for each interval $\{z|h^{m+1}(p) \leq z < h^{m+1}(p+1)\}$, $p = 0, 1, 2, \ldots$, whether g agrees with f_1^{m+1} or with f_2^{m+1} on that interval. Assume g has been computed on $\{z|z < h^{m+1}(p)\}$; then we can find $q \leq 2^p - 1$ such that $\tilde{g}_q \subset g$ and we can compute $\varphi_n^{[s_q]}(z_q)$ and $\varphi_n^{[t_q]}(z_q)$. One and only one of the values must agree with $\varphi_n{}^g(z_q)$. If the former agrees, then g must agree with f_1^{m+1} on the interval in question; otherwise g must agree with f_2^{m+1} on the interval in question.) This completes stage $2n + 2$.

Let $A = \{x|g(x) = 1\}$. It remains to show that A has the desired properties. A is not recursive; otherwise, $g = \varphi_n$ for some n, contrary to the construction at stage $2n + 1$. Assume $B \leq_{\mathrm{T}} A$; that is, $c_B = \varphi_n{}^g$ for some n. If stage $2n + 2$ terminates with substage (*b*), then $\varphi_n{}^g$ is recursive and hence B is recursive. If substage (*c*) is used, then g is recursive in $\varphi_n{}^g$ and $A \leq_{\mathrm{T}} B$. This completes the proof of the theorem. ☒

How difficult is it to go from indices for h^m, $f_1{}^m$, and $f_2{}^m$ to indices for h^{m+1}, f_1^{m+1}, and f_2^{m+1}? For $m = 2n + 1$, we can do this effectively if we can solve instances of the halting problem. For $m = 2n + 2$, however, we must be able, in both substage (*a*) and substage (*b*), to test for membership in $\{x|W_x \textit{ is infinite}\}$ (see Exercise 13-34). (In substage (*c*), if it is reached, the procedure is entirely effective.) Hence we have the following corollary.

Corollary XVII(a) $(\exists \mathbf{a})[\mathbf{0} < \mathbf{a} < \mathbf{0}'' \ \& \ (\forall \mathbf{b})[\mathbf{b} < \mathbf{a} \Rightarrow \mathbf{b} = \mathbf{0}]]$.

Proof. See above. ☒

Note, by Exercise 10-11, that a minimal degree cannot be recursively enumerable.

The theorem and corollary can be relativized as follows.

Corollary XVII(b) $(\forall \mathbf{d})(\exists \mathbf{a})[\mathbf{d} < \mathbf{a} < \mathbf{d}'' \ \& \ (\forall \mathbf{b})[\mathbf{d} \leq \mathbf{b} < \mathbf{a} \Rightarrow \mathbf{b} = \mathbf{d}]]$.

Proof. See Exercise 13-34. ☒

§13.6 PARTIAL DEGREES

Enumeration reducibility ($\leq_{\mathrm{e}}$) and enumeration operators were defined in §9.7. We defined $f \leq_{\mathrm{e}} g$ if $\tau(f) \leq_{\mathrm{e}} \tau(g)$, and $f \equiv_{\mathrm{e}} g$ if $f \leq_{\mathrm{e}} g$ and $g \leq_{\mathrm{e}} f$. Taking equivalence classes with respect to $\equiv_{\mathrm{e}}$ over the collection of all total functions, we obtain a reducibility ordering of e-degrees (of total functions). This ordering is canonically isomorphic to the ordering of degrees; for, by Corollary 9-XXIV, $f \leq_{\mathrm{e}} g \Leftrightarrow \tau(f) \leq_{\mathrm{T}} \tau(g)$. Under this isomorphism, the T-degree of $\tau(f)$ corresponds to the e-degree of f; and conversely, the e-degree of c_A corresponds to the T-degree of A (see Theorem 9-VII). In view of this isomorphism, we sometimes refer to e-degrees of total functions as T-degrees. Equivalence classes with respect to $\equiv_{\mathrm{e}}$ can also be taken over the collection of all partial functions. These e-degrees (of partial functions) are called *partial degrees*, and their reducibility ordering under $\leq_e$ is called the *ordering of partial degrees*. There is, for example, a minimum partial degree which consists of all the partial recursive functions.

Every T-degree (of total functions) is a subcollection of some partial degree. If a partial degree contains a T-degree (of total functions), we call it a *total degree*. (Thus a partial degree is total if it contains at least one total function.) The total degrees therefore constitute a subordering of the partial degrees which is isomorphic to the ordering of degrees (as studied in preceding sections). Is every partial degree a total degree, or is the ordering of partial degrees a more extensive structure? We answer this question in the following theorem announced by Medvedev (see §13.7).

Theorem XVIII $(\exists\psi)[\psi$ *is not partial recursive and* $(\forall f)[f \leq_e \psi \Rightarrow f$ *recursive*]].

Proof. Recall that Φ_n is the enumeration operator of index n. We use the following notation in the proof.

$\Phi_n(\psi)$ abbreviates $\Phi_n(\tau(\psi))$. (Thus $\Phi_n(\psi)$ is a set which may not be single-valued.)

If $\Phi_n(\psi)$ is single-valued, we also abbreviate $\tau^{-1}(\Phi_n(\psi))$ as $\Phi_n(\psi)$. (Thus, for example, we can write $f \leq_e \psi \Leftrightarrow (\exists n)[f = \Phi_n(\psi)]$.)

ψ will be called a *finite segment* if (as before) *domain* ψ is finite. A finite segment $\tilde{\psi}$ will be called a *monotone extension* of ψ if $\psi \subset \tilde{\psi}$ and $(\forall x)(\forall y)[[x \in$ *domain* ψ and $y \in$ *domain* $(\tilde{\psi} - \psi)] \Rightarrow x < y]$.

We prove the theorem by obtaining the desired ψ as the union of a sequence of finite segments $\psi_0, \psi_1, \ldots$, where $m < n \Rightarrow \psi_n$ is a monotone extension of ψ_m.

Stage 0. Define $\psi_0 = \emptyset$.

Stage $2n + 1$. Let $m = 2n$. Let z be the least integer greater than all members of *domain* ψ_m. *See whether* $\varphi_n(z)$ is defined. *If not*, let

$$\psi_{m+1} = \psi_m \cup \{<z,0>\}.$$

If so, let $\psi_{m+1} = \psi_m \cup \{<z, \varphi_n(z) + 1>\}$. (Stage $2n + 1$ guarantees that $\psi \neq \varphi_n$; hence ψ will not be a partial recursive function.)

Stage $2n + 2$. Let $m = 2n + 1$.

Substage (a). *See whether* there exists a monotone extension $\tilde{\psi}$ of ψ_m such that $\Phi_n(\tilde{\psi})$ is not single-valued. *If so*, set $\psi_{m+1} = \tilde{\psi}$, and go to stage $2n + 3$. *If not*, go to substage (b).

Substage (b). *See whether* there exist a w and monotone extensions $\tilde{\psi}^1$ and $\tilde{\psi}^2$ of ψ_m such that $\Phi_n(\tilde{\psi}^1)$ and $\Phi_n(\tilde{\psi}^2)$, as partial functions, are defined and unequal for argument w. *If not*, let $\psi_{m+1} = \psi_m$. (In this case, $\Phi_n(\psi)$ must be recursive, if total, for it can be effectively computed by enumerating all monotone extensions of ψ_m and putting them through Φ_n.) *If so*, let $\psi_{m+1} = \psi_m \cup \{<z,0>\}$, where z is the least integer greater than all members of *domain* $\tilde{\psi}^1 \cup$ *domain* $\tilde{\psi}^2$. (In this case, $\Phi_n(\psi)$ must be undefined at w, for otherwise ψ could be used, together with $\tilde{\psi}^1$ or else $\tilde{\psi}^2$, to provide a monotone extension $\tilde{\psi}$ of ψ_m for which $\Phi_n(\tilde{\psi})$ is not single-valued, contrary to the result of substage (a).)

Define $\psi = \bigcup_i \psi_i$. ψ cannot be partial recursive (see stage $2n + 1$), and for any f, $f \leq_e \psi \Rightarrow f$ is recursive (see stage $2n + 2$).☒

Thus the ordering of partial degrees is more extensive than the ordering of total degrees. In §13.7, we discuss a still more extensive generalization of the notion of degree of unsolvability.

Corollary XVIII $(\exists\psi)[\psi$ *is recursively enumerable in* $\mathbf{0}'$ *and* ψ *is not partial recursive and* $(\forall f)[f \leq_e \psi \Rightarrow f$ *recursive*$]]$.

Proof. See Exercise 13-37.☒

In Exercise 13-38, we give a simple category proof for Theorem XVIII (but not for Corollary XVIII).

If $f \leq_e \psi$, then, by definition, $f = \mathbf{\Psi}(\psi)$ for some partial recursive operator $\mathbf{\Psi}$ (see §9.8). We ask the following question: if $f \leq_e \psi$, must $f = \mathbf{\Psi}(\psi)$ for some recursive operator $\mathbf{\Psi}$? The following theorem gives a negative answer to this question. The first proof of this result has been attributed to Myhill and Shepherdson. The proof given below is due to Rabin.

Theorem XIX $(\exists f)(\exists\psi)[f \leq_e \psi$ *but for no recursive operator* $\mathbf{\Psi}$ *is* $f = \mathbf{\Psi}(\psi)]$.

Proof. We use notation from the previous proof. We must show $(\exists f)(\exists\psi)[f \leq_e \psi \;\&\; (\forall m)[f = \Phi_m(\psi) \Rightarrow (\exists\psi')[\Phi_m(\psi')$ is not single-valued$]]]$.

We construct f as the union of a sequence of initial segments $f_0, f_1, \ldots$; and we construct ψ as the union of a sequence of finite segments $\psi_0, \psi_1, \ldots$, where $m < n \Rightarrow \psi_n$ is a monotone extension of ψ_m. f_{n+1} will have $\{0,1,\ldots,n\}$ as domain. There will be an increasing sequence $x_0, x_1, x_2, \ldots$ such that ψ_{n+1} will have $\{x_0,x_1,\ldots,x_n\}$ as domain. Furthermore, for $0 \leq i \leq n$, we shall have $\psi_{n+1}(x_i) = \tau^*(\langle f(0),f(1),\ldots,f(i)\rangle)$ (see end of §5.6). Thus ψ will be uniquely determined by f and the sequence $x_0,x_1, \ldots$. Indeed, given $f_{n+1}(n)$, ψ_n, and x_n, ψ_{n+1} is determined. If $f_0, f_1, \ldots$ and $\psi_0, \psi_1, \ldots$ are defined subject to these conditions, and if $f = \bigcup_i f_i$ and $\psi = \bigcup_i \psi_i$, then it is evident that $f \leq_e \psi$; for, given any enumeration of the ordered pairs of ψ, we can recover longer and longer initial segments of f.

The segments are constructed as follows.

Stage 0. Let $f_0 = \emptyset$ and $\psi_0 = \emptyset$.

Stage $n + 1$. *See whether* there is a y such that $\langle n,y\rangle \in \Phi_n(\tilde{\psi})$ for some finite monotone extension $\tilde{\psi}$ of ψ_n. *If not*, take $f_{n+1}(n) = 0$ and $x_n = x_{n-1} + 1$ (in the case $n = 0$, set $x_0 = 0$). *If so*, set $f_{n+1}(n) = y + 1$ and $x_n =$ the least integer greater than all members of domain $\tilde{\psi}$. In either case, determine ψ_{n+1} from f_{n+1}, ψ_n, and x_n, as described above.

Let $\psi = \bigcup_i \psi_i$ and $f = \bigcup_i f_i$.

As previously noted, $f \leq_e \psi$. Assume $f = \Phi_n(\psi)$. Then by the construction, $\Phi_n(\psi^*)$ must fail to be single-valued for some ψ^* such that $\psi^* \supset \psi$ (take $\psi^* = \psi \cup \tilde{\psi}$ where $\tilde{\psi}$ is as at stage $n + 1$.) This completes the proof. ☒

Corollary XIX $(\exists f)(\exists \psi)$[*f is recursive in* $\mathbf{0}'$ *and* ψ *is recursively enumerable in* $\mathbf{0}'$ *and* $f \leq_e \psi$, *but for no recursive operator* Ψ *is* $f = \Psi(\psi)$].

Proof. See Exercise 13-39. ☒

Other techniques employed in previous sections can be applied to partial degrees. The following theorem is parallel to Theorem VI.

Theorem XX $(\forall \varphi)[\varphi$ *not partial recursive* $\Rightarrow (\exists \psi)[\varphi \not\leq_e \psi \;\&\; \psi \not\leq_e \varphi]$.

Proof. A proof analogous to the proof of Theorem VI can be found (see Exercise 13-40). A category proof analogous to the proof of Theorem XIII can also be found (see Exercise 13-41). ☒

Partial degrees have not been extensively studied. The analogue to Theorem XVII (on minimal degrees), for example, is an open question. Little is known about automorphisms and order-theoretic properties. The following question is open: is the collection of total degrees order-theoretic, i.e., invariant under all automorphisms of the ordering of partial degrees?

§13.7 THE MEDVEDEV LATTICE

We have studied two structures in preceding sections of this chapter, the ordering of T-degrees (the primary object of our attention) and the more extensive ordering of partial degrees. We now consider a still more comprehensive structure, the *lattice of degrees of difficulty*, defined by Medvedev [1955*a*]. This structure gives added insight into some of the results already obtained about T-degrees and partial degrees, and it appears to exhibit greater regularity in its properties than does either the T-degree ordering or the partial-degree ordering. As we shall see, a *degree of difficulty* is an object of higher, i.e., more abstract, logical type than a T-degree. A degree of difficulty will be a collection of sets of total functions, whereas a T-degree can be viewed as a set of total functions (see §13.6).

Definition $\mathcal{A}$ is a *mass problem* if $\mathcal{A}$ is a collection of total functions.

Motivation. Given any informal "problem," a mass problem corresponding to it is, intuitively, a collection of functions any one of which can "solve" the "problem" and at least one of which can be obtained from any "solution" to the "problem." We give several examples.

Example 1. The unit set $\{c_A\}$ corresponds, as a mass problem, to the "problem" of deciding individual membership questions about A.

Example 2. $\{f \mid range\, f = A\}$ corresponds, as a mass problem, to the "problem" of enumerating A (for $A \neq \emptyset$).

Example 3. $\{f \mid A = \{x \mid f(x) = 0\}\}$ is another mass problem corresponding to the membership "problem" of Example 1.

Example 4. $\{f|range\, f = \{2x|x \in A\}\}$ is another mass problem corresponding to the enumeration "problem" of Example 2.

Of course, given any mass problem $\mathcal{A}$, there is at least one corresponding informal "problem," the "problem" of computing at least one member of $\mathcal{A}$.

The mass problems in Examples 1 and 3 are intuitively "equivalent," and so are the problems in Examples 2 and 4. Furthermore the mass problems in Examples 2 and 4 are "reducible" to the mass problems in Examples 1 and 3. Subsequent definitions will make these notions precise.

By Theorem 9-XXIII, there is a recursive function σ such that for any z, if $\mathbf{\Psi}$ is the partial recursive operator determined by Φ_z, then $\Phi_{\sigma(z)}$ defines a recursive operator which coincides with $\mathbf{\Psi}$ on all total functions in the domain of $\mathbf{\Psi}$.

Definition $\mathbf{\Psi}_z$ is the recursive operator defined by $\Phi_{\sigma(z)}$.

Definition Let $\mathcal{A}$ and $\mathcal{B}$ be mass problems. $\mathcal{A}$ is *reducible* to $\mathcal{B}$ if $(\exists z)[\mathbf{\Psi}_z(\mathcal{B}) \subset \mathcal{A}]$.

Motivation. $\mathcal{A}$ is *reducible* to $\mathcal{B}$ if there is an effective procedure such that, given any member of $\mathcal{B}$, we can get a member of $\mathcal{A}$; i.e., given any "solution" to $\mathcal{B}$, we can get some "solution" to $\mathcal{A}$. (Note that, in the definition, the order of $\mathcal{A}$ and $\mathcal{B}$ appears to be reversed when a superficial comparison is made with earlier reducibility definitions.)

Definition $\mathcal{A}$ is *equivalent* to $\mathcal{B}$ if $\mathcal{A}$ is reducible to $\mathcal{B}$ and $\mathcal{B}$ is reducible to $\mathcal{A}$.

The reducibility relation is trivially seen to be reflexive. It is transitive since the family of recursive operators is closed under composition. Hence the relation *equivalent* defined above is an equivalence relation.

Definition The equivalence classes of this equivalence relation are called *degrees of difficulty*. The reducibility relation is well-defined on these equivalence classes and induces a *reducibility ordering* of the degrees of difficulty.

Notation $\mathbf{A}, \mathbf{B}, \ldots$ will be degrees of difficulty. $\mathbf{A} \leq \mathbf{B}$ will mean that $\mathcal{A}$ is reducible to $\mathcal{B}$ for $\mathcal{A} \in \mathbf{A}$ and $\mathcal{B} \in \mathbf{B}$. $\mathbf{A} < \mathbf{B}$ will mean that $\mathbf{A} \leq \mathbf{B}$ and $\mathbf{B} \not\leq \mathbf{A}$.

Examples. Consider the following mass problems:

(1) $\{c_K\}$.
(2) $\{f|range\, f = K\}$.
(3) $\{f|K = \{x|f(x) = 0\}\}$.
(4) $\{f|range\, f = \{2x|x \in K\}\}$.
(5) $\{f|range\, f = K'\}$.

What reducibility relationships hold? It is immediately seen that (1) and (3) are equivalent (and hence belong to the same degree of difficulty).

Similarly (2) and (4) are equivalent. (2) is reducible to (1), but (1) is not reducible to (2) (otherwise c_K would be a recursive function). Hence the degree of (1) and (3) is distinct from, and higher than, the degree of (2) and (4). What about (5)? Given c_K, we can enumerate K'; conversely, given any enumeration of K', we can enumerate both K and $\bar{K}$ (since $K \leq_1 K'$ and $\bar{K} \leq_1 K'$) and hence we can compute c_K. Thus (5) belongs to the degree of (1) and (3). (In Exercise 13-42, we verify the above statements by explicitly defining appropriate recursive operators.) The concept of degree of difficulty thus makes precise our intuitive idea that "ability" to enumerate K' is "equivalent" to "ability" to test for membership in K. (The above remarks all hold when K is replaced by any nonempty set A, except that for certain sets A (e.g., sets of the form $A = B \textit{ join } \bar{B}$) the mass problems corresponding to (1) to (5) all belong to the same degree of difficulty.)

There is a minimum degree of difficulty; for let f be a recursive function, then $\{f\}$ is reducible to every mass problem. We call this minimum degree the *recursive* degree and denote it $\mathbf{0}$. Any mass problem that contains a recursive function must belong to $\mathbf{0}$. Hence $\mathbf{0}$ has $2^{2^{\aleph_0}}$ members. Conversely, every mass problem in $\mathbf{0}$ must contain a recursive function. Examples of mass problems in $\mathbf{0}$ are $\{c_A\}$ for any recursive A; and $\{f|\textit{range } f = B\}$ for any recursively enumerable, nonempty B.

There is a maximum degree of difficulty; for every mass problem is (trivially) reducible to the empty mass problem. We denote this maximum degree (of the empty mass problem) $\mathbf{1}$. (Medvedev has used the symbol "∞" for this degree.)

In order to study the ordering of degrees of difficulty, we make the following definitions.

Definition Given any integer n and function f, $n \circledast f$ will denote the function h such that $h(0) = n$ and $h(x + 1) = f(x)$ for all x.

Given any functions f and g, $f \wedge g$ will denote the function h such that $f = \lambda x[h(2x)]$ and $g = \lambda x[h(2x + 1)]$.

Definition Let $\mathcal{A}$ and $\mathcal{B}$ be any mass problems. We define

$$\mathcal{A} \wedge \mathcal{B} = \{f|(\exists g)(\exists h)[g \in \mathcal{A} \;\&\; h \in \mathcal{B} \;\&\; f = g \wedge h]\};$$

and $$\mathcal{A} \vee \mathcal{B} = \{f|(\exists g)[[g \in \mathcal{A} \;\&\; f = 0 \circledast g] \textit{ or } [g \in \mathcal{B} \;\&\; f = 1 \circledast g]]\}.$$

It is easy to show that the operations of $\wedge$ and $\vee$ are well-defined on degrees (see Exercise 13-43). Hence we have the following.

Definition

$$\mathbf{A} \wedge \mathbf{B} = \text{the degree of } \mathcal{A} \wedge \mathcal{B} \text{ for } \mathcal{A} \in \mathbf{A} \text{ and } \mathcal{B} \in \mathbf{B}.$$
$$\mathbf{A} \vee \mathbf{B} = \text{the degree of } \mathcal{A} \vee \mathcal{B} \text{ for } \mathcal{A} \in \mathbf{A} \text{ and } \mathcal{B} \in \mathbf{B}.$$

$\mathbf{A} \wedge \mathbf{B}$ is called the *join* of $\mathbf{A}$ and $\mathbf{B}$. $\mathbf{A} \vee \mathbf{B}$ is called the *meet* of $\mathbf{A}$ and $\mathbf{B}$. This definition of *join* is compatible with our original notion of *join* for sets;

for, if **A** is the degree of $\{c_A\}$ and **B** is the degree of $\{c_B\}$, then $\mathbf{A} \wedge \mathbf{B}$ is the degree of $\{c_{A\ join\ B}\}$.

Theorem XXI (***Medvedev***) *The operation* $\wedge$ *gives least upper bound and the operation* $\vee$ *gives greatest lower bound in the ordering of degrees of difficulty. Hence this ordering is a lattice. It is a distributive lattice.*

(Note that the symbols $\wedge$ and $\vee$ are in inverted position with respect to $\cup$ and $\cap$, the usual symbols for join and meet in a lattice. $\wedge$ and $\vee$ are chosen because of their common logical use as symbols for conjunction and disjunction. $\mathcal{A} \wedge \mathcal{B}$ represents the "ability" to solve *both* $\mathcal{A}$ *and* $\mathcal{B}$. $\mathcal{A} \vee \mathcal{B}$ represents the "ability" to solve *either* $\mathcal{A}$ *or* $\mathcal{B}$. This notational difficulty would be corrected (and the notation would be made more natural in certain other intuitive respects) if we inverted the lattice and thought of a problem as being *higher* in the ordering if it is more *easily* solved. We do not make this change, however, in order that our discussion may parallel our treatment of T-degrees and partial degrees.)

Proof. Assume **C** is an upper bound to **A** and **B**. Take representatives $\mathcal{C}$, $\mathcal{A}$, and $\mathcal{B}$ from **C**, **A**, and **B**, respectively. Let Ψ_m map $\mathcal{C}$ into $\mathcal{A}$ and Ψ_n map $\mathcal{C}$ into $\mathcal{B}$. Define an operator Φ^1 such that $[\Phi^1(f)](2x) = [\Psi_m(f)](x)$, and $[\Phi^1(f)](2x + 1) = [\Psi_n(f)](x)$. Φ^1 is evidently a recursive operator, and $\Phi^1(\mathcal{C}) \subset \mathcal{A} \wedge \mathcal{B}$. Hence $\mathbf{A} \wedge \mathbf{B}$ is a least upper bound to **A** and **B**.

Assume **D** is a lower bound to **A** and **B**. Take representative $\mathcal{D}$ from **D**. Let Ψ_p map $\mathcal{A}$ into $\mathcal{D}$ and Ψ_q map $\mathcal{B}$ into $\mathcal{D}$. Define an operator Φ^2 such that $\Phi^2(f) = \Psi_p(\lambda x[f(x + 1)])$ if $f(0) = 0$, and $\Phi^2(f) = \Psi_q(\lambda x[f(x + 1)])$ if $f(0) \neq 0$. Φ^2 is evidently a recursive operator, and $\Phi^2(\mathcal{A} \vee \mathcal{B}) \subset \mathcal{D}$. Hence $\mathbf{A} \vee \mathbf{B}$ is a greatest lower bound to **A** and **B**. (In Exercise 13-44, we consider formal definitions for Φ^1 and Φ^2, and the question of uniformity.)

Hence, by definition, the ordering of degrees of difficulty is a lattice. In order to show distributivity, it is enough to show that

$$\mathbf{A} \wedge (\mathbf{B} \vee \mathbf{C}) = (\mathbf{A} \wedge \mathbf{B}) \vee (\mathbf{A} \wedge \mathbf{C})$$

(see Exercise 12-4). Let any **A**, **B**, and **C** be given. Take representatives $\mathcal{A}$, $\mathcal{B}$, and $\mathcal{C}$ from **A**, **B**, and **C**, respectively. Then the mass problem $\mathcal{A} \wedge (\mathcal{B} \vee \mathcal{C})$ consists, by definition, of all functions of the form $f \wedge (0 \circledast g)$ or $f \wedge (1 \circledast h)$ for $f \in \mathcal{A}$, $g \in \mathcal{B}$, and $h \in \mathcal{C}$; while the mass problem $(\mathcal{A} \wedge \mathcal{B}) \vee (\mathcal{A} \wedge \mathcal{C})$ consists of all functions of the form $0 \circledast (f \wedge g)$ or $1 \circledast (f \wedge h)$. Each of these two mass problems is trivially reducible to the other, and hence the distributive identity is established. ⊠

We refer to this lattice as the *Medvedev lattice.* The lattice is easily shown not to be a Boolean algebra (see Exercise 13-45).

Certain kinds of degrees of difficulty are of special interest.

Definition $\mathcal{A}$ is a *problem of solvability* if $\mathcal{A} = \{c_A\}$ for some A.
$\mathcal{B}$ is a *problem of enumerability* if $\mathcal{B} = \{f | \mathit{range}\, f = B\}$ for some B.

A is a *degree of solvability* (not to be confused with the recursive degree **0**) if **A** contains a problem of solvability.

B is a *degree of enumerability* if **B** contains a problem of enumerability.

Notation $\mathbf{S}_A$ is the degree of solvability containing $\{c_A\}$.

$\mathbf{E}_B$ is the degree of enumerability containing $\{f|\mathit{range}\, f = B\}$ if $B \neq \emptyset$, and is **0** if $B = \emptyset$.

For any partial function ψ, we abbreviate $\mathbf{E}_{\tau(\psi)}$ as $\mathbf{E}_\psi$ and $\mathbf{S}_{\tau(\psi)}$ as $\mathbf{S}_\psi$.

Theorem XXII(a) $(\forall A)[\mathbf{S}_A = \mathbf{E}_{c_A}]$ *(and hence every degree of solvability is a degree of enumerability).*

(*b*) $(\forall A)(\exists \psi)[\mathbf{E}_A = \mathbf{E}_\psi]$ (i.e., *every degree of enumerability contains a problem of enumerability for some partial function*).

(*c*) $\mathbf{S}_A \leq \mathbf{S}_B \Leftrightarrow A \leq_{\mathrm{T}} B$ *(and hence the previously studied ordering of* T*-degrees is isomorphic to the ordering of degrees of solvability in the Medvedev lattice).*

(*d*) $\mathbf{E}_\psi \leq \mathbf{E}_\varphi \Leftrightarrow \psi \leq_{\mathrm{e}} \varphi$ *(and hence the previously studied ordering of partial degrees is isomorphic to the ordering of degrees of enumerability in the Medvedev lattice, and (by (a)) the composition of this isomorphism with the isomorphism from* T*-degrees to total degrees described at the beginning of* §13.6 *yields the isomorphism of* (c)*).*

Proof. (*a*) It is easily verified that $\mathbf{E}_{c_A} = \mathbf{S}_{c_A} = \mathbf{S}_A$.

(*b*) Trivially, $\mathbf{E}_{\iota_A} = \mathbf{E}_A$, where $\iota_A = \{<x,x>|x \in A\}$.

(*c*) This follows from Theorem 9-XXV.

(*d*) $\Leftarrow$: Assume $\psi \leq_{\mathrm{e}} \varphi$. If both ψ and φ are nonempty, then there is a partial recursive operator which carries any function with $\tau(\varphi)$ as range to a function with $\tau(\psi)$ as range. By Theorem 9-XXIII, there must be a recursive operator which does so, and hence $\mathbf{E}_\psi \leq \mathbf{E}_\varphi$. If ψ is empty, then $\mathbf{E}_\psi = \mathbf{0}$, and the conclusion is trivial. If φ is empty, then ψ is partial recursive and again $\mathbf{E}_\psi = \mathbf{0}$.

$\Rightarrow$: Assume $\mathbf{E}_\psi \leq \mathbf{E}_\varphi$. If both ψ and φ are nonempty, then there is a recursive operator mapping $\{f|\mathit{range}\, f = \tau(\varphi)\}$ into $\{f|\mathit{range}\, f = \tau(\psi)\}$. Hence, given any enumeration of $\tau(\varphi)$ (in any order), we can enumerate the function enumerating $\tau(\varphi)$ (in that order), apply the recursive operator, and get an enumeration of a function enumerating $\tau(\psi)$. Hence $\psi \leq_{\mathrm{e}} \varphi$. If ψ is empty, then $\psi \leq_{\mathrm{e}} \varphi$ holds trivially. If ψ is not empty and φ is empty, then $\mathbf{E}_\psi$ must be **0** and $\{f|\mathit{range}\, f = \tau(\psi)\}$ must contain a recursive function. But then ψ is partial recursive and $\psi \leq_{\mathrm{e}} \varphi$ holds trivially. ☒

Corollary XXII(a) $(\forall A)[A$ *is recursive* $\Leftrightarrow \mathbf{S}_A = \mathbf{0}]$.

(*b*) $(\forall \psi)[\psi$ *is partial recursive* $\Leftrightarrow \mathbf{E}_\psi = \mathbf{0}]$.

(*c*) $\mathbf{E}_A \leq \mathbf{E}_B \Leftrightarrow A \leq_{\mathrm{e}} B$.

Proof. Immediate. ☒

By the above theorem, we may view the T-degrees and partial degrees (studied in preceding sections) as embedded in the Medvedev lattice. Hence we can use our previous results on T-degrees and partial degrees to obtain information about the Medvedev lattice. (The existence of incomparable degrees of difficulty, for example, follows immediately.) We list a few results of particular interest in the following theorem.

Theorem XXIII (Medvedev) (*a*) $(\forall \mathbf{A})[\mathbf{A} \neq \mathbf{1} \Rightarrow (\exists B)[\mathbf{A} \leq \mathbf{S}_B]]$ (*i.e., every degree* $\neq \mathbf{1}$ *is reducible to some degree of solvability*).

(*b*) $(\forall A)(\exists \mathbf{B})[\mathbf{S}_A < \mathbf{B}\ \&\ (\forall \mathbf{C})[\mathbf{S}_A < \mathbf{C} \Rightarrow \mathbf{B} \leq \mathbf{C}]]$ (*i.e., every degree of solvability has a degree that is minimum among all degrees above it*).

(*c*) $(\exists \mathbf{B})[\mathbf{0} < \mathbf{B}\ \&\ (\forall \mathbf{C})[\mathbf{0} < \mathbf{C} \Rightarrow \mathbf{B} \leq \mathbf{C}]]$ (*i.e., there is a minimum nonrecursive degree*).

(*d*) $(\exists \mathbf{A})(\forall B)[\mathbf{A} \neq \mathbf{1}\ \&\ \mathbf{A} \neq \mathbf{E}_B]$ (*i.e., there exists a degree* $\neq \mathbf{1}$ *which is not a degree of enumerability*).

(*e*) $(\exists A)[\mathbf{E}_A \neq \mathbf{0}\ \&\ (\forall B)[\mathbf{S}_B \neq \mathbf{0} \Rightarrow \mathbf{S}_B \nleq \mathbf{E}_A]$ (*i.e., there is a nonrecursive degree of enumerability with no nonrecursive degree of solvability reducible to it*).

(*f*) $(\exists A)(\forall B)[\mathbf{E}_A \neq \mathbf{S}_B]$ (*i.e., there is a degree of enumerability which is not a degree of solvability*).

Proof. (*a*) Let $\mathcal{A} \in \mathbf{A}$ and $f \in \mathcal{A}$. Set $B = \tau(f)$. Then $\mathbf{A} \leq \mathbf{S}_B$.

(*b*) Let A be given. Form

$$\mathcal{B} = \{f|(\exists g)(\exists n)[f = n \circledast g\ \&\ c_A = \Psi_n(g)\ \&\ (\forall m)[g \neq \Psi_m(c_A)]]\}.$$

The result is immediate (see Exercise 13-48).

(*c*) This is a corollary to (*b*).

(*d*) Let $\mathbf{A}$ be the minimum degree obtained in (*c*). If $\mathbf{A}$ were a degree of enumerability, then by Theorems XX and XXII(*d*) there would be a degree of enumerability incomparable with it, and this would contradict (*c*).

(*e*) Take ψ as in Theorem XVIII, and let $A = \tau(\psi)$.

(*f*) This is a corollary to (*e*).☒

As we have noted, numerous other results carry over directly; for example, by Theorem XVII, there is a degree of solvability that is minimal among the nonrecursive degrees of solvability.

We have singled out degrees of solvability and enumerability for special attention. Other interesting families of degrees of difficulty can be defined. Consider, for example, the following definition.

Definition $\mathcal{A}$ is a *problem of separability* if, for some A and B,

$$\mathcal{A} = \{f|f(A) = 0 \text{ and } f(B) = 1\}.$$

$\mathbf{A}$ is a *degree of separability* if $\mathcal{A} \in \mathbf{A}$ for some problem of separability $\mathcal{A}$.

Using this concept, a theory of disjoint pairs of sets (as suggested at the end of §7.7) can be obtained within the general theory of degrees of difficulty.

The following definition gives another example.

Definition $\mathcal{A}$ is a *problem of one-one reduction* if, for some A and B, $\mathcal{A} = \{f|f \text{ is one-one } \&\ f^{-1}(B) = A\}$.

A is a *degree of one-one reduction* if $\mathcal{A} \in \mathbf{A}$ for some problem of one-one reduction $\mathcal{A}$.

In the Medvedev lattice, *relativization* (see §9.3) is both generalized and simplified. Given any degree **D**, $\{\mathbf{A} | \mathbf{A} \leq \mathbf{D}\}$ is an ideal in the lattice. A quotient lattice can be formed, and it is natural to call the elements of the quotient lattice *degrees of difficulty relative to* **D**. To form the elements of the quotient lattice, each degree **A** in the original lattice is identified with $\mathbf{A} \wedge \mathbf{D}$ (see Exercise 13-50); hence, in the special case when **D** is a degree of solvability, the *degrees of solvability relative to* **D** in the quotient lattice will correspond to the degrees of unsolvability relative to D in the sense of relativization defined in Chapter 9 (where $\mathbf{D} = \mathbf{S}_D$).

The Medvedev lattice has not been extensively studied. Whether or not the lattice has cardinality $2^{2^{\aleph_0}}$ is an open question. Little is known about nonprincipal ideals, automorphisms, and lattice-theoretic properties. It is not known whether the property of being a degree of solvability or the property of being a degree of enumerability is lattice-theoretic. It is not known whether the jump operation on degrees of solvability is lattice-theoretic or lattice-theoretic with respect to the degrees of solvability, i.e., invariant under all automorphisms of the lattice that carry the degrees of solvability onto the degrees of solvability.

In conclusion, we describe a relation between the Medvedev lattice and certain nonstandard logics.

Theorem XXIV **(*Medvedev*)** *For any* **A** *and* **B**, *the collection* $\{\mathbf{C} | \mathbf{B} \leq \mathbf{A} \wedge \mathbf{C}\}$ *has a least member.*

($\{\mathbf{C} | \mathbf{B} \leq \mathbf{A} \wedge \mathbf{C}\}$ is a dual ideal; the theorem asserts that this dual ideal is principal.)

Proof. Let $\mathcal{A} \in \mathbf{A}$ and $\mathcal{B} \in \mathbf{B}$. Define

$$\mathcal{D} = \{h | (\exists g)(\exists n)[h = n \circledast g \ \& \ (\forall f)[f \in \mathcal{A} \Rightarrow \Psi_n(f \wedge g) \in \mathcal{B}]]\}.$$

Let **D** be the degree of $\mathcal{D}$. Then **D** is the desired degree. To show this, observe first that $\mathcal{B}$ is reducible to $\mathcal{A} \wedge \mathcal{D}$, since $\Phi^1(\mathcal{A} \wedge \mathcal{D}) \subset \mathcal{B}$, where Φ^1 is defined so that $\Phi^1(f \wedge (n \circledast g)) = \Psi_n(f \wedge g)$. Second, take any $\mathcal{C}$ such that $\mathcal{B}$ is reducible to $\mathcal{A} \wedge \mathcal{C}$. Then for some n, $\Psi_n(\mathcal{A} \wedge \mathcal{C}) \subset \mathcal{B}$; but then $\Phi^2(\mathcal{C}) \subset \mathcal{D}$, where Φ^2 is defined so that $\Phi^2(g) = n \circledast g$. Hence $\mathcal{D}$ is reducible to $\mathcal{C}$. It can be immediately verified that Φ^1 and Φ^2 are recursive operators, and the proof is complete.☒

Definition Given **A** and **B**, let $(\mathbf{A} \to \mathbf{B})$ denote the least member of $\{\mathbf{C} | \mathbf{B} \leq \mathbf{A} \wedge \mathbf{C}\}$ (whose existence was proved in Theorem XXIV). We call $(\mathbf{A} \to \mathbf{B})$ the *degree of the problem of reducing* **B** *to* **A**. If $\mathbf{C} = \mathbf{A} \to \mathbf{B}$ for some **A** and **B**, we call **C** a *degree of reducibility* (note that $(\mathbf{A} \to \mathbf{B}) = \mathbf{0} \Rightarrow \mathbf{B} \leq \mathbf{A}$).

Definition Let $(\mathbf{A} \leftrightarrow \mathbf{B}) = (\mathbf{A} \rightarrow \mathbf{B}) \wedge (\mathbf{B} \rightarrow \mathbf{A})$.

Consider expressions formed from the variables "**A**," "**B**," "**C**," . . . and the symbols "$\wedge$", "$\vee$," "$\rightarrow$," and "$\leftrightarrow$" in the usual manner of sentential calculus. We call such an expression an *identity* if, for any assignment of degrees of difficulty to variables, the resulting degree denoted by the entire expression (when $\wedge$, $\vee$, $\rightarrow$, and $\leftrightarrow$ are taken as defined above) is $\mathbf{0}$.

Example 1. "$\mathbf{A} \rightarrow \mathbf{A}$" is an identity, since for any degree $\mathbf{A}$, $(\mathbf{A} \rightarrow \mathbf{A}) = \mathbf{0}$.

Example 2. "$\mathbf{A} \leftrightarrow \mathbf{A}$" is an identity, since $(\mathbf{A} \rightarrow \mathbf{A}) = \mathbf{0}$, and $\mathbf{0} \wedge \mathbf{0} = \mathbf{0}$.

Example 3. The *distributive law* "$(\mathbf{A} \wedge (\mathbf{B} \vee \mathbf{C})) \leftrightarrow ((\mathbf{A} \wedge \mathbf{B}) \vee (\mathbf{A} \wedge \mathbf{C}))$" is an identity, since the lattice is distributive and "$\mathbf{A} \leftrightarrow \mathbf{A}$" is an identity.

Example 4. "$((\mathbf{A} \vee \mathbf{B}) \rightarrow \mathbf{C}) \leftrightarrow ((\mathbf{A} \rightarrow \mathbf{C}) \wedge (\mathbf{B} \rightarrow \mathbf{C}))$" is an identity, as the reader can verify.

Example 5. "$(((\mathbf{A} \rightarrow \mathbf{B}) \rightarrow \mathbf{A}) \rightarrow \mathbf{A})$" is not an identity (see Exercise 13-51).

The expressions considered above are the formulas of sentential logic (in a commonly used symbolism). It can be shown that every identity (in the above sense) is a tautology of sentential logic, i.e., a theorem of classical two-valued sentential logic. The converse does not hold, since Example 5 is a tautology but not an identity. It can be shown (Medvedev) that the identities are just the theorems of the *positive propositional calculus* (see Hilbert and Bernays [1939, vol. 2, suppl. 3]).

Definition $\neg \mathbf{A} = (\mathbf{A} \rightarrow \mathbf{1})$. (Thus $(\neg \mathbf{A}) = \mathbf{1}$ if $\mathbf{A} \neq \mathbf{1}$; and $(\neg \mathbf{A}) = \mathbf{0}$ if $\mathbf{A} = \mathbf{1}$.)

Consider expressions formed as before, but with the additional symbol "$\neg$" allowed. Medvedev has shown that the identities are now just the theorems of the *intuitionistic propositional calculus* (see also Exercise 13-52).

§13.8 FURTHER RESULTS

We now return to the ordering of T-degrees considered in §§13.1 to 13.5. The results proved about the degree ordering in the present chapter have not used the priority methods of Chapter 10. If priority methods are combined with the methods of the present chapter, a number of stronger results can be obtained. These include both theorems about recursively enumerable degrees, e.g., the two theorems of Sacks quoted in §10.3, and stronger forms of theorems about degrees. (For example, Sacks has introduced priority methods into the proof of Theorem XVII and improved the upper bound in Corollary XVII(a) from $\mathbf{0}''$ to $\mathbf{0}'$. Similarly, Sacks has used priority methods to obtain $\mathbf{a}$ and $\mathbf{b}$ as recursively enumerable degrees in Theorem X.)

As an example of the use of priority methods, we present a special case of one of the chief theorems of Sacks [1963*b*] and use it to answer several questions left open in preceding sections. Sacks's proof introduces a fundamental extension of the Friedberg priority idea. The proof achieves priority results while, in effect, allowing individual priority markers to be moved infinitely often. This extended priority method has been used to obtain various recent results (such as density of recursively enumerable degrees) and improvements of earlier results.

***Theorem XXV* (*Sacks*)** $[\mathbf{0}' \leq \mathbf{c}\ \&\ \mathbf{c}$ *recursively enumerable in* $\mathbf{0}'] \Rightarrow (\exists \mathbf{a})[\mathbf{a}$ *recursively enumerable* & $\mathbf{a}' = \mathbf{c}]$.

Proof. (This proof is based in part on a formulation by Yates.) We assume that we have a set C recursively enumerable in K. It clearly suffices to obtain a recursively enumerable set A such that $A' \equiv_T C$ *join* K. Since $C \leq_1 K'$, we have, by §13.2, a recursive f such that $C = \{x | W_{f(x)}$ *is finite*$\}$. In our procedure for enumerating A, we take the integers as given in a doubly infinite matrix *array* where the integer $<i, j>$ occurs in the ith row and jth column. (We call the initial row and initial column the 0th row and 0th column respectively.) As the procedure is carried out, certain integers in the array are marked with a *plus;* these are the members of A. The procedure is arranged so that the xth row contributes finitely many of its members to A if $W_{f(x)}$ is finite and all but finitely many of its members to A if $W_{f(x)}$ is infinite. This yields $x \in \bar{C} \Leftrightarrow x$th row is co-finitely in A. Since the xth row nonmembers of A are obviously uniformly recursively enumerable in A, we have a recursive g such that $\bar{C} = \{x | W^A_{g(x)}$ *is finite*$\}$. By §13.2, this gives $\bar{C} \leq_1 A''$ and hence $\bar{C}$ recursively enumerable in A'. Since C is recursively enumerable in A' (because C is recursively enumerable in K and $K \leq_T A'$), we have $C \leq_T A'$ and hence C *join* $K \leq_T A'$.

Note that we could achieve the above merely by simultaneously enumerating $W_{f(0)}, W_{f(1)}, \ldots$, and, as the rth member of $W_{f(p)}$ appears, putting a *plus* beside $<p, r - 1>$ in the array. This simple procedure would not yield $A' \leq_T C$ *join* K, however; the procedure given below will therefore be a modified and more complex form of this simple procedure.

Two special features of our procedure are as follows. First, for each integer z, there will be a finite collection of integers *restrained* by z. The integers restrained by z will come from rows below the zth row. These collections can be thought of as ancillary lists kept by the person carrying out the enumeration of A. (At any time, only a finite number of the lists are nonempty.) Second, as $W_{f(0)}, W_{f(1)}, \ldots$ are simultaneously enumerated and as the rth member of $W_{f(p)}$ appears, a *plus* is not automatically placed by $<p, r - 1>$; instead a *circle* is drawn around $<p, r - 1>$, and then a *plus* is placed by $<p, r - 1>$ if and when $<p, r - 1>$ is found to be unrestrained.

In detail, our procedure is as follows. Here A_n is the set of integers which have received *plusses* by the end of stage n. The procedure begins with $A_0 = \emptyset$. The procedure concerns computations which place z in

$W_z^{A_n}$. We say that an integer x is used *negatively* in one of these computations if the computation (of $\varphi_z^{A_n}(z)$) converges and uses the fact that $x \notin A_n$. If and when such an integer x is subsequently placed in A, we say that that computation has become *obsolete*. An integer x is said to be *available* for restraint by z if it occurs below the zth row in the array, i.e., if $\pi_1(x) > z$. The set of integers restrained by z (at a given moment) will be called the *z-list* (at that moment). Integers may be either added to or removed from a z-list.

Stage $n + 1$.

(*a*) Carry out n steps in the enumerations of each of $W_0^{A_n}, \ldots, W_n^{A_n}$.

(*b*) Let $z_0 < z_1 < \cdots < z_p$ be all integers z such that z is placed in $W_z^{A_n}$ by a computation (of $\varphi_z^{A_n}(z)$) occurring in (*a*). Add to the z_0-list all integers that are available for z_0 and used negatively in the computation for $z_0 \in W_{z_0}^{A_n}$ and currently not on the z_0-list.

Add to the z_1-list all integers that are available for z_1 and used negatively in the computation for $z_1 \in W_{z_1}^{A_n}$ and now not on either the z_0- or z_1-lists.

. . . . . .

Add to the z_p-list all integers that are available for z_p and used negatively in the computation for $z_p \in W_{z_p}^{A_n}$ and now not on either the z_0- or z_1- or . . . or z_p-lists.

(*c*) Carry out n steps in the enumeration of each of $W_{f(0)}, \ldots, W_{f(n)}$. For each $x \leq n$, if a new member of $W_{f(x)}$ appears (i.e., a member that did not appear in (*c*) of stage n), then circle the first uncircled integer in the xth row of the array.

(*d*) Write a *plus* beside any circled integer which is unrestrained (i.e., restrained by no z).

(*e*) As a result of these added *plusses*, certain of the computations for z in $W_z^{A_n}$ considered in (*b*) may now be obsolete. For each such z, release all integers from restraint by z (i.e., the z-list becomes empty). Write a *plus* beside any circled integer which is now unrestrained.

(*f*) Repeat (*e*) as many times as needed until no further computation from (*b*) has become obsolete.

(*g*) For each x which became unrestrained during (*e*) and (*f*) but which did not receive a *plus*, add x to the z-list of each z such that (i) x is available for z, (ii) x is used negatively in a computation in (*b*) which places z in $W_z^{A_n}$, and (iii) the computation placing z in $W_z^{A_n}$ has not become obsolete during (*e*) and (*f*).

The integers now having *plusses* constitute A_{n+1}. This concludes stage $n + 1$.

It remains to prove that A has the desired properties. First observe that if there are infinitely many stages at which an integer x is restrained then either (i) there are infinitely many stages during which x becomes unrestrained (during (*f*)) or else (ii) there is a z-list to which x permanently belongs from some stage onward. This observation follows from the evident and key feature of the construction that once x leaves a highest z-list it cannot return to that same list without first becoming unrestrained. Hence the only way a circled integer can stay out of A is to become permanently restrained by some z. Observe next that x becomes permanently restrained by $z \Rightarrow x$ used negatively in a computation for z in W_z^A; and that [x used negatively in a computation for z in W_z^A & $\pi_1(x) > z$] $\Rightarrow x$ becomes permanently restrained by z' for some $z' \leq z$. Thus, for any x where $W_{f(x)}$ is infinite (i.e., where all integers in the xth row are circled), only a finite number of integers in the xth row become permanently restrained (since only a finite number can be used in computations for z in W_z^A, $z < x$), and the remaining integers go into A. Therefore the conditions described in the first paragraph of the proof hold and we have C *join* $K \leq_T A'$.

It remains to show $A' \leq_T C$ *join* K. Assume by induction, for a given z, that (α) we have computed the characteristic function of A' for all arguments $<z$, that (β) we know the integers that become permanently restrained by any $z' < z$, and that (γ) we have determined which members of rows above the zth row are in A. (This latter information is finite since the rows are either finitely or co-finitely in A.) Proceed as follows. Determine, by reference to the C-oracle, whether z is in C. If $z \in C$, then the zth row must be only finitely in A; use the K-oracle to locate the largest integer in the row which belongs to A; all preceding integers in this row which do not become permanently restrained by some $z' < z$ must be in A. If $z \notin C$, then the zth row must be co-finitely in A and all integers in the row which do not become permanently restrained by some $z' < z$ must be in A. This gives (γ) for $z + 1$. Next use the K-oracle to determine *whether* there is a stage n in the enumeration of A at which there is a computation placing $z \in W_z^{A_n}$ for which every negatively used integer is either known from (γ) (for $z + 1$) to be in $\bar{A}$, or restrained by z at stage n, or known from (β) to become permanently restrained by some $z' < z$. *If so*, then $z \in A'$ and the integers restrained by z at stage n are exactly the integers that become permanently restrained by z. *If not*, then $z \notin A'$ and no integers become permanently restrained by z. In either case, we have (α) and (β) for $z + 1$. Thus we have an inductive procedure for computing the characteristic function of A' recursively in C and K. Thus $A' \leq_T C$ *join* K.☒

Theorem XXV can be strengthened to the following.

Corollary XXV(a) $[\mathbf{0}' \leq \mathbf{c}$ & $\mathbf{c}$ *recursively enumerable in* $\mathbf{0}'] \Rightarrow (\exists \mathbf{a})$ [$\mathbf{a}$ *recursively enumerable* & $\mathbf{0} < \mathbf{a} < \mathbf{0}'$ & $\mathbf{a}' = \mathbf{c}$].

Proof. Modify the proof of Theorem XXV as follows. (i) Reserve the first integer in each row for use with movable markers in a simple Friedberg priority scheme to guarantee that for all z, $\bar{A} \neq W_z$. (ii) In the proof of Theorem XXV, restraints were arranged in a way that (in effect) associated the computation for z in W_z^A with the zth row. In the present proof, we associate the computation for z in W_z^A with the $2z$th row and associate computations which seek a member of $B \cap W_z^A$ with the $(2z + 1)$st row, where B is a simple set such that $B \equiv_T K$ (e.g., take B to be the set S^* in §8.4). More specifically, at stage $n + 1$, we enumerate n steps in $W_z^{A_n}$ (if $z \leq n$) and n steps in B. If we find no member of $B \cap W_z^{A_n}$, then we consider (at substage (*b*)) restraints for each of the (computations for) members of $W_z^{A_n}$ that has been enumerated; for any new (at stage $n + 1$) member of $W_z^{A_n}$ we start a new restraint list that ranks (for purposes of substage (*b*)) below all restraint lists for nonobsolete, previously obtained computations for members of $W_z^{A_n}$, that ranks below all restraint lists for computations associated with rows above the $(2z + 1)$st, but that ranks above all restraint lists for computations associated with rows below the $(2z + 1)$st. If, on the other hand, we do find a member of $B \cap W_z^{A_n}$, then we consider (at substage (*b*)) restraints for each of those (computations for) members of $W_z^{A_n}$ that have been enumerated out through the first member known to be in B. These restraint lists are ranked as in the case where no member of $B \cap W_z^{A_n}$ is found. The release procedure (for obsolete computations) in substages (*e*) and (*f*) remains as before.

The basic facts about permanent restraint can now be proved as before, and the final inductive proof now carries through as before when we make the crucial observation that once we know all members of A down through the $(2z + 1)$st row, we can recursively enumerate those members of W_z^A which are obtained and for which restraint lists are instituted (as members of $W_z^{A_n}$) in the course of the main construction. Of course, there may be integers obtained (in the main construction) in $W_z^{A_n}$ which are not in W_z^A, but these can be identified (since the possible causes of their failure to be in W_z^A will already be known) and the effect upon other restraint lists of their computations later becoming obsolete can be calculated.

Our recursive enumeration of members of W_z^A must either intersect B (by the fact that B is simple) or else be finite and exhaustive. (Note for purposes of the induction that, either way, only finitely many integers become permanently restrained by computations associated with the $(2z + 1)$st row.) But this yields that $W_z^A \neq \bar{B}$. Hence we have $B \not\leq_T A$ and the proof is complete.☒

Corollary XXV(*a*) shows that all three alternatives in question 3 (page 267) are possible. (This completely answers question 3. A complete answer was first obtained by Shoenfield.)

The construction in Corollary XXV(*a*) carries us uniformly from an index for C (in K) to an index for A. This uniform construction can be relativized. Hence we have the following.

Corollary XXV(b) *There is a recursive f such that for all x and D,* $D <_T W_{f(x)}^D <_T D'$ *and* $(W_{f(x)}^D)' = W_x^{D'}$ *join* D'.

Proof. The relativization is immediate.☒

The following result gives an application of Corollary XXV(*b*).

***Theorem XXVI* (Martin, Lachlan)** *There is a recursively enumerable* **a** *such that* $(\forall n)[\mathbf{0}^{(n)} < \mathbf{a}^{(n)} < \mathbf{0}^{(n+1)}]$.

Proof. Apply the relativized recursion theorem to the function f in Corollary XXV(b) to obtain an n such that for all X, $W_{f(n)}{}^X = W_n{}^X$. Take **a** to be the degree of $W_n{}^{\emptyset}$. The result is immediate. ☒

We conclude by stating several further results.

1. (Sacks). The ordering of recursively enumerable degrees is not a lattice.

2. (Yates). **0** is the greatest lower bound of two other recursively enumerable degrees.

3. (Yates). There exists a recursively enumerable $\mathbf{a} > \mathbf{0}$ such that for every recursively enumerable $\mathbf{b} < \mathbf{0}'$, $\mathbf{a} \cup \mathbf{b} < \mathbf{0}'$.

4. (Sacks). $[\mathbf{0} < \mathbf{a}$ & **a** recursively enumerable$] \Rightarrow$ there exists a recursively enumerable **d** such that $[\mathbf{0} < \mathbf{d} < \mathbf{a}$ & $\mathbf{d}' = \mathbf{0}']$.

5. (Martin, Lachlan). There exist recursively enumerable **a** and **b** such that $(\forall n)[\mathbf{a}^{(n)}|\mathbf{b}^{(n)}]$. See Exercise 13-54.

6. (Yates). There exists a degree below $\mathbf{0}'$ which is incomparable with all recursively enumerable degrees properly between **0** and $\mathbf{0}'$.

7. (Shoenfield, Sacks). $\mathbf{0} < \mathbf{a} < \mathbf{0}' \Rightarrow$ there exists a recursively enumerable **b** such that $\mathbf{b}|\mathbf{a}$.

8. (Spector). No ascending sequence of degrees has a least upper bound; but (Sacks) every countable set of degrees has a minimal upper bound.

9. (Spector). The arithmetical degrees do not form a lattice.

10. (Sacks). If Π is a partial ordering such that (i) the cardinality of $\Pi \leq \aleph_1$ (the least uncountable cardinal), and (ii) every element of Π has at most $\aleph_0$ predecessors, then there is a subordering of the degree ordering isomorphic to Π.

11. (Sacks). If Π is a partial ordering such that (i) the cardinality of $\Pi \leq 2^{\aleph_0}$, and (ii) every element has at most a finite number of predecessors, then there is a subordering of the degree ordering isomorphic to Π.

12. (Titgemeyer). There exists a degree below $\mathbf{0}''$ with one and only one degree between it and **0**.

13. (Sacks). $(\forall \mathbf{a})[(\forall \mathbf{b})[[\mathbf{b} \neq \mathbf{0}$ & **b** recursively enumerable$] \Rightarrow \mathbf{a} \leq \mathbf{b}] \Rightarrow \mathbf{a} = \mathbf{0}]$.

14. (Yates). Every recursively enumerable degree $>\mathbf{0}$ has a minimal degree below it, but there exists a degree $>\mathbf{0}$ with no minimal degree below it.

15. (Shoenfield). For every $\mathbf{a} < \mathbf{0}'$ there is a minimal $\mathbf{b} < \mathbf{0}'$ such that $\mathbf{b} \not\leq \mathbf{a}$ but (Yates) there is a recursively enumerable $\mathbf{a} < \mathbf{0}'$ such that for every recursively enumerable $\mathbf{b} > \mathbf{0}$, **a** and **b** have a lower bound $>\mathbf{0}$.

16. (Martin). **a** is the degree of a maximal set $\Leftrightarrow$ **a** is the degree of a hyperhypersimple set $\Leftrightarrow$ [**a** is recursively enumerable and $\mathbf{a}' = \mathbf{0}''$].

We mention as a final example a theorem of Sacks from which a number of interesting consequences can be derived (including several of the assertions above). These consequences are listed in Exercise 13-53. The theorem is a further generalization of Exercise 10-11.

Theorem (Sacks) *Let B, C, and D be given so that D is recursively enumerable in B, $C \leq_T D$, and $C \not\leq_T B$. Then there exist D_1 and D_2 such that $D_1 \cup D_2 = D$, $D_1 \cap D_2 = \emptyset$, D_1 recursively enumerable in B, D_2 recursively enumerable in B, $C \not\leq_T B$ join D_1, and $C \not\leq_T B$ join D_2.*

Sacks [1963*b*] raises several further open questions. Among these are the following. Does result 11 hold with "finite" replaced by "countable"? Can every *independent* set of degrees (i.e., no degree in the set is reducible to a finite join of other degrees in the set), where the cardinality of the set is $< 2^{\aleph_0}$, be extended to a larger set? A collection of degrees closed under the operation of going from **a** to $\{\mathbf{b}|\mathbf{b} \leq \mathbf{a}\}$ is called an *initial segment;* do all possible *finite* partial orderings appear as initial segments, subject to the conditions (i) that a minimum element exist, and (ii) that any pair of elements have at most one minimal upper bound? Sacks conjectures an affirmative answer to these questions. In the revised edition of [1963*b*], Sacks raises the further question (among others): Is there a z such that (for all A and B) $A <_T W_z^A <_T A'$ and $[A \equiv_T B \Rightarrow W_z^A \equiv_T W_z^B]$?

§13.9 EXERCISES

§13.1

13-1. Let a set A be given. Take two different *acceptable* Gödel numberings of the partial recursive functions (see Exercise 2-10). From them, by Theorem 9-II, get indexings of the partial A-recursive functions. Let A_1 and A_2 be jumps of A as defined from each of these indexings. Show that $A_1 \equiv A_2$.

13-2. Give formal constructions (similar to that given for $\Rightarrow$ in part (d) of Theorem I) for $\Leftarrow$ in part (d), and for $\Rightarrow$ and $\Leftarrow$ in part (e), of Theorem I.

13-3. For any A, show that

(i) $\overline{A'}$ is productive;

(ii) $K \leq_T A \Rightarrow A'$ is productive;

(iii) $(\bar{A})' \not\equiv \overline{A'}$;

(iv) $\bar{A}$ productive does not necessarily imply $(\exists B)[A = B']$.

For any A and B, show that

(v) $A' = B' \Rightarrow A = B$.

13-4. Show that the jump operation on degrees is definable in terms of the relation *recursively enumerable in* and the relation $\leq$.

13-5. Show that there is a uniform procedure such that given $\emptyset^{(n)}$ for any n, we can calculate n; i.e., show that there is a z such that for all n, $\varphi_z^{\emptyset^{(n)}}(0) = n$.

13-6. Assume that $A \in \mathbf{a}$, $B \in \mathbf{b}$, and A is recursively enumerable in B. Which of the following assertions is true and which is false?

(i) For any $A_1 \in \mathbf{a}$, A_1 is recursively enumerable in B.

(ii) For any $B_1 \in \mathbf{b}$, A is recursively enumerable in B_1.

△13-7. Show that $(\forall A)(\forall B)[A^{(\omega)} \not\equiv_m B']$.

13-8. Show that for any set A there exists an increasing function f such that $A \leq_T \{\langle x,y\rangle|x \in \emptyset^{(f(y))}\}$. (*Hint:* See Exercise 13-5.)

13-9. (*a*) Show that A arithmetical in $B \Leftrightarrow (\exists n)[A \leq_1 B^{(n)}]$.
(*b*) Show that A arithmetical in $B \Rightarrow A^{(\omega)} \leq_1 B^{(\omega)}$.

13-10. Show that the relation *arithmetical in* is transitive.

§*13.2*

13-11. Prove: $[\mathbf{0} < \mathbf{a}\ \&\ \mathbf{a} \not\leq \mathbf{d}] \Rightarrow (\exists \mathbf{b})[\mathbf{b}|\mathbf{a}\ \&\ \mathbf{d} \leq \mathbf{b} < \mathbf{a}' \cup \mathbf{d}]$. (*Hint:* See Theorem VII.)

△***13-12.*** (Kleene, Post). Show that for any $\mathbf{a}_1$, $\mathbf{a}_2$, and $\mathbf{a}_3$, such that $\mathbf{a}_1 < \mathbf{a}_2 \leq \mathbf{a}_3$, there exists a $\mathbf{b}$ such that $\mathbf{b}|\mathbf{a}_2$, $\mathbf{b}|\mathbf{a}_3$, and $\mathbf{a}_1 < \mathbf{b} < \mathbf{a}_3'$. (*Hint:* Generalize the proof of Theorem VII.) Apply this result to show that for any finite interval in the sequence $\mathbf{0}', \mathbf{0}'', \ldots$ there exist degrees incomparable with all members of that interval and comparable with all other members of the sequence.

△***13-13.*** (Kleene, Post). In Exercise 13-12, show that the single degree $\mathbf{b}$ can be replaced by an infinite collection of pairwise incomparable degrees.

13-14. Show that $\emptyset' \equiv K$.

13-15. Let f be as in the proof of Theorem VIII. Verify that for any x, $x \in K' \Leftrightarrow \varphi_{f(x)}$ is not total.

△***13-16.*** Show that $\{x|\varphi_x \textit{ is a recursive permutation}\} \equiv \overline{K'}$.

▲***13-17.*** Show that $\{x|W_x \textit{ is co-finite}\} \equiv K''$.

▲***13-18.*** Show that $\{x|W_x \textit{ is recursive}\} \equiv K''$.

§*13.3*

13-19. Show that $(\forall \mathbf{a})(\forall n)(\exists \mathbf{b})[\mathbf{b}|\mathbf{a}\ \&\ \mathbf{a}^{(n)} \leq \mathbf{b}']$. (*Hint:* See the proof of Corollary IX(*b*).)

△***13-20.*** (Spector). Show that $(\exists \mathbf{a})(\exists \mathbf{b})[\mathbf{a}|\mathbf{b}\ \&\ (\mathbf{a} \cup \mathbf{b})' = \mathbf{a}' \cup \mathbf{b}']$. (*Hint:* Construct f and g by the general method of the proof of Theorem X, but insert additional steps to ensure that $\lambda x[f(2x)]$ and $\lambda x[f(2x+1)]$ are incomparable. Take A_1 and A_2 so that $c_{A_1} = \lambda x[f(2x)]$ and $c_{A_2} = \lambda x[f(2x+1)]$. Observe that $A = A_1 \textit{ join } A_2$.)

13-21. Use Corollary IX(*a*) to construct an example of degrees $\mathbf{a}$ and $\mathbf{b}$ such that $\mathbf{a}' \cup \mathbf{b}' \neq (\mathbf{a} \cup \mathbf{b})'$. (*Hint:* See proof of Corollary IX(*b*).)

△***13-22.*** Prove Corollary X(*b*).

13-23. Prove Theorem XI in the following relativized form: $(\forall \mathbf{d})(\exists \mathbf{b})[\mathbf{d} < \mathbf{b} < \mathbf{d}'\ \&\ \mathbf{b}$ not recursively enumerable in $\mathbf{d}]$. (*Hint:* Use the relativization method of Theorem VII.)

13-24. (*a*) Show that δ is a metric on $\mathcal{C}$; i.e., show that
(i) $\delta(f,g) = 0 \Leftrightarrow f = g$;
(ii) $\delta(f,g) = \delta(g,f)$;
(iii) $\delta(f,h) \leq \delta(f,g) + \delta(g,h)$.

(*b*) Show that the metric topology and the cartesian-product topology described for $\mathcal{C}$ in §13.3 are the same.

(*c*) If we identify every set with its characteristic function, the space $\mathfrak{N}$ (of all subsets of N) is identified with $\mathcal{C}$. Prove that the metric topology on $\mathcal{C}$ is the same as the topology described in part (*c*) of Exercise 11-35.

(*d*) Prove that in $\mathcal{C}$ (under our chosen topology), every point is closed and every spherical neighborhood is both open and closed. (Such a space is sometimes called *totally disconnected.*)

(*e*) Show that $\mathcal{C}$ is compact. (*Hint:* Show that it has the *finite-intersection property;* i.e., that if every finite intersection from a given family of closed sets is nonempty, then the intersection of the entire family must be nonempty.)

△***13-25.***

Definition. A sequence of points $f_0, f_1, \ldots$ (in a metric space) is a *Cauchy sequence* if, for every real number $\epsilon > 0$, $(\exists n)(\forall p)(\forall q)[[n < p\ \&\ n < q] \Rightarrow \delta(f_p,f_q) < \epsilon]$.

Definition. $f_0, f_1, \ldots$ has *limit* g if $\lim_{n\to\infty} \delta(f_n,g) = 0$.

Definition. A metric space is *complete* if every Cauchy sequence has a limit. (The property of completeness is the generalization, for metric spaces, of the least-upper-bound property of the real line.)

Prove: $\mathcal{C}$ is complete.

△***13-26.*** Prove Baire's theorem (Theorem XII). (*Hint:* Let $\mathcal{C}$ be a complete metric space. Assume $\mathcal{C}$ is the union of a denumerable sequence of nondense sets. Let $\mathcal{C}_0, \mathcal{C}_1, \ldots$ be the closures of the members of this sequence. Then $\mathcal{C} = \bigcup_{i=0}^{\infty} \mathcal{C}_i$ and $(\forall i)[\mathcal{C}_i$ contains no spherical neighborhood]. Find a chain of spherical neighborhoods $\mathcal{S}_0 \supset \mathcal{S}_1 \supset \cdots$ such that $\mathcal{S}_n \cap (\bigcup_{i=1}^{n} \mathcal{C}_i) = \emptyset$. Form a Cauchy sequence from members of successive neighborhoods, and show that its limit cannot be in $\bigcup_{i=0}^{\infty} \mathcal{C}_i$.)

△***13-27.*** (*a*) Give an example of a set in $\mathcal{C}$ which is of first category but has positive measure. (*Hint:* For any increasing sequence $x_0, x_1, \ldots$, consider $\{f|(\forall i)(\exists x)[x_i \leq x < x_{i+1} \,\&\, f(x) = 1]\}$. Show that this set must be of first category, but that if the gaps between successive members of the sequence increase sufficiently rapidly, the set will have positive measure.)

(*b*) Give an example of a set in $\mathcal{C}$ which is of second category but has measure zero. (*Hint:* Show that the set of first category constructed in (*a*) can be given measure arbitrarily close to 1. Take the union of a denumerable family of such sets and consider its complement.)

13-28. Define a set in $\mathcal{C}$ to be *dense in* $\mathcal{C}$ if its closure is $\mathcal{C}$.

(*a*) Show that any set whose complement is of first category is dense in $\mathcal{C}$.

(*b*) Deduce from Baire's theorem that the intersection of denumerably many sets each of which is open and dense in $\mathcal{C}$ is a set of second category whose complement is of first category. (*Remark:* The proof of Theorem XIII amounts to showing that, for each z, the *interior* of $\{f|c_K \neq \varphi_z^f\}$ is open and dense in $\mathcal{C}$. (The *interior* of a set is the union of all spherical neighborhoods contained in it.))

△***13-29.*** (Sacks). Define: $\mathbf{a}$ is *minimal* if $\mathbf{0} < \mathbf{a}$ and $(\forall \mathbf{b})[\mathbf{b} < \mathbf{a} \Rightarrow \mathbf{b} = \mathbf{0}]$. (In §13.5, we show that such degrees exist.)

Prove that the collecton of all minimal degrees has measure zero. (*Hint:* For any fixed f, define $f^* = \lambda x[f(2x)]$ and $f^{**} = \lambda x[f(2x+1)]$. Clearly $f^* \leq_T f$. Show that $\Pr[f^*$ is recursive$] = 0$ and that $\Pr[f \leq_T f^*] = \Pr[f^{**} \leq_T f^*] = 0$.)

△***13-30.*** (Sacks). Prove the following generalization of Theorem XIV (and deduce an analogue to Corollary XIV): $\{\mathbf{a}|\mathbf{0}'$ *is recursively enumerable in* $\mathbf{a}\}$ has measure zero. (*Hint:* Show that for any A and z, $\Pr[A = W_z^f] = \epsilon > 0 \Rightarrow A$ is recursively enumerable. Apply the methods of the proof of Theorem XIV, beginning with the identity (for any n): $\Pr[A = W_z^f] = \Pr[(\exists f')[f'$ *is an initial segment of* f *and* $(\forall x)[x \leq n \Rightarrow [x \in A \Leftrightarrow x \in W_z^{[f']}]]]] \cdot \Pr[A = W_z^f|(\exists f')[\cdot\ \cdot\ \cdot]] = \epsilon$; where $W_z^{[f']} = domain\ \varphi_z^{[f']}$.)

§13.4

13-31. (*a*) Prove Corollary XV.

(*b*) Give a formal definition (i.e., a definition in terms of recursively enumerable sets) of $\varphi_z^{[f]}$ for f a segment of two variables.

△***13-32.*** (Kleene, Post). Define: a finite collection of degrees is *independent* if no degree in the collection is T-reducible to the (iterated) join of the remaining degrees. Show that in Exercise 13-12, the single degree $\mathbf{b}$ can be replaced by an arbitrarily large independent collection of degrees.

13-33. (Kleene, Post). Show that Theorem XVI holds when **0** (in the statement of Theorem XVI) is replaced by any degree **a**.

§13.5

13-34. (*a*) Prove Corollary XVII(*a*). (*b*) Prove Corollary XVII(*b*).

△(*c*) (Martin). Modify the construction in the proof of Theorem XVII to yield $(\exists \mathbf{a})[\mathbf{0} < \mathbf{a} \,\&\, \mathbf{a}'' = \mathbf{0}'' \,\&\, (\forall \mathbf{b})[\mathbf{b} < \mathbf{a} \Rightarrow \mathbf{b} = \mathbf{0}]]$. (*Hint:* Arrange the construction so that for every n, the question of whether or not $\varphi_n{}^g$ is total can be reduced to $\mathbf{0}''$. To accomplish this, arrange so that in substages (*b*) and (*c*) (of stage $2n + 2$) $\varphi_n{}^g$ is forced to be total. This latter can be accomplished by augmenting the given construction at substages (*b*) and (*c*) in a way that uses the general technique of substage (*c*) as given.

(*d*) (Martin). Show, in (*c*) above, that **a** is an example of a nonrecursive degree which contains no hyperimmune set. (*Hint:* Show that at stage $2n + 2$ there is a recursive f^* which majorizes $\{x|\varphi_n{}^g(x) = 1\}$ in case $\varphi_n{}^g$ is total.)

△*13-35.* (Lacombe). Show that there are $2^{\aleph_0}$ minimal degrees. (*Hint:* In stage $2n + 1$ of Theorem XVII, use the interval $\{z|h^m(1) \leq z < h^m(2)\}$ in place of $\{z|h^m(0) \leq z < h^m(1)\}$, define $h^{m+1} = \lambda x[h^m(x + 2)]$, and make an arbitrary choice (for f_1^{m+1}) on $\{z|h^m(0) \leq z < h^m(1)\}$. In this way, making $\aleph_0$ arbitrary choices, we obtain $2^{\aleph_0}$ versions of g.)

13-36. Show that any degree is the greatest lower bound of two incomparable degrees. (*Hint:* Use Theorem VI.)

§13.6

13-37. Prove Corollary XVIII.

△*13-38.* (Myhill). Let $N^* = N \cup \{\omega\}$. Let $\mathcal{P} = (N^*)^N$ = the collection of all mappings from N into N^*. We can identify $\mathcal{P}$ with the partial functions by interpreting $\psi(x) = \omega$ to mean that "$\psi(x)$ is divergent" (for any ψ in $\mathcal{P}$). Let δ be defined as follows, for ψ and φ in $\mathcal{P}$:

$$\delta(\psi,\varphi) = \begin{cases} 0, & \text{if } \psi = \varphi; \\ \dfrac{1}{\mu x[\psi(x) \neq \varphi(x)] + 1}, & \text{if } \psi \neq \varphi. \end{cases}$$

(*a*) Show that δ is a metric on $\mathcal{P}$, and hence defines a topology on $\mathcal{P}$.

(*b*) Show that $\mathcal{P}$ is complete (as a topological space).

(*c*) Prove that $\{\psi|(\exists f)[f \leq_e \psi \,\&\, f \textit{ not recursive}]\}$ is of first category, and derive Theorem XVIII from this by Baire's theorem. (*Hint:* See substage (*b*) of stage $2n + 2$ in the proof of Theorem XVIII.) (The reader should note in this case (as in others), that although the concept of category provides heuristic simplification and formal economy, the pivotal combinatorial step is the same in both the category and noncategory proofs.)

13-39. (*a*) Prove Corollary XIX.

△(*b*) Let ψ be given such that ψ is partial recursive and *range* $\psi \subset \{0,1\}$. Assume that there exist a function g and a recursive operator $\mathbf{\Phi}$ such that $(\forall f)[[\psi \subset f \,\&\, \textit{range } f \subset \{0,1\}] \Rightarrow \mathbf{\Phi}(f) = g]$. Show that g must be recursive. (*Hint:* Use the compactness theorem for trees (Exercise 9-40), or equivalently, use the compactness of $\mathcal{C}$ (Exercise 13-24).)

13-40. Give a proof for Theorem XX similar to the proof for Theorem VI.

13-41. Take the topological space $\mathcal{P}$ as defined in Exercise 13-38. For any ψ, show that if ψ is not partial recursive, then $\{\varphi|\varphi \not\leq_e \psi \textit{ and } \psi \not\leq_e \varphi\}$ is of second category; and hence obtain Theorem XX. (*Hint:* Apply the method of Theorem XIII; i.e., show that for any n, if $\{\varphi|\psi = \mathbf{\Phi}_n(\varphi)\}$ is not nondense, then ψ is partial recursive.)

§13.7

13-42. Consider the mass problems (1) to (5) given on page 283. Verify the statements made about these examples by defining appropriate recursive operators.

13-43. Show that the operators $\wedge$ and $\vee$, as defined for mass problems, are well-defined on degrees of difficulty.

13-44. Give formal definitions, i.e., definitions in terms of recursively enumerable sets, for the operators Φ^1 and Φ^2 in the proof of Theorem XXI. Show that an index for Φ^1 (in the indexing of recursive operators) can be found effectively from m and n and that an index for Φ^2 can be found effectively from p and q.

13-45. Show that no element in the Medvedev lattice other than **0** or **1** can have a complement. Hence conclude that the Medvedev lattice is not a Boolean algebra.

13-46. Show, for any f, that the degree of difficulty containing $\{f\}$ is a degree of solvability.

13-47. Show that the assertions $\mathbf{E}_A = \mathbf{E}_B$ and $\mathbf{S}_A = \mathbf{S}_B$ are independent, i.e., that neither statement need imply the other.

13-48. Complete the proof of Theorem XXIII(b).

13-49. (a) Show that there is no maximal element among the elements $\neq \mathbf{1}$ in the Medvedev lattice.

(b) Show that there do not exist incomparable elements **A** and **B** in the lattice such that $\mathbf{A} \vee \mathbf{B} = \mathbf{0}$.

(c) Under what circumstances can the meet of two degrees of solvability be a degree of solvability?

13-50. (a) Let $\mathcal{L}$ be a distributive lattice, let $c \in \mathcal{L}$, and let $\mathcal{I}$ be the principal ideal generated by c. Let a and b be elements of $\mathcal{L}$ and let $[a]$ and $[b]$ be the elements of $\mathcal{L}/\mathcal{I}$ determined by a and b. Show that $[a] \leq [b]$ if and only if $a \leq b \cup c$ (see Exercise 12-7).

(b) What is the corresponding result for principal dual ideals?

$\triangle$***13-51.*** Verify the assertions made in Examples 4 and 5 (page 289).

13-52. Let $\mathfrak{M}$ be the Medvedev lattice, $\mathbf{D} \in \mathfrak{M}$, and $\mathfrak{D}$ be the principal dual ideal generated by **D**. Define $\neg \mathbf{A}$ to be $\mathbf{A} \to \mathbf{D}$. Consider expressions in $\neg$, $\vee$, $\wedge$, $\to$, and $\leftrightarrow$. Show that these expressions have a well-defined interpretation over the quotient lattice $\mathfrak{M}/^*\mathfrak{D}$, i.e., that an assignment of elements of the quotient lattice to variables in an expression determines a unique element of the quotient lattice as value for the expression. Assuming Medvedev's result on intuitionistic propositional calculus (see §13.8), show that every theorem of the intuitionistic propositional calculus is an identity for every such quotient lattice. Show that the converse does not hold. In particular, give examples (of **D**) (i) where all expressions are identities; and (ii) where the identities coincide with the theorems of classical two-valued sentential logic.

§13.8

$\triangle$***13-53.*** (Sacks). Assume the theorem of Sacks stated at the end of §13.8. Derive the following consequences.

(i) $[\mathbf{b} < \mathbf{d}$ & $\mathbf{d}$ recursively enumerable in $\mathbf{b}] \Rightarrow (\exists \mathbf{d}_1)(\exists \mathbf{d}_2)[\mathbf{d} = \mathbf{d}_1 \cup \mathbf{d}_2$ & $\mathbf{b} < \mathbf{d}_1 < \mathbf{d}$ & $\mathbf{b} < \mathbf{d}_2 < \mathbf{d}$ & $\mathbf{d}_1$ recursively enumerable in $\mathbf{b}$ & $\mathbf{d}_2$ recursively enumerable in $\mathbf{b}]$. (*Hint:* Prove the following lemma. If $A \cup B = C$, $A \cap B = \emptyset$, and A and B are recursively enumerable in C', then $C \equiv_T A$ *join* B.)

(ii) A recursively enumerable $\Rightarrow A$ is the union of two disjoint recursively enumerable sets of incomparable degree. (This is the result quoted at the end of Exercise 10-11.)

(iii) $\{\mathbf{b} | \mathbf{b} < \mathbf{0}'\}$ has no maximal member.

(iv) $[\mathbf{b} < \mathbf{c} < \mathbf{d}$ & $\mathbf{d}$ recursively enumerable in $\mathbf{b}] \Rightarrow (\exists \mathbf{a})[\mathbf{b} < \mathbf{a} < \mathbf{d}$ & $\mathbf{a} | \mathbf{c}$ & $\mathbf{a}$ recursively enumerable in $\mathbf{b}]$. (This is result 7 in §13.8.)

(v) A is recursive $\Leftrightarrow (\forall B)[B$ recursively enumerable but not recursive $\Rightarrow A \leq_T B]$. (This is result 13 in §13.8.)

(vi) $(\exists \mathbf{b})[\mathbf{d}$ recursively enumerable in $\mathbf{b}$ & $\mathbf{b} < \mathbf{d}] \Rightarrow \mathbf{d}$ is a minimal upper bound to the collection of all degrees below $\mathbf{d}$.

(vii) $(\exists \mathbf{b})[\mathbf{d}$ recursively enumerable in $\mathbf{b}$ & $\mathbf{b} < \mathbf{d}] \Rightarrow$ there exists an increasing sequence of degrees $\mathbf{a}_0 < \mathbf{a}_1 < \mathbf{a}_2 < \cdots$ such that $\mathbf{d}$ is a minimal upper bound to the sequence. (By result 8 in §13.8, this sequence cannot have a least upper bound.)

$\triangle$***13-54.*** (Martin, Lachlan). Show that there exist recursively enumerable degrees $\mathbf{a}$ and $\mathbf{b}$ such that $(\forall n)[\mathbf{a}^{(n)}|\mathbf{b}^{(n)}]$. (*Hint:* Modify the proof of Theorem XXV and Corollary XXV(*b*) to get recursive f and g such that for any x and y, $(W^D_{f(x,y)})' = W_x{}^{D'}$ *join* D'; $(W^D_{g(x,y)})' = W_y{}^{D'}$ *join* D'; and $W^D_{f(x,y)}$, $W^D_{g(x,y)}$ are incomparable. Then make an iterated application of the recursion theorem (see Theorem 11-X).)

$\triangle$***13-55.*** Show that the m-degrees of the arithmetical sets do not form a lattice under m-reducibility. (*Hint:* Take K' and $\overline{K'}$, consider any lower bound, and get a greater lower bound by the method of Theorem 9-I.)

13-56. Conclude from results in §13.8 that there is a recursively enumerable, nonrecursive set to which no hyperhypersimple set is reducible.

$\triangle$***13-57.*** (Martin). Define: f *majorizes all recursive functions* if $(\forall$ recursive $g)(\exists m)(\forall n)[m \leq n \Rightarrow g(n) < f(n)]$. Prove: $\mathbf{0}'' \leq \mathbf{a}' \Leftrightarrow$ there exists f of degree $\mathbf{a}$ such that f majorizes all recursive functions. (*Hint:* $\Leftarrow$: It is enough to get $B = \{z|\varphi_z$ *total*$\}$ recursively enumerable in $\mathbf{a}'$. But we can enumerate $z \in B$ by looking for m such that for all $n \geq m$, $\varphi_z(n)$ converges in fewer than $f(n)$ steps. $\Rightarrow$: We are given a recursive g such that $z \in B \Leftrightarrow W_{g(z)}{}^A$ is finite, where $A \in \mathbf{a}$. For each z, list (recursively in A) a sequence $z^{(0)}, z^{(1)}, \ldots$ where $z^{(n)}$ is either $\varphi_z(n)$ or else the nth member of $W_{g(z)}{}^A$. Compute f to majorize these sequences and to encode, by the parity of its values, c_A.) (Martin uses this result in obtaining his characterization of the degrees of maximal sets.)

14 The Arithmetical Hierarchy (Part I)

§14.1 THE HIERARCHY OF SETS

In preceding chapters, the concepts of *degree* and *reducibility* have been used to classify sets of integers. In the present chapter, we present a related but coarser classification of sets. This new classification gives added insight into the preceding theory and shows further connections between recursive function theory and logic. It also provides methods of considerable combinatorial power and elegance for demonstrating similarity or dissimilarity between sets of integers. The main theorem of the chapter, given in §14.5, is a generalization of the projection theorems (Theorems 5-X and 5-XI).

Definition A relation R is in the *arithmetical hierarchy* if R is recursive or if there exists a recursive relation S such that R can be obtained from S by some finite sequence of complementation and/or projection operations.†

Every recursively enumerable relation is in the arithmetical hierarchy, since every recursively enumerable relation can be obtained from some recursive relation by a projection (Theorem 5-X). The complement of any recursively enumerable relation is hence in the hierarchy, since it can be obtained from a recursive relation by operations of projection and complementation (in that order).

Definition Let a set A be given. R is in the *arithmetical hierarchy in A* if R is A-recursive or if there exists an A-recursive relation S such that R can

† The arithmetical hierarchy was first studied by Kleene [1943] and by Mostowski [1947]. Analogies to hierarchies of Borel and projective sets of real numbers will be considered in Chapter 15.

be obtained from S by some finite sequence of complementation and/or projection operations.

By the relativized projection theorem, every set recursively enumerable in A is in the arithmetical hierarchy in A.

In discussion of the hierarchies, it is convenient to use notation from logic to indicate complementation and projection. Complementation can be indicated by the negation symbol and projection by the use of an existential quantifier. For example, if R is an n-ary relation, then

$$\{\langle x_1, \ldots, x_n\rangle \mid \neg R(x_1, \ldots, x_n)\}$$

is the complement of R;

$$\{\langle x_1, \ldots, x_{n-1}\rangle \mid (\exists x_n) R(x_1, \ldots, x_n)\}$$

is one of the n possible projections of R;

and $\{\langle x_1, \ldots, x_{n-3}\rangle \mid \neg (\exists x_{n-2}) \neg (\exists x_n)(\exists x_{n-1}) R(x_1, \ldots, x_n)\}$

results from an application of projection, projection, complementation, projection, and complementation, in that order, to R. In such an expression (involving possible negation symbols and quantifiers), all but one of the negation symbols can be eliminated if universal quantifiers are introduced and if the following rules of elementary logic are used: $\neg\exists$ is equivalent to $\forall\neg$; $\neg\forall$ is equivalent to $\exists\neg$; and $\neg\neg$ can be deleted. (The following rules are also useful: $\forall$ is equivalent to $\neg\exists\neg$; and $\exists$ is equivalent to $\neg\forall\neg$.) Thus, for example, $\neg\exists\neg\exists\neg\forall$ can be expressed $\forall\exists\exists\neg$. Since the complement of a recursive relation is also recursive, we have the following.

Theorem I *An n-ary relation R is in the arithmetical hierarchy if and only if it is recursive or, for some m, can be expressed as*

$$\{\langle x_1, \ldots, x_n\rangle \mid (Q_1 y_1) \cdots (Q_m y_m) S(x_1, \ldots, x_n, y_1, \ldots, y_m)\}$$

where Q_i is either $\forall$ or $\exists$ for $1 \leq i \leq m$, and S is an $(n+m)$-ary recursive relation.

Proof. Given any sequence of complementation and projection operations applied to a recursive relation, the previously mentioned logical rules can be used and the coordinates of the recursive relation (or its complement) can be reordered to yield the desired expression with quantifiers.

Conversely, given any sequence of quantifiers, the fact that $\forall$ is equivalent to $\neg\exists\neg$ can be used to replace all universal quantifiers by negation symbols and existential quantifiers. But this yields a sequence of complementation and projection operations. ☒

Definition Given an expression that is formed by applying zero or more quantifiers, on distinct coördinates, to a relation symbol, and given a particular recursive relation associated with that relation symbol, we say that the expression and relation together constitute a *predicate form.*†

Definition A predicate form with m quantifiers applied to an n-ary recursive relation (where $m < n$) defines, in the obvious way, an $(n - m)$-ary relation. We say that this $(n - m)$-ary relation is *expressed* by the given form.

For example, if S is a 4-ary recursive relation, then "$(\exists x_3)(\forall x_1)S(x_1,x_2,x_3,x_4)$" together with the relation S is a predicate form, and $\{<x_2,x_4>|(\exists x_3)(\forall x_1)S(x_1,x_2,x_3,x_4)\}$ is the relation expressed by this form.

For convenience of notation, we shall usually assume that the coordinates in a recursive relation occur in the same order as they are mentioned in quantifiers; for example, $(\exists y_2)(\forall y_3)S(y_1,y_2,y_3)$. This constraint is not part of our basic definition however, and, in general, we permit quantifiers to be applied to any coordinates in any order.

Definition Let a set A be given. If, in the definition of predicate form, we replace the recursive relation by an A-recursive relation, we call the result an *A-form.*

Thus we have the following corollary to Theorem I.

Corollary I(a) *R is in the arithmetical hierarchy if and only if there is a predicate form that expresses R.*

(***b***) *R is in the arithmetical hierarchy in A if and only if there is an A-form that expresses R.*

Proof. Immediate.☒

Definition In a predicate form or A-form, the (possibly empty) sequence of quantifier operations is called the *prefix* of the form.

For example, $(\forall x_1)(\exists x_2)(\exists x_3)R(x_1,x_2,x_3,x_4)$ has the prefix $\forall\exists\exists$.

Definition The *number of alternations* in a prefix is the number of pairs of adjacent but unlike quantifiers.

For example, in the prefix $\exists\exists\forall\exists$, there are two alternations.

We shall classify forms (and the relations they express) according to the number of alternations in their prefixes. This classification will prove to be of fundamental significance.

Definitions For $n > 0$, a *Σ_n-prefix* is a prefix which begins with $\exists$ and has $n - 1$ alternations.

A *Σ_0-prefix* is a prefix which is empty.

For $n > 0$, a *Π_n-prefix* is a prefix which begins with $\forall$ and has $n - 1$ alternations.

† Strictly speaking, this definition is set-theoretic rather than syntactic (i.e., symbolic). A predicate form consists of a recursive relation together with a (possibly empty) finite sequence of quantifier operations and a one-one mapping from members of the sequence into coordinates of the relation.

A Π_0-*prefix* is a prefix which is empty.

(The concepts of Σ_0-*prefix* and Π_0-*prefix* therefore coincide.)

Definition A predicate form is a Σ_n-*form* if it has a Σ_n-prefix.

A predicate form is a Π_n-*form* if it has a Π_n-prefix.

Let A be given. An A-form is a Σ_n^A-*form* if it has a Σ_n-prefix. An A-form is a Π_n^A-*form* if it has a Π_n-prefix.

In the following main definition, we extend the Σ_n and Π_n classification from forms to the relations they express.

Definition Σ_n = the class of all relations expressible by Σ_n-forms.
Π_n = the class of all relations expressible by Π_n-forms.

Let A be given.

Σ_n^A = the class of all relations expressible by Σ_n^A-forms.
Π_n^A = the class of all relations expressible by Π_n^A-forms.

Hence, by Corollary I, a relation is in the arithmetical hierarchy if and only if, for some n, it is a member of Σ_n or of Π_n. Similarly for the arithmetical hierarchy in A.

Σ_0 ($= \Pi_0$) is the class of all recursive relations. By the projection theorem, Σ_1 is the class of all recursively enumerable relations. Similarly, Σ_0^A ($= \Pi_0^A$) is the class of all A-recursive relations, and Σ_1^A is the class of all relations recursively enumerable in A.

The following basic relationships are easily demonstrated.

Theorem II(a) $\Sigma_n \cup \Pi_n \subset \Sigma_{n+1} \cap \Pi_{n+1}$.

(b) *For any* R, $R \in \Sigma_n \Leftrightarrow \bar{R} \in \Pi_n$. (Here, for k-ary R, $\bar{R} = N^k - R$.)

Proof. (*a*) Let a form be given by an $(n + m)$-ary recursive relation R and an expression $(Q_1y_1) \cdots (Q_my_m)R(x_1, \ldots, x_n, y_1, \ldots, y_m)$, where $Q_1, \ldots, Q_m$ is some sequence of quantifier operations. "Dummy" quantifiers can be added to the form as follows. Let T be the relation expressed by the form. Define

$$S = \{\langle x_1, \ldots, y_m, z\rangle \mid R(x_1, \ldots, y_m) \ \& \ z \in N\}$$

Then S is recursive, and each of the following expressions gives a form which also expresses T: $(\forall z)(Q_1y_1) \cdots (Q_my_m)S(x_1, \ldots, y_m, z)$; $(\exists z)(Q_1y_1) \cdots (Q_my_m)S(x_1, \ldots, z)$; $(Q_1y_1) \cdots (Q_my_m)(\forall z)S(x_1, \ldots, z)$; and $(Q_1y_1) \cdots (Q_my_m)(\exists z)S(x_1, \ldots, z)$. Part (*a*) of the theorem follows.

(*b*) If $R \in \Sigma_n$, then R is expressed by a Σ_n-form. Take the negation of that form and move the negation symbol to the right by the rules of elementary logic used for Theorem I. This yields a Π_n-form which expresses the complement of R.

Conversely, if $\bar{R} \in \Pi_n$, negation of the Π_n-form yields $R \in \Sigma_n$. ☒

Corollary II *Let A be given.*
(*a*) $\Sigma_n^A \cup \Pi_n^A \subset \Sigma_{n+1}^A \cap \Pi_{n+1}^A$.
(*b*) *For any R*, $R \in \Sigma_n^A \Leftrightarrow \bar{R} \in \Pi_n^A$.
Proof. Similarly.☒

In Exercise 14-3, we show that $n > 0 \Rightarrow (\Sigma_n - \Pi_n) \neq \emptyset$ (and hence, by Theorem II(*b*), that $n > 0 \Rightarrow (\Pi_n - \Sigma_n) \neq \emptyset$). This result is called the *hierarchy theorem.* In conjunction with Theorem II(*a*), it shows that the classes $\Sigma_0, \Sigma_1, \Sigma_2, \ldots$ form a strictly increasing sequence. Solution of Exercise 14-3 requires Theorem III below. A stronger form of Exercise 14-3 will be stated and proved in §14.5.

§14.2 NORMAL FORMS

A relation in $\Sigma_n \cup \Pi_n$ can be expressed by a predicate form having at most n quantifiers. To show this, it is enough to show that if a form has a pair of adjacent quantifiers of the same kind, then there is an equivalent form in which that pair is replaced by a single quantifier of that kind. This follows by use of the coordinate functions π_1 and π_2 (defined in §5.3). For example, $\cdots (\forall x)(\forall y) \cdots S(\cdots x \cdots y \cdots)$ is equivalent to $\cdots (\forall u) \cdots S^*(\cdots u \cdots)$, where

$$S^* = \{<\cdots z \cdots> | S(\cdots \pi_1(z) \cdots \pi_2(z) \cdots)\}$$

and where, if S is an n-ary recursive relation, S^* must be an $(n-1)$-ary recursive relation. This fact is called the *rule for contraction of quantifiers.* (The second projection theorem was a special case of quantifier contraction.) It follows that every relation in the arithmetical hierarchy can be expressed by a predicate form whose prefix has no pair of adjacent quantifiers of the same kind. Quantifiers in A-forms (for given A) can also be contracted by this construction, and hence every relation in the arithmetical hierarchy in A can be expressed by an A-form whose prefix has no pair of adjacent quantifiers of the same kind.

For each class Σ_n and each class Π_n, $n > 0$, an indexing for the sets in that class can be obtained from the indexing of the recursively enumerable sets. Similarly, for each class Σ_n^A and each class Π_n^A, $n > 0$, an indexing for the sets in that class can be obtained from the indexing of the sets recursively enumerable in A. To obtain these new indexings, we first define the following useful and fundamental relations, which we denote T_n and T_n^A.

Definition For $n > 0$, let

$$T_n = \{<z, x_0, x_1, \ldots, x_n> | <x_0, \ldots, x_{n-1}>$$

occurs by the x_nth *step in the enumeration of* W_z *(from index z)*}.

For $n > 0$ and any given A, let

$$T_n{}^A = \{<z,x_0, \ldots ,x_n>|(\exists y)(\exists u)(\exists v)[<<x_0, \ldots ,x_{n-1}>,y,u,v>$$

occurs by the x_n-th *step in the enumeration of* $W_{\rho(z)}$ *(from index* $\rho(z)$*) and* $D_u \subset A$ *and* $D_v \subset \bar{A}$]}. (The *enumeration of* W_z *from index* z is an enumeration based on Corollary 5-V(d).)

T_n is clearly recursive, and $T_n{}^A$ is clearly A-recursive. $(\exists x_n)T_n(z,x_0, \ldots ,x_n)$ asserts that $\varphi_z(<x_0, \ldots ,x_{n-1}>)$ is convergent, and $(\exists x_n)T_n{}^A(z,x_0, \ldots ,x_n)$ asserts that $\varphi_z{}^A(<x_0, \ldots ,x_{n-1}>)$ is convergent.†

For every recursively enumerable relation R, there is a z such that $R(x_0, \ldots ,x_{n-1}) \Leftrightarrow (\exists x_n)T_n(z,x_0, \ldots ,x_n)$. This follows from the definition of recursively enumerable relation. T_n thus gives an indexing for all n-ary recursively enumerable relations, and, if we take complements, it gives an indexing for all n-ary relations whose complements are recursively enumerable.

Definition If $R = \{<x_0, \ldots ,x_{m-1}>|(\exists x_m)T_m(z,x_0, \ldots ,x_m)\}$, then z is a Σ_1-*index* for R.

If $R = \{<x_0, \ldots ,x_{m-1}>|(\forall x_m) \neg T_m(z,x_0, \ldots ,x_m)\}$, then z is a Π_1-*index* for R.

Every set in Σ_n can be obtained by the application of $n - 1$ quantifiers to an n-ary relation in Σ_1 (if n is odd) or to an n-ary relation in Π_1 (if n is even). Hence we have an indexing for all sets in Σ_n (and, similarly, an indexing for all sets in Π_n). More generally, every m-ary relation in Σ_n can be obtained by the application of $n - 1$ quantifiers to an $(m + n - 1)$-ary relation in Σ_1 (if n is odd) or to a $(m + n - 1)$-ary relation in Π_1 (if n is even). Hence, for each $m > 0$, we have an indexing for all m-ary relations in Σ_n (and, similarly, an indexing for all m-ary relations in Π_n).

Definitions If $B = \{x_0|(\exists x_1)(\forall x_2) \cdots T_n(z,x_0, \ldots ,x_n)\}$, then z is a Σ_n-*index* for B (where "$(\exists x_1)(\forall x_2) \cdots$" denotes either a string of alternating quantifiers or a string of alternating quantifiers followed by a negation symbol, according as n is odd or even).

If $B = \{x_0|(\forall x_1)(\exists x_2) \cdots T_n(z,x_0, \ldots ,x_n)\}$, then z is a Π_n-*index* for B (where T_n is preceded by a negation symbol if n is odd).

Thus, for example, the set $B = \{x_0|(\exists x_1)(\forall x_2) \neg T_2(z,x_0,x_1,x_2)$ has z as a Σ_2-index (and the set $\bar{B}$ has z as a Π_2-index). By Theorem II, B is also in Σ_3. As a member of Σ_3 it will have (in general, different) Σ_3-indices.

Σ_n-*indices* and Π_n-*indices* for m-ary relations are defined similarly. (For example, if

$$R = \{<x_0, \ldots ,x_{m-1}>|(\exists x_m)(\forall x_{m+1}) \cdots T_{m+n-1}(z,x_0, \ldots ,x_{m+n-1})\},$$

† The relation called T_n by Kleene is similar to the T_n given here, except that, with Kleene, $T_n(z, \ldots ,x_n)$ has more nearly the meaning: x_n *is a code number for a convergent computation of* $\varphi_z(x_0, \ldots ,x_{n-1})$. Kleene's relation has the advantage that the output of the computation can be recovered directly from x_n. We do not need this added feature.

then z is a Σ_n-index for R (where T_{m+n-1} is preceded by a negation symbol if n is even).)

The above indexings are analogous, in many respects, to our initial indexing of the recursively enumerable sets. In particular, they have the following *acceptability* property: (i) there is an effective procedure such that given any Σ_n-index of a relation R, we can find an index for the recursive characteristic function of a recursive relation S such that R is expressible in Σ_n-form from S; and (ii) there is an effective procedure such that given any Σ_n-form for a relation R, with the recursive relation S given by a characteristic index (i.e., an index for its characteristic function), we can obtain a Σ_n-index for R. Part (i) holds trivially; (ii) holds since the operation of quantifier contraction is uniform; e.g., in the illustration of quantifier contraction, a characteristic index for S^* can be obtained uniformly from a characteristic index for S. Similarly with Π_n in place of Σ_n.

Note that we do not have an indexing for Σ_0. In general no indexing onto N which is acceptable in the above sense can be found for the recursive sets or for the recursive m-ary relations (see Exercise 14-2).

Given a set A, the notions of Σ_n^A-*index* and Π_n^A-*index* ($n > 0$) are defined in exactly the same way, except that T_n is replaced by $T_n{}^A$. The comments on acceptability remain as above, provided that we replace recursive characteristic functions by A-recursive characteristic functions.

We summarize in the following theorem, which we give in relativized form, i.e., for the arithmetical hierarchy in A. We state the theorem for *sets*. A corresponding version for m-ary relations (for any fixed $m > 0$) is immediate.

Theorem III **(*Kleene*) (*the normal-form and enumeration theorem*)** *Let A be given. For $n > 0$, every B in Σ_n^A has a Σ_n^A-index; that is to say, if $B \in \Sigma_n^A$, then there exists a z such that*

$$B = \{x_0 | (\exists x_1)(\forall x_2) \cdots T_n{}^A(z, x_0, \ldots, x_n)\}$$

(where $T_n{}^A$ is preceded by a negation symbol if n is even). A Σ_n^A-index for B can be found uniformly from any Σ_n^A-form for B if the A-recursive relation of that form is given by an index for its A-recursive characteristic function.

Proof. See preceding discussion. ☒

§14.3 THE TARSKI-KURATOWSKI ALGORITHM

Definition Let an expression $Fa_1 \cdots a_n$ be given such that $Fa_1 \cdots a_n$ is built up within quantificational logic from quantifiers, variables, $=$, sentential connectives, and relation symbols, and such that $Fa_1 \cdots a_n$ has $a_1, \ldots, a_n$ as its free (i.e., unquantified) variables. Let the relation symbols be interpreted as certain fixed relations $S_1, S_2, \ldots$. Then the relation $R = \{<x_1, \ldots, x_n> | Fa_1 \cdots a_n$ *is true when* $a_1, \ldots, a_n$ *are interpreted as* $x_1, \ldots, x_n$ *respectively*$\}$ is said to be *definable within quantificational logic from the relations* $S_1, S_2, \ldots$; and $Fa_1 \cdots a_n$ is called a *definition of R from* $S_1, S_2, \ldots$.

If a relation is definable within quantificational logic from recursive relations, then it is in the arithmetical hierarchy. This follows from a fact about recursive relations (paragraph 1 below) and from certain facts of elementary logic (paragraphs 2, 3, and 4 below).

1. If a relation is defined by a sentential combination of recursive relations, then it is recursive. For example, if R and S are recursive 2-ary relations, then $\{<x,y,z> | R(x,y) \vee \neg S(y,z)\}$ is a recursive 3-ary relation.

2. If F and G are expressions such that G does not contain any unquantified occurrence of the variable symbol a, then the following pairs of expressions are equivalent (and hence if one is substituted for the other as part of a larger expression, the original and altered versions of that larger expression are equivalent):

$$
\begin{array}{rl}
(\exists a)F \vee G, & (\exists a)[F \vee G]; \\
(\forall a)F \vee G, & (\forall a)[F \vee G]; \\
(\exists a)F \,\&\, G, & (\exists a)[F \,\&\, G]; \\
(\forall a)F \,\&\, G, & (\forall a)[F \,\&\, G]; \\
(\exists a)F \Rightarrow G, & (\forall a)[F \Rightarrow G]; \\
(\forall a)F \Rightarrow G, & (\exists a)[F \Rightarrow G]; \\
G \Rightarrow (\exists a)F, & (\exists a)[G \Rightarrow F]; \\
G \Rightarrow (\forall a)F, & (\forall a)[G \Rightarrow F].
\end{array}
$$

3. If $F(a)$ is an expression involving the variable symbol a (among others), and if b is a variable symbol not occurring in $F(a)$, let $F(b)$ be the result of substituting b for a at all unquantified occurrences in $F(a)$. Then the following pairs of expressions are equivalent:

$$
\begin{array}{rl}
(\forall a)F(a), & (\forall b)F(b); \\
(\exists a)F(a), & (\exists b)F(b).
\end{array}
$$

4. If F and G are any expressions, then the following pairs of expressions are equivalent:

$$
\begin{array}{rl}
\neg(\forall a)F, & (\exists a) \neg F; \\
\neg (\exists a)F, & (\forall a) \neg F; \\
\neg \neg F, & F; \\
F \Leftrightarrow G, & [F \Rightarrow G] \,\&\, [G \Rightarrow F].
\end{array}
$$

Fact 1 is immediate by Church's Thesis. Facts 2, 3, and 4 can be verified by the reader.

Theorem IV *If a relation R is definable within quantificational logic from recursive relations, then R is in the arithmetical hierarchy.*

Proof. Take the definition of R. Form a chain of equivalences by facts 2, 3, and 4 above, such that all quantifiers are moved to the left until an expression $(Q_1b_1) \cdots (Q_mb_m)G$ is obtained, where $Q_1, \ldots, Q_m$ are quantifiers and G has no quantifiers. Such an expression is said to be in *prenex form*. By fact 1, G defines a recursive relation. Hence we have a predicate form expressing R, and by Corollary I, R is in the arithmetical hierarchy. ☒

Fact 1 holds with "A-recursive" in place of "recursive"; hence Theorem IV holds in the following relativized form.

Corollary IV *Given a set A, if a relation R is definable within quantificational logic from A-recursive relations, then R is in the arithmetical hierarchy in A.*

Proof. See above. ☒†

Example. Let $B = \{x | W_x \textit{ is recursive}\}$. We show that B is in the arithmetical hierarchy. The "computation" occurs in two stages.

Stage 1. We obtain a definition of B from recursive relations.

$$z \in B \Leftrightarrow W_z \text{ is recursive} \Leftrightarrow (\exists y)[W_z = \bar{W}_y] \Leftrightarrow (\exists y)(\forall x)[x \in W_z \Leftrightarrow x \notin W_y] \Leftrightarrow (\exists y)(\forall x)[(\exists u)T_1(z,x,u) \Leftrightarrow \neg (\exists u)T_1(y,x,u)].$$

Stage 2. We use facts 2, 3, and 4 above to form a chain of equivalences leading to prenex form

$$(\exists y)(\forall x)[(\exists u)T_1(z,x,u) \Leftrightarrow \neg (\exists u)T_1(y,x,u)];$$

$$(\exists y)(\forall x)[[(\exists u)T_1(z,x,u) \Rightarrow \neg (\exists u)T_1(y,x,u)] \;\&\; [\neg (\exists u)T_1(y,x,u) \Rightarrow (\exists u)T_1(z,x,u)]] \quad \text{(by fact 4)};$$

$$(\exists y)(\forall x)[[(\exists u)T_1(z,x,u) \Rightarrow \neg (\exists v)T_1(y,x,v)] \;\&\; [\neg (\exists w)T_1(y,x,w) \Rightarrow (\exists s)T_1(z,x,s)]] \quad \text{(by fact 3)};$$

$$(\exists y)(\forall x)[[(\exists u)T_1(z,x,u) \Rightarrow (\forall v) \neg T_1(y,x,v)] \;\&\; [(\forall w) \neg T_1(y,x,w) \Rightarrow (\exists s)T_1(z,x,s)]] \quad \text{(by fact 4)};$$

$$(\exists y)(\forall x)[(\forall u)(\forall v)[T_1(z,x,u) \Rightarrow \neg T_1(y,x,v)] \;\&\; (\exists w)(\exists s)[\neg T_1(y,x,w) \Rightarrow T_1(z,x,s)]] \quad \text{(by fact 2)};$$

$$(\exists y)(\forall x)(\exists w)(\forall u)(\exists s)(\forall v)[[T_1(z,x,u) \Rightarrow \neg T_1(y,x,v)] \;\&\; [\neg T_1(y,x,w) \Rightarrow T_1(z,x,s)]] \quad \text{(by fact 2)}.$$

Now, applying fact 1, we have that B is in the arithmetical hierarchy and, in fact, that $B \in \Sigma_6$.

It is clear from the above example that the final prefix in such a computation derives only from the nature and relative position of the quantifier symbols and sentential connectives at the beginning of stage 2. Hence we can abbreviate stage 2 by indicating only these symbols and connectives. Thus, for the previous example, we would have

$$\exists\forall[\exists \Leftrightarrow \neg\exists];$$
$$\exists\forall[[\exists \Rightarrow \neg\exists] \;\&\; [\neg\exists \Rightarrow \exists]];$$
$$\exists\forall[[\exists \Rightarrow \forall] \;\&\; [\forall \Rightarrow \exists]];$$
$$\exists\forall[\forall\forall \;\&\; \exists\exists];$$
$$\exists\forall\exists\forall\exists\forall.$$

† The converses of Theorem IV and Corollary IV are immediate by Corollary I.

(Note that, for the purposes of the computation (and by fact 1), sentential connectives can be dropped when they no longer apply to subformulas involving quantifiers.) We shall henceforth use this abbreviated notation.

Such computations are not uniquely determined. In the above illustration, we could also have proceeded, at the final step, as follows:

$$\exists\forall[\forall\forall \ \& \ \exists\exists];$$
$$\exists\forall\forall\forall\exists\exists.$$

This shows that $B \in \Sigma_3$.

In general, we wish to proceed so as to get the lowest possible ultimate classification. Thus the second computation for B would be preferred. The best computation (from a given definition) can always be found, since only a finite number of computations are possible.

It often simplifies computations to use the rule for contraction of quantifiers. Thus in our example we could proceed, at the final step, as follows:

$$\exists\forall[\forall\forall \ \& \ \exists\exists];$$
$$\exists\forall[\forall \ \& \ \exists];$$
$$\exists\forall\forall\exists;$$
$$\exists\forall\exists.$$

We give several further examples below. Note that our computations are *uniform* in the sense that, given characteristic indices for the recursive relations in the initial definition of B, we can effectively find a Σ_n-index (or Π_n-index) for B (where Σ_n (or Π_n) is the class yielded by the computation). The general form of computation described above was first used by Tarski and Kuratowski, to classify sets of real numbers in the hierarchies of Borel and projective sets. We therefore call it the *Tarski-Kuratowski algorithm.*

Given a set A, the Tarski-Kuratowski algorithm can also be applied to a definition from A-recursive relations. Such a computation then yields a classification in the arithmetical hierarchy in A. For example, if $B = \{z|(\exists y)[y \in W_z^A \ \& \ W_y^A \textit{ is infinite}]\}$, we obtain

$$(\exists y)[(\exists x)T_1^A(z,y,x) \ \& \ (\forall u)(\exists v)[v > u \ \& \ (\exists x)T_1^A(y,v,x)]];$$
$$\exists[\exists \ \& \ \forall\exists[\& \ \exists]];$$
$$\exists[\exists \ \& \ \forall\exists\exists];$$
$$\exists[\exists \ \& \ \forall\exists];$$
$$\exists\exists\forall\exists;$$
$$\exists\forall\exists;$$

and hence $B \in \Sigma_3^A$.

Further shortcuts are sometimes possible in applications of the Tarski-Kuratowski algorithm.

Abbreviation

$(\forall a \leq b)F$, for $(\forall a)[a \leq b \Rightarrow F]$;
$(\exists a \leq b)F$, for $(\exists a)[a \leq b \ \& \ F]$;

where a and b are distinct variable symbols.

The abbreviations $(\forall a \leq b)$ and $(\exists a \leq b)$ are called *bounded quantifiers.* We note the following facts about bounded quantifiers.

1. A bounded quantifier can be moved to the right past an ordinary quantifier, if certain appropriate changes are made in the recursive relations acted on by those quantifiers. For example, $(\forall a \leq b)(\exists c)(\forall d)R(a,b,c,d)$ is equivalent to $(\exists c)(\forall a \leq b)(\forall d)R(a,b,\pi_a{}^b(c),d)$ where $\pi_a{}^b$ is the recursive function of §5.3 (see Exercise 14-4 for other cases).

2. Application of a bounded quantifier to a recursive relation yields a recursive relation. For example, if R is a recursive 3-ary relation, then $\{<x,u>|(\forall y \leq u)R(x,y,u)\}$ is a recursive binary relation.

It follows from these facts that if a quantifier can be abbreviated to a bounded quantifier, then it can be dropped in a Tarski-Kuratowski computation. (For a generalization of this principle, see Exercise 14-5.) As is evident, this principle applies also to computations for the arithmetical hierarchy in A (for a given set A).

For example, let $R = \{<u,z>|D_u \subset W_z\}$. Then $<u,z> \in R \Leftrightarrow (\forall x)[x \in D_u \Rightarrow (\exists y)T_1(z,x,y)]$. Computing, we have

$$\forall[\Rightarrow \exists];$$
$$\forall\exists;$$

and hence $R \in \Pi_2$. It follows, however, from the definition of *canonical index* in §5.6 that $(\forall x)(\forall u)[x \in D_u \Rightarrow x \leq u]$. Hence $<u,z> \in R \Leftrightarrow (\forall x \leq u)[x \in D_u \Rightarrow (\exists y)T_1(z,x,y)]$. Dropping the bounded quantifier, we get

$$[\Rightarrow \exists];$$
$$\exists;$$

and $R \in \Sigma_1$; that is, R is recursively enumerable, as is intuitively evident from the initial definition of R.

We conclude with the following example. Further examples will be found in the exercises.

Example. Let $B = \{z|W_z \textit{ is simple}\}$. Then $z \in B \Leftrightarrow W_z$ is simple $\Leftrightarrow [\bar{W}_z$ is infinite & $(\forall y)[W_y$ is infinite $\Rightarrow W_y \cap W_z \neq \emptyset]] \Leftrightarrow [(\forall x)(\exists y)[y > x \mathbin{\&} y \notin W_z] \mathbin{\&} (\forall y)[(\forall u)(\exists v)[v > u \mathbin{\&} v \in W_y] \Rightarrow (\exists w)[w \in W_y \mathbin{\&} w \in W_z]]] \Leftrightarrow [(\forall x)(\exists y)[y > x \mathbin{\&} (\forall x_1) \neg T_1(z,y,x_1)] \mathbin{\&} (\forall y)[(\forall u)(\exists v)[v > u \mathbin{\&} (\exists x_1)T_1(y,v,x_1)] \Rightarrow (\exists w)[(\exists x_1)T_1(y,w,x_1) \mathbin{\&} (\exists x_1)T_1(z,w,x_1)]]]$. Computing, we have

$$\forall\exists[\& \forall] \mathbin{\&} \forall[\forall\exists[\& \exists] \Rightarrow \exists[\exists \mathbin{\&} \exists]];$$
$$\forall\exists\forall \mathbin{\&} \forall[\forall\exists \Rightarrow \exists];$$
$$\forall\exists\forall \mathbin{\&} \forall\exists[\forall\exists \Rightarrow];$$
$$\forall\exists\forall \mathbin{\&} \forall\exists\exists\forall;$$
$$\forall\exists\forall.$$

Hence $B \in \Pi_3$.

How can we be sure that a Tarski-Kuratowski computation has given us a best possible result? That is to say, how can we be sure that there is not some equivalent but entirely different initial definition for our given set or relation which could yield (by a Tarski-Kuratowski computation) a lower classification in the hierarchy? We return to this question in §14.8.

§14.4 ARITHMETICAL REPRESENTATION

Consider the logical system of *elementary arithmetic* as defined in §7.8. The *formulas* are built up as formulas of quantificational logic from the following symbols: $+$, $\times$, 0, 1, 2, 3, . . . , $=$, quantifier symbols, variable symbols, and the sentential connectives $\neg$, &, $\vee$, $\Rightarrow$, $\Leftrightarrow$. The formulas without *free* (i.e., unquantified) variables are called *sentences* (in §7.8, sentences were called *wffs*). Sentences are interpreted as assertions about ordinary addition and multiplication over the nonnegative integers with "0," "1," "2," . . . interpreted as 0, 1, 2, . . . and with variable symbols interpreted as ranging over the nonnegative integers. Every sentence is, accordingly, either true or false. An n-ary relation R is *definable in elementary arithmetic* if there is a formula $Fa_1 \cdots a_n$ with free variables $a_1, \ldots, a_n$ such that

$$R = \{<x_1, \ldots, x_n> | F\mathbf{x}_1 \cdots \mathbf{x}_n \textit{ is true}\}$$

where, for any integers $x_1, \ldots, x_n$, $F\mathbf{x}_1 \cdots \mathbf{x}_n$ is the result of substituting the numeral expressions of $x_1, \ldots, x_n$ for $a_1, \ldots, a_n$, respectively, in $Fa_1 \cdots a_n$.

The following well-known and fundamental theorem, due to Gödel, relates the arithmetical hierarchy to the sentences of elementary arithmetic.

Definition An n-ary relation P is a *polynomial relation* if there is a polynomial $p(a_1, \ldots, a_n)$ in n variables, with integer coefficients, such that $P = \{<x_1, \ldots, x_n> | p(x_1, \ldots, x_n) = 0\}$. We permit p to have negative coefficients but (as usual) we restrict $x_1, \ldots, x_n$ to be nonnegative integers. P is hence a relation on the nonnegative integers.

***Theorem V* (*Gödel*)** *For every n-ary recursively enumerable relation R, there is a polynomial relation P and a prefix $Q_1 \cdots Q_m$ such that*

$$R(x_1, \ldots, x_n) \Leftrightarrow (Q_1 y_1) \cdots (Q_m y_m) P(x_1, \ldots, x_n, y_1, \ldots, y_m).$$

Proof. The proof is given in Davis [1958, chap. 7]. We omit it here. It is based on the formal definition of Turing machine and requires a detailed combinatorial construction. The proof uses the chinese remainder theorem (of number theory). ☒

Theorem V can be proved in the following stronger form.

***Corollary V* (*Davis*)** *For every n there is a recursive function f, an integer k, and a q-ary polynomial relation P, such that $q = n + k + 3$ and,* *for all z and* $x_1, \ldots, x_n$,

$$(\exists w)T_n(z,x_1, \ldots ,x_n,w) \Leftrightarrow (\exists y)(\forall u)(\exists v_1) \cdots (\exists v_k)P(f(z),x_1, \ldots ,x_n,y,u,v_1, \ldots ,v_k),$$

where an index for f, a polynomial for P, and hence a value for k, can be found uniformly from n.

Proof. See Davis [1958, chap. 7].☒

Whether or not the universal quantifier in Corollary V can be dropped is unknown. (If it could be dropped, then, by the existence of nonrecursive, recursively enumerable sets, the recursive unsolvability of Hilbert's tenth problem (see §2.2) would follow.) R. M. Robinson [1956] has shown that k in Corollary V can be taken to be $n + 3$.† In the proofs of Gödel, Davis, and Robinson, the polynomials are explicitly defined.

The following consequence of Theorem V explains why the term *arithmetical* is used in connection with the arithmetical hierarchy.

Theorem VI (***the representation theorem***) *For any relation R, R is in the arithmetical hierarchy* $\Leftrightarrow$ *R is definable in elementary arithmetic.*

Proof. $\Leftarrow$: If R is definable in elementary arithmetic, then R is definable from the recursive relations $\{<x,y,z>|x + y = z\}$, $\{<x,y,z>|x \cdot y = z\}$, and $\{x|x = k\}$ (for any integer k). Hence, by Theorem IV, R is in the arithmetical hierarchy.

$\Rightarrow$: Any polynomial relation is definable in elementary arithmetic. For: (i) by transposition, the relation $p(a_1, \ldots ,a_n) = 0$ can be expressed as $q(a_1, \ldots ,a_n) = r(a_1, \ldots ,a_n)$, where q and r are polynomials with nonnegative coefficients; and (ii) exponents can be replaced by iterations of the multiplication operation, for example, $x^3 = (x \cdot x) \cdot x$. The result follows by Theorem V.☒

The relativized form of Theorem VI is given in Theorem VII.

Theorem VII *For any given A and any relation R*: (1) *R is in the arithmetical hierarchy in A* $\Leftrightarrow$ (2) *R is definable within quantificational logic from recursive relations and the set A*‡ $\Leftrightarrow$ (3) *R is definable within quantificational logic from polynomial relations and the set A* $\Leftrightarrow$ (4) *R is definable within an augmented elementary arithmetic obtained by adding atomic expressions of the form "*$a \in X$*" to elementary arithmetic and interpreting "*$a \in X$*" to mean* $x \in A$ *whenever a is interpreted as the integer x.*

Proof. It is immediate that (3) $\Rightarrow$ (4). It is immediate from Theorem V that (2) $\Rightarrow$ (3). To show that (4) $\Rightarrow$ (1), put the defining expression for R in prenex form (as in the Tarski-Kuratowski algorithm); the part acted on by the quantifiers then defines an A-recursive relation and hence, by Corollary I, R is in the arithmetical hierarchy in A. Finally, to show that

† Thus there is a fixed polynomial p of eight variables such that for any z, $W_z = \{x|(\exists y)(\forall u)(\exists v_1)(\exists v_2)(\exists v_3)(\exists v_4)[p(f(z),x,y,u,v_1,v_2,v_3,v_4) = 0]\}$.

‡ That is to say, from recursive relations and the set A considered as a 1-ary relation.

(1) $\Rightarrow$ (2), assume that R is in the arithmetical hierarchy in A. Then, by Theorem III, R is definable from $T_n{}^A$ for some n. But $T_n{}^A(z,x_0, \ldots, x_n) \Leftrightarrow (\exists y)(\exists u)(\exists v)[<<x_0, \ldots, x_{n-1}>,y,u,v>$ occurs by the x_n-th step in the enumeration of $W_{\rho(z)}$ & $(\forall x)[x \in D_u \Rightarrow x \in A]$ & $(\forall x)[x \in D_v \Rightarrow x \notin A]]$. Hence R is definable from recursive relations and the set A.☒

Corollary VII *If A is in the arithmetical hierarchy in B and B is in the arithmetical hierarchy in C, then A is in the arithmetical hierarchy in C.*

Proof. Immediate by the theorem; for if A is definable from B within elementary arithmetic and B is definable from C within elementary arithmetic, then A is definable from C within elementary arithmetic.☒

§14.5 THE STRONG HIERARCHY THEOREM

The theorem below and its corollaries present one of the main results of the present chapter. We call this result the *strong hierarchy theorem.* We state and prove it in relativized form. It relates the arithmetical hierarchy in A to the structure of degrees studied in Chapter 13.

Recall that $A^{(n)}$ is the result of applying the jump operation n times to A.

Theorem VIII (Kleene, Post) *Let any set A be given. Then for all n and any set B*

(a) $B \in \Sigma^A_{n+1} \Leftrightarrow B$ *is recursively enumerable in* $A^{(n)}$;

(b) $B \in \Sigma^A_{n+1} \cap \Pi^A_{n+1} \Leftrightarrow B \leq_T A^{(n)}$. *Furthermore,*

(c) *there are uniform procedures such that for any n and any* $B \in \Sigma^A_{n+1}$, *an index for B as a set recursively enumerable in* $A^{(n)}$ *can be found from any* Σ^A_{n+1}*-index for B; and a* Σ^A_{n+1}*-index for B can be found from any index for B as a set recursively enumerable in* $A^{(n)}$.

(Result (*b*) is sometimes called *Post's theorem.*)

Proof. By Theorem 9-IV, (*a*) $\Rightarrow$ (*b*). It remains to prove (*a*) and (*c*).

(*a*) The proof is by induction.

$n = 0$. By the relativized projection theorem, $B \in \Sigma^A_1 \Leftrightarrow B$ is recursively enumerable in A $(= A^{(0)})$.

$n = k + 1$. Assuming (*a*) is true for $n = k$, we wish to show that $B \in \Sigma^A_{k+2} \Leftrightarrow B$ is recursively enumerable in $A^{(k+1)}$.

$\Leftarrow$: Assume $B = W^{A^{(k+1)}}_{z_0}$. Then $B = \{x|(\exists y)(\exists u)(\exists v)[<x,y,u,v> \in W_{\rho(z_0)}$ & $D_u \subset A^{(k+1)}$ & $D_v \subset \overline{A^{(k+1)}}]\}$. Consider $\{u|D_u \subset A^{(k+1)}\}$. $A^{(k+1)}$ is recursively enumerable in $A^{(k)}$, and hence, by inductive assumption, $A^{(k+1)} \in \Sigma^A_{k+1}$. Furthermore, $D_u \subset A^{(k+1)} \Leftrightarrow (\forall y \leq u)[y \in D_u \Rightarrow y \in A^{(k+1)}]$. Hence, dropping the bounded quantifier and making a trivial computation, we have $\{u|D_u \subset A^{(k+1)}\} \in \Sigma^A_{k+1}$. Similarly, $D_v \subset \overline{A^{(k+1)}} \Leftrightarrow (\forall y \leq v)[y \in D_v \Rightarrow y \in \overline{A^{(k+1)}}]$; and, by inductive assumption, $\overline{A^{(k+1)}} \in \Pi^A_{k+1}$. Hence $\{v|D_v \subset \overline{A^{(k+1)}}\} \in \Pi^A_{k+1}$.

Noting that $<x,y,u,v> \in W_{\rho(z_0)} \Leftrightarrow (\exists w)T_1(\rho(z_0),<x,y,u,v>,w)$, we have a definition for B from relations that are A-recursive. This definition has the form

$$\exists\exists\exists\ [\exists\ \&\ \underbrace{\exists\forall\ \cdots}_{k+1}\ \&\ \underbrace{\forall\exists\ \cdots}_{k+1}].$$

Applying the Tarski-Kuratowski algorithm, we obtain

$$\underbrace{\exists\forall\exists\ \cdots}_{k+2}.$$

Hence $B \in \Sigma_{k+2}^A$.

$\Rightarrow$: Assume $B \in \Sigma_{k+2}^A$. $B = \{x|(\exists y)S(x,y)\}$ where $S \in \Pi_{k+1}^A$. Then $\bar{S} \in \Sigma_{k+1}^A$, and hence, by induction, $\bar{S}$ is recursively enumerable in $A^{(k)}$. Hence, by Theorem 13-I, $\bar{S}$ is recursive in $A^{(k+1)}$. Hence S is recursive in $A^{(k+1)}$ and, by the relativized projection theorem, B must be recursively enumerable in $A^{(k+1)}$ (see also Exercise 14-9). This proves (*a*).

(*c*) The construction is by induction. For $n = 0$, uniformity is trivial, since the indices are the same (see definition of $T_1{}^A$ in §14.2). Assume uniformity in each direction for $n = k$. Uniformity in each direction for $n = k + 1$ follows from the constructions for part (*a*) together with the uniformities given in Corollary 13-I, in Theorem III, and in the relativized projection theorem. ☒

Corollary VIII(*a*) gives parts (*a*) and (*b*) of the theorem in unrelativized form.

Corollary VIII(a) *For all n and any set B,*

$$B \in \Sigma_{n+1} \Leftrightarrow B \text{ is recursively enumerable in } \emptyset^{(n)};$$

and

$$B \in \Sigma_{n+1} \cap \Pi_{n+1} \Leftrightarrow B \leq_{\mathrm{T}} \emptyset^{(n)}.$$

Proof. Immediate. ☒

Corollary VIII(b) **(*the completeness theorem*)** *For given A, and* $n > 0$,

$$B \in \Sigma_n^A \Leftrightarrow B \leq_1 A^{(n)};$$
$$B \in \Pi_n^A \Leftrightarrow B \leq_1 \overline{A^{(n)}};$$
$$B \in \Sigma_n \Leftrightarrow B \leq_1 \emptyset^{(n)};$$
$$B \in \Pi_n \Leftrightarrow B \leq_1 \overline{\emptyset^{(n)}}.$$

(*c*) **(*the hierarchy theorem*)** *Given A, then for every n there exists a set B such that* $B \in (\Sigma_{n+1}^A - \Pi_{n+1}^A)$ *(and hence* $\bar{B} \in (\Pi_{n+1}^A - \Sigma_{n+1}^A)$*).*

Proof. Part (*b*) is immediate by Theorem 13-I. For (*c*), take

$$B = A^{(n+1)}$$

By (*a*) of the theorem, $A^{(n+1)} \in \Sigma_{n+1}^A$. But $A^{(n+1)} \notin \Pi_{n+1}^A$, for otherwise, by (*b*) of the theorem, $A^{(n+1)} \leq_{\mathrm{T}} A^{(n)}$, contrary to Theorem 13-I. (A shorter and more direct proof of (*c*), due to Kleene, is given in Exercise 14-3.) ☒

If $n = 0$, then $\Sigma_n = \Pi_n$, and $\Sigma_n = \Sigma_{n+1} \cap \Pi_{n+1}$. By ($c$), neither of these identities holds for $n > 0$.

Definition B is Σ_n^A*-complete* (or, equivalently, B is *maximal in* Σ_n^A) if $B \in \Sigma_n^A$ and $(\forall C)[C \in \Sigma_n^A \Rightarrow C \leq_1 B]$. Similarly for Π_n^A*-complete.*

By Corollary VIII(b), B is Σ_n^A*-complete* $\Leftrightarrow B \equiv A^{(n)}$. Note that B is Σ_n^A-complete $\Leftrightarrow \bar{B}$ is Π_n^A-complete.

Recall that the concept B *is arithmetical* and the concept B *is arithmetical in* A were defined in §13.1. Corollary VIII(d) below relates these concepts to the arithmetical hierarchy and the arithmetical hierarchy in A.

Corollary VIII(d) *B is arithmetical $\Leftrightarrow$ B is in the arithmetical hierarchy; and B is arithmetical in A $\Leftrightarrow$ B is in the arithmetical hierarchy in A.*

Proof. B is in the arithmetical hierarchy in $A \Leftrightarrow$ (by Theorems I and III) $(\exists n)[B \in \Sigma_{n+1}^A \cap \Pi_{n+1}^A] \Leftrightarrow$ (by Theorem VIII) $(\exists n)[B \leq_T A^{(n)}] \Leftrightarrow$ (by definition in §13.1) B is arithmetical in A. ☒

Corollary VIII(d), taken together with Theorem VII, explains why the word *arithmetical* was used in §13.1. Note that Exercise 13-10 and Corollary VII give two different proofs that the relation *arithmetical in* is transitive. Henceforth we shall often use the phrases *arithmetical* and *arithmetical in A* in place of, respectively, the phrases *in the arithmetical hierarchy* and *in the arithmetical hierarchy in A*.

Let $B = \{z | W_z \textit{ is infinite}\}$. In the proof of Theorem 13-VIII we went to some lengths to show that $B \leq_1 \overline{\emptyset^{(2)}}$. The Tarski-Kuratowski algorithm, together with the strong hierarchy theorem, now gives a simpler and more efficient method for proving such facts. In the case of B, for example, we have W_z infinite $\Leftrightarrow (\forall x)(\exists y)[y > x \,\&\, y \in W_z] \Leftrightarrow \forall\exists[\&\exists] \Leftrightarrow \forall\exists$. Hence $B \in \Pi_2$. Hence by Corollary VIII(b), $B \leq_1 \overline{\emptyset^{(2)}}$. (The proof of Theorem 13-VIII obtained this result by an *ad hoc* construction of some complexity.)

§14.6 DEGREES

Consider Σ_n^A and Π_n^A as classes of sets. The classes Σ_n^A and Π_n^A are recursively invariant and are, in fact, well-defined with respect to m-reducibility (see Exercise 14-12). For $n > 0$, each class contains a maximum recursive-isomorphism type, the maximum type in Σ_n^A being that of $A^{(n)}$, and the maximum type in Π_n^A being that of $\overline{A^{(n)}}$. $\Sigma_0 = \Pi_0$ has maximum type, that of $\{x | x \textit{ is even}\}$, but, by Theorem 9-XVII, $\Sigma_0^K = \Pi_0^K$ has no maximum type. See Exercise 14-14.

For $n > 0$, Σ_n^A and Π_n^A are not well-defined with respect to tt-reducibility and T-reducibility; for, $A^{(n)} \equiv_{tt} \overline{A^{(n)}}$ but, by Corollary VIII(c), $A^{(n)} \in (\Sigma_n^A - \Pi_n^A)$ and $\overline{A^{(n)}} \in (\Pi_n^A - \Sigma_n^A)$. On the other hand, the class $\Sigma_{n+1}^A \cap \Pi_{n+1}^A$ is well-defined with respect to T-reducibility and, by Theorem VIII, consists of all sets T-reducible to $A^{(n)}$.

By Corollary II, $\Sigma_n^A \cup \Pi_n^A \subset \Sigma_{n+1}^A \cap \Pi_{n+1}^A$. Is this containment strict? For $n = 0$ it is not, since $\Sigma_1 \cap \Pi_1 = \Sigma_0 = \Pi_0$. For $n > 0$ it is strict, since, for example, the set $A^{(n)}$ *join* $\overline{A^{(n)}}$ is in $\Sigma_{n+1}^A \cap \Pi_{n+1}^A$ (by Theorem VIII(b)), but is not in $\Sigma_n^A \cup \Pi_n^A$. (If it were in $\Sigma_n^A \cup \Pi_n^A$, then both $A^{(n)}$ and $\overline{A^{(n)}}$ would be recursively enumerable in $A^{(n-1)}$ contrary to Theorem 13-I.)

$A^{(n)}$ *join* $\overline{A^{(n)}}$ can be obtained as the union of two sets from $\Sigma_n^A \cup \Pi_n^A$ (see Exercise 14-13). By Theorem VIII and Exercise 9-16, $\Sigma_{n+1}^A \cap \Pi_{n+1}^A$ is a Boolean algebra of sets. Is this Boolean algebra generated by $\Sigma_n^A \cup \Pi_n^A$? For $n > 0$, a negative answer to this question is obtained from the following theorem.

Definition Given a set B, let $\mathfrak{M}(B) = \{C | C \leq_m B\}$ and let $\mathcal{B}_m(B)$ be the Boolean algebra of sets generated by $\mathfrak{M}(B)$.

Theorem IX *If B is neither N nor $\emptyset$, then for any set C, $C \leq_{btt} B \Leftrightarrow C \in \mathcal{B}_m(B)$.*

Proof. $\Leftarrow$: For any n-ary Boolean function γ (see §8.3 for definition and discussion of Boolean functions) let $\gamma(B_1, \ldots, B_n)$ denote the set whose characteristic function is $\lambda x[\gamma(c_{B_1}(x), c_{B_2}(x), \ldots, c_{B_n}(x))]$. We then have $C \in \mathcal{B}_m(B) \Leftrightarrow$ there is an n such that for some Boolean function γ and some sets $B_1, \ldots, B_n$ in $\mathfrak{M}(B)$, $C = \gamma(B_1, \ldots, B_n)$. Let $B_1, \ldots, B_n$ be m-reducible to B via recursive functions $f_1, \ldots, f_n$, respectively. Then, for any x, $x \in C \Leftrightarrow$ the tt-condition $<<f_1(x), \ldots, f_n(x)>, \gamma>$ is satisfied by B. This tt-condition is uniformly computable from x and its norm is n. Hence $C \leq_{btt} B$.

$\Rightarrow$: Assume $C \leq_{btt} B$ where $B \neq N$ and $B \neq \emptyset$. Then, by Exercise 8-28, there is a recursive g, an n, and a fixed n-ary Boolean function γ such that for all x, $g(x)$ is a tt-condition having γ as its Boolean function, and $g(x)$ is satisfied by B if and only if $x \in C$. Define $f_1, \ldots, f_n$ so that $<f_1(x), \ldots, f_n(x)>$ is the n-tuple of the tt-condition $g(x)$. Take $B_1, \ldots, B_n$ to be $f_1^{-1}(B), \ldots, f_n^{-1}(B)$, respectively. Then $B_1, \ldots, B_n$ are in $\mathfrak{M}(B)$, and $C = \gamma(B_1, \ldots, B_n)$. ☒

We now return to our question: for $n > 0$, is $\Sigma_{n+1}^A \cap \Pi_{n+1}^A$ the Boolean algebra of sets generated by $\Sigma_n^A \cup \Pi_n^A$? We answer it in the following corollary (taking $n = 1$).

Corollary IX(a) *B is in the Boolean algebra generated by $\Sigma_1^A \cup \Pi_1^A \Leftrightarrow B \in \mathcal{B}_m(A') \Leftrightarrow B \leq_{btt} A'$.*

(***b***) *$\Sigma_2^A \cap \Pi_2^A$ contains a set not in $\mathcal{B}_m(A')$.*

(***c***) *$\Sigma_2^A \cap \Pi_2^A$ contains no maximal recursive-isomorphism type.*

Proof. (a) This is immediate since, by Corollary VIII and the fact that A' is a cylinder, $\Sigma_1^A = \mathfrak{M}(A')$.

(b) The proof of Theorem 9-I can be relativized to show that for any A there is an $\tilde{A}$ such that $\tilde{A} \leq_T A'$ but $\tilde{A} \not\leq_{tt} A'$ (see Exercise 14-14).

(c) See the proof of Theorem 9-I and Exercise 14-14. ☒

Almost all the results on degrees in Chapter 13 can be expressed in terms of hierarchy concepts; for the fundamental relationships of T-*reducibility*, *recursive enumerability*, and *jump* can be so expressed. For example, $A \leq_T B \Leftrightarrow A \in \Sigma_1^B \cap \Pi_1^B$; A recursively enumerable in $B \Leftrightarrow A \in \Sigma_1^B$; and $A \equiv_T B' \Leftrightarrow (\exists C)[C \equiv_T A \ \& \ C \in \Sigma_1^B \ \& \ (\forall D)[D \in \Sigma_1^B \Rightarrow D \leq_T C]]$ (see Exercise 14-15). Is the arithmetical hierarchy "monotone" in the sense that every set in Σ_n is T-reducible to every set in $(\Sigma_{n+1} - \Sigma_n)$? Theorem 10-III gives a negative answer to this question, for it shows that there must be a set in $(\Sigma_2 - \Sigma_1)$ which is incomparable with some set in Σ_1 (see Exercise 14-16). Indeed, Corollary 13-IX yields that for each $n > 0$, there is a set in $(\Sigma_{n+2} - \Sigma_n)$ which is incomparable with some set in Σ_1 (see Exercise 14-17).

§14.7 APPLICATIONS TO LOGIC

Let V be the set of true sentences of elementary arithmetic. In §7.8, we saw that V is a productive set. What is the degree of unsolvability of V? The following theorem answers this question.

Theorem X *V, the set of true sentences of elementary arithmetic, is recursively isomorphic to $\emptyset^{(\omega)}$; V is hence of degree $\mathbf{0}^{(\omega)}$.*

Proof. By Corollary 13-I(c) and Theorem VIII(c), there is a recursive f such that for all n, $f(n)$ is a Σ_n-index for $\emptyset^{(n)}$. Hence $\langle x,n \rangle \in \emptyset^{(\omega)} \Leftrightarrow x \in \emptyset^{(n)} \Leftrightarrow (\exists x_1)(\forall x_2) \cdots T_n(f(n),x,x_1,x_2, \ldots)$. By the uniformity of Corollary V, we can effectively obtain, from x and $f(n)$, a sentence of elementary arithmetic which is true if and only if $(\exists x_1)(\forall x_2) \cdots T_n(f(n), x,x_1,x_2, \ldots)$. This gives $\emptyset^{(\omega)} \leq_m V$.

Conversely, let F be a sentence of elementary arithmetic. Let $(Q_1 a_1) \cdots (Q_m a_m)G$ be a sentence in prenex form equivalent to F. Form the expression $(\exists b)(Q_1 a_1) \cdots (Q_m a_m)[c = c \ \& \ b = b \ \& \ G]$, where b and c are variables not occurring in $(Q_1 a_1) \cdots (Q_m a_m)G$. This expression determines a Σ_n-form (for some $n > 0$) since G defines a recursive relation. (G is built up from variables, numerals, $+$, $\times$, $=$, and sentential connectives.) This Σ_n-form expresses a set B such that $B = N$ if F is true and $B = \emptyset$ if F is false. From F, the Σ_n-form can be determined uniformly. From the Σ_n-form, by Theorem III, a Σ_n-index for B can be found uniformly. From this Σ_n-index for B, by Theorem VIII(c), an index for B as a set recursively enumerable in $\emptyset^{(n-1)}$ can be found uniformly. From this index, by Corollary 13-I(c), an index for a recursive function g can be found uniformly such that $B \leq_1 \emptyset^{(n)}$ via g. Hence F is true $\Leftrightarrow 0 \in B \Leftrightarrow g(0) \in \emptyset^{(n)} \Leftrightarrow \langle g(0),n \rangle \in \emptyset^{(\omega)}$. Hence $V \leq_m \emptyset^{(\omega)}$.

Both V and $\emptyset^{(\omega)}$ are easily shown to be cylinders (see Exercise 14-18). Hence $V \equiv \emptyset^{(\omega)}$. ⊠

Sentences of elementary arithmetic can be classified according to prefix.

Definition A sentence of elementary arithmetic in prenex form is a Σ_n-*sentence* if its prefix is a Σ_n-prefix, and a Π_n-*sentence* if its prefix is a Π_n-prefix.

Definition Let V_n be the set of all true Σ_n-sentences.

Corollary X *For all* $n > 0$, $V_n \leq_1 \emptyset^{(n)}$.

Proof. Immediate from the second part of the proof of the theorem and from the fact that $\emptyset^{(n)}$ is a cylinder.☒†

Whether or not $\emptyset^{(n)} \leq_1 V_n$ is an open question (as we have noted in connection with Hilbert's tenth problem).

We assume a coding from sentences of elementary arithmetic onto the integers. (Such a coding has been tacitly assumed in preceding discussion.)

Definitions An *axiomatization* is a recursively enumerable set of sentences. A sentence x is *provable* in an axiomatization if x is deducible by elementary logic from some finite subset of the axiomatization. A sentence y is deducible from a sentence x *under* an axiomatization if $[x \Rightarrow y]$ is provable in the axiomatization, where "$x \Rightarrow y$" denotes the sentence obtained by placing "$\Rightarrow$" between sentences x and y. Sentences x and y are *demonstrably equivalent* under an axiomatization, if each is deducible from the other under that axiomatization.

If an axiomatization is given, $\{\langle x,y\rangle \mid y$ *is deducible from* x *under the axiomatization*$\}$ is recursively enumerable (since we can enumerate all proofs in elementary logic) and an r.e. index for this set can be obtained uniformly from an r.e. index for the axiomatization.

Let some axiomatization be given. (This axiomatization might, for example, consist of Peano's axioms. The provable sentences would then constitute Peano arithmetic as defined in §7.8.) Given a sentence F, what is the minimum n such that there exists a Σ_n-sentence G which is demonstrably equivalent to F under the axiomatization? The following generalization of the Gödel incompleteness theorem shows that if the axiomatization is sound (see §7.8), then the value of n can be made arbitrarily large (by appropriate choice of F).

Theorem XI *Let a sound axiomatization of elementary arithmetic be given. Then for every* n *there exists a true sentence* F_n *such that for no* $m \leq n$ *does there exist a* Σ_m-*sentence demonstrably equivalent to* F_n. *Furthermore, a uniform procedure exists by which, for each* n, *such an* F_n *can be exhibited.*

Proof. Let n be given. Let a be an r.e. index for the given axiomatization. Take V_n as defined above, and form $D(V_n) = \{y \mid (\exists x)[x \in V_n$ & y *is deducible from* x *under the axiomatization*$]\}$.‡ By Corollary X, $V_n \leq_1 \emptyset^{(n)}$.

† It follows that although V is not definable in elementary arithmetic, V_n, for each n, *is* definable.

‡ Note that any finite conjunction of sentences in V_n is equivalent to some sentence in V_n (go to prenex form). Hence $D(V_n)$ consists of all sentences deducible (under the axiomatization) from finite sets of sentences in V_n. This fact is not needed for the proof.

Hence V_n is recursively enumerable in $\emptyset^{(n-1)}$, and hence (by the strong hierarchy theorem and Tarski-Kuratowski algorithm) $D(V_n)$ is recursively enumerable in $\emptyset^{(n-1)}$. Define $\neg D(V_n) = \{x | \textit{the negation of } x \textit{ is in } D(V_n)\}$; then $\neg D(V_n)$ is recursively enumerable in $\emptyset^{(n-1)}$ and $\neg D(V_n) = W_{z_1}^{\emptyset^{(n-1)}}$, where z_1 can be obtained uniformly from n and a. By soundness, $D(V_n) \subset V$ and $\neg D(V_n) \subset \bar{V}$. Let h be the recursive function (constructed in the first part of the proof of Theorem X) such that $\emptyset^{(\omega)} \leq_m V$ via h. Then $h^{-1}(\neg D(V_n)) \subset \overline{\emptyset^{(\omega)}}$. By definition, $\emptyset^{(\omega)} = \{<x,y> | x \in \emptyset^{(y)}\}$. Let

$$B = \pi_1(h^{-1}(\neg D(V_n)) \cap (N \times \{n\}))$$

We then have that $B \subset \overline{\emptyset^{(n)}}$ and that $B = W_{z_2}^{\emptyset^{(n-1)}}$ where z_2 can be obtained uniformly from z_1. Since $\emptyset^{(n)} = \{z | z \in W_z^{\emptyset^{(n-1)}}\}$, we have that $z_2 \in \overline{\emptyset^{(n)}} \cap \bar{B}$. Hence $<z_2,n> \notin \emptyset^{(\omega)}$ and $h(<z_2,n>) \notin V$. Take F_n to be the negation of (the sentence) $h(<z_2,n>)$. Then $F_n \in V$. If $F_n \in D(V_n)$, then $h(<z_2,n>) \in \neg D(V_n)$ and $z_2 \in B$ (by definition of B and h), contrary to our conclusion that $z_2 \notin B$. Hence $F_n \notin D(V_n)$. Thus F_n is a true sentence not deducible from, and hence not demonstrably equivalent to, any true Σ_n-sentence. By soundness, F_n cannot be demonstrably equivalent to any false sentence. Hence F_n is a true sentence not demonstrably equivalent to any Σ_n-sentence.

It follows (by dummy quantifiers) that F_n is not demonstrably equivalent to any Σ_m-sentence, $m < n$. F_n is evidently obtained uniformly from z_2, and hence uniformly from a and n.☒

Corollary XI *F_n (in the theorem) can be taken to be a Π_{n+2}-sentence.*

Proof. From the first part of the proof of Theorem X, we see that $h(<z_2,n>)$ is obtained by way of Theorem VI (the representation theorem) from an assertion $(\exists x_1)(\forall x_2) \cdots T_n(y,z_2,x_1, \ldots)$. By Corollary V, $h(<z_2,n>)$ can be obtained with a Σ_{n+2}-prefix. Hence its negation, F_n, has a Π_{n+2}-prefix.☒

Hierarchy notions can be used to classify sentences of set theory.† For the remainder of our discussion, we assume one of the standard axiomatizations of set theory within quantificational logic; more specifically, we assume the axiomatization given in Exercise 11-23. We call this axiomatization *ZF*. Virtually all of known mathematics can be formulated within set theory, and virtually all the accepted arguments and proofs can be made under *ZF*. In particular, the sentences of elementary arithmetic can be identified with certain sentences of set theory; we therefore speak of the sentences of elementary arithmetic as *embedded* in set theory.‡ The sen-

† By *sentences of set theory* we mean all sentences of a certain formal system, not just those that are demonstrable under a chosen axiomatization for the system. Our usage here is somewhat inconsistent with our usage elsewhere, where "theory" refers to a subset of the sentences of a system.

‡ This embedding is obtained by identifying the integer 0 with the empty set and the integer $n + 1$ with $\alpha_n \cup \{\alpha_n\}$, where α_n is the set identified with n (see introduction to §11.7), and by then identifying the addition and multiplication operations on integers with corresponding relations.

tences of elementary arithmetic that are now provable constitute *set-theory arithmetic* (first defined in §7.8). Peano's axioms are included within set-theory arithmetic (as can be shown without difficulty).† Set-theory arithmetic can be viewed, intuitively, as the collection of all sentences of elementary arithmetic that are provable by known and accepted mathematical means. The Gödel incompleteness theorem applied to Peano arithmetic yields a sentence which is in set-theory arithmetic but not in Peano arithmetic (see the remark on the special assumption on page 98).‡

Certain sentences of set theory are demonstrably equivalent (within set theory under ZF) to sentences of elementary arithmetic. We say that such sentences of set theory are *arithmetically expressible.* We could, if we wished, use the notions of Σ_n-sentence and Π_n-sentence (of elementary arithmetic) to classify the arithmetically expressible sentences of set theory. A more useful classification, however, is obtained as follows. We give it first in rough form, then make the definition more precise. Roughly, we say that a sentence of set theory is in Σ_n^{ZF} if it is demonstrably equivalent (under ZF) to a sentence of the form $(Q_1x_1) \cdots (Q_nx_n)R(x_1, \ldots, x_n)$, where $Q_1 \cdots Q_n$ is a Σ_n-prefix and R is a recursive relation. To make this precise, we must use a name, i.e., index, for the recursive relation. Consider sentences of elementary arithmetic obtained by Corollary V from assertions of the special form $(\exists x_1)(\forall x_2) \cdots T_n(z,0,x_1,x_2, \ldots)$ for various fixed z and n. For given z and n, we denote the special sentence of elementary arithmetic obtained from the assertion $(\exists x_1)(\forall x_2) \cdots T_n(z,0,x_1,x_2, \ldots)$ as $(\exists a_1)(\forall a_2) \cdots T_{\mathbf{n}}(\mathbf{z},0,a_1,a_2, \ldots)$. It will be enough to consider these special sentences and their negations since: (i) any n-ary recursive relation R is

$$\{\langle x_1, \ldots ,x_n\rangle | (\exists x_{n+1}) T_{n+1}(z_1,0,x_1, \ldots ,x_{n+1})\}$$
$$= \{\langle x_1, \ldots ,x_n\rangle | (\forall x_{n+1}) \neg T_{n+1}(z_2,0,x_1, \ldots ,x_{n+1})\}$$

for appropriate z_1 and z_2; and (ii) any sentence of elementary arithmetic is demonstrably equivalent to one of these special sentences by the methods of the Tarski-Kuratowski algorithm together with quantifier contraction and Theorem III. (This is demonstrable equivalence under ZF since ZF includes all accepted forms of mathematical argument.)§

† By Peano's axioms we continue to mean the formulation within elementary arithmetic described in §7.8 rather than the (stronger) formulation that can be made by giving the induction axiom as a single assertion of set theory. (This stronger axiom of induction is also provable in our axiomatization of set theory.)

‡ We can apply the Gödel-Rosser construction (see Exercise 7-65) to set-theory arithmetic and obtain a sentence which is neither provable nor disprovable (provided that the axiomatization of set theory is consistent); however, we cannot establish the truth or falsity of this sentence since we have not proved that the axiomatization of set theory is consistent.

§ A general "metatheorem" establishing all instances of this demonstrable equivalence can be proved in formal detail.

Definition For $n > 0$, Σ_n^{ZF} is the collection of all sentences F of set theory such that for some z, F is demonstrably equivalent to $(\exists a_1)(\forall a_2) \cdots T_{\mathbf{n}}(\mathbf{z},0,a_1, \ldots ,a_n)$. For $n > 0$, Π_n^{ZF} is defined similarly. $\Sigma_0^{ZF} = \Pi_0^{ZF}$ is the set of sentences demonstrably equivalent either to the sentence "$0 = 0$" or to the sentence "$0 = 1$."

It follows by use of dummy quantifiers (as in §14.1) that $\Sigma_n^{ZF} \cup \Pi_n^{ZF} \subset \Sigma_{n+1}^{ZF} \cap \Pi_{n+1}^{ZF}$. The following theorem can be proved by adapting (and simplifying) the proof of Theorem XI.

Theorem XII *If set-theory arithmetic is sound, then, for all n, there exist sentences of set theory in* $(\Sigma_{n+1}^{ZF} - \Pi_{n+1}^{ZF})$ *(and hence sentences in* $(\Pi_{n+1}^{ZF} - \Sigma_{n+1}^{ZF})$*)*.

Proof. The proof is outlined in Exercise 14-21. Several related results are also given in this exercise.☒

This theorem can also be proved for Peano arithmetic (P) if Σ_n^P is defined in a similar way to Σ_n^{ZF}. The hypothesis of soundness can be dropped, since Peano arithmetic is demonstrably sound; the fact that every sentence of elementary arithmetic is demonstrably equivalent (under P) to one of the special sentences is not obvious, however, and an appropriate detailed proof must be given.

A sentence of set theory in $(\Sigma_n^{ZF} - \Pi_n^{ZF})$ cannot be proved or disproved (in set theory), otherwise it would be demonstrably equivalent to "$0 = 0$" or to "$0 = 1$" and would be in Σ_0^{ZF}.

What is known about the classification of various unsettled conjectures of mathematics? *Fermat's last theorem* is known to be in Π_1^{ZF} (since it asserts that $(\forall x)(\forall y)(\forall z)(\forall w)[w > 2 \Rightarrow x^w + y^w \neq z^w]$). The *Riemann hypothesis* (that the *Riemann zeta function* $w = \sum_{n=1}^{\infty} 1/n^z$ (over complex z) has no zeros with real part between 0 and 1 other than those with real part $= \frac{1}{2}$) can also be shown to be in Π_1^{ZF} (by the use of finite approximations). The *prime-number theorem* (that $\lim_{n\to\infty} \frac{P(n) \log n}{n} = 1$, where $P(n)$ is the number of primes $\leq n$) is known to be in Σ_0^{ZF} since it has been proved. Superficial examination, however (without consideration of the proof), shows that the prime-number theorem is in Π_3^{ZF}, for it asserts that $(\forall x)(\exists y)(\forall z)$

$$\left[[x > 0 \;\&\; z > y] \Rightarrow \left|\frac{P(z) \log z}{z}\right| < \frac{1}{x}\right].$$

Almost all statements which (i) have been extensively studied by mathematicians and (ii) are known to be arithmetically expressible can be seen, from a relatively superficial examination, to have quite low level in the Σ_n^{ZF} classification. As has been occasionally remarked, the human mind seems limited in its ability to understand and visualize beyond four or five alternations of quantifier. Indeed, it can be argued that the inventions, subtheories, and central lemmas of various parts of mathematics are devices for assisting the mind in dealing with one or two additional alternations of quantifier. Of course, in Theorem XII above, we have discovered sentences whose minimum arithmetical level is arbitrarily high; however, we have done so by diagonalization arguments rather than by direct insight. Much of recursive function theory can be described as the study of quantifier alternations.

Are all sentences of set theory arithmetically expressible? We give a negative answer to this question in the following theorem.

Theorem XIII *If the axiomatization ZF of set theory is consistent, then there is a sentence of set theory which is not arithmetically expressible.*

Proof. We use a coding that assigns code numbers to formulas with free variables as well as to sentences. We indicate our construction by using the *Gödel substitution function* (see §11.6). Note that $\{x|x$ *is a sentence of elementary arithmetic*$\}$ is recursive, and hence definable in elementary arithmetic; and that $\{x|x$ *is a true sentence of elementary arithmetic*$\}$ is definable in set theory. Choose a fixed z_0 such that $W_{z_0} = \{<x,y>|x$ *is demonstrably equivalent to* $y\}$. z_0 yields, for each y, an effective enumeration of $\{x|x$ *is demonstrably equivalent to* $y\}$ (see Corollary 5-V). We call this the *standard* enumeration of $\{x|x$ *is demonstrably equivalent to* $y\}$. Consider the following formula of set theory: $(\forall b)[[b$ *is the first sentence of elementary arithmetic in the standard enumeration of all formulas demonstrably equivalent to* $\sigma(a,a)] \Rightarrow b$ *is not a true sentence of elementary arithmetic*], where σ is the Gödel substitution function. Let u be a code number for this formula, and let F be the result of substituting the numeral expression of u for the variable a in the formula. Then $\sigma(u,u)$ is the code number for F, and F is a "self-referential" sentence which asserts *"If F is demonstrably equivalent to a sentence of elementary arithmetic, then the first such sentence (in standard enumeration) is false."*

Assume F is demonstrably equivalent to some G, where G is a sentence of elementary arithmetic. Take the first such G, in standard enumeration, and call it H. We can argue in set theory as follows. "Assume H; from this deduce F; from this deduce 'H is false' (by checking that H is the first sentence of elementary arithmetic in the standard enumeration from F); from this deduce $\neg H$. On the other hand, assume $\neg H$; from this deduce $\neg F$; from this deduce 'H is true' (by checking the standard enumeration); from this deduce H." Hence we have a contradiction in set theory. ☒

In Exercise 14-23, we deduce *Gödel's second incompleteness theorem* (for *ZF*) from Theorem XIII.

§14.8 COMPUTING DEGREES OF UNSOLVABILITY

Given a description of a set, it is often possible to use the Tarski-Kuratowski algorithm to get an upper bound to the level of that set in the arithmetical hierarchy. What methods are available for getting lower bounds? We give several results and techniques in the present section.

Definition For any given class $\mathcal{C}$ of recursively enumerable sets, let $P_\mathcal{C} = \{z|W_z \in \mathcal{C}\}$. If $A = P_\mathcal{C}$ for some $\mathcal{C}$, A is called an *index set.*

In our examples, we limit ourselves to index sets; most of the techniques, however, can be applied more generally.

One approach to the problem of lower bounds is to seek a simple structural criterion (on $\mathcal{C}$) for $P_\mathcal{C}$ to fall into various classes of the arithmetical hierarchy. Such criteria can be found for Σ_0 and Σ_1 (and hence for Π_1). The criteria are given by the theorems of Rice and Shapiro mentioned in §2.1 and in Exercise 5-37. We restate and prove these theorems here.

Theorem XIV(a) (Rice) *$P_\mathcal{C} \in \Sigma_0 \Leftrightarrow \mathcal{C}$ is empty or $\mathcal{C}$ is the class of all recursively enumerable sets.*

(b) (Rice, Shapiro). *$P_\mathcal{C} \in \Sigma_1 \Leftrightarrow [\mathcal{C} = \emptyset$ or $(\exists$ recursive $f)$ $[\mathcal{C} = \{A|A$ is recursively enumerable and $(\exists u)$ $[D_{f(u)} \subset A]\}]]$.*

Proof. (*a*) $\Leftarrow$: $P_\mathcal{C} = \emptyset$ or $P_\mathcal{C} = N$, and the result is immediate.

$\Rightarrow$: Assume that $P_\mathcal{C} \in \Sigma_0$ (i.e., that $P_\mathcal{C}$ is recursive), that $A \in \mathcal{C}$, and that $B \notin \mathcal{C}$. Assume further that $A \cup B \in \mathcal{C}$. (The proof for $A \cup B \notin \mathcal{C}$ is parallel.) Then $B \subset A \cup B$, and $B \neq A \cup B$. Define a recursive h such that for any x,

$$W_{h(x)} = \begin{cases} B, & \text{if } \varphi_x(x) \text{ divergent;} \\ A \cup B, & \text{if } \varphi_x(x) \text{ convergent.} \end{cases}$$

It follows from the recursiveness of $P_\mathcal{C}$ that $\{x|\varphi_x(x)$ *convergent*$\} = h^{-1}(P_\mathcal{C})$ is recursive and that the halting problem is solvable. This is a contradiction. (For another proof, see Exercise 11-10.)

(*b*) $\Leftarrow$: If $\mathcal{C} = \emptyset$, then $P_\mathcal{C} = \emptyset \in \Sigma_0 \subset \Sigma_1$. If $\mathcal{C} \neq \emptyset$, let a recursive f be given such that $\mathcal{C} = \{A|A$ *is recursively enumerable and* $(\exists u)[D_{f(u)} \subset A]\}$. Then $x \in P_\mathcal{C} \Leftrightarrow (\exists u)[D_{f(u)} \subset W_x]$. By the Tarski-Kuratowski algorithm, $P_\mathcal{C} \in \Sigma_1$.

$\Rightarrow$: Assume $P_\mathcal{C} \in \Sigma_1$ and $\mathcal{C} \neq \emptyset$. We obtain two lemmas about $\mathcal{C}$.

First lemma *Assume $P_\mathcal{C} \in \Sigma_1$. Then $[A \in \mathcal{C}$ & $A \subset B$ & B is recursively enumerable$] \Rightarrow B \in \mathcal{C}$.*

Second lemma *Assume $P_\mathcal{C} \in \Sigma_1$. Then $A \in \mathcal{C} \Rightarrow (\exists B)$ $[B$ is finite & $B \subset A$ & $B \in \mathcal{C}]$.*

Proof. The lemmas are proved by showing that failure of either lemma would yield solvability of the halting problem (see Exercise 14-24).

Continuing the proof of the theorem, let g be a recursive function such that for all x, $W_{g(x)} = D_x$. Then $g^{-1}(P_\mathcal{C})$ is a recursively enumerable set and, by the second lemma, is not empty. Take f to be a recursive function with $g^{-1}(P_{\mathcal{C}})$ as range. It follows directly from the two lemmas that f has the desired properties with respect to $P_{\mathcal{C}}$. ☒†

We give several examples of the use of Theorem XIV to compute level in the arithmetical hierarchy.

Example 1. Let $\mathcal{C} = \{K\}$. It is immediate that neither $\mathcal{C}$ nor $\bar{\mathcal{C}}$ satisfies the criterion for Σ_1. By the Tarski-Kuratowski algorithm (see Exercise 14-6), $P_{\mathcal{C}} \in \Pi_2$. Hence $P_{\mathcal{C}} \in (\Pi_2 - (\Sigma_1 \cup \Pi_1))$.

Example 2. Let $\mathcal{C} = \{A | A \text{ is recursively enumerable and } A \neq \emptyset\}$. Then $\mathcal{C}$ satisfies the criterion for Σ_1 (take $f = \lambda x[x + 1]$, and recall that $D_0 = \emptyset$). Hence $P_{\mathcal{C}} \in \Sigma_1$. $\mathcal{C}$ does not satisfy the criterion for Σ_0. Hence $P_{\mathcal{C}} \in (\Sigma_1 - \Sigma_0)$.

Example 3. Let $\mathcal{C} = \{A | A \text{ is finite}\}$. By the Tarski-Kuratowski algorithm, $P_{\mathcal{C}} \in \Sigma_2$. Neither $\mathcal{C}$ nor $\bar{\mathcal{C}}$ satisfies the criterion for Σ_1. Hence $P_{\mathcal{C}} \in (\Sigma_2 - (\Sigma_1 \cup \Pi_1))$.

Simple structural criteria like those for Σ_0 and Σ_1 have not been found for Σ_n, $n \geq 2$. Discussion in §§15.1 and 15.2 will give further insight into this matter.‡

We now consider another approach to the problem of lower bounds. This approach, which we call the *reducibility approach,* is to take certain distinguished sets as standard "reference points" and to obtain bounds on the level (and degree) of any other given set by establishing reducibility relationships between it and one or more of the reference sets. In most cases, we shall use sets complete in Σ_n or Π_n $(n > 0)$ as reference sets, and we shall use m-reducibility. The reducibility approach is particularly useful for getting lower bounds on level (and degree). In conjunction with the Tarski-Kuratowski algorithm (and the strong hierarchy theorem), it sometimes enables us to identify not only the level but, indeed, the recursive-isomorphism type of a given set.

Example 1. If we know by the Tarski-Kuratowski algorithm that a given set B is in Σ_3, and if we can show that $\overline{\emptyset^{(2)}} \leq_m B$, then we can conclude that $B \in (\Sigma_3 - \Sigma_2)$.

Example 2. If we know by the Tarski-Kuratowski algorithm that a given set B is in Σ_3, and if we can show that $\emptyset^{(3)} \leq_m B$, then we can conclude that $B \in (\Sigma_3 - (\Sigma_2 \cup \Pi_2))$ and, moreover, we can conclude that B is Σ_3-complete (and $\equiv \emptyset^{(3)}$) (see Exercise 14-10).

Sets with intuitively simple descriptions are especially useful as reference

† The condition in (a) asserts that $\mathcal{C}$ is both open and closed (in the topology of Exercise 11-35(a)). The condition in (b) asserts that $\mathcal{C}$ is open in that topology and is, moreover, a "recursive union" of basic neighborhoods. Topological aspects of the hierarchy will be considered further in Chapter 15.

‡ The author does not know the answer to the following question. If $P_{\mathcal{C}} \in \Sigma_3$, must $\mathcal{C}$ be the intersection of classes which are open in the topology of Exercise 11-35(a)? See preceding footnote.

sets.† For example, $\{x|W_x \textit{ is finite}\}$ is a useful reference set in Σ_2 and $\{x|W_x \textit{ is infinite}\}$ is a useful reference set in Π_2. (By Theorem 13-VIII, $\{x|W_x \textit{ is finite}\}$ is Σ_2-complete, and $\{x|W_x \textit{ is infinite}\}$ is Π_2-complete.) In §13.2 we discussed the relation between certain intuitively simple descriptions and certain degrees of unsolvability.

Example. Let $A = \{x|W_x = K\}$. What is the degree of A? By the Tarski-Kuratowski algorithm, as we have noted, $A \in \Pi_2$. Let

$$B = \{x|W_x \textit{ is infinite}\}$$

Lemma *$B \leq_m A$, where A and B are as defined above.*

Proof of lemma. Let $W_{x_0} = K$. Take a recursive g such that for any z,

$$\varphi_{g(z)}(x) = \begin{cases} \varphi_{x_0}(x), & \text{if at least } x \text{ integers occur in the enumeration of } W_z \text{ (from index } z\text{);} \\ \text{divergent}, & \text{otherwise.} \end{cases}$$

Then $g(z) \in A \Leftrightarrow z \in B$, and $B \leq_m A$.☒

It follows that A must be Π_2-complete. We have established not only the level but also the recursive-isomorphism type of A.

The following theorem provides us with a useful, Σ_3-complete reference set (and hence, incidentally, extends the discussion of §13.2 to $\mathbf{0}^{(3)}$).‡

Theorem XV *$\{x|(\exists y)[y \in W_x \;\&\; W_y \textit{ is infinite}]\}$ is Σ_3-complete, that is, $\equiv \emptyset^{(3)}$.*

Proof. Let $B = \{x|(\exists y)[y \in W_x \;\&\; W_y \textit{ is infinite}]\}$. By Exercise 14-10, it is enough to show $B \equiv_m \emptyset^{(3)}$. By the Tarski-Kuratowski algorithm, $B \in \Sigma_3$ (see Exercise 14-26). Hence $B \leq_m \emptyset^{(3)}$. It remains to show $\emptyset^{(3)} \leq_m B$. Let $\emptyset^{(3)}$ be defined as $\{z|(\exists x_1)(\forall x_2)(\exists x_3)R(z,x_1,x_2,x_3)\}$. Let x_1 and z be fixed. Define ψ by

$$\psi(0) = 0, \quad \text{if } (\exists x_3)R(z,x_1,0,x_3);$$
$$\psi(n+1) = \begin{cases} 0, & \text{if } \psi(n) \text{ is defined and } (\exists x_3)R(z, x_1, n+1, x_3); \\ \text{divergent}, & \text{otherwise.} \end{cases}$$

Let g be a recursive function such that $\psi = \varphi_{g(x_1,z)}$. Then $W_{g(x_1,z)}$ is infinite $\Leftrightarrow$ $(\forall x_2)(\exists x_3)R(z,x_1,x_2,x_3)$. Take a recursive function h such that

$$W_{h(z)} = \textit{range } (\lambda x_1[g(x_1,z)]).$$

Then $z \in \emptyset^{(3)} \Leftrightarrow (\exists x_1)(\forall x_2)(\exists x_3)R(z,x_1,x_2,x_3) \Leftrightarrow (\exists x_1)[W_{g(x_1,z)}$ is infinite$] \Leftrightarrow$ $(\exists y)[y \in W_{h(z)}$ & W_y is infinite$] \Leftrightarrow h(z) \in B$. Hence $\emptyset^{(3)} \leq_m B$.☒

We use this reference set in the following theorem.

† If A is Σ_n-complete, then $A \leq_m C \Leftrightarrow (\forall B)[B \in \Sigma_n \Rightarrow B \leq_m C]$. Thus mention of the special reference set A can be avoided. Use of reference set is both a technical and an intuitive convenience, however, and we describe the reducibility approach in these terms.

‡ The reader should note that the aspects of §13.2 which presented the greatest combinatorial difficulty were the aspects here covered by the Tarski-Kuratowski algorithm.

Theorem XVI *$\{x|W_x$ is recursive$\}$ is Σ_3-complete, that is, $\equiv \emptyset^{(3)}$.*†

Proof. Let $C = \{x|W_x$ *is recursive*$\}$. In the first example in §14.3, we showed that $C \in \Sigma_3$.

We get a lower bound by reducibility. Let

$$B = \{x|(\exists y)[y \in W_x \;\&\; W_y \textit{ is infinite}]\}.$$

By Theorem XV, it will be enough to show that $B \leq_m C$. We use a priority construction similar to that in Theorem 10-III.

Let z be fixed. We take a single *main list* of all the integers. We associate markers $\boxed{0}$, $\boxed{1}$, . . . with members of this main list, and occasionally we write a *plus* beside an integer in the list, or we *move* a marker, or both. Let A be the set of all integers receiving *plus* marks. We shall arrange the construction so that $[z \in B \Rightarrow A$ is co-finite$]$ and $[z \notin B \Rightarrow A$ is not recursive$]$. The instructions for A will depend effectively on z. Hence we shall have a recursive function h such that $A = W_{h(z)}$ and such that $z \in B \Leftrightarrow h(z) \in C$. Thus we shall have $B \leq_m C$.

We first describe an *auxiliary procedure* (which does not involve the main list).

Stage k. Perform k steps in the enumeration (from index z) of W_z. Take each $y \in W_z$ that is obtained and perform k steps in the enumeration (from index y) of W_y.

We now describe the *main procedure.* We say that an integer in the main list is *free* if it does not have and has not had a marker associated with it and that an integer in the main list is *vacant* if it has no *plus* associated with it.

Stage 2n. Associate $\boxed{n}$ with the least free integer in the main list. Let $x_0^{(n)}, \ldots, x_n^{(n)}$ be the current positions of $\boxed{0}, \ldots, \boxed{n}$. Compute n steps in the enumeration of each of $W_0, W_1, \ldots, W_n$. For each i such that $0 \leq i \leq n$ and $x_i^{(n)}$ is vacant and $x_i^{(n)}$ appears in this enumeration of W_i, put a *plus* beside $x_i^{(n)}$.

Stage 2n + 1. Perform the first n stages of the *auxiliary procedure.* See whether there is some y such that: $y \leq n$, y appears in W_z on or before stage n of the auxiliary procedure, and there is a u which appears in W_y for the first time at stage n of the auxiliary procedure. If *not,* go to stage $2n + 3$. *If so,* take the least such y and call it y_n. For all i, $y_n \leq i \leq n$, put a *plus* beside $x_i^{(n)}$ if $x_i^{(n)}$ is vacant, and move the markers $\boxed{y_n}$, $\boxed{y_n + 1}$, . . . , $\boxed{n}$ down to the least $n - y_n + 1$ free integers in the main list.

This defines the set A. Clearly the instructions for A depend uniformly on z. If $(\forall y)[y \in W_z \Rightarrow W_y$ is finite$]$ then (by the odd-numbered stages of the main procedure) each marker moves only finitely often. Let x_n be the final position of $\boxed{n}$. Then (by the even-numbered stages) $x_n \in A \Leftrightarrow x_n \in W_n$. Hence $\bar{A}$ cannot be recursively enumerable and A cannot be recursive.

† Proofs of this theorem were obtained independently by Mostowski and the author. Mostowski's proof is simpler but does not yield Corollary XVI.

If $(\exists y)[y \in W_z \,\&\, W_y$ is infinite], then some marker moves infinitely often and all integers below its initial position must receive *plus* marks. Hence A must be co-finite. This completes the proof.☒

Corollary XVI *$\{x|W_x$ is co-finite$\}$ is Σ_3-complete.*

Proof. By the Tarski-Kuratowski algorithm, $\{x|W_x$ *is co-finite*$\} \in \Sigma_3$. The construction of the theorem yields that $B \leq_m \{x|W_x$ *is co-finite*$\}$, and the result follows.☒

Can we find reference sets for higher values of n that are Σ_n-complete but that have intuitively simpler descriptions than $\emptyset^{(n)}$? In order to answer this question, we introduce the notion of *infinite quantifier* (see Markwald [1954]).

Notation $(\mathrm{U}x)[\cdot\cdot\cdot x \cdot\cdot\cdot]$ asserts that "for infinitely many x, $\cdot\cdot\cdot x \cdot\cdot\cdot$."

The infinite quantifier U is analogous in many respects to the quantifiers $\exists$ and $\forall$. The rules for moving a quantifier to the left are the same as for $\exists$; that is to say, if a is not free in G, the following pairs of expressions are equivalent:

$$\begin{array}{rl} (\mathrm{U}a)F \vee G, & (\mathrm{U}a)[F \vee G]; \\ (\mathrm{U}a)F \,\&\, G, & (\mathrm{U}a)[F \,\&\, G]; \\ G \Rightarrow (\mathrm{U}a)F, & (\mathrm{U}a)[G \Rightarrow F]; \\ (\mathrm{U}a)F \Rightarrow G, & \neg(\mathrm{U}a)\neg[F \Rightarrow G]. \end{array}$$

Unlike $\exists$ and $\forall$, however, the infinite quantifier does not commute with itself. Thus $\mathrm{U}x\mathrm{U}y[x < y]$ is a true assertion about the nonnegative integers, but $\mathrm{U}y\mathrm{U}x[x < y]$ is false.

Definition We shall say that a relation R is *expressible in* $Q_1 \cdots Q_m$-*form* (where for each $i \leq m$, Q_i is either $\exists$, $\forall$, or U) if, for some recursive relation S,

$$R = \{\langle x_1, \ldots, x_n\rangle | (Q_1 y_1) \cdots (Q_m y_m) S(x_1, \ldots, x_n, y_1, \ldots, y_m)\}.$$

Theorem XVII *$R \in \Pi_2 \Leftrightarrow R$ is expressible in U-form.*

Proof. $\Leftarrow$: Let $R = \{\langle x_1, \ldots, x_n\rangle | (\mathrm{U}y)S(x_1, \ldots, x_n, y)\}$. Then $(\mathrm{U}y)S(x_1, \ldots, x_n, y) \Leftrightarrow (\forall u)(\exists y)[y > u \,\&\, S(x_1, \ldots, x_n, y)]$. Hence $R \in \Pi_2$.

$\Rightarrow$: Assume that $R \in \Pi_2$ and that R is n-ary. Then, by Theorem 13-VIII, $\tau^n(R) \leq_1 \{z|W_z$ *is infinite*$\}$ via f for some recursive f. But W_z is infinite $\Leftrightarrow$ $(\mathrm{U}y)S(z,y)$, where S is the recursive relation $\{\langle z,y\rangle | \varphi_z(\pi_1(y))$ *is convergent in exactly* $\pi_2(y)$ *steps*$\}$. Hence

$$R = \{\langle x_1, \ldots, x_n\rangle | (\mathrm{U}y)S(f\tau^n(x_1, \ldots, x_n), y)\}.☒$$

We abbreviate the result of this theorem as $\Pi_2 \Leftrightarrow \mathrm{U}$. By a similar argument, it follows from Theorem XV above that $\Sigma_3 \Leftrightarrow \exists\mathrm{U}$ (see Exercise 14-27). Generalization to higher levels is a matter of some subtlety. The following characterization is due to Kreisel, Shoenfield, and Wang [1960]. We abbreviate $\underbrace{\mathrm{UU}\cdots\mathrm{U}}_{n}$ as $\mathrm{U}^{(n)}$.

Theorem XVIII (Kreisel, Shoenfield, Wang) *For all n,*

$$\Pi_{2n} \Leftrightarrow \mathrm{U}^{(n)};$$
$$\Pi_{2n+1} \Leftrightarrow \mathrm{U}^{(n)}\forall.$$

Proof. The cases Π_0 and Π_1 hold by definition. The case Π_2 holds by Theorem XVII. We prove a relativized version of the case Π_3 in the following lemma.

Lemma *Let A be given. Then for any n-ary R,*

$$R \in \Pi_3^A \Leftrightarrow R = \{\langle x_1, \ldots, x_n\rangle \mid (\mathrm{U}y)(\forall z)S(x_1, \ldots, x_n, y, z)\}$$

for some A-recursive relation S.

Proof of lemma. $\Leftarrow$: This is immediate since $(\mathrm{U}y)(\forall z)S(\cdots) \Leftrightarrow (\forall x)(\exists y)(\forall z)[y \geq x \,\&\, S(\cdots)]$.

$\Rightarrow$: Assume $R \in \Pi_3^A$. Theorem XV holds in the following relativized form: $\{z \mid (\exists y)[y \in W_z{}^A \,\&\, W_y{}^A \textit{ is infinite}]\}$ is Σ_3^A-complete. This form is obtained by a straightforward modification of the earlier proof (see Exercise 14-28). Using this result, a similarly straightforward modification yields the following relativized version of Corollary XVI: $\{z \mid W_z{}^A \textit{ is co-finite}\}$ is Σ_3^A-complete. Hence $\{z \mid W_z{}^A \textit{ is co-infinite}\}$ is Π_3^A-complete (see Exercise 14-29). Hence $\tau^n(R) \leq_1 \{z \mid W_z{}^A \textit{ is co-infinite}\}$ via f for some recursive f. But $W_z{}^A$ is co-infinite $\Leftrightarrow (\mathrm{U}y)[y \notin W_z{}^A] \Leftrightarrow (\mathrm{U}y)(\forall x) \neg T_1{}^A(z,y,x)$. Hence $R = \{\langle x_1, \ldots, x_n\rangle \mid (\mathrm{U}y)(\forall x) \neg T_1{}^A(f\tau^n(x_1, \ldots, x_n), y, x)$ and the lemma is proved.

It remains to prove the theorem. The proof is by induction. The cases of Π_0, Π_1, Π_2, and Π_3 have been established.

$\Leftarrow$, in all cases, is immediate by a Tarski-Kuratowski computation.

Assume $\Pi_{2n} \Leftrightarrow \mathrm{U}^{(n)}$, and assume $R \in \Pi_{2n+2}$, for some $n > 0$. Then $\Pi_{2n+2} = \forall\exists\forall\Sigma_{2n-1}$ (where the notation indicates, in the obvious way, that $\forall\exists\forall$ is followed by a Σ_{2n-1}-prefix). Hence, by the strong hierarchy theorem (Theorem VIII), $R \in \Pi_3^A$, where $A = \emptyset^{(2n-1)}$. By the lemma,

$$R(\ldots) = \{\langle\ldots\rangle \mid (\mathrm{U}y)(\forall x)S(\ldots, y, x)\}$$

where S is recursive in $\emptyset^{(2n-1)}$. By Theorem VIII, $S \in \Sigma_{2n} \cap \Pi_{2n}$. Hence R is expressible in $\mathrm{U}\forall\Pi_{2n}$-form. But $\mathrm{U}\forall\Pi_{2n} = \mathrm{U}\forall\forall\exists\Pi_{2n-2} \Leftrightarrow$ (by contraction) $\mathrm{U}\forall\exists\Pi_{2n-2} = \mathrm{U}\Pi_{2n}$. Hence R is expressible in $\mathrm{U}\Pi_{2n}$-form. Since $\Pi_{2n} \Leftrightarrow \mathrm{U}^{(n)}$, by inductive assumption, we have R expressible in $\mathrm{U}^{(n+1)}$-form. Hence $\Pi_{2n+2} \Leftrightarrow \mathrm{U}^{(n+1)}$.

Similarly, assume $\Pi_{2n+1} \Leftrightarrow \mathrm{U}^{(n)}\forall$ and assume $R \in \Pi_{2n+3}$, for some $n > 0$. We have $\Pi_{2n+3} = \forall\exists\forall\Sigma_{2n}$. Applying the lemma as before, we get

$\forall\exists\forall\Sigma_{2n} \Leftrightarrow \mathrm{U}\forall\Pi_{2n+1}$. Then

$\mathrm{U}\forall\Pi_{2n+1} = \mathrm{U}\forall\forall\exists\Pi_{2n-1} \Leftrightarrow \mathrm{U}\forall\exists\Pi_{2n-1} = \mathrm{U}\Pi_{2n+1} \Leftrightarrow$ (by induction) $\mathrm{U}^{(n+1)}\forall$.

Hence $\Pi_{2n+3} \Leftrightarrow \mathrm{U}^{(n+1)}\forall$. This completes the proof.☒

We can now characterize certain Π_n-complete reference sets as follows.

Corollary XVIII *For* $n > 1$,

$$\{z|(\mathrm{U}x_1) \cdots (\mathrm{U}x_{n-1})[W_{x_{n-1}} \textit{ is infinite } \& \ x_{n-1} \in W_{x_{n-2}} \& \cdots \& \ x_1 \in W_z]\}$$

is Π_{2n}*-complete; and for* $n > 0$,

$$\{z|(\mathrm{U}x_1) \cdots (\mathrm{U}x_n)[W_{x_n} = \emptyset \ \& \ x_n \in W_{x_{n-1}} \& \cdots \& \ x_1 \in W_z]\}$$

is Π_{2n+1}*-complete.*

Proof. These are, respectively, in Π_{2n} and Π_{2n+1}, as can be verified by a Tarski-Kuratowski computation. It remains to show that any set expressible in $\mathrm{U}^{(n)}$-form is m-reducible to the first set and that any set in $\mathrm{U}^{(n)}\forall$-form is m-reducible to the second set. The proof is straightforward and is completed in Exercise 14-30.☒

Computations of degree of unsolvability by the above methods can sometimes be used to show that the sets determined by two different given definitions are, in fact, different sets.† In some cases, knowledge of the degree of one set combined with a Tarski-Kuratowski computation for the other set is enough to show that the sets are distinct. For example, by the result of Theorem XVI, $\{z|W_z \textit{ is not recursive}\}$ is Π_3-complete. It is easy to show, by a Tarski-Kuratowski computation, that $\{z|W_z \textit{ is creative}\} \in \Sigma_3$ (see Exercise 14-7). Hence we have (by the hierarchy theorem) that $\{z|W_z \textit{ is creative}\} \neq \{z|W_z \textit{ is not recursive}\}$ and hence (by the nonrecursiveness of creative sets) we can conclude that there are recursively enumerable sets which are neither recursive nor creative. (We first proved this in Chapter 8 by constructing a simple set.)

The solution to Post's problem (Theorem 10-III) can also be obtained in this way. In Rogers [1959] it is shown that $\emptyset^{(3)} \leq_1 \overline{\{x|W_x \equiv_{\mathrm{T}} K\}}$. Since $\{x|W_x \textit{ is not recursive}\} \equiv \overline{\emptyset^{(3)}}$, and since $\emptyset^{(3)} \not\leq_1 \overline{\emptyset^{(3)}}$ (see Theorem 13-I), we have that $\{x|W_x \textit{ is not recursive}\} \neq \{x|W_x \equiv_{\mathrm{T}} K\}$. (Yates has shown that $\{x|W_x \equiv_{\mathrm{T}} K\}$ is Σ_4-complete, see below.) We shall see another application of this technique in §15.3 (Theorem 15-XXXVI).

Almost all arithmetical sets with intuitively simple definitions that have been studied by the above methods have proved to be Σ_n-complete or Π_n-complete (for some n). The reason for this (i.e., the relation between intuitive simplicity of definition and completeness or noncompleteness) is not fully understood. (See also the last paragraph below.)

In virtually all cases for which best possible results are known, the initial classification yielded by reasonably skillful use of the Tarski-Kuratowski

† This was first pointed out to the author by Norman Shapiro.

algorithm (with recognition and deletion of bounded quantifiers) coincides with the best possible classification. This phenomenon is not fully understood.

General results on index sets have been obtained by Yates, who has shown that for any recursively enumerable A, (i) $\{z|W_z \equiv_T A\}$ is Σ_3^A-complete; and (ii) if $A \neq \emptyset$ and $\bar{A} \neq \emptyset$, then $\{z|W_z \equiv_m A\}$ is Σ_3-complete. (These yield (iii) of Exercise 14-31 and the degree of $\{z|W_z \equiv_T K\}$ as special cases.) Yates has also shown that the sets $\{z|W_z$ *is maximal*$\}$ and $\{z|W_z$ *is hyperhypersimple*$\}$ are Π_4-complete. In connection with (i), note that by applying the uniformity of Corollary 13-XXV(*b*) three times in succession, we have that for any $\mathbf{b}$ such that $\mathbf{b}$ is recursively enumerable in $\mathbf{0}^{(3)}$ and $\mathbf{0}^{(3)} < \mathbf{b} < \mathbf{0}^{(4)}$, there must exist a recursively enumerable $\mathbf{a}$ such that $\mathbf{a}^{(3)} = \mathbf{b}$. In particular, for such a $\mathbf{b}$, if $A \in \mathbf{a}$, then $A^{(3)}$ is Σ_3^A-complete and $K^{(2)} <_T A^{(3)} <_T K^{(3)}$. Then $\{x|W_x \equiv_T A\}$ is an example of an index set which is in $\Sigma_4 - \Pi_4$ but not Σ_4-complete.

§14.9 EXERCISES

§14.1

14-1. Let P and C represent operations of projection and complementation respectively. Let a sequence of such symbols denote a corresponding sequence of operations (in reverse order of application). Thus PPC represents a complementation followed by two projections.

(*a*) Show that $CPPCP$ applied to a relation R can be expressed by a prefix $\forall\forall\exists$ applied to R.

(*b*) Show that $CPPCPC$ can be expressed by a prefix $\forall\forall\exists$ applied to the complement of R.

Because of (*a*) and (*b*), we might say that $CPPCP$ and $CPPCPC$ both correspond to the prefix $\forall\forall\exists$.

(*c*) Find prefixes corresponding to PCP; $CCCP$; $PCPCPCP$.

(*d*) Find sequences of projections and complementations corresponding to: $\exists\forall\exists$; $\exists\forall\exists\forall$; $\forall$.

§14.2

14-2. Show that there is no indexing of the recursive sets in Σ_0 which is *acceptable* in the sense of §14.2. (*Hint:* Use a diagonal argument.)

$\triangle$***14-3.*** (Kleene) (the hierarchy theorem). Prove that if $n > 0$, then $(\Sigma_n^A - \Pi_n^A) \neq \emptyset$ and $(\Pi_n^A - \Sigma_n^A) \neq \emptyset$. (*Hint:* Make a diagonal argument as follows. Let $B = \{z|(\exists x_1)(\forall x_2) \cdots T_n{}^A(z,z,x_1, \ldots ,x_n)\}$. Then $B \in \Sigma_n^A$. Assume $B \in \Pi_n^A$; take a corresponding Π_n^A-index for B, and obtain a contradiction.)

§14.3

14-4. Show that (i) a bounded universal quantifier can be moved to the right past a universal quantifier; (ii) a bounded existential quantifier can be moved to the right past a universal quantifier; and (iii) a bounded existential quantifier can be moved to the right past an existential quantifier.

14-5. Let $(\forall a \leq f(b))F$ abbreviate $(\forall a)[a \leq f(b) \Rightarrow F]$, and let $(\exists a \leq f(b))F$ abbreviate $(\exists a)[a \leq f(b) \ \&\ F]$. Show that if f is interpreted as some given recursive function, then these *recursively bounded quantifiers* can be treated in the same way as bounded quantifiers in Tarski-Kuratowski computations.

14-6. Use the Tarski-Kuratowski algorithm to obtain a classification for

$$\{x|W_x = K\}.$$

△***14-7.*** Obtain a classification for each of the following:

(i) $\{x|W_x$ *is hypersimple*$\}$;
(ii) $\{x|W_x$ *is hyperhypersimple*$\}$;
(iii) $\{x|W_x$ *is creative*$\}$;
(iv) $\{x|W_x$ *is* T-*complete*$\} = \{x|W_x \equiv_T K\}$;
(v) $\{x|W_x$ *is maximal*$\}$.

(*Hint:* Best possible results are Π_3 for (i), Π_4 for (ii), Σ_3 for (iii), Σ_4 for (iv), and Π_4 for (v)).

§*14.4*

14-8. Show that B is in the arithmetical hierarchy in A if and only if B is definable within an augmented elementary arithmetic obtained by adding "relativized quantifiers" of the form $(\forall x)_A$ and $(\exists x)_A$, where $(\forall x)_A$ is interpreted to mean "for all x in A" and $(\exists x)_A$ is interpreted to mean "there exists an x in A such that." (*Hint:* Show that A itself is definable in this augmented system, and use Theorem VII.)

§*14.5*

14-9. Prove the relativized projection theorem (Theorem 9-IX). What uniformity holds?

14-10. Let A and $n > 0$ be given.

(*a*) Show that every Σ_n^A-complete set is a cylinder.
(*b*) Show that the Σ_n^A-complete sets form a 1-degree which is also an m-degree.

▲***14-11.*** Consider sentences of *pure* quantificational logic, i.e., of pure first-order functional calculus with identity (see Church [1956] for terminology). An *integer model* for such a sentence is (i) a nonempty set of integers called the *domain* of the model together with (ii) a finite sequence of relations (on that set) which are associated with the predicate symbols of the sentence, such that if quantifiers and variable symbols are interpreted as ranging over the set, and if predicate symbols are interpreted as the corresponding relations, then the sentence is true. Show that if a sentence is true of some integer model, then it has an integer model for which the domain is recursive and all relations are in $\Sigma_2 \cap \Pi_2$. (*Hint:* Consider Henkin's proof (in Church [1956]) that every consistent sentence has a model and show that it yields an integer model recursive in K.)

§*14.6*

14-12. Show that Σ_n^A and Π_n^A are well-defined with respect to $\leq_m$ (see 14-10).

14-13. (*a*) Show that $A^{(n)}$ *join* $\overline{A^{(n)}}$ is the union of two sets in $\Sigma_n^A \cup \Pi_n^A$.
(*b*) Show that $A^{(n)} \times \overline{A^{(n)}}$ is the intersection of two sets in $\Sigma_n^A \cup \Pi_n^A$.

14-14. Generalize the proof of Theorem 9-I to show that for any A such that $K \leq_T A$, the T-degree of A contains no maximal tt-degree and hence no maximal 1-degree. (*Note:* Martin has shown the existence of a nonrecursive T-degree which consists of a single tt-degree. This exercise shows that such degrees cannot lie above $\mathbf{0}'$.)

14-15. Verify that $A \equiv_T B' \Leftrightarrow (\exists C)[C \in \Sigma_1^A \cap \Pi_1^A \ \&\ A \in \Sigma_1^C \cap \Pi_1^C \ \&\ C \in \Sigma_1^B \ \&\ (\forall D)[D \in \Sigma_1^B \Rightarrow D \in \Sigma_1^C \cap \Pi_1^C]]$.

14-16. Use Theorem 10-III to show that there is a set in $(\Sigma_2 - \Sigma_1)$ which is incomparable (under T-reducibility) with some set in Σ_1.

14-17. Use Corollary 13-IX(*a*) to show that for each $n > 0$, there is a set in $(\Sigma_{n+2} - \Sigma_n)$ which is incomparable with some set in Σ_1.

§14.7

14-18. Show that V and $\emptyset^{(\omega)}$ are cylinders.

14-19. Let A be given. Consider the augmented elementary arithmetic described in Theorem VII. Let V^A be the set of true sentences of this arithmetic. Show that $V^A \equiv A^{(\omega)}$.

14-20. Show that under any axiomatization, every true Σ_1-sentence (of elementary arithmetic) is demonstrably equivalent to some Σ_0-sentence.

△14-21. (*a*) Prove Theorem XII. (*Hint:* For $n \geq 0$, define S_n^Σ = the set of all sentences (of elementary arithmetic) of the special form $(\exists a_1)(\forall a_2) \cdots T_{\mathbf{n}}(\mathbf{z},0,a_1, \ldots ,a_n)$, and S_n^Π = the set of all sentences of the special form $(\forall a_1)(\exists a_2) \cdots T_{\mathbf{n}}(\mathbf{z},0, a_1, \ldots ,a_n)$. Define $S_0^\Sigma = S_0^\Pi$ = the set consisting of the sentences "$0 = 0$" and "$0 = 1$." Let V be the set of all true sentences of elementary arithmetic. Define $V_n^\Sigma = V \cap S_n^\Sigma$, and $V_n^\Pi = V \cap S_n^\Pi$, $n \geq 0$. For any set of sentences A, define $D(A)$ to be the set of all sentences of elementary arithmetic deducible (under ZF) from (finite sets of) sentences in A. Then do the following:

(i) Show $V_n^\Sigma \equiv \emptyset^{(n)}$ and $V_n^\Pi \equiv \overline{\emptyset^{(n)}}$.

(ii) Conclude that $D(V_n^\Pi)$ is recursively enumerable in $\emptyset^{(n)}$.

(iii) Show that $V_{n+1}^\Sigma \subset D(V_n^\Pi)$, and hence that $D(V_{n+1}^\Sigma) \subset D(V_n^\Pi)$.

(iv) Show that $S_{n+1}^\Pi \cap D(V_n^\Pi)$ is recursively enumerable in $\emptyset^{(n)}$, and, by soundness, that $S_{n+1}^\Pi \cap D(V_n^\Pi) \subset V_{n+1}^\Pi$.

(v) Argue that since $V_{n+1}^\Pi \equiv \overline{\emptyset^{(n+1)}}$ and $\overline{\emptyset^{(n+1)}}$ is not recursively enumerable in $\emptyset^{(n)}$, there is a sentence x such that $x \in (V_{n+1}^\Pi - D(V_n^\Pi))$ and hence such that $x \in (V_{n+1}^\Pi - D(V_{n+1}^\Sigma))$.

(vi) Conclude that $x \in \Pi_{n+1}^{ZF}$ by definition, and that $x \notin \Sigma_{n+1}^{ZF}$ by soundness.

(vii) Hence conclude that $x \in (\Pi_{n+1}^{ZF} - \Sigma_{n+1}^{ZF})$, and that the negation of x is in $(\Sigma_{n+1}^{ZF} - \Pi_{n+1}^{ZF})$.)

(*b*) Show that $\Sigma_1^{ZF} \cap \Pi_1^{ZF} = \Sigma_0^{ZF} = \Pi_0^{ZF}$.

(*c*) Take $n > 0$. Assuming soundness, *disprove* the following assertion: any sentence in $\Sigma_{n+1}^{ZF} \cap \Pi_{n+1}^{ZF}$ is demonstrably equivalent (under ZF) to a sentential combination of sentences in $\Sigma_n^{ZF} \cup \Pi_n^{ZF}$. (*Hint:* Use the construction of Theorem 9-I to get a set $\tilde{A}$ such that $\tilde{A} \leq_{\text{T}} \emptyset^{(n)}$ but $\tilde{A} \not\leq_{\text{tt}} \emptyset^{(n)}$. By Theorem VIII, $\tilde{A} \in \Sigma_{n+1} \cap \Pi_{n+1}$. Take normal forms for $\tilde{A}$ in Σ_{n+1} and in Π_{n+1}; use them to get recursive functions f and g such that *range* $f \subset S_{n+1}^\Sigma$; *range* $g \subset S_{n+1}^\Pi$; for all n, $[n \in \tilde{A} \Leftrightarrow f(n)$ is true]; and for each n, $[f(n) \Leftrightarrow g(n)]$ is provable. Then, for each n, $f(n) \in \Sigma_{n+1}^{ZF} \cap \Pi_{n+1}^{ZF}$. Assume that for each n, $f(n)$ is demonstrably equivalent to a sentential combination of sentences in $\Sigma_n^{ZF} \cup \Pi_n^{ZF}$. Conclude from this that $A \leq_{\text{tt}} \emptyset^{(n)}$, a contradiction.)

(*d*) Consider sentences in pure quantificational logic. We say that a *sentence* is a Σ_n-sentence if it is in prenex form and has a Σ_n-prefix. Similarly for Π_n-sentences. Define Σ_n^Q to be the class of all sentences demonstrably equivalent (under elementary logic) to Σ_n-sentences. Similarly for Π_n^Q. The following result in elementary logic is due to Shoenfield: *each sentence in* $\Sigma_{n+1}^Q \cap \Pi_{n+1}^Q$ *is demonstrably equivalent to a sentential combination of sentences in* $\Sigma_n^Q \cup \Pi_n^Q$.

Assume that (*c*) above has been proved with P in place of ZF, i.e., for Peano arithmetic. Conclude from Shoenfield's result that Peano arithmetic is not finitely axiomatizable. (This result was first obtained by Mostowski and Ryll-Nardzewski.) (*Hint:* Show that if Peano arithmetic were finitely axiomatizable, then, taking a large enough value of n, we could obtain a contradiction between (*c*) and Shoenfield's result.)

△14-22. Let S be the assertion, in set theory, that set-theory arithmetic is sound. Show that if set-theory arithmetic is sound, then S is not arithmetically expressible. (This is a weaker form of Theorem XIII.) (*Hint:* The proof of Exercise 14-21 exhibits, for each $n > 0$, a sentence in $(\Pi_n^{ZF} - \Sigma_n^{ZF})$ which is deducible from S.)

△14-23. (Gödel's second incompleteness theorem (for ZF)). Let C be the assertion that ZF is consistent. Show that C is arithmetically expressible. Deduce from Theorem XIII that if ZF is consistent, then C is not provable in ZF. (*Hint:* If C is provable in ZF, then, by Theorem XIII, we have a proof that F (in Theorem XIII) *is not* arithmetically expressible. But this immediately gives a proof of F (since F is of the form "if F is arithmetically expressible then . . . "). Hence we have $F \in \Sigma_0^{ZF}$. Hence we have a proof that F *is* arithmetically expressible. Hence we have a contradiction, and hence ZF is inconsistent. (In this proof we assume and use the fact that all our results, including Theorem XIII, are themselves expressible and provable in ZF).) (The Gödel second incompleteness theorem can be proved for Peano arithmetic by first showing, in detail, that the Gödel-Rosser theorem for Peano arithmetic (see Exercise 7-65) can be expressed within elementary arithmetic and proved within Peano arithmetic. See Feferman [1960].)

§*14.8*

14-24. Prove the lemmas for Theorem XIV, and complete the proof of the theorem.

14-25. Show that every index set is a cylinder.

14-26. Show that $\{x|(\exists y)[y \in W_x \ \&\ W_y \textit{ is infinite}]\} \in \Sigma_3$.

14-27. Show that $\Sigma_3 \Leftrightarrow \exists U$.

14-28. Prove a relativized version of Theorem XV.

14-29. Prove relativized versions of Theorem XVI and Corollary XVI. (*Hint:* Use Exercise 14-28.)

14-30. Complete the proof of Corollary XVIII. (*Hint:* Use the *s-m-n* theorem and Corollary 5-V.)

14-31. Show that:

△ (i) $\{x|W_x \textit{ is simple}\}$ is Π_3-complete;

△ (ii) $\{x|W_x \textit{ is hypersimple}\}$ is Π_3-complete;

▲ (iii) $\{x|W_x \textit{ is creative}\}$ is Σ_3-complete.

(*Hint:* For (i) and (ii) use the construction of Theorem 9-XVI and Exercise 9-25. (iii) is proved in Rogers [1959].)

15 The Arithmetical Hierarchy (Part 2)

§15.1 THE HIERARCHY OF SETS OF SETS

Let $\mathfrak{N}$ ($= 2^N$) be the set of all subsets of N. Let $\mathfrak{N}^k \times N^l$ be the cartesian product $\mathfrak{N} \times \cdots \times \mathfrak{N} \times N \times \cdots \times N$, where $\mathfrak{N}$ occurs k times and N occurs l times, $k,l \geq 0$. For $l \geq 1$, we say that a relation $R \subset \mathfrak{N}^k \times N^l$ is a *single-valued relation* if for every $<X_1, \ldots, X_k, x_1, \ldots, x_{l-1}>$ there exists at most one x_l such that $<X_1, \ldots, x_{l-1}, x_l>$ is in R.

Definition ψ is a *partial function of k set variables and l number variables* if ψ is a single-valued relation in $\mathfrak{N}^k \times N^{l+1}$.

Domain and *range* are defined as for partial functions of k (number) variables in Chapter 1. A partial function of k set variables and l number variables is a (*total*) *function* if its domain is $\mathfrak{N}^k \times N^l$.†

We wish to define the notion of *partial recursive function of k set variables and l number variables*, $k > 0$. Informally, the definition is as follows. Let ψ be a partial function of k set variables and l number variables. ψ is a *partial recursive function of k set variables and l number variables* if there is a set of instructions P such that, for any y and any $<X_1, \ldots, X_k, x_1, \ldots, x_l>$, the following holds: $\psi(X_1, \ldots, x_l) = y \Leftrightarrow$ [P, when supplied with "oracles" for $X_1, \ldots, X_k$, and with $<x_1, \ldots, x_l>$ as input, determines a computation that terminates and yields y as output]. Thus P proceeds algorithmically from the input $<x_1, \ldots, x_n>$ except that (i) from time to time, questions of the form "*is z in X_j?*" (for some j such that $1 \leq j \leq k$) may be asked; and (ii) if and when such a question is asked, the correct answer is supplied by some external agency (the oracle), and the computation continues.

We obtain an appropriate formal definition by means of the definitions in Chapter 9.

† When $k > 0$, a partial function of k set variables and l number variables is sometimes called a *functional*. This terminology and certain related matters are considered further in §15.3.

Definition For $k > 0$ and $l > 0$, ψ is a *partial recursive function of k set variables and l number variables* if, for some z,

$$\psi = \lambda X_1 \cdots X_k x_1 \cdots x_l[\varphi_z^{X_1,\ldots,X_k}(x_1, \ldots, x_l)].$$

For $k > 0$ (and $l = 0$), ψ is a *partial recursive function of k set variables* if, for some z, $\psi = \lambda X_1 \cdots X_k[\varphi_z^{X_1,\ldots,X_k}(0)]$ (see Exercise 15-1).

z is called a *Gödel number* or *index* for ψ.

A number of definitions from the theory of partial recursive functions (of number variables) can be carried over to partial recursive functions of set variables and number variables.

Definition $R \subset \mathfrak{N}^k \times N^l$ is *recursive* if the characteristic function of R is a (total) recursive function of k set variables and l number variables.

Example. $\{<X,x>|x \in X\}$ is a recursive relation (in $\mathfrak{N} \times N$).

We could go on to define a *recursively enumerable* relation to be the domain of a partial recursive function.

Example. $\{X|X \neq \emptyset\} = \textit{domain}\ \psi$, where $\psi(X) = \varphi_z{}^X(0)$, and z is taken so that $\varphi_z{}^X(x) = 1$ if $X \neq \emptyset$ and $\varphi_z{}^X(x)$ is divergent otherwise.

Since the domain of a partial recursive function (of set variables) is not, in general, denumerable, the phrase "recursively enumerable" is inappropriate. We use, instead, the terminology of the arithmetical hierarchy. The discussion that follows will closely parallel the treatment of the arithmetical hierarchy in Chapter 14. For purposes of expository simplicity, however, we omit relativized definitions and results.

Projections

We consider projections, i.e., quantifiers, with respect to number variables only. The more general case (of projections with respect to set variables as well as number variables) is considered in Chapter 16.

The Hierarchy

Definition A relation $R \subset \mathfrak{N}^k \times N^l$ is *arithmetical* if R is recursive or if there exists a recursive relation $S \subset \mathfrak{N}^k \times N^m$ (for some $m > l$) such that R can be obtained from S by some finite sequence of complementation and/or projection operations, where the projection operations occur on number coordinates only.

A counterpart to Theorem 14-I is immediate.

Theorem I *$R \subset \mathfrak{N}^k \times N^l$ is arithmetical $\Leftrightarrow$ [R is recursive or, for some $n > 0$, $R = \{<X_1, \ldots, x_l>|(Q_1x_{l+1}) \cdots (Q_nx_{l+n})S(X_1, \ldots, x_{l+n})\}$ where Q_i is either $\forall$ or $\exists$ for $1 \leq i \leq n$ and $S \subset \mathfrak{N}^k \times N^{l+n}$ is recursive].*

Proof. As for Theorem 14-I. ☒

Σ_n-prefixes and *Π_n-prefixes* are defined as in §14.1. A *$\Sigma_n^{(s)}$-form* is obtained by applying a *Σ_n-prefix* to a recursive relation in $\mathfrak{N}^k \times N^l$; similarly for

$\Pi_n^{(s)}$*-forms.*† (The superscript (s) indicates the possible presence of set variables.) A $\Sigma_n^{(s)}$-form (or $\Pi_n^{(s)}$-form) with recursive relation in $\mathfrak{N}^k \times N^l$ determines a relation in $\mathfrak{N}^k \times N^m$, for some $m \leq l$. We say that the latter relation is *expressed* by the given form.

Definition $\Sigma_n^{(s)}$ is the class of all relations expressible in $\Sigma_n^{(s)}$-form. $\Pi_n^{(s)}$ is the class of all relations expressible in $\Pi_n^{(s)}$-form.‡

Corollary I *$R \subset \mathfrak{N}^k \times N^l$ is arithmetical* $\Leftrightarrow (\exists n)[R \in \Sigma_n^{(s)}$ *or* $R \in \Pi_n^{(s)}]$.
Proof. Immediate.☒
Using "dummy" quantifiers, as in Theorem 14-II, we also have the following.

Theorem II(a) $\Sigma_n^{(s)} \cup \Pi_n^{(s)} \subset \Sigma_{n+1}^{(s)} \cap \Pi_{n+1}^{(s)}$.
(*b*) $R \in \Sigma_n^{(s)} \Leftrightarrow \bar{R} \in \Pi_n^{(s)}$.
Proof. As for Theorem 14-II.☒

Normal Forms

The rule for contraction of quantifiers holds as in §14.2.

Definition For $k > 0$ and $l > 0$,

$$T_{k,l} = \{<X_1, \ldots, X_k, z, x_0, \ldots, x_l>|\varphi_z^{X_1,\ldots,X_k}(x_0, \ldots, x_{l-1})$$

is convergent by the x_l-th *step in the enumeration of* $W_{\rho(z)}$ *(from* $\rho(z)$*)*$\}$. More formally,

$$T_{k,l}(X_1, \ldots, X_k, z, x_0, \ldots, x_l)$$
$$\Leftrightarrow (\exists y)(\exists u)(\exists v)[<<x_0, \ldots, x_{l-1}>, y, u, v>$$

occurs by the x_l-th step in the enumeration of $W_{\rho(z)}$ (from index $\rho(z)$) & $D_u \subset (X_1\ join\ X_2\ join\ \cdots\ X_k)$ & $D_v \subset \overline{(X_1\ join\ \cdots\ X_k)}$].
For $k > 0$ and $l = 0$, $T_{k,l} = \{<X_1, \ldots, X_k, z, x_0>|\varphi_z^{X_1,\ldots,X_k}(0)$ *is convergent by the* x_0-th *step in the enumeration of* $W_{\rho(z)}$ *(from* $\rho(z)$*)*$\}$.
The relation $T_{k,l}$, for $k > 0$, and $l \geq 0$, is evidently recursive.

Convention Henceforth, to parallel previous notation, we write $T_{k,l}(X_1, \ldots, X_k, z, x_0, \ldots, x_l)$ as $T_{k,l}(z, X_1, \ldots, X_k, x_0, \ldots, x_l)$.
We next show that a relation is the domain of a partial recursive function if and only if it is in $\Sigma_1^{(s)}$.

Theorem III Let $R \subset \mathfrak{N}^k \times N^l$. *Then* $R \in \Sigma_1^{(s)} \Leftrightarrow R =$ *domain* ψ *for some partial recursive* ψ *(of k set variables and l number variables).*
Proof. $\Leftarrow$: Assume $R =$ domain ψ, and ψ is partial recursive. Then by definition, for some z, $R = \{<X_1, \ldots, x_l>|(\exists w)T_{k,l}(z, X_1, \ldots, X_k, x_1, \ldots, x_l, w)\}$, and hence $R \in \Sigma_1^{(s)}$.

†A $\Sigma_n^{(s)}$-form thus consists of (i) a Σ_n-prefix, (ii) a recursive relation, and (iii) a one-one mapping from the quantifiers of the prefix into number coordinates of the relation.
‡ The superscript (s) is usually omitted in the literature.

⇒: Conversely, assume $R = \{<X_1, \ldots, x_l> | (\exists w) S(X_1, \ldots, x_l, w)\}$, where S is recursive. Take z so that $\varphi_z^{X_1,\ldots,X_k}(x_1, \ldots, x_l)$ is computed by the following instructions: given $x_1, \ldots, x_l$, test $S(X_1, \ldots, x_l, w)$ for $w = 0, 1, 2, \ldots$ in turn; if and when an affirmative answer is obtained give output zero. Let $\psi = \lambda X_1 \cdots x_l[\varphi_z^{X_1,\ldots,X_k}(x_1, \ldots, x_l)]$. Then ψ has R as domain.☒

A normal-form and enumeration theorem follows directly.

Theorem IV **(Kleene)** *For $n > 0$ and any $R \in \Sigma_n^{(s)}$ (where $R \subset \mathfrak{N}^k \times N^l$) there exists a z such that*

$$R = \{<X_1, \ldots, X_k, x_0, \ldots, x_{l-1}> | (\exists x_l)(\forall x_{l+1})(\exists x_{l+2}) \cdots T_{k,l+n-1}(z, X_1, \ldots, X_k, x_0, \ldots, x_{l+n-1})\}$$

*(where a negation symbol precedes T if n is even). We call z a $\Sigma_n^{(s)}$-*index *for R. Given any $\Sigma_n^{(s)}$-form for R (i.e., given the prefix, given an index for the characteristic function of the recursive relation, and given the mapping from prefix quantifiers into coordinates), a $\Sigma_n^{(s)}$-index for R can be found uniformly. Similarly for $n > 0$ and $R \in \Pi_n^{(s)}$.*

Proof. Immediate.☒

Note that z is a $\Sigma_n^{(s)}$-index for $R \Leftrightarrow z$ is a $\Pi_n^{(s)}$-index for $\bar{R}$.

Tarski-Kuratowski Algorithm

Definition Let $F(\hat{A}_1, \ldots, \hat{A}_k, a_1, \ldots, a_l)$ be an expression which is built up within quantifier logic from sentential connectives, $=$, variables for numbers, quantifiers over variables for numbers, variables $\hat{A}_1, \ldots, \hat{A}_k$ for sets, and symbols for relations (on sets and integers) and which has $a_1, \ldots, a_l$ as its free number variables. Let the relation symbols be interpreted as certain fixed relations $S_1, S_2, \ldots$. Then the relation

$$R = \{<X_1, \ldots, X_k, x_1, \ldots, x_l> | F(\hat{A}_1, \ldots, \hat{A}_k, a_1, \ldots, a_l) \text{ is true when } \hat{A}_1, \ldots, a_l \text{ are interpreted, respectively, as } X_1, \ldots, x_l\}$$

is said to be *definable within quantifier logic from the relations $S_1, S_2, \ldots$, with quantifiers over number variables only,* and $F(\hat{A}_1, \ldots, a_l)$ is called a *definition of R from $S_1, S_2, \ldots$.*

The following analogue to Theorem 14-IV holds.

Theorem V *If $R \subset \mathfrak{N}^k \times N^l$ is definable within quantifier logic from recursive relations, with quantifiers over number variables only, then R is arithmetical.*

Proof. Facts 1 to 4 of §14.3 hold for present case. The theorem follows.☒†

Hence the Tarski-Kuratowski algorithm can be used just as before.

Example 1. Consider $\{X | X \neq \emptyset\}$. $A \neq \emptyset \Leftrightarrow (\exists x)[x \in A] \Leftrightarrow \exists$; and hence this set of sets is in $\Sigma_1^{(s)}$.

† The converse of Theorem V is immediate by Theorem I.

Example 2. Consider $\{X|X \textit{ is finite}\}$. A is finite $\Leftrightarrow (\exists x)(\forall y)[y > x \Rightarrow y \notin A] \Leftrightarrow \exists\forall$; and hence this set of sets is in $\Sigma_2^{(s)}$.

Example 3. Consider $\{X|X \textit{ is recursive}\}$. A is recursive

$$\Leftrightarrow (\exists y)(\exists z)[A = W_y \,\&\, \bar{A} = W_z]$$

$$\Leftrightarrow (\exists y)(\exists z)[(\forall x)[x \in A \Leftrightarrow x \in W_y] \,\&\, (\forall x)[x \notin A \Leftrightarrow x \in W_z]]$$

$$\Leftrightarrow \exists\exists[\forall[\Leftrightarrow \exists] \,\&\, \forall[\Leftrightarrow\exists]] \Leftrightarrow \exists\forall\exists.\dagger$$

Hence this set of sets is seen to be in $\Sigma_3^{(s)}$.

Representation

Theorem 14-VI generalizes to the following.

Theorem VI *Let* $R \subset \mathfrak{N}^k \times N^l$.

(*a*) *R is arithmetical* $\Leftrightarrow$ *R is definable within quantificational logic from recursive relations on numbers and from the relation* $\{<x,X>|x \in X\}$ *with quantifiers on number variables only.*

(*b*) *Let elementary arithmetic be augmented by adding atomic expressions of the form* $a \in \hat{A}_i$, i = 1, 2, . . . , *where a is any number variable and* $\hat{A}_1$, $\hat{A}_2$, . . . *are set variables. Then R is arithmetical* $\Leftrightarrow$ *R is definable within this augmented elementary arithmetic.*

Proof. (*a*) $\Leftarrow$: Immediate by Theorem V.

$\Rightarrow$: By Theorem IV, R is definable from $T_{k,m}$ for some m. $T_{k,m}(z,X_1, \ldots ,x_m) \Leftrightarrow (\exists y)(\exists u)(\exists v)[<<x_0, \ldots ,x_{m-1}>,y,u,v>$ occurs by the x_m-th step in the enumeration of $W_{\rho(z)}$ & $(\forall w)[w \in D_u \Rightarrow w \in (X_1 \textit{ join } \cdots X_k)]$ & $(\forall w)[w \in D_v \Rightarrow w \notin (X_1 \textit{ join } \cdots X_k)]]$; and

$$y \in (X_1 \textit{ join } \cdots X_k) \Leftrightarrow [(\exists y_1)[y = 2y_1 + 1 \,\&\, y_1 \in X_k] \vee (\exists y_1)(\exists y_2)[y = 2y_1 \,\&\, y_1 = 2y_2 + 1 \,\&\, y_2 \in X_{k-1}] \vee \cdots].$$

Hence we have the desired definition of R.

(*b*) $\Leftarrow$: Immediate.

$\Rightarrow$: Immediate by (*a*) and Theorem 14-VI.☒

Topology

In what follows, we shall confine ourselves, for simplicity, to relations of one set variable and no number variables, i.e., to sets of sets of integers. We sometimes (but not always) refer to these as *classes* of sets. We use $\mathcal{A}$, $\mathcal{B}$, $\mathcal{C}$, . . . to denote such classes. Does a general hierarchy theorem hold? That is, does there exist, for each $n > 0$, a class $\mathcal{A}$ such that $\mathcal{A} \in (\Sigma_n^{(s)} - \Pi_n^{(s)})$ (and hence $\bar{\mathcal{A}} \in (\Pi_n^{(s)} - \Sigma_n^{(s)})$)? Before answering this question, we return to the Cantor-set topology on $\mathfrak{N}$ discussed in §13.3 and in Exercise 11-35.

Members of $\mathfrak{N}$, i.e., sets of integers, can be identified with their character-

† $x \in W_y$ is represented as $\exists$ since $x \in W_y \Leftrightarrow (\exists u)T(y,x,u)$. Similarly for $x \in W_z$.

istic functions. The Baire metric defined in §13.3 gives a topology on $\mathfrak{N}$. With respect to this metric, $\mathfrak{N}$ is complete (see Exercise 13-25). The same topology can be obtained by considering $I = \{0,1\}$, taking the discrete topology on I, observing that $\mathfrak{N} = I^N$, and forming the product topology on $\mathfrak{N}$ (see Exercise 13-24). In the latter approach, the *basic neighborhoods* are classes $\mathfrak{D}$ such that $\mathfrak{D} = \{X | D_u \subset X \,\&\, D_v \subset \bar{X}\}$ for some u and v (see §13.3). The space $\mathfrak{N}$ is compact (see Exercise 9-40).

$\Sigma_0^{(s)}$-classes

Classes in $\Sigma_0^{(s)}$ $(= \Pi_0^{(s)})$ are characterized in the following theorem.

Theorem VII (1) $\mathcal{A} \in \Sigma_0^{(s)} \Leftrightarrow$ (2) $\mathcal{A}$ *is both open and closed* (*in the Cantor-set topology on* $\mathfrak{N}$) $\Leftrightarrow$ (3) *there exists a* tt-*condition* x *such that*

$$\mathcal{A} = \{X | X \textit{ satisfies } x\}$$

Proof. (1) $\Rightarrow$ (3). Assume $\mathcal{A} \in \Sigma_0^{(s)}$. Let ψ be the characteristic function (of one set variable) of $\mathcal{A}$. By definition ψ is recursive. Hence $\psi = \lambda X[\varphi_z{}^X(0)]$ for some z. We can generate a computation diagram for $\varphi_z{}^X(0)$ (as X varies), as in §9.2. Since $\varphi_z{}^X(0)$ is defined for all X, all branches must terminate. Hence, by the compactness theorem for trees (Exercise 9-40), the entire diagram is finite. Hence (as in the proof of Theorem 9-XIX) there is a tt-condition x such that $\psi(A) = 1 \Leftrightarrow A$ satisfies x.

(3) $\Rightarrow$ (2). Let x be the given tt-condition. Let B be its associated set. Let $D_{u_0}, \ldots, D_{u_k}$ be the finite subsets of B satisfied by x. Let

$$\mathfrak{D}_i = \{A | D_{u_i} \subset A \,\&\, (B - D_{u_i}) \subset \bar{A}\}, \qquad i \leq k.$$

Then $A \in \mathcal{A} \Leftrightarrow A \in \mathfrak{D}_i$ for some $i \leq k$. Hence

$$\mathcal{A} = \mathfrak{D}_0 \cup \mathfrak{D}_1 \cup \cdots \cup \mathfrak{D}_k.$$

Since $\mathcal{A}$ is a (possibly empty) union of basic neighborhoods, $\mathcal{A}$ is open. Let $D_{v_0}, \ldots, D_{v_l}$ be the finite subsets of B not satisfied by x. Let

$$\mathcal{E}_i = \{A | D_{v_i} \subset A \,\&\, (B - D_{v_i}) \subset \bar{A}\}, \qquad i \leq l.$$

Then $A \in \bar{\mathcal{A}} \Leftrightarrow A \in \mathcal{E}_i$ for some $i \leq l$. Hence

$$\bar{\mathcal{A}} = \mathcal{E}_0 \cup \mathcal{E}_1 \cup \cdots \cup \mathcal{E}_l.$$

Since $\bar{\mathcal{A}}$ is a (possibly empty) union of basic neighborhoods, $\bar{\mathcal{A}}$ is open. Hence $\mathcal{A}$ is both open and closed.

(2) $\Rightarrow$ (1). Assume $\mathcal{A}$ is open and closed. Then $\mathcal{A}$ and $\bar{\mathcal{A}}$ are each a union of basic neighborhoods. The collection of neighborhoods used in the two unions forms an open covering of $\mathfrak{N}$. By compactness, this collection contains a finite subcovering of $\mathfrak{N}$. This implies that $\mathcal{A}$, in particular, is a finite union of basic neighborhoods. But a list of these basic neighborhoods gives a finite set of instructions for computing the characteristic function of $\mathcal{A}$. Hence $\mathcal{A}$ is recursive and $\mathcal{A} \in \Sigma_0^{(s)}$.⊠

Corollary VII $\mathcal{A} \in \Sigma_0^{(s)} \Leftrightarrow \mathcal{A} = \mathfrak{D}_0 \cup \cdots \cup \mathfrak{D}_k$ *for some finite collection of basic neighborhoods* $\mathfrak{D}_0, \ldots, \mathfrak{D}_k$.

Proof. Immediate. ☒

Example. Let $\mathcal{A} = \{X | X \neq \emptyset\}$. As we have noted, $\mathcal{A} \in \Sigma_1^{(s)}$. $\mathcal{A}$ is not a closed set since $\bar{\mathcal{A}} = \{X | X = \emptyset\} = \{\emptyset\}$ is not an open set. Hence $\mathcal{A} \in (\Sigma_1^{(s)} - \Sigma_0^{(s)})$.

$\Sigma_1^{(s)}$-classes and $\Pi_1^{(s)}$-classes

The preceding theorem yields a necessary condition for a class to be in $\Sigma_1^{(s)}$ and a corresponding necessary condition for a class to be in $\Pi_1^{(s)}$.

Theorem VIII(a) $\mathcal{A} \in \Sigma_1^{(s)} \Rightarrow \mathcal{A}$ *is open (in the Cantor-set topology on* $\mathfrak{N}$*).*

(b) $\mathcal{A} \in \Pi_1^{(s)} \Rightarrow \mathcal{A}$ *is closed (in the Cantor-set topology on* $\mathfrak{N}$*).*

Proof. $\mathcal{A} \in \Sigma_1^{(s)} \Rightarrow \mathcal{A} = \{X | (\exists x)R(X,x)\}$ for some recursive R

$$\Rightarrow \mathcal{A} = \bigcup_{x=0}^{\infty} \{X | R(X,x)\}$$

$\Rightarrow \mathcal{A}$ is a union of open and closed sets $\Rightarrow \mathcal{A}$ is open. Similarly, $\mathcal{A} \in \Pi_1^{(s)} \Rightarrow \mathcal{A}$ is an intersection of open and closed sets $\Rightarrow \mathcal{A}$ is closed. ☒

The conditions in Theorem VIII are clearly not sufficient, for there are uncountably many open classes and uncountably many closed classes (e.g., for every A, $\{A\}$ is closed), while there are only countably many classes in $\Sigma_1^{(s)}$ and $\Pi_1^{(s)}$, by Theorem IV.

Corollary VIII $\mathcal{A} \in \Sigma_0^{(s)} \Leftrightarrow \mathcal{A} \in \Pi_0^{(s)} \Leftrightarrow \mathcal{A} \in \Sigma_1^{(s)} \cap \Pi_1^{(s)}$.

Proof. $\mathcal{A} \in \Sigma_0^{(s)} \Rightarrow \mathcal{A} \in \Pi_0^{(s)} \Rightarrow \mathcal{A} \in \Sigma_1^{(s)} \cap \Pi_1^{(s)}$ by Theorem II. $\mathcal{A} \in \Sigma_1^{(s)} \cap \Pi_1^{(s)} \Rightarrow \mathcal{A} \in \Sigma_0^{(s)}$ by Theorems VIII and VII. ☒

$\Sigma_2^{(s)}$-classes and $\Pi_2^{(s)}$-classes

The necessary conditions for $\Sigma_1^{(s)}$ and $\Pi_1^{(s)}$ yield necessary conditions for $\Sigma_2^{(s)}$ and $\Pi_2^{(s)}$.

Theorem IX(a) $\mathcal{A} \in \Sigma_2^{(s)} \Rightarrow \mathcal{A}$ *is a union of closed sets.*

(b) $\mathcal{A} \in \Pi_2^{(s)} \Rightarrow \mathcal{A}$ *is an intersection of open sets.*

Proof. As in the proof of Theorem VIII, an existential quantifier yields a union and a universal quantifier yields an intersection. The result follows. ☒

Category arguments are occasionally helpful in finding the level of a given class in the hierarchy.

Example. Consider $\mathcal{A} = \{X | X$ *is finite*$\}$. By a Tarski-Kuratowski computation, as we have noted, $\mathcal{A} \in \Sigma_2^{(s)}$. From the proof of Theorem IX, $\mathcal{A}$ is a denumerable union of closed sets. Since every basic neighborhood contains an infinite set, each of these closed sets must be nondense. Hence $\mathcal{A}$ must be of first category. Now assume $\bar{\mathcal{A}} \in \Sigma_2^{(s)}$. Then $\bar{\mathcal{A}}$ is a denumerable union of closed sets. Since every basic neighborhood contains a finite set, each of these closed sets must be nondense. Hence $\bar{\mathcal{A}}$ must be of first category. But this is a contradiction since $\mathfrak{N} = \mathcal{A} \cup \bar{\mathcal{A}}$ is of second category. We thus have that $\mathcal{A} \in (\Sigma_2^{(s)} - \Pi_2^{(s)})$ and $\bar{\mathcal{A}} \in (\Pi_2^{(s)} - \Sigma_2^{(s)})$.

The above examples are evidently parallel to certain results in §14.8. For instance, in §14.8, we saw that $\{x|W_x \neq \emptyset\} \in (\Sigma_1 - \Pi_1)$ and $\{x|W_x$ *is finite*$\} \in (\Sigma_2 - \Pi_2)$; while, above, we saw that $\{X|X \neq \emptyset\} \in (\Sigma_1^{(s)} - \Pi_1^{(s)})$ and $\{X|X$ *is finite*$\} \in (\Sigma_2^{(s)} - \Pi_2^{(s)})$. Can we find a general theorem to the effect that there exists a family of "simply defined" (in some sense) classes such that for all classes $\mathcal{C}$ in the family, $\{x|W_x \in \mathcal{C}\}$ and $\{X|X \in \mathcal{C}\}$ must occupy similar hierarchy positions? Apparently we cannot; for, in §14.6, we saw that $\{x|W_x$ *is co-finite*$\} \in (\Sigma_3 - \Pi_3)$, while a Tarski-Kuratowski computation shows that $\{X|X$ *is co-finite*$\} \in \Sigma_2^{(s)}$ (and $\Sigma_2^{(s)} \subset \Pi_3^{(s)}$). In this regard, see Exercise 15-4.

The Classical Borel Hierarchy

Define Σ_n^* and Π_n^* as follows (for classes of sets).

$\Sigma_0^* = \Pi_0^* = \Sigma_0^{(s)} = \Pi_0^{(s)} =$ the family of all classes that are both open and closed.

$\Sigma_{n+1}^* =$ the family of all denumerable unions of classes in Π_n^*.

$\Pi_{n+1}^* =$ the family of all denumerable intersections of classes in Σ_n^*.

Σ_1^* consists of the open sets and Π_1^* of the closed sets.

Σ_2^* consists of what are usually called the F_σ sets, Π_2^* of what are usually called the G_δ sets, etc. These classes form the *finite Borel hierarchy on the Cantor set.* Theorem IX generalizes to the following.

Theorem X *For all n*, $\mathcal{A} \in \Sigma_n^{(s)} \Rightarrow \mathcal{A} \in \Sigma_n^*$, *and* $\mathcal{A} \in \Pi_n^{(s)} \Rightarrow \mathcal{A} \in \Pi_n^*$.

Proof. Immediate.☒

By a cardinality argument there are classes in Σ_n^* but not in $\Sigma_n^{(s)}$ and in Π_n^* but not in $\Pi_n^{(s)}$, for all $n > 0$. The classes $\Sigma_n^{(s)}$ and $\Pi_n^{(s)}$ are sometimes said to form the *effective finite Borel hierarchy on* $\mathfrak{N}$. The effective hierarchy may be viewed as the Borel hierarchy with unions and intersections limited to be "recursive" (see Exercise 15-8). We discuss this matter further in §15.2, where we show that both the classical and the effective hierarchies can be studied in a common theoretical framework.

The Hierarchy Theorem

We now prove a hierarchy theorem for classes of sets.

Theorem XI *For every* $n > 0$, *there exists a class* $\mathcal{A}$ *such that* $\mathcal{A} \in (\Sigma_n^{(s)} - \Pi_n^{(s)})$ *(and hence* $\bar{\mathcal{A}} \in (\Pi_n^{(s)} - \Sigma_n^{(s)})$*).*

Proof. $n = 1$. We showed above that $\{X|X = \emptyset\} \in (\Pi_1^{(s)} - \Sigma_1^{(s)})$ and that $\{X|X \neq \emptyset\} \in (\Sigma_1^{(s)} - \Pi_1^{(s)})$.

$n = 2$. We showed above that $\{X|X \textit{ is finite}\} \in (\Sigma_2^{(s)} - \Pi_2^{(s)})$ and that $\{X|X \textit{ is infinite}\} \in (\Pi_2^{(s)} - \Sigma_2^{(s)})$.

$n \geq 3$. For any B, let $\mathcal{A}_B = \{\{x\}|x \in B\}$.

Lemma *For* $n \geq 3$, (i) $\mathcal{A}_B \in \Sigma_n^{(s)} \Leftrightarrow B \in \Sigma_n$, *and* (ii) $\mathcal{A}_B \in \Pi_n^{(s)} \Leftrightarrow B \in \Pi_n$.

Proof of lemma. (i) $\Rightarrow$: Assume $\mathcal{A}_B \in \Sigma_n^{(s)}$. Then

$$\mathcal{A}_B = \{X|(\exists x_1)(\forall x_2) \cdots R(X,x_1, \ldots ,x_n)\}$$

where $R \subset \mathfrak{N} \times N^n$ is recursive. Define

$$S = \{<x,x_1, \ldots ,x_n>|R(\{x\},x_1, \ldots ,x_n)\}$$

Then $B = \{x|(\exists x_1)(\forall x_2) \cdots S(x,x_1, \ldots ,x_n)\}$ and S is recursive. Hence $B \in \Sigma_n$.

(ii) $\Rightarrow$: Similarly.

(i) $\Leftarrow$: Assume $B \in \Sigma_n$. Let $B = \{x|(\exists x_1) \cdots S(x,x_1, \ldots ,x_n)\}$. Then $\mathcal{A}_B = \{X|(\exists x)[(\forall y)[y \in X \Leftrightarrow y = x] \& (\exists x_1) \cdots S(x,x_1, \ldots ,x_n)]\}$. Since $n \geq 2$, we have, by the Tarski-Kuratowski algorithm, that $\mathcal{A}_B \in \Sigma_n^{(s)}$.

(ii) $\Leftarrow$: Assume $B \in \Pi_n$. Let $B = \{x|(\forall x_1) \cdots S(x,x_1, \ldots ,x_n)\}$. Then

$$\mathcal{A}_B = \{X|(\exists x)(\forall y)[y \in X \Leftrightarrow y = x] \\ \& (\forall x)[x \in X \Rightarrow (\forall x_1) \cdots S(x,x_1, \ldots ,x_n)]\}$$

Since $n \geq 3$, we have, by the Tarski-Kuratowski algorithm, that $\mathcal{A}_B \in \Pi_n^{(s)}$. This proves the lemma.

The desired result now follows from the hierarchy theorem for sets of integers (Corollary 14-VIII(c)).⊠

Degrees

Concepts of reducibility and degree provided a fine-structure for the arithmetical hierarchy of sets. An analogous fine-structure for the arithmetical hierarchy of classes has not been studied, nor have results corresponding to the strong hierarchy theorem been found. The Medvedev lattice may yield appropriate tools for such a study.

Implicit Definability

An apparently more natural way to derive the hierarchy theorem for classes of sets from the hierarchy theorem for sets would be to show (if possible) that for all B and $n \geq 3$, (i) $\{B\} \in \Sigma_n^{(s)} \Leftrightarrow B \in \Sigma_n$, and (ii) $\{B\} \in \Pi_n^{(s)} \Leftrightarrow B \in \Pi_n$. This approach is shown to be impossible by the following remarkable result, which can be found in work of Hilbert and Bernays [1939] and in work of Kuznecov and Trahtenbrot [1955]: there exists a set B such that $\{B\}$ is an arithmetical class but B is not an arithmetical set. We prove this fact in the following theorem.

Theorem XII (Hilbert, Bernays, Kuznecov, Trahtenbrot) *Let V be the set of true sentences of elementary arithmetic. Then* $\{V\} \in \Pi_2^{(s)}$.

We say that A is *explicitly definable in elementary arithmetic* if A is *definable in elementary arithmetic* in the sense of §14.4. We say that A is *implicitly definable in elementary arithmetic* if there is an expression $F(\hat{A})$ in the augmented elementary arithmetic of Theorem VI(b) with $\hat{A}$ as a free set variable such that $F(\hat{A})$ is true if and only if $\hat{A}$ is interpreted as A. Hence, by Theorem VI, A is implicitly definably in elementary arithmetic $\Leftrightarrow \{A\}$ is arithmetical. By Tarski's theorem (Exercise 11-45), or by Theorems 14-X, 14-VIII, and 14-VI, V $(\equiv \emptyset^{(\omega)})$ is not *explicitly* definable in elementary arithmetic. Theorem XII shows that V is implicitly definable in elementary arithmetic, since $\{V\} = \{X|(\forall y)(\exists z)R(X,y,z)\}$ for some recursive R, and R is definable in the augmented elementary arithmetic of Theorem VI(b) (by Theorem VI(b)).

Proof of Theorem XII. If $R \subset \mathfrak{N} \times N^2$ is a recursive relation and h is a recursive permutation, then $S = \{<X,y,z>|R(h(X),y,z)\}$ is a recursive relation. Hence if $\{A\} = \{X|(\forall y)(\exists z)R(X,y,z)\} \in \Pi_2^{(s)}$, then

$$\{h^{-1}(A)\} = \{X|(\forall y)(\exists z)S(X,y,z)\} \in \Pi_2^{(s)}$$

Thus it will be enough to show that $\{\emptyset^{(\omega)}\} \in \Pi_2^{(s)}$, since, by Theorem 14-X, $\emptyset^{(\omega)} \equiv V$.

Lemma *There exists a recursive f such that for all n, f(n) is a* $\Pi_2^{(s)}$*-index for* $\{\emptyset^{(n)}\}$*; i.e., for all n,* $\{\emptyset^{(n)}\} = \{X|(\forall y)(\exists z)T_{1,1}(f(n),X,y,z)\}$.

Proof of lemma. The function f is defined inductively. Set $f(0) = z_0$, where z_0 is a $\Pi_2^{(s)}$-index for $\{\emptyset\}$. z_0 exists since, as we have previously noted, $\{\emptyset\} \in \Pi_1^{(s)}$.

Assume that $\{\emptyset^{(n)}\} = \{X|(\forall y)(\exists z)T_{1,1}(f(n),X,y,z)\}$. We show how to compute $f(n+1)$. By Corollary 13-I, there exists a recursive g such that for all A, $A \leq_1 A'$ via g, and there exists a w_0 such that for all A, $A' = W_{w_0}^A$. Hence

$$A = \emptyset^{(n+1)} = (\emptyset^{(n)})' \Leftrightarrow [g^{-1}(A) = \emptyset^{(n)} \ \&\ A = (g^{-1}(A))']$$

$$\Leftrightarrow [g^{-1}(A) = \emptyset^{(n)} \ \&\ (\forall x)[x \in A \Leftrightarrow (\exists y)(\exists u)(\exists v)[<x,y,u,v> \in W_{\rho(w_0)} \ \&\ D_u \subset g^{-1}(A) \ \&\ D_v \subset \overline{g^{-1}(A)}]]]$$

$$\Leftrightarrow (\forall y)(\exists z)T_{1,1}(f(n),g^{-1}(A),y,z) \ \&\ (\forall x)[x \in A \Leftrightarrow (\exists y)(\exists u)(\exists v)[<x,y,u,v> \in W_{\rho(w_0)} \ \&\ g(D_u) \subset A \ \&\ g(D_v) \subset \bar{A}]]$$

By a Tarski-Kuratowski computation and Theorem IV, this final assertion can be expressed as $(\forall y)(\exists z)T_{1,1}(z_{n+1},A,y,z)$, where z_{n+1} is obtained from $f(n)$ by the uniformity of Theorem IV. Set $f(n+1) = z_{n+1}$. Then $f(n+1)$ is a $\Pi_2^{(s)}$-index for $\{\emptyset^{(n+1)}\}$. This completes the proof of the lemma.

By definition, $\emptyset^{(\omega)} = \{<u,v>|u \in \emptyset^{(v)}\}$. Taking f as in the lemma, we have

$$X = \emptyset^{(\omega)} \Leftrightarrow (\forall v)(\forall y)(\exists z)T_{1,1}(f(v),\{u|<u,v> \in X\},y,z).$$

$\{<X,v,y,z>|T_{1,1}(f(v),\{u|<u,v> \in X\},y,z)\}$ is evidently recursive since, for any fixed v, individual membership questions about $\{u|<u,v> \in X\}$ are equivalent to corresponding individual membership questions about X. Hence $\{\emptyset^{(\omega)}\} \in \Pi_2^{(s)}$. This completes the proof of the theorem.☒

Computing Levels in the Hierarchy

We conclude with a computation for the hierarchy of classes that is similar (in purpose) to examples given in §14.8 for the hierarchy of sets.

Theorem XIII **(*Shoenfield* [*1958*])** $\{X|X$ *is recursively enumerable*$\}$ $\in (\Sigma_3^{(s)} - \Pi_3^{(s)})$.

Proof. A is recursively enumerable

$$\Leftrightarrow (\exists x)[A = W_x] \Leftrightarrow (\exists x)[(\forall y)[y \in A \Leftrightarrow y \in W_x]]$$
$$\Leftrightarrow \exists\forall[\Leftrightarrow \exists] \Leftrightarrow \exists\forall\exists$$

Hence $\{X|X$ *is recursively enumerable*$\} \in \Sigma_3^{(s)}$.

Assume $\{X|X$ *is recursively enumerable*$\} \in \Pi_3^{(s)}$. The proof of the lemma in Theorem 14-XVIII carries over directly to yield that there exists a recursive R such that for all A, A is recursively enumerable $\Leftrightarrow$ $(\mathrm{U}x)(\forall y)R(A,x,y)$. For any X, define $C_X = \{x|(\forall y)R(X,x,y)\}$. Then X is recursively enumerable $\Leftrightarrow C_X$ is infinite.

For any Y, define $\mathcal{S}_Y = \{X|C_X = C_Y\}$. Since there are at most denumerably many C_Y when Y is recursively enumerable (because there are at most denumerably many recursively enumerable sets), and since there are at most denumerably many C_Y when Y is not recursively enumerable (because there are at most denumerably many finite sets), there are at most denumerably many classes $\mathcal{S}_Y$ as Y varies over $\mathfrak{N}$. Furthermore, these classes cover $\mathfrak{N}$. Since the union of these classes is of second category, at least one of the classes is not nondense. Take Y_0 such that $\mathcal{S}_{Y_0}$ is not nondense; i.e., the closure of $\mathcal{S}_{Y_0}$ contains a nonempty open class.

For any x, $\{X|x \in C_X\} = \{X|(\forall y)R(X,x,y)\} \in \Pi_1^{(s)}$ and is hence closed. It follows that $\{Y|C_{Y_0} \subset C_Y\} = \bigcap_{x \in C_{Y_0}}\{X|x \in C_X\}$ is an intersection of closed classes and hence is itself closed. Hence $\{Y|C_{Y_0} \subset C_Y\}$ contains the closure of $\mathcal{S}_{Y_0}(= \{Y|C_{Y_0} = C_Y\})$ and hence contains a neighborhood. But every neighborhood contains a set that is not recursively enumerable, and hence $\{Y|C_{Y_0} \subset C_Y\}$ contains a set that is not recursively enumerable. But this implies that C_{Y_0} must be finite. Let $x_0, x_1, x_2, \ldots$ be the members of the co-finite set $\bar{C}_{Y_0}$ in increasing order.

For the remainder of the proof, we identify sets with characteristic functions. The open class contained in the closure of $\mathcal{S}_{Y_0}$ must contain a neighborhood of the form $\{f|f_0 \subset f\}$ for some initial segment f_0 (of a characteristic function). We define a recursive characteristic function g by successive finite extensions, as follows. Let z be chosen so that $\varphi_z{}^X(x,y) = 1$ if $R(X,x,y)$ holds and $\varphi_z{}^X(x,y) = 0$ if $R(X,x,y)$ does not hold. (As in Chapter 13, we use $f_0, g_0, g_1, \ldots$ to denote finite initial segments of functions; and for any segment h, $\varphi_z^{[h]}(x,y) = w$ indicates that $\varphi_z{}^h(x,y)$ is convergent with output w by a computation whose questions about $\tau(h)$ involve only arguments for which the segment h is defined.)

Stage 0. Let $g_0 = f_0$.

Stage $n + 1$. Consider $\{g|g_n \subset g\}$. This neighborhood is contained in $\{g|g_0 \subset g|$ and hence contains (the characteristic function of) a set X such that $C_X = C_{Y_0}$ and hence such that $(\forall y)R(X,x_n,y)$ is not true. The finite extensions $\tilde{g}$ of g_n such that $(\exists y)[\varphi_z^{[\tilde{g}]}(x_n,y) = 0]$ can be effectively listed. Let g_{n+1} be the first extension in this list.

Let $g = \bigcup_n g_n$. g is recursive, and hence is the characteristic function of a recursive set B. By the construction, $C_B \subset C_{Y_0}$ and C_{Y_0} is finite. This contradicts our earlier conclusion that X recursively enumerable $\Rightarrow C_X$ infinite. Hence $\{X|X$ *is recursively enumerable*$\}$ cannot be in $\Pi_3^{(s)}$.☒

***Corollary XIII* (*Shoenfield*)** $\{X|X$ *is recursive*$\} \in \Sigma_3^{(s)} - \Pi_3^{(s)}$.

Proof. By a Tarski-Kuratowski computation, as we have noted, this class is in $\Sigma_3^{(s)}$. That it is not in $\Pi_3^{(s)}$ follows from the proof of the theorem (see Exercise 15-5).☒

Relativization

Material in this section has not been given in relativized form. Relativization is easily carried through. We define $R \subset \mathfrak{N}^k \times N^l$ to be *A-recursive* if there is a recursive $S \subset \mathfrak{N}^{k+1} \times N^l$ such that

$$R = \{<X_1, \ldots, x_l>|S(A,X_1, \ldots, x_l)\}$$

A normal-form and enumeration theorem follows directly. For example, $\mathfrak{A} \in \Sigma_3^{(s)A} \Leftrightarrow$ for some z, $\mathfrak{A} = \{X|(\exists x_0)(\forall x_1)(\exists x_2)T_{2,2}(z,A,X,x_0,x_1,x_2)\}$. The topological comments for classes of sets remain as before, except that "recursive" unions and intersections are replaced by unions and intersections "recursive in A." In particular, $\Sigma_0^{(s)} = \Sigma_0^{(s)A}$. For nonrecursive A and $n > 0$, however, $\Sigma_n^{(s)A}$ contains classes which are not in $\Sigma_n^{(s)}$ (see Exercise 15-6). The hierarchy theorem is proved as before (see Exercise 15-7).

§15.2 THE HIERARCHY OF SETS OF FUNCTIONS

An arithmetical hierarchy of sets of functions can also be defined. We carry out a development parallel to that in §15.1.

Let $\mathfrak{F}$ be the set of all functions of one number variable, i.e., the set of all total functions mapping N into N. Define $\mathfrak{F}^k \times N^l$ in the same way as $\mathfrak{N}^k \times N^l$ (with $\mathfrak{F}$ in place of $\mathfrak{N}$). Define *partial function of k function variables and l number variables* in the same way as partial function of k set variables and l number variables (with $\mathfrak{F}$ in place of $\mathfrak{N}$). These partial functions are sometimes called *functionals* (see §15.3). ψ is a *(total) function of k function variables and l number variables* if it is a partial function (of k function variables and l number variables) whose domain is $\mathfrak{F}^k \times N^l$.

The informal definition of *partial recursive function of k function variables and l number variables* is like that for partial recursive function of k set variables and l number variables (in §15.1) except that oracles for $\tau(f_1)$, $\ldots$, $\tau(f_k)$ (where $f_1, \ldots, f_k$ are the function arguments) replace oracles for $X_1, \ldots, X_k$. Note the following distinctive feature of the function-variable case: given an oracle for $\tau(f)$, not only can we answer questions of the form "*is* $\langle x,y\rangle$ *in* f?", but we can also "compute" f. That is to say, given any x, we can find $f(x)$ by asking, in turn, the questions "*is* $\langle x,0\rangle$ *in* f?", "*is* $\langle x,1\rangle$ *in* f?", $\ldots$. Since f is a (total) function, an affirmative answer must eventually occur. The fact that computations for function variables always have this feature gives the hierarchy of sets of functions certain special properties, as we shall see.

The formal definition of *partial recursive function of k function variables and l number variables* is also analogous to that given for set variables in §15.1.

Definition For $k > 0$ and $l > 0$, ψ is a *partial recursive function of k function variables and l number variables* if, for some z,

$$\psi = \lambda f_1 \ldots f_k x_1 \ldots x_l[\varphi_z^{f_1,\ldots,f_k}(x_1, \ldots, x_l)].$$

For $k > 0$ (and $l = 0$), ψ is a *partial recursive function of k function variables* if, for some z, $\psi = \lambda f_1 \cdots f_k[\varphi_z^{f_1,\ldots,f_k}(0)]$.

z is called a *Gödel number* or *index* for ψ.

The above concepts are closely related to those defined in §15.1, since $\varphi_z^{f_1,\ldots,f_k} = \varphi_z^{\tau(f_1),\ldots,\tau(f_k)}$ by definition. Thus, for example, ψ is a partial recursive function of one function variable if and only if there is a partial recursive function φ of one set variable such that $\psi = \lambda f[\varphi(\tau(f))]$. Note (by a trivial construction) that it is possible for φ_z^f to be total for all f, even though φ_z^X is not total for all X.

Definition $R \subset \mathfrak{F}^k \times N^l$ is *recursive* if the characteristic function of R is a (total) recursive function of k function variables and l number variables.

Example 1. $\{f|f(5) \neq 0\}$ is recursive (although $\{X|(\exists y)[y \neq 0 \;\&\; \langle 5,y\rangle \in X]\}$ is not a recursive class of sets).

Example 2. If A is recursive, $\{f|f(5) \in A\}$ is recursive.

Example 3. $\{\langle f,g\rangle|f(g(5)) = 0\}$ is recursive, and $\{f|f(f(5)) = 0\}$ is recursive.

Example 4. $\{f|f(0) = \mu y[0 < y \;\&\; 0 < f(y)]\}$ is recursive. This example suggests that, in general, recursive sets of functions may be more complex in structure than recursive sets of sets.

Example 5. $\{f|5 \in range\, f\}$ is the domain of a partial recursive function of one function variable (see Exercise 15-10).

The Hierarchy

We define a hierarchy by applying complementations and number-variable projections to recursive relations. Application of function-variable projections will be considered in Chapter 16 (see also Exercise 15-19).

Definition A relation $R \subset \mathfrak{F}^k \times N^l$ is *arithmetical* if R is recursive or if there exists a recursive relation $S \subset \mathfrak{F}^k \times N^m$ such that R can be obtained from S by some finite sequence of complementation and/or projection operations, where the projection operations occur on number coordinates only.

A counterpart to Theorem I is immediate.

Theorem XIV *$R \subset \mathfrak{F}^k \times N^l$ is arithmetical $\Leftrightarrow$ [R is recursive or, for some $m > 0$, $R = \{<f_1, \ldots, x_l>|(Q_1x_{l+1}) \cdots (Q_mx_{l+m})S(f_1, \ldots, x_{l+m})\}$, where Q_i is either $\forall$ or $\exists$ for $1 \leq i \leq m$ and $S \subset \mathfrak{F}^k \times N^{l+m}$ is recursive].*

Proof. As for Theorem 14-I. ☒

$\Sigma_n^{(fn)}$*-forms* and $\Pi_n^{(fn)}$*-forms* are defined in the same way as $\Sigma_n^{(s)}$-forms and $\Pi_n^{(s)}$-forms. (The superscript (fn) indicates the possible presence of function variables.) *Expressibility* by a form is defined as before.

Definition $\Sigma_n^{(fn)}$ is the class of all relations expressible in $\Sigma_n^{(fn)}$-form. $\Pi_n^{(fn)}$ is the class of all relations expressible in $\Pi_n^{(fn)}$-form.

Corollary XIV *$R \subset \mathfrak{F}^k \times N^l$ is arithmetical $\Leftrightarrow (\exists n)[R \in \Sigma_n^{(fn)}$ or $R \in \Pi_n^{(fn)}]$.*

Proof. Immediate. ☒

Using dummy quantifiers, as before, we have the following.

Theorem XV(a) $\Sigma_n^{(fn)} \cup \Pi_n^{(fn)} \subset \Sigma_{n+1}^{(fn)} \cap \Pi_{n+1}^{(fn)}$.

(b) $R \in \Sigma_n^{(fn)} \Leftrightarrow \bar{R} \in \Pi_n^{(fn)}$.

Proof. As for Theorem 14-II. ☒

Normal Forms

The rule for contraction of quantifiers holds as before.

Definition For $k > 0$ and $l \geq 0$,

$$T'_{k,l} = \{<f_1, \ldots, f_k, z, x_0, \ldots, x_l>| <\tau(f_1), \ldots, \tau(f_k), z, x_0, \ldots, x_l> \in T_{k,l}\}$$

Convention As with $T_{k,l}$, we write

$$T'_{k,l}(f_1, \ldots, f_k, z, x_0, \ldots, x_l) \quad \text{as} \quad T'_{k,l}(z, f_1, \ldots, f_k, x_0, \ldots, x_l).$$

The following theorems are parallel to Theorems III and IV.

Theorem XVI *Let $R \subset \mathfrak{F}^k \times N^l$. Then $R \in \Sigma_1^{(\mathrm{fn})} \Leftrightarrow R = domain\ \psi$ for some partial recursive ψ (of k function variables and l number variables.)*

Proof. As for Theorem III. ☒

Theorem XVII **(Kleene)** *For $n > 0$ and any $R \in \Sigma_n^{(\mathrm{fn})}$ (where $R \subset \mathfrak{F}^k \times N^l$), there exists a z such that*

$$R = \{<f_1, \ldots, f_k, x_0, \ldots, x_{l-1}>|(\exists x_l)(\forall x_{l+1})(\exists x_{l+2}) \cdots T'_{k,l+n-1}(z, f_1, \ldots, f_k, x_0, \ldots, x_{l+n-1})\}$$

*(where a negation symbol precedes T if n is even). We call z a $\Sigma_n^{(\mathrm{fn})}$-*index *for R. Given any $\Sigma_n^{(\mathrm{fn})}$-form for R (see Theorem IV), a $\Sigma_n^{(\mathrm{fn})}$-index for R can be found uniformly. Similarly for $n > 0$ and $R \in \Pi_n^{(\mathrm{fn})}$.*

Proof. Immediate. ☒

Tarski-Kuratowski Algorithm

Definition Let $F(p_1, \ldots, p_k, a_1, \ldots, a_l)$ be an expression which is built up within quantifier logic from sentential connectives, $=$, variables for numbers, quantifiers over variables for numbers, variables $p_1, \ldots, p_k$ for functions (of one variable), and symbols for relations (on functions and integers), and which has $a_1, \ldots, a_l$ as its free number variables. Let the relation symbols be interpreted as certain fixed relations $S_1, S_2, \ldots$. Then the relation

$$R = \{<f_1, \ldots, f_k, x_1, \ldots, x_l>|F(p_1, \ldots, p_k, a_1, \ldots, a_l) \text{ is true when } p_1, \ldots, a_l \text{ are interpreted, respectively, as } f_1, \ldots, x_l\}$$

is said to be *definable within quantifier logic from the relations $S_1, S_2, \ldots$ with quantifiers over number variables only*, and $F(p_1, \ldots, a_l)$ is called a *definition of R from $S_1, S_2, \ldots$*. Note that *nesting* of variables for functions is permitted; for example, $(\exists b)[p(p(b)) = a]$ defines $\{<f,x>|x \in range\ \lambda y[ff(y)]\}$.

The following analogue to Theorem V holds.

Theorem XVIII *If $R \subset \mathfrak{F}^k \times N^l$ is definable within quantifier logic from recursive relations, with quantifiers over number variables only, then R is arithmetical.*

Proof. As for Theorem V. ☒†

Hence the Tarski-Kuratowski algorithm can be used as before.

Example 1. Consider $\{f|range\ f \text{ is infinite}\}$. *range f* is infinite

$$\Leftrightarrow (\forall x)(\exists y)[y > x \ \&\ y \in range\ f]$$
$$\Leftrightarrow (\forall x)(\exists y)[y > x \ \&\ (\exists z)[f(z) = y]] \Leftrightarrow \forall\exists\exists \Leftrightarrow \forall\exists;$$

and hence this set of functions is in $\Pi_2^{(\mathrm{fn})}$.

† The converse of Theorem XVIII is immediate by Theorem XIV.

Example 2. Consider $\{f|f \textit{ is recursive}\}$. f is recursive

$$\Leftrightarrow (\exists z)[f = \varphi_z] \Leftrightarrow (\exists z)(\forall x)[f(x) = \varphi_z(x)]$$
$$\Leftrightarrow (\exists z)(\forall x)(\exists y)(\exists w)[\varphi_z(x) = y \text{ in } w \text{ steps } \& \ y = f(x)]$$

$\Leftrightarrow \exists\forall\exists$; and this set of functions is in $\Sigma_3^{(\text{fn})}$.

Representation

Theorem XIX Let $R \subset \mathfrak{F}^k \times N^l$.

(*a*) *R is arithmetical $\Leftrightarrow$ R is definable within quantificational logic from recursive relations on numbers and from the relation $\{\langle f,x,y\rangle | f(x) = y\}$ with quantifiers on number variables only.*

(*b*) *Let elementary arithmetic be augmented by adding* 1-*ary operator symbols (i.e., function variables) $p_1, p_2, \ldots$. Then R is arithmetical $\Leftrightarrow$ R is definable within this augmented elementary arithmetic.*

Proof. (*a*) $\Leftarrow$: Immediate by Theorem XVIII.

$\Rightarrow$: As for Theorem VI; note that

$$D_u \subset \tau(f) \Leftrightarrow (\forall x)(\forall y)[\langle x,y\rangle \in D_u \Rightarrow f(x) = y].$$

(*b*) $\Leftarrow$: By Theorem XVIII. Operator symbols may be nested, but any such nesting defines a recursive relation.

$\Rightarrow$: By (*a*) and Theorem 14-VI. ☒

Topology

Henceforth we confine ourselves, for simplicity, to relations of one function variable and no set variables, i.e., to sets of functions. We sometimes (but not always) refer to these as *classes* of functions. We use $\mathcal{A}, \mathcal{B}, \mathcal{C}, \ldots$ to denote such classes. Before obtaining a hierarchy theorem, we introduce a topology on $\mathfrak{F}$ analogous to that on $\mathfrak{N}$. As we shall see, $\mathfrak{F}$ and $\mathfrak{N}$ are significantly different as topological spaces.

Definition The *Baire metric* is defined on $\mathfrak{F}$ as follows. Let

$$d(f,g) = \begin{cases} 0, & \text{if } f = g; \\ \dfrac{1}{\mu x[f(x) \neq g(x)] + 1}, & \text{if } f \neq g. \end{cases}$$

Then d is a metric and, under this metric, $\mathfrak{F}$ is a complete metric space (see Exercise 15-11). Hence Theorem 13-XII applies, and category arguments can be used. $\mathfrak{F}$ is commonly called *Baire space.*

The same topology on $\mathfrak{F}$ can be obtained by taking the discrete topology on N, observing that $\mathfrak{F} = N^N$, and forming the product topology on $N^N = N \times N \times \cdots$ (see Exercise 15-12).

Elements of $\mathfrak{F}$ can be identified with the irrational real numbers between 0 and 1 by means of continued fraction representations (see Exercise 15-13). The topology on $\mathfrak{F}$ as Baire space is the same as the topology on the irrationals induced by the standard topology on the reals. (Of course, under the *metric* of the reals, $\mathfrak{F}$ is not complete.)† For intuitive purposes, Baire space can be viewed as the space of all infinite paths in the function tree (see below and Exercise 9-42), and the topology can then be viewed as the topology generated by taking, as a basis of neighborhoods, classes of the form $\{f|\tilde{f} \subset f\}$, where $\tilde{f}$ is a finite initial segment.

Definition A class $\mathfrak{A}$ of functions is a *basic neighborhood* if $\mathfrak{A} = \{f|\tilde{f} \subset f\}$ for some finite initial segment $\tilde{f}$ (see Exercise 15-14).

The space $\mathfrak{F}$ is not compact (see Exercise 15-15). Several important differences between the arithmetical hierarchy of classes of functions and the arithmetical hierarchy of classes of sets can be associated with this fact.

Classes That Are Both Open and Closed

Let $\mathfrak{T}^*$ be the *function tree* as defined in Exercise 9-42. Define *subtree* and *branch* as in 9-42. Each branch of $\mathfrak{T}^*$ can be identified with a member of $\mathfrak{F}$ (and conversely). A *subbranch* is an initial segment of a branch. Each member, i.e., vertex, of $\mathfrak{T}^*$ determines a finite subbranch of $\mathfrak{T}^*$ (and conversely); and each finite subbranch of $\mathfrak{T}^*$ can be identified with a finite initial segment of a function (and conversely). Thus each finite subbranch of $\mathfrak{T}^*$ determines a basic neighborhood in $\mathfrak{F}$ (and conversely), where the neighborhood determined by a given finite subbranch consists of those functions which correspond to branches of $\mathfrak{T}^*$ that contain the given subbranch.

Definition $\mathfrak{T}$ is a *finite-path tree* if $\mathfrak{T}$ is a subtree of $\mathfrak{T}^*$ and every branch of $\mathfrak{T}$ is finite.

Theorem XX $\mathfrak{A}(\subset \mathfrak{F})$ *is both open and closed* $\Leftrightarrow$ *there is a finite-path tree* $\mathfrak{T}$ *such that* $\mathfrak{A}$ *is the union of the neighborhoods determined by the branches of* $\mathfrak{T}$.

Proof. $\Leftarrow$: For any nonempty finite subbranch f of $\mathfrak{T}^*$, define f^- to be the finite subbranch obtained by deleting from f the last element of f. Let a finite-path tree $\mathfrak{T}$ be given and let $\mathfrak{A}$ be the union of the neighborhoods determined by the branches of $\mathfrak{T}$. Take the collection of all finite subbranches f of $\mathfrak{T}^*$ such that (i) f is not a subbranch of $\mathfrak{T}$; (ii) if f^- exists (i.e., if $f \neq \emptyset$), then f^- is a subbranch of $\mathfrak{T}$ but f^- is not a branch of $\mathfrak{T}$. Let $\mathfrak{T}'$ be the union of the subbranches in this collection. Then $\mathfrak{T}'$ is a finite-path tree, since any infinite branch of $\mathfrak{T}'$ would have to be a branch of $\mathfrak{T}$, contrary to the assumption that $\mathfrak{T}$ is finite-path. By the definition of $\mathfrak{T}'$, no branch of $\mathfrak{T}'$ can be contained in a branch of $\mathfrak{T}$ and no branch of $\mathfrak{T}$ can be contained in a branch of $\mathfrak{T}'$. Furthermore, every branch of $\mathfrak{T}^*$ contains either a branch of $\mathfrak{T}$ or a branch of $\mathfrak{T}'$. Thus $\mathfrak{A}$ is the union of the neighborhoods determined by the

† As is well known, a space may be complete under one metric and not complete under another, where both metrics yield the same topology.

branches of $\mathfrak{I}$, and $\bar{\mathcal{A}}$ is the union of the neighborhoods determined by the branches of $\mathfrak{I}'$. Hence $\mathcal{A}$ is both open and closed.†

$\Rightarrow$: Let $\mathcal{A}$ be both open and closed. With each f in $\mathfrak{F}$ associate the *shortest* finite initial segment $\tilde{f}$ of f such that either $\{f|\tilde{f} \subset f\} \subset \mathcal{A}$ or $\{f|\tilde{f} \subset f\} \subset \bar{\mathcal{A}}$. Such an $\tilde{f}$ must exist, since both $\mathcal{A}$ and $\bar{\mathcal{A}}$ are open. The finite subbranches of $\mathfrak{I}^*$ identified with these finite segments constitute a finite-path tree (since for every branch f of $\mathfrak{I}^*$, only the finite subbranch $\tilde{f}$ can appear). A subtree of this finite-path tree is determined by those branches $\tilde{f}$ for which $\{f|\tilde{f} \subset f\} \subset \mathcal{A}$. This subtree is finite-path (since it is contained in a finite-path tree), and $\mathcal{A}$ is the union of the neighborhoods determined by the branches of this subtree. ☒

Corollary XX *There are $2^{\aleph_0}$ classes that are both open and closed.*

Proof. Every set of integers determines a finite-path tree, each of whose branches has length $=1$. Any two distinct such trees determine distinct open and closed classes. There are $2^{\aleph_0}$ sets of integers, hence there are at least $2^{\aleph_0}$ open and closed sets.

There are $\aleph_0$ basic neighborhoods, and every open class is a union of basic neighborhoods. Hence there are at most $2^{\aleph_0}$ open and closed classes of functions. ☒

$\Sigma_0^{(\mathrm{fn})}$-classes

Since there are only a denumerable number of classes in $\Sigma_0^{(\mathrm{fn})}$, Theorem VII fails for $\Sigma_0^{(\mathrm{fn})}$. We have, however, the following necessary condition.

Theorem XXI $\mathcal{A} \in \Sigma_0^{(\mathrm{fn})} \Rightarrow \mathcal{A}$ *is both open and closed (in the Baire-space topology on $\mathfrak{F}$).*

Proof. Let $\mathcal{A} \in \Sigma_0^{(\mathrm{fn})}$. Let ψ be the characteristic function (of one function variable) of $\mathcal{A}$. By definition, ψ is recursive. Hence $\psi = \lambda f[\varphi_z^{\tau(f)}(0)]$ for some z. Consider the collection of all finite initial segments $\tilde{f}$ such that (i) $\varphi_z^{\tau(\tilde{f})}(0)$ is convergent by a computation whose questions about $\tau(\tilde{f})$ involve only arguments occurring in the domain of $\tilde{f}$, and (ii) $\varphi_z^{\tau(\tilde{f})}(0)$ is not convergent by a computation whose questions about $\tau(\tilde{f})$ involve only arguments occurring in the domain of some shorter initial subsegment of $\tilde{f}$; i.e. (in the notation of Chapter 13), $\varphi_z^{[\tilde{f}]}(0)$ is convergent, but $\varphi_z^{[\tilde{g}]}(0)$ is not convergent for any proper initial subsegment $\tilde{g}$ of $\tilde{f}$. By the recursiveness of $\mathcal{A}$, every f in $\mathfrak{F}$ contains such an initial segment. As in the proof of Theorem XX, these segments determine a finite-path tree, and $\mathcal{A}$ is the union of the neighborhoods determined by branches of some subtree of this finite-path tree. Hence the subtree must be finite-path and $\mathcal{A}$ is open and closed. ☒

(The recursive classes are thus associated with a certain denumerable

† Note that the preceding argument covers the special cases $\mathfrak{I} = \{\emptyset\}$ ($\mathfrak{I}$ consists of the empty sequence) and $\mathfrak{I} = \emptyset$ ($\mathfrak{I}$ is empty). If $\mathfrak{I} = \{\emptyset\}$, then $\mathfrak{I}' = \emptyset$, $\mathcal{A} = \mathfrak{F}$, and $\bar{\mathcal{A}} = \emptyset$. If $\mathfrak{I} = \emptyset$, then $\mathfrak{I}' = \{\emptyset\}$, $\mathcal{A} = \emptyset$, and $\bar{\mathcal{A}} = \mathfrak{F}$.

subfamily of the family of all finite-path trees. We further consider this "recursive" subfamily, and certain related subfamilies, in Chapter 16.)

Corollary XXI *Let $\mathfrak{A}$ be the domain of a partial recursive function of one function variable. Then $\mathfrak{A}$ is open.*

Proof. $\mathfrak{A}$ is a union of neighborhoods determined by initial segments $\tilde{f}$ such that $\varphi_z^{[\tilde{f}]}(0)$ is convergent.☒

Example. Let $\mathfrak{A} = \{f|0 \in \textit{range } f\}$. As we have previously noted, $\mathfrak{A} \in \Sigma_1^{(\text{fn})}$. Given any initial segment $\tilde{f}$, there exists an $f \in \mathfrak{F}$ such that $\tilde{f} \subset f$ and $0 \in \textit{range } f$; hence $\bar{\mathfrak{A}}$ contains no basic neighborhood and $\bar{\mathfrak{A}}$ is not open. Hence $\mathfrak{A} \in (\Sigma_1^{(\text{fn})} - \Sigma_0^{(\text{fn})})$.

$\Sigma_1^{(\text{fn})}$-classes and $\Pi_1^{(\text{fn})}$-classes

Theorem VIII carries over directly to classes of functions.

Theorem XXII (*a*) $\mathfrak{A} \in \Sigma_1^{(\text{fn})} \Rightarrow \mathfrak{A}$ *is open.*
(*b*) $\mathfrak{A} \in \Pi_1^{(\text{fn})} \Rightarrow \mathfrak{A}$ *is closed.*

Proof. By Theorem XVI and Corollary XXI.☒

Corollary XXII $\mathfrak{A} \in \Sigma_0^{(\text{fn})} \Leftrightarrow \mathfrak{A} \in \Pi_0^{(\text{fn})} \Leftrightarrow \mathfrak{A} \in \Sigma_1^{(\text{fn})} \cap \Pi_1^{(\text{fn})}$.

Proof. $\mathfrak{A} \in \Sigma_0^{(\text{fn})} \Rightarrow \mathfrak{A} \in \Pi_0^{(\text{fn})} \Rightarrow \mathfrak{A} \in \Sigma_1^{(\text{fn})} \cap \Pi_1^{(\text{fn})}$ by Theorem XV. Assume $\mathfrak{A} \in \Sigma_1^{(\text{fn})} \cap \Pi_1^{(\text{fn})}$. Then $\mathfrak{A} = \textit{domain } \psi_1$ and $\bar{\mathfrak{A}} = \textit{domain } \psi_2$, for partial recursive functions (of one function variable) ψ_1 and ψ_2. Define a partial recursive function ψ (of one function variable) by the following instructions. Given (an oracle for) f, begin computing $\psi_1(f)$ and $\psi_2(f)$; if $\psi_1(f)$ converges before $\psi_2(f)$ give output 1, and if $\psi_2(f)$ converges before $\psi_1(f)$ give output 0. Then ψ is a total recursive function of one function variable and is, in fact, the characteristic function of $\mathfrak{A}$. Hence $\mathfrak{A} \in \Sigma_0^{(\text{fn})}$.☒

The Classical Borel Hierarchy

A hierarchy of Borel classes can be defined on $\mathfrak{F}$ similar to that defined on $\mathfrak{N}$ in §15.1.

Definition $\Sigma_0^{**} = \Pi_0^{**} =$ the family of all classes contained in $\mathfrak{F}$ which are both open and closed.

$\Sigma_{n+1}^{**} =$ the family of all denumerable unions of classes in Π_n^{**}.
$\Pi_{n+1}^{**} =$ the family of all denumerable intersections of classes in Σ_n^{**}.

Theorem XXIII *For all n, $\mathfrak{A} \in \Sigma_n^{(\text{fn})} \Rightarrow \mathfrak{A} \in \Sigma_n^{**}$ and $\mathfrak{A} \in \Pi_n^{(\text{fn})} \Rightarrow \mathfrak{A} \in \Pi_n^{**}$.*

Proof. Immediate.☒

In particular, all classes in $\Sigma_2^{(\text{fn})}$ are F_σ and all classes in $\Pi_2^{(\text{fn})}$ are G_δ. (F_σ and G_δ are defined as for the hierarchy on $\mathfrak{N}$.) (In view of Exercise 15-13, the (finite-level) Borel hierarchy on $\mathfrak{F}$ is sometimes called the *finite hierarchy of Borel sets on the irrationals.*)

Example. Let $\mathcal{A} = \{f|f^{-1}(0) \text{ is finite}\}$. A Tarski-Kuratowski computation and a category argument can be combined to show that $\mathcal{A} \in (\Sigma_2^{(\text{fn})} - \Pi_2^{(\text{fn})})$. (The proof is similar to that for the corresponding example following Theorem IX in §15.1; see Exercise 15-17.)

The Hierarchy Theorem

A hierarchy theorem for classes of functions is easier to derive (from Theorem 14-VIII) than was the hierarchy theorem for classes of sets.

Theorem XXIV *For every $n > 0$, there exists a class $\mathcal{A}$ such that $\mathcal{A} \in (\Sigma_n^{(\text{fn})} - \Pi_n^{(\text{fn})})$ (and hence $\bar{\mathcal{A}} \in (\Pi_n^{(\text{fn})} - \Sigma_n^{(\text{fn})})$).*

Proof. For any B, let $\mathcal{A}_B = \{f|f(0) \in B\}$. We show that for all n: (i) $\mathcal{A}_B \in \Sigma_n^{(\text{fn})} \Leftrightarrow B \in \Sigma_n$; and (ii) $\mathcal{A}_B \in \Pi_n^{(\text{fn})} \Leftrightarrow B \in \Pi_n$.

$\Leftarrow$: Let $B = \{x|(Q_1x_1) \cdots (Q_nx_n)R(x,x_1, \ldots ,x_n)\}$, where $R \subset N^{n+1}$ is recursive. Then $\mathcal{A}_B = \{f|(Q_1x_1) \cdots (Q_nx_n)R(f(0),x_1, \ldots ,x_n)\}$. $\{<f,x_1, \ldots ,x_n>|R(f(0),x_1, \ldots ,x_n)\}$ is evidently a recursive relation in $\mathfrak{F} \times N^n$. Hence $B \in \Sigma_n \Rightarrow \mathcal{A}_B \in \Sigma_n^{(\text{fn})}$, and $B \in \Pi_n \Rightarrow \mathcal{A}_B \in \Pi_n^{(\text{fn})}$.

$\Rightarrow$: Assume $\mathcal{A}_B = \{f|(Q_1x_1) \cdots (Q_nx_n)S(f,x_1, \ldots ,x_n)\}$, where $S \subset \mathfrak{F} \times N^n$ is recursive. Define $R = \{<x,x_1, \ldots ,x_n>|S(\lambda y[x],x_1, \ldots ,x_n)\}$. Then R is evidently a recursive relation in N^{n+1} and

$$B = \{x|(Q_1x_1) \cdots (Q_nx_n)R(x,x_1, \ldots ,x_n)\}.$$

Hence $\mathcal{A}_B \in \Sigma_n^{(\text{fn})} \Rightarrow B \in \Sigma_n$ and $\mathcal{A}_B \in \Pi_n^{(\text{fn})} \Rightarrow B \in \Pi_n$.

The theorem is now immediate by Theorem 14-VIII. ☒

Note that, in the above proof, $\mathcal{A}_B$ is both open and closed. We thus have, at all levels of the arithmetical hierarchy, classes which occur at the Σ_0^{**} level of the classical Borel hierarchy.

What can be said about connections between the hierarchy of sets of functions and the hierarchy of sets of sets? In particular, what relationships must hold between the classification of a class $\mathcal{A} \subset \mathfrak{F}$ and the classification of the corresponding class $\tau(\mathcal{A}) \subset \mathfrak{N}$?

Theorem XXV *Let $\mathcal{A} \subset \mathfrak{F}$.*
For all n, $\tau(\mathcal{A}) \in \Sigma_n^{(\text{s})} \Rightarrow \mathcal{A} \in \Sigma_n^{(\text{fn})}$.
For all n, $\tau(\mathcal{A}) \in \Pi_n^{(\text{s})} \Rightarrow \mathcal{A} \in \Pi_n^{(\text{fn})}$.
For $n \geq 3$, $\mathcal{A} \in \Sigma_n^{(\text{fn})} \Rightarrow \tau(\mathcal{A}) \in \Sigma_n^{(\text{s})}$.
For $n \geq 2$, $\mathcal{A} \in \Pi_n^{(\text{fn})} \Rightarrow \tau(\mathcal{A}) \in \Pi_n^{(\text{s})}$.

Proof. Assume $\tau(\mathcal{A}) = \{X|(Q_1x_1) \cdots (Q_nx_n)R(X,x_1, \ldots ,x_n)\}$, where $R \subset \mathfrak{N} \times N^n$ is recursive. Take $S = \{<f,x_1, \ldots ,x_n>|R(\tau(f),x_1, \ldots ,x_n)\}$. Then $S \subset \mathfrak{F} \times N^n$ is recursive and

$$\mathcal{A} = \{f|(Q_1x_1) \cdots (Q_nx_n)S(f,x_1, \ldots ,x_n)\}.$$

Hence $\tau(\mathcal{A}) \in \Sigma_n^{(\text{s})} \Rightarrow \mathcal{A} \in \Sigma_n^{(\text{fn})}$, and $\tau(\mathcal{A}) \in \Pi_n^{(\text{s})} \Rightarrow \mathcal{A} \in \Pi_n^{(\text{fn})}$.

Conversely, assume $\mathcal{A} \in \Sigma_n^{(\mathrm{fn})}$ with $n \geq 3$. Then for some z,

$$\mathcal{A} = \{f|(\exists x_0)(\forall x_1) \cdots T'_{1,n-1}(z,f,x_0, \ldots ,x_{n-1})\}.$$

By the definition of T', $\mathcal{A} = \{f|(\exists x_0)(\forall x_1) \cdots T_{1,n-1}(z,\tau(f),x_0, \ldots ,x_{n-1})\}$. Then

$$\tau(\mathcal{A}) = \{X|(\forall x)(\forall y)(\forall u)[<x,y> \in X \,\&\, <x,u> \in X \Rightarrow y = u] \\ \& \,(\forall x)(\exists y)[<x,y> \in X] \,\&\, (\exists x_0)(\forall x_1) \cdots T_{1,n-1}(z,X,x_0, \ldots ,x_{n-1})\}.$$

Applying the Tarski-Kuratowski algorithm, we have $[\forall \,\&\, \forall\exists \,\&\, \exists\forall\exists \cdots] \Leftrightarrow \exists\forall\exists \cdots$. Hence $\tau(\mathcal{A}) \in \Sigma_n^{(\mathrm{s})}$.

Similarly, if $\mathcal{A} \in \Pi_n^{(\mathrm{fn})}$ with $n \geq 2$, we obtain $[\forall \,\&\, \forall\exists \,\&\, \forall\exists \cdots] \Leftrightarrow \forall\exists \cdots$, and hence $\tau(\mathcal{A}) \in \Pi_n^{(\mathrm{s})}$.☒

Corollary XXV *Let $\mathcal{A} \subset \mathfrak{F}$. $\mathcal{A}$ is arithmetical $\Leftrightarrow \tau(\mathcal{A})$, as a subclass of $\mathfrak{N}$, is arithmetical.*

Proof. Immediate.☒

We can also consider relationships between the classification of a class $\mathcal{A} \subset \mathfrak{N}$ and the classification of the corresponding class

$$\mathcal{C}(\mathcal{A}) = \{f|(\exists X)[X \in \mathcal{A} \,\&\, f = c_X]\} \subset \mathfrak{F}.$$

Theorem XXVI *Let $\mathcal{A} \subset \mathfrak{N}$.*
For all n, $\mathcal{C}(\mathcal{A}) \in \Sigma_n^{(\mathrm{fn})} \Rightarrow \mathcal{A} \in \Sigma_n^{(\mathrm{s})}$.
For all n, $\mathcal{C}(\mathcal{A}) \in \Pi_n^{(\mathrm{fn})} \Rightarrow \mathcal{A} \in \Pi_n^{(\mathrm{s})}$.
For $n \geq 2$, $\mathcal{A} \in \Sigma_n^{(\mathrm{s})} \Rightarrow \mathcal{C}(\mathcal{A}) \in \Sigma_n^{(\mathrm{fn})}$.
For $n \geq 1$, $\mathcal{A} \in \Pi_n^{(\mathrm{s})} \Rightarrow \mathcal{C}(\mathcal{A}) \in \Pi_n^{(\mathrm{fn})}$.

Proof. The proof is similar, in general outline, to the proof for Theorem XXV (see Exercise 15-18).☒

Corollary XXVI *Let $\mathcal{A} \subset \mathfrak{N}$. $\mathcal{A}$ is arithmetical $\Leftrightarrow \mathcal{C}(\mathcal{A})$, as a subclass of $\mathfrak{F}$, is arithmetical.*

Proof. Immediate.☒

Definability and Implicit Definability

Definition f is an *arithmetical* function if f, as a relation, is arithmetical.

The reader can easily verify that for any set A, A is arithmetical $\Leftrightarrow c_A$ is arithmetical.

Definition *f is explicitly definable in elementary arithmetic* if f, as a relation, is *definable in elementary arithmetic* in the sense of §14.4.

By Theorem 14-VI, f is arithmetical $\Leftrightarrow f$ is explicitly definable in elementary arithmetic. Hence for any A, A is arithmetical $\Leftrightarrow c_A$ is explicitly definable in elementary arithmetic. (The reader can also easily verify: f is arithmetical $\Leftrightarrow \tau(f)$ is arithmetical $\Leftrightarrow \tau(f)$ is explicitly definable in elementary arithmetic.)

Definition *f is implicitly definable in elementary arithmetic* if there is an expression $F(p)$ in the augmented elementary arithmetic of Theorem XIX(*b*), with p as free function variable, such that $F(p)$ is true if and only if p is interpreted as f.

By Theorem XIX, f is implicitly definable in elementary arithmetic $\Leftrightarrow \{f\}$ is an arithmetical class. Let V be as in Theorem XII. By Theorem XXVI, $\{c_V\} \in \Pi_2^{(\text{fn})}$. Hence we have that nonarithmetical functions are implicitly definable in elementary arithmetic and, in particular, that the characteristic function of the set of true sentences of elementary arithmetic is implicitly definable in elementary arithmetic.

Computing Levels in the Hierarchy

Techniques used in the hierarchy of classes of sets can be used in the hierarchy of classes of functions. For example, an argument parallel to that of Theorem XIII can be given to show that $\{f | f \textit{ is recursive}\} \in (\Sigma_3^{(\text{fn})} - \Pi_3^{(\text{fn})})$ (see Exercise 15-24).

Relativization

Material in this section has not been given in relativized form. Relativization is easily carried through by first defining the notion of *partial A-recursive function of k function variables and l number variables* (where, for example, a partial function ψ of one function variable is A-recursive if $\psi = \lambda f[\varphi_z^{A,\tau(f)}(0)]$ for some z). Normal forms can be obtained in terms of $T'^A_{m,n}$, where for example, $T'^A_{1,n} = \{<f,z,x_0, \ldots ,x_n> | T_{2,n}(z,A,\tau(f),x_0, \ldots ,x_n)\}$, and $T_{2,n}$ is as in §15.1. In §15.1 we noted that relativization leaves the lowest level of the hierarchy of classes of sets unaltered; i.e., for all A, $\Sigma_0^{(\text{s})} = \Sigma_0^{(\text{s})A}$. In contrast to this, it is possible for $\Sigma_0^{(\text{fn})}$ to differ from $\Sigma_0^{(\text{fn})A}$ (see Exercise 15-20).

A somewhat different kind of relativization is discussed immediately below.

The Generalized Hierarchy Theorem

Addison [1955] shows that there is a common theory for the (finite) classical Borel hierarchy on $\mathfrak{F}$ and the arithmetical hierarchy on $\mathfrak{F}$. More specifically, let $\mathcal{C}$ be any class of sets. We consider relations $R \subset \mathfrak{F}^k \times N^l$.

Definition R is *recursive in* $\mathcal{C}$ if R is recursive in A for some A in $\mathcal{C}$. $\Sigma_n^{(\text{fn})}[\mathcal{C}]$*-forms* and $\Pi_n^{(\text{fn})}[\mathcal{C}]$*-forms* are defined in the same way as $\Sigma_n^{(\text{fn})}$-forms and $\Pi_n^{(\text{fn})}$-forms, except that recursive relations are replaced by relations recursive in $\mathcal{C}$.

Definition $\Sigma_n^{(\text{fn})}[\mathcal{C}]$ = the class of relations expressible in $\Sigma_n^{(\text{fn})}[\mathcal{C}]$-form. $\Pi_n^{(\text{fn})}[\mathcal{C}]$ = the class of relations expressible in $\Pi_n^{(\text{fn})}[\mathcal{C}]$-form.

Note that the classes (of relations) $\Sigma_n^{(\text{fn})}[\mathcal{C}]$ and $\Pi_n^{(\text{fn})}[\mathcal{C}]$ may now be nondenumerable. In particular, if $\mathcal{C} = \mathfrak{N}$, we have that $\mathcal{A} \in \Sigma_0^{(\text{fn})}[\mathcal{C}] \Leftrightarrow \mathcal{A}$ is both open and closed (see Exercise 15-21). In spite of this possible uncountability, we can obtain a generalized normal-form and enumeration theorem as follows.

Definition

$$T_n^* = \{<g,f,x_1, \ldots ,x_n>|T'_{2,n-1}(g(0),\lambda x[g(x+1)],f,x_1, \ldots ,x_n)\}$$
$$= \{<g,f,x_1, \ldots ,x_n>|T_{2,n-1}(g(0),\tau(\lambda x[g(x+1)]),\tau(f),x_1, \ldots ,x_n)\}.$$

***Theorem XXVII* (*normal-form and enumeration theorem*)** *For $n > 0$: $\mathcal{A} \in \Sigma_n^{(\text{fn})}[\mathcal{C}] \Leftrightarrow$ for some A in $\mathcal{C}$ and for some g recursive in A, $\mathcal{A} = \{f|(\exists x_1)(\forall x_2) \cdots T_n^*(g,f,x_1, \ldots ,x_n)\}$ (where T_n^* is preceded by a negation symbol if n is even). Similarly for $\Pi_n^{(\text{fn})}[\mathcal{C}]$.*

Proof. $\Leftarrow$: Immediate, since [g recursive in A and $A \in \mathcal{C}$] implies that $\{<f,x_1, \ldots ,x_n>|T_n^*(g,f,x_1, \ldots ,x_n)\}$ is recursive in $\mathcal{C}$.

$\Rightarrow$: Assume $\mathcal{A} \in \Sigma_n^{(\text{fn})}[\mathcal{C}]$.

Case (i): n is odd. Then for some A in $\mathcal{C}$,

$$\mathcal{A} = \{f|(\exists x_1)(\forall x_2) \cdots (\exists x_n)R(f,x_1, \ldots ,x_n)\},$$

where R is recursive in A. By a relativized version of Theorem XVI, $\{<f,x_1, \ldots ,x_{n-1}>|(\exists x_n)R(f,x_1, \ldots ,x_n)\}$ is the domain of a partial function ψ of 1 function variable and $n-1$ number variables, where $\psi = \lambda f x_1 \cdots x_{n-1}[\varphi_z^{A,\tau(f)}(x_1, \cdots ,x_{n-1})]$ for some z. Hence,

$$(\exists x_n)R(f,x_1, \ldots ,x_n) \Leftrightarrow (\exists x_n)T_{2,n-1}(z,A,\tau(f),x_1, \ldots ,x_n).$$

Take z' so that $T_{2,n-1}(z,A,\tau(f),x_1, \ldots ,x_n) \Leftrightarrow T_{2,n-1}(z',\tau(c_A),\tau(f),x_1, \ldots ,x_n)$. Then $\mathcal{A} = \{f|(\exists x_1)(\forall x_2) \cdots (\exists x_n)T'_{2,n-1}(z',c_A,f,x_1, \ldots ,x_n)\}$. Take g such that $g(0) = z'$, $\lambda x[g(x+1)] = c_A$. Then g is recursive in A and

$$\mathcal{A} = \{f|(\exists x_1)(\forall x_2) \cdots (\exists x_n)T_n^*(g,f,x_1, \ldots ,x_n)\}.$$

Case (ii): n is even. Make a construction for $\bar{\mathcal{A}}$ similar to that in case (i) for $\mathcal{A}$, then take the negation.

Similarly for $\Pi_n^{(\text{fn})}[\mathcal{C}]$. ☒

The functions recursive in sets in $\mathcal{C}$ thus serve as Gödel numbers at each level of the [$\mathcal{C}$]-arithmetical hierarchy. In particular, as g ranges over the recursive functions, we obtain, for a given n, all classes in $\Sigma_n^{(\text{fn})}$ and all classes in $\Pi_n^{(\text{fn})}$.

We now obtain, by simple diagonalization, a generalized hierarchy theorem.

Theorem XXVIII *For each $n > 0$, there exists an $\mathcal{A}$ such that for any $\mathcal{C}$, $\mathcal{A} \in (\Sigma_n^{(\text{fn})}[\mathcal{C}] - \Pi_n^{(\text{fn})}[\mathcal{C}])$.*

Proof. Take $\mathcal{A} = \{f|(\exists x_1)(\forall x_2) \cdots T_n^*(f,f,x_1, \ldots ,x_n)\}$. Then $\mathcal{A} \in \Sigma_n^{(\text{fn})} \subset \Sigma_n^{(\text{fn})}[\mathcal{C}]$. But

$$\mathcal{A} \in \Pi_n^{(\text{fn})}[\mathcal{C}] \Rightarrow \mathcal{A} = \{f|(\forall x_1)(\exists x_2) \cdots T_n^*(g,f,x_1, \ldots ,x_n)\}$$

for some g recursive in $\mathcal{C}$. Hence

$$\bar{\mathcal{A}} \in \{f|(\exists x_1)(\forall x_2) \cdots T_n^*(g,f,x_1, \ldots ,x_n)\}.$$

Then $g \in \mathcal{A} \Leftrightarrow (\exists x_1)(\forall x_2) \cdots T_n^*(g,g,x_1, \ldots ,x_n) \Leftrightarrow g \in \bar{\mathcal{A}}$, and this is a contradiction.☒

In Theorem XXIV, we exhibited, at all levels of the arithmetical hierarchy, classes which were both open and closed, i.e., of level Σ_0^{**} in the classical Borel hierarchy. In contrast to this, the present theorem exhibits, for all levels, classes which occur at the same level in both the arithmetical and classical hierarchies.

§15.3 FUNCTIONALS

Consider functions and partial functions of one (number) variable.

Definition $\mathcal{P}$ = the class of all partial functions.

$\mathcal{PR}$ = the class of all partial recursive functions.

$\mathcal{F}$ = the class of all (total) functions.

$\mathcal{R}$ = the class of all recursive functions.

Then $\mathcal{R} \subset \mathcal{PR} \subset \mathcal{P}$, and $\mathcal{R} \subset \mathcal{F} \subset \mathcal{P}$.

We first briefly consider the concept of functional on $\mathcal{P}$, then consider in more detail the somewhat simpler concept of functional on $\mathcal{F}$.†

Functionals on $\mathcal{P}$

Definition A *functional on* $\mathcal{P}$ is a single-valued subset of $\mathcal{P} \times (N \cup \{\omega\})$. We use **F**, **G**, . . . to denote functionals. If $x \in N$ and $\langle\varphi,x\rangle \in \mathbf{F}$, we say that **F** is *strongly defined* at φ and that $\mathbf{F}(\varphi) = x$. If $\langle\varphi,\omega\rangle \in \mathbf{F}$, we say that **F** is *weakly defined* at φ and that $\mathbf{F}(\varphi)$ is *divergent*. If $\langle\varphi,\omega\rangle \notin \mathbf{F}$ and $(\forall x)[\langle\varphi,x\rangle \notin \mathbf{F}]$, we say that **F** is *strongly undefined* at φ and that $\mathbf{F}(\varphi)$ is *strongly undefined.* We call $\mathbf{F}^{-1}(N \cup \{\omega\})$ the *weak domain* of **F**, and we call $\mathbf{F}^{-1}(N)$ the *strong domain* of **F**.

Every enumeration operator Φ determines a functional **F** on $\mathcal{P}$ as follows. The *weak domain* of **F** is $\{\varphi|\Phi(\tau(\varphi))$ *has at most one member*$\}$. The *strong domain* of **F** is $\{\varphi|\Phi(\tau(\varphi))$ *has exactly one member*$\}$. For φ in the strong domain of **F**, $\mathbf{F}(\varphi)$ is the unique member of $\Phi(\tau(\varphi))$.

Definition If **F** is determined (as above) by an enumeration operator, **F** is called a *partial recursive functional on* $\mathcal{P}$.

Definition Let **F** be a partial recursive functional on $\mathcal{P}$.

F is a *recursive functional* on $\mathcal{P}$ if the weak domain of $\mathbf{F} = \mathcal{P}$ (or, equivalently, if $\mathcal{F}$ is contained in the weak domain of **F**; see Theorem 9-XXII and Exercise 15-26).

† In §15.3, we use boldface symbols to denote certain classes of partial functions. In particular, the class denoted $\mathcal{F}$ in §15.2 is now denoted $\mathcal{F}$.

F is a *general recursive functional on* $\mathcal{P}$ if $\mathfrak{F}$ is contained in the strong domain of **F**. (Hence, if **F** is a general recursive functional on $\mathcal{P}$, then (by Exercise 15-26) **F** is a recursive functional on $\mathcal{P}$.)

The concepts defined above are closely related (and parallel) to the concepts of partial recursive operator, recursive operator, and general recursive operator, respectively, as defined in Chapter 9. For example, the notion of partial recursive functional can be generalized from the case of functionals on $\mathcal{P}$ to the case of functionals on $\mathcal{P} \times N$ in a straightforward way. There is then a canonical one-one correspondence between partial recursive functionals of this latter kind and partial recursive operators as defined in Chapter 9. For let **F** be such a functional; $\mathbf{\Phi}(\varphi) = \lambda x[\mathbf{F}(\varphi,x)]$ determines a corresponding partial recursive operator if we make the convention that $\mathbf{\Phi}(\varphi)(x)$ is undefined when $\mathbf{F}(\varphi,x) = \omega$ and that $\mathbf{\Phi}$ is undefined (as an operator) at φ if $(\exists x)[<\varphi,x> \not\in$ weak domain of **F**]. Conversely, let $\mathbf{\Phi}$ be a given partial recursive operator; then $\mathbf{F}(\varphi,x) = \mathbf{\Phi}(\varphi)(x)$ determines a corresponding partial recursive functional if we make the convention that $\mathbf{F}(\varphi,x) = \omega$ when $\mathbf{\Phi}(\varphi)$ is defined and $\mathbf{\Phi}(\varphi)(x)$ is undefined. Note that this canonical correspondence carries recursive functionals onto recursive operators and general recursive functionals onto general recursive operators.

Functionals on $\mathcal{PR}$

If a recursive functional on $\mathcal{P}$ is restricted to $\mathcal{PR}$, then it can be represented by a partial recursive function on Gödel numbers, and conversely. This is the content of the following theorem.

Definition A mapping **F** from $\mathcal{PR}$ into $N \cup \{\omega\}$ is an *effective operation* on $\mathcal{PR}$ if there is a partial recursive ψ such that: (i) $\mathbf{F}(\varphi_x) = \psi(x)$ if $\psi(x)$ is convergent, and (ii) $\mathbf{F}(\varphi_x) = \omega$ if $\psi(x)$ is divergent.

***Theorem XXIX* (*Myhill and Shepherdson*) (*a*)** *Every recursive functional on* $\mathcal{P}$ *determines an effective operation on* $\mathcal{PR}$.

(*b*) *Every effective operation on* $\mathcal{PR}$ *is the restriction to* $\mathcal{PR}$ *of a recursive functional on* $\mathcal{P}$.

Proof. (*a*) Let **F** be a given recursive functional on $\mathcal{P}$. Then **F** is determined by some enumeration operator Φ. Define $\psi(x) =$ the unique member of $\Phi(\tau(\varphi_x))$ if $\Phi(\tau(\varphi_x))$ is nonempty; $\psi(x)$ divergent otherwise. ψ is partial recursive since $\Phi(\tau(\varphi_x))$ can be enumerated uniformly in x. Thus ψ gives the desired effective operation on $\mathcal{PR}$.

(*b*) Take f to be a one-one, increasing recursive function such that for any u, if D_u is single-valued, then $\varphi_{f(u)} = \tau^{-1}(D_u)$. Let

$$B = \{y|(\exists u)[D_u \text{ is single-valued and } y = f(u)]\}.$$

Then B is a recursive set which contains one Gödel number for each finite function. Take g to be a recursive function such that for all u, $gf(u) = u$. Then g takes us from any Gödel number in B to a corresponding canonical index.

Let ψ be given, where ψ determines an effective operation on $\mathcal{PR}$. Define an operator Φ as follows.

$$\Phi(A) = \{x|(\exists y)[y \in B \ \& \ D_{g(y)}(=\tau(\varphi_y)) \subset A \ \& \ x = \psi(y)]\}.$$

Φ is evidently an enumeration operator. Let **F** be the partial recursive functional on $\mathcal{P}$ determined by Φ. Then (i) **F** is a recursive functional, and (ii) **F** coincides, on $\mathcal{PR}$, with the effective operation determined by ψ.

To prove (i), assume **F** is not a recursive functional. Then there is a φ such that $\Phi(\tau(\varphi))$ has at least two members. But then there are $y_1, y_2 \in B$ such that $\varphi_{y_1} \subset \varphi$, $\varphi_{y_2} \subset \varphi$, $\psi(y_1)$ and $\psi(y_2)$ are convergent, and $\psi(y_1) \neq \psi(y_2)$. This gives a solution to the halting problem (see Exercise 15-27).

To prove (ii), assume **F** differs at φ_x from the effective operation determined by ψ. This can occur only if either $\mathbf{F}(\varphi_x) \neq \omega$ and $\psi(x)$ is divergent or $\mathbf{F}(\varphi_x) = \omega$ and $\psi(x)$ is convergent. Each case leads to solvability of the halting problem (see Exercise 15-27; the proof is similar to the proof of Theorem 14-XIV).☒

Functionals on $\mathfrak{F}$

We now confine ourselves to mappings from total functions to integers. Under this limitation, it is possible to redefine the concepts of recursive functional and general recursive functional in a simpler (but equivalent) way. We first prove the following theorem, which shows that recursive functionals on $\mathcal{P}$ correspond to partial recursive functions of one function variable (on $\mathfrak{F}$) as defined in §15.2.

Theorem XXX *Let* **F** *be a recursive functional on* $\mathcal{P}$. *For all* $f \in \mathfrak{F}$, *define* $\psi(f) = \mathbf{F}(f)$ if $\mathbf{F}(f) \in N$, *and* $\psi(f)$ *divergent if* $\mathbf{F}(f) = \omega$. *Then* ψ *is a partial recursive function of one function variable.*

Conversely, let ψ *be a partial recursive function of one function variable. Define* $\mathbf{F}(f) = \psi(f)$ if $\psi(f)$ *convergent, and* $\mathbf{F}(f) = \omega$ *if* $\psi(f)$ *divergent. Then* **F** *is the restriction to* $\mathfrak{F}$ *of a recursive functional on* $\mathcal{P}$.

Proof. Given a recursive functional **F**, and given f, we can compute $\psi(f)$ by listing the members of $\tau(f)$ and applying the enumeration operator that determines **F**. But if we can test for membership in $\tau(f)$, we can list $\tau(f)$; hence $\psi(f) = \varphi_z^{\tau(f)}(0)$ for some z. Hence ψ is a partial recursive function.

Conversely, given ψ such that $\psi(f) = \varphi_z^{\tau(f)}(0)$, define an enumeration operator Φ as follows. For any A, $\Phi(A) = \{y|y = \varphi_z^{|\tilde{f}|}(0)$ *for some finite initial segment* $\tilde{f}$ *such that* $\tau(\tilde{f}) \subset A\}$. Then the functional determined by Φ is the desired recursive functional, as is easily verified (see Exercise 15-28).☒

Corollary XXX *Let* **F** *be a general recursive functional on* $\mathcal{P}$ *(as defined above), then the restriction of* **F** *to* $\mathfrak{F}$ *is a recursive function of one function variable. Conversely, let* ψ *be a recursive function of one function variable, then* ψ *is the restriction to* $\mathfrak{F}$ *of a general recursive functional on* $\mathcal{P}$.

Proof. Immediate.☒

We note the following result in passing. Let **F** be a partial recursive functional on $\mathcal{P}$. For any $f \in \mathfrak{F}$, let $\psi(f) = \mathbf{F}(f)$ for $\mathbf{F}(f) \in N$, and let $\psi(f)$ be divergent otherwise. Then ψ need not be a partial recursive function of one function variable. On the other hand, by the construction of Theorem 9-XXIII, every such ψ can be extended to a partial recursive function of one function variable (see Exercise 15-29).

In view of the above theorem, we define the notion of *functional on* $\mathfrak{F}$ as follows.

Definition A *functional on* $\mathfrak{F}$ is a partial function of one function variable.

A *recursive functional on* $\mathfrak{F}$ is a partial recursive function of one function variable.

A *general recursive functional on* $\mathfrak{F}$ is a recursive function of one function variable.

Henceforth, unless we indicate otherwise, we use "functional" to mean *functional on* $\mathfrak{F}$, and we dispense with the use of ω to represent divergence of a functional.

Note on terminology. Functionals considered in the literature are usually taken over $\mathfrak{F}$. Results about functionals over $\mathcal{P}$ are often obtained as results about functional operators (in order to avoid explicit consideration of weak and strong domains). Theorem XXIX was first obtained in this form (see the formulation in §11.5). Terminology for functionals over $\mathfrak{F}$ varies. What we call *recursive functionals* are sometimes called "partial recursive functionals," and what we call *general recursive functionals* are sometimes called "recursive functionals." (This alternative usage is not unnatural, since, under it, "partial recursive functionals" correspond to partial recursive functions of one function variable, and "recursive functionals" correspond to recursive functions of one function variable.)

Observe that if the Baire topology is taken on $\mathfrak{F}$ and the discrete topology is taken on N, then every recursive functional **F** is a continuous mapping on its domain (since, for any x, $\mathbf{F}^{-1}(x) \in \Sigma_1^{(\text{fn})}$ and hence $\mathbf{F}^{-1}(x)$ is open).

Functionals on $\mathcal{R}$

We consider the restriction to $\mathcal{R}$ of functionals on $\mathfrak{F}$.

The recursive functions are dense in $\mathfrak{F}$. Is it true that if the domain of a recursive functional contains $\mathcal{R}$, then the domain contains $\mathfrak{F}$ (that is, **F** is general recursive)? A simple example, presented in Exercise 15-30, gives a negative answer to this question; a partial recursive function of one function variable may be total on $\mathcal{R}$ without being total on $\mathfrak{F}$.

The notion of effective operation on $\mathcal{R}$ can be defined as follows.

Definition Let **F** be the restriction to $\mathcal{R}$ of a functional on $\mathfrak{F}$. **F** is called an *effective operation* on $\mathcal{R}$ if there is a partial recursive function (of one number variable) ψ such that for all x, $\varphi_x \in \mathfrak{R} \Rightarrow [[\mathbf{F}(\varphi_x)$ convergent $\Leftrightarrow$ $\psi(x)$ convergent] & $[\mathbf{F}(\varphi_x)$ convergent $\Rightarrow \mathbf{F}(\varphi_x) = \psi(x)]]$.

Does a counterpart to Theorem XXIX hold? A negative answer has been obtained by Friedberg [1958*a*].

***Theorem XXXI* (Friedberg)** *There exists an effective operation on* $\mathfrak{R}$ *which is not the restriction to* $\mathfrak{R}$ *of a recursive functional on* $\mathfrak{F}$.

Proof. Define a partial recursive function ψ as follows:

$$\psi(x) = \begin{cases} 0, & \text{if either } (\forall y)[y \leq x \Rightarrow \varphi_x(y) = 0] \text{ or } (\exists z)[\varphi_x(z) \neq 0 \\ & \quad \& \ (\forall y)[y < z \Rightarrow \varphi_x(y) = 0] \ \& \ (\exists x')[x' < z \ \& \\ & \quad (\forall u)[u \leq z \Rightarrow \varphi_{x'}(u) = \varphi_x(u)]]]; \\ \text{divergent}, & \text{otherwise.} \end{cases}$$

It is easily verified that ψ defines an effective operation $\mathbf{F}$ on $\mathfrak{R}$. Assume $\mathbf{G}$ is a recursive functional whose restriction to $\mathfrak{R}$ is $\mathbf{F}$. Then $\mathbf{G}(\lambda x[0]) = 0$. Since $\mathbf{G}$ is a recursive functional, any value of $\mathbf{G}$ is determined by answers to a finite number of questions about the argument. Hence there is an n such that $(\forall f)[(\forall x)[x < n \Rightarrow f(x) = 0] \Rightarrow \mathbf{G}(f) = 0]$. Define a recursive function g as follows:

$$g(x) = 0, \qquad \text{for } x \neq n;$$
$$g(n) = k,$$

where k is chosen so that there is no index for g less than n. Then, by the definition of ψ, $\mathbf{F}(g)$ is undefined. But $\mathbf{G}(g)$ must equal 0. This is a contradiction. ☒

If we restrict our attention to operations and functionals which are convergent on all of $\mathfrak{R}$, however, a counterpart to Theorem XXIX can be obtained.

Definition Let $\mathbf{F}$ be the restriction to $\mathfrak{R}$ of a functional on $\mathfrak{F}$. $\mathbf{F}$ is called a *total effective operation* on $\mathfrak{R}$ if there is a partial recursive ψ such that for all x, $\varphi_x \in \mathfrak{R} \Rightarrow [\psi(x)$ is defined & $\mathbf{F}(\varphi_x) = \psi(x)]$.

***Theorem XXXII* (Kreisel, Lacombe, Shoenfield [*1957*])** $\mathbf{F}$ *is a total effective operation on* $\mathfrak{R} \Leftrightarrow \mathbf{F}$ *is the restriction to* $\mathfrak{R}$ *of a recursive functional which is total on* $\mathfrak{R}$.

(Note, by Exercise 15-30, that we cannot obtain, as an immediate corollary, that every total effective operation on $\mathfrak{R}$ is the restriction to $\mathfrak{R}$ of a general recursive functional.)

Proof. $\Leftarrow$: Assume $\mathbf{F}$ is the restriction to $\mathfrak{R}$ of a recursive functional total on $\mathfrak{R}$. Hence, for $f \in \mathfrak{R}$, we need ask only finitely many questions about f in order to compute $\mathbf{F}(f)$. Define a partial recursive function ψ as follows. To compute $\psi(y)$, compute values for $\varphi_y(0)$, $\varphi_y(1)$, . . . until enough values are obtained so that $\mathbf{F}(f)$ can be computed for any function f agreeing with φ_y for those arguments and values. If and when this occurs, set $\psi(y) = \mathbf{F}(f)$. Then ψ gives the desired total effective operation on $\mathfrak{R}$.

$\Rightarrow$: Let ψ be a partial recursive function which determines a total effective operation on $\mathfrak{R}$. Let a particular set of instructions for ψ be given. We introduce the following notation for the remainder of this proof. If $\tilde{f}$ is a finite initial segment, then $\tilde{f}^0$ is the function which coincides with $\tilde{f}$ on *domain* $\tilde{f}$ and is zero elsewhere. Note that a Gödel number for $\tilde{f}^0$ can be obtained uniformly from an explicit description of $\tilde{f}$, i.e., from a canonical index for $\tau(\tilde{f})$. If φ is a partial function, $\varphi^{[w]}$ is the (finite) restriction of φ to arguments in $\{0,1, \ldots ,w\}$.

Let y and z be fixed. We give instructions for a partial recursive function η. The instructions for computing $\eta(x)$ are as follows.

Compute $\psi(y)$. If $\psi(y)$ is divergent, then $\eta(x)$ is divergent. If $\psi(y)$ converges, then *see whether* $\psi(z)$ converges (when computed from the given instructions for ψ) in fewer than x steps. *If not*, set $\eta(x) = \varphi_y(x)$. *If so, see whether* $\psi(z) = \psi(y)$; *if not* set $\eta(x) = \varphi_y(x)$. *If so*, let w be the number of steps in which $\psi(z)$ converges ($w < x$); compute values of φ_y until $\varphi_y(u)$ is defined for all $u \leq w$ (if this does not occur, then $\eta(x)$ is undefined). If and when this occurs, let f_w be the finite initial segment determined by $\varphi_y(0), \ldots, \varphi_y(w)$. Examine, in turn, all finite extensions $\tilde{f} \supset f_w$. For each such extension $\tilde{f}$, get a Gödel number $v_{\tilde{f}}$ for $\tilde{f}^0$, and see whether $\psi(v_{\tilde{f}}) \neq \psi(y)$. If no such extension $\tilde{f}$ (with $\psi(v_{\tilde{f}}) \neq \psi(y)$) is found, $\eta(x)$ is divergent; if such an extension $\tilde{f}$ is found, set $\eta(x) = \tilde{f}^0(x)$.

Summarizing, we have

$$\eta = \begin{cases} \text{everywhere divergent}, & \text{if } \psi(y) \text{ is divergent;} \\ \varphi_y, & \text{if } \psi(y) \text{ is convergent and } \psi(z) \text{ is divergent;} \\ \varphi_y, & \text{if } \psi(y) \text{ is convergent, } \psi(z) \text{ is convergent, and } \psi(y) \neq \psi(z); \\ \tilde{f}^0, & \text{if } \psi(y) \text{ is convergent, } \psi(z) \text{ is convergent in exactly } w \text{ steps, } \psi(y) = \psi(z), \\ & \{0, \ldots ,w\} \subset \textit{domain } \varphi_y, \text{ and } \tilde{f} \text{ is an extension of } \varphi_y^{[w]} \text{ such that } \mathbf{F}(\tilde{f}^0) \neq \psi(y); \\ \varphi_y^{[w]}, & \text{otherwise } (w \text{ is the number of steps in which } \psi(z) \text{ converges).} \end{cases}$$

The instructions for η depend uniformly on y and z. Hence, applying the recursion theorem, we obtain an n such that

$$\varphi_n = \begin{cases} \text{everywhere divergent}, & \text{if } \psi(y) \text{ is divergent;} \\ \varphi_y, & \text{if } \psi(y) \text{ is convergent and } \psi(n) \text{ is either divergent or } \neq \psi(y); \\ \tilde{f}^0, & \text{if } \psi(y) \text{ is convergent, } \psi(n) \text{ is convergent in } w \text{ steps, } \psi(y) = \psi(n), \{0, \ldots ,w\} \subset \\ & \textit{domain } \varphi_y, \text{ and } \tilde{f} \text{ is an extension of } \varphi_y^{[w]} \text{ such that } \mathbf{F}(\tilde{f}^0) \neq \psi(y); \\ \varphi_y^{[w]}, & \text{otherwise;} \end{cases}$$

where n depends uniformly on y.

Now if φ_y is total, then $\psi(y)$ is defined. Hence $\psi(n)$ must be defined, $\psi(n) = \psi(y)$, and there can be no extension $\tilde{f}$ of $\varphi_y^{[w]}$ such that $\mathbf{F}(\tilde{f}^0) \neq \psi(y)$ (otherwise φ_n would be $\tilde{f}^0$, and we would have $\psi(n) \neq \psi(y)$.) Hence, given any total φ_y, $\psi(n(y))$ must be defined, and if w is the number of steps necessary to compute $\psi(n(y))$, then every extension of $\varphi_y^{[w]}$ of the form $\tilde{f}^0$ must be mapped into $\psi(y)$ by $\mathbf{F}$.

We now give instructions for computing a recursive functional $\mathbf{G}$. Given any input f, look for a y and a w such that (i) $\psi(y)$ is convergent; (ii) $\psi(n(y))$ is convergent in w steps; (iii) φ_y is defined on $\{0, \ldots, w\}$; and (iv) f coincides with φ_y on $\{0, \ldots, w\}$. If and when such a y and w are found, take a Gödel number u for $(f^{[w]})^0$ and compute $\psi(u)$. Set $\mathbf{G}(f) = \psi(u)$. If no such y and w are found, $\mathbf{G}(f)$ is undefined. The reader can verify that $\mathbf{G}$ is well-defined and that $\mathbf{G}$ coincides with $\mathbf{F}$ on $\mathcal{R}$ (see Exercise 15-31). ☒

Comment. The above proof shows, in effect, that $\mathbf{F}$ is continuous on $\mathcal{R}$ and that, for any recursive function φ_y, a *modulus of continuity* w can be computed from y. (We define w to be a modulus of continuity for $\mathbf{F}$ at f if $\mathbf{F}(g) = \mathbf{F}(f)$ for all g in the domain of $\mathbf{F}$ such that $g \supset f^{[w]}$.)

We thus see that the class of total effective operations on $\mathcal{R}$ coincides (on $\mathcal{R}$) with the class of recursive functionals that are total on $\mathcal{R}$. Two other, more general, classes of functionals on $\mathcal{R}$ have been studied. We define them and discuss them briefly.

Notation Let $\tilde{f}$ be a finite partial function, then $[\tilde{f}]$ denotes the canonical index of $\tau(\tilde{f})$, and (as in the proof of Theorem XXXII) $\tilde{f}^0$ is the function which coincides with $\tilde{f}$ on *domain* $\tilde{f}$ and is zero elsewhere. Let f be a function, then (as in the proof of Theorem XXXII) $f^{[x]}$ denotes the restriction of f to arguments in $\{0, \ldots, x\}$.

Definition $\mathbf{F}$ is a *total functional* on $\mathcal{R}$ if $\mathcal{R} \subset$ *domain* $\mathbf{F}$. Let $\mathbf{F}$ be a total functional on $\mathcal{R}$. $\mathbf{F}$ is a *limit functional* if there is a partial recursive ψ such that (i) $\psi([\tilde{f}])$ is defined for all finite initial segments $\tilde{f}$; and (ii) $\lim_{x \to \infty} \psi([f^{[x]}])$ exists and equals $\mathbf{F}(f)$, for every recursive f.

Definition Let $\mathbf{F}$ be a total functional on $\mathcal{R}$. $\mathbf{F}$ is a *Banach-Mazur functional* if, for every recursive function f of two variables, there is a recursive function g such that for all x, $\mathbf{F}(\lambda y[f(x,y)]) = g(x)$.

We show below that the limit functionals contain the Banach-Mazur functionals and that the Banach-Mazur functionals contain (on $\mathcal{R}$) the recursive functionals total on $\mathcal{R}$; and we show that these containments are proper.

Theorem XXXIII *The restriction to $\mathcal{R}$ of any recursive functional total on $\mathcal{R}$ is a Banach-Mazur functional.*

Proof. Let $\mathbf{F}$ be a recursive functional total on $\mathcal{R}$. By Theorem XXXII, $\mathbf{F}$ is a total effective operation. The result is immediate by the *s-m-n* theorem. ☒

***Theorem XXXIV* (Mazur, Kreisel, Pour-El)** *Every Banach-Mazur functional is a limit functional.*

Proof. We first prove the following lemma.

Lemma *Let* $\mathbf{F}$ *be a Banach-Mazur functional. Then if f is recursive,* $\mathbf{F}(f) = \lim_{x\to\infty} \mathbf{F}((f^{[x]})^0)$.

Proof of lemma. Let j be defined as follows:

$$j(x,y) = \begin{cases} \tilde{f}^0(y), & \text{if } x = [\tilde{f}] \text{ for some finite initial segment } \tilde{f}; \\ 0, & \text{otherwise.} \end{cases}$$

Since $\mathbf{F}$ is a Banach-Mazur functional, there must be a recursive g such that $\mathbf{F}(\lambda y[j(x,y)]) = g(x)$. Hence, for any initial segment $\tilde{f}$, $\mathbf{F}(\tilde{f}^0) = g([\tilde{f}])$. Assume, for some recursive f, that $\mathbf{F}(f) \neq \lim_{x\to\infty} \mathbf{F}((f^{[x]})^0)$. For every z, define h_z as follows:

$$h_z(y) = \begin{cases} f(y), & \text{if } \varphi_z(z) \text{ does not converge in fewer than } y \text{ steps;} \\ (f^{[x]})^0(y), & \text{if } \varphi_z(z) \text{ converges in } w \text{ steps, where } w < y \text{ and} \\ & \quad x = \mu u[w \leq u \;\&\; g([f^{[u]}]) \neq \mathbf{F}(f)]. \end{cases}$$

$\lambda zy[h_z(y)]$ is evidently a recursive function of two variables. Since $\mathbf{F}$ is Banach-Mazur, there is a recursive k such that $\mathbf{F}(\lambda y[h_z(y)]) = k(z)$. But, for any z, $k(z) = \mathbf{F}(f) \Leftrightarrow z \notin K$, and hence we have a solution to the halting problem. This is a contradiction, and hence the lemma is proved.

Using g from the proof of the lemma, we have the theorem. ☒.

***Theorem XXXV* (Pour-El [*1960*])** *There exists a limit functional which is not a Banach-Mazur functional.*

Proof. Define ψ as follows:

$$\psi([\tilde{f}]) = \begin{cases} 1, & \text{if } \varphi_z(z) \text{ is convergent in fewer than } w \text{ steps, where } z = \tilde{f}(0) \\ & \quad \text{and } w \text{ is the length of } \tilde{f}; \\ 0, & \text{otherwise.} \end{cases}$$

ψ determines a limit functional $\mathbf{F}$. Assume $\mathbf{F}$ is a Banach-Mazur functional. Let $h(z,y) = z$ for all z and y. Then there must be a recursive k such that for all z, $\mathbf{F}(\lambda y[h(z,y)]) = k(z)$. But k must be the characteristic function of K, and this is a contradiction. ☒

***Theorem XXXVI* (Friedberg [*1958*])** *There exists a Banach-Mazur functional which does not coincide (on* $\mathcal{R}$*) with a recursive functional total on* $\mathcal{R}$.

Proof. The proof uses a computation of degree of unsolvability. We omit details. In general outline, the proof is as follows. Consider partial recursive functions which define limit functionals. Consider, in particular,

$A_1 = \{z|\varphi_z$ *defines a limit functional which is Banach-Mazur*$\}$ and
$A_2 = \{z|\varphi_z$ *defines a limit functional which coincides* (*on* $\mathfrak{R}$) *with a recursive functional that is total on* $\mathfrak{R}\}$.

Friedberg uses the methods of §14.8 to show that A_2 is Σ_4-complete; while a Tarski-Kuratowski computation shows that A_1 is in Π_4 (see Exercise 15-33). By the hierarchy theorem (14-VIII) the result follows. (The computation for A_2 is more complex than for any of the examples given in §14.8).☒

Relativization

In a straightforward manner, functionals can be relativized with respect to a given set or family of sets. In this way one can develop, for example, a theory of arithmetical functionals. Generalizations to higher types (such as mappings that take functionals themselves as arguments) have also been studied. The notion of effective operation can be applied to functionals of higher type. We do not consider these matters further.

Applications of Functionals

Functionals have been an important tool in foundational investigations of classical mathematics. One area of application has been the study of effective objects within classical mathematics. For example, if the reals are identified with sequences of nested intervals with rational end-points, then a mapping from reals to reals becomes a functional operator (and an integral becomes a functional of higher type). If we restrict our attention to recursive operators, then a corresponding theory of "recursive analysis" can be developed. (See Exercise 15-35, where we show that every such recursive operator gives a continuous mapping from the reals to the reals.)

A second area of application has been the study of denumerable structures *analogous* to the nondenumerable structures of classical mathematics, with mappings on these denumerable structures given by effective operations. For example, as "recursive reals" we can take those rational nested-interval sequences which are recursive, and as "functions" we can take effective operators on indices (of these "recursive reals"). (See Exercise 15-36; discontinuous "functions" now appear.) (An integral now becomes an effective operation on indices of "functions.")

A third area of application has been the study of the "effective content" of certain statements and proofs of classical mathematics. We now describe an application of this kind due to Kreisel [1951].

The No-counterexample Interpretation

Let any sentence of elementary arithmetic be given in prenex form. This sentence can be transformed into an equivalent assertion about recursive functionals as follows. (We illustrate for an $\exists\forall\exists\forall$ prefix; however, the method is general.)

$(\exists x)(\forall y)(\exists z)(\forall w)R(x,y,z,w)$ is true $\Leftrightarrow$
$(\forall x)(\exists y)(\forall z)(\exists w) \neg R(x,y,z,w)$ is false $\Leftrightarrow$
$(\exists f^{(1)})(\exists g^{(2)})(\forall x)(\forall z) \neg R(x,f(x),z,g(x,z))$ is false $\Leftrightarrow$
$(\forall f^{(1)})(\forall g^{(2)})(\exists x)(\exists z)R(x,f(x),z,g(x,z))$ is true $\Leftrightarrow$
$(\exists \mathbf{F})(\exists \mathbf{G})(\forall f^{(1)})(\forall g^{(2)})R(\mathbf{F}(f),f(\mathbf{F}(f)),\mathbf{G}(f,g),g(\mathbf{F}(f),\mathbf{G}(f,g)))$ is true $\Leftrightarrow$
$(\exists$ recursive $\mathbf{F})(\exists$ recursive $\mathbf{G})(\forall f^{(1)})(\forall g^{(2)})$
$R(\mathbf{F}(f),f(\mathbf{F}(f)),\mathbf{G}(f,g),g(\mathbf{F}(f),\mathbf{G}(f,g)))$ is true.

The last step holds since, in the third from last statement, for any f and g, the values of x and z (i.e., the values of $\mathbf{F}(f)$ and $\mathbf{G}(f,g)$) can be found by testing all pairs of values for x and z, using oracles for f and g; hence $\mathbf{F}$ and $\mathbf{G}$ must be recursive functionals.†

The logical transformations indicated are called the *no-counterexample interpretation* since, in the example given, f and g may be thought of as possible counterexamples to the original assertion. The recursive functionals $\mathbf{F}$ and $\mathbf{G}$ compute, from any alleged counterexamples f and g, an instance for which f and g fail to be counterexamples.

Kreisel has shown that the proof of a sentence under certain standard axiomatizations, e.g., Peano's axioms, yields additional information about the corresponding no-counterexample functionals. Indeed, provability of a sentence under certain axiomatizations can be characterized in terms of corresponding properties of the functionals. As might be expected, more elementary axiomatizations correspond to more "effective" functionals (in a sense analogous to that in which a primitive recursive function is more "effective" than a recursive function which is not primitive recursive).

§15.4 EXERCISES

§15.1

15-1. Justify (on the basis of the informal discussion) the formal definition for *partial recursive function of k set variables.*

15-2. Formulate and prove analogues to Theorems 1-IX and 1-X for partial recursive functions of one set variable.

15-3. Prove the following characterizations of tt-reducibility and T-reducibility.
(i) $A \leq_{tt} B \Leftrightarrow A = \{x|R(B,x)\}$ for some $R \in \Sigma_0^{(s)}$.
(ii) $A \leq_{T} B \Leftrightarrow A = \{x|R(B,x)\} = \{x|S(B,x)\}$ for some $R \in \Sigma_1^{(s)}$ and $S \in \Pi_1^{(s)}$.
(*Hint:* Show A recursively enumerable in $B \Leftrightarrow A = \{x|R(B,x)\}$ for some $R \in \Sigma_1^{(s)}$.)

† In case there are initial universal quantifiers, these are not replaced by function quantifiers at the third step, but remain as number quantifiers (a number may be thought of as a "function of zero variables"). In this case, the functionals that appear at the next-to-last step will have numbers as well as functions as arguments.

15-4. (Jockusch). For any class $\mathcal{C}$ of sets, define $P_{\mathcal{C}} = \{x | W_x \in \mathcal{C}\}$. Show the following for any $\mathcal{C}$.

(i) $\mathcal{C} \in \Sigma_n^{(s)} \Rightarrow P_{\mathcal{C}} \in \Sigma_{n+1}$, $n \geq 1$.

(ii) $\mathcal{C} \in \Pi_n^{(s)} \Rightarrow P_{\mathcal{C}} \in \Pi_{n+1}$, $n \geq 1$.

(iii) $\mathcal{C} \in \Sigma_0^{(s)} \Rightarrow P_{\mathcal{C}} \in \Sigma_2 \cap \Pi_2$.

Show the following for any $\mathcal{C}$ all of whose members are recursively enumerable.

(iv) $P_{\mathcal{C}} \in \Sigma_n \Rightarrow \mathcal{C} \in \Sigma_n^{(s)}$, $n \geq 3$.

(v) $P_{\mathcal{C}} \in \Pi_n \Rightarrow \mathcal{C} \in \Sigma_{n+1}^{(s)}$, $n \geq 2$.

Give examples showing that (iii) and (iv) are best possible. (*Hint:* For (iii), take $\mathcal{C} = \{X | 0 \in X \;\&\; 1 \notin X\}$; for (iv), use Theorem XIII.)

15-5. (*a*) In the proof of Theorem XII, justify in more detail the construction leading to $f(n + 1)$.

(*b*) Prove Corollary XIII.

15-6. Let A be a nonrecursive set. Show that $\Sigma_0^{(s)} = \Sigma_0^{(s)A}$ but that there is a class $\mathcal{A} \in (\Sigma_1^{(s)A} - \Sigma_1^{(s)})$.

15-7. Let A be given. Prove a hierarchy theorem for the $\Sigma_n^{(s)A}$, $\Pi_n^{(s)A}$ hierarchy.

15-8. (*a*) Consider classes that are both open and closed. Introduce a notion of canonical index for such classes, and call these indices *0-indices*. Let $\mathcal{D}_x$ be the class with 0-index x. Show that $\mathcal{A} \in \Sigma_1^{(s)} \Leftrightarrow (\exists \text{ recursive } f)\, [\mathcal{A} = \bigcup_{x=0}^{\infty} \mathcal{D}_{f(x)}]$. Similarly for $\Pi_1^{(s)}$.

(*b*) Define: z is a *1-index* for $\mathcal{A}$ if φ_z is total and $\mathcal{A} = \bigcup_{x=0}^{\infty} \mathcal{D}_{f(x)}$, where $f = \varphi_z$. Let $\mathcal{G}_z$ be the $\Sigma_1^{(s)}$ class with 1-index z. Show that $\mathcal{A} \in \Pi_2^{(s)} \Leftrightarrow (\exists \text{ recursive } f)[\mathcal{A} = \bigcap_{x=0}^{\infty} \mathcal{G}_{f(x)}]$.

(*c*) Extend the above to all levels of the arithmetical hierarchy.

△***15-9.*** (Hilbert, Bernays). (*a*) Define V_n as in §14.6. Let $V^* = \bigcup_{n=0}^{\infty} V_n$. Formulate an inductive definition of V_n, $n = 0, 1, 2, \ldots$, and from this conclude that $\{V^*\} \in \Pi_2^{(s)}$. (*Hint:* Observe that V_0 is recursive and that for any n, $V_{n+1} = \{x | x$ *is a* Σ_{n+1}-*sentence and some substitution instance of the "*Π_n*-part" of* x *is true*$\}$.

(*b*) Deduce directly from the above that $\{V\} \in \Pi_2^{(s)}$.†

§15.2

15-10. Show that $\{f | 5 \in \textit{range } f\}$ is the domain of a partial recursive function of one function variable.

15-11. Show that d, as defined in §15.2, is a metric; and that $\mathcal{F}$ is complete under this metric.

15-12. Show that the product topology defined for $\mathcal{F}$ as in §15.2 is the same as the topology yielded by the Baire metric.

△***15-13.*** For any real number β, let $[\beta]$ = the greatest integer $\leq \beta$. Let α be a real number between 0 and 1. Define two sequences, $n_0, n_1, n_2, \ldots$, and $r_0, r_1, \ldots$, as follows:

$$r_0 = \alpha;$$
$$n_i = \left[\frac{1}{r_i}\right], \quad \text{if } r_i \neq 0;$$
$$r_{i+1} = \frac{1}{r_i} - n_i.$$

† The author is indebted to Hao Wang for calling this proof of Theorem XII to his attention.

The (possibly finite) sequence $n_0, n_1, \ldots$ is called the *continued fraction expansion* of α, since the sequence

$$\alpha_0 = \frac{1}{n_0}, \qquad \alpha_1 = \frac{1}{n_0 + \frac{1}{n_1}}, \qquad \alpha_2 = \frac{1}{n_0 + \frac{1}{n_1 + \frac{1}{n_2}}}, \qquad \ldots$$

either terminates with α (if the expansion is finite) or has α as its limit. Note that if the expansion is infinite, all members are positive.

(*a*) Show that every real α, such that $0 < \alpha < 1$, has an expansion.

(*b*) Show that the expansion is infinite if and only if α is irrational.

(*c*) Show that every infinite sequence of positive integers occurs as the expansion for some irrational between 0 and 1.

(*d*) Identify every member f of $\mathfrak{F}$ with the expansion $f(0) + 1$, $f(1) + 1$, ... By (*a*) and (*c*) this yields a one-one correspondence between $\mathfrak{F}$ and the irrationals between 0 and 1. Show that every basic neighborhood in $\mathfrak{F}$ determines an open set of irrationals (in the natural induced topology on the irrationals) and that every open interval of irrationals determines an open class in $\mathfrak{F}$. Hence conclude that the correspondence defined between $\mathfrak{F}$ and the irrationals is a homeomorphism.

15-14. Explain why the basic neighborhoods form a *basis* of neighborhoods for $\mathfrak{F}$.

15-15. Prove that $\mathfrak{F}$ is not compact. (*Hint:* Give a covering of $\mathfrak{F}$ by basic neighborhoods that has no finite subcovering.)

15-16. Prove or disprove the following assertion. $\mathfrak{J}$ is a finite-path tree if and only if $\mathfrak{J}$ is a union of pairwise incomparable finite subbranches of $\mathfrak{J}^*$. (Two subbranches are incomparable if neither is contained in the other.)

15-17. Use a category argument to prove that $\{f|f^{-1}(0) \textit{ is finite}\} \in (\Sigma_2^{(\mathrm{fn})} - \Pi_2^{(\mathrm{fn})})$.

15-18. Prove Theorem XXVI.

$\triangle$*15-19.* Consider all sets of integers obtainable from recursive relations in $\mathfrak{N}^k \times N^l$ (for all $k \geq 0$, $l \geq 0$) by operations of complementation and projection where projection on set coordinates as well as on number coordinates is permitted. Let $\mathfrak{A}_1$ be the class of all such sets. Consider all sets of integers obtainable from recursive relations in $\mathfrak{F}^k \times N^l$ (for all $k \geq 0$, $l \geq 0$) by operations of complementation and projection, where projection on function coordinates as well as on number coordinates is permitted. Let $\mathfrak{A}_2$ be the class of all such sets. Show that $\mathfrak{A}_1 = \mathfrak{A}_2$. (*Hint:* See the proofs of Theorems XXV and XXVI.) The sets in this class are called the *analytical* sets. These sets are studied further in Chapter 16.

15-20. Show that for appropriate A, $\Sigma_0^{(\mathrm{fn})A}$ will contain classes not in $\Sigma_0^{(\mathrm{fn})}$. (*Hint:* Show that for every open and closed class $\mathfrak{A}$, there is a set A such that $\mathfrak{A} \in \Sigma_0^{(\mathrm{fn})A}$. Take A so that it encodes appropriate information about the finite-path tree determining $\mathfrak{A}$.)

15-21. Show that $\mathfrak{A} \in \Sigma_0^{(\mathrm{fn})}[\mathfrak{N}] \Leftrightarrow \mathfrak{A}$ is both open and closed. (*Hint:* See hint for 15-20.)

$\triangle$*15-22.* Show that the assertions in Theorems XXV and XXVI are best possible.

15-23. If $\mathfrak{A} \in \Sigma_0^{(\mathrm{fn})}$, must there exist a $\mathfrak{B} \in \Sigma_0^{(\mathrm{s})}$ such that $\mathfrak{A} = \{f|\tau(f) \in \mathfrak{B}\}$?

15-24. Adapt the proof of Theorem XIII to show that $\{f|f \textit{ is recursive}\} \in (\Sigma_3^{(\mathrm{fn})} - \Pi_3^{(\mathrm{fn})})$.

$\triangle$*15-25.* Define a partial function ψ to be *arithmetical* if $\tau(\psi)$ is arithmetical. Give an appropriate *indexing* for the arithmetical partial functions. Develop a theory (based on arithmetical partial functions) analogous to the theory (based on partial recursive functions) developed in Chapters 1, 4, and 5. Pay special attention to divergences between the two theories. (The notion of arithmetical computability is a natural generalization of the notion of recursive computability. We consider still a further generalization in Chapter 16.)

§15.3

15-26. Let **F** be a partial recursive functional on $\mathcal{P}$. (*a*) Show that **F** is a recursive functional on $\mathcal{P}$ if and only if $\mathcal{F} \subset$ weak domain of **F**.

(*b*) Show that if $\mathcal{P}$ = strong domain of **F**, then **F** is a constant.

15-27. (*a*) Complete the proof of part (*b*) of Theorem XXIX by showing that failure of either (i) or (ii) gives solvability of the halting problem.

(*b*) Deduce the theorem of Exercise 11-43 from Theorem XXIX.

15-28. Show that Φ, in the proof of Theorem XXX, gives the functional desired.

$\triangle$***15-29.*** (*a*) Find a partial recursive functional on $\mathcal{P}$, call it **F**, and a partial function of one function variable, call it ψ, such that (i) $\psi(f) = \mathbf{F}(f)$ if $\mathbf{F}(f)$ is strongly defined at f; (ii) $\psi(f)$ is divergent, otherwise; and (iii) ψ is not a partial recursive function of one function variable. (*Hint:* Define **F** by an enumeration operator Φ such that

$$\Phi(\tau(f)) = \begin{cases} \{0\}, & \text{if } f(0) \notin K; \\ \{0,1\}, & \text{if } f(0) \in K.) \end{cases}$$

(*b*) Let **F** be a partial recursive functional on $\mathcal{P}$. Let ψ be obtained from **F** by (i) and (ii) in (*a*). Show that ψ can be extended to a partial recursive function of one function variable. (*Hint:* See Theorem 9-XXIII.)

15-30. (*a*) Give a partial recursive function of one function variable which is total on $\mathcal{R}$ but not on $\mathcal{F}$. (*Hint:* Define $\psi(f) = 0$, if $(\exists x)[f(x) = \varphi_x(x)]$; $\psi(f)$ divergent otherwise.)

$\triangle$(*b*) Define a partial recursive function of one function variable which is total on $\mathcal{R}$ but cannot be extended to a total function on $\mathcal{F}$. (*Hint:* Define $\psi(f)$ = the first x in an effective enumeration of K such that $f(x) = \varphi_x(x)$, if such an x exists; $\psi(f)$ divergent, otherwise.)

(*c*) Deduce that not every total effective operation on $\mathcal{R}$ is the restriction to $\mathcal{R}$ of a general recursive functional.

15-31. In the proof of Theorem XXXII, verify that **G** is well-defined (i.e., at most one value $\psi(u)$ can be found), and that **G** coincides with **F** on $\mathcal{R}$.

15-32. (Kreisel, Lacombe, Shoenfield). Let R be the set of Gödel numbers of total recursive functions. Let $\mathcal{A}$ be a class of recursive functions. Define $P_{\mathcal{A}} = \{x|\varphi_x \in \mathcal{A}\}$. We say (for the purposes of this problem) that $\mathcal{A}$ is *recursive* if there are disjoint recursively enumerable sets A_1 and A_2 such that $R \subset A_1 \cup A_2$ and $P_{\mathcal{A}} = A_1 \cap R$, and that $\mathcal{A}$ is *recursively enumerable* if there is a recursively enumerable set A such that $P_{\mathcal{A}} = A \cap R$.

(*a*) Show that $\mathcal{A}$ is recursive if and only if there exist disjoint recursively enumerable sets B and C such that

$$f \in \mathcal{A} \Leftrightarrow (\exists \text{ initial segment } \tilde{f})[f \supset \tilde{f} \ \&\ [\tilde{f}] \in B],$$

and

$$f \notin \mathcal{A} \Leftrightarrow (\exists \text{ initial segment } \tilde{f})[f \supset \tilde{f} \ \&\ [\tilde{f}] \in C],$$

where $[\tilde{f}]$ is the canonical index of $\tau[\tilde{f}]$. (*Hint:* $\Leftarrow$ is immediate. For $\Rightarrow$, use Theorem XXXII to get from an effective operation to a recursive functional.)

(*b*) Show that if $\mathcal{A}$ is recursively enumerable, it need not follow that there is a recursively enumerable set B such that

$$f \in \mathcal{A} \Leftrightarrow (\exists \text{ initial segment } \tilde{f})[f \supset \tilde{f} \ \&\ [\tilde{f}] \in B].$$

(*Hint:* Use the construction from Theorem XXXI.)

15-33. Let A_1 and A_2 be as in the outlined proof of Theorem XXXVI.

(*a*) Show that $A_1 \in \Pi_4$.

(*b*) Show that $A_2 \in \Sigma_4$.

$\blacktriangle$(*c*) Show that A_2 is Σ_4-complete. (*Hint:* Use Theorem 14-XVIII.)

△*15-34.* Identify the real numbers with a subclass of $\mathfrak{F}$ as follows. Choose a fixed effective coding of pairs of rationals onto N. Let f be a function such that $f(0), f(1), \ldots$ gives (code numbers for) a convergent nested sequence of these rational intervals. If α is the (real number) limit of this sequence, we say that f *represents* α. Clearly every real is represented by an infinite number of functions. A real is called *recursive* if it is represented by a recursive function. z is called an *index* for the recursive real α if φ_z represents α.

(*a*) Show that a real is recursive if and only if its decimal expansion can be given effectively.

(*b*) Show that the operations of addition and multiplication on recursive reals can be given by effective operations on indices of recursive reals.

(*c*) (Rice [1954]). Define the *recursive complex numbers* to be those complex numbers whose real and imaginary parts are recursive reals. Show that the recursive complex numbers constitute an algebraically closed field. (*Hint:* Use an appropriate effective approximation procedure for "solving" any given polynomial equation.)

(*d*) Show that one can go uniformly from (instructions for) decimal expansions to (indices for) nested interval representations for recursive reals. Show, however, that one cannot go uniformly in the converse direction. (Thus certain effective mappings into the recursive reals can be defined if we use the more inclusive nested interval representation but not if we use decimal expansions. It is for this reason that the effective decimal representation is not taken as basic.)

△*15-35.* Let $\mathfrak{F}^*$ consist of those functions which represent reals (see Exercise 15-34). Let Φ be a recursive operator which maps $\mathfrak{F}^*$ into $\mathfrak{F}^*$ and determines a corresponding well-defined mapping from the reals into the reals. Show that Φ determines a continuous mapping from the reals into the reals. (*Hint:* Represent Φ as a functional on $\mathfrak{F}^* \times N$, and use the continuity of that functional on its domain.)

△*15-36.* (Moschovakis). Let R^* be the set of all indices for recursive reals (as defined in 15-34). Let ψ be partial recursive function which maps R^* into R^* and determines a corresponding well-defined mapping from the recursive reals into the recursive reals. Show that this mapping is continuous at every recursive real. (*Hint:* See Theorem XXXII and Exercise 15-35.) Show that this mapping need not be the restriction to the recursive reals of a continuous mapping on the reals. (*Hint:* Define an effective sequence of open intervals with rational end-points which covers all recursive reals yet has total length ≤ 1. Do this by computing, for each z, the values $\varphi_z(0), \varphi_z(1), \ldots$. If and when, for any z, we find an n such that (i) $\varphi_z(y)$ is defined for $y \leq n$, (ii) $\varphi_z(0), \ldots, \varphi_z(n)$ give (code numbers for) a finite nested sequence of rational intervals, and (iii) $\varphi_z(n)$ gives an interval of length $<1/2^{z+1}$; then add to the main list of open intervals an open interval of length $1/2^z$ containing the interval given by $\varphi_z(n)$. The main list, so generated, constitutes the desired open covering. Let $I_0, I_1, \ldots$ be the intervals in this covering in effectively given order. For each k, define h_k to be a continuous function of one real variable which $= 0$ on $I_0 \cup I_1 \cup \cdots \cup I_k$, and $= 1$ elsewhere except for straight-line interpolation on a recursively given set of total length $1/2^k$. For any recursive real α, define $g(\alpha) = \sum_{k=0}^{\infty} h_k(\alpha)$. Show that g is finite (for every recursive real); that g can be given by a partial recursive function ψ (of the kind described at the beginning of the problem), and that g is not the restriction to the recursive reals of a continuous mapping on the reals. To do the last, show that g is unbounded in any interval of length 3. (This shows, incidentally, that the Heine-Borel theorem fails for the recursive reals on a closed interval with a recursively given infinite open covering.)) For further results in "recursive analysis" see Moschovakis [1963] and Klaua [1961].

△*15-37.* (Hodes [1962]). The notion of recursive real given in Exercise 15-34 can be generalized to *arithmetical real* if arithmetical partial functions are used in place of partial recursive functions (see Exercise 15-25). Show that every arithmetical open covering (i.e., arithmetical sequence of open intervals with rational end points) of a closed interval of arithmetical reals has a finite subcovering. (*Hint:* Show that an arithmetical version of the Bolzano-Weierstrass theorem holds. (A recursive version fails; see hint for 15-36).) (*Comment:* The Heine-Borel theorem can be formulated, and proved, in a somewhat more general form for the arithmetical reals. The theory of "arithmetical analysis" is more similar, in this and other respects, to classical analysis than is the theory of "recursive analysis" of 15-35 or 15-36. See also Ritter [1962].)

15-38. Give the no-counterexample interpretation for the prime-number theorem (see §14.7) and describe how to compute the recursive functional in question.

16 *The Analytical Hierarchy*

§16.1 THE ANALYTICAL HIERARCHY

In §15.2, we considered relations obtained from recursive relations in $\mathfrak{F}^k \times N^l$ (for all $k \geq 0, l \geq 0$) by operations of complementation and number-coordinate projection. In the present chapter, we consider the larger class of relations obtained by allowing projection on function coordinates as well. These relations are called *analytical*. They fall into a natural hierarchy which was first studied by Kleene [1955, and 1955*a*].

The Hierarchy

Recursive relations in $\mathfrak{F}^k \times N^l$(for all $k \geq 0, l \geq 0$) were defined in §15.2.

Definition A relation $R \subset \mathfrak{F}^k \times N^l$ is *analytical* if R is recursive or if there exists a recursive relation $S \subset \mathfrak{F}^m \times N^n$ such that R can be obtained from S by some finite sequence of complementation and/or projection operations.

A counterpart of Theorem 14-I is immediate.

Theorem I *$R \subset \mathfrak{F}^k \times N^l$ is analytical $\Leftrightarrow$ R is recursive or, for some m and n,*

$$R = \{<f_1, \ldots ,f_k,x_1, \ldots ,x_l>|(\mathrm{Q}_1\xi_1)(\mathrm{Q}_2\xi_2) \cdots (\mathrm{Q}_{m+n-k-l}\xi_{m+n-k-l})S(f_1, \ldots ,f_m,x_1, \ldots ,x_n)\},$$

where $S \subset \mathfrak{F}^m \times N^n$ is recursive, where Q_i is either $\forall$ or $\exists$ for $1 \leq i \leq$

$m + n - k - l$, and where $\xi_1, \xi_2, \ldots, \xi_{m+n-k-l}$ are $f_{k+1}, \ldots, f_m, x_{l+1}, \ldots, x_n$ in some order.†

Proof. As for Theorem 14-I. ☒

Definition A quantifier operation is of *type* 1 if it applies to a function coordinate. A quantifier operation is of *type* 0 if it applies to a number coordinate. For certain purposes, e.g., the Tarski-Kuratowski algorithm, we use the symbols "$\forall^1$" and "$\exists^1$" for type-1 quantifier operations and the symbols "$\forall^0$" and "$\exists^0$" for type-0 quantifier operations.‡

Predicate form is defined as in §14.1 (except that function coordinates and quantifiers of type 1 are now allowed). Henceforth "predicate form" will be used in this new generalized sense, rather than in the sense of §14.1. A predicate form determines, in the obvious way, a corresponding analytical relation. We say that this relation is *expressed* by the given form. Two predicate forms are equivalent if they express the same relation.

Definition In a predicate form, the (possibly empty) sequence of quantifier operations, indexed by type, is called the *prefix* of the form. (Thus, for example, $(\forall x_2)(\exists f_1)(\exists x_1)(\forall f_2)R(f_1,f_2,x_1,x_2,x_3)$ has the prefix $\forall^0\exists^1\exists^0\forall^1$). The *reduced prefix* is the (possibly empty) sequence of quantifier operations obtained by deleting all operations of type 0 from the prefix. (Thus, for example, a form with prefix $\forall^0\exists^1\exists^0\forall^1$ has reduced prefix $\exists^1\forall^1$.)

We obtain the analytical hierarchy by classifying forms according to the number of alternations in the reduced prefix.

Definition For $n > 0$, a *Σ_n^1-prefix* is a prefix whose reduced prefix begins with $\exists^1$ and has $n - 1$ alternations of quantifier.

A *Σ_0^1-prefix* is a prefix whose reduced prefix is empty.

For $n > 0$, a *Π_n^1-prefix* is a prefix whose reduced prefix begins with $\forall^1$ and has $n - 1$ alternations of quantifier.

A *Π_0^1-prefix* is a prefix whose reduced prefix is empty. (The concepts of Σ_0^1-prefix and Π_0^1-prefix therefore coincide.)

Definition A predicate form is a *Σ_n^1-form* if it has a Σ_n^1-prefix and a *Π_n^1-form* if it has a Π_n^1-prefix.

Definition Σ_n^1 = the class of all relations (in $\mathfrak{F}^k \times N^l$, $k \geq 0$, $l \geq 0$) expressible in Σ_n^1-form.

† To accord with previous usage in this book and to avoid confusion with our other notations (such as $\alpha, \beta, \ldots$ for ordinals) we continue to use $f, g, h, \ldots$ for functions.

‡ More generally, in the theory of hierarchies, integers are often called *objects of type* 0, sets and functions of integers are called *objects of type* 1, sets and functionals of objects of type 1 are called objects of type 2, etc. The functionals considered in §15.3, for example, are objects of type 2. The study of recursive relations and functionals for objects of type higher than 1 has been initiated by Kleene [1959, 1963] (see also Clarke [1964]).

Π_n^1 = the class of all relations (in $\mathfrak{F}^k \times N^l$, $k \geq 0$, $l \geq 0$) expressible in Π_n^1-form.

The class Σ_0^1 ($=\Pi_0^1$) thus consists of the arithmetical relations defined and studied in §15.2.

At the end of the present section, we shall extend this notation to relations in $\mathfrak{F}^k \times \mathfrak{N}^l \times N^m$, $k \geq 0$, $l \geq 0$, $m \geq 0$.

Corollary I *R ($\subset \mathfrak{F}^k \times N^l$) is analytical $\Leftrightarrow$ $(\exists n)[R \in \Sigma_n^1$ or $R \in \Pi_n^1$.]*

Proof. Immediate.☒

Using dummy quantifiers, as in Theorem 14-II, we have the following.

Theorem II(a) $\Sigma_n^1 \cup \Pi_n^1 \subset \Sigma_{n+1}^1 \cap \Pi_{n+1}^1$.

(*b*) $R \in \Sigma_n^1 \Leftrightarrow \bar{R} \in \Pi_n^1$.

Proof. As for Theorem 14-II.☒

The following theorem gives rules which generalize and extend the rule for contraction of quantifiers given in §14.2. It shows that for $n > 0$, any relation in $\Sigma_n^1 \cup \Pi_n^1$ can be expressed by a predicate form having at most n quantifiers of type 1 and at most one quantifier of type 0, where the quantifier of type 0 is the last quantifier in the prefix.

Theorem III (Kleene) *The following prefix transformations are permissible; i.e., in each case, for any predicate form with the given prefix, an equivalent predicate form with the new prefix can be obtained.*

$$
\begin{array}{lll}
\text{(i)} & \cdots \forall^0\forall^0 \cdots \rightarrow \cdots \forall^0 \cdots & \\
 & \cdots \exists^0\exists^0 \cdots \rightarrow \cdots \exists^0 \cdots & , \\
\text{(ii)} & \cdots \forall^1\forall^1 \cdots \rightarrow \cdots \forall^1 \cdots & \\
 & \cdots \exists^1\exists^1 \cdots \rightarrow \cdots \exists^1 \cdots & , \\
\text{(iii)} & \cdots \forall^0 \cdots \rightarrow \cdots \forall^1 \cdots & \\
 & \cdots \exists^0 \cdots \rightarrow \cdots \exists^1 \cdots & , \\
\text{(iv)} & \cdots \forall^0\exists^1 \cdots \rightarrow \cdots \exists^1\forall^0 \cdots & \\
 & \cdots \exists^0\forall^1 \cdots \rightarrow \cdots \forall^1\exists^0 \cdots & .
\end{array}
$$

(It will follow from the hierarchy theorem that converses to (iii) and (iv) fail; see Exercise 16-1).

Proof. We indicate the appropriate changes in the recursive relation of the given predicate form. The recursiveness of the new relation and the equivalence of the predicate forms is easily demonstrated (see Exercise 16-2).

(i) As in §14.2.

(ii) Given $\cdots (\forall f_1)(\forall f_2) \cdots R(—,f_1,—,f_2,—)$,
use $\{\langle f,———\rangle | R(—,\pi_1 f,—,\pi_2 f,—)\}$. Similarly for $\cdots (\exists f_1)(\exists f_2) \cdots$.

(iii) Given $\cdots (\forall x) \cdots R(—,x,—)$,
use $\{\langle f,——\rangle | R(—,f(0),—)\}$. Similarly for $\cdots (\exists x) \cdots$.

(iv) Given $\cdots (\forall x)(\exists f) \cdots R(—,f,—,x,—)$,
use $\{\langle f,x,———\rangle | R(—,\lambda y[f(\langle x,y\rangle)],—,x,—)\}$. Similarly for $\cdots (\exists x)(\forall f) \cdots$.☒

Corollary III *For $n > 0$, any relation R in Σ_n^1 can be expressed by a predicate form with a prefix consisting of $n + 1$ alternating quantifiers, of which the first n are of type 1 and the last is of type 0, and in which the initial quantifier is $\exists^1$. Similarly for Π_n^1 with the initial quantifier $\forall^1$.*

Proof. A systematic procedure (with use of dummy quantifiers if necessary) is evident from the theorem (see Exercise 16-3). We limit ourselves to an example. Let the prefix $\forall^0\exists^1\exists^0\forall^1$ be given. Then,

$$\begin{aligned} \forall^0\exists^1\exists^0\forall^1 &\to \forall^0\exists^1\exists^1\forall^1 && \text{by (iii)},\\ &\to \forall^0\exists^1\forall^1 && \text{by (ii)},\\ &\to \exists^1\forall^0\forall^1 && \text{by (iv)},\\ &\to \exists^1\forall^1\forall^1 && \text{by (iii)},\\ &\to \exists^1\forall^1 && \text{by (ii)},\\ &\to \exists^1\forall^1\exists^0 && \text{adding dummy.} \boxtimes \end{aligned}$$

A predicate form can be described by giving (i) the prefix, (ii) an index for the characteristic function of the recursive relation, and (iii) the mapping from prefix quantifiers into coordinates of the relation. Call this a *description* of the form. The quantifier rules of Theorem III are evidently uniform in the sense that a description of the result of applying a quantifier rule to a predicate form can be obtained uniformly from a description of the original form.

Normal Form and Enumeration

Define $T'_{k,l}$ as in §15.2. (Then, for $k > 0$ and $l > 0$, $T'_{k,l}(z,f_1, \ldots ,f_k,x_1, \ldots ,x_l,w) \Leftrightarrow \varphi_z^{\tau(f_1),\ldots,\tau(f_k)}(x_1, \ldots ,x_l)$ is convergent by the wth step in the enumeration of $W_{\rho(z)}$ (from $\rho(z)$).) By Theorem 15-XVII, for any predicate form $(\exists w)R(f_1, \ldots ,f_k,x_1, \ldots ,x_l,w)$, we can uniformly obtain a z such that $(\exists w)T'_{k,l}(z,f_1, \ldots ,f_k,x_1, \ldots ,x_l,w)$ expresses the same relation. This is a special case of the following theorem.

Theorem IV **(Kleene)** *Let n be even and > 0. Then for any $R \in \Sigma_n^1$ (where $R \subset \mathfrak{F}^k \times N^l$) there exists a z such that*

$$R = \{<f_1, \ldots ,f_k,x_1, \ldots ,x_l>|(\exists g_1)(\forall g_2) \cdots (\forall g_n)(\exists w)T'_{k+n,l}(z,f_1, \ldots ,f_k,g_1, \ldots ,g_n,x_1, \ldots ,x_l,w)\}.$$

Let n be odd. Then for any $R \in \Sigma_n^1$ (where $R \subset \mathfrak{F}^k \times N^l$) there exists a z such that

$$R = \{<f_1, \ldots ,f_k,x_1 \cdots x_l>|(\exists g_1)(\forall g_2) \cdots (\exists g_n)(\forall w) \neg T'_{k+n,l}(z,f_1, \ldots ,f_k,g_1, \ldots ,g_n,x_1, \ldots ,x_l,w)\}.$$

*In each case, z is called a Σ_n^1-*index *for R. Given any Σ_n^1-form for R, a Σ_n^1-index for R can be found uniformly (in the sense of Theorem 15-IV).*
Similarly for Π_n^1.

Proof. Immediate by Theorem 15-XVII and the proof of Corollary III. ☒

Let a fixed coding from finite sequences of integers (including the empty sequence) onto N be chosen. More specifically, take the coding to be that given by τ^* in §5.6. We introduce the following notation.

Definition

$$\bar{f}(x) = \begin{cases} \tau^*(<f(0),f(1), \ldots ,f(x-1)>), & \text{if } x > 0; \\ \tau^*(\emptyset) = 0, & \text{if } x = 0. \end{cases}$$

Thus for any function f, $\bar{f}(x)$ is the code number for the finite initial segment of f of length x. $\bar{f}(x)$ is called the *sequence number of length x determined by f.*†

An especially useful normal form, based on recursive relations over integers (rather than over both functions and integers) can be obtained as follows.

Definition For $k > 0$ and $l \geq 0$, define

$T^*_{k,l} = \{<z,y_1, \ldots ,y_k,x_1, \ldots ,x_l> \mid$ *there exist* $f_1, \ldots ,f_k$ *and there exists a* w *such that for all* i, $1 \leq i \leq k$, $y_i = \bar{f}_i(w)$, *and* $T'_{k,l}(z,f_1, \ldots ,f_k,x_1, \ldots ,x_l,w)$, *where, in the computation for* $T'_{k,l}$, *all questions asked about* $f_1, \ldots ,f_k$ *involve arguments less than* $w\}$.‡

$T^*_{k,l}$ is evidently a recursive relation, since the length of any sequence number can be effectively obtained from that sequence number. The following result is immediate.

Theorem V (Kleene) (alternative normal-form theorem) *Let* $n > 0$, *let* R $(\subset \mathfrak{F}^k \times N^l)$ *be a relation in* Σ^1_n, *and let* z *be a* Σ^1_n*-index for* R. *If* n *is even, then*

$$R = \{<f_1, \ldots ,f_k,x_1, \ldots ,x_l> \mid (\exists g_1)(\forall g_2) \cdots (\forall g_n)(\exists w)T^*_{k+n,l}(z,\bar{f}_1(w), \ldots ,\bar{f}_k(w),\bar{g}_1(w), \ldots ,\bar{g}_n(w),x_1, \ldots ,x_l)\}.$$

If n *is odd, then*

$$R = \{<f_1, \ldots ,f_k,x_1, \ldots ,x_l> \mid (\exists g_1)(\forall g_2) \cdots (\exists g_n)(\forall w) \neg T^*_{k+n,l}(z,\bar{f}_1(w), \ldots ,\bar{f}_k(w),\bar{g}_1(w), \ldots ,\bar{g}_n(w),x_1, \ldots ,x_l)\}.$$

Similarly for Π^1_n.

Proof. Immediate by definition. ☒

As a special case, we have the following corollary.

† Other notations for such code numbers have been used in Chapters 13 and 15. The present notation is common in discussion of the analytical hierarchy and will be used throughout Chapter 16.

‡ In the notation of Chapter 13 (and for the case $l > 0$), $T^*_{k,l}(z,y_1, \ldots ,y_k,x_1, \ldots ,x_l) \Leftrightarrow (\exists w)[y_1, \ldots , y_k$ are sequence numbers for initial segments $f_1, \ldots , f_k$ each of length w, and $\varphi_z^{[f_1], \ldots ,[f_k]}(x_1, \ldots ,x_l)$ is convergent by the wth step in the enumeration of $W_{\rho(z)}]$.

Corollary V

$$A \in \Pi_1^1 \Leftrightarrow (\exists R)[R \text{ is recursive } \& \ R \subset N^2 \ \& \ A = \{x|(\forall f)(\exists w)R(\bar{f}(w),x)\}].$$

Proof. Take $R = \{\langle y,x\rangle | T^*_{1,1}(z,y,x)\}$, where z is a Π_1^1-index for A. ☒

The Tarski-Kuratowski Algorithm

Let $F(p_1, \ldots, p_k, a_1, \ldots, a_l)$ be an expression which is built up within quantificational logic from sentential connectives, $=$, variables for numbers, variables for functions (of one variable), quantifiers over variables for numbers, quantifiers over variables for functions, and symbols for relations (on functions and integers), and which has $p_1, \ldots, p_k$ as its free function variables and $a_1, \ldots, a_l$ as its free number variables. Let the relation symbols be interpreted as certain fixed relations $S_1, \ldots, S_m$. Then the relation

$$R = \{\langle f_1, \ldots, f_k, x_1, \ldots, x_l\rangle | F(p_1, \ldots, p_k, a_1, \ldots, a_l) \text{ is true when } p_1, \ldots, a_l \text{ are interpreted respectively as } f_1, \ldots, x_l\}$$

is said to be *definable within quantificational logic from the relations* $S_1, \ldots, S_m$. $F(p_1, \ldots, a_l)$ is called a *definition of* R *from* $S_1, \ldots, S_m$.

The following analogue to Theorem 14-IV holds.

Theorem VI *If* $R \subset \mathfrak{F}^k \times N^l$ *is definable within quantificational logic from recursive relations, then* R *is analytical.*

Proof. As for Theorem 14-IV. ☒

(The converse of Theorem VI holds by Theorem I.)

Thus the Tarski-Kuratowski algorithm can be used as before, provided that we maintain a distinction between type-1 and type-0 quantifiers. In addition to the quantifier rules of Theorem III, several further rules are sometimes useful for obtaining lowest possible classification.

Theorem VII (***Addison and Kleene*** [*1957*]) *The following prefix transformations are permissible* (*in the sense of Theorem III*).

(i) $\cdots \exists^1 \rightarrow \cdots \exists^0,$
$\cdots \forall^1 \rightarrow \cdots \forall^0;$

where, in each case, the quantifier changed is the final quantifier of the prefix.

(ii) $\cdots (\forall f)(\exists x)(\exists g) \cdots \rightarrow \cdots (\exists g)(\forall f)(\exists x) \cdots,$
$\cdots (\exists f)(\forall x)(\forall g) \cdots \rightarrow \cdots (\forall g)(\exists f)(\forall x) \cdots;$

provided that, in each case, the recursive relation of the given predicate form is a relation in $\bar{f}(x)$ *and the coordinates other than* f.

Proof. (i) Let the initial predicate form be $\cdots (\exists f)R(f,\text{—})$. Then applying Theorem V to R, we have a fixed z such that

$$\cdots (\exists f)(\exists w)T^*(z,\bar{f}(w), \ldots)$$

is an equivalent predicate form. But this is equivalent to $\cdots (\exists y)(\exists w)[y$ is a sequence number of length w & $T^*(z,y, \ldots)]$. Contracting the number quantifiers (by (i) of Theorem III) we get the desired result. The proof for $\forall^1$ is similar.

(ii) Given $\cdots (\forall f)(\exists x)(\exists g) \cdots R(\text{—},\bar{f}(x),\text{—},x,\text{—},g,\text{—})$, where there is no occurrence of f other than that explicitly indicated, use

$$\{<f,g,x, \ldots>|R(\text{—},\bar{f}(x),\text{—},x,\text{—},\lambda y[g(<\bar{f}(x),y>)],\text{—})\}.$$

The case for $\cdots (\exists f)(\forall x)(\forall g) \cdots$ is similar (see Exercise 16-4).☒

Example 1. Define W as in Exercise 11-61; that is,

$W = \{z|\varphi_z^{(2)}$ *is the characteristic function of a well-ordering* $(\leq)$ *of some set of integers*$\}$.

Then $z \in W \Leftrightarrow [\varphi_z$ is the characteristic function of a linear ordering $(\leq)$ which has no infinite descending chains$] \Leftrightarrow [\varphi_z$ is the characteristic function of a linear well-ordering $(\leq)$ and $(\forall f)(\exists n)[f(n+1) = f(n)$ or $\varphi_z(f(n+1),f(n)) = 0]] \Leftrightarrow \forall^0\exists^0$ & $\forall^0$ & $\forall^0\forall^0\forall^0$ & $\forall^0\forall^0$ & $\forall^1\exists^0\exists^0$ (see Exercise 16-5), $\Leftrightarrow \forall^1\exists^0$.
Hence $W \in \Pi_1^1$.

Example 2. Let O be the set of constructive ordinal notations defined in §11.7. In Exercise 11-61 we saw that $O \equiv W$. Hence O is also in Π_1^1.

Example 3. Returning to the example given in the proof of Corollary III, we see that a relation with prefix $\forall^0\exists^1\exists^0\forall^1$ is in fact in Σ_1^1, since $\exists^1\forall^1 \rightarrow \exists^1\forall^0$ by Theorem VII.

The Hierarchy Theorem

For the sake of simplicity, we henceforth confine ourselves largely to analytical sets of integers. We thus emphasize, for the analytical hierarchy, the case treated in Chapter 14 for the arithmetical hierarchy. Extension of our theory to relations in $\mathfrak{F}^k \times N^l$, $k > 0$, is analogous to the extended version of the arithmetical hierarchy treated in §15.2. We further consider such extension in §16.7, §16.8, and Exercises 16-39 through 16-41.

Definition For $n > 0$, let

$$E^n = \{z|(\exists f_1)(\forall f_2) \cdots (\forall f_n)(\exists w)T'_{n,1}(z,f_1, \ldots ,f_n,z,w)\}$$

if n is even; and let

$$E^n = \{z|(\exists f_1)(\forall f_2) \cdots (\exists f_n)(\forall w) \neg T'_{n,1}(z,f_1, \ldots ,f_n,z,w)\}$$

if n is odd.

The following hierarchy theorem is analogous to Corollary 14-VIII(c) for the arithmetical hierarchy.

***Theorem VIII* (Kleene)** *For every $n > 0$, $E^n \in (\Sigma_n^1 - \Pi_n^1)$ (and hence $\bar{E}^n \in (\Pi_n^1 - \Sigma_n^1)$).*

Proof. Let $n > 0$ be given. Assume n is even. (The case for n odd is similar.) It is immediate that $E^n \in \Sigma_n^1$. Assume $E^n \in \Pi_n^1$. Then by Theorem II(*b*), $\bar{E}^n \in \Sigma_n^1$, and by Theorem IV, for some z_0,

$$\bar{E}^n = \{z|(\exists f_1) \cdots (\forall f_n)(\exists w)T'_{n,1}(z_0,f_1, \ldots ,f_n,z,w)\}.$$

But then

$$z_0 \in \bar{E}^n \Leftrightarrow (\exists f_1) \cdots (\forall f_n)(\exists w)T'_{n,1}(z_0,f_1, \ldots ,f_n,z_0,w) \Leftrightarrow z_0 \in E^n$$

(by definition of E^n), and we have a contradiction. Hence $E^n \in (\Sigma_n^1 - \Pi_n^1)$, and $\bar{E}^n \in (\Pi_n^1 - \Sigma_n^1)$.☒

The Completeness Theorem

The sets E^n and $\bar{E}^n$ have the following completeness property.

Theorem IX (Kleene) *For every $n > 0$ and for every A,*

$$A \in \Sigma_n^1 \Leftrightarrow A \leq_1 E^n,$$

and

$$A \in \Pi_n^1 \Leftrightarrow A \leq_1 \bar{E}^n.$$

Proof. Assume $A \leq_1 E^n$. Then

$$A = \{z|(\exists f_1) \cdots T'_{n,1}(g(z),f_1, \ldots ,f_n,g(z),w)\}$$

for some recursive g. Hence $A \in \Sigma_n^1$.

Conversely, assume $A \in \Sigma_n^1$. Then

$$A = \{z|(\exists f_1) \cdots T'_{n,1}(z_0,f_1, \ldots ,f_n,z,w)\}.$$

For any given x define

$$A_x = \{y|y = y \ \& \ (\exists f_1) \cdots T'_{n,1}(z_0,f_1, \ldots ,f_n,x,w)\}.$$

Then $A_x = N$ if $x \in A$, and $A_x = \emptyset$ if $x \notin A$. By Theorem IV, $A_x \in \Sigma_n^1$ and a Σ_n^1-index for A_x can be found uniformly from x. Let h be a recursive function such that $h(x)$ is a Σ_n^1-index for A_x. By definition of E^n, $x \in A \Leftrightarrow A_x = N \Leftrightarrow h(x) \in E^n$. Thus $A \leq_1 E^n$.☒

Definition For $n > 0$, if $A \equiv E^n$, then A is said to be Σ_n^1*-complete*, and $\bar{A}$ is said to be Π_n^1*-complete*.

By the normal-form theorem, there are only denumerably many analytical sets. Hence nonanalytical sets exist. We obtain one such set as follows.

Definition $E^\omega = \{<x,n>|x \in E^n\}$.

Corollary IX *E^ω is not analytical.*

Proof. Assume E^ω analytical. Then for some n, $E^\omega \leq_1 E^n$ by Theorem IX. But $E^{n+1} \leq_1 E^\omega$ trivially. Hence $E^{n+1} \leq_1 E^n$, contrary to Theorems VIII and IX.☒

The sets E^n form a strictly increasing sequence of T-degrees of unsolvability, as the following theorem and corollaries show. (A' is the *jump* of A, as defined in §13.1.)

Theorem X *For all n,* $A \in \Sigma_n^1 \cap \Pi_n^1 \Rightarrow A' \in \Sigma_n^1 \cap \Pi_n^1$.

Proof.

$$A' = \{x | x \in W_x{}^A\} = \{x | (\exists y)(\exists u)(\exists v)[<x,y,u,v> \in W_{\rho(x)} \ \& \ (\forall x)[x \in D_u \Rightarrow x \in A] \ \& \ (\forall x)[x \in D_v \Rightarrow x \notin A]]\}.$$

Inserting, in this last, a Σ_n^1-form for the first occurrence of A and a Π_n^1-form for the second occurrence of A, we have

$$\exists^0\exists^0\exists^0[\exists^0 \ \& \ \forall^0[\ \Rightarrow \Sigma_n^1] \ \& \ \forall^0[\ \Rightarrow \Pi_n^1]].$$

By the Tarski-Kuratowski algorithm, this yields a Σ_n^1-form. Similarly, putting a Π_n^1-form for the first occurrence of A and a Σ_n^1-form for the second occurrence of A, we obtain a Π_n^1-form for A'. Hence $A' \in \Sigma_n^1 \cap \Pi_n^1$. ☒

Corollary X(a) *For all n,* $B \in \Sigma_n^1 \cap \Pi_n^1 \ \& \ A \leq_T B \Rightarrow A \in \Sigma_n^1 \cap \Pi_n^1$.

(b) *For* $n > 0$, E^{n+1} *is of higher* T*-degree than* E^n.

Proof. (*a*) Observe that $A \leq_T B \Rightarrow A \leq_1 B'$. The result then follows by the theorem.

(*b*) $E^n \in \Sigma_n^1 \subset \Sigma_{n+1}^1 \cap \Pi_{n+1}^1$. Hence, by the theorem, $(E^n)' \in \Sigma_{n+1}^1 \cap \Pi_{n+1}^1 \subset \Sigma_{n+1}^1$. Hence by Theorem IX, $(E^n)' \leq_1 E^{n+1}$. ☒

Hyperarithmetical Sets

In the arithmetical hierarchy, we had the basic result that

$$\Sigma_1 \cap \Pi_1 = \Sigma_0 (=\Pi_0)$$

(Theorem 15-XXII and Corollary 15-XXII). The corresponding result for the analytical hierarchy fails, as we now show.

Theorem XI *There exists a set A such that* $A \in \Sigma_1^1 \cap \Pi_1^1$ *but* $A \notin \Sigma_0^1$.

Proof. Let $A = V$, where V is defined as in §14.7; that is, V is the set of true sentences of elementary arithmetic. Since sets in Σ_0^1 are arithmetical, and since V is not arithmetical (see proof of Theorem 15-XII), $V \notin \Sigma_0^1$. On the other hand, by Theorem 15-XII (on the implicit definability of V) there is a recursive R ($\subset \mathfrak{N} \times N^2$) such that $(\forall y)(\exists z)R(X,y,z)$ holds if and only if $X = V$. Hence, by Theorem 15-XXVI, there is a recursive S ($\subset \mathfrak{F} \times N^2$) such that $(\forall y)(\exists z)S(f,y,z)$ holds if and only if $f = c_V$. Hence for any x,

$$\begin{aligned} x \in V &\Leftrightarrow (\forall f)[(\forall y)(\exists z)S(f,y,z) \Rightarrow f(x) = 1] \\ &\Leftrightarrow (\exists f)[(\forall y)(\exists z)S(f,y,z) \ \& \ f(x) = 1]. \end{aligned}$$

But this yields that $V \in \Pi_1^1$ and that $V \in \Sigma_1^1$. Hence $V \in \Sigma_1^1 \cap \Pi_1^1$. ☒

Corollary XI *For all n, there exist sets in* $\Sigma_{n+1}^1 \cap \Pi_{n+1}^1$ *which are not in* $\Sigma_n^1 \cup \Pi_n^1$.

Proof. The case $n = 0$ is given by Theorem XI. For $n > 0$, consider $A = E^n$ *join* $\bar{E}^n$; then $A \in \Sigma_{n+1}^1 \cap \Pi_{n+1}^1$ but $A \notin \Sigma_n^1 \cup \Pi_n^1$ (see Exercise 16-7). ☒

The following notations and terminology are common, and will be used below.

Definition For all n, $\Delta_n^1 = \Sigma_n^1 \cap \Pi_n^1$. Relations in Δ_n^1 are said to be in *both n-function-quantifier forms*.

Definition R is *hyperarithmetical* if $R \in \Delta_1^1$.

Thus Δ_0^1 $(=\Sigma_0^1 = \Pi_0^1)$ is the class of arithmetical relations (in $\mathfrak{F}^k \times N^l$), and Δ_1^1 $(=\Sigma_1^1 \cap \Pi_1^1)$ is the class of hyperarithmetical relations. Theorem XI shows that there exist hyperarithmetical sets which are not arithmetical.

Set Variables and Set Quantifiers

We could have defined an analytical hierarchy by beginning with recursive relations in $\mathfrak{N}^k \times N^l$ (rather than in $\mathfrak{F}^k \times N^l$) and by allowing set quantifiers and set variables in place of function quantifiers and function variables in our definitions above. If we had done so, the same relations in N^l would have been obtained, and the same Σ_n^1, Π_n^1 classification of these relations would have been obtained, as before. This is a consequence of the proofs of Theorems 15-XXV and 15-XXVI (see Exercise 16-8).

Rules for set quantifiers are somewhat less convenient than those (of Theorems III and VII) for function quantifiers. In particular, Corollary III does not hold with set quantifiers (see Exercise 16-9). Instead, we have the following: *for* $n > 0$, *any set of integers in* Σ_n^1 *can be expressed by a set-quantifier predicate form with a prefix consisting of* $n + 2$ *alternating quantifiers, of which the first n are set quantifiers and the last two are number quantifiers, and in which the first quantifier is existential; similarly for sets in* Π_n^1 (see Exercise 16-10).

Moreover, function quantifiers appear to be of greater heuristic value than set quantifiers in studies of the analytical hierarchy. The rules (of Theorems III and VII) for manipulating function quantifiers are tools of surprising combinatorial power. Much of the elegance and depth of later results in this chapter will rest on their use. For these reasons, we use function quantifiers for our definitions and normal forms.

We can also define a hierarchy by beginning with recursive relations in $\mathfrak{F}^k \times \mathfrak{N}^l \times N^m$ $(k,l,m \geq 0)$ and allowing both set quantifiers and function quantifiers (as well as number quantifiers). If we do so, and if we take set quantifiers to be of type 1 and to be indistinguishable from function quantifiers for purposes of prefix classification (into the classes Σ_n^1 and Π_n^1), then the relations in $\mathfrak{F}^k \times N^l$ so obtained are the same as before, and the classification of these relations into Σ_n^1 and Π_n^1 is the same as before. This is a consequence of the proofs of Theorems 15-XXV and 15-XXVI. But relations in $\mathfrak{F}^k \times \mathfrak{N}^l \times N^m$ $(k,l,m \geq 0)$ are also obtained, and a Σ_n^1, Π_n^1 classification of these relations is provided. Moreover, the same classification of the same relations is obtained if we use only function and number quantifiers (on our recursive relations in $\mathfrak{F}^k \times \mathfrak{N}^l \times N^m$) and not set quantifiers (see Exercise 16-8).

Similarly, we can define an arithmetical hierarchy on relations in $\mathfrak{F}^k \times \mathfrak{N}^l \times N^m$ by beginning with recursive relations in $\mathfrak{F}^k \times \mathfrak{N}^l \times N^m$ and allowing number quantifiers only.

Henceforth, we adopt the following convention.

Convention on notation *Arithmetical hierarchy.* Σ_n^0 = the class of all relations in $\mathfrak{F}^k \times \mathfrak{N}^l \times N^m$ $(k,l,m \geq 0)$ definable from recursive relations by a Σ_n-prefix.

Π_n^0 = the class of all relations in $\mathfrak{F}^k \times \mathfrak{N}^l \times N^m$ $(k,l,m \geq 0)$ definable from recursive relations by a Π_n-prefix.

$\Delta_n^0 = \Sigma_n^0 \cap \Pi_n^0$.

(The class Σ_n^0 thus includes the classes Σ_n, $\Sigma_n^{(s)}$, and $\Sigma_n^{(fn)}$ studied in Chapters 14 and 15. We no longer use the superscripts (s) and (fn), which were introduced for expository purposes in Chapter 15.)

Analytical hierarchy. Σ_n^1 = the class of all relations in $\mathfrak{F}^k \times \mathfrak{N}^l \times N^m$ $(k,l,m \geq 0)$ definable from recursive relations by a Σ_n^1-prefix.

Π_n^1 = the class of all relations in $\mathfrak{F}^k \times \mathfrak{N}^l \times N^m$ $(k,l,m \geq 0)$ definable from recursive relations by a Π_n^1-prefix.

$\Delta_n^1 = \Sigma_n^1 \cap \Pi_n^1$.

As we previously noted, we shall be chiefly concerned with analytical sets of integers. We shall sometimes refer to sets in Σ_n^1 as Σ_n^1-sets. Similarly for Π_n^1, Δ_n^1, Σ_n^0, Π_n^0, and Δ_n^0.

Summary of Chapter

In §16.2, we prove a representation theorem for analytical sets analogous to the representation theorem of §14.4 for the arithmetical sets. In particular, we augment elementary arithmetic by adding variables and quantifiers over the nonnegative real numbers, and we show that the sets of integers definable in this augmented system (which we call *elementary analysis*) are exactly the analytical sets. It is for this reason that the hierarchy considered in the present chapter is called the *analytical* hierarchy.

In the remainder of the chapter, we shall be chiefly concerned with hyperarithmetical sets and Π_1^1-sets; the theory of these sets is related to a variety of problems and results in logic and the foundations of mathematics. To a lesser extent, we shall be concerned with the Δ_2^1-sets. This class has been the subject of considerable recent interest. Virtually all sets of integers that are known to be definable by iterative "constructive" procedures† have

† We do not attempt to make this notion more precise.

been shown to be in Δ_2^1.† As has been remarked, the human mind seems severely limited in its ability to understand and visualize beyond a single alternation of function quantifiers (see the related discussion following Theorem 14-XII in §14.7). Higher levels of the analytical hierarchy have not been as extensively studied as the levels at Δ_2^1 and below.

Analogies

We now comment briefly on the analytical hierarchy of classes of functions (subclasses of $\mathfrak{F}$ determined by predicate forms with one unquantified function variable and no unquantified number variables). We saw in §15.2 that the arithmetical classes of functions were analogous to the Borel sets of finite level in Baire space. In Theorems 15-XXVII and 15-XXVIII, we saw that the arithmetical classes and the Borel sets could be put in a common theoretical framework. This theoretical framework can be extended to the analytical hierarchy (although we do not make this extension here, see Addison [1955]). The following analogies then appear: (1) the hyperarithmetical classes correspond to the Borel sets (of finite and transfinite level); (2) the Σ_1^1-classes correspond to the sets known in descriptive set theory as the *analytic sets;*‡ (3) the hierarchy of analytical classes corresponds to the hierarchy known in descriptive set theory as the *hierarchy of projective sets.*§ The reader should note the unfortunate conflict in terminology between "analytical" in recursive function theory and "analytic" in descriptive set theory. (It is for the purpose of making this distinction that the suffix "-al" has come into use in recursive function theory.)

In recent studies of this common framework, the boldface symbols $\mathbf{\Sigma}_n^1$, $\mathbf{\Pi}_n^1$, $\mathbf{\Delta}_n^1$ are used to indicate levels in the "classical" projective hierarchy of descriptive set theory, and the symbols $\mathbf{\Sigma}_n^0$, $\mathbf{\Pi}_n^0$, $\mathbf{\Delta}_n^0$ (sometimes $\mathbf{\Sigma}_n$, $\mathbf{\Pi}_n$, $\mathbf{\Delta}_n$) are used to indicate finite levels in the "classical" hierarchy of Borel sets.

§16.2 ANALYTICAL REPRESENTATION; APPLICATIONS TO LOGIC

The logical system known as *second-order arithmetic with set variables* is defined as follows. The basic symbols are the symbols of elementary arith-

† This "absorptive" quality of Δ_2^1 is illustrated, on a small scale (and for Δ_1^1 and Δ_0^1 as well), by Theorem X above.

‡ These are sometimes defined as the sets obtainable from Borel sets by the operation A. See Kuratowski [1950]. (The *operation* A is a "generalized projection" defined as follows. If Γ is a class of subsets of $\mathfrak{F}$, and if R is any relation in $N \times \mathfrak{F}$ such that for all x, $\{g|R(x,g)\} \in \Gamma$, then the class $\{g|(\exists f)(\forall x)R(\bar{f}(x),g)\}$ is said to be obtained from Γ by an "application of operation A." In applications of this operation, the relation R is sometimes spoken of as a *sieve*.

§ Historically, the analogy first noted was between the arithmetical hierarchy of recursive function theory and the projective hierarchy of descriptive set theory. The analogy between the analytical hierarchy and the projective hierarchy, however, proves to be more natural (see Addison [1955]).

metic together with set variables and the symbol $\in$. Formulas are built up by allowing, in addition to the atomic expressions and terms of elementary arithmetic, atomic expressions of the form $a \in \hat{A}$ where a is any number variable and $\hat{A}$ is any set variable, and by allowing quantifiers over set variables as well as over number variables. Formulas in which all variables (set or number) are acted on by quantifiers are called *sentences*. Sentences are interpreted as assertions about ordinary addition and multiplication over the integers, with number variables ranging over N and set variables ranging over the collection of all subsets of N. Every sentence is, accordingly, either true or false. For example,

$$(\forall \hat{A})[[0 \in \hat{A} \;\&\; (\forall a)[a \in \hat{A} \Rightarrow a + 1 \in \hat{A}]] \Rightarrow (\forall a)[a \in \hat{A}]]$$

is a true sentence (expressing the law of mathematical induction over N).

The logical system known as *second-order arithmetic with function variables* is defined as follows. Basic symbols are the symbols of elementary arithmetic together with function variables of one argument. Formulas are built up as in elementary arithmetic but with function variables of one argument allowed as additional operator symbols and with quantification over function variables, as well as number variables, permitted. (Nesting of function variables can, of course, occur.) Formulas in which all variables (function variables or number variables) are acted on by quantifiers are called *sentences*. Sentences are interpreted as assertions about ordinary addition and multiplication over the integers, with number variables ranging over N and function variables ranging over the collection of all functions from N into N. Every sentence is, accordingly, either true or false. For example,

$$(\forall p)[[p(0) = 1 \;\&\; (\forall a)[p(a) = 1 \Rightarrow p(a + 1) = 1]] \Rightarrow (\forall a)[p(a) = 1]]$$

is a true sentence (also expressing the law of mathematical induction over N).

The following theorem is immediate from previous results.

Theorem XII *A is analytical $\Leftrightarrow$ A is definable in second-order arithmetic with set variables $\Leftrightarrow$ A is definable in second-order arithmetic with function variables.*

Proof. Immediate from the definition of analytical set, from Exercise 15-19 (see Theorems 15-XXV and 15-XXVI), and from the representation theorems of §§15.1 and 15.2 (Theorems 15-VI(b) and 15-XIX(b)). ⊠†

Henceforth, by *second-order arithmetic* we shall mean second-order arithmetic with function variables. In the literature, second-order arithmetic is sometimes called *second-order number theory*.

† Note that if symbols and quantifiers for k-ary relations on N are allowed in second-order arithmetic with set variables, or if symbols and quantifiers for functions of more than one variable are allowed in second-order number theory with function variables, Theorem XII remains true; for the recursive pairing function τ can be represented in elementary arithmetic, and by the use of τ, symbols for relations (or functions of more than one variable) can be replaced by symbols for sets (or functions of one variable).

The logical system known as *elementary analysis* is defined as follows. The basic symbols are the symbols of elementary arithmetic together with variables for real numbers ("real variables") in addition to, and distinct from, the variables for integers. We use $\eta, \eta_1, \eta_2, \ldots$ as variables for real numbers. Formulas are built up by allowing quantifiers over real variables as well as over number, i.e., integer, variables. Formulas in which all variables are acted on by quantifiers are called *sentences*. Sentences are interpreted as assertions about ordinary addition and multiplication over the nonnegative real numbers, with number variables ranging over the nonnegative integers and real variables ranging over the nonnegative real numbers. Every sentence is, accordingly, either true or false. For example, $(\forall a)(\exists \eta)[\eta \times \eta = a]$ is a true sentence (expressing the fact that every nonnegative integer has a nonnegative real square root).

We say that the set A is *definable in elementary analysis* if there is a formula of elementary analysis Fa with number variable a as its only free variable, such that $A = \{x|F\mathbf{x} \text{ is true}\}$, where, for any integer x, $F\mathbf{x}$ is the sentence which results by substitution of the numeral expression of x for the variable a in Fa. We obtain the following representation theorem.

Theorem XIII *A is analytical* $\Leftrightarrow$ *A is definable in elementary analysis.*

Proof. Consider the following five logical systems.

S_1: elementary arithmetic with variables for nonnegative real numbers $(\eta, \eta_1, \eta_2, \ldots)$, i.e., elementary analysis.

S_2: elementary arithmetic with variables for nonnegative rational numbers $(r, r_1, r_2, \ldots)$ and variables for nonnegative irrational numbers $(\zeta, \zeta_1, \zeta_2, \ldots)$.

S_3: elementary arithmetic with variables for nonnegative irrational numbers $(\zeta, \zeta_1, \zeta_2, \ldots)$.

S_4: elementary arithmetic with function variables $(p, p_1, p_2, \ldots)$ but with no nesting of function variables permitted.

S_5: elementary arithmetic with function variables $(p, p_1, p_2, \ldots)$ and nesting of function variables permitted, i.e., second-order arithmetic.

(More detailed definitions of S_2, S_3, and S_4 would parallel the definitions previously given for S_1 and S_5.)

We show that for any set A, [A is definable in S_1] $\Leftrightarrow$ [A is definable in S_2] $\Leftrightarrow$ [A is definable in S_3] $\Leftrightarrow$ [A is definable in S_4] $\Leftrightarrow$ [A is definable in S_5]. The result then follows by Theorem XII.

We show, for each successive pair of systems, how free-number-variable formulas of one system can be translated into equivalent (with respect to interpretation) free-number-variable formulas of the other. (A *free-number-variable formula* is a formula with no free variables other than number variables. In particular, sentences are free-number-variable formulas.) In each case we briefly indicate the method of translation. In the first six cases (i.e., the first three equivalences) we assume that an expression is put in prenex form and that all universal quantifiers ($\forall$) are then replaced by existential quantifiers and negations; that is, $\forall$ goes to $\neg\exists\neg$; we then give an inductive procedure for dealing with the rightmost existential quantifier on a variable to be eliminated. (At intermediate stages, the expression being operated on will be a "mixed" expression involving symbols from both systems.) Justification is given in Exercise 16-11.

(1) $S_1 \rightarrow S_2$:

$$(\exists\eta)M(\eta) \rightarrow (\exists r)M(r) \vee (\exists\zeta)M(\zeta).$$

$S_2 \rightarrow S_1$:

$$(\exists\zeta)M(\zeta) \rightarrow (\exists\eta)[M(\eta) \;\&\; (\forall a)(\forall b)[b \neq 0 \Rightarrow a \neq b \times \eta]],$$
$$(\exists r)M(r) \rightarrow (\exists\eta)[M(\eta) \;\&\; (\exists a)(\exists b)[b \neq 0 \;\&\; a = b \times \eta]].$$

(2) $S_2 \rightarrow S_3$:

$$(\exists r)M(r) \rightarrow (\exists a)(\exists b)[b \neq 0 \;\&\; M(a/b)],$$

where $M(a/b)$ is the formula obtained by putting a/b for r in $M(r)$ and then clearing each atomic subformula (i.e., equation) of fractions by the usual rules of elementary algebra.

$S_3 \rightarrow S_2$: trivial.

(3) We associate nonnegative irrationals α with functions f as follows: $[\alpha]$ (the greatest integer in α) is $f(0)$ and $\alpha - [\alpha]$ is given by the continued fraction expansion $f(1) + 1$, $f(2) + 1$, . . . (see Exercise 15-13). (We speak of the successive approximations to α given by successive initial segments of f as the *continued fraction approximations given by f*.) Thus we have a one-one correspondence between $\mathfrak{F}$ and the collection of all nonnegative irrationals.

$S_3 \rightarrow S_4$:

$$(\exists\zeta)M(\zeta) \rightarrow (\exists p)M'(p),$$

where $M'(p)$ is obtained by transforming each atomic subformula (i.e., equation) of $M(\zeta)$ as follows. Let $R(p,a,b,c)$ be the formula of S_4 which defines (by Theorem 15-XIX) the recursive relation $\{\langle f,x,y,z\rangle | x/y$ *is the* zth *continued fraction approximation given by* $f\}$. Let $E_1(\zeta) = E_2(\zeta)$ be an atomic subformula of $M(\zeta)$. We replace $E_1(\zeta) = E_2(\zeta)$ by

$$\begin{aligned}(\forall d)[d \neq 0 \Rightarrow (\forall c_1)(\exists c_2)(\exists c)[c = c_1 + c_2 \\ \& (\exists a)(\exists b)[R(p,a,b,c) \;\&\; (\exists a_1)(\exists a_2)[a_2 \neq 0 \\ \& \left(\left(E_1\left(\frac{a}{b}\right) - E_2\left(\frac{a}{b}\right)\right) \times \left(E_1\left(\frac{a}{b}\right) - E_2\left(\frac{a}{b}\right)\right)\right) + \frac{a_1}{a_2} = \frac{1}{d}\Big]\Big]\Big]\Big],\end{aligned}$$

where $E_1(a/b)$ and $E_2(a/b)$ are obtained by formally substituting a/b for ζ. We then use the rules of elementary algebra to clear of fractions and to eliminate the symbol "$-$" by transposition.

$S_4 \rightarrow S_3$:

$$(\exists p)M(p) \rightarrow (\exists\zeta)M''(\zeta),$$

where $M''(\zeta)$ is obtained by transforming atomic subformulas as follows. Let $S(\zeta,a,b)$ assert that if $b = 0$ then $a = [\zeta]$, and that if $b > 0$ then $a + 1$ is the bth term in the continued fraction expansion of $\zeta - [\zeta]$. It is easy to show, on the basis of the arithmetical representation of recursive relations (Theorem 14-VI) and the definitions in Exercise 15-13, that $S(\zeta,a,b)$ can be taken as a formula of S_3 (see Exercise 16-12). Let $E(p(t_1), \ldots, p(t_k))$ be an atomic subformula of $M(p)$ with k occurrences of p as indicated ($t_1, \ldots t_k$ are terms). Replace $E(p(t_1), \ldots, p(t_k))$ by

$$(\exists a_1) \cdots (\exists a_k)[E(a_1, \ldots, a_k) \mathbin{\&} S(\zeta,a_1,t_1) \mathbin{\&} \cdots \mathbin{\&} S(\zeta,a_k,t_k)]$$

(where substitutions in S and E are as indicated).

(4) $S_4 \to S_5$: trivial.

$S_5 \to S_4$: nesting of function variables in an atomic subformula is eliminated as in the following example:

$$p(p_1(a) + p(b)) = c \to$$

$$(\exists a_1)(\exists b_1)[a_1 = p_1(a) \mathbin{\&} b_1 = p(b) \mathbin{\&} p(a_1 + b_1) = c].$$

In the case of deeper nesting, we iterate the procedure beginning with outermost occurrences of function variables.†

This concludes the proof. ☒

Assume an appropriate fixed coding from sentences of elementary analysis onto N and an appropriate fixed coding from sentences of second-order arithmetic onto N.

Corollary XIII(a) *The set of true sentences of elementary analysis is recursively isomorphic to the set of true sentences of second-order arithmetic.*

Proof. The translations in the proof of the theorem evidently take true sentences to true sentences and false sentences to false sentences. This yields that the true sentences of each system are 1-reducible to the true sentences of the other. Recursive isomorphism follows by Theorem 7-VI. ☒

Definition Consider a sentence of *second-order arithmetic* in prenex form. The sentence is said to be a Σ_n^1-*sentence* if its prefix is a Σ_n^1-prefix in the sense of §16.1. Similarly for Π_n^1.

Consider a sentence of *elementary analysis* in prenex form. The sentence is said to be a Σ_n^1-*sentence* if its prefix is a Σ_n^1-prefix in the sense of §16.1 when real-variable quantifiers are taken to be of type 1 and number quantifiers are taken to be of type 0. Similarly for Π_n^1.

Corollary XIII(b) *There is a recursive permutation which maps the true sentences of elementary analysis onto the true sentences of second-order arithmetic and which also, for all n, maps the Σ_n^1-sentences of elementary analysis onto the Σ_n^1-sentences of second-order arithmetic and the Π_n^1-sentences of elementary analysis onto the Π_n^1-sentences of second-order arithmetic.*

† In logic, this procedure, combined with a procedure for replacing function symbols by relation symbols, is known as the method of *eliminating descriptions*.

Proof. The translations in the proof of the theorem are easily seen to take Σ_n^1-sentences into sentences which can become Σ_n^1-sentences when put into prenex form. Appropriate modifications of the translation process in connection with the construction of Theorem 7-VI now yield the result (see Exercise 16-13).☒

Results in the remainder of the present section are stated for elementary analysis. By virtue of Theorem XIII and Corollary XIII(*b*), these results hold, *mutatis mutandis*, for second-order arithmetic. Indeed, the "equivalence" to second-order arithmetic given by Theorem XIII and its corollaries will be used to supply most of the proofs.†

We note in passing that if we take elementary arithmetic and *replace* number variables by real variables, we obtain a system that is much weaker than elementary analysis; in fact, this new system (which we call *elementary real-number theory*) is decidable (as we noted in §2.2). It follows, by the undecidability of elementary arithmetic, and hence of S_5, and hence of S_1, that the set N cannot be defined in elementary real-number theory. We shall see below that the set of true sentences of elementary analysis is not even analytical.

In connection with Theorem XII, we note also that the phenomenon of Theorem 15-XII does not occur for analytical sets of integers. If a set is implicitly definable, then it is explicitly definable by the construction used in the proof of Theorem XI (see Exercise 16-14).

Applications to Logic

Definition Let V^* be the set of true sentences of elementary analysis (under an appropriate fixed coding from the sentences of elementary analysis onto N).

The set V^* is not an analytical set. In fact, we have the following na-logue to Theorem 14-X.

† We do not consider questions of formal deducibility in these systems (except as subsystems of some standard set theory; see below). If an appropriate version of Peano's axioms is adopted within second-order arithmetic, together with axioms of comprehension (all axioms of the form $(\forall \cdots)(\exists p)(\forall a)[p(a) = 0 \Leftrightarrow S]$, where S is any formula not containing p as a free variable, and where $(\forall \cdots)$ indicates universal quantifiers over all free variables in S other than a), and if counterparts to these axioms are chosen in elementary analysis, then, in the presence of these (respective) axiomatizations, the translations of Theorem XIII (and hence the recursive permutation of Corollary XIII) preserve quantificational deducibility. The phrases "second-order arithmetic" and "second-order number theory" are sometimes used in the literature to refer to the collection of sentences of elementary arithmetic provable from these axioms (this collection is the same for either system). We, of course, use "second-order arithmetic" to refer to the general system called S_5 in the proof of Theorem XIII. Moreover, we are concerned with sentences that involve function variables (as well as sentences of elementary arithmetic), and we are concerned with true sentences rather than sentences provable from a particular axiomatization.

Theorem XIV $V^* \equiv E^\omega$.

Proof. We prove the result for second-order arithmetic, then use Corollary XIII. The proof is as for Theorem 14-X (see Exercise 16-15).☒

Definition Let V_n^* be the set of all true Σ_n^1-sentences of elementary analysis. (In particular, $V_0^* \equiv V$.†)

Corollary XIV *For* $n > 0$, $V_n^* \equiv E^n$.

Proof. In the first part of the proof of the theorem (paralleling the proof of Theorem 14-X), the quantifier rules of Theorem III give, immediately, that for any $<x,n>$ there is a corresponding Σ_n^1-sentence which is true if and only if $x \in E^n$. It follows that $E^n \leq_1 V_n^*$. The second part of the proof gives that $V_n^* \leq_1 E^n$. Hence $V_n^* \equiv E^n$ (see Exercise 16-16).☒

Corollary XIV is stronger than the analogous result for the arithmetical hierarchy (Corollary 14-X). As a result, analogues to Theorems 14-XI, 14-XII, and 14-XIII can be proved more easily (and, in the case of 14-XI, in a stronger form). We state these results and leave the proofs for exercises. The results amount to a further generalization of the Gödel incompleteness theorem. For motivation and explanation of terminology, see §14.7.

Theorem XV *Let a sound axiomatization of elementary analysis be given. Then, for every n, there exists a* Σ_n^1*-sentence* F_n *such that for no* $m < n$ *does there exist a* Σ_m^1*-sentence demonstrably equivalent to* F_n*. Furthermore a uniform procedure exists by which, for each n, such an* F_n *can be exhibited.*

Proof. See Exercise 16-17.☒

Corollary XV *Let a sound axiomatization of elementary analysis be given and let A be a set of true sentences such that* $A \in \Sigma_n^1$*. Then there is a true* Σ_{n+1}^1*-sentence which is not deducible (under the axiomatization) from A.*

Proof. See Exercise 16-17.☒

As before, we take the axiomatization ZF of set theory. The sentences of elementary analysis can be identified with certain sentences of set theory. Under this embedding, the sentences of elementary analysis that are now provable constitute what we shall call *set-theory analysis.*‡ The soundness of set-theory analysis can be expressed within set theory, since the set E^ω is definable in set theory.

Definition A sentence of set theory is *analytically expressible* if it is demonstrably equivalent (under ZF) to a sentence of elementary analysis.

† These sets are not identical since V was defined under a coding from the sentences of elementary arithmetic onto N and V_0^* under a coding from the sentences of elementary analysis onto N.

‡ We adopt this terminology for its brevity. More specific and less ambiguous terminology would be "set-theory elementary analysis" or "set-theory second-order arithmetic." Our shorter phrase may suggest an unwarranted degree of generality; one cannot, for example, speak of *sets* of real numbers in elementary analysis as we define it.

Definition $\Sigma_n^{1,ZF}$ = the set of all sentences of set theory, which are demonstrably equivalent, under *ZF*, to Σ_n^1-sentences of elementary analysis. (Thus, in the notation of Chapter 14, $\Sigma_0^{1,ZF} = \bigcup_{n=0}^{\infty} \Sigma_n^{ZF}$.) Similarly for $\Pi_n^{1,ZF}$.

Using dummy quantifiers, we have that for all n,

$$\Sigma_n^{1,ZF} \cup \Pi_n^{1,ZF} \subset \Sigma_{n+1}^{1,ZF} \cap \Pi_{n+1}^{1,ZF}.$$

The following analogue to Theorem 14-XII now follows.

Theorem XVI *If set-theory analysis is sound, then, for all n, there exist sentences of set theory in* $(\Sigma_{n+1}^{1,ZF} - \Pi_{n+1}^{1,ZF})$ *(and hence sentences in* $(\Pi_{n+1}^{1,ZF} - \Sigma_{n+1}^{1,ZF})$*).*

Proof. Immediate from Theorem VIII and Corollary XIV.☒

In Theorem 14-XIII we obtained a sentence of set theory which is not arithmetically expressible (provided *ZF* is consistent). Since V is analytical, it is easy to show that the particular sentence given in that proof is analytically expressible. Using exactly the same method of proof, however, we can obtain the following.

Theorem XVII *If the axiomatization ZF of set theory is consistent, then there is a sentence of set theory which is not analytically expressible.*

Proof. See Exercise 16-18.☒

The incompleteness of analysis under any analytical axiomatization (see Corollary XV above) is of special interest in view of the highly nonconstructive nature of analytical sets beyond the level Δ_2^1 (see comments on Δ_2^1, §16.1).

Results about elementary analysis yield results about pure logic. For if we treat "+" and "×" as unspecified (and unquantified) function symbols, if we treat "0" and "1" as unspecified (and unquantified) number symbols, if we express any other numeral **n** as $(\cdot\ ((1 + 1) + 1) \cdots + 1)$ (n times), and if we adopt the semantical conventions of pure second-order logic (i.e., if we insist that for any nonempty domain $\mathfrak{D}$ over which the number variables are to range, the function variables range over the collection of all mappings from $\mathfrak{D}$ into $\mathfrak{D}$), then a single formula P can be found such that for any sentence F of second-order arithmetic, $[P \Rightarrow F]$ is a universally true formula of second-order logic if and only if F is a true sentence of second-order arithmetic. (It suffices to take P to be the conjunction of an appropriate version of Peano's axioms; the conjunction can be made finite since the induction assertion given at the beginning of this section (as an example of a sentence of second-order arithmetic) is now available.) An incompleteness theorem for pure second-order logic now follows from Theorem XV. In fact we have that the true sentences of pure second-order logic do not constitute an analytical set.†

† Second-order logic is defined and discussed in Church [1956, Chap. 5].

If we do not adopt the semantical conventions of second-order logic but instead allow a model (i.e., semantical interpretation) to include both (i) an arbitrary nonempty collection of objects $\mathfrak{D}$ over which the number variables are to range and (ii) an arbitrary nonempty collection $\mathfrak{D}'$ of mappings from $\mathfrak{D}$ into $\mathfrak{D}$ over which the function variables are to range,† then it is possible to obtain interpretations for which all true sentences of second-order arithmetic are true but in which both $\mathfrak{D}$ and $\mathfrak{D}'$ are denumerable collections. Such an interpretation is called a *denumerable model* (for the true sentences of second-order arithmetic). (It includes a particular assignment to "+" and "×" of mappings from $\mathfrak{D} \times \mathfrak{D}$ into $\mathfrak{D}$ and a particular assignment to "0" and "1" of members of $\mathfrak{D}$. As before, we express a numeral $\mathbf{n}$ ($n > 1$) as $(\cdot\ (1 + 1) + 1) \cdot \cdot \cdot + 1)$ (n times).)

An interpretation which satisfies all true sentences of second-order arithmetic and in which $\mathfrak{D}$ is the set N and "0," "1," "+," and "×" are given their standard interpretation, is called an *ω-model* (for the true sentences of second-order arithmetic). In Exercise 16-20 we show the existence of denumerable ω-models for the true sentences of second-order arithmetic.

§16.3 FINITE-PATH TREES‡

The *function tree* $\mathfrak{J}^*$ was defined in Exercise 9-42 and considered further in the discussion of *open and closed classes* in §15.2. In that discussion, the concepts of *subtree, branch, subbranch,* and *finite-path tree* were also used. Finite-path trees will play an important role in our study of the analytical hierarchy.

Recall that $\mathfrak{J}^*$, the function tree, is the collection of all finite sequences of integers partially ordered under the relation *is an initial segment of*. Henceforth (and in contrast to previous usage), we shall take this ordering in the sense of $\geq$; that is, $a \geq b \Leftrightarrow a$ is an initial segment of $b \Leftrightarrow a$ is "above" $b \Leftrightarrow b$ is "below" a. (We thus visualize a tree as "growing downwards.")

In what follows, a *tree* will be a subtree of $\mathfrak{J}^*$. Its members will be called *elements* or *vertices.* A vertex is thus a finite sequence of integers. Every nonempty tree has $\emptyset$, the empty sequence, as its maximum vertex. If a tree possesses minimal vertices, we sometimes speak of these minimal vertices as *terminal* vertices.

† This means (in effect) that we are adopting the semantical conventions of first-order (quantificational) logic and treating the sentences of second-order arithmetic as sentences in a *two-sorted* first-order system.

‡ Some familiarity with the notion of constructive ordinal and with the system of notations O will occasionally be assumed in the remaining sections of this chapter (see §§11.7 and 11.8).

Our terminology is illustrated in the following examples.

Let $\mathcal{A}_1$ consist of all finite sequences of integers.

Let $\mathcal{A}_2$ consist of the empty sequence, $\emptyset$, together with all finite sequences formed from the integer 0; i.e., the sequences $\emptyset$, $<0>$, $<0,0>$, $<0,0,0>$, ...

Let $\mathcal{A}_3$ consist of $\emptyset$, $<0>$, $<0,0>$, and $<0,0,0>$.

Let $\mathcal{A}_4$ consist of $\emptyset$, $<0,0>$, and $<0,0,0>$.

Let $\mathcal{A}_5$ consist of $\emptyset$ together with all sequences $<n, n+1, \ldots, m>$, $0 \leq n$, $n \leq m \leq 2n$.

Let $\mathcal{A}_6$ have $\emptyset$ as its only member.

Let $\mathcal{A}_7$ be empty.

Then $\mathcal{A}_1$, $\mathcal{A}_2$, $\mathcal{A}_3$, $\mathcal{A}_5$, $\mathcal{A}_6$, and $\mathcal{A}_7$ are trees. $\mathcal{A}_4$ is not a tree (since the vertex $<0>$ is missing). $\mathcal{A}_1$ is the tree $\mathfrak{J}^*$. Each tree is a subtree of itself. Furthermore, $\mathcal{A}_2$, $\mathcal{A}_3$, $\mathcal{A}_5$, $\mathcal{A}_6$, and $\mathcal{A}_7$ are subtrees of $\mathcal{A}_1$; $\mathcal{A}_3$, $\mathcal{A}_6$, and $\mathcal{A}_7$ are subtrees of $\mathcal{A}_2$; $\mathcal{A}_6$ and $\mathcal{A}_7$ are subtrees of $\mathcal{A}_5$; and $\mathcal{A}_7$ is a subtree of $\mathcal{A}_6$. $\mathcal{A}_1$ has $2^{\aleph_0}$ branches, all of which are infinite—one branch for each infinite sequence of integers (i.e., one branch for each member of $\mathfrak{F}$). $\mathcal{A}_2$ has a single branch which is infinite and corresponds to the function $\lambda x[0]$. $\mathcal{A}_3$ has a single branch which is finite and is a subbranch of the branch of $\mathcal{A}_2$. $\mathcal{A}_5$ has $\aleph_0$ branches, all of which are finite. $\mathcal{A}_6$ has a single branch; this branch is a subbranch of all branches $\mathcal{A}_1$, $\mathcal{A}_2$, $\mathcal{A}_3$, $\mathcal{A}_5$, and $\mathcal{A}_6$. $\mathcal{A}_7$ has a single branch; this branch is empty and is a subbranch of all branches of all trees. $\mathcal{A}_3$, $\mathcal{A}_5$, $\mathcal{A}_6$, and $\mathcal{A}_7$ are finite-path trees. $\mathcal{A}_1$ and $\mathcal{A}_2$ are not finite-path trees.

Notation $\tau, \tau_0, \tau_1, \ldots$ will be trees. Whether the symbol "τ" denotes a tree or the pairing function of §5.3 will be clear from context on all occasions below.

We shall use sequence numbers as a coding for vertices. Thus, for any f and x, $\bar{f}(x)$ is the code number for the $(x+1)$st vertex along the branch of the function tree determined by f. In particular, $\bar{f}(0)$ is a code number for the maximum vertex $\emptyset$. (Recall that for all f, $\bar{f}(0) = 0$.) We shall, for many purposes, identify vertices with sequence numbers, and trees with certain sets of sequence numbers. We remark that a tree τ is finite-path if and only if $(\forall f)(\exists x)$[the vertex (encoded by) $\bar{f}(x)$ is not in τ].

Ordinal numbers can be used to classify finite-path trees. Let a finite-path tree τ be given. If τ is nonempty, we define an ordinal-valued function o_τ on the vertices of τ as follows.

If a is a minimal vertex of τ, let $o_\tau(a) = 1$.

If a is not a minimal vertex of τ, let $o_\tau(a) =$ the least ordinal greater than all ordinals in $\{o_\tau(b) | b \in \mathcal{B}\}$, where $\mathcal{B} = \{b | b \textit{ is below } a \textit{ in } \tau\}$.

It is easily shown (using the finite-path property) that these inductive conditions determine a unique function o_τ on the tree τ. The ordinal which o_τ associates with the maximum element of τ is called the *ordinal* of τ, and we denote it $o(\tau)$. If τ is empty, we take its ordinal to be 0. Note that if

$\tau_1 \subset \tau_2$, then $o(\tau_1) \leq o(\tau_2)$. In the examples above, $o(\mathcal{A}_3) = 4$, $o(\mathcal{A}_5) = \omega$, $o(\mathcal{A}_6) = 1$, and $o(\mathcal{A}_7) = 0$.†

What ordinals can be associated with trees in this way? Note first that for a fixed finite-path tree τ, every nonzero ordinal less than $o(\tau)$ appears as a value of o_τ (see Exercise 16-23). Hence, since there are at most denumerably many vertices in τ, $o(\tau)$ must be a countable ordinal. Conversely, every countable ordinal occurs as the ordinal of some finite-path tree. To show this, let α be a countable ordinal. For each limit number β, $\beta \leq \alpha$, choose a single fixed fundamental sequence having β as limit. We associate ordinals with vertices in the function tree as follows. If $\alpha \neq 0$, we associate α with the maximum vertex in the function tree. Assume that an ordinal γ has been associated with a vertex a. If $\gamma = \beta + 1$ and $\beta \neq 0$, we associate β with each of the (infinitely many) vertices immediately below a. If γ is a limit number, we associate the nonzero members of the fundamental sequence chosen for γ with the (infinite) succession of vertices that are immediately below a. Consider those vertices which have ordinals associated with them by this inductive procedure. They are easily seen to constitute a finite-path tree τ such that $o(\tau) = \alpha$ (see Exercise 16-24). We thus have the following.

Theorem XVIII(a) *For every finite-path tree τ there is a countable ordinal α such that $o(\tau) = \alpha$.*

(***b***) *For every countable ordinal α there is a finite-path tree τ such that $\alpha = o(\tau)$.*

Proof. See discussion above.☒‡

Natural operations of addition and multiplication can be defined for finite-path trees. We summarize the definitions informally. Let τ_1 and τ_2 be finite-path trees. (1) $\tau_1 + \tau_2$ is the finite-path tree obtained by "affixing" a replica of τ_1 to every terminal element of τ_2. (2) $\tau_1 \cdot \tau_2$ is the finite-path tree obtained by "inserting" a replica of τ_1 at every non-limit vertex of τ_2. These definitions accord with the usual addition and multiplication operations for ordinals in that $o(\tau_1 + \tau_2) = o(\tau_1) + o(\tau_2)$ and $o(\tau_1 \cdot \tau_2) = o(\tau_1) \cdot o(\tau_2)$ (see Exercise 16-25).

Definition Let n be given and let y be the sequence number of the finite sequence $\langle y_1, \ldots, y_k \rangle$. Then $n \circledast y$ is the sequence number of the finite sequence $\langle n, y_1, \ldots, y_k \rangle$.§

† For certain purposes (in treating the notion of finite-path tree as a generalization of the notion of ordinal number), it is more convenient to limit our attention to nonempty trees and to define o_τ by letting $o_\tau(a) = 0$ if a is a minimal vertex. We do not do this here, since we wish to include the empty tree as a finite-path tree.

‡ Finite-path trees give a useful insight into the nature of ordinals. In particular, it is sometimes easier to visualize and work with ordinals like ω, ω^ω, and ϵ_0 in terms of finite-path trees than in terms of linear well-orderings.

§ In particular, if y is the sequence number for the empty sequence, then $n \circledast y$ is the sequence number for $\langle n \rangle$. In §13.7, $n \circledast f$ was defined for integer n and function f. The two definitions are related: $n \circledast (\bar{f}(x)) = \overline{(n \circledast f)}(x + 1)$.

Definition If τ is a tree, then the nth *branch tree of* τ is the tree consisting of all sequence numbers y such that $n \circledast y$ is in τ. (In the examples at the beginning of this section every branch tree of $\mathcal{A}_1$ is $\mathcal{A}_1$ itself; all branch trees of $\mathcal{A}_2$ are empty except for the 0th which is $\mathcal{A}_2$ itself; every branch tree of $\mathcal{A}_3$ is empty except for the 0th which is $\{\emptyset, <0>, <0,0>\}$; all branch trees of $\mathcal{A}_6$ and $\mathcal{A}_7$ are empty.)

It is useful for certain kinds of inductive argument to reduce questions about finite-path trees to questions about their branch trees. In the following combinatorial lemma we show how to compare the ordinal size of finite-path trees by comparing, in an appropriate way, the ordinal size of their branch trees.

Basic tree lemma *Let τ_1 and τ_2 be finite-path trees. Then*

$$o(\tau_1) < o(\tau_2) \Leftrightarrow [[o(\tau_1) = 0 \;\&\; o(\tau_2) = 1] \vee (\exists \textit{ branch tree } \sigma_2 \textit{ of } \tau_2)(\forall \textit{ branch tree } \sigma_1 \textit{ of } \tau_1)[o(\sigma_1) < o(\sigma_2)]]$$
$$\Leftrightarrow [o(\tau_2) \neq 0 \;\&\; [o(\tau_1) = 0 \vee (\exists \textit{ branch tree } \sigma_2 \textit{ of } \tau_2)(\forall \textit{ branch tree } \sigma_1 \textit{ of } \tau_1)[o(\sigma_1) < o(\sigma_2)]]]$$

Proof. See Exercise 16-26.☒

A tree can be considered as a collection of sequence numbers. The characteristic function of this set of sequence numbers will be called the *characteristic function of the tree.*

Definition Let τ be a tree. τ is a *recursive tree* if τ has a recursive characteristic function.

If φ_z is the characteristic function of a tree τ, then $\lambda y[\varphi_z(n \circledast y)]$ is evidently the characteristic function of the nth branch tree of τ, and an index for this characteristic function can be found uniformly from z and n.

Definition b is a (chosen and fixed) recursive function such that for any n and z, $\varphi_{b(n,z)} = \lambda y[\varphi_z(n \circledast y)]$.

Definition $T = \{z|\varphi_z$ *is the characteristic function of a finite-path tree*$\}$.

The members of T may be thought of as "notations" for recursive finite-path trees (analogous to the "notations" for recursive ordinals considered in Exercise 11-61.)

Definition For $z \in T$, let τ_z be the tree determined by φ_z, and let $\|z\| = o(\tau_z)$.

What ordinals are represented by recursive finite-path trees? We answer this in the following theorem.

Theorem XIX(a) *For every recursive finite-path tree τ there is a constructive ordinal α such that $o(\tau) = \alpha$.*

(b) *For every constructive ordinal α there is a recursive finite-path tree τ such that $\alpha = o(\tau)$.*

Proof. (a) Let τ be any tree. Each vertex of τ is a finite sequence.

Linearly order these finite sequences under the relation: $a < b$ if b is a proper initial segment of a or (in the case that neither a nor b is an initial segment of the other) if the first member of a different from a corresponding member of b is smaller than that member of b. For example, $\langle 0,4,7,9\rangle$ is smaller than $\langle 0,4\rangle$, and $\langle 0,3\rangle$ is smaller than $\langle 0,4,7,9\rangle$. This ordering is known as the *Kleene-Brouwer ordering* of τ. It is easily shown that the Kleene-Brouwer ordering is a well-ordering if and only if τ is finite-path, and that the ordinal of the Kleene-Brouwer ordering is at least as great as $o(\tau)$ (see Exercise 16-27). Assume τ is a recursive finite-path tree. Then the Kleene-Brouwer ordering of τ is a recursive well-ordering (see Exercise 16-27). Hence its order type is a recursive ordinal, and hence, by Theorem 11-XX, a constructive ordinal. Hence $o(\tau)$ is a constructive ordinal.

(*b*) If $\alpha = 0$, take τ to be the empty tree, which is recursive.

Let α be a constructive ordinal $\neq 0$. Take a notation for α in the notation system O of §11.7. In the system O, each limit notation determines a fundamental sequence of notations uniquely and effectively. If we now construct the finite-path tree used to prove (*b*) in Theorem XVIII, we see that this finite-path tree is recursive (see Exercise 16-28).☒

Other kinds of "effective" finite-path trees can be defined. We mention two such: (1) call a finite-path tree *recursively enumerable* if its vertices (i.e., sequence numbers) constitute a recursively enumerable set; (2) call a finite-path tree τ *strongly recursive* if there is a recursive function h such that $h(x) = 2$ if x is (the sequence number for) a nonterminal vertex of τ, $h(x) = 1$ if x is a terminal vertex of τ, and $h(x) = 0$ if x is not a vertex of τ. The "recursive" finite-path trees mentioned in §15.2 were, in fact, strongly recursive trees. It is easy to show that there are recursively enumerable finite-path trees which are not recursive and that there are recursive finite-path trees which are not strongly recursive. It is also easy to show that all three definitions yield the same ordinals, namely, the constructive ordinals (see Exercise 16-29).

What is the degree of the set T? Where does it fall in the analytical hierarchy? We answer these questions in the following theorem.

Theorem XX *T is a complete Π^1_1-set. That is to say,* $(\forall A)[A \in \Pi^1_1 \Leftrightarrow A \leq_1 T]$ *(and hence $\bar{T} \equiv E^1$).*

Proof. We first show that T is a Π^1_1-set. Observe that φ_z is a characteristic function for a tree $\Leftrightarrow [(\forall x)[\varphi_z(x) = 0 \vee \varphi_z(x) = 1] \;\&\; (\forall x)(\forall y)[[\varphi_z(y) = 1$ & the vertex (with sequence number) x is an initial segment of (i.e., lies above) the vertex (with sequence number) $y] \Rightarrow \varphi_z(x) = 1]]$. Hence, by the Tarski-Kuratowski algorithm, there is a recursive relation C such that φ_z is the characteristic function of a tree $\Leftrightarrow (\forall x)(\exists y)C(x,y,z)$. Thus we have $z \in T \Leftrightarrow (\forall x)(\exists y)C(x,y,z) \;\&\; (\forall f)(\exists x)[\varphi_z(\bar{f}(x)) = 0]$. Hence, by the Tarski-Kuratowski algorithm, $T \in \Pi^1_1$.

It remains to show that for any $A \in \Pi^1_1$, $A \leq_1 T$. Let $A \in \Pi^1_1$. Then by Theorem V, $A = \{x|(\forall f)(\exists w)T^*_{1,1}(z,\bar{f}(w),x)\}$, for some z. Let x be given. Define ψ as follows:

$$\psi(y) = \begin{cases} 1, & \text{if } \neg T^*_{1,1}(z,y,x); \\ 0, & \text{otherwise.} \end{cases}$$

It follows from the definition of the relation T^* that ψ is the characteristic function of a tree. Furthermore ψ is recursive with index depending effectively on x. Take h to be a recursive function such that h is one-one and $\psi = \varphi_{h(x)}$. Then $x \in A \Leftrightarrow \psi$ is the characteristic function of a finite-path tree $\Leftrightarrow h(x) \in T$. Thus $A \leq_1 T$ via h.☒

Corollary XX(a) *There exists a recursive function f^* of two variables such that for any z, the set with Π_1^1-index z is 1-reducible to T via $\lambda x[f^*(z,x)]$.*

Proof. In the proof of the theorem, the index of ψ depends uniformly on both x and z.☒

Corollary XX(b) *The sets O and W defined in* §11.7 *and Exercise* 11-61 *are Π_1^1-complete. Hence $T \equiv O \equiv W$.*

Proof. In §16.1 we showed that $W \in \Pi_1^1$. Hence $W \leq_1 T$. In Exercise 16-27 we use the Kleene-Brouwer ordering to show that $T \leq_1 W$. In Exercise 11-61, we showed that $O \equiv W$. Hence $T \equiv O \equiv W$, and all three sets are Π_1^1-complete.☒

It follows by the hierarchy theorem that none of these three sets is hyperarithmetical (and a fortiori that none is arithmetical).

We shall use recursive finite-path trees to study Π_1^1-sets, hyperarithmetical sets, and Δ_2^1-sets. In the proof of Theorem XX we have seen how any given predicate form for a Π_1^1-set associates recursive finite-path trees with the members of that Π_1^1-set. In the next section we shall see that the degree of unsolvability of a Π_1^1-set is related to the ordinals given by these finite-path trees.

We conclude with the following definition.

Definition For any ordinal α, $T_\alpha = \{z|z \in T \;\&\; \|z\| < \alpha\}$.

Note that $\alpha < \beta \Rightarrow T_\alpha \subset T_\beta$; that α is nonconstructive if and only if $T_\alpha = T$; and that for each α, T_α is a cylinder.

§16.4 Π_1^1-SETS AND Δ_1^1-SETS

In this section we obtain certain fundamental results concerning Δ_1^1-sets and Π_1^1-sets. The results were first obtained by Spector (in a form based on recursive ordinals rather than on recursive finite-path trees).

Definition Let $A \in \Delta_1^1$. If u is a Σ_1^1-index for A and v is a Π_1^1-index for A, then $w = \langle u,v \rangle$ will be called a Δ_1^1-*index* for A.

Theorem XXI *For any constructive ordinal* α, $T_\alpha \in \Delta_1^1$. *Moreover, this result holds uniformly in the following sense:* $(\exists$ *recursive* $g)(\forall z)[z \in T \Rightarrow$ $g(z)$ *is a* Δ_1^1*-index* for $T_{\|z\|}]$.

The proof of Theorem XXI uses the recursion theorem. The application of the recursion theorem in this proof, and in several later proofs below, is summarized in the following lemma, which we call the *recursion lemma.* It generalizes Exercise 11-33. One form of the lemma was indicated in Rogers [1959*a*]; the specific form given here is found in Enderton [1964].

Definition A partial ordering is *well-founded* if every nonempty subset has a minimal element (or, equivalently (under the axiom of choice), if there are no infinite descending chains).

Recursion lemma *Let* $S \subset N$ *be partially ordered by* $<_S$, *where* $<_S$ *is well-founded. For any* $y \in S$, *we define* $S_y = \{x|x <_S y\}$. *Let* $R \subset N^2$ *be a given relation. For any* $S' \subset S$, *we say that a partial function* φ *is* satisfactory *on* S' *(with respect to* R*) if* $S' \subset$ *domain* φ *and* $(\forall x)[x \in S' \Rightarrow R(x,\varphi(x))]$. *Assume that there is a partial recursive function* η *of two variables such that for any* $y \in S$ *and any* z, $[\varphi_z$ *satisfactory on* $S_y \Rightarrow [\eta(z,y)$ *is convergent and* $R(y,\eta(z,y))]]$. *It follows that there is a partial recursive function* ψ *which is satisfactory on* S *(and, a fortiori, that* $(\forall x)[x \in S \Rightarrow (\exists w)R(x,w)]$).

Comment on motivation. In applications, R will be a relation for which we wish to show that $(\forall x)[x \in S \Rightarrow (\exists w)R(x,w)]$. In Theorem XXI, for example, S will be T and R will be $\{<x,w>|x \in T$ & $T_{\|x\|} \in \Delta_1^1$ & w *is a* Δ_1^1*-index for* $T_{\|x\|}\}$.

Proof of recursion lemma. Take a recursive g such that for all z,

$$\varphi_{g(z)} = \lambda y[\eta(z,y)].$$

By the recursion theorem, find a z_0 such that $\varphi_{z_0} = \varphi_{g(z_0)} = \lambda y[\eta(z_0,y)]$. We claim that φ_{z_0} is the desired ψ.

Assume otherwise. Then $\{x|x \in S$ & $[\varphi_{z_0}(x)$ *is divergent or* $R(x,\varphi_{z_0}(x))$ *fails*$]\}$ must have a minimal element y_0. But φ_{z_0} is satisfactory on S_{y_0}, and by the property of η, $\eta(z_0,y_0)$ is convergent and $R(y_0,\eta(z_0,y_0))$. Since $\varphi_{z_0}(y_0) = \eta(z_0,y_0)$, this is a contradiction.☒

The following corollary is immediate.

Corollary to recursion lemma *If* η *(in the lemma) is a total recursive function, then* ψ *can be chosen to be a total recursive function.*

Proof. Immediate.☒

Note, in the recursion lemma, that an index for ψ depends uniformly on an index for η. It follows, trivially, that the lemma holds uniformly in the presence of integer parameters provided that η depends effectively on those parameters (see Exercise 16-30).

Proof of Theorem XXI. The recursion lemma is applied as follows. Let S be T, and let $<_S$ be $\{<x,y> | x \in T \ \& \ y \in T \ \& \ \|x\| < \|y\|\}$. (Then for any $y \in T$, $S_y = T_{\|y\|}$.) $<_S$ is immediately seen to be well-founded. Take R (as noted above) to be $\{<x,w> | x \in T \ \& \ T_{\|x\|} \in \Delta_1^1 \ \& \ w \textit{ is a } \Delta_1^1\textit{-index for } T_{\|x\|}\}$. It remains to demonstrate the inductive assumption of the recursion lemma, i.e., to supply the partial recursive function η. Recall that for any n and y, if φ_y is the characteristic function of a tree τ, then $\varphi_{b(n,y)}$ is the characteristic function of the nth branch tree of τ.

Let $y \in T$ be given, and assume that we have a z such that for every $x \in S_y$ $(= T_{\|y\|})$, $\varphi_z(x)$ is a Δ_1^1-index for $T_{\|x\|}$ (and hence $\pi_1\varphi_z(x)$ is a Σ_1^1-index for $T_{\|x\|}$ and $\pi_2\varphi_z(x)$ is a Π_1^1-index for $T_{\|x\|}$). (This is the assumption that φ_z is *satisfactory* on S_y.)

By the basic tree lemma, for $y \in T$,

$$\begin{aligned} x \in T_{\|y\|} &\Leftrightarrow [x \in T \ \& \ \|x\| < \|y\|] \\ &\Leftrightarrow [\neg [\|y\| = 0] \ \& \ [[x \in T \ \& \ \|x\| = 0] \vee [x \in T \ \& \\ &\qquad\qquad (\exists m)(\forall n)[b(n,x) \in T_{\|b(m,y)\|}]]]] \\ &\Leftrightarrow [\varphi_y(0) = 1 \ \& \ [(\forall u)[\varphi_x(u) = 0] \vee [\varphi_x(0) = 1 \ \& \\ &\qquad\qquad (\exists m)(\forall n)[b(n,x) \in T_{\|b(m,y)\|}]]]]. \end{aligned}$$

(Recall that 0 is the sequence number for the maximum vertex of the function tree.)

Now if $\|y\| \neq 0$, then $(\forall m)[b(m,y) \in T_{\|y\|}]$; hence, by assumption, for any n, m, and x,

$$\begin{aligned} b(n,x) \in T_{\|b(m,y)\|} &\Leftrightarrow (\exists f)(\forall w) \neg T_{1,1}^*(\pi_1\varphi_z(b(m,y)), \bar{f}(w), b(n,x)) \\ &\Leftrightarrow (\forall f)(\exists w) T_{1,1}^*(\pi_2\varphi_z(b(m,y)), \bar{f}(w), b(n,x)). \end{aligned}$$

Since either of these last expressions can be substituted in the final expression above for $x \in T_{\|y\|}$, we have, by Tarski-Kuratowski computations, that $T_{\|y\|} \in \Delta_1^1$.

If $\|y\| = 0$, then the same substitutions yield $T_{\|y\|} \in \Delta_1^1$, since (i) the Σ_1^1-form and Π_1^1-form obtained by the Tarski-Kuratowski computations each express the empty set (because of the condition that $\varphi_y(0) = 1$), and (ii) $T_{\|y\|}$ is, in fact, the empty set.

Hence, by the uniformity of Theorem IV, we have a Δ_1^1-index for $T_{\|y\|}$ uniformly from z and y. Let η be a partial recursive function giving this index as $\eta(z,y)$. η now satisfies the assumption of the recursion lemma. In fact, as is easily seen, η can be chosen a total function, and the corollary to the recursion lemma applies. Theorem XXI now follows. ☒

A converse result, that a set A is Δ_1^1 if and only if ($\exists$ constructive α) $[A \leq_T T_\alpha]$, follows from Theorem XXII. Theorem XXII is the main result of the present section. Recall from Theorem XX that for any A, $A \in \Pi_1^1 \Leftrightarrow A \leq_1 T$. Theorem XXII asserts that if $A \leq_1 T$, then $A \in \Delta_1^1$ if and only if its image in T is bounded (by a constructive ordinal).

Theorem XXII **(*the boundedness theorem*)** *Let h be a recursive function such that* $(\forall x)[x \in A \Leftrightarrow h(x) \in T]$. *Then* $A \in \Delta_1^1 \Leftrightarrow (\exists$ *constructive* $\alpha)$ $(\forall x)[x \in A \Rightarrow \|h(x)\| < \alpha]$.

Proof. $\Leftarrow$: If the ordinal α exists, then $A = \{x|h(x) \in T_\alpha\}$. But $T_\alpha \in \Delta_1^1$ by Theorem XXI; hence $A \in \Delta_1^1$.

$\Rightarrow$: Assume that no such constructive α exists. It follows that $T = \{z|(\exists x)[x \in A \;\&\; z \in T_{\|h(x)\|}]\}$. Assume $A \in \Delta_1^1$; then $A \in \Sigma_1^1$, and by Theorem XXI and a Tarski-Kuratowski computation we have $T \in \Sigma_1^1$, contrary to Theorem XX.⊠

Corollary XXII(a) $A \in \Delta_1^1 \Leftrightarrow (\exists$ *constructive* $\alpha)[A \leq_1 T_\alpha] \Leftrightarrow (\exists$ *constructive* $\alpha)$ $[A \leq_T T_\alpha]$.

Proof. The first equivalence is immediate from Theorems XXI and XXII. In the second equivalence, $\Rightarrow$ is trivial. To show $\Leftarrow$, assume $A \leq_T T_\alpha$. Then $A \leq_1 (T_\alpha)'$. By Theorem X, $(T_\alpha)' \in \Delta_1^1$. Hence $(T_\alpha)' \leq_1 T_\beta$ for some constructive β, and $A \leq_1 T_\beta$.⊠

In the following corollary, we show that for $A \in \Delta_1^1$, the bound of Theorem XXII can be found uniformly (from a Δ_1^1-index for A).

Corollary XXII(b) $(\exists$ *recursive* $g)(\forall A)(\forall w)[[A \in \Delta_1^1 \;\&\; A$ *has* Δ_1^1*-index* $w] \Rightarrow [g(w) \in T \;\&\; A \leq_1 T_{\|g(w)\|}]]$.

Proof. Let $A \in \Delta_1^1$, and let $w = \langle u,v \rangle$ be a Δ_1^1-index for A.

For any z, let W_z^1 be the Π_1^1-set with Π_1^1-index z. Then the set E^1, as defined in §16.1, is $\{z|z \notin W_z^1\}$. Let p be a recursive permutation such that $p(\bar{T}) = E^1$. (p exists since both $\bar{T}$ and E^1 are Σ_1^1-complete.)

Since $A \in \Delta_1^1$, there is a recursive f such that $A \leq_1 T$ via f, where an index for f can be found uniformly from v (see Corollary XX(a)). Consider $B = \{z|(\exists x)[x \in A \;\&\; z \in T_{\|f(x)\|}]\}$. By assumption, $A \in \Sigma_1^1$. Hence by Theorem XXI, $B \in \Sigma_1^1$ with Σ_1^1-index uniform in u and v. Hence $p(\bar{B}) \in \Pi_1^1$ with Π_1^1-index q uniform in u and v. Thus $p(\bar{B}) = W_q^1$. Since $B \subset T$, we have $p(\bar{B}) \supset E^1$. Hence, by the definition of E^1, $q \in W_q^1$ and $q \notin E^1$. Hence $p^{-1}(q) \notin B \;\&\; p^{-1}(q) \in T$. Now $p^{-1}(q)$ is uniform in u and v. Let g' be a recursive function giving $p^{-1}(q)$ uniformly from $w = \langle u,v \rangle$. Then our construction yields $A \leq_1 T_{\|g'(w)\|+1}$ via f.

For any y, define

$$\psi(r) = \begin{cases} \varphi_y(s), & \text{if } r = n \circledast s \text{ for some } n; \\ 1, & \text{otherwise, i.e., if } r = 0. \end{cases}$$

Let h be a recursive function such that for all y, $\psi = \varphi_{h(y)}$. If $y \in T$, then $h(y) \in T$ and $\|h(y)\| = \|y\| + 1$, as is easily verified.

We now set $g = hg'$, and the proof is complete.⊠

In proving the above corollary, we have turned the proof by contradiction of Theorem XXII into a constructive proof by using the "productive" property of E^1 (and hence of $\bar{T}$) implicit in the definition of E^1. Analogous versions of Theorem XXII hold for W and O (see Exercise 16-32).

The main combinatorial tools used to obtain Theorem XXII were the quantifier rules of Theorem III. The reader will see the power of these rules more clearly if, for example, he tries to prove, by a direct argument about trees, that $(\forall \text{ constructive } \alpha)(\exists \text{ constructive } \beta)[\bar{T}_\alpha \leq_1 T_\beta]$. (This result is a corollary to Theorems XXI and XXII via Theorem X.)

Is it true that $\alpha \leq \beta \Rightarrow T_\alpha \leq_T T_\beta$? Singleterry [1965] shows that this is the case.

Definition $T_\alpha^* = \{<x,y>|x \in T \ \&\ y \in T \ \&\ \|x\| < \|y\| < \alpha\}$.

Evidently $\alpha \leq \beta \Rightarrow T_\alpha^* \leq_{\text{btt}} T_\beta^*$. Taking tt-cylinders, we have a well-ordered chain of 1-degrees such that for any set A, $A \in \Delta_1^1 \Leftrightarrow$ the 1-degree of A lies below some member of this chain (see Exercise 16-33). In §16.8 we consider the problem of finding, for the Δ_1^1-sets, the most natural and invariant analogue to the chain of degrees $\mathbf{0}$, $\mathbf{0}'$, $\mathbf{0}''$, . . . for the arithmetical sets.

The following theorem is a restatement and extension of Theorem XXI.

Theorem XXIII(a) *For* $y \in T$, $\{x|x \in T \ \&\ \|x\| < \|y\|\} \in \Delta_1^1$ *uniformly in* y *by a total function. (That is,* $(\exists \text{ recursive } g)(\forall y)[y \in T \Rightarrow g(y)$ *is a* Δ_1^1*-index for* $\{x|x \in T \ \&\ \|x\| < \|y\|\}]$.)

(*b*) *For* $y \in T$, $\{x|x \in T \ \&\ \|x\| = \|y\|\} \in \Delta_1^1$ *uniformly in* y *by a total function.*

(*c*) $\{<x,y>|x \in T \ \&\ y \in T \ \&\ \|x\| < \|y\|\} \in (\Pi_1^1 - \Sigma_1^1)$.

(*d*) $\{<x,y>|x \in T \ \&\ y \in T \ \&\ \|x\| = \|y\|\} \in (\Pi_1^1 - \Sigma_1^1)$.

Proof. (*a*) This is Theorem XXI.

(*b*) Take the recursive function h as defined at the end of the proof of Corollary XXII(*b*). Then for $y \in T$,

$$\{x|x \in T \ \&\ \|x\| = \|y\|\} = \{x|x \in T \ \&\ \|x\| < \|y\| + 1 \ \&\ \|y\| < \|x\| + 1\}$$
$$= \{x|x \in T_{\|h(y)\|} \ \&\ y \in T_{\|h(x)\|}\}.$$

Inserting the Π_1^1-forms and Σ_1^1-forms given (uniformly from $h(y)$ and $h(x)$) by Theorem XXI, we have the desired result by Tarski-Kuratowski computations. Note that the forms inserted for $T_{\|h(x)\|}$ express $T_{\|h(x)\|}$ in case $h(x) \in T$. What they express when $h(x) \notin T$ is irrelevant, since the forms for $T_{\|h(y)\|}$ will cause the final forms to be false in that case.

(*c*) Introduce the Π_1^1-form for $y \in T$ given by Theorem XX and the Π_1^1-form for $x \in T \ \&\ \|x\| < \|y\|$ given (uniformly from y) by (*a*) above. (What the latter expresses when $y \notin T$ is irrelevant.) By a Tarski-Kuratowski computation, we have the desired Π_1^1-form. If $\{<x,y>|x \in T \ \&\ y \in T \ \&\ \|x\| < \|y\|\}$ were in Σ_1^1, then

$$T = \{x|(\exists y)[x \in T \ \&\ y \in T \ \&\ \|x\| < \|y\|]\}$$

would be in Σ_1^1, contrary to Theorem XX.

(*d*) Similarly from Theorem XX and (*b*) above (see Exercise 16-34). ☒

It is occasionally convenient to introduce a Π_1^1 linear ordering on T. This can be done as follows.

Definition $x <^T y$ holds if

$$[x \in T \mathbin{\&} y \in T \mathbin{\&} [\|x\| < \|y\| \vee [\|x\| = \|y\| \mathbin{\&} x < y]]].$$

This is easily seen to be a well-ordering of order type of the least nonconstructive ordinal (see Exercise 16-35).

Corollary XXIII(a) *For* $y \in T$, $\{x|x <^T y\} \in \Delta_1^1$ *uniformly in* y *by a total function.*

(b) $\{\langle x,y\rangle|x <^T y\} \in (\Pi_1^1 - \Sigma_1^1)$.

Proof. (*a*) This is immediate by introduction of the appropriate predicate forms from parts (*a*) and (*b*) of the theorem.

(*b*) $\{\langle x,y\rangle|x <^T y\} \in \Pi_1^1$ by introduction of appropriate predicate forms from parts (*c*) and (*d*) of the theorem. If $\{\langle x,y\rangle|x <^T y\}$ were in Σ_1^1, then $T = \{x|(\exists y)[x <^T y]\}$ would be in Σ_1^1, contrary to Theorem XX. ☒

The corollary shows that there is a Π_1^1 linear ordering of order type of the least nonconstructive ordinal.

Theorem XXII can be formulated and proved in terms of the linear ordering $<^T$. Modifications of the earlier proof are trivial. We give the result in recursively invariant form.

Theorem XXIV *For every set A which is* Π_1^1*-complete, there exists a* Π_1^1 *well-ordering of A such that for every* Π_1^1*-set B and every recursive f, if* $B \leq_1 A$ *via f, then* $B \in \Delta_1^1 \Leftrightarrow f(B)$ *is bounded above in the linear ordering of A.*

Proof. As for Theorem XXII (see Exercise 16-36). ☒†

§16.5 GENERALIZED COMPUTABILITY

The theory of §16.1 was analogous in several respects to the theory of the arithmetical hierarchy developed in Chapter 14. We now explore further analogies between the analytical hierarchy and the arithmetical hierarchy. More specifically, we consider an analogy under which hyperarithmetical sets correspond to recursive sets and Π_1^1-sets correspond to recursively enumerable sets. The fact that the class Σ_1 consists of the recursively enumerable sets might at first suggest that an analogy between Σ_1^1-sets and recursively enumerable sets would be more natural. This is not the case,

† The reader with knowledge of set theory will note several analogies between Theorem XXIV and certain results in set theory (see especially the last part of Exercise 16-36).

1. Let a set of cardinality $\aleph_1$ be analogous to A (in Theorem XXIV); let sets of cardinality $\aleph_0$ or less be analogous to Δ_1^1-sets; and take an ordering of order type of the least uncountable ordinal as analogous to the Π_1^1 well-ordering of A.

2. In Gödel-Bernays set theory, let the universe be analogous to A, let *classes* be analogous to Π_1^1 subsets of A, let *sets* be analogous to Δ_1^1 subsets of A, and take a well-ordering of the universe as analogous to the Π_1^1 well-ordering of A. (Indeed, the definition of the usual well-ordering of the universe obtained through the concept of *rank* of a set closely parallels the definition of $<^T$.)

however, as we shall see. (The reduction principle, for example (see Theorem 5-XVII), will hold for Π_1^1-sets but not for Σ_1^1-sets.)

Recalling the results and discussion of §5.7, we begin our exploration of this analogy by proving a single-valuedness theorem.

Definition For each z, let $W_z^1 = \{x|(\forall f)(\exists w)T_{1,1}^*(z,\bar{f}(w),x)\}$. Thus W_z^1 is the Π_1^1-set whose Π_1^1-index is z.

In §5.7, we defined *domain* A to be $\{x|(\exists y)[<x,y> \in A]\}$, for any set A.

Theorem XXV ***(the single-valuedness theorem for*** Π_1^1***-sets)*** *There exists a recursive k such that for all z,*

(i) $W_{k(z)}^1$ *is single-valued;*

(ii) $W_{k(z)}^1 \subset W_z^1$*;*

(iii) *domain* $W_{k(z)}^1 =$ *domain* W_z^1.

(Hence W_z^1 *single-valued* $\Rightarrow W_{k(z)}^1 = W_z^1$.*)*

Proof. In Corollary XX(a) we obtained a recursive f^* such that for any z, $W_z^1 \leq_1 T$ via $\lambda x[f^*(z,x)]$. Define

$$A_z = \{<x,y>|<x,y> \in W_z^1 \,\&\, (\forall y')[[y' \neq y \,\&\, <x,y'> \in W_z^1] \Rightarrow f^*(z,<x,y>) <^T f^*(z,<x,y'>)]\}.$$

But this can be expressed

$$A_z = \{<x,y>|<x,y> \in W_z^1 \,\&\, (\forall y') \neg [f^*(z,<x,y'>) <^T f^*(z,<x,y>)]\}.$$

Now for any $<x,y>$, Corollary XXIII(a) gives a Σ_1^1-form which expresses $\{u|u <^T f^*(z,<x,y>)\}$ when $<x,y> \in W_z^1$. Using this form (it does not matter what the form expresses when $<x,y> \notin W_z^1$), and using the Π_1^1-form for W_z^1, we have, by a Tarski-Kuratowski computation, that $A_z \in \Pi_1^1$ with a Π_1^1-index obtainable uniformly from z. Take a recursive k such that $A_z = W_{k(z)}^1$. The desired result is now immediate. ☒

Definition For each z, let $\varphi_z^1 = \{<x,y>|<x,y> \in W_{k(z)}^1\}$. The partial functions φ_z^1 are called the *partial* Π_1^1 *functions.*

Note that the domain of a partial Π_1^1 function is a Π_1^1-set (by a simple Tarski-Kuratowski computation), and that every Π_1^1-set is the domain of some partial Π_1^1 function. (Given the Π_1^1-set A, take the function $\{<x,x>|x \in A\}$.) Similarly for the range of a partial Π_1^1 function.

We now have the sets W_0^1, W_1^1, . . . analogous to the recursively enumerable sets W_0, W_1, . . . ; the partial functions φ_0^1, φ_1^1, . . . analogous to the partial recursive functions; and the hyperarithmetical sets (i.e., the Π_1^1-sets with Π_1^1 complements) analogous to the recursive sets.

Definition If a partial Π_1^1 function is total, we call it a *hyperarithmetical function.*

Note that if a partial Π_1^1 function ψ is total, then $\tau(\psi) = \{<x,y>|\psi(x) = y\}$ is a hyperarithmetical set. ($\tau(\psi) \in \Pi_1^1$ by definition. If ψ is total, we also have $\tau(\psi) = \{<x,y>|(\forall z)[\psi(x) = z \Rightarrow z = y]\}$ and this yields $\tau(\psi) \in \Sigma_1^1$.

In fact, for φ_z^1 total, a Δ_1^1-index for $\tau(\varphi_z^1)$ can be obtained uniformly from z by a total function. More generally, by the same argument, for any n and any f, $f \in \Pi_n^1 \Leftrightarrow f \in \Sigma_n^1 \Leftrightarrow f \in \Delta_n^1$.) Thus our terminology is consistent. A hyperarithmetical function is a function which is, as a relation, hyperarithmetical. Note further that a set is hyperarithmetical if and only if its characteristic function is hyperarithmetical.

A basic theorem of recursive function theory fails under our analogy. In recursive function theory, the range of a recursive function can be nonrecursive. Here we have the following (pointed out to the author by Hodes).

Theorem XXVI *The range of a hyperarithmetical function is a hyperarithmetical set.*

Proof. Let f be a hyperarithmetical function. Then

$$\textit{range}\, f = \{y | (\exists x)[f(x) = y]\}.$$

Since $f \in \Delta_1^1$, the result is immediate by a Tarski-Kuratowski computation. ☒

An analogue to Theorem 5-IV can, however, be obtained from this result.

Corollary XXVI **(Luckham)** *Every infinite* Π_1^1*-set has an infinite hyperarithmetical subset.*

Proof. Let A be an infinite Π_1^1-set. Form the Π_1^1-set

$$B = \{<x,y> | x < y \;\&\; y \in A\}.$$

Apply the single-valuedness theorem to B. This yields a total Π_1^1 function whose range is an infinite subset of A. By the theorem, this subset is Δ_1^1. ☒

Other theorems of §5.7 go through under the analogy.

Theorem XXVII **(reduction principle)** *Given any two* Π_1^1*-sets* A *and* B, *there exist* Π_1^1*-sets* A' *and* B' *such that* $A' \subset A$, $B' \subset B$,

$$A \cup B = A' \cup B',$$

and $A' \cap B' = \emptyset$.

Proof. From the single-valuedness theorem exactly as in Theorem 5-XVII (in §5.7). ☒

Theorem XXVIII **(selection theorem)** *There is a partial* Π_1^1 *function* ψ *such that for any* z,

(i) $\psi(z)$ *convergent* $\Leftrightarrow (\exists w)[w \in W_z^1 \;\&\; W_w^1 \neq \emptyset]$;

(ii) $\psi(z)$ *convergent* $\Rightarrow [\psi(z) \in W_z^1 \;\&\; W_{\psi(z)}^1 \neq \emptyset]$.

Proof. As in Theorem 5-XVIII (in §5.7). ☒

A generalized reduction principle follows as in Exercise 5-36.

Theorem 7-XII showed that there exists a disjoint pair of recursively enumerable sets which are not recursively separable. Does an analogue to this result go through; i.e., does there exist a disjoint pair of "hyperarithmetically inseparable" Π_1^1-sets?

Theorem XXIX *There exist disjoint Π_1^1-sets A and B such that for no Δ_1^1-set C is it true that $A \subset C$ and $B \subset \bar{C}$.*

Proof. Define $A = \{x|\varphi_x^1(x) = 0\}$ and $B = \{x|\varphi_x^1(x) = 1\}$. The result follows as in Theorem 7-XII(*c*). ☒

It follows from Theorem XXIX that the Σ_1^1-sets do not satisfy a reduction principle, just as the Π_1^0-sets did not satisfy a reduction principle (by Theorem 7-XII) (see Exercise 5-3). For this reason (among others), the analogy between Π_1^1-sets and Σ_1^0-sets is closer than the analogy between Σ_1^1-sets and Σ_1^0-sets.

The present discussion is, of course, concerned with sets of integers. Questions of reduction and separation can also be raised for sets of functions (see Exercises 16-39 to 16-41). As Addison [1959, 1959*a*] has shown, questions of reduction and separation can be viewed as questions concerning the permutability of quantifiers in certain predicate forms (see Exercise 16-41).

In Exercises 16-42 to 16-45, we give several further examples of results that parallel earlier results of recursive function theory. In Exercise 16-42, we use the single-valuedness theorem to construct a hyperarithmetically simple set (and conclude that there are sets in $(\Pi_1^1 - \Sigma_1^1)$ which are not recursively isomorphic to T (and in fact not hyperarithmetically isomorphic to T)). Exercise 16-43 gives an analogue to the recursion theorem. Exercise 16-44 uses this to obtain analogues to Myhill's theorem that every creative set is 1-complete. Exercise 16-45 gives an analogue to Rice's theorem (Theorem 14-XIV).

Given a definition or result of recursive function theory, formulation of the hyperarithmetical analogue is not, in general, unique. A *full analogue* is obtained when all partial recursive functions become partial Π_1^1 functions. Various *partial analogues* may be possible in which some partial recursive functions become partial Π_1^1 functions and some remain partial recursive functions. In a notion of "Π_1^1-productive" set, for example, we might or might not require the productive partial function to be recursive (see Exercises 16-44 and 16-46). Similarly, in results on 1-reducibility, we might or might not allow the reducibility functions to be hyperarithmetical. (The matter of full and partial relativization, discussed in §9.3, is similar.)

An interesting divergence between our new hyperarithmetical theory and the earlier recursive theory appears in Exercise 16-45 on Rice's theorem. Here (and occasionally but not always elsewhere) we find the role of finite sets (in the earlier theory) taken over by hyperarithmetical sets and the role of canonical indices taken over by Δ_1^1-indices. (See Exercise 16-47, where we show that given any Δ_1^1-index, a unique "canonical" Δ_1^1-index for the same set can be hyperarithmetically computed.)

In §16.6 we shall continue our development of this analogue theory by introducing counterparts to T-reducibility and to the jump operation. We shall see that the analogue to Post's problem has a different answer from before, and that, in the analytical hierarchy, a strong hierarchy theorem (analogous to Theorem 14-VIII) fails.

We remark that the similarity between our analogue theory and recursive function theory can be traced back to the use of the relation $<^T$ and the function f^* in the proof of Theorem XXV. In effect, all convergent "Π_1^1-computations" are placed in a linear ordering such that we can tell hyperarithmetically (by §16.4) which of any two given convergent computations "comes first." This transfinite ordering is thus a counterpart to the ordering of recursive computations given by Corollary 5-V(*d*) and used in the proof of Theorem 5-XVI.

Generalized Machines

The concept of Turing machine can be generalized as follows (see also Kleene [1962]). The set of "instructions" defining a *generalized machine* remains finite, but certain *special steps* are permitted to occur in a computation, and appropriate symbols and conventions are introduced for indicating these special steps. The concept of *admissible* special step is defined inductively. An admissible special step tells whether or not an ordinary Turing-machine computation (instructions for which are currently given on the tape of the generalized machine) halts; or, more generally, an admissible special step tells whether or not a certain generalized computation (instructions for which are currently given on the tape) halts, provided that the generalized computation named on the tape involves no steps other than ordinary Turing-machine steps or special steps already defined to be admissible. Given an input and a generalized machine, the resulting computation is said to be *divergent* if either (i) it does not terminate, or (ii) a nonadmissible special step occurs. We do not go further into formal details.

With any convergent computation (on a generalized machine) can be associated a *tree diagram.* This diagram indicates, at its first level, the steps of the computation; at its second level, substeps of special steps occurring at the first level; at its third level, substeps of special steps occurring at the second level; etc. The *tree diagram* is isomorphic to a finite-path tree, and this finite-path tree has a hyperarithmetical characteristic function. (In §16.6, we shall see that the ordinal of a hyperarithmetical finite-path tree is constructive.) The Kleene-Brouwer ordering corresponding to this finite-path tree suggests a "physical" realization for such a computation. Let the unit interval (of real numbers) be thought of as a finite interval of time. Take an order-preserving mapping of the Kleene-Brouwer ordering (of our computation) into this interval. The resulting points may be thought of as moments of time at which steps of the generalized computation occur. Thus, for example, the generalized machine which solves the ordinary halting problem might, for a given input, perform its steps at times $\frac{1}{2}$, $\frac{3}{4}$, $\frac{7}{8}$, . . . , 1.

What partial functions can be computed by such generalized machines?

It is possible to show that exactly the partial Π_1^1 functions are obtained. We do not give details of this result.

The notion of generalized machine thus throws considerable light on the notion of hyperarithmetical function (and of partial Π_1^1 function), and suggests that these functions constitute a class that is even more intuitively natural than we might previously have supposed. To put it loosely, a hyperarithmetical function is a function which we can compute deterministically, in discrete steps, from a finite set of instructions, if we have the ability to carry out a search through an infinite sequence of steps whenever we wish, i.e., in any given time interval of positive length. Spector and Addison have referred to this level of computational ability as the "$\aleph_0$-mind," and Kreisel has suggested a counterpart to Church's Thesis for the hyperarithmetical functions. This intuitive view of hyperarithmetical computability can be of considerable heuristic value in connection with certain general properties of the hyperarithmetical sets (e.g., some of the strong closure properties of hyperarithmetical sets considered in the next section), and in a variety of specific problems (Exercise 16-46, for example). It has also been used to support "constructivist" claims that the hyperarithmetical sets are, in some sense, the only sets of integers that exist.

This intuitive view also suggests that hyperarithmetical computability of functions and functionals may be an appropriate framework within which to develop a constructive version of real-variable analysis. Kreisel and Feferman have initiated such a development as a study of effective objects within classical mathematics; Hodes and Ritter, following work of Moschovakis for the recursive case, have initiated such a development as a study of denumerable structures that are separate from, but analogous to, the nondenumerable structures of classical mathematics (see discussion in §15.3, and Exercises 15-36 and 15-37).

Unfortunately, this intuitive view of hyperarithmetical computability is, by itself, of less value in obtaining specific mathematical results about hyperarithmetical functions than Church's Thesis is in obtaining results about recursive functions. The main results that we have so far obtained have depended heavily on the combinatorial power of the quantifier rules of Theorem III. It is difficult, for example, to decide on the basis of intuitive considerations alone whether or not the selection theorem (Theorem XXVIII) is true; whereas the recursive counterpart (Theorem 5-XVIII) is easily seen to be true on the basis of Church's Thesis and the enumeration theorem.†

Another Analogue to Recursive Function Theory

An analogue (to recursive function theory) like that described at the beginning of this section suggests looking for a more abstract theory that

† Of course, once the selection theorem is proved, we can (and should) use it to enrich our intuitive view of hyperarithmetical computability.

would apply to both the recursive and hyperarithmetical cases. With this in mind, Kreisel and Sacks have studied a somewhat different analogue to recursive function theory based on Π_1^1-sets and Δ_1^1-sets. The Kreisel-Sacks analogue proves to be closer to the recursive case than the analogue discussed earlier in this section. In the Kreisel-Sacks analogue, T plays the role of N. Let ψ be a partial function such that $\psi \subset T \times T$. ψ is "partial recursive" if $\psi \in \Pi_1^1$. ψ is "recursive" if ψ is "partial recursive" and *domain* $\psi = T$. Functions of more than one variable are introduced by a recursive "pairing" function that maps $T \times T$ onto T. (Such a function exists since $T \times T$ is easily seen to be Π_1^1-complete, and hence $T \times T \equiv T$.) Let $A \subset T$. A is "recursively enumerable" if A is Π_1^1. A is "recursive" if A and $T - A$ are both Π_1^1. A "Gödel numbering" of the "partial recursive" functions (*onto* T) is obtained by taking any fixed one-one recursive map f of N into T, letting $f(z)$, for any z, be the "Gödel number" of $\varphi_z^1 \cap (T \times T)$, and letting the remaining members of T (i.e., the set $T - f(N)$) be Gödel numbers for the empty partial function. (Some members of $f(N)$ will also be Gödel numbers for the empty partial function.) See Exercise 16-48. This gives, via domains, an indexing (*onto* T) for the "recursively enumerable" sets. The existence of "recursively enumerable but not recursive" sets is now proved by an exact formal analogue to Theorem 5-VI.

Interestingly, as in the previous analogy, the Δ_1^1-subsets of T have many of the properties that finite sets have in the recursive theory. (Note that "recursive" sets need not be Δ_1^1; for instance, T is "recursive.") For example the following holds: if A is the range of a one-one "recursive" function, then A is not in Δ_1^1 (see Exercise 16-50).

The difficulty of Theorem XXVI no longer arises, and we have the result that [A is "recursively enumerable" and $A \neq \emptyset$] $\Leftrightarrow$ A is the range of some "recursive" function. ($\Leftarrow$ is immediate; to prove $\Rightarrow$, consider separately the cases A "finite" (i.e., in Δ_1^1) and A not "finite." Assume A not "finite" and let z be a Π_1^1-index for A. Then a Tarski-Kuratowski computation, using Corollary XXIII(a), shows that

$$\{\langle x,y\rangle | x \in T \;\&\; y \in A \;\&\; [x = f^*(z,y) \vee x <^T f^*(z,y)] \\ \& \; (\forall u)[f^*(z,u) <^T f^*(z,y) \Rightarrow f^*(z,u) <^T x]\}$$

is the desired "recursive" function. For the case A "finite," see Exercise 16-51.)

A variety of further results can be obtained, including the existence of "recursively enumerable" sets which are, in an analogous sense to that of §12.4, "maximal." We give one further illustration. Define a "recursive" function to be "strictly increasing" if it is strictly increasing with respect to $<^T$. Then the range of a "strictly increasing" function must be "recursive." For let A be the range of "strictly increasing" ψ. Then

$$A = \{y | (\exists z)[[z <^T y \vee z = y] \;\&\; y = \psi(z)]\}$$

and $T - A = \{y | y \in T \ \& \ (\forall z)[[z <^T y \vee z = y] \Rightarrow (\exists w)[w \neq y \ \& \ w = \psi(z)]]\}$. Using Corollary XXIII($a$), we have both A and $T - A$ in Π_1^1.

§16.6 HYPERDEGREES AND THE HYPERJUMP; Σ_2^1-SETS AND Δ_2^1-SETS

Pursuing the analogy discussed in the first part of the preceding section, we seek a concept analogous to Turing reducibility. In Chapter 14, we saw that $B \leq_T A \Leftrightarrow B \in \Pi_1^A \cap \Sigma_1^A$. This suggests that we define an analogue to Turing reducibility by relativization in the analytical hierarchy.

Relativization

The definitions and results of §16.1 can be relativized as follows. Let a set A be given.

Definitions ψ is a *partial A-recursive function of k function variables and l number variables* ($k \geq 0$, $l > 0$) if, for some z,

$$\psi = \lambda f_1 \cdots f_k x_1 \cdots x_l[\varphi_z^{A,\tau(f_1),\ldots,\tau(f_k)}(x_1, \ldots, x_l)].$$

ψ is a *partial A-recursive function of k function variables and* 0 *number variables* ($k \geq 0$) if, for some z, $\psi = \lambda f_1 \cdots f_k[\varphi_z^{A,\tau(f_1),\ldots,\tau(f_k)}(0)]$.

ψ is a *(total) A-recursive function of k function variables and l number variables* ($k \geq 0$, $l \geq 0$) if ψ is a partial A-recursive function of k function variables and l number variables and *domain* $\psi = \mathfrak{F}^k \times N^l$.

R ($\subset \mathfrak{F}^k \times N^l$) is an *A-recursive* relation if its characteristic function is a (total) A-recursive function of k function variables and l number variables.

R ($\subset \mathfrak{F}^k \times N^l$) is *analytical in A* if R is either A-recursive or is obtained from an A-recursive relation by projections and/or complementations.

A-forms are defined in the same way as predicate forms, except that the recursive relation is replaced by an A-recursive relation.

A relativized version of Theorem I holds. $\Sigma_n^{1,A}$-forms, $\Pi_n^{1,A}$-forms, and the classes $\Sigma_n^{1,A}$, $\Pi_n^{1,A}$, and $\Delta_n^{1,A}$ are now defined in the same way as before. Relativized versions of Theorems II and III hold.

Definition

$$T_{k,l}'^A = \{\langle z, f_1, \ldots, f_k, x_1, \ldots, x_l, w\rangle | T_{k+1,l}(z, A, \tau(f_1), \ldots, \tau(f_k), x_1, \ldots, x_l, w)\}.$$

Using $T_{k,l}'^A$ in place of $T_{k,l}'$, Theorem IV now holds in relativized form, and the notions of $\Sigma_n^{1,A}$*-index* and $\Pi_n^{1,A}$*-index* are defined.

Definition For $k > 0$, $l \geq 0$, $T_{k,l}^{*A} = \{\langle z, y_1, \ldots, y_k, x_1, \ldots, x_l\rangle |$ *there exist* $f_1, \ldots, f_k$ *and there exists a* w *such that for all* i, $1 \leq i \leq k$,

$$y_i = \bar{f}_i(w), \text{ and } T_{k,l}'^A(z, f_1, \ldots, f_k, x_1, \ldots, x_l, w),$$

where, in the computation for $T'^{A}_{k,l}$, all questions asked about $f_1, \ldots, f_k$ involve arguments less than w}.

We now have a relativized version of Theorem V.

Theorem XXX *Let $n > 0$, let R ($\subset \mathfrak{F}^k \times N^l$) be a relation in $\Sigma_n^{1,A}$, and let z be a $\Sigma_n^{1,A}$-index for R. If n is even, then*

$$R = \{<f_1, \ldots, f_k, x_1, \ldots, x_l>|(\exists g_1)(\forall g_2) \cdots (\forall g_n)(\exists w) T^{*A}_{k+n,l}(z, \bar{f}_1(w), \ldots, \bar{f}_k(w), \bar{g}_1(w), \ldots, \bar{g}_n(w), x_1, \ldots, x_l)\}.$$

Similarly for n odd and for $\Pi_n^{1,A}$ (see statement of Theorem V).

Proof. Immediate by definition. ☒

Relativization of Theorems VI and VII is now straightforward. Letting $E^{n,A}$ be the relativized counterpart to E^n (for example, $E^{1,A} = \{z|(\exists f)(\forall w) \neg T^{*A}_{1,1}(z, \bar{f}(w), z)\}$) we have immediate relativizations of Theorems VIII and IX.

Relativization with respect to a given function h is obtained by using $\tau(h)$ in place of A in the preceding definitions and discussion. This yields, in particular, that $T'^{\tau(h)}_{k,l}(z, \ldots) \Leftrightarrow T'_{k+1,l}(z, h, \ldots)$ and that Theorem XXX holds in the following form. (We write $\Sigma_n^{1,\tau(h)}$, $T'^{\tau(h)}_{k,l}$, ... as $\Sigma_n^{1,h}$, $T'^{h}_{k,l}$, ...)

Corollary XXX *Let $n > 0$, let R ($\subset \mathfrak{F}^k \times N^l$) be a relation in $\Sigma_n^{1,h}$, and let z be a $\Sigma_n^{1,h}$-index for R. If n is even, then*

$$R = \{<f_1, \ldots, f_k, x_1, \ldots, x_l>|(\exists g_1)(\forall g_2) \cdots (\forall g_n)(\exists w) T^{*}_{k+n+1,l}(z, \bar{h}(w), \bar{f}_1(w), \ldots, \bar{f}_k(w), \bar{g}_1(w), \ldots, \bar{g}_n(w), x_1, \ldots, x_l)\}.$$

Similarly for n odd and for $\Pi_n^{1,h}$.

Proof. Immediate from definitions (see Exercise 16-52). ☒

Simultaneous relativization with respect to several sets (or several functions) is also straightforward. $\Sigma_n^{1,A_1,\ldots,A_m}$, $\Pi_n^{1,A_1,\ldots,A_m}$, and $\Delta_n^{1,A_1,\ldots,A_m}$ are defined in the obvious way, where we take $T'^{A_1,\ldots,A_m}_{k,l}$ to be

$$\{<z, f_1, \ldots, f_k, x_1, \ldots, x_l, w>|T_{k+m,l}(z, A_1, \ldots, A_m, \tau(f_1), \ldots, \tau(f_k), x_1, \ldots, x_l, w)\}.$$

We shall write $\Sigma_n^{1,\tau(h_1),\ldots,\tau(h_m)}$, ... as $\Sigma_n^{1,h_1,\ldots,h_m}$, ...

Hyperdegrees and the Hyperjump

We now define our counterpart to T-reducibility.

Definition B is *hyperarithmetical in A* (notation: $B \leq_h A$) if $B \in \Delta_1^{1,A}$ $(= \Sigma_1^{1,A} \cap \Pi_1^{1,A})$. If $\tau(f) \leq_h \tau(g)$, we abbreviate this $f \leq_h g$.

Note the parallel in terminology to the notions *recursive in*, *arithmetical in*, and *analytical in*. If $B \leq_h A$, we sometimes say that B is *hyperarithmetically reducible* to A.

The relation $\leq_h$ is clearly reflexive. Is it transitive?

Theorem XXXI *If $C \leq_h B$ and $B \leq_h A$, then $C \leq_h A$.*

Proof. If $C \leq_h B$, then, in particular, $C \in \Sigma_1^{1,B}$. Hence, for some z, $C = \{x|(\exists f)(\forall w) \neg T'^B_{1,1}(z,f,x,w)\}$. Therefore, $x \in C \Leftrightarrow (\exists f)(\forall w)[\varphi_z^{B,\tau(f)}(x)$ is not convergent in w steps] $\Leftrightarrow (\exists f)(\forall w) \neg (\exists y)(\exists u)(\exists v)[<x,y,u,v>$ occurs by the wth step in the enumeration of $W_{\rho(z)}$ (from index $\rho(z)$) & $D_u \subset B \text{ join } \tau(f)$ & $D_v \subset \overline{B \text{ join } \tau(f)}$]. [$D_u \subset B \text{ join } \tau(f)$] can be expressed $(\forall x)[[2x \in D_u \Rightarrow x \in B]$ & $[2x + 1 \in D_u \Rightarrow x \in \tau(f)]]$, and $D_v \subset \overline{B \text{ join } \tau(f)}$ can be expressed $(\forall x)[[2x \in D_v \Rightarrow x \notin B]$ & $[2x + 1 \in D_v \Rightarrow x \notin \tau(f)]]$. By assumption, B can be expressed in both $\Sigma_1^{1,A}$-form and $\Pi_1^{1,A}$-form. Substitute the $\Pi_1^{1,A}$-form at the first occurrence of B (in the final expression for C above); and substitute the $\Sigma_1^{1,A}$-form at the second occurrence of B. A Tarski-Kuratowski computation now yields $C \in \Sigma_1^{1,A}$. By a similar argument, beginning with a $\Pi_1^{1,B}$-form for C, and substituting the $\Sigma_1^{1,A}$-form and $\Pi_1^{1,A}$-form for B in an appropriate way, we obtain $C \in \Pi_1^{1,A}$.⊠

Corollary XXXI(a) *If $B \leq_h A$ & $A \in \Delta_1^1$, then $B \in \Delta_1^1$.*

Proof. Immediate from proof of the theorem.⊠

Corollary XXXI(b) *If C is analytical in B and B is analytical in A, then C is analytical in A.*

Proof. By the same technique as for the theorem (see Exercise 16-53).⊠

Definition $A \equiv_h B$ if $A \leq_h B$ and $B \leq_h A$.

By Theorem XXXI, $\equiv_h$ is an equivalence relation. The equivalence classes of $\equiv_h$ are called *hyperdegrees* (and are the counterparts, in our analogue, to Turing degrees).

By Corollary XXXI(*a*), there is a minimum hyperdegree, which consists of the hyperarithmetical sets. By the hierarchy theorem (Theorem VIII) there must be other hyperdegrees containing Π_1^1-sets. We ask the counterpart to Post's problem (Chapter 10): are there more than two hyperdegrees containing Π_1^1-sets? The answer proves to be negative, as we now show.†

Let T be the Π_1^1-complete set defined in §16.3.

Theorem XXXII $A \in \Pi_1^1 \Rightarrow [A \in \Delta_1^1$ *or* $A \equiv_h T]$.

Proof. Let $A \in \Pi_1^1$. Then $A \leq_1 T$ and it is enough to show that either $A \in \Delta_1^1$ or $T \leq_h A$. Assume $A \notin \Delta_1^1$ and let $A \leq_1 T$ via f. Then by the boundedness theorem (Theorem XXII), we have

$$T = \{y|(\exists x)[x \in A \ \&\ y \in T_{\|f(x)\|}]\}.$$

Introducing the Σ_1^1-form or Π_1^1-form for $y \in T_{\|f(x)\|}$ (by Theorem XXI), and making a Tarski-Kuratowski computation, we obtain, respectively, a $\Sigma_1^{1,A}$-form or $\Pi_1^{1,A}$-form for T. Hence $T \in \Delta_1^{1,A}$ and $T \leq_h A$.⊠

† In the Kreisel-Sacks analogue (see discussion at end of §16.5), the counterpart to Post's problem has the same affirmative answer (by way of a similar proof) as Post's problem for recursively enumerable sets.

As a counterpart to *recursively enumerable in*, we make the following definition.

Definition A is Π_1^1 *in* B if $A \in \Pi_1^{1,B}$.

We define a counterpart to the *jump* (of §13.1) by choosing a particular set in $\Pi_1^{1,A}$ that is complete with respect to $\Pi_1^{1,A}$.

Definition $T^A = \{z | \varphi_z{}^A$ *is the characteristic function of a finite-path tree*$\}$.

A relativized version of Theorem XX is immediate, and we have that $T^A \in \Pi_1^{1,A}$, and that for all B, $B \in \Pi_1^{1,A} \Rightarrow B \leq_1 T^A$. By the relativized hierarchy theorem, $T^A \not\leq_h A$. We call T^A the *hyperjump* of A. (By analogy with §13.1, we could have defined a hyperjump by first defining $W_z^{1,A} = \{x | (\forall f)(\exists w) T_{1,1}'^A(z,f,x,w)\}$, and then taking $K^{1,A} = \{x | x \in W_x^{1,A}\}$ as the "hyperjump" of A. In Exercise 16-54, we show that $K^{1,A} \equiv T^A$.)

Theorem XXXIII $A \leq_h B \Leftrightarrow T^A \leq_1 T^B$.

Proof. The proof is parallel to that for Theorem 13-I(*e*).

$\Rightarrow$: Assume $A \leq_h B$. Since $T^A \in \Pi_1^{1,A}$, and $A \in \Delta_1^{1,B}$, we have, by a predicate form substitution as in the proof of Theorem XXXI, that $T^A \in \Pi_1^{1,B}$. Since T^B is $\Pi_1^{1,B}$-complete, we have $T^A \leq_1 T^B$.

$\Leftarrow$: Conversely, assume $T^A \leq_1 T^B$. Since $T^B \in \Pi_1^{1,B}$, we have $T^A \in \Pi_1^{1,B}$. But $A \leq_1 T^A$ and $\bar{A} \leq_1 T^A$ (since, trivially, $A \in \Pi_1^{1,A}$ and $\bar{A} \in \Pi_1^{1,A}$). Hence $A \in \Pi_1^{1,B}$ and $\bar{A} \in \Pi_1^{1,B}$. But this yields $A \in \Delta_1^{1,B}$.⊠

We thus have, from Theorem XXXIII and the remarks preceding it, a full counterpart to Theorem 13-I.

It follows from Theorem XXXIII that the hyperjump operation is well defined on hyperdegrees. We define one hyperdegree to be Π_1^1 *in* another if there is a member of the first which is Π_1^1 in a member of the second. We can ask questions about the structure of the ordering of hyperdegrees under the *hyperjump* and the relation Π_1^1 *in*, analogous to the questions studied in Chapter 13 about the T-degrees under the *jump* and the relation *recursively enumerable in*. This study was begun by Spector. Several results analogous to results in Chapter 13 have been found, including a proof of the existence of incomparable hyperdegrees (see Exercise 16-55). Are there hyperdegrees below the hyperdegree of T other than the minimum hyperdegree? (By Theorem XXXII, such degrees cannot contain any Π_1^1-sets.) In §16.7, we show that such hyperdegrees exist.

Does a counterpart to Theorem 14-VIII (the strong hierarchy theorem) hold? A negative answer is obtained as a corollary to the following theorem, a special case of which was first proved by Addison and Kleene [1957]. The theorem is a strong generalization of Theorem X.

Theorem XXXIV **(*Shoenfield* [*1962*])** *For all* n, $[A \in \Delta_n^{1,B}$ & $B \in \Delta_n^{1,C}] \Rightarrow A \in \Delta_n^{1,C}$.

Proof. The case $n = 0$ was proved in Corollary 14-VII, and the case

$n = 1$ was proved above in Theorem XXXI. We give the general proof for n odd; the proof for n even is similar.

Assume $A \in \Delta_n^{1,B}$ and $B \in \Delta_n^{1,C}$. Then

$$A = \{x|(\exists f_1)(\forall f_2) \cdots (\exists f_n)(\forall w)R(B,f_1, \ldots ,f_n,x,w)\}$$

and
$$A = \{x|(\forall f_1) \cdots (\forall f_n)(\exists w)S(B,f_1, \ldots ,f_n,x,w)\}$$

for some recursive R and S.

Hence

$$A = \{x|(\exists D)[D = B \,\&\, (\exists f_1) \cdots (\exists f_n)(\forall w)R(D,f_1, \ldots ,f_n,x,w)]\}.$$

But $D = B$ can be expressed $(\forall u)[u \in D \Rightarrow u \in B] \,\&\, (\forall u)[u \in B \Rightarrow u \in D]$. Introducing the $\Sigma_n^{1,C}$-form for B at the first occurrence and the $\Pi_n^{1,C}$-form for B at the second occurrence, we have a $\Sigma_n^{1,C}$-form for $\{D|D = B\}$. Introducing this into the expression for A, we obtain, by an appropriate Tarski-Kuratowski computation, that $A \in \Sigma_n^{1,C}$.

Similarly,

$$A = \{x|(\forall D)[D = B \Rightarrow (\forall f_1) \cdots (\forall f_n)(\exists w)S(D,f_1, \ldots ,f_n,x,w)]\}.$$

Introducing the same $\Sigma_n^{1,C}$-form for $\{D|D = B\}$ in this expression, we obtain $A \in \Pi_n^{1,C}$.⊠

Corollary XXXIV **(*Addison, Kleene*)** $A \in \Delta_2^1 \Rightarrow T^A \in \Delta_2^1$ *(and hence, in particular, $T^T \in \Delta_2^1$ and cannot be Σ_2^1-complete or Π_2^1-complete).*

Proof. Immediate, since $T^A \in \Pi_1^{1,A} \subset \Delta_2^{1,A}$.⊠

Other results and concepts in §§16.3 to 16.5 can be relativized. We make the following definitions. Let a set A be given.

Definition For $z \in T^A$, let $\tau_z{}^A$ be the finite-path tree of which $\varphi_z{}^A$ is the characteristic function, and let $\|z\|^A = o(\tau_z{}^A)$.

Definition α is an *A-constructive ordinal* if $\alpha = \|z\|^A$ for some $z \in T^A$. (The class of A-constructive ordinals can also be obtained by relativizing the notions of §11.8 or of Exercise 11-61.)

Definition $T_\alpha{}^A = \{x|x \in T^A \,\&\, \|x\|^A < \alpha\}$.

Definition Let b' be a fixed recursive function such that for any n, z, and X, $\varphi^X_{b'(n,z)} = \lambda y[\varphi_z{}^X(n \circledast y)]$. Hence, for any A and n, if $z \in T^A$, then $b'(n,z) \in T^A$ and $\tau^A_{b'(n,z)}$ is the nth branch tree of $\tau_z{}^A$. (b' is thus a version, appropriate to the relativized case, of the function b defined in §16.3.)

Relativized versions of Theorems XXI, XXII, and XXIII can now be proved. A discussion of A-hyperarithmetical computability can be carried out parallel to the discussion in §16.5; $W_z^{1,A}$ can be defined as the set with $\Pi_1^{1,A}$-index z, a single-valuedness theorem can be proved, and a concept of *partial function Π_1^1 in A* can be obtained. It is possible to show that the partial functions Π_1^1 in A are exactly the partial functions obtained if we equip the generalized machines of §16.5 with oracles in the obvious way.

We do not give this relativized development here. Instead we prove a somewhat stronger relativization of Theorem XXIII. Its usefulness in subsequent work makes it one of the chief results of the present section.

Theorem XXXV *For* $y \in T^B$, $\{x | x \in T^A \,\&\, \|x\|^A < \|y\|^B\} \in \Delta_1^{1,A,B}$ *uniformly in* y *by a total function. That is to say,* ($\exists$ *recursive* g)

$$(\forall A)(\forall B)(\forall y)[y \in T^B \Rightarrow g(y) \text{ is a } \Delta_1^{1,A,B}\text{-index for } \{x | x \in T^A \,\&\, \|x\|^A < \|y\|^B\} \; (= \{x | x \in T^A_{\|y\|^B}\})].$$

Proof. We apply the recursion lemma in a proof that parallels the proof of Theorem XXI. Let S (for the recursion lemma) be T^B, and let $<_S$ be $\{\langle x,y\rangle | x \in T^B \,\&\, y \in T^B \,\&\, \|x\|^B < \|y\|^B\}$. Let $y \in T^B$ be given. We assume that we have a z such that for every $x \in S_y$ $(= T^B_{\|y\|^B})$, $\varphi_z(x)$ is a $\Delta_1^{1,A,B}$-index for $T^A_{\|x\|^B}$. It will suffice to find a $\Delta_1^{1,A,B}$-index for $T^A_{\|y\|^B}$ uniformly from z and y.

Applying the basic tree lemma, we have

$$x \in T^A_{\|y\|^B} \Leftrightarrow [\varphi_y{}^B(0) = 1 \,\&\, [(\forall u)[\varphi_x{}^A(u) = 0] \vee [\varphi_x{}^A(0) = 1 \,\&\, (\exists m)(\forall n)[b'(n,x) \in T^A_{\|b'(m,y)\|^B}]]]].$$

Now if $\|y\|^B \neq 0$, then by assumption, for any n, m, and x,

$$\begin{aligned} b'(n,x) \in T^A_{\|b'(m,y)\|^B} &\Leftrightarrow (\exists f)(\forall w) \neg T_{1,1}^{*A,B}(\pi_1\varphi_z(b'(m,y)),\bar{f}(w),b'(n,x)) \\ &\Leftrightarrow (\forall f)(\exists w) T_{1,1}^{*A,B}(\pi_2\varphi_z(b'(m,y)),\bar{f}(w),b'(n,x)). \end{aligned}$$

Since either of these last expressions can be substituted in the expression above of $x \in T^A_{\|y\|^B}$, we have by Tarski-Kuratowski computations that $T^A_{\|y\|^B} \in \Delta_1^{1,A,B}$.

If $\|y\|^B = 0$, the same substitutions yield $T^A_{\|y\|^B} \in \Delta_1^{1,A,B}$, as explained in the proof of Theorem XXI.

This yields the desired result, as in the proof of Theorem XXI.☒

Corollary XXXV *For* $y \in T^B$, $\{x | x \in T^A \,\&\, \|x\|^A = \|y\|^B\} \in \Delta_1^{1,A,B}$ *uniformly in* y *by a total function.*

Proof. Observe that

$$\{x | x \in T^A \,\&\, \|x\|^A = \|y\|^B\} = \{x | x \in T^A \,\&\, \|x\|^A < \|y\|^B + 1 \,\&\, \|y\|^B < \|x\|^A + 1\},$$

and proceed as for Theorem XXIII(*b*).☒

What can be said about the class of A-constructive ordinals as A varies? In particular, what are the *hyperarithmetical ordinals*, i.e., the ordinals obtained from finite-path trees with hyperarithmetical characteristic functions? As a first application of Theorem XXXV, we prove the following theorem which yields as a corollary that the hyperarithmetical ordinals are just the constructive ordinals.

Definition Let λ = the least nonconstructive ordinal. For any A, let λ^A = the least ordinal that is not A-constructive.

Theorem XXXVI **(*Spector* [*1955*])** $B \leq_h A \Rightarrow \lambda^B \leq \lambda^A$.

Proof. Assume $B \leq_h A$, but $\lambda^A < \lambda^B$. Then there is a $y \in T^B$ such that $||y||^B = \lambda^A$. But then $T^A = \{x|x \in T^A \,\&\, ||x||^A < ||y||^B\}$. By Theorem XXXV, $T^A \in \Delta_1^{1,A,B}$. But $B \leq_h A \Rightarrow \Delta_1^{1,A,B} = \Delta_1^{1,A}$ by a predicate form substitution similar to that in the proof of Theorem XXXI. Hence $T^A \in \Delta_1^{1,A}$ and this contradicts the fact that T^A is a $\Pi_1^{1,A}$-complete set.☒

Corollary XXXVI **(*Spector*)** **(*a*)** $[x \in T^B \,\&\, B \in \Delta_1^1] \Rightarrow ||x||^B$ *constructive.*

(*b*) $\lambda < \lambda^A \Leftrightarrow T \leq_h A$.

(*c*) $B \leq_h A \Rightarrow [\lambda^B < \lambda^A \Leftrightarrow T^B \leq_h A]$.

Proof. (*a*) Note that for A recursive, $\lambda^A = \lambda$. Apply the theorem with A recursive and $B \in \Delta_1^1$.

(*b*) $\Rightarrow$: Assume $\lambda < \lambda^A$. Then $||y||^A = \lambda$ for some $y \in T^A$. Hence $T = \{x|x \in T \,\&\, ||x|| < ||y||^A\}$. By Theorem XXXV, $T \in \Delta_1^{1,A}$; that is, $T \leq_h A$ (by definition).

$\Leftarrow$: Assume $T \leq_h A$. By the theorem, $\lambda^T \leq \lambda^A$. It remains to show that $\lambda < \lambda^T$. Define ψ, a function recursive in T, as follows:

$$\psi(0) = 1;$$

$$\psi(n \circledast \tilde{f}(x)) = \begin{cases} 0, & \text{if } n \notin T; \\ \varphi_n(\tilde{f}(x)), & \text{if } n \in T. \end{cases}$$

Then ψ is the characteristic function of a finite-path tree τ, and every recursive finite-path tree occurs as some branch tree of τ. Thus $o(\tau) = \lambda$. Let $\psi = \varphi_z^T$. Then $||z||^T = \lambda$, and hence $\lambda < \lambda^T$.

(*c*) This is a generalization of (*b*) (see Exercise 16-57).☒

In §16.7, we shall show that there exist hyperdegrees of nonhyperarithmetical sets below the hyperdegree of T. This implies that the converse to Theorem XXXVI fails (see Exercise 16-58).

The result of Corollary XXXVI(*a*) can be viewed as a rather strong closure property of the hyperarithmetical sets. Given any hyperarithmetical set A, we might attempt to get more and more sets of higher degree by first forming all sets $T_\alpha{}^A$, as α ranges over the A-constructive ordinals, and then repeating this process on the sets so obtained. The corollary shows us that we cannot "escape" from the hyperarithmetical sets in this way; indeed, a single application of this operation to any recursive set gives all the sets T_α (up to recursive isomorphism) and all further sets obtained by repeating the operation must be reducible to these. This closure property accords with our informal discussion of generalized computability in §16.5.

Σ_2^1-sets and Δ_2^1-sets

Using Theorem XXXV, we obtain a theory of Σ_2^1-sets and Δ_2^1-sets parallel to the theory of Π_1^1-sets and Δ_1^1-sets given in §16.4. The results are due to Moschovakis and are related to results of Suzuki (see Theorems XLVI, XLVII, and XLVIII).

We define certain concepts analogous to T, $\|x\|$, and T_α. We use a superscript to distinguish the notations for these new concepts.

Definition Let $T^2 = \bigcup_{A \subset N} T^A$.

For $x \in T^2$, let $\mathfrak{A} = \{A | x \in T^A\}$, and let $\|x\|^2 = \min_{A \in \mathfrak{A}} \|x\|^A$.

For any ordinal α, let $T_\alpha{}^2 = \{x | x \in T^2 \ \& \ \|x\|^2 < \alpha\}$.

We first prove the analogue of Theorem XX, then obtain analogues to Theorems XXI and XXII.

Theorem XXXVII *T^2 is Σ_2^1-complete.*

Proof. By definition, $T^2 = \{z | (\exists A)[z \in T^A]\}$. We have seen that $T^A \in \Pi_1^{1,A}$. Hence $T^2 \in \Sigma_2^1$ (see comment on set variables and set quantifiers at end of §16.1).

Take any $B \in \Sigma_2^1$. Then $B = \{x | (\exists f)(\forall g)(\exists w) T_{2,1}^*(z, \bar{f}(w), \bar{g}(w), x)\}$, for some z. Given any x and f, define an f-recursive function ψ as follows. For all g and w,

$$\psi(\bar{g}(w)) = \begin{cases} 1, & \text{if } \neg\, T^*(z, \bar{f}(w), \bar{g}(w), x); \\ 0, & \text{otherwise.} \end{cases}$$

An index for ψ as an f-recursive function can be found uniformly from x (and independent of f). Let h be a one-one recursive function giving this index. Then $x \in B \Leftrightarrow (\exists f)$ [$\varphi^f_{h(x)}$ is the characteristic function of a finite-path tree] $\Leftrightarrow (\exists f)$ $[h(x) \in T^{\tau(f)}] \Leftrightarrow h(x) \in T^2$ (see Exercise 16-60). Hence $B \leq_1 T^2$ via h. ☒

Definition For any set A, $w = \langle u, v \rangle$ is a *Δ_2^1-index* for A if u is a Σ_2^1-index for A and v is a Π_2^1-index for A.

Theorem XXXVIII *For $y \in T^2$, $T^2_{\|y\|^2} \in \Delta_2^1$ uniformly in y by a total function; that is to say, ($\exists$ recursive g)$(\forall y)[y \in T^2 \Rightarrow g(y)$ is a Δ_2^1-index for $T^2_{\|y\|^2}$].*

Proof. Assume $y \in T^2$. Then

$$\begin{aligned} x \in T^2_{\|y\|^2} &\Leftrightarrow (\forall B)[y \in T^B \Rightarrow (\exists A)[x \in T^A \ \& \ \|x\|^A < \|y\|^B]] \\ &\Leftrightarrow (\exists A)[x \in T^A \ \& \ \neg\, (\exists B)[y \in T^B \ \& \ \|y\|^B < \|x\|^A + 1]]. \end{aligned}$$

Theorem XXXV yields a Π_2^1-form from the first expression and a Σ_2^1-form from the second, with indices uniform in y (see Exercise 16-61). ☒

Theorem XXXIX (boundedness theorem) *Let $A \in \Sigma_2^1$ and let h be a recursive function such that $(\forall x)[x \in A \Leftrightarrow h(x) \in T^2]$. Then $A \in \Delta_2^1 \Leftrightarrow (\exists y)[y \in T^2 \;\&\; (\forall x)[x \in A \Rightarrow \|h(x)\|^2 < \|y\|^2]]$.*

Proof. $\Leftarrow$: If y exists, then $A \leq_1 T^2_{\|y\|^2}$. But then $A \in \Delta_2^1$ by Theorem XXXVIII.

$\Rightarrow$: If no such y exists, then $T^2 = \{z|(\exists x)[x \in A \;\&\; z \in T^2_{\|h(x)\|^2}]\}$. If $A \in \Pi_2^1$, then, by Theorem XXXVIII, $T^2 \in \Pi_2^1$, contrary to Theorem XXXVII.☒

Corollary XXXIX(a)

$$A \in \Delta_2^1 \Leftrightarrow (\exists y)[y \in T^2 \;\&\; A \leq_1 T^2_{\|y\|^2}] \Leftrightarrow (\exists y)[y \in T^2 \;\&\; A \leq_T T^2_{\|y\|^2}].$$

(b) *$(\exists$ recursive $g)(\forall A)(\forall w)[$if $A \in \Delta_2^1$ with Δ_2^1-index w, then $g(w) \in T^2$ and $A \leq_1 T^2_{\|g(w)\|^2}]$.*

Proof. Proofs for (*a*) and (*b*) are parallel to the proofs for Corollary XXII (see Exercise 16-62).☒

Definition Let $\mathfrak{A} = \{\alpha|(\exists y)[y \in T^2 \;\&\; \alpha = \|y\|^2]\}$.

Definition α is a *Δ_2^1-ordinal* if there is a Δ_2^1-set A and a z such that $z \in T^A$ and $\alpha = \|z\|^A$.

What ordinals are in $\mathfrak{A}$? Every constructive ordinal is evidently in $\mathfrak{A}$. In §16.7, we shall show that every ordinal in $\mathfrak{A}$ is a Δ_2^1-ordinal and that there exist Δ_2^1-ordinals not in $\mathfrak{A}$ (Corollaries XLV(*d*) and XLV(*e*)). In the following theorem, we show that $\mathfrak{A}$ is not bounded above by a Δ_2^1-ordinal. It follows from this and the mentioned results of §16.7 that the ordinals of $\mathfrak{A}$ do not constitute an initial segment of ordinals. In Exercise 16-63 we see that the order type of $\mathfrak{A}$ is not a Δ_2^1-ordinal. $\mathfrak{A}$ thus consists of a proper cofinal subset of the Δ_2^1-ordinals with the same order type as the set of all Δ_2^1-ordinals.

Theorem XL $(\exists x)(\exists B)[B \in \Delta_2^1 \;\&\; x \in T^B \;\&\; \alpha = \|x\|^B] \Rightarrow (\exists y)[y \in T^2 \;\&\; \alpha < \|y\|^2]$.

Proof. Assume not. Then $B \in \Delta_2^1$, $x \in T^B$, and $(\forall y)[y \in T^2 \Rightarrow \|y\|^2 < \|x\|^B + 1]$. By the technique used for the end of the proof for Corollary XXII(*b*), find an x' such that $\|x'\|^B = \|x\|^B + 1$. Then

$$y \in T^2 \Leftrightarrow (\exists A)[\|y\|^A < \|x'\|^B].$$

But by Theorems XXXV and XXXIV, this yields $T^2 \in \Delta_2^1$, contrary to Theorem XXXVII.☒

Definition

$$x <^{T_2} y \text{ if } x \in T^2 \;\&\; y \in T^2 \;\&\; [\|x\|^2 < \|y\|^2 \vee [\|x\|^2 = \|y\|^2 \;\&\; x < y]].$$

$<^{T_2}$ is a Σ_2^1 relation and every bounded initial segment of $<^{T_2}$ is a Δ_2^1 relation (proof as for Corollary XXIII). This yields, incidentally, and without using the result of §16.7, that every bounded initial segment of $\mathfrak{A}$ is isomorphic (as a well-ordering) to a Δ_2^1-ordinal.

A theory of Σ_2^1-computability can now be developed parallel to the theory of Π_1^1-computability given in §16.5. In particular, a single-valuedness theorem, a reduction principle, and the existence of a disjoint pair of Σ_2^1-sets inseparable by any Δ_2^1-set can be proved. Similarly, a reduction principle for Σ_2^1-sets of functions can be shown (see Exercises 16-64 and 16-65).

At first glance, an extension of the above theory to Π_3^1 appears promising. Define $T^{A,B}$ in the obvious way. Let

$$T^3 = \bigcap_{B \subset N} \bigcup_{A \subset N} T^{A,B} = \{z|(\forall B)(\exists A)[z \in T^{A,B}]\}.$$

Let $||z||^3 = \sup_{B \subset N} \min_{A \in \mathfrak{a}} ||x||^{A,B}$, where $\mathfrak{a} = \{A|z \in T^{A,B}\}$. Let

$$T_\alpha{}^3 = \{z|z \in T^3 \;\&\; ||z||^3 < \alpha\}.$$

T^3 is easily seen to be Π_3^1-complete. $T^3_{||y||^3} \in \Sigma_3^1$ uniformly in y (if $y \in T^3$, then $[x \in T^3_{||y||^3} \Leftrightarrow (\exists D)(\forall C)[y \in T^{C,D} \Rightarrow (\forall B)(\exists A)[||x||^{A,B} < ||y||^{C,D}]]]$). However, we cannot show that $T^3_{||y||^3} \in \Pi_3^1$ uniformly. (Why does the expression $(\forall B)(\exists A)[x \in T^{A,B} \;\&\; (\exists D)(\forall C) \neg [||y||^{C,D} < ||x||^{A,B} + 1]]$ fail? See Exercise 16-66.) Indeed, if we could show this, we would have a proof of the reduction principle for Π_3^1-sets, and it is known, from work of Gödel and Addison, that such a proof would yield an inconsistency in *ZF* set theory (see Addison [1959]).

§16.7 BASIS RESULTS AND IMPLICIT DEFINABILITY

Definition Let $\mathfrak{K}$ be the class of hyperarithmetical functions.

Notation We abbreviate $(\exists f)[f \in \mathfrak{K} \;\&\; \cdots]$ as $(\exists f)_{\mathfrak{K}}[\cdots]$.

Theorem XLI (Kleene) *Let R be a recursive relation. Let*

$$A = \{x|(\exists f)_{\mathfrak{K}}(\forall w)R(\bar{f}(w),x)\}.$$

Then $A \in \Pi_1^1$.

Proof. Let φ_z^1 be as defined in §16.5. Recall that a Π_1^1-index for φ_z^1 can be obtained uniformly from z, and that for φ_z^1 total, a Σ_1^1-index for φ_z^1 can be found uniformly from z by a total function.

Now $(\exists f)_{\mathfrak{K}}(\forall w)R(\bar{f}(w),x) \Leftrightarrow (\exists z)[\varphi_z^1$ is total $\& (\forall w)R(\bar{\varphi}_z^1(w),x)] \Leftrightarrow (\exists z)[(\forall u)(\exists v)[<u,v> \in \varphi_z^1] \;\&\; (\forall w)(\forall y)[[y$ is a sequence number of length $w \;\&\; (\forall u)(\forall n)[[n < w \;\&\; u$ is the $(n + 1)$st value in the sequence represented by $y] \Rightarrow <n,u> \in \varphi_z^1]] \Rightarrow R(y,x)]]$.

Putting in a Π_1^1-form for φ_z^1 at the first occurrence, and the uniform Σ_1^1-form at the second occurrence, we have $A \in \Pi_1^1$. ☒

Corollary XLI(a) *There exists a recursive set S such that*

$$(\exists f)(\forall w)S(\bar{f}(w))$$

even though $\neg(\exists f)_{\mathcal{H}}(\forall w)S(\bar{f}(w))$.

Proof. Let $B = E^1 = \{z | (\exists f)(\forall w) \neg T^*_{1,1}(z,\bar{f}(w),z)\}$. Take

$$A = \{z | (\exists f)_{\mathcal{H}}(\forall w) \neg T^*_{1,1}(z,\bar{f}(w),z)\}.$$

Then $A \subset B$. Let z_0 be a Π^1_1-index for A. Then, by the definition of B, $z_0 \in B \;\&\; z_0 \not\in A$. Hence $(\exists f)(\forall w) \neg T^*_{1,1}(z_0,\bar{f}(w),z_0)$ but $\neg(\exists f)_{\mathcal{H}}(\forall w) \neg T^*_{1,1}(z_0,\bar{f}(w),z_0)$. Letting $S = \{y | \neg T^*_{1,1}(z_0,y,z_0)\}$ we have the result.☒

This corollary can be restated in the following interesting form.

Corollary XLI(b) *There is a recursive tree which has infinite branches but which has no infinite hyperarithmetical branches.*

Proof. Immediate.☒

The converse of Theorem XLI will be proved as Theorem XLIV.

Definition Let $\mathbf{R}$ be a family of classes of functions. Let $\mathcal{C}$ be a class of functions. $\mathcal{C}$ is called a *basis* for $\mathbf{R}$ if $(\forall \mathcal{A})[\mathcal{A} \in \mathbf{R} \Rightarrow [(\exists f)[f \in \mathcal{A}] \Leftrightarrow (\exists f)[f \in \mathcal{C} \;\&\; f \in \mathcal{A}]]$.

For example, it is easy to show that the class of recursive functions is a basis for the family of recursive classes of functions (see Exercise 16-67).

In basis problems we sometimes use hierarchy notation to refer to classes of functions or to families of classes of functions. Thus the result just mentioned can be stated: Σ^0_0 is a basis for Σ^0_0 (i.e., the class of functions each definable in Σ^0_0-form is a basis for the family of classes each definable in Σ^0_0-form).

Consider the family of classes $\mathcal{A}$ such that for some recursive S, $\mathcal{A} = \{f | (\forall w)S(\bar{f}(w))\}$. Corollary XLI shows that the class of hyperarithmetical functions does not constitute a basis for this family. In view of Theorem V, this family is the family of all Π^0_1-classes. Hence we can abbreviate this result: Δ^1_1 is not a basis for Π^0_1.

Do the Π^1_1-functions form a basis for Π^0_1? Trivially not, since by §16.5, the Π^1_1-functions coincide with the hyperarithmetical functions. (Indeed, for any n, the Σ^1_n-functions, the Π^1_n-functions, and the Δ^1_n-functions all coincide.)

Does the class of all functions recursive in Π^1_1-sets form a basis for Π^0_1? It does form a basis, as we now show.

Theorem XLII *Let S be recursive.* $(\exists f)(\forall w)S(\bar{f}(w)) \Rightarrow (\exists f)$ [*f recursive in T* & $(\forall w)S(\bar{f}(w))$].

Proof. Let S be given. Then S determines a recursive tree τ. (A vertex is in τ if S holds for it and for all vertices above it.) Assume that $(\exists f)(\forall w)S(\bar{f}(w))$. We give instructions for computing such an f recursively in T.

To compute $f(0)$, examine, in turn, the successive branch trees of τ, and use T to test, for each one, whether or not it is finite-path. Some branch tree must fail to be finite-path. Let x_0 be the least u such that the uth branch tree of τ is not finite-path. Set $f(0) = x_0$.

To compute $f(n)$, assume $f(0), \ldots, f(n-1)$ have been computed. Hence $\bar{f}(n)$ is known. Examine in turn the successive branch trees of the tree obtained from τ by taking the vertex $\bar{f}(n)$ together with all vertices of τ that lie below the vertex $\bar{f}(n)$.☒

Corollary XLII(a) *The class of functions recursive in T is a basis for Π_1^0.*

Proof. Immediate.☒

Corollary XLII(b) *The class of functions recursive in T is a basis for Σ_1^1.*

Proof. Let $\mathcal{A}$ be a Σ_1^1 class. Then $\mathcal{A} = \{f|(\exists g)(\forall w)R(\bar{f}(w),\bar{g}(w))\}$ for some recursive R.

But

$$\begin{aligned}
(\exists f)[f \in \mathcal{A}] &\Rightarrow (\exists f)(\exists g)(\forall w)R(\bar{f}(w),\bar{g}(w)\\
&\Rightarrow (\exists h)(\forall w)R(\overline{\pi_1 h}(w),\overline{\pi_2 h}(w))\\
&\Rightarrow (\exists h)[h \text{ is recursive in } T \ \&\ (\forall w)R(\pi_1 h(w),\pi_2 h(w))]\\
&\qquad\qquad\qquad\qquad\qquad\qquad\text{(by the theorem)}\\
&\Rightarrow (\exists f)[f \text{ recursive in } T \ \&\ (\exists g)[g \text{ recursive in } T\\
&\qquad\qquad\qquad\qquad\&\ (\forall w)R(\bar{f}(w),\bar{g}(w))]]\\
&\Rightarrow (\exists f)[f \text{ recursive in } T \ \&\ (\exists g)(\forall w)R(\bar{f}(w),\bar{g}(w))]\\
&\Rightarrow (\exists f)[f \in \mathcal{A}].
\end{aligned}$$

Hence the above assertions are all equivalent, and $(\exists f)[f \in \mathcal{A}] \Leftrightarrow (\exists f)[f \text{ recursive in } T \ \&\ f \in \mathcal{A}]$.☒

Must we use functions of degree as high as T (or even of hyperdegree as high as T) to get a basis for Π_1^0? Gandy has obtained the following result.

Theorem XLIII **(*Gandy* [*1960*])** *Let $\mathcal{C}$ be the class of all functions of hyperdegree less than the hyperdegree of T;* that is, $\mathcal{C} = \{f|f \leq_h T \ \&\ T \not\leq_h f\}$. *Then $\mathcal{C}$ forms a basis for Π_1^0.*

Proof. Let a function g be given. As indicated in §16.6, the treatment of hyperarithmetical computability in §16.5 can be relativized to yield $\Pi_1^{1,g}$-sets $W_0^{1,g}$, $W_1^{1,g}$, . . . , and partial $\Pi_1^{1,g}$ functions $\varphi_0^{1,g}$, $\varphi_1^{1,g}$, ... The proof of Theorem XLI can then be relativized directly to yield: if R is recursive in g, then $\{x|(\exists f)[f \leq_h g \ \&\ (\forall w)R(\bar{f}(w),x)]\} \in \Pi_1^{1,g}$.

The proof of Corollary XLI(a) can then be relativized directly to yield z_0 such that $(\exists f)(\forall w) \neg T_{1,1}^{*g}(z_0,\bar{f}(w),z_0)$ $(\Leftrightarrow (\exists f)(\forall w) \neg T_{2,1}^{*}(z_0,\bar{g}(w),\bar{f}(w),z_0))$, but such that $\neg (\exists f)[f \leq_h g \ \&\ (\forall w) \neg T_{1,1}^{*g}(z_0,\bar{f}(w),z_0)]$ $(\Leftrightarrow \neg (\exists f)[f \leq_h g \ \&\ (\forall w) \neg T_{2,1}^{*}(z_0,\bar{g}(w),\bar{f}(w),z_0)])$. (See Exercise 16-68.) This z_0 is independent of the choice of g. Hence we have a recursive S such that

$$(\forall g)(\exists f)(\forall w)S(\bar{g}(w),\bar{f}(w))$$

but such that $(\forall g)(\forall f)[f \leq_h g \Rightarrow \neg (\forall w)S(\bar{g}(w),\bar{f}(w))]$.

Let R be any recursive set. We wish to show that

$$(\exists g)(\forall w)R(\bar{g}(w)) \Rightarrow (\exists g)[g \leq_h T \;\&\; T \not\leq_h g \;\&\; (\forall w)R(\bar{g}(w))].$$

Define $Q = \{\langle u,v\rangle | R(u) \;\&\; S(u,v)\}$. Then

$$\begin{aligned}(\exists g)(\forall w)R(\bar{g}(w)) &\Rightarrow (\exists h)(\forall w)Q(\overline{\pi_1 h}(w),\overline{\pi_2 h}(w)) \qquad \text{(by the definition of } S)\\ &\Rightarrow (\exists h)[h \text{ is recursive in } T \;\&\; (\forall w)Q(\overline{\pi_1 h}(w),\overline{\pi_2 h}(w))]\\ &\qquad\qquad \text{(by Theorem XLII)}\\ &\Rightarrow (\exists g)(\exists f)[f \text{ and } g \text{ are recursive in } T \;\&\; (\forall w)R(\bar{g}(w))\\ &\qquad\qquad \&\; (\forall w)S(\bar{g}(w),\bar{f}(w))]\\ &\Rightarrow (\exists g)(\exists f)[f \text{ and } g \text{ are recursive in } T \;\&\; (\forall w)R(\bar{g}(w))\\ &\qquad\qquad \&\; f \not\leq_h g] \qquad \text{(by the construction of } S)\\ &\Rightarrow (\exists g)[g \text{ recursive in } T \;\&\; T \not\leq_h g \;\&\; (\forall w)R(\bar{g}(w))]\\ &\Rightarrow (\exists g)[g \leq_h T \;\&\; T \not\leq_h g \;\&\; (\forall w)R(\bar{g}(w))].\boxtimes\end{aligned}$$

Corollary XLIII(a) *The class of all functions of hyperdegree less than the hyperdegree of T forms a basis for* Σ_1^1.

Proof. As in Corollary XLII(*b*).$\boxtimes$

Corollary XLIII(b) *There exists a hyperdegree between* Δ_1^1 *(the minimum hyperdegree) and the hyperdegree of T.*

Proof. Immediate by Corollary XLI(*a*).$\boxtimes$

In Exercise 16-69, we show that there are infinitely many such hyperdegrees.

A number of basis results have been obtained in connection with hyperarithmetical sets. Let us say that A is *definable in* Σ_1^1*-form with* $\mathcal{C}$ *as a basis* if there is a recursive R such that: (i) $A = \{z | (\exists f)(\forall w)R(\bar{f}(w),z)\}$, and (ii) $(\forall z)[(\exists f)(\forall w)R(\bar{f}(w),z) \Rightarrow (\exists f)[f \in \mathcal{C} \;\&\; (\forall w)R(\bar{f}(w),z)]]$. It is immediate from Theorem XLI that a set definable in Σ_1^1-form with the hyperarithmetical functions as a basis is a hyperarithmetical set. Kleene [1959*a*] has shown that the hyperarithmetical sets can be inductively generated as follows: every hyperarithmetical set is definable in Σ_1^1-form with functions recursive in previously obtained T-degrees as basis. See Exercise 16-99. These results are another aspect of the closure property of hyperarithmetical sets discussed in §16.6.

We now get a converse to Theorem XLI by means of the basic tree lemma.

Theorem XLIV (Spector) *If* $A \in \Pi_1^1$, *then there is a recursive R such that* $A = \{x | (\exists f)_{\mathcal{HC}}(\forall w)R(\bar{f}(w),x)\}$.

Proof. Our main goal in the proof is to obtain an arithmetical $P \subset \mathfrak{F} \times N$ with the following two properties. (i) $z \in T \Rightarrow [P(f,z) \Leftrightarrow f$ is the characteristic function of $\{\langle u,v\rangle | u \in T \;\&\; v \in T \;\&\; \|u\| < \|v\| \leq \|z\|\}]$. (ii) $[z \notin T \;\&\; P(f,z)] \Rightarrow f$ coincides on $N \times T$ with the characteristic function of $\{<u,v> | u \in T \,\&\, v \in T \,\&\, \|u\| < \|v\|\}$. Once P is obtained, the theorem follows by straightforward argument.

For any x, we say that *x is a tree* if φ_x is the characteristic function for a tree, i.e., if $(\forall u)[\varphi_x(u) = 0 \vee \varphi_x(u) = 1] \,\&\, (\forall u)(\forall v)[[\varphi_x(u) = 1 \,\&$ the sequence (with sequence number) u is an extension of the sequence (with sequence number) $v] \Rightarrow \varphi_x(v) = 1]$. Note that this last gives a Π_2^0-form for $\{x | x \textit{ is a tree}\}$.

Define $C = \{f | (\forall u)[f(u) = 0 \vee f(u) = 1]\}$.

Define

$$\begin{aligned} M = \{<f,x,y> | x \textit{ is a tree} \,\&\, y \textit{ is a tree} \,\&\, [[\|x\| = 0 \,\&\, \|y\| = 1] \\ \vee [\neg [\|y\| = 0] \,\&\, (\exists m)(\forall n)[f(<b(n,x),b(m,y)>) = 1] \\ \&\, (\forall m)[\|b(m,y)\| = 0 \vee (\exists s)[f(s,b(m,y)) = 1]]]]\} \end{aligned}$$

where b is the recursive function for obtaining branch trees defined in §16.3.

Recalling the basic tree lemma, we note several facts about M.

1. Let f be the characteristic function of $\{<x,y> | x \in T \,\&\, y \in T \,\&\, \|x\| < \|y\|\}$. Then f satisfies the relationship

$$C(f) \,\&\, (\forall x)(\forall y)[f(<x,y>) = 1 \Leftrightarrow M(f,x,y)].$$

2. This arithmetical relationship can be satisfied by other functions f. Were this not so, then, by the argument used to prove Theorem XI, the characteristic function in fact 1 would be hyperarithmetical, contrary to Theorem XXIII(*c*) (see Exercise 16-71).

3. Let $y_0 \in T$ and assume that f coincides with the characteristic function of $\{<u,v> | u \in T_{\|y_0\|} \,\&\, v \in T_{\|y_0\|} \,\&\, \|u\| < \|v\|\}$ on $N \times T_{\|y_0\|}$. Assume further that $C(f) \,\&\, (\forall x)(\forall y)[f(<x,y>) = 1 \Leftrightarrow M(f,x,y)]$. Then

$$C(f) \,\&\, (\forall x)[f(<x,y_0>) = 1 \Leftrightarrow x \in T \,\&\, \|x\| < \|y_0\|]$$

and hence f coincides with the characteristic function of $\{<u,v> | u \in T_{\|y_0\|+1} \,\&\, v \in T_{\|y_0\|+1} \,\&\, \|u\| < \|v\|\}$ on $N \times T_{\|y_0\|+1}$. This is immediate by the basic tree lemma.

4. Indeed, for any f, if $C(f) \,\&\, (\forall x)(\forall y)[f(<x,y>) = 1 \Leftrightarrow M(f,x,y)]$, then f coincides with the characteristic function of $\{<u,v> | u \in T \,\&\, v \in T \,\&\, \|u\| < \|v\|\}$ on $N \times T$. This follows from fact 3 by induction on $\|y_0\|$.

Now define

$$\begin{aligned} P = \{<f,z> | C(f) \\ \&\, (\forall x)(\forall y)[f(<x,y>) = 1 \Leftrightarrow [M(f,x,y) \\ \&\, (\forall m)(\exists n)[f(<b(n,z),b(m,y)>) = 0]]] \\ \&\, (\forall x)(\forall y)[f(<x,y>) = 1 \Rightarrow (\forall n)[f(<b(n,y),z>) = 1]\}. \end{aligned}$$

We prove several facts about P.

5. If $z \in T$ and f is the characteristic function for $\{<u,v> | u \in T \,\&\, v \in T \,\&\, \|u\| < \|v\| \leq \|z\|\}$, then $P(f,z)$ holds. The first and third clauses

of P are immediate. In the second clause, $\Rightarrow$ is immediate (see fact 1), and $\Leftarrow$ follows from the basic tree lemma (see fact 3).

6. If $z \in T$ and $P(f,z)$ holds, then f is the characteristic function for $\{<u,v>|u \in T \& v \in T \& \|u\| < \|v\| \leq \|z\|\}$. We prove this as follows. Assume that $z \in T$ and $P(f,z)$. We first show that f coincides with the characteristic function of $\{<u,v>|u \in T \& v \in T \& \|u\| < \|v\| \leq \|w\|\}$ on $N \times T_{\|w\|+1}$ for any w such that $w \in T$ and $0 \leq \|w\| \leq \|z\|$. We do this by general induction on $\|w\|$, using the second clause of P. The proof is straightforward (see Exercise 16-71 and fact 4). It is then immediate from the third clause of P that $f = 0$ on $N \times \bar{T}_{\|z\|+1}$.

7. If $z \in T$, then $P(f,z)$ holds for one and only one f, and this f is hyperarithmetical. Uniqueness of f is given by facts 5 and 6. But if f is unique, then

$$f = \{<x,y>|(\exists g)[P(g,z) \& g(x) = y]\}$$
$$= \{<x,y>|(\forall g)[P(g,z) \Rightarrow g(x) = y]\}.$$

As P is arithmetical, this gives $f \in \Delta_1^1$.†

8. If $z \notin T$ and $P(f,z)$, then f coincides with the characteristic function of $\{<u,v>|u \in T \& v \in T \& \|u\| < \|v\|\}$ on $N \times T$. This is proved by an induction parallel to that in fact 6, but with $\|w\|$ ranging over all constructive ordinals. As in fact 6, the induction uses the second clause of P. The third clause is not used (see Exercise 16-71). (For certain $z \notin T$, there may be no f such that $P(f,z)$.)

9. If $z \notin T$ and $P(f,z)$, then f is not hyperarithmetical. Take any $y \in T$. Then by fact 8, $T_{\|y\|} = \{x|f(<x,y>) = 1\}$. Hence, by Corollary XXII, every hyperarithmetical set is recursive in f. Hence f is not hyperarithmetical (see Theorem X).

From facts 7 and 9 we can immediately conclude

$$T = \{z|(\exists f)_{\mathfrak{H}}P(f,z)\}.$$

Applying the Tarski-Kuratowski algorithm to the arithmetical relations C, M, and P, we obtain first that $C \in \Pi_1^0$ and $M \in \Delta_3^0$, and then that $P \in \Pi_3^0$. Hence we have:

$$z \in T \Leftrightarrow (\exists f)_{\mathfrak{H}}(\forall x)(\exists y)(\forall w)R(f,x,y,w,z),$$

for some recursive R, where, for $z \in T$, f is unique. We now modify the number quantifiers by introducing appropriate function quantifiers. Thus,

$$z \in T \Leftrightarrow (\exists f)_{\mathfrak{H}}(\exists g)(\forall x)[(\forall w)R(f,x,g(x),w,z) \& (\forall u)[u < g(x) \Rightarrow (\exists w) \neg R(f,x,u,w,z)]].$$

For $z \in T$, g is evidently unique and arithmetical in the unique hyperarith-

† Facts 6 and 7 give an independent proof of Theorem XXI, i.e., that $z \in T \Rightarrow T_{\|z\|} \in \Delta_1^1$ (see Exercise 16-71).

metical f. Hence we have

$$z \in T \Leftrightarrow (\exists f)_{\mathcal{H}}(\exists g)_{\mathcal{H}}(\forall x)(\exists w)S(f,g,x,w),$$

for some recursive S, where f and g are unique if $z \in T$; hence we have

$$z \in T \Leftrightarrow (\exists f)_{\mathcal{H}}(\exists g)_{\mathcal{H}}(\exists h)(\forall x)[S(f,g,x,h(x)) \ \& \ (\forall u)[u < h(x) \Rightarrow \neg S(f,g,x,u)]].$$

For $z \in T$, h is evidently unique and arithmetical in the unique hyperarithmetical f and g. Hence we have

$$z \in T \Leftrightarrow (\exists f)_{\mathcal{H}}(\exists g)_{\mathcal{H}}(\exists h)_{\mathcal{H}}(\forall w)Q(\bar{f}(w),\bar{g}(w),\bar{h}(w),z),$$

for some recursive Q where f, g, and h are unique for $z \in T$. Hence we have

$$z \in T \Leftrightarrow (\exists h)_{\mathcal{H}}(\forall w)Q(\overline{\pi_1{}^3h}(w),\overline{\pi_2{}^3h}(w),\overline{\pi_3{}^3h}(w),z),$$

where h is unique for $z \in T$.

Thus T is expressed in the desired form. Since any Π_1^1-set is 1-reducible to T, any Π_1^1-set can be expressed in the desired form.☒

In Exercise 16-101, a somewhat shorter proof of Theorem XLIV will be given. The present proof, however, leads to the following characterization of hyperarithmetical sets: every set implicitly definable in elementary arithmetic is hyperarithmetical, and every hyperarithmetical set is 1-reducible to some set implicitly definable in elementary arithmetic. We state this in the following corollary.

Corollary XLIV(a) $\{A\} \in \Delta_0^1 \Rightarrow A \in \Delta_1^1$.

(*b*) $A \in \Delta_1^1 \Leftrightarrow (\exists B)[A \leq_1 B \ \& \ \{B\} \in \Delta_0^1]$.

Proof. Part (*a*) is immediate by the method used for fact 7.

To show (*b*), it is enough to show that for every $y \in T$, $(\exists B)[T_{\|y\|} \leq_1 B \ \& \ \{B\} \in \Delta_0^1]$. By fact 7, $\{f\} \in \Delta_0^1$, where f is the characteristic function for $\{<u,v> \mid \|u\| < \|v\| \leq \|y\|\}$. Then $T_{\|y\|} \leq_1 \tau(f)$ and, by the methods of 15-XXV, $\{\tau(f)\} \in \Delta_0^1$ (see Exercise 16-74). Conversely, assume $\{B\} \in \Delta_0^1$ and $A \leq_1 B$. Then $B \in \Delta_1^1$ by (*a*), and hence $A \in \Delta_1^1$.☒

Assertion (*b*) can be strengthened to the following.

Corollary XLIV(c) $A \in \Delta_1^1 \Leftrightarrow (\exists B)[A \leq_1 B \ \& \ \{B\} \in \Pi_3^0]$.

Proof. Immediate from the proof of the theorem and from Theorem 15-XXV.☒

The following also hold.

Corollary XLIV(d) $A \in \Delta_1^1 \Leftrightarrow (\exists f)[A \leq_T f \ \& \ \{f\} \in \Pi_1^0]$.

(*e*) $A \in \Delta_1^1 \Leftrightarrow (\exists B)[A \leq_T B \ \& \ \{B\} \in \Pi_2^0]$.

Proof. Part (*d*) follows from the final form for T in the proof of the theorem, since $\pi_1{}^3h \leq_T h$.

Part (*e*) then follows by Theorem 15-XXV.☒

Corollaries XLIV(*d*) and XLIV(*e*) are strengthened in Theorem LII and Exercise 16-98.

Is it true that every hyperarithmetical set is implicitly definable in elementary arithmetic; i.e., that $A \in \Delta_1^1 \Rightarrow \{A\} \in \Delta_0^1$? Feferman has obtained a negative answer to this question using "forcing-generic" methods of Cohen. See Exercise 16-72. We note, however, the following.

Corollary XLIV(f) $A \in \Delta_1^1 \Leftrightarrow \{A\} \in \Delta_1^1 \Leftrightarrow \{A\} \in \Sigma_1^1$.
(*g*) $A \in \Delta_1^1 \Leftrightarrow (\exists B)[B \in \Delta_1^1 \;\&\; \{<A,B>\} \in \Delta_0^1]$.
(*h*) $A \in \Delta_1^1 \Leftrightarrow (\exists B)[B \in \Delta_1^1 \;\&\; \{<A,B>\} \in \Pi_2^0]$.

Part (*g*) asserts that any hyperarithmetical set is "almost implicitly definable" in elementary arithmetic, in the sense that it and another hyperarithmetical set can be "simultaneously implicitly defined." Implicit definability results for hyperarithmetical sets were first obtained in this form.

Proof. (*f*) (i) $A \in \Delta_1^1 \Rightarrow \{A\} \in \Delta_1^1$. Assume $A \in \Delta_1^1$. Observe $\{A\} = \{X|(\forall x)[x \in X \Rightarrow x \in A] \;\&\; (\forall x)[x \in A \Rightarrow x \in X]\}$. Inserting appropriate forms for A on the right, we have the result.

(ii) $\{A\} \in \Delta_1^1 \Rightarrow \{A\} \in \Sigma_1^1$. This is trivial.

(iii) $\{A\} \in \Sigma_1^1 \Rightarrow A \in \Delta_1^1$. Assume $\{A\} \in \Sigma_1^1$. Let $R(A)$ be a Σ_1^1-form expressing $\{A\}$. Then

$$\begin{aligned} A &= \{x|(\exists X)[R(X) \;\&\; x \in X]\} \\ &= \{x|(\forall X)[R(X) \Rightarrow x \in X]\}, \end{aligned}$$

and this yields the desired Σ_1^1-form and Π_1^1-form for A.

(*g*) $\Leftarrow$: Assume $(\exists B)[B \in \Delta_1^1 \;\&\; \{<A,B>\} \in \Delta_0^1]$. Let $R(A,B)$ be an arithmetical form for $\{<A,B>\}$. Then $\{A\} = \{X|(\exists B)R(X,B)\}$, and we have $\{A\} \in \Sigma_1^1$. Hence $A \in \Delta_1^1$ by (*f*).

$\Rightarrow$: Assume $A \in \Delta_1^1$. By (*b*), take B such that $\{B\} \in \Delta_0^1$ and $A \leq_1 B$. Let $S(B)$ be an arithmetical form for B. Let $A \leq_1 B$ via f. Then $\{<A,B>\} = \{<X,Y>|S(Y) \;\&\; (\forall x)[x \in X \Leftrightarrow f(x) \in Y]\}$, and we have the desired arithmetical form.

(*h*) See Exercise 16-75. ☒

In connection with (*f*) Benson has announced the result that for every level of the hyperarithmetical hierarchy, there are hyperarithmetical sets possessing implicit "hyperarithmetical" definitions that lie above that level but possessing no implicit definitions that fall below it. (The *levels* of the hyperarithmetical hierarchy are defined in §16.8.)

Is $\mathcal{H}$, the class of hyperarithmetical functions, a hyperarithmetical class? In Exercise 16-76, we show that this is not the case and that $\mathcal{H} \in (\Pi_1^1 - \Sigma_1^1)$. Is the class of arithmetical functions an arithmetical class? Addison has shown, by Cohen's methods, that this is not true. See Exercise 16-73.

We now turn to the Δ_2^1-sets. We obtain a basis result opposite in form to that obtained in Corollary XLI for Δ_1^1, but we obtain results on implicit definability parallel in form to those obtained in Corollary XLIV for Δ_1^1. Our results follow from a fundamental fact first proved for the classical hierarchy of projective sets by Kondo, and later demonstrated for the analytical hierarchy by Addison.

Let $\mathcal{P} \subset \mathcal{N}$ and $\mathcal{P} \in \Pi_1^1$. Then $\mathcal{P} = \{A | (\forall f)(\exists y)R(A,f,y)\}$ for some recursive $R \subset \mathcal{N} \times \mathcal{F} \times N$.

Definition z is a Π_1^1-*index* for $\mathcal{P}$ if

$$\mathcal{P} = \{A | (\forall f)(\exists y) T_{2,0}(z,A,\tau(f),y)\}.$$

Note that

$$(\forall f)(\exists y) T_{2,0}(z,A,\tau(f),y) \Leftrightarrow (\forall f)(\exists y) T'^{A}_{1,0}(z,f,y) \Leftrightarrow (\forall f)(\exists y) T^{*A}_{1,0}(z,\bar{f}(y)),$$

by definition. (See definitions preceding Theorem XXX.) Thus, z is a Π_1^1-index for $\mathcal{P} \Leftrightarrow \mathcal{P} = \{A | (\forall f)(\exists y) T^{*A}_{1,0}(z,\bar{f}(y))\}$. Note also that

$$\{<A,z,u> | T^{*A}_{1,0}(z,u)\}$$

is a recursive relation.

Theorem XLV (Kondo, Addison) *If $\mathcal{P} \subset \mathcal{N}$ and $\mathcal{P} \in \Pi_1^1$, then there is an $\mathcal{Q}$ such that $\mathcal{Q} \subset \mathcal{N}$ and $\mathcal{Q} \in \Pi_1^1$, and such that*

(i) $\mathcal{P} = \emptyset \Rightarrow \mathcal{Q} = \emptyset$,

(ii) $\mathcal{P} \neq \emptyset \Rightarrow (\exists A)[\mathcal{Q} = \{A\} \ \& \ A \in \mathcal{P}]$,

where a Π_1^1-index for $\mathcal{Q}$ can be found uniformly, by a total function, from a Π_1^1-index for $\mathcal{P}$.

Proof. We make several definitions related to our chosen coding from finite sequences into sequence numbers.

Definition We say that *y is an extension of x* if the sequence represented by (sequence number) y is a proper extension of the sequence represented by (sequence number) x.

Note the following property of the coding into sequence numbers: $(\forall x)(\forall y)[y$ is an extension of $x \Rightarrow y > x]$ (see Exercise 16-77).

Definition If x is the sequence number for the sequence $x_1, \ldots, x_k$ and y is the sequence number for the sequence $y_1, \ldots, y_m$, let $x * y$ be the sequence number for the sequence $x_1, \ldots, x_k, y_1, \ldots, y_m$. $\lambda xy[x * y]$ is clearly recursive. Note that $0 * y = y$ for any y.

Definition Take t to be a recursive function such that for any X,

$$\varphi^X_{t(x,z)} = \lambda y[\varphi_z{}^X(x * y)].$$

Thus if $\varphi_z{}^X$ is the characteristic function for a tree τ, then $\varphi^X_{t(x,z)}$ is the characteristic function for the tree obtained from τ by taking all vertices in τ that lie at or below the vertex (with sequence number) x. (The function t is evidently a generalization of the branch-tree function b defined in §16.3.)

Throughout the proof below we adopt the convention that a given finite-path tree not only associates a nonzero ordinal with each of its own vertices, as in §16.3, but also associates the ordinal 0 with all other vertices (of the function tree).

Let $\mathcal{P}$ be given and let z be a Π_1^1-index for $\mathcal{P}$. Then

$$\mathcal{P} = \{A|(\forall f)(\exists y)T_{1,0}^{*A}(z,\bar{f}(y))\}.$$

Take w so that for all A,

$$\varphi_w{}^A(x) = \begin{cases} 1, & \text{if } \neg\, T_{1,0}^{*A}(z,x); \\ 0, & \text{otherwise.} \end{cases}$$

w is evidently uniform in z by a total function. By definition, $\varphi_w{}^A$ is the characteristic function of a finite-path tree if and only if $A \in \mathcal{P}$; that is, $w \in T^A \Leftrightarrow A \in \mathcal{P}$. (Recall that T^A, $\|z\|^A$, and $\tau_z{}^A$ were defined in §16.6.)

We now define $\mathcal{Q}$. In the expression for $\mathcal{Q}$, we abbreviate $[u \in T^B \;\&\; \|u\|^B < \|v\|^A]$ as $\|u\|^B < \|v\|^A$, and we abbreviate $[u \in T^B \;\&\; \|u\|^B = \|v\|^A]$ as $\|u\|^B = \|v\|^A$ (for any terms u and v).

$$\begin{aligned}\mathcal{Q} = \{A|&w \in T^A \\ &\& (\forall B)\neg\,[\|w\|^B < \|w\|^A] \\ &\& (\forall B)[\|w\|^B = \|w\|^A \Rightarrow (\forall y)[(\forall x)[x < y \Rightarrow \|t(x,w)\|^B = \|t(x,w)\|^A] \\ &\qquad\qquad \Rightarrow \neg\,[\|t(y,w)\|^B < \|t(y,w)\|^A]]] \\ &\& (\forall B)[(\forall x)[\|t(x,w)\|^B = \|t(x,w)\|^A] \\ &\qquad \Rightarrow (\forall y)[(\forall x)[x < y \Rightarrow c_B(x) = c_A(x)] \Rightarrow c_A(y) \le c_B(y)]]\}.\end{aligned}$$

Note that the definition occurs in four clauses.

It remains to show that $\mathcal{Q}$ has the desired properties.

$\mathcal{Q} \in \Pi_1^1$. Use a $\Pi_1^{1,A}$-form for T^A, and use the appropriate $\Pi_1^{1,A,B}$-forms and $\Sigma_1^{1,A,B}$-forms given by Theorem XXXV and Corollary XXXV. Then apply the Tarski-Kuratowski algorithm (see Exercise 16-8). (If $w \in T^A$, then $t(x,w) \in T^A$ for all x, and the forms used from Theorem XXXV and Corollary XXXV have the desired meaning. If $w \notin T^A$, then the meaning of these forms is irrelevant.) It follows that a Π_1^1-index for $\mathcal{Q}$ is uniform in z.

(i) $\mathcal{P} = \emptyset \Rightarrow \mathcal{Q} = \emptyset$. This is immediate since $w \in T^A \Leftrightarrow A \in \mathcal{P}$, as we have noted.

(ii) $\mathcal{P} \neq \emptyset \Rightarrow (\exists A)[\mathcal{Q} = \{A\} \;\&\; A \in \mathcal{P}]$. We first show that $\mathcal{Q}$ contains at most one A. Consider the clauses in order.

The first clause says that $A \in \mathcal{P}$.

The second clause says that for any B, if $B \in \mathcal{P}$, then $\|w\|^A \le \|w\|^B$. Hence for any A_1 and A_2 in $\mathcal{Q}$, $\|w\|^{A_1} = \|w\|^{A_2}$.

The third clause says that for any B, if $B \in \mathcal{P}$ and $\|w\|^A = \|w\|^B$, then $\tau_w{}^B$ must be at least as high as $\tau_w{}^A$ in a certain "lexicographical" linear ordering of the finite-path trees. Hence for any A_1 and A_2 in $\mathcal{Q}$, $\tau_w{}^{A_1} = \tau_w{}^{A_2}$.

The fourth clause says that for any B, if $\tau_w{}^A = \tau_w{}^B$, then B must be at least at high as A in a certain "lexicographical" linear ordering of $\mathfrak{N}$.† Hence, for any A_1 and A_2 in $\mathfrak{N}$, $A_1 = A_2$.

† This is the usual linear ordering of the Cantor set.

It remains to show that $\mathcal{P} \neq \emptyset \Rightarrow \mathcal{Q} \neq \emptyset$. (This is not immediate since the "lexicographical" linear ordering of finite-path trees used is not (as a linear ordering of all finite-path trees) a well-ordering; and the "lexicographical" linear ordering of $\mathfrak{N}$ used is not a well-ordering.)

Assume $\mathcal{P} \neq \emptyset$.

1. $\{A|A$ *satisfies the first clause*$\} \neq \emptyset$. This is immediate, since $A \in \mathcal{P} \Leftrightarrow w \in T^A$.

2. $\{A|A$ *satisfies the first and second clauses*$\} \neq \emptyset$. This is immediate, since the ordering of the ordinals is a well-ordering.

3. $\{A|A$ *satisfies the first, second, and third clauses*$\} \neq \emptyset$. We prove this as follows. Define the classes $\mathcal{G}_i$, $i \geq 0$, as below.

$\mathcal{G}_0 = \{A|A \in \mathcal{P}$ & *the ordinal associated with vertex* 0 *by* A (*i.e., by* $\tau_w{}^A$) *is a minimum among all ordinals associated with vertex* 0 *by members of* $\mathcal{P}\}$. (Hence $\mathcal{G}_0$ is just the class named in statement 2 immediately above.)

$\mathcal{G}_{n+1} = \{A|A \in \mathcal{G}_n$ & *the ordinal associated with vertex* $n+1$ *by* A (*i.e., by* $\tau_w{}^A$) *is a minimum among all ordinals associated with vertex* $n+1$ *by members of* $\mathcal{G}_n\}$.

It is immediate that $\mathcal{G}_{n+1} \neq \emptyset$ by the well-ordering of the ordinals. To show statement 3, we must show $\bigcap_n \mathcal{G}_n \neq \emptyset$.

For each n, we define $o(n)$ to be the (uniquely defined) ordinal associated with vertex n by the members of $\mathcal{G}_n$.

Lemma *If n is an extension of m, and if $o(m) \neq 0$, then $o(n) < o(m)$.*

Proof. Assume n an extension of m, and $o(m) \neq 0$. By Exercise 16-77, $m < n$. $o(n)$ is the ordinal associated with vertex n by some $A \in \mathcal{G}_n$. But $\mathcal{G}_n \subset \mathcal{G}_m$. Hence A must also associate $o(m)$ with m. Hence $o(m)$ is assigned to m and $o(n)$ is assigned to n by the same finite-path tree ($\tau_w{}^A$), and since $o(m) \neq 0$, we have $o(n) < o(m)$. This completes the proof of the lemma.

Consider $\{n|o(n) = 0\}$. By the lemma and the well-ordering of the ordinals, every branch of the function tree must contain a member of this set; hence this set is infinite. Let $q_0, q_1, \ldots$ be the members of this set in increasing order. Define the classes $\mathcal{G}_i^*$, $i \geq 0$, as follows

$$\mathcal{G}_0^* = \{A|T_{1,0}^{*A}(z,q_0)\},$$

(that is, $$\mathcal{G}_0^* = \{A|t(q_0,w) \in T^A \;\&\; \|t(q_0,w)\|^A = 0\}).$$

$$\mathcal{G}_{n+1}^* = \{A|A \in \mathcal{G}_n^* \;\&\; T_{1,0}^{*A}(z,q_{n+1})\}.$$

It is immediate that $\mathcal{G}_{n+1}^* \subset \mathcal{G}_n^*$, for all n. $\mathcal{G}_{q_0} \subset \mathcal{G}_0^*$, since all sets in $\mathcal{G}_{q_0}$ assign the ordinal 0 to vertex q_0. Assume that $\mathcal{G}_{q_n} \subset \mathcal{G}_n^*$. Then $\mathcal{G}_{q_{n+1}} \subset \mathcal{G}_{n+1}^*$; for $\mathcal{G}_{q_{n+1}} \subset \mathcal{G}_{q_n} \cap \{A|T_{1,0}^{*A}(z,q_{n+1})\} \subset \mathcal{G}_n^* \cap \{A|T_{1,0}^{*A}(z,q_{n+1})\} = \mathcal{G}_{n+1}^*$. Hence $\mathcal{G}_{q_n} \subset \mathcal{G}_n^*$, for all n; and hence $\bigcap_n \mathcal{G}_{q_n} \subset \bigcap_n \mathcal{G}_n^*$. Note that $\mathcal{G}_n^* \neq \emptyset$ for all n, since $\mathcal{G}_n \neq \emptyset$ for all n.

For each n, $\mathcal{G}_n^*$ is a recursive class and hence is open-and-closed in the

Cantor-set topology on $\mathfrak{N}$ (Theorem 15-VII). We can therefore conclude, from the compactness of $\mathfrak{N}$ under this topology (Exercise 13-24(*e*)), that $\bigcap_n \mathcal{G}_n^* \neq \emptyset$.

It remains to show that $\bigcap_n \mathcal{G}_n^* \subset \bigcap_n \mathcal{G}_n$; for then $\bigcap_n \mathcal{G}_n \neq \emptyset$, and statement 3 follows.

Take $A \in \bigcap_n \mathcal{G}_n^*$. Observe that $\varphi_w{}^A$ is the characteristic function of a tree (by the definition of w). Call this tree τ_A. Then τ_A is finite-path. For assume $k_0, k_1, \ldots$ are the successive vertices on an infinite branch of τ_A. Then for all n, $o(k_n) \neq 0$. (If $o(k_n) = 0$, then $k_n = q_m$, for some m; but since $A \in \mathcal{G}_m^*$, q_m is not a vertex of τ_A, by the definition of $\mathcal{G}_m^*$.) But then, by the lemma above, $o(k_1) > o(k_2) > \cdots$, and this contradicts the well-ordering of the ordinals.

For any n, let $o_A(n)$ be the ordinal associated with vertex n by the finite-path tree τ_A. (If $n \notin \tau_A$, then $o_A(n) = 0$.) Assume $A \notin \bigcap_n \mathcal{G}_n$. We obtain a contradiction as follows. Let n_0 be the least n such that $A \notin \mathcal{G}_n$. Then by the definition of $\mathcal{G}_{n_0}$, $o_A(n_0) > o(n_0)$. We consider two cases.

Case (*a*): $o(n_0) = 0$. Then $n_0 = q_j$ for some j. Hence $o_A(q_j) \neq 0$. But $A \in \mathcal{G}_j^*$; and $o_A(q_j) = 0$ by the definition of $\mathcal{G}_j^*$. This is a contradiction.

Case (*b*): $o(n_0) > 0$. Then by the definition of o_A (see §16.3), there is a proper extension n_1 of n_0 such that $o_A(n_1) = o(n_0)$. But by the lemma above, $o(n_0) > o(n_1)$. Hence $o_A(n_1) > o(n_1)$. Taking successive extensions $n_2, n_3, \ldots$ in this manner, we must eventually reach (by the well-ordering of the ordinals) an extension n_k such that $o_A(n_k) > o(n_k)$ and $o(n_k) = 0$. But this is case (*a*), and we have a contradiction.

We have thus shown that $A \in \bigcap_n \mathcal{G}_n^* \Rightarrow A \in \bigcap_n \mathcal{G}_n$. Hence $\bigcap_n \mathcal{G}_n^* \subset \bigcap_n \mathcal{G}_n$, and in fact $\bigcap_n \mathcal{G}_n^* = \bigcap_n \mathcal{G}_n$. Hence $\bigcap_n \mathcal{G}_n \neq \emptyset$, and the proof of statement 3 is complete.

4. $\{A|A$ *satisfies the first, second, third, and fourth clauses*$\} \neq \emptyset$. We prove this as follows. Any nonempty subset of $\mathfrak{N}$ has a greatest lower bound in the "lexicographical" linear ordering of $\mathfrak{N}$, as is easily demonstrated. Let A^* be the greatest lower bound of $\bigcap_n \mathcal{G}_n$. It is enough to show that $A^* \in \bigcap_n \mathcal{G}_n$. This is an immediate consequence of the following facts: (*a*) $\mathcal{G}_n^*$ is closed for all n (as remarked above); (*b*) $\bigcap_n \mathcal{G}_n^*$ is closed (since the intersection of closed sets is closed); (*c*) $\bigcap_n \mathcal{G}_n^* = \bigcap_n \mathcal{G}_n$ (this was shown in the proof of statement 3); (*d*) either $A^* \in \bigcap_n \mathcal{G}_n$, or A^* is a limit point of $\bigcap_n \mathcal{G}_n$ in the topology of $\mathfrak{N}$ (this follows from the definition of the linear ordering on $\mathfrak{N}$); (*e*) a closed set contains all its limit points.

This completes the proof of statement 4, and the proof of the theorem is now complete. ☒

Definition Let P and Q be relations in $\mathfrak{N}^k \times N^l$. We say that Q *uniformizes* P if, for every $B_1, \ldots, B_{k-1}, x_1, \ldots, x_l$,

(i) $(\forall A)[Q(A,B_1, \ldots ,x_l) \Rightarrow P(A,B_1, \ldots ,x_l)]$,

(ii) $(\exists A)P(A,B_1, \ldots ,x_l) \Rightarrow (\exists A)Q(A,B_1, \ldots ,x_l)$,

(iii) $(\forall A_1)(\forall A_2)[[Q(A_1,B_1, \ldots, x_l) \mathbin{\&} Q(A_2,B_1, \ldots, x_l)] \Rightarrow A_1 = A_2]$.

We now give our theorem in the following more general form.

Corollary XLV(a) *Every Π_1^1-relation* (in $\mathfrak{N}^k \times N^l$) *can be uniformized by a Π_1^1-relation.*

Proof. For given P and fixed $x_1, \ldots, x_l$, we relativize the theorem with respect to $B_1, \ldots, B_{k-1}$ (by introducing $T^{A,B_1,\ldots,B_{k-1}}$ in place of T^A, and $\|z\|^{A,B_1,\ldots,B_{k-1}}$ in place of $\|z\|^A$, where T^A and $\|z\|^A$ are as defined in §16.6.) Theorem XXXV and Corollary XXXV apply directly. A Π_1^1-index for the resulting Q is uniform in $x_1, \ldots, x_l$.☒

For this reason, Theorem XLV is sometimes called the *Kondo-Addison uniformization theorem* (see footnote on page 72).

Next we observe that Theorem XLV holds for classes of functions as well as for classes of sets.

Corollary XLV(b) *If $\mathcal{P} \subset \mathcal{F}$ and $\mathcal{P} \in \Pi_1^1$, then there is a $\mathcal{Q}$ such that $\mathcal{Q} \subset \mathcal{F}$ and $\mathcal{Q} \in \Pi_1^1$, and such that*

(i) $\mathcal{P} = \emptyset \Rightarrow \mathcal{Q} = \emptyset$,

(ii) $\mathcal{P} \neq \emptyset \Rightarrow (\exists g)[\mathcal{Q} = \{g\} \mathbin{\&} g \in \mathcal{P}]$,

where a Π_1^1-index for $\mathcal{Q}$ can be found uniformly, by a total function, from a Π_1^1-index for $\mathcal{P}$.

Proof. Immediate by Exercise 16-74.☒

(Corollary XLV(a) also holds with functions in place of sets.)

Our basis result for Δ_2^1 now follows.

Corollary XLV(c) *The Δ_2^1 functions form a basis for Π_1^1, and hence for Σ_2^1.*

Proof. Assume $\mathcal{P} \in \Pi_1^1$ and $\mathcal{P} \neq \emptyset$. Obtain $\mathcal{Q}$ as in Corollary XLV(b). Then $\mathcal{Q} \in \Pi_1^1$, and $\mathcal{Q} = \{g\}$ for some g.

Now

$$\begin{aligned} g(x) = y &\Leftrightarrow (\forall f)[f \in \mathcal{Q} \Rightarrow f(x) = y] \\ &\Leftrightarrow (\exists f)[f \in \mathcal{Q} \mathbin{\&} f(x) = y], \end{aligned}$$

and these expressions yield, from the Π_1^1-form for $\mathcal{Q}$, that $g \in \Delta_2^1$. The case for Σ_2^1 follows trivially by the argument used for Corollary XLII(b).☒

This result is in contrast to Corollary XLI(a), where we saw that the Δ_1^1 functions do not even form a basis for Π_1^0.

Consider the set of ordinals associated with T^2 in the last part of §16.6 (see Theorem XL).

Corollary XLV(d) $(\forall z)[z \in T^2 \Rightarrow \|z\|^2$ *is a Δ_2^1-ordinal*].

Proof. $z \in T^2 \Leftrightarrow (\exists B)[z \in T^B]$. But by Corollary XLV($c$) (in set-variable form), $(\exists B)[z \in T^B] \Leftrightarrow (\exists B)[B \in \Delta_2^1 \mathbin{\&} z \in T^B]$.☒

Corollary XLV(e) (Gandy) $(\exists \alpha)[\alpha$ *is a Δ_2^1-ordinal* & $(\forall z)[z \in T^2 \Rightarrow \|z\|^2 \neq \alpha]]$.

Proof. As in §16.6, define $\mathfrak{A} = \{\alpha | (\exists y)[y \in T^2 \mathbin{\&} \alpha = \|y\|^2]\}$. We wish to show that there is a Δ_2^1-ordinal not in $\mathfrak{A}$. (In conjunction with Theorem XL, this will show that $\mathfrak{A}$ is not an initial segment of ordinals.)

Define f to be *finite-path* if f is the characteristic function of a finite-path tree. If f is finite-path, define $\|f\| = o(\tau)$ where τ is the finite-path tree with characteristic function f. Let x_0 be a Gödel number such that for all f, $\varphi_{x_0}^{\tau(f)} = f$. Then for finite-path f, $\|f\| = \|x_0\|^{\tau(f)}$.

Consider the assertion:

$$f \text{ is finite-path } \& \ (\forall z)[(\exists B)[\|z\|^B < \|f\| + 1] \Rightarrow (\exists B)[\|z\|^B < \|f\|]].$$

Clearly, for any f for which the assertion holds, $\|f\| \notin \mathfrak{A}$. The assertion can be written equivalently as:

$$f \text{ is finite-path } \& \ (\forall z)[(\exists B)[\|z\|^B < \|x_0\|^{\tau(f)} + 1] \Rightarrow (\exists B)[\|z\|^B < \|x_0\|^{\tau(f)}]].$$

By Theorem XXXV, the assertion can be written:

$$R(f) \ \& \ (\forall z)[(\exists B)S_1(z,f,B) \Rightarrow (\exists B)S_2(z,f,B)],$$

where $R \in \Pi_1^1$, $S_1 \in \Sigma_1^1$, and $S_2 \in \Sigma_1^1$. Hence the assertion can be written: $P(f)$, where $P \in \Sigma_2^1$. By Corollary XLV(d), the assertion is true for any f such that f is finite-path and $\|f\|$ is larger than all members of $\mathfrak{A}$. Hence $(\exists f)P(f)$. Applying Corollary XLV(c), we have an $f_0 \in \Delta_2^1$ such that $P(f_0)$. Thus $\|f_0\|$ is a Δ_2^1-ordinal which does not belong to $\mathfrak{A}$.☒

We now use the preceding theorem to obtain an implicit definability result parallel to Corollary XLIV(b).

***Theorem XLVI* (*Suzuki*)** $A \in \Delta_2^1 \Leftrightarrow (\exists B)[A \leq_1 B \ \& \ \{B\} \in \Pi_1^1]$.

Proof. $\Leftarrow$: If $\{B\} \in \Pi_1^1$, then $B \in \Delta_2^1$ by Corollary XLV(c) (in set-variable form). The result is immediate.

$\Rightarrow$: Let A be a fixed Δ_2^1-set. We may assume $A \neq \emptyset$. (If $A = \emptyset$, then $\{A\} \in \Pi_1^1$ by the construction: $\{A\} = \{X|(\forall x)[x \notin X]\}$.) Then, $\{A\} = \{X|(\forall x)[x \in X \Rightarrow x \in A] \ \& \ (\forall x)[x \in A \Rightarrow x \in X]\}$. Insert a Σ_2^1-form for A at the first occurrence of A on the right, and a Π_2^1-form for A at the second occurrence. This yields $\{A\} \in \Sigma_2^1$. Hence

$$\begin{aligned}\{A\} &= \{X|(\exists f)(\forall g)(\exists w)R(X,f,g,w)\} \quad \text{for some recursive } R\\ &= \{X|(\exists f)(\forall g)(\exists w)R'(X,\tau(f),g,w)\} \quad \text{for some recursive } R'\\ &= \{X|(\exists Y)[(\forall x)(\exists y)(\forall u)[\langle x,u\rangle \in Y \Leftrightarrow u = y]\\ &\qquad \& \ (\forall g)(\exists w)R'(X,Y,g,w)]\}.\end{aligned}$$

The quantifier transformations of Theorem III can be applied in the presence of a free set variable, as is easily verified. Hence we obtain

$$\{A\} = \{X|(\exists Y)(\forall g)(\exists w)R''(X,Y,g,w)\} \text{ for some recursive } R''.$$

We may assume that

$$\{A\} = \{X|(\exists Y)[Y \neq \emptyset \ \& \ (\forall g)(\exists w)R''(X,Y,g,w)]\}$$

(otherwise $\{A\} = \{X|(\forall g)(\exists w)R''(X,\emptyset,g,w)\}$, and $\{A\} \in \Pi_1^1$).

Let $\mathcal{P} = \{Z | (\forall g)(\exists w)R''(\pi_1(Z),\pi_2(Z),g,w)\}$. Then $\mathcal{P} \in \Pi_1^1$. Observing that $[X \neq \emptyset \ \& \ Y \neq \emptyset] \Leftrightarrow X \times Y \neq \emptyset$, and that

$$X \times Y \neq \emptyset \Rightarrow [\pi_1(X \times Y) = X \ \& \ \pi_2(X \times Y) = Y],$$

and recalling our assumption that $A \neq \emptyset$, we have $\mathcal{P} \neq \emptyset$. Applying Theorem XLV, we obtain $\mathcal{Q} \in \Pi_1^1$ and $\mathcal{Q} = \{B\}$, where $B \in \mathcal{P}$. Then $A = \pi_1(B)$ and $\pi_2(B) \neq \emptyset$, by the construction. Let n be a member of $\pi_2(B)$. Then $A \leq_1 B$ via $\lambda x[\tau(x,n)]$, and the proof is complete. ☒

We next observe that Corollary XLIV(f) generalizes to Δ_2^1 and, indeed, to Δ_n^1 for all n.

Corollary XLVI(a) $A \in \Delta_n^1 \Leftrightarrow \{A\} \in \Delta_n^1 \Leftrightarrow \{A\} \in \Sigma_n^1$.
The proof of Corollary XLIV(f) generalizes directly.

Corollary XLVI(b) $A \in \Delta_2^1 \Leftrightarrow (\exists B)[B \in \Delta_2^1 \ \& \ \{\langle A,B\rangle\} \in \Pi_1^1]$.

Proof. As for Corollary XLIV(g). ☒

Is it true that $A \in \Delta_2^1 \Rightarrow \{A\} \in \Pi_1^1$? In the following theorems of Suzuki we show that this is not the case, and we obtain further information about the sets implicitly definable in Π_1^1, i.e., about $\{A | \{A\} \in \Pi_1^1\}$.

Theorem XLVII (Suzuki) $(\exists B)[B \in \Delta_2^1 \ \& \ \{B\} \notin \Pi_1^1]$.

Proof. If $\{A\} \in \Pi_1^1$ with Π_1^1-index z, then

$$\{A\} = \{X | (\forall f)(\exists y) T_{1,0}^{*X}(z,\bar{f}(y))\}.$$

Take w such that for all X,

$$\varphi_w{}^X(x) = \begin{cases} 1, & \text{if } \neg\, T_{1,0}^{*X}(z,x); \\ 0, & \text{otherwise.} \end{cases}$$

(w is uniform in z.) Then $(\forall X)[w \in T^X \Leftrightarrow X = A]$. We call w a *tree for* A.

Lemma *If* $\{A_1\}$ *and* $\{A_2\} \in \Pi_1^1$, *and if* w_1 *and* w_2 *are trees for* A_1 *and* A_2, *respectively, then*

$$\|w_1\|^{A_1} \leq \|w_2\|^{A_2} \Rightarrow A_1 \leq_h A_2.$$

Proof of lemma. Assume $\|w_1\|^{A_1} \leq \|w_2\|^{A_2}$. Then

$$\begin{aligned} A_1 &= \{x | (\exists X)[w_1 \in T^X \ \& \ \|w_1\|^X \leq \|w_2\|^{A_2} \ \& \ x \in X]\} \\ &= \{x | (\forall X)[[w_1 \in T^X \ \& \ \|w_1\|^X \leq \|w_2\|^{A_2}] \Rightarrow x \in X]\}. \end{aligned}$$

From Theorem XXXV, Corollary XXXV, and a Tarski-Kuratowski computation, the first expression yields $A_1 \in \Sigma_1^{1,A_2}$ and the second yields $A_1 \in \Pi_1^{1,A_2}$. Hence $A_1 \leq_h A_2$ and the lemma is proved.

It follows from the above lemma that the sets implicitly definable in Π_1^1 lie in a linear ordering of hyperdegrees.

Lemma *If* $\{A_1\}$ *and* $\{A_2\} \in \Pi_1^1$, *then*

$$A_1 \leq_h A_2 \Rightarrow [A_2 \leq_h A_1 \vee T^{A_1} \leq_h A_2].$$

Proof of lemma. Assume $A_1 \leq_h A_2$. Let w_2 be a tree for A_2. Consider the set $C = \{x | x \in T^{A_1} \& \|x\|^{A_1} < \|w_2\|^{A_2}\}$. By Theorem XXXV, $C \leq_h A_2$. Now either $C = T^{A_1}$ or for some x_0, $[x_0 \in T^{A_1} \& \|x_0\|^{A_1} = \|w_2\|^{A_2}]$. In the first case, we have $T^{A_1} \leq_h A_2$. In the second case, we can define

$$\begin{aligned} A_2 &= \{x | (\exists X)[w_2 \in T^X \& \|x_0\|^{A_1} = \|w_2\|^X \& x \in X]\} \\ &= \{x | (\forall X)[[w_2 \in T^X \& \|x_0\|^{A_1} = \|w_2\|^X] \Rightarrow x \in X]\}. \end{aligned}$$

Hence, by Corollary XXXV and Tarski-Kuratowski computations, we have $A_2 \in \Sigma_1^{1,A_1}$ and $A_2 \in \Pi_1^{1,A_1}$; and hence $A_2 \leq_h A_1$. This proves the lemma.

Our second lemma shows that the sets implicitly definable in Π_1^1 are, in a certain sense, sparse, for if A is implicitly definable, no set with hyperdegree properly between the hyperdegrees of A and the hyperjump of A is implicitly definable.

The theorem is now immediate from Corollary XLIII(b); for let B have hyperdegree properly between the hyperdegrees of $\emptyset$ and $T(\equiv T^{\emptyset})$. Since $\{\emptyset\} \in \Pi_1^1$, we have $\{B\} \notin \Pi_1^1$ by the second lemma.☒

In the lemma of Theorem 15-XII, we proved, in effect, that if A is implicitly definable in arithmetic, then A' (the jump of A) is implicitly definable in arithmetic. The following theorem gives the analogous result for sets implicitly definable in Π_1^1 and the hyperjump. It hence shows that the sets implicitly definable in Π_1^1 are, in a certain sense, not too sparse.

Theorem XLVIII (Suzuki) $\{A\} \in \Pi_1^1 \Rightarrow \{T^A\} \in \Pi_1^1$.

Proof. We first prove a lemma.

Lemma $[\{T^A\} \in \Pi_1^{1,A} \& \{A\} \in \Pi_1^1] \Rightarrow \{T^A\} \in \Pi_1^1$.

Proof of lemma. Define a one-one recursive f such that for all X and x,

$$\varphi_{f(x)}^X = \begin{cases} \lambda y[0], & \text{if } x \in X; \\ \lambda y[1], & \text{if } x \notin X. \end{cases}$$

Then, trivially, for any A, $A \leq_1 T^A$ via f.

Let

$$\{T^A\} = \{X | (\forall g)(\exists y)R_1(X,A,g,y)\},$$

and

$$\{A\} = \{X | (\forall g)(\exists y)R_2(X,g,y)\},$$

where R_1 and R_2 are recursive.

Consider $\{X | (\forall g)(\exists y)R_2(f^{-1}(X),g,y) \& (\forall g)(\exists y)R_1(X,f^{-1}(X),g,y)\}$. Evidently this is $\{T^A\}$, and hence $\{T^A\} \in \Pi_1^1$. This proves the lemma.

To prove our theorem, it will now suffice to show that for any A, $\{T^A\} \in \Pi_1^{1,A}$. We begin with the following lemma.

Lemma $\{T\} \in \Pi_1^1$.

Proof of lemma. Let $B = \{\langle x,y\rangle | x \in T \& y \in T \& \|x\| < \|y\|\}$. By Exercise 16-34, $B \equiv T$. By Exercise 16-74, it will be enough to show that $\{c_B\} \in \Pi_1^1$.

Define M and C as in the proof of Theorem XLIV. It is immediate from facts 1 to 4 in that proof that

$$\{c_B\} = \{g|C(g) \ \&\ (\forall x)(\forall y)[M(g,x,y) \Leftrightarrow g(<x,y>) = 1] \ \&\ (\forall u)(\forall v)[g(<u,v>) = 1 \Rightarrow (\forall h)[[C(h) \ \&\ (\forall x)(\forall y)[M(h,x,y) \Leftrightarrow h(<x,y>) = 1]] \Rightarrow h(<u,v>) = 1]]\},$$

and by a Tarski-Kuratowski computation, this gives the lemma.

The proof of the above lemma can be relativized to a set A directly, using $\varphi_x{}^A$ and $\varphi_y{}^A$ in place of φ_x and φ_y, and using b' in place of b, in the definition of M. (b' was defined after Corollary XXXIV in §16.6.) This gives $\{T^A\} \in \Pi_1^{1,A}$ for any A, and the proof of the theorem is complete. ☒

§16.8 THE HYPERARITHMETICAL HIERARCHY

Hyperarithmetical sets and hyperarithmetical computability were considered in §16.4 and §16.5. Results on hyperarithmetical sets and implicit definability were given in §16.7. We now look further at the structure of the class of hyperarithmetical sets.

We can approach this class in three different ways: (1) We can look for a classification that extends the Σ_n^0, Π_n^0 classification of the arithmetical hierarchy. Such a classification would be based on an appropriate generalized notion of predicate form and would yield a classification of sets of sets and sets of functions as well as of sets of integers. We would hope to obtain results analogous to the basic results of Chapters 14 and 15 about the arithmetical hierarchy. (2) We can study the Turing degrees of hyperarithmetical sets. We would hope to obtain results analogous to the basic results of Chapter 13 for the arithmetical hierarchy. In particular, we would seek a "natural" chain of degrees for the hyperarithmetical sets analogous to the chain $\mathbf{0}$, $\mathbf{0}'$, . . . for the arithmetical sets. (3) We can look at ways of generating the hyperarithmetical sets "from below" by inductive procedures that define (in some sense) new sets from sets already obtained. We would expect to study the "levels" of hyperarithmetical sets obtained at successive stages of such an inductive procedure.

All three approaches are closely interrelated, and the formal development given below is important for each approach. We shall motivate it in terms of the first approach. We give this motivation in a *heuristic discussion*, where we consider the classification of sets of integers and outline a method for extending the arithmetical hierarchy to nonarithmetical hyperarithmetical sets. We then give a *formal development*, in which this method is made precise and certain questions of invariance are answered. We then turn to the classification of *sets of sets* and *sets of functions*. Here we obtain a second method for extending the hierarchy that provides a uniform treatment of sets of sets, sets of functions, and sets of integers. The classification which it yields coincides with that of the first method on sets of integers.

We then conclude with *comments* that relate our results to approaches 2 and 3 above.

Heuristic Discussion

For the time being, we consider only sets of integers (the case studied in Chapter 14). In keeping with this limitation, we use the notation of Chapter 14, and take $\Sigma_0, \Sigma_1, \ldots$ to be the classes of sets of integers in the arithmetical hierarchy.

In the arithmetical hierarchy, we have the classes $\Sigma_0, \Sigma_1, \ldots$ associated with the finite ordinals $0, 1, \ldots$. We wish to extend this classification in a way that will associate appropriate classes (which we might call Σ_β) with infinite ordinals β. Assume that a classification has been obtained up to a given limit ordinal α. If we can associate a particular new (i.e., as yet unclassified) set of integers A with α (we shall call A the *basic set* at α), then we can take the relativized classes $\Sigma_0^A, \Sigma_1^A, \ldots$ as the classes $\Sigma_\alpha, \Sigma_{\alpha+1}, \ldots$, and thus obtain a classification up to the next limit ordinal. Moreover, for $n > 0$, we can take the Σ_n^A-forms as our new $\Sigma_{\alpha+n}$-forms, and the Σ_n^A-indices as our new $\Sigma_{\alpha+n}$-indices.† For example, given the classes $\Sigma_0, \Sigma_1, \ldots$, we could take $A = \emptyset^{(\omega)}$ as the basic set at ω (see §13.1), and obtain $\Sigma_0^A, \Sigma_1^A, \ldots$ as the classes $\Sigma_\omega, \Sigma_{\omega+1}, \ldots$.

The extended classification obtained in this way clearly depends on the method used for choosing basic sets. More specifically, let α be a limit ordinal; let Σ_β, $\beta < \alpha$, be given; and let A be a basic set at α. The classes $\Sigma_{\alpha+n}$, $n = 0, 1, \ldots$, depend on the T-degree of A; the concept of $\Sigma_{\alpha+n}$-form depends on the T-degree of A; and the $\Sigma_{\alpha+n}$-indices depend on the particular set A itself. One constraint on the T-degree of A is immediate: in order that the classification be monotone (i.e., that $\beta < \gamma \Rightarrow \Sigma_\beta \subset \Sigma_\gamma$) and that a hierarchy theorem hold (i.e., that $\beta < \gamma \Rightarrow (\Sigma_\gamma - \Sigma_\beta) \neq \emptyset$), we must take the T-degree of A to be higher than the T-degree of any set in $\bigcup_{\beta<\alpha}\Sigma_\beta$. Most desirable, in order to give as fine a classification as possible, would be to have the T-degree of A be a minimum among all higher T-degrees. Unfortunately, as we saw in §13.4, such a minimum higher T-degree may not exist. How then shall we choose the set A? Can it be done in a way that is "natural," i.e., invariant in some appropriate sense?

The construction of transfinite levels in the classical hierarchy of Borel sets suggests that we take A to be, in some sense, an "infinite join" of the sets already classified. We do this as follows. At every limit ordinal β below the given limit ordinal α, define A_β to be the basic set already chosen at β. For every successor number β below α, define A_β to be some Σ_β-complete set. For $\beta = 0$, take A_β to be some recursive set. (We take $A_0 = \emptyset$; our results will not depend on this choice). Let ν be a numerical indexing

† The notation "Σ_α" is introduced temporarily for the purposes of the present heuristic discussion. For $\alpha \geq \omega$, we shall later follow the common practice in the literature and use the notation "Σ_α^0" for the class here called $\Sigma_{\alpha+1}$.

(i.e., a one-one map from integers to ordinals) that assigns the set of indices D_α to $\{\beta|\beta < \alpha\}$. Then define

$$A = A_\alpha = \{<x,y>|y \in D_\alpha \;\&\; x \in A_{\nu(y)}\}.$$

Several questions of invariance immediately arise. (i) How does the classification obtained depend upon the numerical indexing of ordinals used, and what constraints should be placed on the choice of such an indexing? (ii) How does the classification obtained depend upon the choice of Σ_β-complete sets, and what constraints, if any, should be placed on this choice? (See also Exercise 16-81.) We consider these questions in turn.

(i) The T-degree of a basic set can be shown to vary with the numerical indexing used (see Exercise 16-78). Some constraint must therefore be placed on the choice of an indexing. The discussion at the beginning of §11.7 suggests the requirement that an indexing be a *univalent system of notation.* We adopt this requirement (which incidentally limits us to the constructive ordinals). In the presence of this requirement, does the degree of a basic set depend on the choice of indexing? We return to this question below.

(ii) It is trivial to show that the T-degree of A_α varies with our choice of the Σ_β-complete sets $A_{\beta+1}$, $\beta < \alpha$ (see Exercise 16-79). Earlier theory suggests the requirement that these sets be chosen in a uniform and effective way. Recall from Theorem 14-VIII (the strong hierarchy theorem) that $\Sigma_{\beta+1}$ must be the collection of all sets recursively enumerable in A_β. We require that there be a fixed z_0 such that for all β, $A_{\beta+1} = W_{z_0}{}^{A_\beta}$ and that there be a fixed recursive function f_0 of two variables such that for any u and β, $W_u{}^{A_\beta} \leq_1 A_{\beta+1}$ via $\lambda x[f_0(u,x)]$. Given this requirement, and given a particular univalent system of notations over a segment of ordinals, it is easy to show (using the recursion lemma) that the classification obtained over that segment of ordinals is independent of the fixed z_0 used (see Exercise 16-82). We make the following specific choice of z_0. We take $A_{\beta+1} = (A_\beta)'$. The required z_0 and f_0 exist by Corollary 13-I(*c*).

The following problem now remains. Given that the above requirements are satisfied, do two univalent systems of notation determine the same classification over the ordinals that they have in common? The affirmative answer to this problem of "notational invariance" is given in Theorem XLIX below. This result shows that we have a well-defined classification over all the constructive ordinals.

How much of the analytical hierarchy does this classification include? In Theorems L and LI we show that it coincides with the hyperarithmetical sets, i.e., that $\bigcup_{\beta \text{ constructive}} \Sigma_\beta$ = the class of Δ_1^1-sets. We call this classification (into the classes Σ_β) the *hyperarithmetical hierarchy.*

To what extent can the restriction (in (i) above) be weakened? Can we, for example, allow all recursive well-orderings and still get notational invariance? This seems doubtful, although Enderton and Luckham [1964] have shown that for a certain z_0 (in (ii) above) notational invariance exists provided that z_0 does not vary as recursive well-orderings vary. If the restriction to systems of notation is weakened, is the hyperarithmetical hierarchy a *minimum* among all hierarchies then obtained (taking the hierarchy $\{\Sigma_\beta\} \leq$ the hierarchy $\{\Sigma'_\beta\}$ if $\Sigma_\beta \subset \Sigma'_\beta$ for all β for which both hierarchies are defined)? Enderton has shown that the hyperarithmetical hierarchy is minimum even when condition (iii) in the definition of system of notation is dropped (see Exercise 16-84).

In the formal development below, we shall use the system of notations O, as defined in §11.7. Every branch of O is a univalent system of notation. Theorem XLIX will show that any two branches of O yield the same Σ_β classification over the ordinals that they have in common. This yields an invariance result over all univalent systems of notation since, by Corollary 11-XVIII(b), any univalent system can be recursively mapped into a branch of O in an ordinal-preserving way (see Exercise 16-80).

Formal Development

Define O and $<_O$ as in §11.7. For any $x \in O$, let $|x|$ be the ordinal associated with x (denoted $|x|_O$ in §11.7). We say that

$$x \leq_O y \text{ if } x \in O \;\&\; y \in O \;\&\; [x <_O y \vee x = y].$$

Definition We associate sets with notations in O as follows:

$$\begin{aligned} H(1) &= \emptyset; \\ H(2^x) &= (H(x))', \qquad \text{for } x \in O; \\ H(3 \cdot 5^y) &= \{\langle u,v\rangle \mid v <_O 3 \cdot 5^y \;\&\; u \in H(v)\}, \qquad \text{for } 3 \cdot 5^y \in O. \end{aligned}$$

These inductive rules evidently associate a unique set, which we call $H(x)$, with each x in O. Along each branch of O, sets are associated with notations according to the general construction described in the heuristic discussion above for extending the arithmetical hierarchy. Our goals, as indicated in that discussion, are as follows: first, to show that at any limit number α, $[[x \in O \;\&\; y \in O \;\&\; |x| = |y| = \alpha] \Rightarrow H(x) \equiv_T H(y)]$; and second, to show that for any B, $B \in \Delta_1^1 \Leftrightarrow (\exists x)[x \in O \;\&\; B \leq_1 H(x)]$. We achieve these goals in Theorems XLIX, L, and LI. We first prove two lemmas.

First lemma $[x \in O \;\&\; y \in O \;\&\; x \leq_O y] \Rightarrow H(x) \leq_1 H(y)$ *uniformly in x and y (that is to say, if $x \leq_O y$, then $H(x) \leq_1 H(y)$ via h, where an index for h can be found uniformly in x and y.)*

Proof. By Corollary 13-I(c), there is a recursive function f such that for any A, $A \leq_1 A'$ via f.

Let x and y be given such that $x \leq_O y$. We show how to compute h so that $H(x) \leq_1 H(y)$ via h. We consider two cases.

$|x| + n = |y|$ for some n. Take $h = \lambda u[f^n(u)]$ (where $f^0(u) = u$ and $f^{n+1}(u) = f(f^n(u))$).

$|x| + \omega \leq |y|$. Take z such that $z \in O$, $|z|$ is a limit ordinal, and $|y| = |z| + n$. Then $u \in H(x) \Leftrightarrow <u,x> \in H(z) \Leftrightarrow f^n(<u,x>) \in H(y)$. Take $h = \lambda u[f^n(<u,x>)]$.

The two cases can be effectively distinguished, and n in each case can be effectively computed from x and y. Hence the desired uniformity holds.☒

***Second lemma* (Spector)** *For $y \in O$, $\{u|u \in O \ \&\ |u| < |y|\}$ is recursive in $H(2^y)$ uniformly in y (that is to say, an index for the characteristic function of $\{u|u \in O \ \&\ |u| < |y|\}$ as a set recursive in $H(2^y)$ can be found uniformly in y).*

Proof. We use the recursion lemma over the well-founded relation $<_O$. Take R (for the recursion lemma) to be

$$\{<y,w>|y \in O \ \&\ \varphi_w^{H(2^y)} \textit{ is the characteristic function of } \{u|u \in O \ \&\ |u| < |y|\}\}.$$

Given a z such that $(\forall x)[x <_O y \Rightarrow R(x,\varphi_z(x))]$, we must show how to compute a w, uniformly in z and y, such that $R(y,w)$. (w was called $\eta(z,y)$ in the statement of the recursion lemma.) For any ordinal α, define

$$O_\alpha = \{u|u \in O \ \&\ |u| < \alpha\}.$$

We consider four cases.

$y = 1$. In this case $O_{|y|} = \emptyset$. Choose w so that $(\forall u)[\varphi_w{}^{H(2)}(u) = 0]$.

$y = 2^s$ and $s = 2^t$. Then $O_{|y|} = O_{|s|} \cup \{2^v|v \in O_{|s|}\}$. But, by assumption, $O_{|s|}$ is recursive in $H(y)$ with index (for characteristic function) $\varphi_z(s)$. Therefore $O_{|y|}$ is recursive in $H(y)$ with an index obtainable uniformly from $\varphi_z(s)$. Hence $O_{|y|}$ is recursive in $H(2^y)$ ($= (H(y))'$) with index w obtainable uniformly from $\varphi_z(s)$.

$y = 2^s$ and $s = 3 \cdot 5^t$. Then

$O_{|y|} = O_{|s|} \cup \{3 \cdot 5^r|\varphi_r \textit{ is total } \& (\forall u)(\forall v)[\varphi_r(u) \subset W_{f\varphi_r(v)}] \ \&\ (\forall u)[\varphi_r(u) \in O_{|s|}]\}$, where f is the recursive function of Exercise 11-55.

$\{r|\varphi_r \textit{ is total}\}$ is recursive in $H(4)$ ($= \emptyset''$) (by §13.2); $\{r|(\forall u)(\forall v)|\varphi_r(u) \subset W_{f\varphi_r(v)}\}$ is recursive in $H(4)$ directly; and $H(4)$ is recursive in $H(2^y)$ by the first lemma. Since $O_{|s|}$ is recursive in $H(y)$ with index $\varphi_z(s)$, $\{r|(\forall u)[\varphi_r(u) \in O_{|s|}]\}$ is recursive in $H(2^y)$ with index obtainable uniformly from $\varphi_z(s)$. Hence we have $O_{|y|}$ recursive in $H(2^y)$ with index w obtainable uniformly from $\varphi_z(s)$.

$y = 3 \cdot 5^t$. Then $O_{|y|} = \{u|(\exists v)[u \in O_{|\varphi_t(v)|}]\}$. But for any v, $O_{|\varphi_t(v)|}$ is recursive in $H(2^{\varphi_t(v)})$ with index $\varphi_z(\varphi_t(v))$ by inductive assumption, and $H(2^{\varphi_t(v)})$ is recursive in $H(y)$ uniformly by the first lemma. Hence $O_{|y|}$ is recursively enumerable in $H(y)$ and therefore recursive in $H(2^y)$ with index w obtainable uniformly from z and t.

The four cases can be effectively distinguished, and w is uniform in y in each case. Hence we have our result.☒

***Corollary to second lemma* (*Spector*)** *For* $y \in O$,

$$\{u|u \in O \,\&\, |u| = |y|\}$$

is recursive in $H(2^{2^y})$, *uniformly in* y.

Proof.

$$\{u|u \in O \,\&\, |u| = |y|\} = \{u|u \in O \,\&\, |u| < |2^y| \,\&\, \neg[u \in O \,\&\, |u| < |y|]\}$$
$$= \{u|u \in O \,\&\, |u| < |2^y|\} \cap \{u| \neg[u \in O \,\&\, |u| < |y|]\}.$$

By the lemma, the set $\{u|u \in O \,\&\, |u| < |y|\}$ is recursive in $H(2^y)$ uniformly in y. Also by the lemma, the set $\{u|u \in O \,\&\, |u| < |2^y|\}$ is recursive in $H(2^{2^y})$ uniformly in y. It follows that the intersection of these two sets is recursive in $H(2^{2^y})$ uniformly in y. ☒

***Theorem XLIX* (*Spector*)**

$$[x \in O \,\&\, y \in O \,\&\, |x| = |y|] \Rightarrow H(x) \leq_T H(y),$$

uniformly in x *and* y. (*That is to say, if* $|x| = |y|$, *then an index for* $c_{H(x)}$ *as a* $H(y)$*-recursive function can be found uniformly in* x *and* y.)

Proof. Let $S = \{<x,y>|x \in O \,\&\, y \in O \,\&\, |x| = |y|\}$. Define the partial ordering $<_S$ over S as follows: $<u,v> \, <_S \, <x,y>$ if $|u| < |x|$. We use the recursion lemma over the well-founded relation $<_S$, where R (for the recursion lemma) is taken to be

$$\{<<x,y>,w>|x \in O \,\&\, y \in O \,\&\, |x| = |y| \,\&\, c_{H(x)} = \varphi_w^{H(y)}\}.$$

Given a z such that $(\forall u)(\forall v)[<u,v> \, <_S \, <x,y> \Rightarrow R(<u,v>,\varphi_z(<u,v>))]$, we must show how to compute a w uniformly in x, y, and z such that $R(<x,y>,w)$. We consider three cases.

Case (*a*): $<x,y> = <1,1>$. Take w so that $\varphi_w^{\emptyset} = c_{\emptyset}$.

Case (*b*): $<x,y> \in S \,\&\, <x,y> = <2^u,2^v>$. By assumption on z, we can find an r such that $c_{H(u)} = \varphi_r^{H(v)}$. But $H(x) = (H(u))'$, and $H(y) = (H(v))'$. Applying Corollary 13-I(*c*), we can obtain the desired w.

Case (*c*): $<x,y> \in S \,\&\, <x,y> = <3 \cdot 5^u, 3 \cdot 5^v>$. We seek a way to reduce $H(x)$ to $H(y)$. Let any $<s,t>$ be given. We must show how to test whether $<s,t> \in H(x)$, given an oracle for $H(y)$. By the definition of $H(x)$ $(= H(3 \cdot 5^u))$, we wish to test whether $[t <_O x \,\&\, s \in H(t)]$. We proceed as follows. First, see whether $t <_O x$. By Exercise 11-55, this is an instance of the halting problem and can be tested by reference to $H(2)$ $(\equiv K)$. Since $H(2) \leq_1 H(y)$ by the first lemma, this can be tested by reference to $H(y)$.

If $t <_O x$, then see whether $s \in H(t)$ as follows. Enumerate $\{p|p <_O y\}$ (see Exercise 11-55). As each member p of this set appears, see whether $|t| = |p|$. We can do this, since

$$\{q|q \in O \,\&\, |q| = |p|\}$$

is recursive in $H(2^{2^p})$, by the second lemma, and $H(2^{2^p})$ is uniformly reducible to $H(y)$ by the first lemma. Eventually we must identify a p such that $|t| = |p|$ and $p <_O y$. By assumption on z, $\varphi_z(<t,p>)$ gives us a way to reduce $H(t)$ to $H(p)$. The first lemma gives us a way to reduce $H(p)$ to $H(y)$. Hence we have a way to reduce $H(t)$ to $H(y)$ and we can use it to test whether $s \in H(t)$. Thus we have a procedure for testing whether $<s,t> \in H(x)$.

Instructions for the above procedure are evidently uniform in x, y, and z, and this concludes the proof.☒

Corollary XLIX(a) $[x \in O \;\&\; y \in O \;\&\; |x| = |y|] \Rightarrow H(x) \equiv_T H(y)$ *uniformly in x and y.*

(**b**) $[x \in O \;\&\; y \in O \;\&\; |x| = |y| = \beta \;\&\; \beta$ *is a successor ordinal*$] \Rightarrow H(x) \equiv H(y)$ *uniformly in x and y.*

(**c**) $[x \in O \;\&\; y \in O \;\&\; |x| \leq |y|] \Rightarrow H(x) \leq_T H(y)$ *uniformly in x and y.*

Proof. (*a*) Immediate from the theorem.

(*b*) Immediate by Corollary 13-I(*a*). (Moschovakis has shown that (*b*) can fail when β is a limit ordinal. See Exercise 16-97.)

(*c*) That $H(x) \leq_T H(y)$ is immediate from (*a*) and the first lemma. This proof does not yield uniformity, however, since we have no apparent way for deciding whether $|x| = |y|$ or $|x| < |y|$. In Exercise 16-96, we modify the proof of Theorem XLIX to yield the desired uniformity.☒

Theorem L (Kleene) $y \in O \Rightarrow H(y) \in \Delta_1^1$ *uniformly in y.*

Proof. We use the recursion lemma over $<_O$ where R is taken to be $\{<y,w>|y \in O \;\&\; w$ *is a* Δ_1^1*-index for* $H(y)\}$. Given a z such that $(\forall x)[x <_O y \Rightarrow R(x,\varphi_z(x))]$, we must show how to compute a w, uniformly in y and z, such that $R(y,w)$. We consider three cases.

$y = 1$. Let w be a Δ_1^1-index for $\emptyset$.

$y = 2^s$. Then $\varphi_z(s)$ is a Δ_1^1-index for $H(s)$. Since $H(y) = (H(s))'$ we can apply the construction of Theorem X and obtain a Δ_1^1-index w for $H(y)$.

$y = 3 \cdot 5^t$. Then

$$H(y) = \{<u,v>|v <_O y \;\&\; u \in H(v)\} = \{<u,v>|v <_O y \;\&\; (\exists f)(\forall x) \neg T'_{1,1}(\pi_1\varphi_z(v),f,u,x)\} = \{<u,v>|v <_O y \;\&\; (\forall f)(\exists x)T'_{1,1}(\pi_2\varphi_z(v),f,u,x)\}$$

Since $\{v|v \leq_O y\}$ is recursively enumerable uniformly in y (Exercise 11-55) and is hence in Σ_1^0, a Tarski-Kuratowski computation yields that $H(y) \in \Delta_1^1$, and a Δ_1^1-index w can be obtained uniformly in y.☒

Theorem LI (Kleene) $B \in \Delta_1^1 \Rightarrow (\exists y)[y \in O \;\&\; B \leq_1 H(y)]$.

Proof. It is enough to show that $B \in \Delta_1^1 \Leftrightarrow (\exists y)[y \in O \;\&\; B \leq_T H(y)]$ (since $A \leq_T H(y) \Rightarrow A \leq_1 H(2^y)$). By Corollary XXII(*b*), it will be enough to show that for all $y \in O$, $T_{|y|} \leq_T H(2^{2^y})$. We abbreviate 2^{2^y} as $e(y)$.

W~. use the recursion lemma over $<_O$, where R is taken to be

$$\{<y,w>|y \in O \ \&\ c_{T_{|y|}} = \varphi_w^{H(e(y))}\}.$$

Given a z such that $(\forall x)[x <_O y \Rightarrow R(x,\varphi_z(x))]$, we must show how to compute a w uniformly in y and z such that $R(y,w)$. We consider three cases.

$y = 1$. Take w so that $\varphi_w^{H(4)} = \lambda x[0]$.

$y = 2^s$. Observe that

$$T_{|y|} = \{x|[x \in T \ \&\ \|x\| = 0] \vee [\varphi_x(0) = 1 \ \&\ (\forall n)[b(n,x) \in T_{|s|}]]\}.$$

Now

$$\{x|x \in T \ \&\ \|x\| = 0\} = \{x|(\forall u)[\varphi_x(u) = 0]\} \in \Pi_2^0.$$

Hence $\{x|x \in T \ \&\ \|x\| = 0\} \leq_T H(2^2) \leq_T H(e(y))$ by Theorem 14-VIII and the first lemma above.

Similarly $\{x|\varphi_x(0) = 1\} \in \Sigma_1^0$, and hence

$$\{x|\varphi_x(0) = 1\} \leq_T H(2) \leq_T H(e(y)).$$

Finally, $\{x|(\forall n)[b(n,x) \in T_{|s|}]\} \in \Pi_1^{0,T_{|s|}}$. Hence

$$\{x|(\forall n)[b(n,x) \in T_{|s|}]\} \leq_T (T_{|s|})'$$

(by Theorem 14-VIII). But $T_{|s|} \leq_T H(e(s))$ by the assumption on z, and hence $(T_{|s|})' \leq_T (H(e(s)))' = H(e(y))$.

Thus $T_{|y|} \leq_T H(e(y))$.

$y = 3 \cdot 5^t$. Observe that

$$T_{|y|} = \{x|[x \in T \ \&\ \|x\| = 0] \vee [\varphi_x(0) = 1 \ \&\ (\exists m)(\forall n)[b(n,x) \in T_{|\varphi_t(m)|}]]\}$$

(by the basic tree lemma). By assumption on z, we can reduce $T_{|\varphi_t(m)|}$ to $H(e(\varphi_t(m)))$ uniformly. By the first lemma, we can reduce $H(e(\varphi_t(m)))$ to $H(y)$ uniformly. Hence, making a Tarski-Kuratowski computation, we have $T_{|y|} \in \Delta_3^{0,H(y)}$ uniformly in y. But by Theorem 14-VIII, this yields $T_{|y|} \leq_T (H(y))'' = H(e(y))$. We can obtain our desired w, and the proof is complete.☒

Corollary LI *There is a recursive g such that for any w, if w is a* Δ_1^1*-index for B, then* $g(w) \in O$ *and* $B \leq_1 H(g(w))$.

Proof. By Corollary XXII(*b*) and the proof of the theorem.☒

The preceding lemmas and theorems were initially obtained under a somewhat different assignment of sets to notations in O. The assignment used here is due to Enderton, and is more convenient for our purposes. In Exercise 16-83 we show that the earlier assignment, due to Kleene, is equivalent (with respect to T-reducibility) to the assignment used by us.

We now introduce notation for the degrees of unsolvability determined by the sets $H(x)$.

Definition Let $x \in O$ and $|x| = \alpha$. Define $\mathbf{h}_\alpha$ to be the (uniquely defined) T-degree of $H(x)$.

Observe that $\mathbf{h}_{\alpha+1} = \mathbf{h}'_\alpha$; that $\alpha \leq \beta \Leftrightarrow \mathbf{h}_\alpha \leq \mathbf{h}_\beta$; and that $\mathbf{a}$ is a degree of hyperarithmetical sets if and only if $\mathbf{a} \leq \mathbf{h}_\alpha$ for some (constructive) α.

We next introduce notation for the classes of the hierarchy (as classes of sets of integers).

Definition Let $\beta = \alpha + n$ where α is a limit ordinal, and let $y \in O$ and $|y| = \alpha$. Define

$$\Sigma^0_\beta = \Sigma^{H(y)}_{n+1};$$
$$\Pi^0_\beta = \Pi^{H(y)}_{n+1};$$
$$\Delta^0_\beta = \Sigma^0_\beta \cap \Pi^0_\beta.$$

Let $\beta = n < \omega$. Define

$$\Sigma^0_\beta = \Sigma_n;$$
$$\Pi^0_\beta = \Pi_n;$$
$$\Delta^0_\beta = \Sigma^0_\beta \cap \Pi^0_\beta.\dagger$$

By Theorem XLIX, the classes Σ^0_β and Π^0_β depend only on β and not on the particular y in O such that $|y| + n = \beta$. It follows from Theorem 14-VIII that for any x such that $x \in O$ and $|x| = \beta + 1$, $H(x)$ is a Σ^0_β-complete set; that for $\beta \geq \omega$, $A \in \Sigma^0_\beta \Leftrightarrow A$ is recursively enumerable in $\mathbf{h}_\beta$; and that for $\beta \geq \omega$, $A \in \Delta^0_\beta \Leftrightarrow \mathbf{a} \leq \mathbf{h}_\beta$, where $\mathbf{a}$ is the degree of A. (The reader should note the difference from the Σ_β notation temporarily used in the heuristic discussion. For $\beta \geq \omega$, what was there called $\Sigma_{\beta+1}$ is here called Σ^0_β.)

The following results are now immediate from Chapter 14. For all $\beta > 0$:

$$A \in \Sigma^0_\beta \Leftrightarrow \bar{A} \in \Pi^0_\beta;$$
$$\Sigma^0_\beta \cup \Pi^0_\beta \subset \Sigma^0_{\beta+1} \cap \Pi^0_{\beta+1};$$
$$(\Sigma^0_\beta - \Pi^0_\beta) \neq \emptyset.$$

Note that in our Σ^0_β scheme of notation for the hyperarithmetical hierarchy, two rather natural classes are omitted at each limit number α. These are (i) $\bigcup_{\beta<\alpha} \Sigma^0_\beta$; and (ii) $\Sigma^0_\alpha \cap \Pi^0_\alpha$. All lower classes are properly contained in (i); (i) is properly contained in (ii); and (ii) is properly contained in Σ^0_α. Our temporary notation in the heuristic discussion also gave a name to (ii) (it was called Σ_α). We now have the notation Δ^0_α for (ii). We shall see that the definitions which we introduce below for sets of sets and sets of functions give a name to (i) and provide notions of predicate form and index for (i).

Sets of Sets and Sets of Functions

Can Σ_β-forms, as defined at the beginning of the heuristic discussion, be relativized to yield a hierarchy of sets of sets? Direct relativization of such

† The classes Σ^0_β and Π^0_β are also known as P_β and Q_β, respectively, in the literature.

forms unfortunately leads to a classification that is neither monotone nor sufficiently inclusive (see Exercise 16-86). To obtain a satisfactory hierarchy, we must make a "deeper" relativization as follows.

Definition For any set X, define

$$H^X(1) = X;$$
$$H^X(2^x) = (H^X(x))', \qquad \text{for } x \in O;$$
$$H^X(3 \cdot 5^y) = \{<u,v>|v <_O 3 \cdot 5^y \;\&\; u \in H^X(v)\}, \qquad \text{for } 3 \cdot 5^y \in O.$$

This associates a unique set $H^X(x)$ with each $x \in O$.

Definition Let $\mathcal{A} \subset \mathfrak{N}$, let $y \in O$, $|y| > 0$, and let z be an integer such that $\mathcal{A} = \{X|z \in H^X(y)\}$. We call the assertion "$z \in H^X(y)$" a *y-form* for $\mathcal{A}$, and we call z a *y-index* for $\mathcal{A}$. y itself plays a role analogous to that of the prefix in arithmetical and analytical forms.

Assume that $u \in O$, $v \in O$, and $|u| \leq |v|$; and let $\mathcal{A}$ be a set of sets expressed by a u-form. Must $\mathcal{A}$ also be expressed by a v-form? In Exercise 16-88 we use Theorem XLIX to show that this is the case. We go on to make the following definition.

Definition If α is a successor ordinal, $y \in O$, $|y| = \alpha$ and $\mathcal{A}$ is expressed by a y-form, then we say that

$$\mathcal{A} \in \Sigma^0_\alpha, \qquad \text{if } \alpha < \omega;$$
$$\mathcal{A} \in \Sigma^0_{\alpha-1}, \qquad \text{if } \alpha > \omega;$$
$$\mathcal{A} \in \Pi^0_\alpha, \qquad \text{if } \bar{\mathcal{A}} \in \Sigma^0_\alpha.$$

This defines a hierarchy for sets of sets. A simple inductive argument shows that this classification coincides with the arithmetical hierarchy of §15.1 for $0 < \alpha < \omega$ (see Exercise 16-89). In Exercise 16-85 we show that it includes all hyperarithmetical sets of sets and no other sets of sets. We therefore call it the *hyperarithmetical hierarchy* of sets of sets.

Note that we also have y-forms when $|y|$ is a limit number. In this case, $\mathcal{A}$ is expressed by a y-form if and only if $\mathcal{A} \in \bigcup_{\beta<\alpha} \Sigma^0_\beta$, as is easily shown.

Can the notion of y-form be generalized to include y-forms for sets of integers as well as sets of sets? The method of Theorem 15-XI suggests a way to do this. We say that a y-form *expresses a set A with y-index z* if $A = \{x|z \in H^{\{x\}}(y)\}$. The hyperarithmetical hierarchy of sets of integers obtained in this way coincides with that defined in the preceding formal development (see Exercise 16-90).

More general y-forms can now be defined as $\mathfrak{H}_y(z,A_1, \ldots ,x_1, \ldots)$ if $z \in H^{A_1 \, join \cdots \{x_1\} \, join \cdots}(y)$.

Does a hierarchy theorem hold for the hyperarithmetical hierarchy of sets of sets? Such a theorem is easily obtained by a proof similar to that for Theorem 15-XI.

Sets of functions can also be incorporated by saying that a y-form expresses a set of functions $\mathcal{A}$ if $\mathcal{A} = \{f|z \in H^{\tau(f)}(y)\}$. The hierarchy obtained can be shown to coincide at positive finite levels with the hierarchy of §15.2, and a hierarchy theorem can be proved as in Theorem 15-XXIV.

The definitions and notations immediately above lead to a generalized formulation that extends the formulation used for Theorems 15-XXVII and 15-XXVIII. For such a formulation, the system of notations O must be replaced by a system $O[\mathcal{C}]$ in which functions rather than integers serve as notations, and in which fundamental sequences (of notations) recursive in $\mathcal{C}$ (rather than just recursive fundamental sequences) are allowed to determine limit notations. In addition, the jump operation must be based on the relation T_n^* of Theorem 15-XXVII (not the same as the relation $T_{k,l}^*$ defined in §16.1). If $\mathcal{C} = \emptyset$, the resulting hierarchy classification of subsets of $\mathfrak{F}$ extends over the constructive ordinals and coincides with the hyperarithmetical hierarchy on $\mathfrak{F}$. If $\mathcal{C} = \mathfrak{N}$, the resulting hierarchy classification extends over all countable ordinals and coincides with the classical hierarchy of Borel sets on $\mathfrak{F}$.

Comments

Let us again consider the three approaches mentioned at the beginning of this section.

Approach 1. Several aspects of our theories of the arithmetical and analytical hierarchies are missing in our theory of the hyperarithmetical hierarchy. We have not given a representation theorem, nor have we given an analogue to the Tarski-Kuratowski algorithm. In Exercise 16-91, a form of representation theorem is proved. In Exercise 16-92, rules for a form of Tarski-Kuratowski algorithm are described. These do not provide a fully integrated treatment, however. Problems of uniformity and notational invariance appear difficult.

Approach 2. The chain of degrees $\mathbf{h}_0, \mathbf{h}_1, \ldots, \mathbf{h}_\omega, \ldots$ appears to be a satisfactory hyperarithmetical counterpart to the chain of degrees $\mathbf{0}, \mathbf{0}'$, ... of arithmetical sets. In view of §13.4, however, the question can be raised: is this chain *minimal* among all chains in some invariantly defined collection of "reasonable" chains (where we say that chain $\{\mathbf{h}_\alpha\} \leq$ chain $\{\mathbf{g}_\alpha\}$ if $(\forall\alpha)[\mathbf{h}_\alpha \leq \mathbf{g}_\alpha]$)? The result of Exercise 16-84 yields one affirmative answer to this question. Considerations of minimality are important in the examination of other possible ways of generating chains of hyperarithmetical degrees, for such other ways may omit intervals (usually at limit ordinals) from the chain $\{\mathbf{h}_\alpha\}$. This proves to be the case, for example, in Exercise 16-93, where we consider a "notation free" method (due to Shoenfield) for generating degrees in the chain $\{\mathbf{h}_\alpha\}$. This is also the case with the sets T_α.

Can we compute degrees of unsolvability in the hyperarithmetical hierarchy? Methods which generalize the methods of §14.8, and in which the recursion lemma plays a fundamental role, can be developed. Spector [1958*a*] has announced the results of such a computation for the degrees O_α

and W_α, where $O_\alpha = \{x | x \in O \ \& \ |x| < \alpha\}$ and $W_\alpha = \{x | \varphi_x$ *is the characteristic function of a well-ordering of order type* $< \alpha\}$.

We note in passing that the word "hierarchy" has come to be applied not only to classifications based on predicate forms but also to more general orderings of degrees obtained by constructions analogous to the construction of the sets $H(x)$. Other (and possibly less effective) numerical indexings of ordinals are used, and sets are associated with "notations" in other (and possibly less effective) inductive ways. Several studies of such "hierarchies" have been begun, and work on notational invariance in such hierarchies has been done. See Kreider and Rogers [1961], Richter [1965], Enderton [1964], and Putnam and Hensel [1965].

Approach 3. The hyperarithmetical sets of integers can be obtained by iterating the operations of jump and "infinite join," and taking all sets T-reducible to the sets thus obtained. This is a consequence of Theorem L, Theorem LI, and the definition of the sets $H(x)$ (see also Exercise 16-94). This inductive characterization of the hyperarithmetical sets "from below" can be used to prove a strengthened version of Corollaries XLIV(c) and XLIV(e) on implicit definability.

Theorem LII $A \in \Delta_1^1 \Leftrightarrow (\exists B)[A \leq_1 B \ \& \ \{B\} \in \Pi_2^0]$.

Proof. $\Leftarrow$: Immediate by Corollary XLIV.

$\Rightarrow$: By Theorem LI, it is enough to show that $x \in O \Rightarrow \{H(x)\} \in \Pi_2^0$. The proof generalizes the construction in Theorem 15-XII. We use the recursion lemma over $<_O$, and consider three cases.

$x = 1$. Then $\{H(1)\} \in \Pi_2^0$, trivially.

$x = 2^s$. If $\{H(s)\} \in \Pi_2^0$, then $\{H(2^s)\} \in \Pi_2^0$ by the argument of Theorem 15-XII.

$x = 3 \cdot 5^t$. Assume that we have a z such that $u <_O 3 \cdot 5^t \Rightarrow \varphi_z(u)$ is a Π_2^0-index for $\{H(u)\}$. Then

$$\{H(3 \cdot 5^t)\} = \{X | (\forall u)(\forall v)[<u,v> \in X \Rightarrow v <_O 3 \cdot 5^t] \ \& \ (\forall v)[v <_O 3 \cdot 5^t \Rightarrow (\forall y_1)(\exists y_2) T_{1,1}(\varphi_z(v), \{x | <x,v> \in X\}, y_1, y_2)]\}.$$

By a Tarski-Kuratowski computation, we have $\{H(3 \cdot 5^t)\} \in \Pi_2^0$ with Π_2^0-index uniform in z and t.

The recursion lemma therefore applies, and the proof is complete. ☒

In Exercise 16-98, we see that for each $x \in O$ there is a function f such that $H(x) = \{u | f(u) \neq 0\}$ and $\{f\} \in \Pi_1^0$.

§16.9 EXERCISES

§16.1

16-1. Use Theorem VIII (the hierarchy theorem) to show that the converses to rules (iii) and (iv) in Theorem III do not hold.

16-2. Complete the proof of Theorem III by showing, in (ii), (iii), and (iv), that the new relations are recursive, and that the new predicate forms are equivalent to the given predicate forms. (Note that the proof of equivalence for (iv) uses the axiom of choice.)

16-3. Give a general proof of Corollary III.

16-4. Complete the proof of Theorem VII by showing that in (ii) the new relations are recursive and the new predicate forms are equivalent to the given predicate forms.

16-5. Give details for Examples 1 and 2 after Theorem VII.

△***16-6.*** (*a*) The set O was defined (and temporarily denoted D_O) in §11.7. Give a direct definition of O that yields $O \in \Pi_1^1$ (without using the fact that $O \equiv W$). (*Hint:* A definition without reference to ordinals must be found. First define $<_O$ as the intersection of all binary relations R closed under the conditions

$$<1,2> \in R;$$
$$<x,y> \in R \Rightarrow <y,2^y> \in R;$$
$$[<x,y> \in R \;\&\; <y,z> \in R] \Rightarrow <x,z> \in R;$$
$$(\forall m)(\forall n)[m < n \Rightarrow <\varphi_z(m),\varphi_z(n)> \in R] \Rightarrow (\forall n)[<\varphi_z(n),3 \cdot 5^z> \in R].$$

Then express this intersection by a universal function quantifier. Then define O as $\{x|(\exists y)[x <_O y]\}$.)

(*b*) Let the system of ordinal notations S_1 be as defined in §11.7. Show that $D_{S_1} \in \Pi_1^1$. (*Hint:* Obtain an analytical definition for the relation $\leq_{S_1} = \{<x,y>|x \in D_{S_1} \;\&\; y \in D_{S_1} \;\&\; |x|_{S_1} \leq |y|_{S_1}\}$.)

16-7. Use Theorem VIII to complete the proof of Corollary XI.

16-8. (*a*) Show that if set quantifiers are allowed (and treated in the same way as function quantifiers for purposes of prefix classification) and if recursive relations in $\mathfrak{N}^k \times N^l$ are allowed, then the same classes of relations on integers are obtained as with function quantifiers.

(*b*) Show that if set quantifiers are allowed (and treated in the same way as function quantifiers for purposes of prefix classification) and if recursive relations in $\mathfrak{F}^k \times \mathfrak{N}^l \times N^m$ $(k,l,m \geq 0)$ are allowed, then the same classes of relations in $\mathfrak{F}^k \times \mathfrak{N}^l \times N^m$ $(k,l,m \geq 0)$ are obtained as would be obtained from these recursive relations by the use of function and number quantifiers alone.

△***16-9.*** Let $R \subset \mathfrak{N} \times N^2$ be recursive. Let $A = \{x|(\forall X)(\exists y)R(X,x,y)\}$. Show that A is recursively enumerable and hence in Σ_0^1 $(=\Pi_0^1)$. (*Hint:* Use the compactness theorem for trees in Exercise 9-40.)

16-10. Prove that for $n > 0$, any set of integers in Σ_n^1 can be expressed by a set-quantifier predicate form whose prefix consists of $n + 2$ alternating quantifiers of which the first n are set quantifiers, the last two are number quantifiers, and the initial quantifier is existential.

§*16.2*

16-11. Show that the translation procedures indicated in the proof of Theorem XIII do, in fact, lead to equivalent formulas.

16-12. In part (3) of the proof of Theorem XIII (for the case $S_4 \rightarrow S_3$) show that $S(\zeta,a,b)$ can be taken as a formula of S_3. (*Hint:* Use Theorem 14-VI and the definitions of Exercise 15-13. First show that $S^*(\zeta,c,b)$ can be taken as a formula of S_3, where $S^*(\zeta,c,b)$ asserts that c is a sequence number of length b giving the first b terms in the continued fraction expansion of $\zeta - [\zeta]$.)

16-13. Complete the proof of Corollary XIII(*b*).

16-14. Show that if A is implicitly definable in second-order arithmetic, then A is explicitly definable in second-order arithmetic. (*Hint:* See proof of Theorem XI.)

Conclude that if A is implicitly definable in elementary analysis, then A is explicitly definable in elementary analysis.

16-15. Prove Theorem XIV.

16-16. Give the proof of Corollary XIV in more detail.

16-17. Prove Theorem XV and Corollary XV. (*Hint:* See proof of Theorem 14-XI.)

16-18. Outline the proof of Theorem XVII.

△***16-19.*** Take P as suggested in the discussion at the end of §16.2, and take F to be a sentence of second-order arithmetic. Show that the formula $[P \Rightarrow F]$ is universally valid (under the semantical conventions of second-order logic) $\Leftrightarrow$ F is a true sentence of second-order arithmetic.

△***16-20.*** (Skolem). Show the existence of denumerable ω-models for the true sentences of elementary arithmetic. (*Hint:* Take $\mathfrak{D}$ to be the integers and "+" and "×" to have their standard interpretation. Consider the denumerable collection of all true sentences of second-order arithmetic in prenex form. Call an existential quantifier in such a sentence *initial* if it is preceded by no universal quantifier. With each initial existential quantifier of each true prenex sentence associate (using the axiom of choice) a corresponding integer or function. Call the denumerable collection of all such chosen functions $\mathcal{E}$. With each noninitial existential quantifier of each true prenex sentence associate (using the axiom of choice) a corresponding (not necessarily recursive) functional (mapping from $\mathfrak{F}^k \times N^l$ into N) or functional operator (mapping from $\mathfrak{F}^k \times N^l$ into $\mathfrak{F}$). (See the second step of the illustration for the no-counterexample interpretation in §15.3.) Call the denumerable collection of all these functional operators $\mathcal{E}^*$. Take $\mathfrak{D}'$ to be the closure (in the obvious sense) of $\mathcal{E}$ under all functional operators in $\mathcal{E}^*$.)

§16.3

16-21. Show that a subtree τ of $\mathfrak{J}^*$ is finite path $\Leftrightarrow$ $(\forall f)(\exists x)$[the vertex $\bar{f}(x)$ is not in τ].

16-22. Define the concepts of *tree, subtree, branch, subbranch, finite-path tree,* and *tree ordering* directly in terms of sequence numbers. (For example, A is a *tree* if $(\forall f)(\forall x)[\bar{f}(x) \in A \Rightarrow (\forall y)[y < x \Rightarrow \bar{f}(y) \in A]]$).

16-23. Let τ be a finite-path tree. Show that every nonzero ordinal less than $o(\tau)$ appears as a value of o_τ.

16-24. Let τ and α be as in the discussion immediately preceding the statement of Theorem XVIII. Complete the proof of Theorem XVIII by showing that τ is finite-path and that $o(\tau) = \alpha$.

△***16-25.*** Give more precise definitions for $\tau_1 + \tau_2$ and $\tau_1 \cdot \tau_2$, and show that $o(\tau_1 + \tau_2) = o(\tau_1) + o(\tau_2)$ and that $o(\tau_1 \cdot \tau_2) = o(\tau_1) \cdot o(\tau_2)$.

16-26. Prove the basic tree lemma.

16-27. (*a*) Show that for a tree τ, the Kleene-Brouwer ordering of τ is a well-ordering if and only if τ is finite-path.

(*b*) Show that for a finite-path tree τ, the Kleene-Brouwer ordering has order type at least as great as $o(\tau)$.

(*c*) Show that the Kleene-Brouwer ordering of a recursive finite-path tree is a recursive well-ordering.

(*d*) Show that $T \leq_1 W$, where W is defined as in Exercise 11-61.

16-28. Complete the proof of Theorem XIX by showing that the finite-path tree constructed from the notation for α is recursive.

△***16-29.*** (*a*) Give an example of a recursively enumerable finite-path tree which is not recursive, and an example of a recursive finite-path tree which is not strongly recursive.

(*b*) Show that the ordinal associated with a recursively enumerable finite-path tree is constructive and that every constructive ordinal is given by a strongly recursive finite-path tree.

§16.4

16-30. Let S, $<_S$, and R in the recursion lemma vary (not necessarily effectively) with a parameter q which ranges over some set A. Assume that for any $q \in A$ and any $y \in S^{(q)}$, if φ_z is satisfactory on $S_y^{(q)}$, then $\eta(z,y,q)$ is defined and $R^{(q)}(y,\eta(z,y,q))$, where η is partial recursive. Conclude that there is a partial recursive ψ of two variables such that for any $q \in A$, $\lambda x[\psi(x,q)]$ is satisfactory on $S^{(q)}$ with respect to $R^{(q)}$.

16-31. Show that the recursion lemma remains true if the fifth sentence in its statement is replaced by the following. "Assume that there is a recursive function f of two variables such that for any z and any $y \in S$, if φ_z is satisfactory on S_y, then $\varphi_{f(z,y)}$ is satisfactory on $S_y \cup \{y\}$."

$\triangle$***16-32.*** Recall that for $x \in O$, $|x|_O$ is the constructive ordinal associated with notation x. For x in W, let $|x|_W$ be the constructive ordinal associated with the recursive well-ordering given by x. Prove that

(i) If $A \leq_1 O$ via h, then $A \in \Delta_1^1 \Leftrightarrow (\exists$ constructive $\alpha)$ $(\forall x)[x \in A \Rightarrow |h(x)|_O < \alpha]$.

(ii) If $A \leq_1 W$ via h, then $A \in \Delta_1^1 \Leftrightarrow (\exists$ constructive $\alpha)$ $(\forall x)[x \in A \Rightarrow |h(x)|_W < \alpha]$.

(*Hint:* Consider the mappings from T to W and from O to T provided by Theorem XIX, and the mapping from W to O provided by Theorem 11-XX. Show that bounded sets go to bounded sets and unbounded sets to unbounded sets under these mappings.)

16-33. Show that $T_\alpha \leq_1 T^*_{\alpha+1}$.

16-34. (*a*) Show that $\{<x,y>|x \in T \;\&\; y \in T \;\&\; \|x\| < \|y\|\}$ is, in fact, Π_1^1-complete. (*Hint:* Use the function h defined at the end of the proof of Corollary XXII(*b*), and observe that $x \in T \Leftrightarrow [<x,h(x)>$ belongs to this set].)

(*b*) Prove (*d*) in Theorem XXIII.

16-35. Show that the linear ordering $<^T$ is of the order type of the least nonconstructive ordinal. (*Hint:* Let λ be the least nonconstructive ordinal. Then $<^T$ has order type $\omega \cdot \lambda = \lambda$, by continuity of $\lambda\beta[\omega \cdot \beta]$.)

16-36. Complete the proof of Theorem XXIV. Show that the Π_1^1 ordering can be taken so that every proper initial segment of A, in this ordering, is a Δ_1^1-set.

16-37. Show that in Theorem XXIV not every Π_1^1 well-ordering of A can be used.

§16.5

16-38. Show that a set is in Δ_1^1 if and only if its characteristic function is a hyperarithmetical function.

$\triangle$***16-39.*** (Addison). (*a*) Prove the reduction principle for the class of all Σ_1^0-subsets of $\mathfrak{F}$; i.e., in the notation of §15.2, prove the reduction principle for the class $\Sigma_1^{(\text{fn})}$. (*Hint:* Let

$$\mathfrak{A} = \{f|(\exists x)T'_{1,0}(z_1,f,x)\} \text{ and } \mathfrak{B} = \{f|(\exists x)T'_{1,0}(z_2,f,x)\}.$$

Define

$$\mathfrak{A}_1 = \{f|(\exists x)[T'_{1,0}(z_1,f,x) \;\&\; (\forall y < x) \neg T'_{1,0}(z_2,f,y)]\},$$

and

$$\mathfrak{B}_1 = \{f|(\exists y)[T'(z_2,f,y) \;\&\; (\forall x \leq y) \neg T'_{1,0}(z_1,f,x)\}.)$$

(*b*) Prove that there exist disjoint Σ_1^0-subsets of $\mathfrak{F}$, call them $\mathfrak{A}$ and $\mathfrak{B}$, such that for no Δ_0^0-subset $\mathfrak{C}$ of $\mathfrak{F}$ is it true that $\mathfrak{A} \subset \mathfrak{C}$ and $\mathfrak{B} \subset \bar{\mathfrak{C}}$. Conclude that the Π_1^0-subsets of $\mathfrak{F}$ do not satisfy the reduction principle. (*Hint:* Let A and B be recursively inseparable recursively enumerable sets. Take $\mathfrak{A} = \{f|f(0) \in A\}$ and $\mathfrak{B} = \{f|f(0) \in B\}$, and apply the techniques of the proof of Theorem 15-XXIV.) (The "Gödel numbering" by functions of Theorem 15-XXVII can also be used to give a proof more directly analogous to that of Theorem 7-XII(*c*).)

$\triangle$***16-40.*** (Addison). (*a*) For any $n \geq 1$, prove the reduction principle for the class of all Σ_n^0-subsets of $\mathfrak{F}$. (*Hint:* See Exercise 16-39(a), and recall the rules for bounded quantifiers in §14.3.)

(b) For any $n \geq 1$, find a disjoint pair of Σ_n^0-subsets of $\mathfrak{F}$ not separable by any Δ_n^0-subset of $\mathfrak{F}$. Conclude that the Π_n^0-subsets of $\mathfrak{F}$ do not satisfy the reduction principle.

△16-41. (Addison). (a) Prove the reduction principle for the class of all Π_1^1-subsets of $\mathfrak{F}$. (*Hint:* As in Exercise 16-39(a), except that the bounded quantifiers use the ordering $<^T$ and range over values of the function f^* (in Corollary XX(a))).

(b) Find a disjoint pair of Π_1^1-subsets of $\mathfrak{F}$ which are not separable by any Δ_1^1-subset of $\mathfrak{F}$. Conclude that the class of Σ_1^1-subsets of $\mathfrak{F}$ does not satisfy the reduction principle.

16-42. (a) Define: A is *hyperarithmetically simple* if, (i) A is Π_1^1, (ii) $\bar{A}$ is infinite, and (iii) $(\forall B)[B$ is infinite and in $\Pi_1^1 \Rightarrow B \cap A \neq \emptyset]$. Show that hyperarithmetically simple sets exist. (*Hint:* See the proof of Theorem 8-II.)

(b) Define: A is h-*isomorphic* to B if there is a permutation f such that f is a hyperarithmetical function and $A = f(B)$. Show that there are sets in $(\Pi_1^1 - \Sigma_1^1)$ which are not h-isomorphic to a Π_1^1-complete set.

16-43. Let f be a hyperarithmetical function. Show that there is an n such that $\varphi_n^1 = \varphi_{f(n)}^1$, and that n can be found uniformly recursively from a Π_1^1-index for f.

△16-44. (Luckham). (a) Define: A is *recursively* Π_1^1*-productive* if there exists a partial recursive ψ such that for any z, $W_z^1 \subset A \Rightarrow [\psi(z)$ convergent & $\psi(z) \in A - W_z^1]$.

Define: A is *recursively* Π_1^1*-creative* if A is Π_1^1 and $\bar{A}$ is recursively Π_1^1-productive.

Prove: A is recursively Π_1^1-creative $\Leftrightarrow$ A is Π_1^1-complete. (*Hint:* Proof parallels the p oof of Theorem 11-V.)

(b) Define: A is Π_1^1*-productive* if there exists a partial Π_1^1 function ψ such that for any z, $W_z^1 \subset A \Rightarrow [\psi(z)$ convergent & $\psi(z) \in A - W_z^1]$.

Define: A is Π_1^1*-creative* if A is Π_1^1 and $\bar{A}$ is Π_1^1-productive.

Define: B is h-1-*reducible* to A if there is a hyperarithmetical one-one function f such that $(\forall x)[x \in B \Leftrightarrow f(x) \in A]$.

Prove: A is Π_1^1-creative $\Leftrightarrow [A \in \Pi_1^1$ & $(\forall B)[B \in \Pi_1^1 \Leftrightarrow B$ is h-1-reducible to $A]]$.

(c) Prove that any two Π_1^1-creative sets are h-isomorphic. (*Hint:* Prove a hyperarithmetical version of Theorem 7-VI.) Note Exercise 16-46 below.

△16-45. (Luckham). Let $\mathfrak{C}$ be a class of Π_1^1-sets. Define: $P_{\mathfrak{C}}^1 = \{x|W_x^1 \in \mathfrak{C}\}$. If z is a Δ_1^1-index for some set, let E_z be that set. Prove the following theorems.

(a) $P_{\mathfrak{C}}^1 \in \Sigma_0^1 \Leftrightarrow [\mathfrak{C}$ is empty or $\mathfrak{C}$ is the class of all Π_1^1-sets].

(b) $P_{\mathfrak{C}}^1 \in \Pi_1^1 \Leftrightarrow (\exists B)[B \in \Pi_1^1$ & $(\forall z)[z \in B \Rightarrow z$ is a Δ_1^1-index for some set] & $\mathfrak{C} = \{A|A \in \Pi_1^1$ & $(\exists z)[z \in B$ & $E_z \subset A]\}]$. (*Hint:* Proof is parallel to the proof of Theorem 14-XIV.)

▲16-46. (a) Show that there is a Π_1^1-set A which is h-isomorphic to T but not recursively isomorphic to T. (*Hint:* Carry out the following hyperarithmetical computation. Let B and C be infinite recursive subsets of T and $\bar{T}$ respectively. Enumerate the recursive permutations $p_0, p_1, \ldots$. Define a hyperarithmetical permutation f such that for each x, $(\exists y \in B)(\exists z \in C)[f(y) \neq p_x(z)]$.)

(b) Hence conclude, by Exercise 16-44, that a Π_1^1-productive set need not be recursively Π_1^1-productive.

△16-47. Show that there exist a set C of Δ_1^1-indices such that every Δ_1^1-set has one and only one index in C, and a partial Π_1^1 function ψ such that for any Δ_1^1-set A, if z is a Δ_1^1-index for A, then $\psi(z)$ is convergent, $\psi(z) \in C$, and $\psi(z)$ is a Δ_1^1-index for A. (*Hint:* Set $B = \{z|\varphi_z^1$ *is empty or* $[\varphi_z^1$ *is total and* φ_z^1 *is either strictly increasing, or strictly increasing and then constant*$]\}$. Then B is Π_1^1. Use $<^T$ to form a Π_1^1-set C consisting of one and only one Δ_1^1-index for the range of each partial Π_1^1-function having an index in B. Obtain a recursive h such that $(\forall A)(\forall z)[z$ is a Δ_1^1-index for $A \Rightarrow [h(z) \in B$ & *range* $\varphi_{h(z)}^1 = A]]$. Define $\varphi = \{\langle x,y\rangle|x \in B$ & $y \in C$ & y *is a* Δ_1^1*-index for range* $\varphi_x^1\}$. Set $\psi = \varphi h$.)

△*16-48.* For any "partial recursive" function ψ, take any Π_1^1-index for $\tau(\psi)$ to be "instructions" for computing ψ. Show that the "Gödel numbering" of "partial recursive" functions given for the Kreisel-Sacks analogy is "acceptable" in a reasonable sense (see Exercise 2-10). More specifically, show that an enumeration theorem (see Theorem 1-IV) and an *s-m-n* theorem (see Theorem 1-V) hold.

16-49. In the Kreisel-Sacks analogy, show the existence of sets which are "recursively enumerable" but not "recursive."

△*16-50.* In the Kreisel-Sacks analogy, show that if A is the range of a one-one "recursive" function, then A is not in Δ_1^1. (*Hint:* Consider the inverse mapping and use Theorem XXVI.)

16-51. In the discussion of the Kreisel-Sacks analogy, (i) show that the construction given to show that [A "recursively enumerable" and not "finite" $\Rightarrow$ A is the range of a "recursive" function] is satisfactory; and (ii) show that if A is "finite" and $\neq \emptyset$, then A is the range of a "recursive" function. (*Hint:* For (ii), let the "recursive" function map A onto itself and $T - A$ onto some fixed member of A.)

§16.6

16-52. Prove Corollary XXX.

16-53. Prove Corollary XXXI(*b*).

16-54. For any A, define $K^{1,A}$ as in the comment preceding Theorem XXXIII. Show that $K^{1,A}$ is $\Pi_1^{1,A}$-complete.

▲*16-55.* (Spector [1958]). Show that incomparable hyperdegrees exist. (*Hint:* A measure-theoretic method which generalizes some of the results in §13.3 is due to Spector. It is enough to show that for ome A, $\mu(\{B|A \leq_h B\}) \neq 1$. To obtain this, it is enough to show that $\mathcal{R} = \{<A,B>|A \leq_h B\}$ is a measurable relation. For $(\forall B)[\{A|A \leq_h B\}$ is denumerable], and hence $(\forall B)[\mu(\{A|A \leq_h B\}) = 0]$. Hence, by Fubini's theorem (see Halmos [1950]), $\mu(\{<A,B>|A \leq_h B\}) = 0$. Hence, again by Fubini's theorem, $\mu(\{B|A \leq_h B\}) = 0$ for almost all A. To show that $\mathcal{R}$ is measurable, it is enough to show that every Σ_1^1 relation is measurable (since $\mathcal{R}$ is the intersection of a Σ_1^1 relation and the complement of a Σ_1^1 relation). To show this, it is enough to show that the application of operation **A** to a collection of open and closed sets yields a measurable set. See footnote on *analytic* sets, page 384, and see Kuratowski [1950].)

△*16-56.* Let $\{<x,y>|<x,y> \in A\}$ be the ordering relation of a well-ordering $(<)$ of order type α.

(*a*) (Spector). Show that $A \in \Delta_1^1 \Rightarrow \alpha$ constructive. (*Hint:* Obtain $z \in T^A$ such that $\alpha \leq \|z\|^A$, and apply Corollary XXXVI. To obtain z, associate members of $A_1 = \{x|(\exists y)[<x,y> \in A]\}$ with vertices of the function tree as follows. With the vertices immediately below the maximum vertex associate (in order of numerical size) the members of A_1. More generally, given a member x of A_1 associated with a vertex, with the vertices immediately below that vertex associate (in order of numerical size) the members of A_1 that fall below x in the well-ordering A. This determines a finite-path tree with characteristic function recursive in A. Let $\varphi_z{}^A$ be this characteristic function.)

(*b*) (Tanaka, Kreisel). Show that, in fact, $A \in \Sigma_1^1 \Rightarrow \alpha$ constructive. (*Hint:* For y in the well-ordering, define $|y|^A$ to be the order type of $\{x|<x,y> \in A\}$ and show that $\{x|x \in T \;\&\; \|x\| < |y|^A\} \in \Sigma_1^1$. To show this, use the recursion lemma as in Theorem XXXV. Apply the lemma to the well-ordering given by A, and take as basic expression: $x \in T_{|y|^A} \Leftrightarrow [(\exists u)[<u,y> \in A] \;\&\; [(\forall u)[\varphi_x(u) = 0] \vee [\varphi_x(0) = 1 \;\&\; (\exists z)(\forall n)[<z,y> \in A \;\&\; b(n,x) \in T_{|z|^A}]]]]$. *Alternative hint:* show $T_\alpha \in \Sigma_1^1$ by considering possible existence of order-preserving maps from trees into A.)

16-57. Prove Corollary XXXVI(*c*).

16-58. (*a*) Assume that $B \notin \Delta_1^1$, that $B \leq_h T$, and that $T \nleq_h B$. What can be concluded about the value of λ^B?

(*b*) Assume that $B \in \Pi_1^1$ and $\lambda^B = \lambda$. What can be concluded about the hyperdegree of B?

$\triangle$**16-59.** Show that for every f there exists a z such that $\{g|\tau(f) = W_z^{1,g}\}$ is dense in $\mathfrak{F}$. (*Hint:* Let z define f in terms of the "infinite" behavior of g; e.g., let $f(x) = \limsup g(y_i)$ for an appropriate sequence y_i depending on x.) Note how this result contrasts with the recursive case, where, for nonrecursive f, $\{g|\tau(f) = W_z^g\}$ cannot be dense in $\mathfrak{F}$.

16-60. Verify from basic definitions that $T^2 = \{z|(\exists f)[z \in T^{\tau(f)}]\}$.

16-61. Show in more detail that the expressions in the proof of Theorem XXXVIII yield the desired Σ_2^1-form and Π_2^1-form. (*Hint:* See Exercise 16-8.)

$\triangle$**16-62.** Complete the proof of Corollary XXXIX. (*Hint:* See proofs for Corollary XXII.)

$\triangle$**16-63.** Show that the order type of $\mathfrak{A}$ is not a Δ_2^1-ordinal. (*Hint:* We obtain the result by showing that every Δ_2^1-ordinal can be mapped in a one-one order-preserving way into $\mathfrak{A}$. Let α be a Δ_2^1-ordinal. Then it is easy to show from the definition of Δ_2^1-ordinal that there is a Δ_2^1 linear ordering of integers of order type α. Choose such an ordering and call it S. We construct a partial recursive φ which gives a map from S into T^2 such that [x precedes y in $S \Leftrightarrow \|\varphi(x)\|^2 < \|\varphi(y)\|^2$]. We do this by showing that there is a partial recursive η such that for any y in S, if φ_z restricted to the predecessors (in S) of y is an order-preserving map from the predecessors of y into T^2, then the ordered pair $<y,\eta(z,y)>$ extends this map to y (as an order-preserving map). The existence of φ then follows by a recursion theorem argument closely parallel to the proof of the recursion lemma in §16.4. The existence of the desired η follows from two observations: first that $[A \subset T^2 \,\&\, A \in \Delta_2^1] \Leftrightarrow \{x|(\exists y)[y \in A \,\&\, x \in T^2_{\|y\|^2}]\} \in \Pi_2^1$ uniformly in a Δ_2^1-index for A by Theorem XXXVIII; and second, that a Σ_2^1-complete set is "productive" with respect to Π_2^1 subsets (for example, E^2 is "productive" with $\lambda x[x]$ as "productive function."))

16-64. Formulate and prove analogues for Theorems XXV to XXIX, with Σ_2^1 in place of Π_1^1 and Δ_2^1 in place of Δ_1^1.

16-65. (*a*) Prove the reduction principle for Σ_2^1-subsets of $\mathfrak{F}$.

(*b*) Find a disjoint pair of Σ_2^1-subsets of $\mathfrak{F}$ which are not separable by any Δ_2^1-subset of $\mathfrak{F}$. Conclude that the Π_2^1-subsets of $\mathfrak{F}$ do not satisfy the reduction principle.

16-66. Explain why the expression

$$(\forall B)(\exists A)[x \in T^{A,B} \,\&\, (\exists D)(\forall C) \neg [\|y\|^{C,D} < \|x\|^{A,B} + 1]]$$

does not give a satisfactory definition of $T^3_{\|y\|^3}$, for $y \in T^3$.

§16.7

16-67. Show that Δ_0^0 is a basis for Δ_0^0. (*Hint:* See the discussion of recursive classes (that is, $\Sigma_0^{(\mathrm{fn})}$-classes) in §15.2.)

16-68. Verify that the relativized proofs of Theorem XLI and Corollary XLI(*a*), needed for the proof of Theorem XLIII, go through, and that the z_0 obtained is independent of g.

$\triangle$**16-69.** (Gandy). Generalize the proof of Theorem XLIII to show that infinitely many hyperdegrees exist between the minimum hyperdegree and the hyperdegree of T.

16-70. Let A be a nonhyperarithmetical set with hyperdegree below T. What are the A-constructive ordinals?

16-71. (*a*) Give, in more detail, the proofs for facts 2, 3, and 4 in the proof of Theorem XLIV.

(*b*) Give, in more detail, the proofs for facts 6, 7, and 8 in the proof of Theorem XLIV.

(*c*) Show that if $z \notin T$ and $P(f,z)$, then $f = 1$ on $T \times \{z\}$.

(*d*) Deduce Theorem XXI from facts 6 and 7 in the proof of Theorem XLIV.

$\triangle$***16-72.*** (Cohen, Feferman). Show that there exists a set $A \in \Delta_1^1$ such that $\{A\} \notin \Delta_0^1$ as follows.

Let elementary arithmetic be augmented by adding atomic expressions of the form $t \in \hat{A}$, where $\hat{A}$ is a single, fixed, free, set variable and t is a number variable or numeral. Consider sentences which contain no quantifier symbols other than $\exists$ and no propositional connectives other than $\vee$ and $\neg$. Let $F, G, \ldots$ denote such sentences. Let the *rank* of a sentence F be the total number of occurrences of $\exists$, $\vee$, and $\neg$ in F. Let $f, g, \ldots$ be characteristic functions and let $\tilde{f}, \tilde{g}, \ldots$ be finite initial segments of such functions. For any such f, let S_f be the set with characteristic function f. $\tilde{f}$ is an *extension* of $\tilde{g}$ if $\tilde{g} \subset \tilde{f}$.

Define the concept $\tilde{f}$ *forces* F inductively (on rank of F) as follows: $\tilde{f}$ forces $\mathbf{n} \in \hat{A}$ if $n \in$ *domain* $\tilde{f}$ and $\tilde{f}(n) = 1$; $\tilde{f}$ forces other atomic expressions (e.g., $\mathbf{m} + \mathbf{n} = \mathbf{p}$) if these expressions are true (under the usual interpretation); $\tilde{f}$ forces $(\exists a)[\ldots a \ldots]$ if, for some n, $\tilde{f}$ forces $\ldots \mathbf{n} \ldots$; $\tilde{f}$ forces $F \vee G$ if either $\tilde{f}$ forces F or $\tilde{f}$ forces G; and $\tilde{f}$ forces $\neg F$ if no extension of $\tilde{f}$ forces F. Define f *forces* F if some finite initial segment of f forces F

(*a*) Show that if F does not contain variable $\hat{A}$, then f forces $F \Leftrightarrow F$ is true.

(*b*) Show that $\tilde{f}$ forces $\neg \mathbf{n} \in \hat{A} \Leftrightarrow n \in$ *domain* $\tilde{f}$ and $\tilde{f}(n) = 0$.

(*c*) Show that $\tilde{f}$ forces $F \Leftrightarrow$ every extension of $\tilde{f}$ forces F. (*Hint:* Use induction on rank of F.)

We say that F is *true* of f if F is true when $\hat{A}$ is taken to represent S_f.

(*d*) Show that no f forces $\neg(\exists a)\neg a \in \hat{A}$. Conclude that it is possible to have F true of f with neither F nor $\neg F$ forced by f, and (taking a further negation) that it is possible to have F false of f but forced by a finite segment of f.

Define: f is *generic* if, for all F, F true of $f \Rightarrow F$ forced by f (and hence F true of $f \Leftrightarrow F$ forced by f). If f is generic, the set S_f is called a *generic set*.

(*e*) Show that f is generic $\Leftrightarrow$ for every F, f forces either F or $\neg F$. (*Hint:* $\Rightarrow$ is immediate, $\Leftarrow$ is proved by induction on rank.)

(*f*) Show that generic sets exist. (*Hint:* Use (*e*). Construct generic f by making successive extensions to guarantee that for every F, f forces either F or $\neg F$.)

(*g*) Show that there exists a generic f in Δ_1^1. (*Hint:* Show that the relation $\tilde{f}$ *forces* F (under appropriate Gödel numbering) is a Δ_1^1 relation (in fact, it is recursive in $\mathbf{0}^{(\omega)}$), and that the construction in (*f*) can therefore be made Δ_1^1.

(*h*) Let f and an integer n be given. Define (characteristic function) g by the condition: $g(x) = f(x)$ for $x \neq n$, and $g(n) = 1 \Leftrightarrow f(n) = 0$. We denote g as $n(f)$. Show that for any f and n, f is generic $\Leftrightarrow n(f)$ is generic. (*Hint:* For any sentence F, define $n(F)$ to be the sentence obtained by replacing each atomic expression $t \in \hat{A}$ by the expression $(\neg(t = \mathbf{n} \vee \neg t \in \hat{A}) \vee \neg(\neg t = \mathbf{n} \vee t \in \hat{A}))$. Observe that $n(\neg F) = \neg n(F)$. Show that for any f and F: (i) f forces $n(n(F)) \Leftrightarrow f$ forces F; and (ii) f forces $F \Leftrightarrow n(f)$ forces $n(F)$. Observe that $n(n(f)) = f$, and use (*e*).)

(*i*) Show that f generic $\Rightarrow \{f\} \notin \Delta_0^1$. (*Hint:* Assume f generic and $\{f\} \in \Delta_0^1$. Let F be an implicit definition for S_f. Since F is true of f, $\tilde{f}$ forces F for some $\tilde{f} \subset f$. Take $n \notin$ *domain* $\tilde{f}$. Then $n(f)$ forces F (since $\tilde{f} \subset n(f)$). But $n(f)$ is generic (by (*h*)). Hence F is true of $n(f)$ contrary to our assumption that F defines the unique function f.)

$\triangle$***16-73.*** (*a*) Define generic set as in 16-72. Show that the collection of all non-generic sets is a set of first category in Cantor space. (*Hint:* For each F, $\{f \mid (\exists\tilde{f})[\tilde{f} \subset f \;\&\; \tilde{f}$ *forces either* F *or* $\neg F]\}$ is an open dense set. It is open because it is a union of neighborhoods given by the segments forcing F or $\neg F$. It is dense because any neighborhood disjoint from the set would be given by a segment forcing $\neg F$. The generic sets are thus the denumerable intersection of open dense sets. Their complement is hence of first category.)

(*b*) (Addison). Show that any arithmetical class of first category has an arithmetical set in its complement; hence conclude that the class of arithmetical sets is not arithmetical. (*Hint:* Define: f is n-generic if, for any F of rank n or less, F true of $f \Rightarrow F$ forced by f. Show that for each n there exists an arithmetical f which is n-generic. Then let $\mathcal{C}$ be an arithmetical class of first category with definition F of rank n (see Theorem 15-VI). Let f be n-generic and arithmetical. If $S_f \notin \mathcal{C}$, we are done. If $S_f \in \mathcal{C}$, then F is true of f and some $\tilde{f} \subset f$ forces F; by argument as in (*a*), the n-generic extensions of $\tilde{f}$ yield a collection of second category lying in $\mathcal{C}$, and this is a contradiction.)

16-74. Extend Theorem 15-XXV and Theorem 15-XXVI to the analytical hierarchy. More specifically, prove the following.

(*a*) Let $\mathcal{A} \subset \mathfrak{F}$ and let $\tau(\mathcal{A}) = \{X|(\exists f)[f \in \mathcal{A}\ \&\ X = \{<x,y>|f(x) = y\}]\}$. Show that for all n, $\mathcal{A} \in \Sigma_n^1 \Leftrightarrow \tau(\mathcal{A}) \in \Sigma_n^1$, and $\mathcal{A} \in \Pi_n^1 \Leftrightarrow \tau(\mathcal{A}) \in \Pi_n^1$. (*Hint:* See proof of Theorem 15-XXV. Corollary 15-XXV gives the case $n = 0$.)

(*b*) Let $\mathcal{A} \subset \mathfrak{N}$ and let $\mathcal{C}(\mathcal{A}) = \{f|(\exists X)[X \in \mathcal{A}\ \&\ f = c_X]\}$. Show that for all n, $\mathcal{A} \in \Sigma_n^1 \Leftrightarrow \mathcal{C}(\mathcal{A}) \in \Sigma_n^1$, and $\mathcal{A} \in \Pi_n^1 \Leftrightarrow \mathcal{C}(\mathcal{A}) \in \Pi_n^1$. (*Hint:* See proof of Theorem 15-XXVI. Corollary 15-XXVI gives the case $n = 0$.)

(*c*) For $\mathcal{A} \subset \mathfrak{F}$ and $n > 0$, the notions of *Σ_n^1-index* of $\mathcal{A}$ and *Π_n^1-index* of $\mathcal{A}$ were defined in §16.1. For $\mathcal{A} \subset \mathfrak{N}$ and $n > 0$, define z to be a *Σ_n^1-index* for $\mathcal{A}$ if $\mathcal{A} = \{X|(\exists f_1) \cdots T_{n+1,0}(z,X,\tau(f_1), \ldots)\}$. Similarly for *$\Pi_n^1$-indices.*

Show that for $n > 0$, (*a*) and (*b*) above hold uniformly in the appropriate indices.

16-75. Prove Corollary XLIV(h). (*Hint:* Use Corollary XLIV(e).)

16-76. Show that $\mathcal{H} \in (\Pi_1^1 - \Sigma_1^1)$. (*Hint:* $f \in \mathcal{H} \Leftrightarrow (\exists z)(\forall x)[f(x) = \varphi_z^1(x)]$. This yields $\mathcal{H}$ in Π_1^1. Assume $\mathcal{H} \in \Sigma_1^1$ and apply Theorem XLIV to get $T \in \Sigma_1^1$, a contradiction.)

16-77. Show that if y is an extension of x (as defined in the proof of Theorem XLV), then $y > x$. (*Hint:* Consider the definition of sequence numbers from the function τ^*, the definition of τ^* from τ, and the definition of τ.)

§*16.8*

$\triangle$***16-78.*** Let f be any one-one function. Define

$$K(f(0)) = \emptyset;$$
$$K(f(n+1)) = (K(f(n)))'.$$

Let $A = \{<x,y>|y \in \textit{range}\, f\ \&\ x \in K(y)\}$. Show that the T-degree of A can be made arbitrarily high by appropriate choice of f.

16-79. Let $A_0, A_1, \ldots$ be a sequence of sets such that $A_0 = \emptyset$, and A_n is Σ_n-complete for all $n > 0$. Let $A = \{<x,y>|x \in A_y\}$. Show that the T-degree of A can be made arbitrarily high by an appropriate choice of the sets $A_0, A_1, \ldots$.

16-80. Let ν be any univalent system of notation. Define

$$K(\nu^{-1}(0)) = \emptyset;$$
$$K(\nu^{-1}(\beta+1)) = (K(\nu^{-1}(\beta)))';$$
$$K(\nu^{-1}(\alpha)) = \{<u,v>|\nu(v) < \alpha\ \&\ u \in K(v)\}, \qquad \text{for } \alpha \text{ a limit ordinal.}$$

Show that for all x in the domain of ν, there is a $y \in O$ such that $\nu(x) = |y|$ and $K(x) \equiv_T H(y)$. (*Hint:* Use Corollary 11-XVIII(b) and the recursion lemma.)

16-81. Let σ be any one-one recursive map of $N \times N$ into N, range recursive. Define the sets $K(x)$ like the sets $H(x)$, except that σ is used as a pairing function in place of τ. Show that for all x in O, $H(x) \equiv K(x)$. (*Hint:* Use the recursion lemma.)

16-82. Take any z_0 such that (i) $(\forall A)[W_{z_0}{}^A \equiv A']$, and (ii) there is a recursive f_0 such that for all A and all u, $W_u{}^A \leq_1 W_{z_0}{}^A$ via $\lambda x[f_0(u,x)]$. Define $A^* = W_{z_0}{}^A$. Define the sets $K(x)$ in the same way as the sets $H(x)$, except that the operation* is used in place of the jump. Show that $x \in O \Rightarrow H(x) \equiv_T K(x)$. (*Hint:* Use the recursion lemma.)

$\triangle$***16-83.*** (Enderton). Associate sets with notations in O as follows:

$$\begin{aligned} H_1 &= N; \\ H_{2^x} &= (H_x)'; \\ H_{3\cdot 5^y} &= \{<u,v>|u \in H_{\varphi_y(v)}\}. \end{aligned}$$

Theorems XLIX, L, and LI were first formulated in terms of this association, which is due to Kleene (but see the footnotes on pages 207 and 209). Show that for $x \in O$, $H_x \equiv_{\mathrm{T}} H(x)$. (*Hint:* Use the recursion lemma to prove that $x \leq_O y \Rightarrow H_x \leq_1 H_y$ uniformly in x and y. Then use the recursion lemma to show that $H(x) \leq_{\mathrm{T}} H_x$ and to show that $H_x \leq_{\mathrm{T}} H(x)$.)

$\blacktriangle$***16-84.*** (Enderton). Let ν_S be a one-one mapping from a set of integers D_S onto a segment of ordinal numbers such that conditions (i) and (ii) hold in the definition of *system of notation* at the beginning of §11.7. For $x \in D_S$, define $K(x)$ by the following rules:

$$\begin{aligned} K(\nu^{-1}(0)) &= \emptyset; \\ K(\nu^{-1}(\beta + 1)) &= (K(\nu^{-1}(\beta)))'; \\ K(\nu^{-1}(\alpha)) &= \{<u,v>|v \in D_S \;\&\; \nu(v) < \alpha \;\&\; u \in K(v)\}, \qquad \text{for } \alpha \text{ a limit ordinal.} \end{aligned}$$

Show that if $x \in D_S$, $y \in O$, and $\nu(x) = |y|$, then $H(y) \leq_{\mathrm{T}} K(x)$. (*Hint:* First use the recursion lemma to show that, given $v \in D_S$, $y \in O$, and $\nu(v) \leq |y|$, the integer u such that $u \leq_O y$ and $\nu(v) = |u|$ can be computed recursively in $K(2^{2^v})$ uniformly in y and v. Then prove the desired result by an application of the recursion lemma.)

$\blacktriangle$***16-85.*** (Kleene, Souslin). Let $\mathfrak{A} \subset \mathfrak{N}$. Show that $\mathfrak{A} \in \Delta_1^1 \Leftrightarrow (\exists$ constructive $\beta)$ $[\mathfrak{A} \in \Sigma_\beta^0]$. (*Hint:* To prove $\Rightarrow$, proceed as follows. Use the Π_1^1-form of $\mathfrak{A}$ to obtain z_0 such that $\mathfrak{A} = \{A|z_0 \in T^A\}$. Prove that for any fixed z_0, if $\{A|z_0 \in T^A\} \in \Sigma_1^1$, then $(\exists x_0)[x_0 \in T \;\&\; (\forall A)[z_0 \in T^A \Rightarrow \|z_0\|^A < \|x_0\|]]$ (otherwise $T = \{x|(\exists B)[z_0 \in T^B \;\&\; \|x\| < \|z_0\|]\}$, and $T \in \Sigma_1^1$ by Theorem XXXV). Conclude that $\mathfrak{A} = \{A|z_0 \in T^A_{\|x_0\|}\}$ for some $x_0 \in T$. Relativize the proof of Theorem LI to show that $T^A_{|y|} \leq_1 H^A(2^{e(y)})$. (This proof can be made to yield a notation in O for β uniformly in the Δ_1^1-index for $\mathfrak{A}$.) To prove $\Leftarrow$, use the recursion lemma in a proof parallel to that for Theorem L.)

16-86. Show that there is a set of functions in $\Sigma_2^{(\mathrm{fn})}$ which is not in $\Sigma_1^{(\mathrm{fn})B}$, where $B = H(y)$, $y \in O$, and $|y| = \omega$. (*Hint:* See final comment in §15.1 and final comment in §15.2.)

$\triangle$***16-87.*** For any X and all $y \in O$, define $H^X(y)$ as in §16.8. Show that it is not in general true that $A \in \Delta_1^{1,B} \Leftrightarrow (\exists y)[y \in O \;\&\; A \leq_{\mathrm{T}} H^B(y)]$. (*Hint:* Take $B = T$, show that there is a set in $\Delta_1^{1,T}$ of higher T-degree than any $H^T(y)$, $y \in O$.)

$\triangle$***16-88.*** Show that if $u \in O$, $v \in O$, $|u| \leq |v|$, $\mathfrak{A} \subset \mathfrak{N}$, and $\mathfrak{A}$ is expressed by a u-form, then $\mathfrak{A}$ is expressed by a v-form. (*Hint:* Do this in two steps. First, show for $u <_O v$, then for $|u| = |v|$. For the latter, first prove that Theorem XLIX holds for the sets $H^X(y)$ "uniformly in" X, i.e., with uniformity independent of X. Then use the recursion lemma. ($u = 2^s$ is the key case.))

16-89. Define $\Sigma_n^{(s)}$ as in §15.2 and Σ_n^0 as in §16.8. Show that $\mathfrak{A} \in \Sigma_n^{(s)} \Leftrightarrow \mathfrak{A} \in \Sigma_n^0$.

16-90. Let $y = 2^s \in O$. Show that $A \leq_1 H(y) \Leftrightarrow (\exists z)[A = \{x|z \in H^{\{x\}}(y)\}]$. (*Hint:* Use the recursion lemma over O.)

$\triangle$***16-91.*** To elementary arithmetic add atomic formulas of the form $D(\mathbf{n},a,b)$, where $\mathbf{n}$ is a numeral and a and b are variables. Let a fixed Gödel numbering of the formulas of this augmented system be chosen. A formula containing formulas $D(\mathbf{n},a,b)$ is *admissible* only if each n is the index of a recursive function all of whose values are Gödel numbers of formulas with one free variable that are already defined to be admissible. In an admissible formula, $D(\mathbf{n},a,b)$ is interpreted to mean "b satisfies the formula with Gödel number $\varphi_n(a)$." Make the notion of admissible formula precise and prove that a set is hyperarithmetical if and only if it is definable by an admissible formula. (What is the degree of unsolvability of all true admissible sentences?)

16-92. Let "Σ_α^0" stand for an x-form and "Π_β^0" stand for the negation of a y-form, where $|x| = \alpha > |y| = \beta$. Show that the following "quantifier rules" hold:

(i) $\exists\Sigma_\alpha^0 \rightarrow \Sigma_\alpha^0$;
(ii) $\forall\Sigma_\alpha^0 \rightarrow \Pi_{\alpha+1}^0$;
(iii) $[\Sigma_\alpha^0 \,\&\, \Pi_\beta^0] \rightarrow \Sigma_\alpha^0$;
(iv) $[\Sigma_\alpha^0 \rightarrow \Pi_\beta^0] \rightarrow \Pi_\alpha^0$.

$\triangle$***16-93.*** The following definition yields a chain of hyperarithmetical sets such that every Δ_1^1-set is reducible to some member of the chain. The definition avoids the use of an indexing of ordinals. It is due to Shoenfield. Define

$$D_0 = N;$$
$$D_{\beta+1} = D_\beta \cap (D_\beta)';$$
$$D_\alpha = \bigcap_{\beta<\alpha} D_\beta, \qquad \text{for } \alpha \text{ a limit ordinal.}$$

Show that $D_n \in \mathbf{h}_n$, for $n < \omega$, but that $D_\omega \in \mathbf{h}_{\omega+1}$. (*Hint:* Show that D_ω is Π_ω^0-complete.)

16-94. Let f be a recursive function such that for all x, $f(x)$ is a Δ_1^1-index for some Δ_1^1-set. For all x, let A_x be the set with Δ_1^1-index $f(x)$. Show that the "infinite join" $\{<u,v>|u \in A_v\}$ is a Δ_1^1-set. (*Hint:* This is immediate with the use of appropriate forms.)

16-95. (Geiser). Let h and A be given such that $h \in \Delta_1^1$, $A \in \Sigma_1^1$, and $A \subset \bigcup_{y\in T} W_{h(y)}^1$. Show that for some constructive α, $A \subset \bigcup_{y\in T_\alpha} W_{h(y)}^1$. (*Hint:* Take P as in Theorem XLIV and use Exercise 16-72 to show that if the desired result does not hold, then $z \in T \Leftrightarrow (\exists f)[P(f,z) \,\&\, (\exists x)[x \in A \,\&\, (\forall u)(\forall y)[[f(<u,z>) = 1 \,\&\, u \in T \,\&\, \|y\| < \|u\|] \Rightarrow x \notin W_{h(y)}^1]]]$. This yields $T \in \Sigma_1^1$).

16-96. (Enderton). Prove the uniformity asserted in Corollary XLIX(c). (*Hint:* Generalize the statement and proof of Theorem XLIX to yield Corollary XLIX(c) directly. Take $S = \{<x,y>|x \in O \,\&\, y \in O \,\&\, |x| < |y|\}$, and define an appropriate relation $<_S$. Methods are as in Theorem XLIX. The key case is: $x = 2^u$, $y = 3 \cdot 5^v$.)

$\triangle$***16-97.*** (Moschovakis). Definition: Let x and y be members of O such that $|x| = |y|$. We say that x is *majorized* by y if there exists a partial recursive ψ such that $u <_O x \Rightarrow [\psi(u) \text{ defined} \,\&\, \psi(u) \leq_O y \,\&\, |u| \leq |\psi(u)|]$.

(a) Show that there exist x_0 and y_0 such that $|x_0| = |y_0| = \omega^2$, but x_0 is not majorized by y_0. (*Hint:* Take any y_0 such that $|y_0| = \omega^2$. Construct $x_0 = 3 \cdot 5^v$ so that

$$|\varphi_v(n+1)| = \begin{cases} |\varphi_v(n)| + |\varphi_n(n+1)| + \omega, \text{ if } \varphi_n(n+1) <_O y_0; \\ |\varphi_v(n)| + \omega, \text{ otherwise.} \end{cases}$$

Assume x_0 majorized by y_0 via ψ. Take n so that $\varphi_n = \psi\varphi_v$ and get a contradiction.)

(b) Show that there is a uniform procedure such that given a (Gödel number for a) recursive f, one can find an n such that for all A, $n \in A' \Leftrightarrow f(n) \notin A$. (*Hint:* Use the recursion theorem to get n such that $W_n{}^A = N$ if $f(n) \notin A$, and $W_n{}^A = \emptyset$ if $f(n) \in A$.)

(c) Show that $[x \in O \,\&\, y \in O \,\&\, |x| = |y| = \text{a limit number}] \Rightarrow [x \text{ majorized by } y \Leftrightarrow H(x) \leq_1 H(y)]$. Hence conclude that for x_0, y_0 as in (a), $|x_0| = |y_0|$ but $H(x_0) \not\equiv H(y_0)$. (*Hint:* $\Rightarrow$ follows by the use of Corollary XLIX(c). To show $\Leftarrow$, assume $H(x) \leq_1 H(y)$. Take $t <_O x$. We show how to compute $\psi(t) <_O y$ so that $|t| \leq |\psi(t)|$. Let $e(t) = 2^{2^t}$, and let any s be given. Then $s \in H(e(t)) \Leftrightarrow <s,e(t)> \in H(x) \Leftrightarrow <p(s),q(s)> \in H(y)$ (where $p(s)$ and $q(s)$ are got by the 1-reducibility of $H(x)$ to $H(y)$) $\Leftrightarrow q(s) <_O y \,\&\, p(s) \in H_{q(s)}$. Assume for the moment that $|q(s)| < |t|$ for all s such that $q(s) <_O y$. Under this assumption we can use $H(t)$ to test $s \in H(e(t))$ as follows: see whether $q(s) <_O y$ (this uses the uniform reducibility of K to $H(t)$), then use the reduction of $H(q(s))$ to $H(t)$ given by Corollary XLIX(c). This T-reduction of $H(e(t))$ to $H(t)$ yields 1-reducibility of $H(e(t))$ to $H(2^t)$ via a recursive f (with index uniform in t) defined so that: $f(s) \in H(2^t) \Leftrightarrow$ the T-reduction procedure places s in $H(e(t))$. Of course $H(e(t))$ is not in fact 1-reducible to $H(2^t)$ since $H(e(t)) = (H(2^t))'$. Hence for some s, $|q(s)| \geq |t|$. We find such a $q(s)$ as follows. Take the f defined above (its definition was independent of our false assumption; only the assertion that $H(e(t)) \leq_1 H(2^t)$ via f depended on that assumption.) Apply (b) to f to get an n such that $n \in H(e(t)) \Leftrightarrow f(n) \notin H(2^t)$. It follows from the definition of f and the structure of the alleged T-reduction of $H(e(t))$ to $H(t)$ that $q(n) <_O y$ & $|q(n)| \geq |t|$. We therefore take $\psi(t) = q(n)$.)

△***16-98.*** (a) For $x \in O$ let H_x be the set defined as in 16-83. Show that for each $x \in O$, there exists a function f_x such that (i) $H_x = \{u | f_x(u) \neq 0\}$, (ii) $\{f_x\} \in \Pi_1^0$ uniformly in x, and (iii) $f_x \equiv_T H_x$ uniformly in x. (The existence of f_x with properties (ii) and (iii) has been remarked by Kreisel in [1962].) (This exercise gives a proof of Corollary XLIV(d) that does not depend on Theorem XLIV.) (*Hint:* Let g_0 be a recursive function such that for all A, $A \leq_1 A'$ via g_0 and *range* g_0 is recursive. For $x \in O$, define f_x as follows.

$x = 1$: $f_x = \lambda u[1]$.

$x = 2^y$ & $u \in$ *range* g_0: $f_x(u) = f_y g_0^{-1}(u)$. (Thus $f_y = f_x g_0$.)

$x = 2^y$ & $u \notin$ *range* g_0: $f_x(u) = w \Leftrightarrow [w = <p,q,r,s>$ & $<u,p,q,r> \in W_{\rho(u)}$ in exactly s steps & $D_q \subset \{z | f_x g_0(z) \neq 0\}$ & $D_r \subset \{z | f_x g_0(z) = 0\}]$. (Note in this case that the possibility $f_x(u) = 0$ is automatically covered by the assertion that f_x is a function.)

$x = 3\cdot5^t$: $f_x(<u,v>) = f_{\varphi_t(v)}(u)$.)

(b) Prove the assertion of (a) with $H(x)$ in place of H_x. (*Hint:* In the definition of $f_{3\cdot5^t}$ in (a), replace $\varphi_t(v)$ by $\varphi(t,v)$ where φ is a partial recursive function such that for $x = 3\cdot5^t \in O$, $\lambda v[\varphi(t,v)]$ is a function whose range is $\{z | z <_O x\}$.)

16-99. (Kleene). For any $\mathcal{C} \subset \mathfrak{F}$, define f (or A) $\in \Sigma^1_{n,\mathcal{C}}$ if f (or A) is expressible by a Σ_n^1-form with all function quantifiers restricted to range over $\mathcal{C}$.

(a) Show that $f \in \Sigma^1_{1,\mathcal{H}} \Rightarrow f \in \mathcal{H}$

△(b) Show that for each $x \in O$, H_x is definable in Σ_1^1-form with $\mathcal{C}_x$ as basis, where $\mathcal{C}_x$ is $\{f | (\exists u)[u <_O x$ & $f \leq_T H_u]\}$. (*Hint:* Use the Π_1^0-form constructed in 16-98. Let $S(x,f)$ be the Π_1^0-form such that for $x \in O$, $g = f_x \Leftrightarrow S(x,g)$. In the case $x = 2^y$, define $H_x = \{u | (\exists f)[S(y,f)$ & $[[u \in$ *range* g_0 & $fg_0^{-1}(u) \neq 0] \vee [u \notin$ *range* g_0 & $(\exists w)[w \neq 0$ & $w = <p,q,r,s>$ & . . . & $D_r \subset \{z | f(z) = 0\}]]]]\}$. The case $x = 3\cdot5^t$ is similar.)

16-100. (a) Show that for all n, $\Sigma^1_{n,\mathcal{H}} \subset \Delta_2^1$. (*Hint:* Observe that $(\exists f)_{\mathcal{H}}[. . . f . . .]$ can be written either as $(\exists f)[f \in \mathcal{H}$ & $. . . f . . .]$ or as $(\exists z)[\varphi_z^1$ is total & $(\forall f)[f = \varphi_z^1 \Rightarrow . . . f . . .]].)$

(b) Show that $\bigcup_{n=0}^{\infty} \Sigma^1_{n,\mathcal{H}} \subset \Delta_2^1$.

▲(c) Define $\mathcal{C}_0 = \mathcal{H}$;

$$\mathcal{C}_{\beta+1} = \bigcup_{n=0}^{\infty} \Sigma^1_{n,\mathcal{C}_\beta};$$

$$\mathcal{C}_\lambda = \bigcup_{\beta<\lambda} \mathcal{C}_\beta, \text{ for } \lambda \text{ a limit number.}$$

Let β_0 be the least ordinal β such that $\mathcal{C}_{\beta+1} = \mathcal{C}_\beta$. Show that β_0 is countable and that $\mathcal{C}_{\beta_0} \subset \Delta_2^1$. (The classes $\Sigma^1_{n,\mathcal{C}_\beta}$, $\beta \leq \beta_0$, form what is called, in the literature, the *ramified analytical hierarchy*. Result (c) has been obtained using set theory; the author does not know of a purely recursive-function-theoretic proof.)

△16-101. (*a*) (Kreisel). Let $R \in \Pi_1^1$. Show that

$$(\forall x)(\exists f)_{\mathcal{H}} R(f,x) \Leftrightarrow (\exists g)_{\mathcal{H}}(\forall x)R(\lambda y[g(<x,y>)],x).$$

(*Hint:* Consider the relation $\{<f,x> | f \in \mathcal{H} \ \&\ R(f,x)\} = P$. Then $P \in \Pi_1^1$. Applying Theorem XLV (Kondo-Addison) to P, we obtain P^* such that $P^* \subset P$, $P^* \in \Pi_1^1$, and $(\forall x)$ ($\exists$ unique f) $P^*(f,x)$. For each x, let g_x be the unique f such that $P^*(f,x)$. Define $g(<x,y>) = g_x(y)$. Evidently $(\forall x)R(\lambda y[g(<x,y>)],x)$. It remains to show $g \in \mathcal{H}$. But $g(<x,y>) = z \Leftrightarrow (\exists f)_{\mathcal{H}}[P^*(f,x) \ \&\ f(y) = z] \Leftrightarrow (\forall f)_{\mathcal{H}}[P^*(f,x) \Rightarrow f(y) = z]$. By the method of Theorem XLI (and 16-100(*a*)), this gives $g \in \Delta_1^1$.)

(The result can be obtained without the use of Theorem XLV as follows. $(\forall x)(\exists f)_{\mathcal{H}}R(f,x) \Rightarrow (\forall x)(\exists z)[\varphi_z^1$ is total $\&\ (\forall f)[(\forall y)(\forall w)[\varphi_z^1(y) = w \Rightarrow f(y) = w] \Rightarrow R(f,x)]]$. Abbreviate this last as $(\forall x)(\exists z)S(z,x)$. Then $S \in \Pi_1^1$. Applying Theorem XXV (single-valuedness for Π_1^1), we obtain: $(\exists h)_{\mathcal{H}}(\forall x)S(h(x),x)$. Let $g(<x,y>) = \varphi_{h(x)}^1(y)$. Evidently $g \in \mathcal{H}$ and $(\forall x)R(\lambda y[g(<x,y>)],x)$.)

(*b*) Define $\Delta_{1,\mathcal{H}}^1$ (for sets) as in 16-99. Show, without using Theorem XLIV, that $\Delta_{1,\mathcal{H}}^1 = \Delta_1^1$. (*Hint:* $\Delta_{1,\mathcal{H}}^1 \subset \Delta_1^1$ by Theorem XLI. To show $\Delta_1^1 \subset \Delta_{1,\mathcal{H}}^1$, use the quantifier rule from (*a*) above in the proof of Theorem XXI to get $T_\alpha \in \Delta_{1,\mathcal{H}}^1$ for all constructive α.

(*c*) (Moschovakis). Show that $\mathcal{H} \in (\Pi_1^1 - \Delta_1^1)$ without the use of Theorem XLIV (used in 16-76). (*Hint:* Define $P(f,z)$ to be the assertion: $[f \in \mathcal{H} \ \&\ z \in T \ \&\ f \leq_{\mathrm{T}} T_{\|z\|}] \vee [f \notin \mathcal{H} \ \&\ (\forall y)[\varphi_z(y) = 0]]$. Assume $\mathcal{H} \in \Delta_1^1$. Then we have $P \in \Pi_1^1$. Applying a modified version of Theorem XXV (a Π_1^1 single-valuedness theorem for partial functions of one function variable), we obtain P^* such that $P^* \subset P$, $P^* \in \Pi_1^1$, $(\forall f)(\forall z)[P^*(f,z) \Rightarrow z \in T]$, and $(\forall f)(\exists$ unique $z)P^*(f,z)$. Evidently, from the definition of P, $x \in T \Leftrightarrow (\exists f)(\exists z)[P^*(f,z) \ \&\ \|x\| < \|z\|]$. But this yields $x \in T \Leftrightarrow (\exists f)(\exists z)[(\forall y)[y \neq z \Rightarrow \neg P^*(f,y)] \ \&\ \|x\| < \|z\|]$. By Theorem XXI, this gives $T \in \Sigma_1^1$, and we have a contradiction.)

(*d*) (Moschovakis). Deduce $\Pi_1^1 \subset \Sigma_{1,\mathcal{H}}^1$ (Theorem XLIV) from (*b*) and (*c*) above. (*Hint:* Since $\mathcal{H} \in \Pi_1^1$, take z_0 so that $f \in \mathcal{H} \Leftrightarrow z_0 \in T^f$. By (*c*), $\{\|z_0\|^f | f \in \mathcal{H}\}$ is unbounded in the constructive ordinals. Hence we have $x \in T \Leftrightarrow (\exists f)_{\mathcal{H}}[\|x\| < \|z_0\|^f]$. Take the proof of Theorem XXI as modified in (*b*) above, and relativize it to f to obtain $\Delta_{1,\mathcal{H}}^1$ forms in f and x which express $\{x | \|x\| < \|z_0\|^f\}$ when f is in $\mathcal{H}$. Using the $\Sigma_{1,\mathcal{H}}^1$-form, we get $T \in \Sigma_{1,\mathcal{H}}^1$. Hence $\Pi_1^1 \subset \Sigma_{1,\mathcal{H}}^1$.) (While possibly simpler than the proof of Theorem XLIV given in §16.7, this proof does not yield the implicit definability corollaries to Theorem XLIV that are obtained in §16.7.)

△16-102. (Kreisel). Show that $\mathcal{H}$ is the smallest class $\mathcal{C}$ such that: (i) $(\forall f)(\forall g)[[g \in \mathcal{C} \ \&\ f \leq_{\mathrm{T}} g] \Rightarrow f \in \mathcal{C}]$, and (ii) for any $R \in \Pi_1^0$, $(\forall x)(\exists g)_{\mathcal{C}}R(g,x) \Leftrightarrow (\exists h)_{\mathcal{C}}(\forall x)R(\lambda y[h(<x,y>)],x)$. (That $\mathcal{H}$ has properties (i) & (ii) follows from Corollary X(*a*) and from (*a*) in 16-101.) (*Hint:* For $z \in O$, define $\mathcal{C}_z = \{g | (\exists y)[y <_O z \ \&\ g \leq_{\mathrm{T}} H(y)]\}$. It will be enough to show that for every $z \in O$, $|z| \neq 0$, there is an $R \in \Pi_1^0$ such that $(\forall v)(\exists g)_{\mathcal{C}_z}R(g,v)$ but $\neg(\exists h)_{\mathcal{C}_z}(\forall v)R(\lambda y[h(<v,y>)],v)$. For $z \in O$, let f_z be as defined in 16-98, and let S be the Π_1^0 relation constructed in 16-98 such that $g = f_z \Leftrightarrow S(z,g)$. For $z = 3 \cdot 5^t$, take $R(g,v) \Leftrightarrow S(\varphi_t(v),g)$. For $z = 2^y$, take $R(g,v) \Leftrightarrow [S(y,\lambda n[g(n+1)]) \ \&\ [g(0) = w \Leftrightarrow [[v \in range\ g_0 \ \&\ w = g(g_0^{-1}(v) + 1)] \vee [v \notin range\ g_0 \ \&\ w = <p,q,r,s> \ \&\ <v,p,q,r> \in W_{\rho(v)}$ in exactly s steps $\&\ D_q \subset \{z | g(z+1) \neq 0\} \ \&\ D_r \subset \{z | g(z+1) = 0\}]]]]$, where g_0 is as in 16-98.)

Bibliography

Addison, John W.

[1955] Analogies in the Borel, Lusin, and Kleene hierarchies, I, *Bulletin of the American Mathematical Society*, vol. 61, p. 75 (abstract).

[1959] Some consequences of the axiom of constructibility, *Fundamenta mathematicae*, vol. 46, pp. 337–357.

[1959*a*] Separation principles in the hierarchies of classical and effective descriptive set theory, *Fundamenta mathematicae*, vol. 46, pp. 123–135.

[1962] Some problems of hierarchy theory, *Proceedings of symposia in pure mathematics*, vol. V, *Recursive function theory*, pp. 123–130, American Mathematical Society, Providence, R.I.

[1962*a*] The theory of hierarchies, *Logic, methodology, and philosophy of science* (*Proceedings of the 1960 international congress*), pp. 26–37, Stanford University Press, Stanford, Calif.

Addison, John W., and Stephen C. Kleene

[1957] A note on function quantification, *Proceedings of the American Mathematical Society*, vol. 8, pp. 1002–1006.

Appel, Kenneth I., and Thomas G. McLaughlin

[1965] On properties of regressive sets, *Transactions of the American Mathematical Society*, vol. 115, pp. 83–93.

Asser, Günter

[1960] Rekursive Wortfunktionen, *Zeitschrift für mathematische Logik und Grundlagen der mathematik*, vol. 6, pp. 258–278.

Axt, Paul

[1959] On a subrecursive hierarchy and primitive recursive degrees, *Transactions of American Mathematical Society*, vol. 92, pp. 85–105.

Bernays, P., and A. A. Fraenkel

[1958] *Axiomatic set theory*, North-Holland Publishing Company, Amsterdam.

Birkhoff, Garrett

[1940] *Lattice theory*, American Mathematical Society colloquium publications, vol. 25, New York.

Boone, William W.

[1957] Certain simple, unsolvable problems of group theory, V, VI, *Indagationes mathematicae,* vol. 19, pp. 22–27, 227–232. (*Koninklijke Nederlandse Akademie van Wetenschappen, Proceedings, series A,* vol. 60.)

Church, Alonzo

[1936] An unsolvable problem of elementary number theory, *American journal of mathematics,* vol. 58, pp. 345–363.

[1936*a*] A note on the Entscheidungsproblem, *The journal of symbolic logic,* vol. 1, pp. 40–41, 101–102.

[1938] The constructive second number class, *Bulletin of the American Mathematical Society,* vol. 44, pp. 224–232.

[1956] *Introduction to mathematical logic,* Princeton University Press, Princeton, N.J.

Church, Alonzo, and Stephen C. Kleene

[1937] Formal definitions in the theory of ordinal numbers, *Fundamenta mathematicae,* vol. 28, pp. 11–21.

Clarke, Douglas A.

[1964] *Hierarchies of predicates of finite types,* Memoirs of the American Mathematical Society, no. 51, Providence, R.I.

Cleave, J. P.

[1961] Creative functions, *Zeitschrift für mathematische Logik und Grundlagen der Mathematik,* vol. 7, pp. 205–212.

Craig, William

[1953] On axiomatizability within a system, *The journal of symbolic logic,* vol. 18, pp. 30–32.

Crossley, John N.

[1965] Constructive order Types I, *Formal systems and recursive functions,* J. N. Crossley and M. A. E. Dummett (eds.), pp. 189–264, Amsterdam, 1965.

Curry, Haskell B.

[1963] *Foundations of mathematical logic,* McGraw-Hill Book Company, New York.

Davis, Martin

[1956] A note on universal Turing machines, *Automata studies, Annals of mathematics studies,* no. 34, pp. 167–175, Princeton, N.J.

[1957] The definition of universal Turing machine, *Proceedings of the American Mathematical Society,* vol. 8, pp. 1125–1126.

[1958] *Computability and unsolvability,* McGraw-Hill Book Company, New York.

Davis, Martin, Hilary Putnam, and Julia Robinson

[1961] The decision problem for exponential diophantine equations, *Annals of mathematics,* vol. 74, pp. 425–436.

Dekker, James C. E.

[1953] Two notes on recursively enumerable sets, *Proceedings of the American Mathematical Society,* vol. 4, pp. 495–501.

[1954] A theorem on hypersimple sets, *Proceedings of the American Mathematical Society,* vol. 5, pp. 791–796.

[1955] Productive sets, *Transactions of the American Mathematical Society,* vol. 78, pp. 129–149.

[1962] Infinite series of isols, *Proceedings of the symposia in pure mathematics,* vol. V, *Recursive function theory,* pp. 77–96, American Mathematical Society, Providence, R.I.

Dekker, James C. E., and John Myhill

[1958] Some theorems on classes of recursively enumerable sets, *Transactions of the American Mathematical Society,* vol. 89, pp. 25–59.

[1958*a*] Retraceable sets, *Canadian journal of mathematics*, vol. 10, pp. 357–373.

[1960] Recursive equivalence types, *University of California publications in mathematics*, n.s., vol. 3, pp. 67–213.

Dreben, Burton S.

[1962] Solvable Suranyi subclasses: an introduction to the Herbrand theory, *Annals of the Computation Laboratory of Harvard University*, vol. 31, pp. 32–47.

Enderton, Herbert B.

[1964] Hierarchies in recursive function theory, *Transactions of the American Mathematical Society*, vol. 111, pp. 457–471.

Enderton, Herbert B., and David Luckham

[1964] Hierarchies over recursive well-orderings, *Journal of symbolic logic*, vol. 29, pp. 183–190.

Feferman, Solomon

[1957] Degrees of unsolvability associated with classes of formalized theories, *The journal of symbolic logic*, vol. 22, pp. 161–175.

[1960] Arithmetization of metamathematics in a general setting, *Fundamenta mathematicae*, vol. 49, pp. 35–92.

[1961] Classifications of recursive functions by means of hierarchies, *Office of Ordnance Research technical report no.* 4 (DA-04-200-ORD), pp. 1–43.

Feferman, Solomon, and Clifford Spector

[1962] Incompleteness along paths in progressions of theories, *The journal of symbolic logic*, vol. 27, pp. 383–390.

Fischer, Patrick C.

[1962] *Theory of provable recursive functions*, Ph.D. Dissertation, Massachusetts Institute of Technology, Cambridge, Mass.

[1963] A note on bounded-truth-table reducibility, *Proceedings of the American Mathematical Society*, vol. 14, pp. 875–877.

Fraenkel, A. A.

[1928] *Einleitung in die Mengenlehre*, 3rd ed., Springer-Verlag OHG, Berlin.

Friedberg, Richard M.

[1957] Two recursively enumerable sets of incomparable degrees of unsolvability (solution of Post's problem 1944), *Proceedings of the National Academy of Sciences*, vol. 43, pp. 236–238.

[1957*a*] A criterion for completeness of degrees of unsolvability, *The journal of symbolic logic*, vol. 22, pp. 159–160.

[1958] Four quantifier completeness: a Banach-Mazur functional not uniformly partial recursive, *Bulletin de l'Académie Polonaise des Sciences, Série des sciences mathématiques, astronomiques et physiques*, vol. 6, pp. 1–5.

[1958*a*] Un contre-example relatif aux fonctionnelles récursives, *Comptes rendus hebdomadaires des séances de l'Académie des Sciences (Paris)*, vol. 247, pp. 852–854.

[1958*b*] Three theorems on recursive enumeration: I, Decomposition, II, Maximal set, III, Enumeration without duplication, *The journal of symbolic logic*, vol. 23, pp. 309–316.

Gandy, Robin O.

[1960] On a problem of Kleene's, *Bulletin of the American Mathematical Society*, vol. 66, pp. 501–502.

[1960*a*] Proof of Mostowski's conjecture, *Bulletin de l'Academie Polonaise des Sciences, série des sciences mathématiques, astronomiques et physiques*, vol. 8, pp. 571–575.

Gödel, Kurt

[1931] Über formal unentscheidbare Sätze der Principia Mathematica und verwandter Systeme, I, *Monatshefte für Mathematik und Physik*, vol. 38, pp. 173–198.

Grzegorczyk, Andrzej

[1953] *Some classes of recursive functions*, Rozprawy matematyczne no. 4, Instytut Matematyczny Polskiej Akademie Nauk, Warsaw.

Halmos, Paul

[1950] *Measure theory*, D. Van Nostrand Company, Inc., Princeton, N.J.

Hanf, William

[1962] Degrees of finitely axiomatizable theories, *Notices of the American Mathematical Society*, vol. 9, pp. 127–128 (abstract).

Henkin, Leon

[1949] The completeness of the first-order functional calculus, *The journal of symbolic logic*, vol. 14, pp. 159–166.

Hensel, Gustav, and Hilary Putnam

[1965] On the notational independence of various hierarchies of degrees of unsolvability, *Journal of symbolic logic*, vol. 30, pp. 69–86.

Herbrand, Jacques

[1931] Sur le problème fondamental de la logique mathématique, *Comptes rendus des séances de la Société des Sciences et des Lettres de Varsovie, classe III*, vol. 24, pp. 12–56.

[1931*a*] Sur la non-contradiction de l'arithmétique, *Journal für die reine und angewandte Mathematik*, vol. 166, pp. 1–8.

Higman, Graham

[1961] Subgroups of finitely presented groups, *Proceedings of the Royal Society of London, series A*, vol. 262, pp. 455–475.

Hilbert, D., and P. Bernays

[1939] *Grundlagen der Mathematik*, Springer-Verlag OHG, Berlin.

Hodes, Louis

[1962] *Hyperarithmetical real numbers and hyperarithmetical analysis*, Ph.D. Dissertation, Massachusetts Institute of Technology, Cambridge, Mass.

Hooper, Philip K.

[1966] The undecidability of the Turing machine immortality problem, *Journal of symbolic logic*, vol. 31, pp. 219–234.

Jockusch, Carl G., Jr.

[1966] *Reducibilities in recursive function theory*, Ph.D. Dissertation, Massachusetts Institute of Technology, Cambridge, Mass.

Kalmár, László

[1955] Über ein Problem, betreffend die Definition des Begriffes der allgemeinrekursivenFunktion, *Zeitschrift für mathematische Logik und Grundlagen der Mathematik*, vol. 1, pp. 93–96.

Kent, Clement F.

[1962] Constructive analogues of the group of permutations of the natural numbers, *Transactions of the American Mathematical Society*, vol. 104, pp. 347–362.

Klaua, Dieter

[1961] *Konstruktive Analysis*, Mathematische Forschungsberichte, XI, Deutscher Verlag der Wissenschaften, Berlin.

Kleene, Stephen C.

[1936] General recursive functions of natural numbers, *Mathematische Annalen*, vol. 112, pp. 727–742.

[1936*a*] λ-definability and recursiveness, *Duke mathematical journal*, vol. 2, pp. 340–353.

[1938] On notation for ordinal numbers, *The journal of symbolic logic*, vol. 3, pp. 150–155.

[1943] Recursive predicates and quantifiers, *Transactions of the American Mathematical Society*, vol. 53, pp. 41–73.

[1950] A symmetric form of Gödel's theorem, *Indagationes mathematicae*, vol. 12, pp. 244–246. (*Koninklijke Nederlandse Akademie van Wetenschappen, Proceedings, series A*, vol. 53.)

[1952] *Introduction to metamathematics*, D. Van Nostrand Company, Inc., Princeton, N.J.

[1955] Arithmetical predicates and function quantifiers, *Transactions of the American Mathematical Society*, vol. 79, pp. 312–340.

[1955*a*] Hierarchies of number theoretic predicates, *Bulletin of the American Mathematical Society*, vol. 61, pp. 193–213.

[1955*b*] On the forms of the predicates in the theory of constructive ordinals, II, *American journal of mathematics*, vol. 77, pp. 405–428.

[1959] Recursive functionals and quantifiers of finite types, I, *Transactions of the American Mathematical Society*, vol. 91 pp. 1–52.

[1959*a*] Quantification of number theoretic functions, *Compositio mathematica*, vol. 14, pp. 23–40.

[1962] Turing machine computable functionals of finite types, I, *Logic, methodology, and philosophy of science* (*Proceedings of the 1960 international congress*), pp. 38–45, Stanford University Press, Stanford, Calif.

[1963] Recursive functionals and quantifiers of finite types, II, *Transactions of the American Mathematical Society*, vol. 108, pp. 106–142.

Kleene, Stephen C., and Emil L. Post

[1954] The upper semi-lattice of degrees of recursive unsolvability, *Annals of mathematics*, ser. 2, vol. 59, pp. 379–407.

Kondo, Motokiti

[1938] Sur l'uniformization des complémentaires analytiques et les ensembles projectifs de la seconde classe, *Japanese journal of mathematics*, vol. 15, pp. 197–230.

Kreider, D. L., and Hartley Rogers, Jr.

[1961] Constructive versions of ordinal number classes, *Transactions of the American Mathematical Society*, vol. 100, pp. 325–369.

Kreisel, Georg

[1951] Some remarks on the foundations of mathematics: an expository article, *The mathematical gazette*, vol. 35, pp. 23–28.

[1962] The axiom of choice and the class of hyperarithmetic functions, *Indagationes mathematicae*, vol. 24, pp. 307–319. (*Koninklijke Nederlandse Akademie van Wetenschappen, Proceedings, series A*, vol. 65, pp. 307–319.)

Kreisel, Georg, Daniel Lacombe, and Joseph R. Shoenfield

[1957] Partial recursive functionals and effective operations, in A. Heyting (ed.), *Constructivity in mathematics: proceedings of the colloquium held at Amsterdam*, 1957, pp. 195–207, North-Holland Publishing Company, Amsterdam.

[1957*a*] Fonctionnelles récursivement définissables et fonctionelles récursives, *Comptes rendus hebdomadaires des séances de l'Académie des Sciences* (*Paris*), vol. 245 pp. 399–402.

Kreisel, Georg, and Gerald Sacks

[1965] Metarecursive sets, *Journal of symbolic logic*, vol. 30, pp. 318–338.

Kreisel, Georg, Joseph R. Shoenfield, and Hao Wang

[1960] Number theoretic concepts and recursive well-orderings, *Archive für mathematische Logik und Grundlagenforschung*, vol. 5, pp. 42–64.

Kuratowski, Kazimierz

[1950] *Topologie*, vol. 1, 2d ed. (1948); vol. 2 (1950), Warsaw.

Kuznecov, A. V.

[1950] On primitive recursive functions of large oscillation (Russian), *Doklady Akademii Nauk SSSR*, n.s., vol. 71, pp. 233–236.

Kuznecov, A. V., and B. A. Trahtenbrot

[1955] Investigation of partially recursive operators by means of the theory of Baire Space, *Doklady Akademii Nauk SSSR*, n.s., vol. 105, pp. 897–900.

Lachlan, A. H.

[1965] Some notions of reducibility and productiveness, *Zeitschrift für mathematische Logik und Grundlagen der Mathematik*, vol. 1, pp. 97–108.

[1966] A note on universal sets, *Journal of symbolic logic*, vol. 31, pp. 573, 574.

Lacombe, Daniel

[1954] Sur le semi-réseau constitué par les degrés d'indécidabilité récursive, *Comptes rendus hebdomadaires des séances de l'Académie des Sciences* (*Paris*), vol. 239, pp. 1108–1109.

[1959] Quelques procédés de définition en topologie récursive, in A. Heyting (ed.), *Constructivity in mathematics: proceedings of the colloquium held at Amsterdam*, 1957, pp. 129–158, North-Holland Publishing Company, Amsterdam.

[1960] La théorie des fonctions récursives et ses applications (Exposé d'information générale), *Bulletin de la Société Mathématique de France*, vol. 88, pp. 393–468.

Loève, Michel

[1955] *Probability theory*, D. Van Nostrand Company, Inc., Princeton, N.J.

Markov, A. A.

[1947] On the representation of recursive functions (Russian). *Doklady Akademii Nauk SSSR*, n.s., vol. 58, pp. 1891–1892.

[1951] The theory of algorithms (Russian), *Trudy Mathematicheskogo Instituta imeni V.A. Steklova*, vol. 38, pp. 176–189.

[1954] The theory of algorithms (Russian), *Trudy Mathematicheskogo Instituta imeni V.A. Steklova*, vol. 42 (English translation, 1961, National Science Foundation, Washington, D.C.).

[1958] The problem of homeomorphy (Russian), *Proceedings of the international congress of mathematicians, 1958*, Cambridge University Press, London, pp. 300–306 (1960).

Markwald, Werner

[1954] Zur Theorie der konstruktiven Wohlordnungen, *Mathematische Annalen*, vol. 127, pp. 135–149.

Martin, Donald A.

[1963] A theorem on hyperhypersimple sets, *The journal of symbolic logic*, vol. 28, pp. 273–278.

McLaughlin, Thomas G.

[1962] On an extension of a theorem of Friedberg, *Notre Dame journal of formal logic*, vol. 3, pp. 270–273.

Medvedev, Yu. T.

[1955] On nonisomorphic recursively enumerable sets (Russian), *Doklady Academii Nauk SSSR*, n.s., vol. 102, pp. 211–214.

[1955*a*] Degrees of difficulty of the mass problem (Russian), *Doklady Academii Nauk SSSR*, n.s., vol. 104, pp. 501–504.

Minsky, Marvin L.

[1961] Recursive unsolvability of Post's problem of "tag" and other topics in the theory of Turing machines, *Annals of mathematics*, vol. 74, pp. 437–455.

Moschovakis, Yiannis N.

[1963] *Recursive analysis*, Ph.D. Dissertation, University of Wisconsin, Madison, Wis.

Mostowski, Andrzej

[1947] On definable sets of positive integers, *Fundamenta mathematicae*, vol. 34, pp. 81–112.

[1948] On a set of integers not definable by means of one-quantifier predicates, *Annales de la Société Polonaise de Mathématique*, vol. 21, pp. 114–119.

Muchnik, A. A.

[1956] On the unsolvability of the problem of reducibility in the theory of algorithms (Russian), *Doklady Akademii Nauk SSSR*, n.s., vol. 108, pp. 194–197.

[1956*a*] On the separability of recursively enumerable sets (Russian), *Doklady Akademii Nauk SSSR*, n.s., vol. 109, pp. 29–32.

[1958] Isomorphism of systems of recursively enumerable sets with effective properties (Russian), *Trudy Moskovskogo Mathematicheskogo Obshchestva*, vol. 7, pp. 407–412 (English translation in *American Mathematical Society translations*, vol. 23).

Myhill, John

[1953] Three contributions to recursive function theory, *Actes du XIème congrès international de philosophie, Bruxelles*, 20–26 *Août* 1953, vol. XIV, pp. 50–59, North-Holland Publishing Company, Amsterdam.

[1955] Creative sets, *Zeitschrift für mathematische Logik und Grundlagen der Mathematik*, vol. 1, pp. 97–108.

[1956] The lattice of recursively enumerable sets, *The journal of symbolic logic*, vol. 21, p. 220 (abstract).

[1959] Finitely representable functions, in A. Heyting (ed.), *Constructivity in mathematics: proceedings of the colloquium held at Amsterdam, 1957*, pp. 195-207, North-Holland Publishing Company, Amsterdam.

[1959*a*] Recursive digraphs, splinters and cylinders, *Mathematische Annalen*, vol. 138, pp. 211–218.

[1961] Note on degrees of partial functions, *Proceedings of the American Mathematical Society*, vol. 12, pp. 519–521.

[1961*a*] Category methods in recursion theory, *Pacific journal of mathematics*, vol. 11, pp. 1479–1486.

Myhill, J., and J. C. Shepherdson

[1955] Effective operations on partial recursive functions, *Zeitschrift für mathematische Logik und Grundlagen der Mathematik*, vol. 1, pp. 310–317.

Nerode, Anil

[1957] General topology and partial recursive functionals, *Summaries of talks presented at the Summer Institute for Symbolic Logic, Cornell University*, 1957, pp. 247–251.

Novikov, P. S.

[1955] On the algorithmic unsolvability of the word problem in group theory (Russian), *Trudy Matematicheskogo Institutaimeni V.A. Steklova*, no. 44.

Péter, Rósza

[1951] *Rekursive funktionen*, Académiai Kiadó, Budapest.

Post, Emil L.

[1936] Finite combinatory processes—formulation, I, *The journal of symbolic logic*, vol. 1, pp. 103–105.

[1943] Formal reductions of the general combinatorial decision problem, *American journal of mathematics*, vol. 65, pp. 197–215.

[1944] Recursively enumerable sets of positive integers and their decision problems, *Bulletin of the American Mathematical Society*, vol. 50, pp. 284–316.

[1946] Note on a conjecture of Skolem, *The journal of symbolic logic*, vol. 11, pp. 73–74.

[1947] Recursive unsolvability of a problem of Thue, *The journal of symbolic logic*, vol. 12, pp. 1–11.

Poúr-El, Marian B.

[1960] A comparison of five "computable" operators, *Zeitschrift für mathematische Logik und Grundlagen der Mathematik*, vol. 6, pp. 325–340.

Putnam, Hilary

[1964] On hierarchies and systems of notations, *Proceedings of the American Mathematical Society*, vol. 15, pp. 44–50.

Quine, Willard V.

[1959] *Methods of logic*, Holt, Rinehart and Winston, Inc., New York.

[1963] *Set theory and its logic*, Harvard University Press, Cambridge, Mass.

Rabin, Michael O.

[1960] Computable algebra, general theory and theory of computable fields, *Transactions of the American Mathematical Society*, vol. 95, pp. 341–360.

Rabin, Michael O., and D. Scott

[1959] Finite automata and their decision problems, *IBM journal of research and development*, vol. 3, pp. 114–125.

Rice, H. Gordon

[1953] Classes of recursively enumerable sets and their decision problems, *Transactions of the American Mathematical Society*, vol. 74, pp. 358–366.

[1954] Recursive real numbers, *Proceedings of the American Mathematical Society*, vol. 5, pp. 784–791.

[1956] On completely recursively enumerable classes and their key arrays, *The journal of symbolic logic*, vol. 21, pp. 304–308.

Richter, Wayne

[1965] Extensions of the constructive ordinals, *Journal of symbolic logic*, vol. 30, pp. 193–211.

Ritchie, Robert W.

[1963] Classes of predictably computable functions, *Transactions of the American Mathematical Society*, vol. 106, pp. 139–173.

Ritter, William

[1962] *Some results in hyperarithmetical analysis*, Ph.D. Dissertation, Massachusetts Institute of Technology, Cambridge, Mass.

Robinson, Julia

[1949] Definability and decision problems in arithmetic, *The journal of symbolic logic*, vol. 14, pp. 98–114.

Robinson, Raphael M.

[1956] Arithmetical representation of recursively enumerable sets, *The journal of symbolic logic*, vol. 21, pp. 162–186.

Rogers, Hartley, Jr.

[1959] Computing degrees of unsolvability, *Mathematische Annalen*, vol. 138, pp. 125–140.

[1959*a*] Recursive functions over partial well orderings, *Proceedings of the American Mathematical Society*, vol. 10, pp. 847–853.

Rose, G. F., and Joseph S. Ullian

[1963] Approximations of functions on the integers, *Pacific journal of mathematics*, vol. 13, pp. 693–701.

Rosenbloom, Paul C.

[1950] *The elements of mathematical logic*, Dover Publications, Inc., New York.

Rosser, J. Barkley

[1936] Extensions of some theorems of Gödel and Church, *The journal of symbolic logic*, vol. 1, pp. 87–91.

Sacks, Gerald E.

[1961] A minimal degree less than 0′, *Bulletin of the American Mathematical Society*, vol. 67, pp. 416–419.

[1961*a*] On suborderings of degrees of recursive unsolvability, *Zeitschrift für mathematische Logik und Grundlagen der Mathematik*, vol. 7, pp. 46–56.

[1963] On the degrees less than 0′, *Annals of mathematics*, vol. 77, pp. 211–231.

[1963*a*] Recursive enumerability and the jump operator, *Transactions of the American Mathematical Society*, vol. 108, pp. 223–239.

[1963*b*] *Degrees of unsolvability, Annals of mathematics studies*, no. 55, Princeton, N.J.

[1964] A simple set which is not effectively simple, *Proceedings of the American Mathematical Society*, vol. 15, pp. 51–55.

[1964*a*] The recursively enumerable degrees are dense, *Annals of mathematics*, vol. 80, pp. 300–312.

[1964*b*] A maximal set which is not complete, *Michigan mathematical journal*, vol. 11, pp. 193–205.

Shapiro, Norman

[1956] Degrees of computability, *Transactions of the American Mathematical Society*, vol. 82, pp. 281–299.

Shepherdson, J. C., and H. E. Sturgis

[1963] Computability of recursive functions, *Journal of the Association for Computing Machinery*, vol. 10, pp. 217–255.

Shoenfield, Joseph R.

[1957] Quasicreative sets, *Proceedings of the American Mathematical Society*, vol. 8, pp. 964–967.

[1958] The class of recursive functions, *Proceedings of the American Mathematical Society*, vol. 9, pp. 690–692.

[1958*a*] Degrees of formal systems, *The journal of symbolic logic*, vol. 23, pp. 389–392.

[1959] On degrees of unsolvability, *Annals of mathematics*, ser. 2, vol. 69, pp. 644–653.

[1960] An uncountable set of incomparable degrees, *Proceedings of the American Mathematical Society*, vol. 11, pp. 61–62.

[1962] The form of the negation of a predicate, *Proceedings of symposia in pure mathematics*, vol. V, *Recursive function theory*, pp. 131–134, American Mathematical Society, Providence, R.I.

Singleterry, Ann M.

[1965] *Degrees of unsolvability in the hyperarithmetical hierarchy*, Ph.D. Dissertation, Massachusetts Institute of Technology, Cambridge, Mass.

Skolem, Thoralf

[1944] Some remarks on recursive arithmetic, *Det Kongelige Norske Videnskabers Selskabs Forhandlinger*, vol. 17, pp. 103–106.

Smullyan, Raymond M.

[1961] *Theory of formal systems, Annals of mathematics studies*, no. 47, Princeton, N.J.

Spector, Clifford

[1955] Recursive well-orderings, *The journal of symbolic logic*, vol. 20, pp. 151–163.

[1956] On degrees of recursive unsolvability, *Annals of mathematics*, vol. 64, pp. 581–592.

[1958] Measure theoretic construction of incomparable hyperdegrees, *The journal of symbolic logic*, vol. 23, pp. 280–288.

[1958*a*] Strongly invariant hierarchies, *Notices of the American Mathematical Society*, vol. 5, p. 851 (abstract).

[1959] Hyperarithmetical quantifiers, *Fundamenta mathematicae*, vol. 48, pp. 313–320.

Suppes, Patrick

[1957] *Introduction to logic*, D. Van Nostrand Company, Inc., Princeton, N.J.

Suzuki, Yoshindo

[1964] A complete classification of the Δ_2^1-functions, *Bulletin of the American Mathematical Society*, vol. 70, pp. 246–253.

Tarski, Alfred

[1932] Der Wahrheitsbegriff in den Sprachen der deduktiven Disziplinen, *Akademie der Wissenschaften in Wien, Mathematisch-naturwissenschaftliche Klasse, Anzeiger*, vol. 69, pp. 24–25.

[1936] Der Wahrheitsbegriff in den formalisierten Sprachen, *Studia philosophica*, vol. 1, pp. 261–405. (English translation in [1956].)

[1948] *A decision method for elementary algebra and geometry*, RAND Corporation, Santa Monica, Calif.

[1956] *Logic, semantics, metamathematics*, Oxford University Press, New York.

Tarski, Alfred, Andrzej Mostowski, and Raphael M. Robinson

[1953] *Undecidable theories*, North-Holland Publishing Company, Amsterdam.

Trahtenbrot, B. A.

[1950] The impossibility of an algorithm for the decision problem for finite domains (Russian), *Doklady Academii Nauk SSSR*, n.s., vol. 70, pp. 569–572.

[1955] Tabular representation of recursive operators (Russian), *Doklady Akademii Nauk SSSR*, n.s., vol. 101, pp. 417–420.

Turing, Alan M.

[1936] On computable numbers, with an application to the Entscheidungsproblem, *Proceedings of the London Mathematical Society*, ser. 2, vol. 42, pp. 230–265; vol. 43, pp. 544–546.

[1937] Computability and λ-definability, *The journal of symbolic logic*, vol. 2, pp. 153–163.

Ullian, Joseph S.

[1960] Splinters of recursive functions, *The journal of symbolic logic*, vol. 25, pp. 33–38.

[1961] A theorem on maximal sets, *Notre Dame journal of formal logic*, vol. 2, pp. 222–223.

Uspenskii, V. A.

[1955] On computable operations (Russian), *Doklady Akademii Nauk SSSR*, n.s., vol. 103, pp. 773–776.

[1955*a*] Systems of denumerable sets and their enumeration (Russian), *Doklady Akademii Nauk SSSR*, n.s., vol. 105, pp. 1155–1158.

[1957] Some notes on recursively enumerable sets (Russian), *Zeitschrift für mathematische Logik und Grundlagen der Mathematik*, vol. 3, pp. 157–170. (English translation in *American Mathematical Society translations*, vol. 23.)

[1960] *Lectures on computable functions* (Russian), Gosudarstvennoye Izdat. Fiz.-Mat. Lit., Moscow.

Whitehead, Alfred N., and Bertrand Russell

[1910] *Principia mathematica*, vol. 1 (1910); vol. 2 (1912); vol. 3 (1913), Cambridge University Press, London.

Yates, C. E. M.

[1962] Recursively enumerable sets and retracing functions, *Zeitschrift für mathematische Logik und Grundlagen der Mathematik*, vol. 8, pp. 331–345.

[1965] Three theorems on the degree of recursively enumerable sets, *Duke mathematical journal*, vol. 32, pp. 461–468.

Young, Paul R.

[1963] *On the structure of recursively enumerable sets*, Ph.D. Dissertation, Massachusetts Institute of Technology, Cambridge, Mass.

[1966] Linear orderings under one-one reducibility, *Journal of symbolic logic*, vol. 31, pp. 70–85.

Index of Notations†

† Entries are listed in order of occurrence in text.

V	318
V_n	319
ZF	320
$T_n(z, \ldots)$	321
Σ_n^{ZF}, Π_n^{ZF}	322
$P_{\mathfrak{C}}$	324
$(\mathrm{U}x)$	328
$\mathrm{U}^{(n)}$	329
$\Sigma_n^{(s)}$, $\Pi_n^{(s)}$	337
$T_{k,l}$	337
Σ_n^*, Π_n^*	342
$\mathfrak{F}$ (Baire space)	346
$\Sigma_n^{(s)A}$	346
$\Sigma_n^{(\mathrm{fn})}$, $\Pi_n^{(\mathrm{fn})}$	348
$T'_{k,l}$	348
Σ_n^{**}, Π_n^{**}	353
$\Sigma_n^{(\mathrm{fn})A}$	356
$\Sigma_n^{(\mathrm{fn})}[\mathfrak{C}]$, $\Pi_n^{(\mathrm{fn})}[\mathfrak{C}]$	356
T_n^*	357
$\mathcal{PR}$, $\mathcal{R}$	358
$[f]$	364
Σ_n^1, Π_n^1	374, 375, 383
$\bar{f}(x)$	377
$T_{k,l}^*$	377
E^n	379
E^ω	380
Δ_n^1	382, 383
Σ_n^0, Π_n^0, Δ_n^0	383
$\Sigma_n^{1,ZF}$	391
$o(\tau)$	393
$b(n,z)$	395
T	395
$\|z\|$	395
T_α	397
W_z^1	400, 403
T_α^*	401
$<^T$	402
φ_z^1	403
$\Sigma_n^{1,A}$, $\Pi_n^{1,A}$, $\Delta_n^{1,A}$	409
$T'^A_{k,l}$	409
$T^{*A}_{k,l}$	409
$\Sigma_n^{1,h}, \ldots, T'^{h}_{k,l} \ldots$	410
$\leq_{\mathrm{h}}$	410
$\equiv_{\mathrm{h}}$	411
T^A	412
$\|z\|^A$	413
$T_\alpha{}^A$	413
$b'(n,z)$	413
λ, λ^A	415
T^2, $\|z\|^2$, $T_\alpha{}^2$	416
$\mathfrak{A}$	417

Subject Index